T4-AKW-720

OXIDATIVE STRESS

To Nancy Jaye

OXIDATIVE STRESS:
Oxidants and Antioxidants

Edited by

Helmut Sies

Universität Düsseldorf, Düsseldorf, Germany

Academic Press

Harcourt Brace Jovanovich, Publishers

London San Diego New York

Boston Sydney Tokyo Toronto

ACADEMIC PRESS LIMITED
24–28 Oval Road
London NW1 7DX

United States Edition published by
ACADEMIC PRESS INC.
San Diego, CA 92101

This book is printed on acid-free paper

A catalogue record for this book is available from the British Library
ISBN 0-12-642762-3

Filmset by Bath Typesetting Limited
and printed in Great Britain by St Edmundsbury Press Ltd,
Bury St. Edmunds, Suffolk

Contents

Contributors ix

Preface xiii

Oxidative Stress: introduction xv

OCCURRENCE OF OXIDANTS

1 Oxygen radicals and air pollution
S. Hippeli and E. F. Elstner 3

2 UVA (320–380 nm) radiation as an oxidative stress
R. M. Tyrrell 57

3 Non-radical mechanisms for the oxidation of glutathione
D. M. Ziegler 85

4 Formation of 8-hydroxydeoxyguanosine in DNA by oxygen radicals and its biological significance
H. Kasai and S. Nishimura 99

DEFENCE AGAINST OXIDANTS

5 Repair systems for radical-damaged DNA
B. Demple and J. D. Levin 119

6 The bacterial adaptation to hydrogen peroxide stress
L. A. Tartaglia, G. Storz, S. B. Farr and B. N. Ames 155

7 Glutathione transferases and products of reactive oxygen
B. Ketterer and B. Coles 171

8 The role of NAD(P)H: quinone reductase in protection against the toxicity of quinones
H. J. Prochaska and P. Talalay 195

9 Endogenous antioxidant defences in human blood plasma
R. Stocker and B. Frei 213

10 Protective and adverse biological actions of phenolic antioxidants
R. Kahl 245

PROCESSES AND CELL RESPONSES

11 Role of free radicals in carcinogen activation
M. A. Trush and T. W. Kensler 277

12 Membrane hydroperoxides
F. Ursini, M. Maiorino and A. Sevanian 319

13 Biological activities of 4-hydroxyalkenals
H. Zollner, R. J. Schaur and H. Esterbauer 337

14 Oxidation of low-density lipoprotein in vitro
M. Gebicki, G. Jürgens and H. Esterbauer 371

15 The phagocytes and the respiratory burst
M. Baggiolini and M. Thelen 399

16 Oxidative stress in platelets
B. Brüne, F. von Appen and V. Ullrich 421

17 Vasodilation and oxygen radical scavenging by nitric oxide/EDRF and organic nitrovasodilators
E. Noack and M. Murphy 445

TOWARDS CLINICAL MEDICINE

18 Ischaemia–reperfusion
B. Omar, J. McCord and J. Downey 493

19 The lens and oxidative stress
A Spector 529

20 Photooxidative stress in the skin
J. Fuchs and L. Packer 559

21 Interactions between reactive oxygen and mediators of sepsis and shock
A. Wendel, M. Niehörster and G. Tiegs 585

22 Involvement of oxygen radicals in shock and organ failure
H. Redl, H. Gasser, S. Hallström, E. Paul, S. Bahrami, G. Schlag and R. Spragg 595

Abbreviations 617

Index 621

Contributors

Bruce N. Ames Division of Biochemistry and Molecular Biology, University of California, Berkeley, CA 94720, USA

Frank von Appen Fakultät für Biologie, Universität Konstanz, Postfach 5560, D-7750 Konstanz, Germany

Marco Baggiolini Theodor-Kocher Institute, University of Bern, P.O. Box 99, CH-3000 Bern 9, Switzerland

Soheyl Bahrami Ludwig Boltzmann Institute for Experimental and Clinical Traumatology, Donaueschingenstrasse 13, A-1200 Vienna, Austria

Bernhard Brüne Fakultät für Biologie, Universität Konstanz, Postfach 5560, D-7750 Konstanz, Germany

Brian Coles Cancer Research Campaign, Molecular Toxicology Research Group, Department of Biochemistry, University College and Middlesex School of Medicine, Cleveland Street, London W1P 6DB, UK

Bruce Demple Laboratory of Toxicology, Harvard School of Public Health, 665 Huntington Avenue, Boston, Massachusetts 02115, USA

James Downey Department of Physiology, The University of South Alabama, Mobile, Alabama 36688, USA

Erich F. Elstner Institut für Botanik und Mikrobiologie der Technischen Universität München, Biochemisches Labor, Arcisstrasse 21, D-8000 München 2, Germany

Hermann Esterbauer Institute of Biochemistry, University of Graz, Schubertstrasse 1, A-8010 Graz, Austria

Spencer B. Farr Division of Biochemistry and Molecular Biology, University of California, Berkeley, CA 94720, USA

Balz Frei Division of Biochemistry and Molecular Biology, Harvard School of Public Health, 665 Huntington Avenue, Boston, MA 02115, USA

Jürgen Fuchs Zentrum der Dermatologie und Venerologie, Klinikum der Johann Wolfgang, Goethe Universität, 6000 Frankfurt/M 70, Germany

Harald Gasser Ludwig Boltzmann Institute for Experimental and Clinical Traumatology, Donaueschingenstrasse 13, A-1200 Vienna, Austria

Janusz M. Gebicki School of Biological Sciences, Macquarie University, Sydney, Australia

Seth Hallström Ludwig Boltzmann Institute for Experimental and Clinical Traumatology, Donaueschingenstrasse 13, A-1200 Vienna, Austria

Susanne Hippeli Institut für Botanik und Mikrobiologie der Technischen Universität München, Biochemisches Labor, Arcisstrasse 21, D-8000 München 2, Germany

Günther Jürgens Institute of Biochemistry, University of Graz, Schubertstrasse 1, A-8010 Graz, Austria

Regine Kahl Abteilung Klinische Pharmakologie, Universität Göttingen, Robert-Koch-Strasse 40, D-3400 Göttingen, Germany

Hiroshi Kasai Biology Division, National Cancer Center Research Institute, Tsukiji 5-1-1, Chuo-ku, Tokyo 104, Japan

Thomas W. Kensler Department of Environmental Health Sciences, Johns Hopkins School of Hygiene and Public Health, 615 N. Wolfe Street, Baltimore, Maryland 21205, USA

Brian Ketterer Cancer Research Campaign Molecular Toxicology Research Group, Department of Biochemistry, University College and Middlesex School of Medicine, Cleveland Street, London W1P 6DB, UK

J. D. Levin Laboratory of Toxicology, Harvard School of Public Health, 665 Huntington Avenue, Boston, Massachusetts 02115, USA

Matilde Maiorino Department of Biological Chemistry, University of Padova, Via F. Trieste 75, I-35100 Padova, Italy

Joe McCord Webb-Waring Lung Institute, University of Colorado, Denver, Colorado 80262, USA

Michael Murphy Institut für Physiologische Chemie I, Heinrich-Heine-Universität Düsseldorf, Moorenstrasse 5, D-4000 Düsseldorf 1, Germany

Marcus Niehörster Faculty of Biology, University of Konstanz, Postfach 5560, D-7750 Konstanz, Germany

Susumo Nishimura National Cancer Center Research Institute, Biology Division, Tsukiji 5-1-1, Chuo-ku, Tokyo 104, Japan

Eike Noack Institut für Pharmakologie, Heinrich-Heine-Universität Düsseldorf, Moorenstrasse 5, D-4000 Düsseldorf 1, Germany

Bassam Omar Webb-Waring Lung Institute, University of Colorado, Denver, Colorado 80262, USA

Lester Packer Department of Molecular and Cell Biology, University of California, Berkeley and Membrane Bioenergetics Group, Lawrence Berkeley Laboratory, Berkeley, CA 94720, USA

Eva Paul Ludwig Boltzmann Institute for Experimental and Clinical Traumatology, Donaueschingenstrasse 13, A-1200 Vienna, Austria

Hans J. Prochaska Department of Pharmacology and Molecular Sciences and Department of Medicine, The Johns Hopkins University School of Medicine, Baltimore, Maryland 21205, USA

H. Redl Ludwig Boltzmann Institute for Experimental and Clinical Traumatology, Donaueschingenstrasse 13, A-1200 Vienna, Austria

Rudolf Jörg Schaur Institute of Biochemistry, University of Graz, Schubertstrasse 1, A-8010 Graz, Austria

Günther Schlag Ludwig Boltzmann Institute for Experimental and Clinical Traumatology, Donaueschingenstrasse 13, A-1200 Vienna, Austria

Alex Sevanian Institute for Toxicology, University of Southern California, Los Angeles, California, USA

Abraham Spector Biochemistry and Molecular Biology Laboratory, Department of Ophthalmology, College of Physicians and Surgeons of Columbia University, New York, NY, USA

Roger Spragg Department of Medicine, Pulmonary and Critical Care Division, University of California, Medical Center, San Diego, CA, USA

Roland Stocker The Heart Research Institute, 145 Missenden Road, Camperdown, Sydney, NSW 2050, Australia

Gisela Storz Division of Biochemistry and Molecular Biology, University of California, Berkeley, CA 94720, USA

Paul Talalay Department of Pharmacology and Molecular Sciences, The Johns Hopkins University School of Medicine, Baltimore, Maryland 21205, USA

Louis A. Tartaglia Division of Biochemistry and Molecular Biology, University of California, Berkeley, CA 94720, USA

M. Thelen Theodor-Kocher Institute, University of Bern, P.O. Box 99, CH-3000 Bern 9, Switzerland

Gisa Tiegs Faculty of Biology, University of Konstanz, Postfach 5560, D-7750 Konstanz, Germany

Michael A. Trush Department of Environmental Health Sciences, Johns Hopkins School of Hygiene and Public Health, 615 N. Wolfe Street, Baltimore, Maryland 21205, USA

R. M. Tyrrell ISREC, Ch. des Boveresses, CH-1066 Epalinges, Switzerland

Volker Ullrich Fakultät für Biologie, Universität Konstanz, Postfach 5560, D-7750 Konstanz, Germany

Fulvio Ursini Department of Biological Chemistry, University of Padova, Via F. Trieste 75, I-35100 Padova, Italy

Albrecht Wendel Faculty of Biology, University of Konstanz, Postfach 5560, D-7750 Konstanz, Germany

D. M. Ziegler The Clayton Foundation, Biochemical Institute, Department of Chemistry and Biochemistry, The University of Texas at Austin, Austin, Texas 78712, USA

Helmward Zollner Institute of Biochemistry, University of Graz, Schubertstrasse 1, A-8010 Graz, Austria

Preface

Oxidants and antioxidants have, in recent years, attracted widespread interest in many scientific disciplines, ranging from free radical chemistry to biochemistry, biology and medicine. Life on this planet utilizes oxygen and oxygen metabolites in energy conversion, and it has become clear that constant generation of pro-oxidants, for example, oxygen free radicals, is an essential attribute of aerobic life. This challenge is met by a system of antioxidants which maintain a steady state; a disturbance in the prooxidant/antioxidant system has been defined as oxidative stress.

Since the publication of *Oxidative Stress* in 1985 there have been fascinating scientific developments. The present book attempts to highlight some of these. Interest now ranges from geochemistry (radicals in the atmosphere) to clinical conditions (sepsis, shock, multiorgan failure). The molecular biology of adaptation to oxidative challenge and the ensuing damage to DNA and other macromolecules has made fast progress. The nature and activity of antioxidants such as vitamins (tocopherols, carotenoids) and micronutrients has come into scientific focus. This is a topic for the future: epidemiological and experimental evidence suggests close links between oxidants and major degenerative diseases as well as cancer and aging.

This treatise focuses on the major topics in this bustling area, and the individual chapters are intended as in-depth accounts of current knowledge. My thanks go to the authors for their excellent contributions and to the scientific community at large as represented under the umbrella of the Society for Free Radical Research.

Helmut Sies
Düsseldorf

Oxidative Stress: Introduction

The exposure of living organisms to reactive oxygen species, notably oxygen free radicals and hydrogen peroxide, is associated with the very fact of aerobic life. Notions that the challenge comes from external noxious sources such as ionizing radiation, toxins, drugs, chemicals, environmental pollutants are correct, but it is equally true that every cell in the living organism can generate reactive oxygen species as well, and some cell-types are even specialized to do so continually or in the form of an 'oxidative burst'. Therefore, there is no qualitative difference in this regard between Claude Bernard's *milieu interne* and the *milieu externe*: our own metabolism produces free radicals even under healthy conditions, and one may be surprised to find that these attacking species have a diversity of useful effects. An interesting result of studies has been that oxygen free radicals can be generated under the control of stimuli and signal molecules; a veritable network of functions is being uncovered here. One recent highlight has been the elucidation of the biology of the nitric oxide radical.

OXIDATIVE STRESS

'Oxidative stress' was defined as a disturbance in the prooxidant-antioxidant balance in favour of the former (Sies, 1985). As there has been a proliferation of publications using this term, a few cautionary words may be appropriate. How can it be defined operationally? The answer to this question is somewhat arbitrary. Considering the normal (healthy) state, oxidative challenge occurs in many cell-types, but this alone does not constitute an oxidative stress. Likewise, a simple loss of antioxidant as resulting from limited nutritional supply is not sufficient. However, when there is an increased formation of prooxidants such as hydrogen peroxide, which is accompanied by a loss of glutathione caused by the formation of glutathione disulphide, we approach a definition. Even a severe loss of antioxidant may, however, still mean that there is no resulting damage. A useful definition of 'oxidative stress' therefore would be: a disturbance in the prooxidant–antioxidant balance in favour of the former, leading to potential damage. Such a definition would incorporate damage products as indicators of oxidative stress (Sies, 1986), and accordingly this area has been the subject of much research with damaged DNA bases, protein oxidation products and products of lipid peroxidation being examined as indicators of oxidative stress (Hayaishi et al, 1989; Pryor and Davies, 1990; Pryor and Godber, 1991).

OCCURRENCE AND NATURE OF REACTIVE OXYGEN SPECIES

The diversity of prooxidants is shown in Table 1 (Pryor, 1986). The vast difference in the half-life, from nanoseconds for the hydroxyl radical to seconds for peroxyl radicals or nitric oxide, underscores the variety of defence problems. There are several levels of defence: prevention, interception and repair. Clearly, defence against the highly reactive hydroxyl radical can only be by prevention because any agent capable of interception would have to be present at very high concentration; experimentally, more than 0.1 M mannitol is required (Czapski, 1984). The superoxide anion radical is dismutated to oxygen and hydrogen peroxide (McCord and Fridovich, 1969). The metabolism of hydroperoxides is taken care of by powerful enzymes, catalases and other hydroperoxidases (Chance et al, 1979). Interception is the domain of chain-breaking antioxidants, such as tocopherols, reacting with peroxyl radicals. Repair, in general, occurs via enzymatic mechanisms and evolution has provided sets of complementary enzyme systems, in some cases, under concerted control. In view of these powerful repair activities, the oxidative stress as detected by the accumulation of products might be underestimated. Therefore, an increased flux through repair pathways should be taken into consideration in assessments of biological oxidative challenge.

These few words already indicate that an adequate assessment of the occurrence and activity of free radicals in biological systems is difficult. In particular, the widespread use of measurements of malondialdehyde and thiobarbituric acid-reactivity as diagnostic indices of lipid peroxidation and peroxidative tissue injury requires caution and correlative data (Janero, 1990); in fact, it is scientifically unsound to equate increased plasma or serum levels of thiobarbituric acid-reactive material alone with the occurrence of a 'free radical disease'.

The occurrence of reactive oxygen species is not limited to biological settings. Atmospheric radicals and their relationship to pollutants and the consequences of the greenhouse effect and global warming are discussed by

Hippeli and Elstner (this volume). Further, the geochemical occurrence of hydrogen peroxide is astounding. Rainwater can contain about 10 μM hydrogen peroxide; several orders of magnitude greater than the concentration in surface seawater (Cooper et al, 1987). If calculated for oceanic surfaces, the upper layers of seawater probably constitute the largest pool of hydrogen peroxide. The principal source of hydrogen peroxide is thought to be the photooxidation of dissolved organic matter (referred to as humic substances) and this explains the large variations in concentrations in space and time. Interestingly, these oceanic reactive species may play a major role in the degradation of many pollutants in surface waters (Cooper et al, 1989).

Table 1 Estimate of the half-lives of oxy-radicals and related species (Modified from Pryor, 1986)

Radical	Substrate[a]	Concentration[b] (mM)	Half-life (s) (at 37°C)
$HO^\bullet$	LH[c]	100	10^{-9}
$RO^\bullet$	LH	100	10^{-6}
$ROO^\bullet$	LH	1	7
$L^{\bullet}$[d]	O_2	2	10^{-8}
H_2O_2	e	—	e
$O_2^{-}\bullet$	e	—	e
$^1O_2$	H_2O	Solvent	10^{-6}
$Q^{-}\bullet$	e, f	—	Days
$NO^\bullet$	g	—	1–10

[a] Substrate chosen as representative of typical reactive target molecules for the species in the first column.

[b] A concentration of the substrate that is meant to approximate the sum of all reactive species in the vicinity of the radical and that is chosen to reflect the selectivity of the radical.

[c] Linoleate.

[d] • is the linolenyl radical. The reversibility of the $L^\bullet$ O_2 reaction has been neglected. An oxygen concentration is used that is typical of moderately oxygenated tissue.

[e] Enzymatic.

[f] The cigarette tar free radical.

[g] Several biological targets.

Photochemical generation of reactive oxygen species is a rapidly developing field, and the chapter by Tyrrell focuses on UVA effects. With the advent of laser techniques applicable to biology, a whole new field of photodynamic therapy has emerged, applying suitable generated reactive species for targeted therapeutic purposes. There have been major advances in the chemistry and photochemistry of nucleic acids (Steenken, 1989; Cadet and Vigny, 1990).

The diversity of reaction products in oxidative stress is illustrated by the results obtained by employing novel techniques (Packer and Glazer, 1990).

In DNA, the occurrence of 8-hydroxydeoxyguanosine has been highlighted (Floyd, 1990; Kasai and Nishimura, this volume), but assays of oxidative DNA damage using selected-ion mass spectrometry reveal a plethora of further products that also include DNA-protein crosslinks (Dizdaroglu, 1991). We can expect interesting novel biological functions in this area. Likewise, novel products of protein oxidation and their relationship to proteolysis during aging and oxidative stress have been discovered (Pacifici and Davies, 1990; Stadtman and Oliver, 1991). The degradation of proteoglycans by hydrogen peroxide (Roberts et al, 1987) and the release of hyaluronate from cells by radical-induced degradation (Prehm, 1990) are phenomena of biological significance in the control of proliferation and in aging. Finally, lipid peroxidation products are now well-established as biologically active compounds, exemplified by the chapter by Zollner et al (this volume) on 4-hydroxyalkenals.

ANTIOXIDANT DEFENCE

The genetic control of antioxidant defence has been elucidated with microorganisms (Demple and Levin; Tartaglia et al, this volume), and the striking relationships to other adaptive responses such as the heat shock response (Donati et al, 1990) underline the central functions in evolution and development. There is evidence for induction of transcription of the protooncogenes, *c-fos* and *c-myc*, in mouse epidermis cells by oxidative stress (Crawford et al, 1988) and for redox regulation of Fos and Jun proteins on DNA-binding activity (Abate et al, 1990). Thus, oxidants are implicated in alterations in gene expression and have potential significance in differentiation and development (Allen, 1991).

Another fascinating aspect of antioxidant defence is related to a number of antioxidant vitamins and micronutrients (Machlin, 1991). The tocopherols in the lipid phase protect membranes and lipoproteins (Diplock et al, 1989; Sies and Murphy, 1991; Gebicki et al, this volume; Stocker and Frei, this volume). Other nutritive antioxidants are the carotenoids (Krinsky, 1990; DiMascio et al, 1991), a topic that deserves attention because of epidemiological evidence in cancer prevention. Thus, leafy green and yellow vegetables are of interest in this respect, but also in plant physiology itself, the carotenoids fulfil essential roles in preventing harmful photooxidation reactions: what keeps plants from getting sunburnt? There would be no photosynthesis were it not for the presence of carotenoids in the photosynthetic reaction center. Also, we should not forget the other side of the coin: plants contain carcinogens, and dietary pesticides are predominantly of natural origin (Ames, 1983; Ames et al, 1990).

The realm of food science and technology encompasses the use of antioxidants. The metabolism of some of the phenolic antioxidants (see Kahl as well as Trush and Kensler, this volume) generates potentially harmful reactive species, so that currently there is an increased search for the employment of natural antioxidants in food processing, food preservation and food colouring.

Health food, to a large extent, focuses on protective functions by biological antioxidants. An adequate assessment of the daily needs (Anonymous, 1989a,b), the definition of a 'prudent diet' and the discussion on the benefits and risks of dietary supplements (Pryor, 1987) all underline the need for appropriate basic research in this area. As presented in the chapter by Stocker and Frei (this volume), there are multiple layers of endogenous antioxidant defence in human blood plasma and similar analyses for cells in tissues and specialized organs are required in the future.

The minds of numerous medicinal chemists and pharmacologists have been applied to research and development of compounds exerting antioxidant effects, in the hope of controlling oxidative stress in a number of human diseases (Rice-Evans, 1989; Emerit et al, 1990). An area of particular interest emerged in the prevention of arteriosclerosis, as studied by the effects of probucol, a phenolic antioxidant (Parthasarathy et al, 1986; Gebicki et al, this volume). Longterm protection is required in modulating degenerative diseases (arthritis, cataract, diabetes, arteriosclerosis, cancer, etc) as well as the aging process itself (Harman, 1981).

PROCESSES AND CELL RESPONSES

Numerous molecular and cellular processes involve reactive radical intermediates (Slater, 1985; Witmer et al, 1991). While the role of free radicals in carcinogen activation (Trush and Kensler, this volume) and the neutrophil respiratory burst (Babior et al, 1973; Baggiolini and Thelen, this volume) have been studied for some time, the role of oxygen radicals in vasodilation has been one of the more recent additions to this field. Starting out from the discovery by Furchgott and Zawadski (1980) that endothelium produces a 'relaxing factor' (EDRF, endothelium-derived relaxing factor) of short biological half-life (see Table 1), and the assignment to NO, nitric oxide, by Palmer et al (1987) to this function, there has been a veritable explosion of knowledge in this area, unravelling a novel biochemistry of arginine and explaining the long-known pharmacological effects of nitrovasodilators (Noack and Murphy, this volume).

Otto Warburg (1908) discovered that sea urchin eggs consume oxygen after fertilization. This respiratory burst and the associated oxidative reactions of cross-linking involving hydrogen peroxide and a secreted peroxidase present an interesting biological example of oxidative stress used for a biological process (Shapiro, 1991). Ovothiols serve as protective agents (Turner et al, 1988).

TOWARDS CLINICAL MEDICINE

From the outset of these introductory remarks, assertions have been made that reactive oxygen species may be critically involved in human health and disease (see Halliwell and Gutteridge, 1989, 1990). Clearly, the numerous attempts in recent years to successfully apply antioxidant therapy have generated less than satisfactory results. One area that was greeted enthusiastically in clinical medicine is that of reoxygenation injury, or 'ischaemia–reperfusion', and the role of xanthine oxidase, as made popular by McCord (1985). As discussed by Omar et al (this volume), the initial euphoria has been dampened but in certain organs and clinical settings the ischaemia–reperfusion problem has potential. One basic problem in the transposition of biochemistry and pharmacology of reactive oxygen to clinical medicine is the fact that diseases under scrutiny are multifactorial. For example, the adult respiratory distress syndrome (ARDS), multi-organ failure (MOF), and traumatic or haemorrhagic shock include components amenable to antioxidant therapy and protection, but other factors such as proteinase activation are also of cardinal importance (Redl et al, this volume). Even an activation of HIV-1 by oxidative stress has been reported (Legrand-Poels et al, 1990) and intracellular thiols regulate activation of nuclear factor kappa-B(NF-κB) and transcription of human immunodeficiency virus (Staal et al, 1990; Schreck et al, 1991). Interleukins, tumour necrosis factor, and other cytokines are key players in this field (see Wendel et al, this volume).

Organs such as the eye lens (Spector, this volume), the skin (Fuchs and Packer, this volume) or the lung apparently have direct relations to oxygen species, but certainly other organs in the body are open to oxygen radical attack, given the activity of the neutrophils and other cell-types, e.g. macrophages. Even tissue cells have recently been shown to give off superoxide in response to cytokines and other types of stimuli, potentially as a metabolic regulatory signal (Meier et al, 1989, 1990).

Coverage of many more fascinating topics relating reactive oxygen species to biology and medicine is possible now and will increase in the future. The present in-depth accounts of some of the areas will be useful for further understanding and will generate enthusiasm for embarking on research projects. One caveat might be appropriate at the end: critical thought in addition to experimental diligence are required in utilizing methods in the analysis of oxidative stress.

Helmut Sies

REFERENCES

Allen RG (1991) *Proc. Soc. Exp. Biol. Med.* **196**: 117–129.

Ames BN (1983) *Science* **221**: 1256–1263.

Ames BN, Profet M & Gold LS (1990) *Proc. Natl. Acad. Sci.* **87**: 7777–7781.

Anonymous (1989a) Diet and Health: Implications for reducing chronic disease risk. Washington: National Academy Press.

Anonymous (1989b) Recommended Daily Allowances, 10th edition. Washington: National Academy Press.

Babior B (1987) *Trends Biochem. Sci.* **12**: 241–243.

Cadet J & Vigny P (1990) In: Morrison, H. (ed.) *Bioorganic Photochemistry* pp. 1–272, New York: Wiley.

Chance B, Sies H & Boveris A (1979) *Physiol. Rev.* **59**: 527–605.

Cooper WJ, Saltzman ES & Zika RG (1987) *J. Geophys. Res.* **92**: 2970–2980.

Cooper WJ, Zika RG, Petasne RG & Fischer AM (1989) In: (Suffet, I.H., MacCarthy, P., eds) *Aquatic Humic Substances: Influence on Fate and Treatment of Pollutants* pp. 333–362, American Chem. Soc.

Crawford D, Zbinden I, Amstad P & Cerutti P (1988) *Oncogene* **3**: 27–32.

Czapski G (1984) *Isr. J. Chem.* **24**: 29–32.

Diplock AT, Machlin LJ, Packer L & Pryor WA (eds) (1989) *Ann. N.Y. Acad. Sci.* **570**.

DiMascio P, Murphy ME & Sies H (1991) *Am. J. Clin. Nutr.* **53**: 194S–200S.

Dizdaroglu M (1991) *Free Rad. Biol. Med.* **10**: 225–242.

Donati YR, Slosman DO & Polla, BS (1990) *Biochem. Pharmacol.* **40**: 2571–2577.

Emerit I, Packer L & Auclair C (eds) (1990) Antioxidants in Therapy and Preventive Medicine, New York; Plenum Press

Floyd RA (1990) *Free Rad. Res. Comm.* **8**: 139 –141.

Furchgott RF & Zawadski JV (1980) *Nature* **288**, 373–376.

Halliwell B, Gutteridge JMC (1989) *Free Radicals in Biology and Medicine*, 2nd ed., Oxford: Clarendon Press.

Halliwell B & Gutteridge JMC (1990) *Meth. Enzymol.* **186**: 1–85.

Harman D (1981) *Proc. Natl. Acad. Sci. USA* **78**: 7124–7128.

Hayaishi O, Niki E, Kondo M & Yoshikawa T (eds) (1989) *Medical, Biochemical and Chemical Aspects of Free Radicals*, Volumes 1 & 2, Amsterdam: Elsevier.

Ignarro LJ, Buga GM, Wood KS, Byrns RE & Chaudhuri (1987) *Proc. Natl. Acad. Sci. USA* **84**: 9265–9269.

Krinsky NI (1989) *Free Rad. Biol. Med.* **7**: 617–635.

Legrand-Peols S, Vaira D, Pincemail J, Van de Vorst A & Piette J (1990) *Aids Res. Hum. Retrovir* **6**: 1389–1397.

Machlin LJ (1991) *Handbook of Vitamins*, 2nd ed., New York: Marcel Dekker.

McCord JM & Fridovich I (1969) *J. Biol. Chem.* **244**: 6049–6055.
Meier B, Radeke HH, Selle S, Younes M, Sies H, Resch K & Habermehl GG (1989) *Biochem. J.* **263**: 539–545.
Meier B, Radeke HH, Selle S, Habermehl GG, Resch K & Sies H (1990) *Biol. Chem. Hoppe-Seyler* **371**: 1021–1025.
Pacifici RE & Davies KJA (1990) *Meth. Enzymol.* **186**: 485–502.
Packer L & Glazer AN (eds) (1990) *Meth. Enzymol.* **186**.
Palmer RMJ, Ferrige AG & Moncada S (1987) *Nature* **327**: 524–526.
Parthasarathy S, Young SG, Witztum JL, Pittman RC & Steinberg D (1986) *J. Clin. Invest.* **7**: 641–644.
Prehm P (1990) *Biochem. J.* **267**: 185–189.
Pryor WA (1986) *Ann. Rev. Physiol.* **48**: 657–667.
Pryor WA (1987) *Free Rad. Biol. Med.* **3**: 189–191.
Pryor WA & Davies KJA (eds) (1990) Oxidative Damage and Repair. *Free Rad. Biol. Med.* **9 (Supplement 1)**.
Pryor WA & Godber SS (1991) *Free Rad. Biol. Med.* **10**: 177–184.
Rice-Evans C (ed.) (1989) *Free Radicals, Disease States and Anti-Radical Interventions*, London: Richelieu Press.
Roberts CR, Mort JS & Roughley PJ (1987) *Biochem. J.* **247**: 349–357.
Schreck R, Rieber P & Baeuerle PA (1991) *EMBO J.* **10**: 2247–2258.
Shapiro BM (1991) *Science* **252**: 533–536.
Sies H (1985) In: Sies H (ed.) *Oxidative Stress* pp. 1–8, London: Academic Press.
Sies H (1986) *Angew. Chem. Int. Ed.* **25**: 1058–1071.
Sies H & Murphy ME (1991) *J. Photochem. Photobiol. B* **8**: 211–224.
Slater TF (1985) *Biochem. J.* **222**: 1–15.
Staal FJ, Roederer M & Herzenberg LA (1990) *Proc. Natl. Acad. Sci. USA* **87**: 9943–9947.
Stadtman ER & Oliver CN (1991) *J. Biol. Chem.* **266**: 2005–2008.
Steenken S (1989) *Chem. Rev.* **89**: 203–520.
Turner E, Hager L & Shapiro BM (1988) *Science* **242**: 939–941.
Warburg O (1908) *Z. Physiol. Chem.* **57**: 1.
Witmer CM, Snyder RR, Jollow DJ, Kalf GF, Kocsis JJ & Sipes IG (eds) (1991) Biological Reactive Intermediates IV, New York: Plenum Press.

OCCURRENCE OF OXIDANTS

1

Oxygen Radicals and Air Pollution

SUSANNE HIPPELI and ERICH F. ELSTNER
Institut für Botanik und Mikrobiologie der Technischen Universität München, Biochemisches Labor, Arcisstr.21, D-8000 München 2

I. Introduction: history and relevant aspects 3
II. Emissions and atmospheric chemistry 4
III. Reactions with biomolecules 17
IV. Symptoms and diseases 29
V. Conclusions and regulatory decisions 44

I. INTRODUCTION: HISTORY AND RELEVANT ASPECTS

Air pollution has been recognized as a worldwide problem as far as direct influences on plants, animals and man are concerned. Global climate changes (Adams et al, 1990) as a result of the so-called greenhouse effect due to anthropogenic gas emissions (Lashof and Ahuja, 1990) and stratospheric ozone depletion as a result of the emissions of chlorofluorocarbons (Prather and Watson, 1990) have been added as another threat. These worrying findings have in common that oxidative processes are the basis for the promotion of emission-driven toxic events. Air pollution research has increased in the past two decades, accompanied by the publication of a wealth of data, e.g. Mudd and Kozlowski (1975), Guthrie and Perry (1980), Smith (1981), Unsworth and Ormrod (1982), Koziol and Whatley (1984), Guderian (1985) and Schulte-Hostede et al (1988). A minireview on oxygen radicals in air pollution has appeared (Elstner and Osswald, 1991).

The general biology of oxygen has attracted much attention, triggered by

Oxidative Stress: Oxidants and Antioxidants
ISBN 0-12-642762-3

the findings of McCord and Fridovich (1969) on the superoxide dismutase activity of erythrocuprein; e.g. Sies (1985), Elstner (1987), Elstner et al (1988) and Rice-Evans (1987).

The fact that aerobic life has to cope with oxygen toxicity was predicted in 1771 by Priestley, who stated that 'pure oxygen has adverse effects on the healthy state of the body'. Likewise, the role of oxidative processes in air pollution was reported in the nineteenth century when the production of acid rain was recognized as an oxidation by atmospheric hydrogen peroxide of sulphur dioxide stemming from coal combustion (Holle 1892). Similarly, chlorophyll bleaching was described as a process catalysed by oxidative enzymes (Woods, 1899). All these reports had been more or less forgotten: we have been greatly surprised in recent times by the overwhelming number of findings that oxygen toxicity is the dominating factor in general toxicology and pathology (Elstner, 1990).

Due to this fact, political activities have launched new research activities, directed by project carriers such as the Health Effects Institute (HEI) in the USA or the Project of the European Research Communities (PEF-Karlsruhe) and the Project on Air Pollution Research (PBWU-München) in Germany. In the present chapter, special attention will be paid to basic redox mechanisms involved in (a) processing of primary air pollutants, (b) chemical interactions between different gaseous and/or particulate pollutants and (c) mechanisms of ultimate toxic effects in living cells from animals and plants.

II. EMISSIONS AND ATMOSPHERIC CHEMISTRY

A. Emissions

Important air pollutants can be subdivided into anthropogenic and non-anthropogenic emissions (Table 1). Primarily emitted compounds react with each other during transport from the site of emission to the site of deposition. Natural and anthropogenic emissions are important for the formation and conversion of pollutants in the atmosphere and may directly damage biological systems and buildings. In contrast, stratospheric reactions are not directly involved in damage of living cells but are indirectly active via influences on the ozone layer and thus the penetration of damaging UV radiation into the biosphere.

1. Natural emissions

Volcanism affects the composition of the atmosphere, causing the emission of dust and gases. Volcanic ashes contain arsenic, mercuric and fluorine compounds, the main gaseous emissions being CO_2, SO_2, HCl and HF. In addition, several organic compounds such as CH_4, aldehydes, ketones, benzene and toluene are emitted.

Table 1. Yearly global emission rates of important air pollutants.

Compound	Emission rate (in 10^9 kg $year^{-1}$)			Main sources
CO_2	830 000	na	800 000	Respiration, biological degradation
		a	30 000	Combustion
CO	3 400	na	—	
		a	3 400	Mainly combustion
Hydrocarbons	1 000	na	930	Trees
(without CH_4)		a	70	Industry, motor vehicles
CH_4	500	na	300	Swamps, rice fields
		a	200	Ruminants
SO_2	400	na	20	Volcanoes
		a	380	Coal and oil combustion
NO_x	160	na	10	Flashes
		a	150	Combustion
NH_3*	35	na	2	Soil, biological degradation
		a	33	Cattle

na, non-anthropogenic; a, anthropogenic.
Data from Fond Chem. Ind. (1987) and *ABEF (1990).

Swamps, sediments and the intestinal contents of animals represent bioreactors forming volatile organic compounds (VOC), especially methane. Blake and Rowland (1986) reported on a worldwide increase in tropospheric methane during 1978–1983. Contributing reasons include increases in the source strengths from cattle and rice fields. The concentrations of CO, CH_4 and $^{\bullet}OH$ in the air are closely coupled, so that higher concentrations of CO or CH_4 should cause decreased concentrations of $^{\bullet}OH$, which in turn should lengthen the atmospheric lifetime of CO or CH_4. In addition, an increased CH_4 source strength may be due to the effect of increasing atmospheric temperatures, partially from the known aqueous biological methane sources such as swamps.

In the USA, natural VOC emissions amount to approximately 28 million metric tons per year, of which 15 million metric tons may come from trees. In comparison, anthropogenic sources emitted approximately 20 million metric tons in 1985 (NAPAP, 1990). Most of these emissions occur during the summer months. The contributions of anthropogenic and natural sources to the total VOC reactivity budget are shown in Figure 1. The mentioned reactivity is based on the rate constants of various compounds with $^{\bullet}OH$; the reactivity budget accounts for the differences in the carbon contents and molecular weights of the compounds. Isoprene and alkenes are the most important reactive VOCs emitted from natural sources, whereas the anthropogenic contribution to the reactivity budget consists of alkanes, aromatics and others. The natural emissions of VOCs are of special interest. Bufalini and Arnts (1981) edited two volumes on atmospheric biogenic hydrocarbons (1981), where the sources of these compounds, their concentration and composition, their photochemical oxidant potential and ozonolysis products have been described.

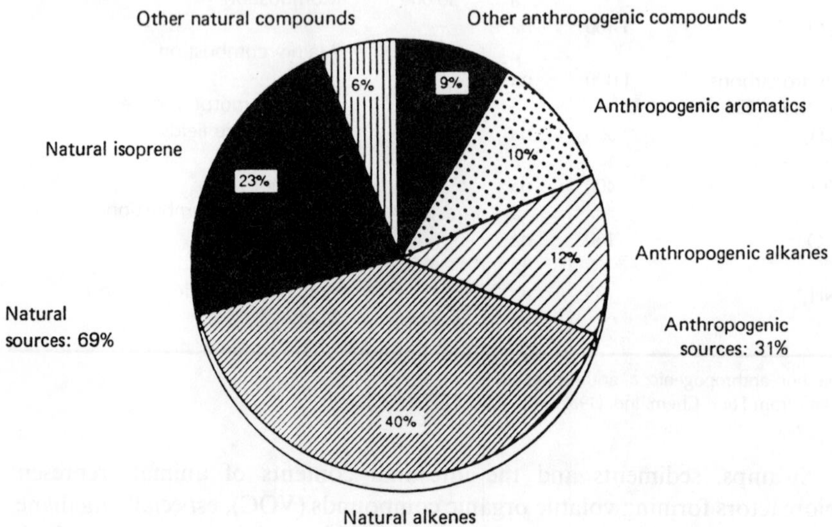

Figure 1. Contributions of anthropogenic and natural sources to total VOC reactivity budget (NAPAP, 1990).

NO_x from natural sources such as soyabeans and other nitrogen-fixing plants may also be evolved to some extent (Klepper, 1990). Natural NO_x emission varies strongly between seasons. During winter, natural sources are not expected to contribute significantly to the atmospheric burden of NO_x, whereas during summer NO_x from natural sources may influence the atmospheric burden in regions remote from the large anthropogenic NO_x sources and, hence, may siginificantly contribute to ozone production in these areas.

2. Anthropogenic emissions

The intensive use of fossil and mineral fuel, mining activities, large man-governed landscapes and cattle are sources of anthropogenic emissions. The

Table 2. Yearly anthropogenic emissions of main air pollutants in West Germany (FRG) (Fond Chem. Industrie, 1987) and the USA (NAPAP, 1990).

	CO		SO_2		NO_x		VOCs	
Producer	FRG	USA	FRG	USA	FRG	USA	FRG	USA
Manufacturing	18.7%		24.0%	14.5%	10.7%	4.6%	49.9%	47.6%
Transportation	59.2%		3.6%	1.9%	57.3%	38.8%	45.2%	42.4%
Utilities	0.6%		62.9%	69.9%	27.7%	33.7%	1.1%	—
Other cumbustion	21.5%		9.5%	14.7%	4.3%	20.9%	3.8%	10.0%
Total emission (million tons)	7.4 = 100%		2.6 = 100%	21 = 100%	3.0 = 100%	18.6 = 100%	1.8 = 100%	20 = 100%

USA, data from 1985; FRG, data from 1984.

yearly emission rates of the main air pollutants (except CO_2) in West Germany and the USA are shown in Table 2.

CO is formed in combustion processes under oxygen-deficient conditions. SO_2 is mainly formed during the combustion of sulphur-containing fuels in power plants. More than 90% of the anthropogenic emitted nitrogen oxides are formed as byproducts of combustion processes. Because of the stable triple bond of nitrogen, NO is mainly formed at high temperatures. Motor vehicles with high compression and melting chamber firing, which is typical for German power plants, have high temperatures in the combustion chambers and therefore have high NO_x emissions. The sources of nitric oxides in the atmosphere at altitudes between 5 and 15 km are tropical thunderstorms and, especially in the Northern hemisphere, emissions from air traffic (Schmailzl, 1990). Manufacturing and transportation are the main contributors to the anthropogenic emission of organic compounds. In Europe, 6.4 million tons of NH_3 were emitted. About 80% of this came from cattle (see ABEF, 1989).

Like the gaseous emissions, particulate air pollutants play an important role. Atmospheric aerosol particles are classified as primary or secondary, depending on their origin. Primary particles are produced by sources such as combustion devices and are imported into the atmosphere in particulate form. Secondary particles are formed in the atmosphere by chemical reactions among primary and secondary gaseous species, primary particles, gases and liquid water. Most of the particulate carbon can be identified as combustion-generated soot (Novakov, 1982). Table 3 shows the yearly soot particle emissions of the main anthropogenic sources in West Germany (Metz, 1989). Especially in the last 15 years, the interest in diesel soot particles has risen, since they have been shown to be carcinogenic (McClellan, 1980; Lewtas, 1982; Schlipköter et al, 1987; Mitchell, 1988). Asbestos fibres as particles with high toxicity play an important role, especially as indoor pollutants.

Table 3. Sources and emissions of soot particles in 1986 (Metz, 1989).

	10^3 t/year	%
Power plants	85	15.4
Industry	360	65.5
Households	44	8.0
Gasoline motor cars	8.8	1.6
Diesel motor cars	10.3	1.9
Lorries	27.7	5.0
Other vehicles	14.2	2.6
Total	550	100

All combustion processes, especially incineration and combustion in motor cars, are possible sources of polychlorinated dioxins, a group of extremely toxic chemicals in the atmosphere. The yearly emission of 2,3,7,8-tetrachlorodibenzo-*p*-dioxin (TCDD) in (West) Germany by incineration amounts to some grams.

B. Reactions in the atmosphere

After emission from natural and anthropogenic sources, primary air pollutants are removed from the atmosphere by sink processes such as wet and dry deposition or chemical and photochemical conversion. The latter processes lead to the formation of secondary air pollutants such as photochemical oxidants and strong acids. The formation of atmospheric acidity is largely driven by photochemistry dependent on the quantum flux densities.

The life cycles of many trace molecules in the troposphere are controlled by emissions of gases from both biogenic sources and man-made activities and reaction with oxidants such as $^{\bullet}OH$ radicals, which in turn originate from ozone photolysis in the moist lower atmosphere.

The main role seen for atmospheric particulates is their function as a medium for physical removal of water-soluble and low-volatility oxidation products. There have been surprises, such as the loss of ozone in the Antarctic lower stratosphere in spring via the reaction with chlorine atoms, originating from photolysis of Cl_2O_2 by near-UV light (Cox, 1990).

1. *Primary photochemical reactions*

The formation of a radical by absorption of a photon by an atom or molecule is called a primary photochemical reaction. Photolysis or photodissociation results in molecule destruction caused by light absorption. The reaction rate of a photochemical process depends on the wavelength and the intensity of light, and the concentrations of the absorbing molecules and their absorption coefficients.

Main primary photochemical reactions are the photolysis of (1) nitrogen dioxide, (2) ozone and (3) formaldehyde by the following reactions:

$$NO_2 + h\nu\ (\leqslant 400\,\text{nm}) \longrightarrow NO^{\bullet} + < O > \tag{1}$$

$$< O > + O_2 \longrightarrow O_3$$

$$O_3 + h\nu\ (\leqslant 310\,\text{nm}) \longrightarrow O_2^*\ (^1\Delta g) + O^*\ (^1D) \tag{2}$$

$$O^*\ (^1D) + H_2O \longrightarrow 2^{\bullet}OH$$

$$H_2CO + h\nu\ (\leqslant 330\,\text{nm}) \longrightarrow H^{\bullet} + HCO^{\bullet} \tag{3}$$

$$H^{\bullet} + O_2 \longrightarrow HO_2^{\bullet}$$

$$HCO^{\bullet} + O_2 \longrightarrow HO_2^{\bullet} + CO$$

$$HO_2^{\bullet} + NO^{\bullet} \longrightarrow {}^{\bullet}OH + NO_2$$

Primary photochemical reactions initiate the atmospheric oxidation of several tracer compounds. The presence of photooxidants in the atmosphere allows the oxidation of compounds which generally in the dark and at low air temperature would react much more slowly.

Main photooxidants are $^{\bullet}OH$ and $HO_2^{\bullet}$. Important processes for the $^{\bullet}OH$/$HO_2^{\bullet}$ radical cycle are the photolytical $^{\bullet}OH$ formation from O_3, the photolytical generation of O_3 deriving from NO_2, and the $^{\bullet}OH$ consumption by oxidation of other compounds such as hydrocarbons and CO.

2. *Chemical lifespan of organic compounds*

The abiotic degradation or reactivity (k) of any compound X is given by the following equation:

$$k = \sum_{i=1}^{n} k_i \cdot c(Y_i) + J_{\text{photo}}$$

where k_i = reaction rate
$c(Y_i)$ = concentrations of the reactants Y_i
J_{photo} = photolysis rate

The reciprocal value k^{-1} describes the chemical lifespan τ_X of the compound X.

The chemical lifespan of X in the troposphere is determined by reactions in the gas phase, by oxidations in clouds and rain droplets and by oxidation on the surface of soil.

Especially in maritime air, reactions with halogenic molecules, atoms and oxides may take place. For most volatile organic compounds degradation processes on the surfaces or within aerosols seem not to play a major role, since gas-phase reactions with $^{\bullet}OH$ radicals, $NO_3^{\bullet}$ radicals and ozone appear to be the main mechanisms of degradation.

Prinn et al (1987) demonstrated that the global rate of loss of methyl chloroform, a long-lived atmospheric man-made pollutant, can be used to deduce appropriately weighted, globally averaged tropospheric concentrations for the major atmospheric oxidant, the hydroxyl radical ($^{\bullet}OH$).

Butane, propene, toluene and dimethylsulphide are typical substances representing alkanes, alkenes, aromatics and organic sulphur compounds, respectively. Table 4 shows representative reaction rates with $^{\bullet}OH$, O_3 and $NO_3^{\bullet}$. The data clearly indicate that butane and toluene are degraded by reaction with $^{\bullet}OH$ radicals, whereas dimethylsulphide mainly reacts with $NO_3^{\bullet}$ radicals and propene with ozone.

Table 4. Contribution of $^{\bullet}OH$, O_3 and $NO^{\bullet}$ to the chemical degradation of organic compounds (Fond d. Chem. Industrie, 1987).

	K^*_{total} $(10^{-6}\,S^{-1})$	Contribution to degradation		
		$^{\bullet}OH$ K_{OH}/K_{total}	O_3 K_{O_3}/K_{total}	$NO_3^{\bullet}$ K_{NO_3}/K_{total}
Butane	1.3	1.00	0.00	0.00
Propene	27	0.56	0.41	0.03
Toluene	3.1	1.00	0.00	0.00
Dimethylsulphide	123	0.02	0.00	0.98

$^*K_{total} = K_{OH} + K_{O_3} + K_{NO_3^{\bullet}}$

3. *Mechanisms of the most important reactions*

The $^{\bullet}OH$ radical seems to play a predominant role in atmospheric chemistry via $H^{\bullet}$ and electron abstraction from various donors. During these reactions primary emissions such as CO, CH_4, SO_2 and NO_2 are oxidized, yielding the corresponding acids. The oxdiation of CO is based on the following reactions:

$$CO + {}^{\bullet}OH \longrightarrow CO_2 + H^{\bullet} \tag{4}$$

$$H^{\bullet} + O_2 \longrightarrow HO_2^{\bullet}$$

$$HO_2^{\bullet} + NO^{\bullet} \longrightarrow {}^{\bullet}OH + NO_2$$

$$\overline{CO + O_2 + NO^{\bullet} \longrightarrow CO_2 + NO_2}$$

In the presence of $NO^{\bullet}$, $^{\bullet}OH$ is regenerated (see also reaction sequence 3). In the absence of $NO^{\bullet}$, the $HO_2^{\bullet}$ radicals will form H_2O_2 and O_2 by disproportionation.

The oxidation of methane is also driven by the $^{\bullet}OH$ radical, and involves a series of reaction steps:

$$CH_4 + {}^{\bullet}OH \longrightarrow CH_3^{\bullet} + H_2O \quad (5)$$

$$CH_3^{\bullet} + O_2 \longrightarrow CH_3O_2^{\bullet}$$

$$CH_3O_2^{\bullet} + NO^{\bullet} \longrightarrow CH_3O^{\bullet} + NO_2$$

$$CH_3O^{\bullet} + O_2 \longrightarrow HCHO + HO_2^{\bullet}$$

$$HCHO + {}^{\bullet}OH \longrightarrow CHO^{\bullet} + H_2O$$

$$CHO^{\bullet} + O_2 \longrightarrow CO + HO_2^{\bullet}$$

$$CO + {}^{\bullet}OH \longrightarrow CO_2 + H^{\bullet}$$

$$H^{\bullet} + O_2 \longrightarrow HO_2^{\bullet}$$

$$3HO_2^{\bullet} + 3NO^{\bullet} \longrightarrow 3{}^{\bullet}OH + 3NO_2$$

$$CH_4 + 4O_2 + 4NO^{\bullet} \longrightarrow CO_2 + 2H_2O + 4NO_2$$

By this sequence, ${}^{\bullet}OH$ radicals are converted into $HO_2^{\bullet}$ radicals, which regenerate ${}^{\bullet}OH$ radicals in the presence of $NO^{\bullet}$. At low $NO^{\bullet}$ concentrations chain-breaking reactions predominate:

$$HO_2^{\bullet} + HO_2^{\bullet} \longrightarrow H_2O_2 + O_2 \quad (6)$$

$$CH_3O_2^{\bullet} + HO_2^{\bullet} \longrightarrow CH_3OOH + O_2$$

If no ${}^{\bullet}OH$ is produced via other processes such as photolysis of ozone, these reactions cause a decrease in the ${}^{\bullet}OH$ concentration and thus the termination of CH_4 oxidation. Therefore, $NO^{\bullet}$ concentrations are decisive for methane degradation.

The oxidation of sulphur dioxide yields sulphuric acid:

$$SO_2 + {}^{\bullet}OH \longrightarrow SO_2\text{—}{}^{\bullet}OH \quad (7)$$

$$SO_2\text{—}{}^{\bullet}OH + O_2 \longrightarrow SO_3 + HO_2^{\bullet}$$

$$SO_3 + H_2O \longrightarrow H_2SO_4$$

$$HO_2^{\bullet} + NO^{\bullet} \longrightarrow {}^{\bullet}OH + NO_2$$

$$SO_2 + O_2 + NO^{\bullet} + H_2O \longrightarrow H_2SO_4 + NO_2$$

Besides this, SO_2 can be oxidized by H_2O_2 in an acid-catalysed reaction. Thus, an alternative to the oxidation of SO_2 in the gas phase is liquid-phase oxidation (Fuhrer, 1985; Mohnen, 1988). It was believed previously that sulphuric acid formation in fog was catalysed by transition metals; however, the uncatalysed oxidation of SO_2 by H_2O_2 or by O_3 seems to be the dominating process, especially outside urban regions in clouds over remote and clean areas (Neftel et al, 1984). The presence of H_2O_2 in cloud water may result from dissolution of gaseous H_2O_2 during cloud formation, or from formation of H_2O_2 inside the cloud droplets (Kelly et al, 1985). Production of H_2O_2 in either the gas or aqueous phase proceeds by reaction of $HO_2^{\bullet}$ radicals (cf. reaction 6).

$$HO_2^{\bullet} + HO_2^{\bullet} \longrightarrow H_2O_2 + O_2 \qquad (8)$$

$$SO_2 + H_2O_2 \longrightarrow H_2SO_4$$

Aqueous-phase chemistry was shown to significantly decrease ozone concentration in the troposphere (Lelieveld and Crutzen, 1990). The efficient scavenging of $HO_2^{\bullet}$ from the gas phase affects O_3, H_2O_2 and HO_x chemistry in the interstitial air. An important reaction under cloudless conditions is the one between $HO_2^{\bullet}$ and $NO^{\bullet}$, which leads to the formation of NO_2 and subsequently of O_3. In clouds, however, this reaction is suppressed.

In contrast to the formation of sulphuric acid via a radical chain, the generation of nitric acid proceeds via a break-off reaction during daytime:

$$NO_2 + {}^{\bullet}OH \longrightarrow HNO_3 \qquad (9)$$

At night NO_2 can react with remaining O_3, generating the nitrate radical, which rapidly reacts with NO_2 to form N_2O_5. The N_2O_5 yields HNO_3 via reaction with water droplets (Fuhrer, 1985).

$$NO_2 + O_3 \longrightarrow NO_3 + O_2 \qquad (10)$$

$$NO_3 + NO_2 \longrightarrow N_2O_5$$

$$N_2O_5 + H_2O \longrightarrow 2HNO_3$$

The neutralization of both acids occurs by reaction with ammonia with the formation of sulphates and nitrates. $(NH_4)_2SO_4$ is one major component of urban hazard dust, markedly decreasing visibility.

4. *Smog*

The word 'smog' was created as a combination of the terms 'smoke' and 'fog', known as 'London'-type smog. Photosmog can build up under strong sunlight irradiation and high temperatures ('inversions'), leading to an increase in photooxidants, typical for 'Los Angeles'. The main differences between 'London' and 'Los Angeles' types of smog are summarized in Table 5.

Table 5. Comparison of London-type and Los Angeles-type smog (Kerr et al, 1976).

London	Los Angeles
Peaks early in morning	Peaks midday
Temperature 30–40°F	Temperature 75–90°F
High relative humidity and fog	Low relative humidity and clear sky
Radiative or surface inversion	Subsidence or overhead inversion
Chemical reducing atmosphere	Chemical oxidizing atmosphere
Bronchial irritation	Eye irritation

Acid smog ('London' smog)

The main reactions of acid smog generation are the catalysed formation of sulphuric acid and, as mentioned above, the uncatalysed oxidation of SO_2 by hydrogen peroxide:

$$SO_2 + H_2O + 0.5O_2 \xrightarrow{\text{catalysts}} H_2SO_4 \tag{11}$$

$$SO_2 + H_2O_2 \longrightarrow H_2SO_4$$

Heavy metal ions present in coal and the surface of soot particles may function as catalysts. Improvements in combustion technology and the use of better-grade fuels have led to the virtual elimination of visible smoke emission and consequently to a considerable decrease of acid smog.

Photosmog ('Los Angeles' smog)

Photosmog was first recognized in the Los Angeles area in the 1940s. The geographical situation of the Los Angeles basin is particularly favourable for temperature inversions and therefore stable air masses. In addition, the exhausts of the intensive motor vehicle traffic in this area are responsible for the accumulation of primary pollutants. Oxy radicals are of outstanding importance for the production of photochemical smog (Kerr et al, 1976). Simulated photoreactions have been conducted to show that reactions between $NO^{\bullet}$ and oxygen seem to proceed via the catalysis of hydrocarbons.

The following equations represent the main reactions of photosmog chemistry:

$$\text{a.}\quad NO_2 + O_2 + h\nu \longrightarrow NO^{\bullet} + O_3 \qquad (12)$$

$$\text{b.}\quad O_3 + NO^{\bullet} \longrightarrow NO_2 + O_2$$

$$\text{c.}\quad RH + {}^{\bullet}OH + O_2 \longrightarrow RO_2^{\bullet} + H_2O$$

$$\text{d.}\quad RO_2^{\bullet} + NO^{\bullet} \longrightarrow RO^{\bullet} + NO_2$$

$$\text{e.}\quad RCH_2O^{\bullet} + O_2 \longrightarrow HO_2^{\bullet} + RCHO$$

$$\text{f.}\quad HO_2^{\bullet} + NO^{\bullet} \longrightarrow {}^{\bullet}OH + NO_2$$

$$\text{g.}\quad NO_2 + {}^{\bullet}OH \longrightarrow HNO_3$$

$$\text{h.}\quad RO_2^{\bullet} + NO_2 \longrightarrow RO_2NO_2$$

where $RO_2^{\bullet}$ may represent the acetic acid radical

$$CH_3COO^{\bullet} + NO_2 \longrightarrow CH_3COONO_2$$

peroxyacetyl nitrate (PAN)

$$\text{i.}\quad HO_2^{\bullet} + HO_2^{\bullet} \longrightarrow H_2O_2 + O_2$$

$$\text{k.}\quad RO_2^{\bullet} + HO_2^{\bullet} \longrightarrow RO_2H + O_2$$

During the irradiation of hydrocarbons, $NO^{\bullet}$ and NO_2 in an air system, the only molecule initially undergoing photodissociation is NO_2 (reaction a). The known reaction rates of oxygen atoms and ozone with hydrocarbons are too slow to account for the rapid removal of the hydrocarbons from the system. Therefore, it is believed that a large part of the hydrocarbon removal is brought about by a free radical chain reaction involving hydroxyl radicals (Kerr et al, 1976). In the presence of $NO^{\bullet}$ this mechanism always leads to the re-formation of the initiating ${}^{\bullet}OH$ radical. The concentrations of the reactants in the photosmog system change with time of day. In the morning ozone photolysis causes the generation of ${}^{\bullet}OH$ radicals. The ${}^{\bullet}OH$ radicals react with hydrocarbons in the ambient air (reaction c). $NO^{\bullet}$ is oxidized to NO_2 with regeneration of ${}^{\bullet}OH$ (reaction f) and photodissociation of NO_2 results in the formation of ozone (reactions a and b), which accumulates since NO_2 photolysis is much faster than ozone photolysis. At midday the ozone concentration is maximal. Most of the $NO^{\bullet}$ is converted into NO_2.

Later on, termination reactions predominate, the most important being the recombination of $^{\bullet}OH$ and NO_2 to form nitric acid (reaction g). If the concentrations of $NO^{\bullet}$ and NO_2 are decreased to a low level, especially in the evening, other break-off reactions may overlap (reactions i and k).

Photochemically produced secondary air pollutants of biological and ecological concern arise from reactions which depend on the presence of nitric oxides and hydrocarbons emitted primarily from motor vehicles. It must be remembered, however, that hydrocarbon emission from vegetation may also be important and initiate photochemical reactions close to the surface of leaves (Figure 2), increasing the O_3 formation. The reaction of ozone with monoterpenes (pinenes, limonene) leads to the formation of H_2O_2. Becker et al (1990) observed an increase in the yields of H_2O_2 in the presence of water vapour as a result of the reaction of the Criegee biradical, the main intermediate of the reaction of ozone with alkenes. The biogenic-hydrocarbon-driven ozone formation and the following ozonolysis of naturally occurring alkenes may be an explanation for the high concentrations of H_2O_2 found in forest areas, especially under conditions of high relative humidity. On the other hand, night-time reactions of monoterpenes with NO_3 radicals lead to very low monoterpene concentrations in ambient air during the early morning (Winer et al, 1984).

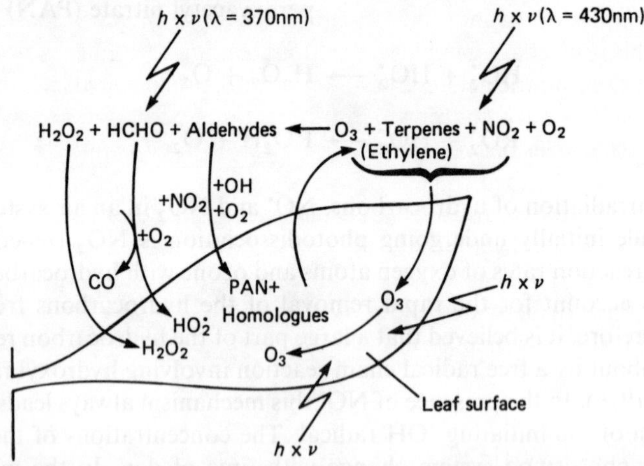

Figure 2. Possible formation of photooxidants (O_3, PAN) and consecutive products near the surface of leaves (Osswald and Elstner, 1986).

Hewitt et al (1990) showed the presence of organic hydroperoxides (ROOH) in leaves of isoprene-emitting plants after ozone exposure. The authors concluded that the reaction of ozone with biogenic alkenes produces toxic ROOH, thus representing one basic mechanism causing damage to plants. This could be particularly important for areas with acidic deposition, where the stability of ROOH is enhanced.

The hydroperoxyl radical plays a key role in stratospheric chemistry through the HO_x catalytic cycle of ozone destruction (Traub et al, 1990). Related to ozone depletion in the stratosphere probably mediated by organohalogens, Blumthaler and Ambach (1990) observed an increased flux of solar UVB radiation (290–330 nm) of about 1% per year (measurements in Switzerland). This might have various consequences, particularly an increased risk of skin cancer by approximately 2% for each 1% of ozone loss (Henriksen et al, 1990).

5. *Heterogeneous atmospheric chemistry*

Heterogeneous or multiphase processes play an important role in the atmosphere (Schryer, 1982). These processes concern: (a) effects of gas-to-particle conversions, (b) development of aerosol size distributions, (c) the role of ions in heteromolecular nucleation, (d) reactions of gases on aerosol particle surfaces, (e) kinetics of reactions between free radicals and surfaces, (f) photoassisted heterogeneous catalysis, and (g) soot-catalysed atmospheric reactions, to list only some.

Considerable work has been carried out to identify the chemical pathways of incorporation of vapour species into atmospheric particles. Of high interest in this respect are gas-phase reactions leading to condensable vapour species such as sulphuric acid, ammonium nitrate, organic acids and nitrates. Additional studies have concentrated on reactions occurring at the surface of, or within, particles between particulate-phase components such as metals and carbon and adsorbed molecules.

As mentioned above, most of the carbon consists of combustion-generated soot. Soot serves as an efficient catalyst for atmospheric reactions such as the oxidation of SO_2 to sulphate (Novakov, 1982). Soot has properties similar to those of activated carbon, which is well known to be both a catalytic and surface-active material.

III. REACTIONS WITH BIOMOLECULES

Since it is almost impossible to differentiate between the individual influences under natural conditions, there is no single influence of just one component of air pollutants on aerobic cells. We now discuss the influences of SO_2, NO_x, ozone, PAN and derivatives and certain particle-bound activities such as soot particles and asbestos fibres.

A. Dark reactions

1. SO_2 and derivatives

SO_2 is produced during combustion of organic materials, especially coal. SO_2 and HSO_3^- are involved in radical-chain processes. HSO_3^- is a reductant which can also accelerate oxidative processes by donating electrons to peroxidic bonds. Principally, SO_2 or HSO_3^- can react with the following types of molecules:

1. Aldehydes and ketones, forming hydroxysulphonates which, in turn, are inhibitors of several enzymes.
2. Alkenes, when sulphonic acids are formed and the double bond is lost.
3. Pyrimidines, with the formation of dihydrosulphonates, which may have mutagenic effects.
4. Disulphides, with the formation of S-sulphonates, where the S—S bridge is split.
5. Superoxide ($O_2^{\cdot-}$) where the very reactive bisulphite radical is formed:

$$SO_2^{\cdot-} + HSO_3^- + 2H^+ \rightleftharpoons HSO_3^{\cdot} + H_2O_2$$

The radical $HSO_3^{\cdot}$ is a powerful initiator of several radical-chain processes such as lipid peroxidation.

Most of the physiological effects can be induced by these primary reactions. None of the above-mentioned (a–e) reactions, however, seems to be solely responsible for the observed toxic effects or symptoms. Of special importance is certainly the reaction with preformed hydroperoxides (see below).

SO_2 is a highly water-soluble gas which is rapidly hydrated, forming sulphite (Peiser and Yang, 1978):

$$SO_2\,(g) \rightleftharpoons SO_2\,(aq) \xrightarrow[pK\,=\,1.76]{H_2O \quad H^+} HSO_3^- \xrightarrow[pK\,=\,7.20]{H^+} SO_3^{2-}$$

Sulphite modifies the cellular energy metabolism in mammalian tissues by decreasing cellular stores of ATP, possibly due to binding and disruption of the function of pyridine and flavin nucleotides (McManus et al, 1989).

Sulphite oxidase, an enzyme absent in human lung tissue, oxidizes sulphite. If not enzymatically metabolized, sulphite will be oxidized via a radical-chain mechanism in the presence of oxygen, generating a series of SO_x^- and oxygen radicals (see below). These radicals can induce lipid peroxidation and subsequent secondary reactions.

In so-called clean air areas, SO_2 is present in concentrations lower than 20 $\mu g\ m^{-3}$; in heavily polluted areas close to coal power plants such as the border between the former DDR and Czechoslovakia, concentrations up to 1200 $\mu g\ m^{-3}$ have been observed.

The conversion factors for SO_2 are as follows: 1 ppm = 2670 $\mu g\ m^{-3}$; ppb = $\mu g\ m^{-3} \times 0.375$; $\mu g\ m^{-3}$ = ppb $\times$ 2.67.

2. NO_x

NO or NO_2 are free radicals by themselves. Due to their unpaired electrons, several typical reactions can be initiated where the attack at alkenic structures (double bonds) are especially characteristic. Since the first reaction seems to be an abstraction of a hydrogen atom, the following reaction can be drawn:

$$NO_2 + \text{—CH=CH—CH}_2\text{—} \longrightarrow HNO_2 + \text{—}\overset{\cdots\cdots\cdots}{\text{CH—CH—CH}}\text{—} \xrightarrow{(O_2)} \text{chain reaction}$$

The resulting acid, HNO_2 may react with amines, forming mutagenic nitrosamines. Also, addition reactions can be observed:

$$NO_2 + \text{—C=C—} \longrightarrow NO_2\text{—}\overset{|}{\underset{|}{C}}\text{—}\overset{|}{\underset{|}{C}}\text{—}^{\bullet}$$

$$NO_2\text{—}\overset{|}{\underset{|}{C}}\text{—}\overset{|}{\underset{|}{C}}\text{—}^{\bullet} + O_2 \longrightarrow NO_2\text{—}\overset{|}{\underset{|}{C}}\text{—}\overset{|}{\underset{|}{C}}\text{—OO}^{\bullet} \longrightarrow \text{chain reaction}$$

The primary product of combustion is NO. Since NO and NO_2 are connected with each other in a very complicated chain of reactions (see above, reactions 1–5), involving ozone and a vast amount of organic molecules (VOCs), referring to concentrations only of NO or NO_2 is not very meaningful (see Section IV).

The conversion factors for NO_2 are as follows: 1 ppm = 1910 $\mu g\ m^{-3}$; ppb = $\mu g \times 0.523$; $\mu g\ m^{-3}$ = ppb $\times$ 1.91.

3. Ozone and peroxides

Ozone is formed from oxygen and NO_x, catalysed by VOCs in a chain of atmospheric events (Section II). Ozone ($^{(-)}O—O^{(+)}=O$) has a standard redox potential of about 2 V and is therefore a very reactive molecule due to its ionic resonance structure. The characteristic chemical reaction of biochemical relevance is the splitting of alkenic bonds and the reaction with thiols. Ozonization of alkenes yields addition products (ozonides), which finally decompose into ketones and the so-called 'Criegee zwitterions', which undergo different intramolecular changes. The resonance structures allow additions of carboxylic acids, alcohols and water, yielding peroxidic products:

$$>C=C< + O_3 \longrightarrow \text{ozonide (C(-O-O-O-)C)} \longrightarrow \left[-C^{(+)}(-O-O^{(-)}) \leftrightarrow >C=\overset{+}{O}-\overset{-}{O} \right] + >C=O$$

ozonide

additions to the 'Criegee zwitterion'

$$>C^{(+)}-O-O^{(-)} + ROH \longrightarrow -C(OOH)(OR)-$$

$$+ RCOOH \longrightarrow -C(OOH)-O-C(=O)-R$$

peroxidic products

$$+ H_2O \longrightarrow >C=O + H_2O_2$$

Thiol oxidation leads to sulphonic acids or disulphides:

$$\underset{\text{cysteine}}{HOOC{-}\underset{\displaystyle NH_2}{\underset{|}{CH}}{-}CH_2{-}SH} \xrightarrow{O_3} \underset{\text{cystin}}{\begin{matrix} CH_2{-}S{-}S{-}CH_2 \\ | \qquad\qquad | \\ H_2N{-}C{-}H \quad H{-}C{-}NH_2 \\ | \qquad\qquad | \\ COOH \qquad COOH \end{matrix}}$$

$$\downarrow O_3$$

$$\underset{\text{cysteine sulphonic acid}}{HOOC{-}\underset{\displaystyle NH_2}{\underset{|}{CH}}{-}CH_2{-}SO_3H}$$

In a similar way to cysteine, glutathione and proteinic SH-groups are oxidized. The oxidation of SH-groups is responsible for inhibition of the lipid synthesis in mitochondria and microsomes: Acyl-CoA-thioesterases and -thiokinases and acyltransferases are blocked; this is detectable as inhibition of the glycerol-3-phosphate acylation (Peters and Mudd, 1982). In a similar way glyceraldehyde-3-phosphate dehydrogenase (GAPDH), a key enzyme of glycolysis, is inactivated, with the oxidation of SH-groups in the active centre being responsible for rapid loss of function. In addition, peripheral SH-groups, tryptophan, histidine and methionine residues are oxidized (Knight and Mudd, 1984). Ozone inactivates α_1-proteinase inhibitor via methionine oxidation (Mohsenin and Gee, 1989), and lysozyme via the conversion of tryptophan to *N*-formyl kynurenine (Dooley and Mudd, 1982).

Since ozone, due to its ionic resonance structure, is more water-soluble than oxygen, hydrophilic compartments may be the main site of reaction. Under acidic conditions ozone is quite stable, while under basic conditions it decomposes into molecular oxygen. At 22°C 100 ppm ozone on a water surface yield micromolar solutions: 10^{-6} moles of ozone are dissolved in 55 moles of H_2O. In the presence of aromatic compounds, O_3 is decomposed into $^{\bullet}OH$-radical-like compounds. Depending on the concentrations of NO, aromatic compounds and high quantum flux density, O_3 can increase up to 400–500 $\mu g\ m^{-3}$. Average concentrations in rural areas during summer days may be 20–80 $\mu g\ m^{-3}$, depending on time of day and altitude. In the summer of 1990 in Germany, episodes with 150–250 $\mu g\ m^{-3}$ were observed quite frequently.

The conversion factors for O_3 are as follows: 1 ppm = 2000 $\mu g\ m^{-3}$, ppb = $\mu g\ m^{-3} \times 0.501$, $\mu g\ m^{-3}$ = ppb $\times$ 2.00.

Among the peroxides relevant to air pollution, PAN and H_2O_2 are of major importance. PAN in some industrial areas and centres of pollution, like the Los Angeles region, may amount to 5–20% of the relevant ozone concentrations. Since PAN is never the single peroxide present in polluted air, other peroxides such as peroxypropionyl, peroxybutyryl and peroxybenzyl nitrate have to be considered. PAN at physiological pH (7.2) has a half life of 4.4 min. It undergoes base-catalysed decay, essentially into acetate and nitrite:

$$CH_3{-}\overset{\displaystyle O}{\overset{\|}{C}}{-}OONO_2 + 2OH^- \longrightarrow CH_3COO^- + NO_2^- + O_2 + H_2O$$

PAN reacts as follows:

1. With thiol groups, especially in enzymes, such as the SH-groups of glucose-6-phosphate dehydrogenase, where the substrates, glucose-6-phosphate and NAD, may prevent SH oxidation. Smaller sulphur-containing molecules are also oxidized, and thus methionine is converted into methionine sulphoxide, and lipoic acid and coenzyme A are converted into the corresponding disulphides. With glutathione one mole of PAN may convert three moles of SH-groups.
2. PAN reacts with NADPH, where the oxidized compounds (NAD^+, $NADP^+$) do not seem to further react with PAN.
3. PAN reacts with alkenes, with the formation of epoxides and further products of lipid peroxidation:

$$\text{PAN} + \text{alkene} \longrightarrow CH_3COO^\bullet + NO_2 + \text{epoxide}$$

$$CH_3COO^\bullet \longrightarrow CO_2 + CH_3^\bullet$$

$$CH_3^\bullet + LH \longrightarrow L^\bullet + CH_4$$

4. Reactions with amines result in acetylation:

$$\text{PAN} + RNH_2 \longrightarrow O_2 + HNO_2 + CH_3{-}\overset{\displaystyle O}{\overset{\|}{C}}{-}NH{-}R$$

The conversion factors for PAN are as follows: 1 ppm PAN = 4370 μg m^{-3}; ppb = μg m^{-3} × 0.223; μg m^{-3} = ppb × 4.37.

An overview of the toxic reactions of the above-mentioned components of air pollution is presented in Table 6.

Table 6. Examples of experimental phytotoxic effects of trace gases and aerosols (Fond. der Chem. Industrie, 1987).

Trace gas aerosols	Reaction partner	Main reaction	Physiological reaction partner	Physiological reaction products
SO_2		To CO_2 competitive inhibition	Carboxylases	Radicals
	Probably thiols	?	Thioredoxine regulation system	
HSO_3^-	Peroxide	Homolytical splitting	Lipid hydroperoxide	Diff. radicals ethane
	Aldehydes	Addition	Pyridoxal phosphate	Addition products (inactivated vitamin)
NO_2	Alkenes	Addition lipids	Unsaturated with nitro group	Peroxides
HNO_2	Amines	Nitrosylation	Amines, amino acids	Alcohols, carboacids
O_3	Alkenes	Addition	Unsaturated lipids	Aldehydes, ketones
	Thiols	Oxidation	Cysteine, glutathione, protein	Cysteine, cysteine-sulphonic acid
Peroxyacetyl nitrate	Thiols	Oxidation	Methionine, cysteine	Methionine-sulphoxide
	Primary amines	Acetylation	Amines, amino acids	O_2, HNO_2, acyl amino compounds
	Aryles	H-abstraction and reaction	Indolacetic acid (growth hormone) $NO_3^{\bullet}$ and O_2	3-Hydroxymethoxyindole (growth inhibitor)

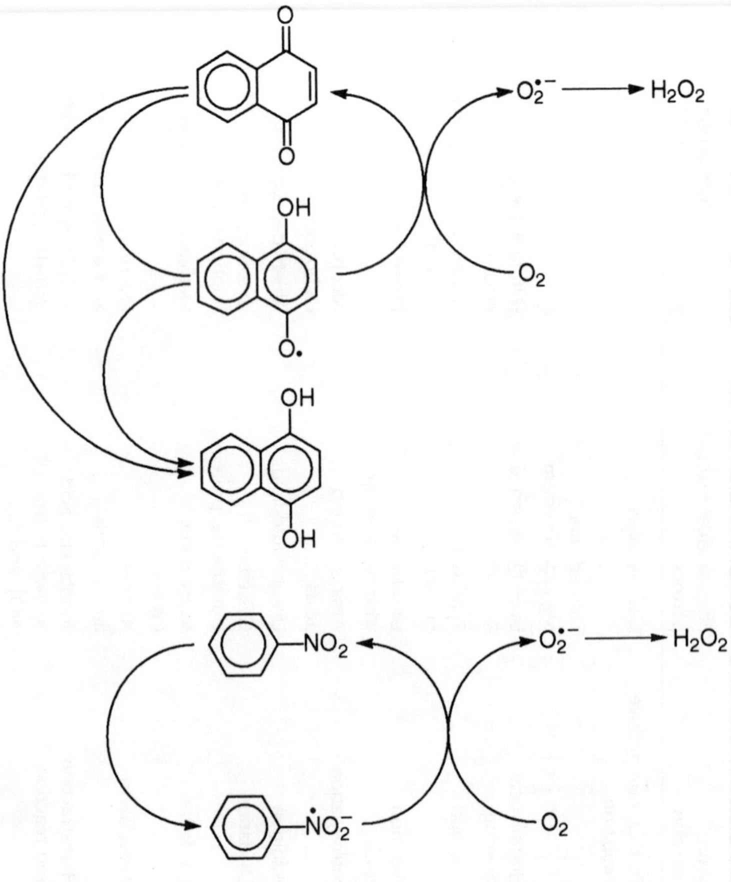

Figure 3. Redox cycling of naphthoquinones and nitroaromatic compounds.

4. Particle catalysis

Airborne particles may be inorganic or organic. A further differentiation may be possible by comparing the reactivities: some particles have no catalytic activity and influence living cells just mechanically by covering surfaces (especially important for plants) or stimulating leukocyte (macrophage) activities in respiratory systems of animals. Particles with catalytic activities such as soot particles or asbestos fibres, in addition to their mechanical or surface-based functions, interfere with metabolism. As catalytically inert particles, quartz particles have been described as activators of alveolar macrophages. While inorganic particles function more or less mechanically or by transition metal catalysis, organic particles may contain both transition metal chelates and redox-cycling organic compounds. In this context the interactions of chelated iron or copper with quinones, such as certain naphthoquinones and nitroaromatic compounds, have been documented by several workers. Diesel soot and other soot particles have been shown to contain naphthoquinones, which undergo redox cycling, as well as nitroaromatic compounds, which, via the nitroxyl radical, are also able to undergo redox cycling, finally forming superoxide and H_2O_2 (Figure 3).

B. Light-driven reactions

Under conditions with limited $NADP^+$ in the stroma of the chloroplasts (i.e. strong light, transport limitations, ATP uncoupling, inhibition of Calvin cycle activities and others), reductive and photodynamic oxygen activation may be expected. In particular, inhibited CO_2-fixation due to stomatal closure or inhibition of the Calvin cycle due to blocked translocation of photosynthetic products can lead to insufficient reoxidation of $NADPH_2$. These conditions result in 'over'-reduction of the electron transport chain and a lack of the potential for charge separation after chlorophyll activation in the light. Thus, inhibited energy transfer supports photodynamic reactions of type II, generating singlet oxygen (Elstner et al, 1985):

$$\begin{aligned} \text{Chl} &\longrightarrow \text{Chl}^* \ (\text{Chl}^* = \text{activated chlorophyll}) \\ \text{Chl}^* &\longrightarrow {}^3\text{Chl} \ ({}^3\text{Chl} = \text{triplet state chlorophyll}) \\ {}^3\text{Chl} + \text{O}_2 &\longrightarrow \text{Chl} + {}^1\text{O}_2 \end{aligned}$$

The electrophilic nature of singlet oxygen results in chemical reactions with double bonds or other electron-rich functions. Singlet oxygen may eventually be quenched by carotenoids or α-tocopherol in the thylakoid membranes or directly react with unsaturated fatty acids (LH), forming hydroperoxides (LOOH) (Knox and Dodge, 1985):

$$\text{LH} + {}^1\text{O}_2 \rightarrow \text{LOOH}$$

Membrane disruption due to lipid peroxidation is a common feature of photodynamic actions. Valenzo (1987) reviewed different photomodifications of biological membranes, with emphasis on singlet oxygen mechanisms. He concluded that effective membrane sensitization usually involves an association of the sensitizer with the membrane. This is especially so in thylakoid membranes of the chloroplasts.

Limited $NADP^+$ availability results also in the formation of $O_2^{\bullet -}$ and H_2O_2 via 'over'-reduction of the electron transport system and subsequent electron channelling to oxygen. Oxygen may be reduced by either the primary acceptor (PA) of photosystem I, by reduced ferredoxin (fd_{red}), or by both (Osswald and Elstner, 1986):

$$fd_{red} + O_2 \longrightarrow fd_{ox} + O_2^{\cdot -}$$

$$fd_{red} + O_2^{\cdot -} + 2H^+ \longrightarrow fd_{ox} + H_2O_2$$

$$PA_{red} + O_2 \longrightarrow PA_{ox} + O_2^{\cdot -}$$

H_2O_2 may be further reduced by reduced ferredoxin, forming oxygen species with properties similar to those of the hydroxyl-radical ($^{\bullet}OH$) (Elstner et al, 1985):

$$fd_{red} + H_2O_2 \longrightarrow fd_{ox} + OH^- + (^{\bullet}OH)$$

Light-dependent generation of reactive oxygen species in the thylakoid membrane is summarized in Figure 4.

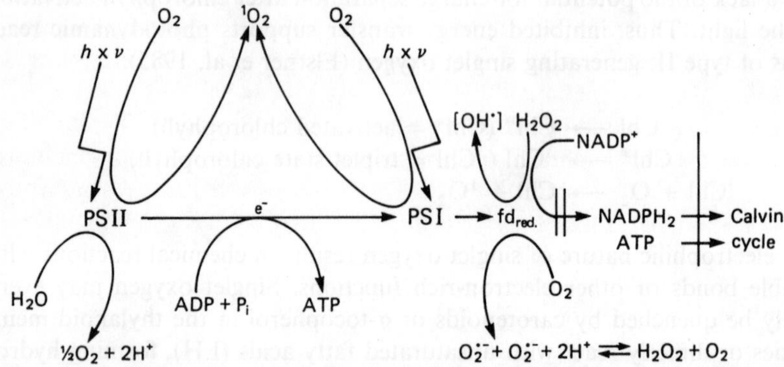

Figure 4. Light-dependent generation of reactive oxygen species in the thylakoid membrane (after Osswald and Elstner 1986).

Reductively and photodynamically generated reactive oxygen species degrade fatty acids in the membranes. The formed hydroperoxides of fatty acids undergo metal-catalysed decay, forming alkoxy radicals ($LO^{\bullet}$), which in turn cooxidize chlorophyll and other pigments, initiating their bleaching (Elstner et al, 1985):

$$LOOH \xrightarrow{Me} LO^{\bullet} + OH^-$$

$$LO^{\bullet} + CHl_{red} \longrightarrow LOH + Chl_{ox}$$

Reactive oxygen species can be detoxified in the chloroplasts. Superoxide dismutation is catalysed by copper–zinc superoxide dismutase (SOD), which is found in the stroma as well as bound to the thylakoids. Since chloroplasts are devoid of catalase, H_2O_2 detoxification is brought about by the ascorbate ($AscH_2$/Asc) and glutathione (2GSH/GSSG) redox cycle at the expense of $NADPH_2$ ('Beck–Halliwell–Asada' cycle; see Elstner et al (1985)):

$$H_2O_2 + AscH_2 \xrightarrow{\text{Ascorbate peroxidase}} Asc + 2H_2O$$

$$Asc + 2GSH \longrightarrow AscH_2 + GSSG$$

$$GSSG + NADPH_2 \xrightarrow{\text{Glutathione reductase}} 2GSH + NADP$$

The whole process not only detoxifies H_2O_2 but also decreases the $NADPH_2$/ATP ratio. Thus, lipid peroxidation and pigment cooxidation due to the generation of reactive oxygen species only take place when the defence system of the chloroplast is overcharged. Furthermore, accumulation of H_2O_2 in the stroma inhibits certain Calvin cycle enzymes (generally the phosphatases). Uncoupling of oxidative phosphorylation results in insufficient reoxidation of $NADPH_2$ and thus causes inhibition of photosynthesis in the light due to a lack of ATP.

Plant damage caused by SO_2 is oxygen- and light-dependent. Typical SO_2 effects are the rapid inhibition of photosynthesis before direct symptoms such as bleaching of chlorophyll can be observed. SO_2 or its aqueous solution, sulphite (HSO_3^-), can be oxidized in the presence of oxygen via a radical-chain mechanism (Peiser and Yang, 1978):

$$\text{Initiation: } SO_3^{2-} + O_2 \longrightarrow SO_3^{\bullet -} + O_2^{\bullet -}$$

$$\begin{aligned} \text{Propagation: } SO_3^{\bullet -} + O_2 &\longrightarrow SO_5^{\bullet -} \\ SO_5^{\bullet -} + SO_3^{2-} &\longrightarrow SO_4^{-} + SO_4^{2-} \\ SO_4^{\bullet -} + SO_3^{2-} &\longrightarrow SO_4^{2-} + SO_3^{\bullet -} \\ SO_4^{\bullet -} + OH^- &\longrightarrow SO_4^{2-} + OH^{\bullet} \\ OH^{\bullet} + SO_3^{2-} &\longrightarrow OH^- + SO_3^{\bullet -} \end{aligned}$$

$$\begin{aligned} \text{Termination: } OH^{\bullet} + SO_3^{\bullet -} &\longrightarrow OH^- + SO_3 \\ SO_3^{\bullet -} + O_2^{\bullet -} &\longrightarrow SO_3 + H_2O_2 \\ O_2^{\bullet -} + O_2^{\bullet -} &\longrightarrow H_2O_2 + O_2 \end{aligned}$$

The initiation can be started by metal ions, UV irradiation or enzymatic reactions. Free radicals generated during sulphite autoxidation attack cellular membranes via lipid peroxidation and subsequent chlorophyll bleaching.

In addition, lipid peroxidation is initiated by sulphite radicals generated during the reaction between sulphite and superoxide (Peiser and Yang, 1978):

$$SO_3^{2-} + O_2^{\cdot -} + 2H^+ \longrightarrow SO_3^{\cdot -} + H_2O_2$$
$$SO_3^{\cdot -} + LH \longrightarrow HSO_3^- + L^\cdot$$
$$L^\cdot + O_2 \longrightarrow LOOL^\cdot$$
$$LOO^\cdot + LH \longrightarrow LOOH + L^\cdot$$

The bleaching of chlorophyll by lipid peroxidation is only observed if HSO_3^- is simultaneously present. Neither LOOH nor HSO_3^- alone can destroy the green colour of chlorophyll (Figure 5). The generated H_2O_2 is thought to be responsible for the inactivation of ascorbate peroxidase, glutathione reductase and enzyme such as the phosphatases of the Calvin cycle. SO_2 inhibits the translocation of photosynthetic products via a disturbed phloem loading. This effect is also observed after ozone exposure of higher plants and is associated with an inactivation of the plasmalemma-bound ATPase (Dominy and Heath, 1985). Secondary events lead to an increased starch accumulation and finally bleaching of the photosynthetic pigments. Ozone further acts as an effective uncoupler of photophosphorylation (Robinson and Wellburn, 1983). Studies with isolated chloroplasts showed that ozone blocks the energy transfer from the primary light acceptors to the reaction centres, initiating photodynamic reactions of type II as mentioned above. The significance of this observation remains to be established in vivo, since ozone can probably not penetrate the outer chloroplast membrane.

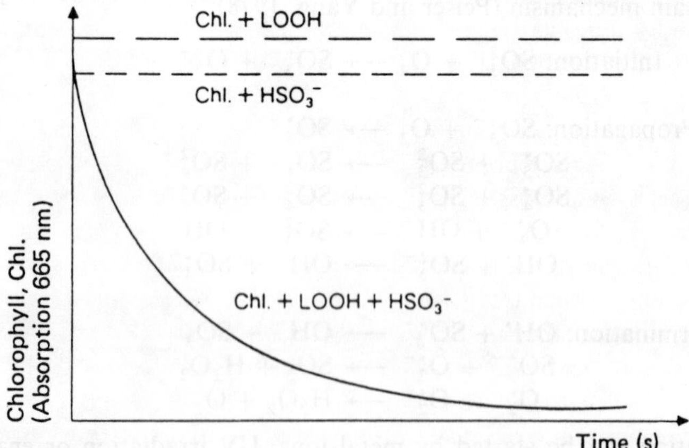

Figure 5. Cooperative effects of sulphite and hydroperoxide in pigment bleaching (Peiser and Yang, 1978).

Aben et al (1990) showed that low-level ozone (120 μg m^{-3} at 8 h per day over two weeks) can have a direct effect on the stomata and photosynthetic system, causing decreased photosynthesis and stomatal conductance.

In contrast, the high toxicity of NH_3 results from early penetration of biomembranes and uncoupling of photophosphorylation (ABEF, 1989). This may explain yellowing and necrosis of the outer leaf surface of plants after treatment with NH_3 (1–3 mg m^{-3}).

IV. SYMPTOMS AND DISEASES

A. Plants

1. *Changes of the surface*

As already mentioned in Section II, natural emissions of unsaturated VOCs from plant surfaces may react with oxidants from the atmosphere. The reaction of the plant stress hormone ethylene with ozone or other components of air pollution may produce H_2O_2 and formaldehyde:

$$C_2H_4 + O_3 + H_2O \rightarrow 2HCHO + H_2O_2$$

These compounds, together with acid mist and aerosols may attack the outer surface of plants, damaging in particular the cuticular waxes (Elstner et al, 1985).

Stomata of young intact needles of *Pinus sylvestris* are covered by tubular wax crystals. Damaged two-year-old needles show only a few crystalline wax fibres. In addition, nearly 50% of the stomata of these damaged needles are completely filled with particles, identified as fly ash (Hafner et al, 1989). Masuch et al (1986) reported a decreased resistance to drought after treatment with H_2O_2-containing acidic fog of young trees.

Leaves from heavily polluted areas exhibit higher levels of transpiration and enhanced water loss. Simultaneously, changes in wax structure have been observed (Poborski, 1988). Application of acidic fog (pH 2.75) resulted in a disintegration of the epicuticular wax layer and of the epistomatal wax props of young spruce trees (Mengel et al, 1990; Magel and Ziegler, 1986), indicating a disturbance in the water status of plants.

SO_2, O_3 and acid rain are known to increase the rate of leaching of mineral ions, such as potassium and calcium. The latter plays an essential role in the buffering of H^+ and the control of cuticular permeability (Poborski, 1988). Chronic injury of leaves via photodynamic processes, photooxidants and/or dry acidic depositions may result in a partial loss of structural resistance. Under these conditions special pathogens, which otherwise would not be successful, cause infections of plants. This 'secondary' effect of air pollution leads to new symptoms such as necrosis, which may overlap with the primary symptoms. The correlation of necrotic events and fungal infection has been demonstrated recently by Osswald and Elstner (1986), using ergosterol, a sterol specific to fungi, as a biochemical indicator for fungal infection. Figure 6 shows an inverse linear correlation between ergosterol and chlorophyll contents of spruce needles collected from different areas.

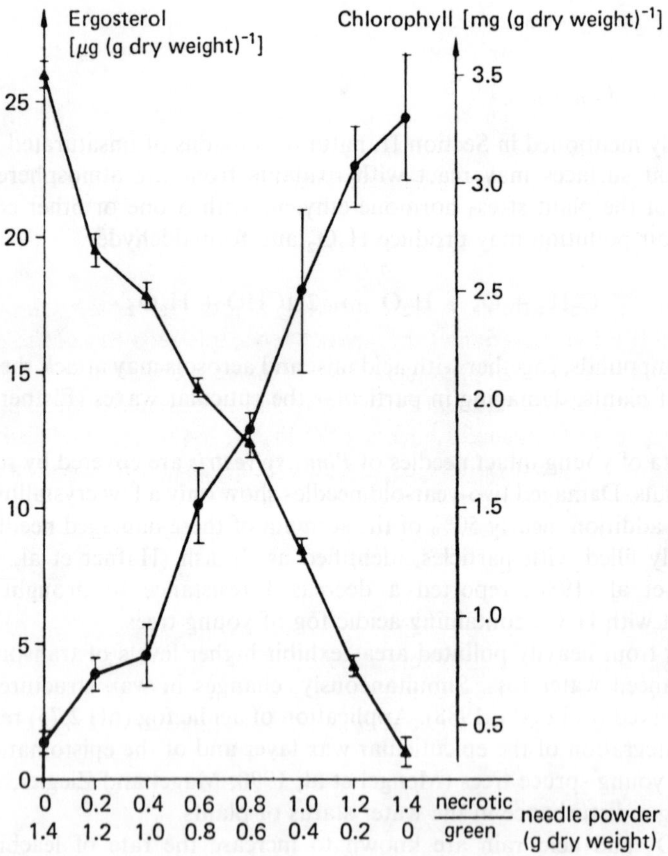

Figure 6. Ergosterol standard curve of differently infected spruce needle samples. The values represent means of six measurements. ●, Ergosterol; ▲, chlorophyll (Osswald et al, 1986).

The change in the outer surface of plants, which is the ultimate defence barrier against pathogenic invaders, may therefore change host–parasite relationships and thus render plants less resistant to infectious diseases.

2. *Effects on physiological processes*

Air pollutants and especially their combinations (review: Darrall, 1989) show different effects on all aspects of plant metabolism, especially concerning photosynthesis, stomatal aperture, carbon partitioning, respiration and antioxidant systems. O_3 in combination with SO_2 and/or NO_2 proved to be particularly noxious. NO_2, which per se is regarded as not being injurious to plants at concentrations measured in remote areas, increased the damage caused by O_3/SO_2. An inhibition of nitrogen-assimilating enzymes is supposed to be responsible for this synergistic effect (Guderian et al, 1988).

In addition to combination effects, a complication in evaluating the physiological mechanism of pollution injury has to be envisaged since such diverse factors as light, water, temperature and mineral nutrition influence the responses of plants to pollutants.

Effects on photosynthesis

The lowest concentrations of SO_2 showing effects on photosynthesis are 200–400 ppb. After cessation of the stress a very rapid recovery of the plants is observed. Depression of photosynthesis by ozone is observed with 100 ppb for 1 h or intermittent treatment for 170 h within three weeks with 35 ppb. In contrast to these influences, effects of NO_2 or NO are only visible at extremely high concentrations of about 500 or 700 ppb. These concentrations, however, are rarely obtained in nature. Combinations of SO_2 and O_3 or O_3 and NO_2 decrease the above-mentioned thresholds; in certain cases they increase the amount of damage. At very low concentrations of SO_2 an increase in photosynthesis is actually observed. This may be due to the reactions of HSO_3^- with superoxide intermittently formed during 'pseudocyclic' electron transport. Under these conditions sulphide is produced, which may reduce sulphur limitations in intermediary metabolism after reduction of sulphate to the cysteine level by photosynthetic electron transport (Elstner, 1987).

Effects on stomatal apertures

Low doses of SO_2, O_3 and NO_x increase stomatal apertures, thus allowing more SO_2 to penetrate the plants. Therefore, besides the deleterious activities of SO_2 and NO_x, increased SO_2 availability may influence plant growth. This does not, however, hold for drought conditions, since increased stomatal apertures cause increased water losses. High doses of SO_2 (close to

2000 ppb) in most cases cause stomatal closure. Again, interactions with water stress induce different responses. Since ozone in plants exclusively penetrates through the stomata, stomatal closure caused by other pollutants may even lower ozone sensitivity. 'The intricate connection between tissue water balance and movement and stomate behaviour plays a key role in the control of pollution injury' (Heath, 1980).

Effects on carbon partitioning

Air pollutants influence the distribution of carbohydrates stemming from photosynthetic events. If less fixed carbon is transported to the roots, in the form of sugars, amino acids or fat precursors, the shoot/root ratio is finally shifted, causing a lack of water supply for the growing plant. Treatment with < 50 ppb O_3 over 40 h decreases carbohydrate transport into the roots within five weeks. This can be measured as a disturbed dry weight balance with decreases in the roots. Assimilates seem to be accumulated in the leaves, and are measurable as a strongly increased concentration of sucrose or starch in the mesophyll cells. This has an influence on the primary events of photosynthesis. Since NADPH can no longer be rapidly reoxidized by Calvin cycle metabolism, the electrons from the photosystems are eventually transferred to oxygen, resulting in an increased hydrogen peroxide content (see Section III.B).

Influence on respiration

Low concentrations of SO_2 and O_3 increase dark respiration in plant cells, while NO_x has little or no effect. Concentrations in the range of 1000 ppb inhibit dark respiration. Altogether, air pollutants in ambient concentrations do not seem to have major effects on mitochondrial respiration in plants.

Effects on antioxidative systems

The antioxidative system of higher plants consists of enzymes, low molecular weight compounds and integrated detoxification chains. Chloroplasts of green plants contain thylakoid-bound as well as soluble (stroma) Cu/Zn SOD and large amounts of catalase in the peroxisomes. Since chloroplasts do not seem to contain catalase, H_2O_2 detoxification of chloroplasts is mainly due to ascorbate peroxidase, coupled to the glutathione system. The so-called 'Beck–Halliwell–Asada' cycle includes peroxidation of ascorbic acid, yielding dehydro- or semidehydroascorbate, which in turn is rereduced by GSH. GSSG is reduced at the expense of NADPH, which is kept in the reduced state by photosynthetic electron-transport (Elstner, 1982, 1987). In addition to this system, α-tocopherol, carotenoids and the xanthophyll cycle cooperate in quenching and scavenging excited states, free radicals and singlet oxygen, whereby lycopene is the most efficient carotenoid singlet oxygen quencher (di Mascio et al, 1989). Experiments by Sakaki et al (1983) showed that gassing of spinach leaves with ozone in the light results in a degradation of chlorophyll and carotenoids as well as in the formation of malondialdehyde. External applications of SOD and ascorbate diminished the toxic effects, indicating the involvement of superoxide.

Treatment of plant material with oxidants such as O_3, H_2O_2 or SO_2 causes an increase in the antioxidative potential. Tanaka and Sugahara (1980) found that SOD activity is increased after SO_2 treatment. Peroxidase (POD) and SOD were increased; ascorbate, however, was decreased when young spruce plants were treated with ozone (Castillo et al, 1987). Similar findings were reported by Mehlhorn et al (1986) and Guderian et al (1988), who treated conifers and other forest plants with SO_2/O_3 combinations. In these cases SOD as well as POD, ascorbate, GSH and α-tocopherol were increased as compared to the controls. In spruce needles ascorbate and GSH inversely correlate with the chlorophyll content (Table 7).

Table 7. Ascorbic acid and glutathione contents correlated with chlorophyll of healthy (h) and bleached (b) spruce needles, collected in the Bavarian Forest (Osswald et al, 1990).

	Year									
	1984		1983		1982		1981		1980	
	h	b	h	b	h	b	h	b	h	b
Chlorophyll	2.2	2.0	2.7	1.7	3.2	0.8	3.3	0.7	2.9	1.4
Ascorbate	2.0	4.4	3.9	5.1	5.6	7.7	4.6	6.2	4.4	7.8
Dehydroascorbate	0.4	1.0	0.3	0.8	0.1	0.3	0.6	0.6	0.2	2.0
GSH	51	84	46	104	84	203	57	120	41	202

Data are given in mg (g dry weight)$^{-1}$.

Ozone damage in the ozone-sensitive tobacco plant mutant Bel W3 may be prevented by treatment with polyamines such as putrescine, spermidine and spermine. Ozone treatment especially seems to inactivate diamine oxidase, while peroxidase is not inhibited (Peters et al, 1989). The function of amines or polyamines is not yet clear, since Bors et al (1989) have shown that these compounds do not seem to act as radical scavengers. It is assumed that diamine oxidase is integrated in the lignification process via formation of H_2O_2, finally yielding a higher degree of resistance. Therefore, a possible function of amines in resistance against air pollutants is indicated. In addition, H_2O_2 seems to be involved as a messenger in the induction of defence reactions in higher plants, yielding, for example, an increase in phytoalexins (Apostol et al, 1989). Keen and Taylor (1975) reported that ozone treatment of soya beans increased the de novo synthesis of the phytoalexins daidzein, cumestrol and soyagole.

Treatment of needles from *Pinus sylvestris* with low ozone doses leads to the induction of phytoalexins. The lignin-biosynthetic enzyme cinnamic alcohol dehydrogenase and several stress proteins in spruce and *pinus* needles were induced by ozone (Sandermann et al, 1990). The authors speculate that ozone functions as a gene inducer for proteins of plant defence reactions.

Ozone exposure of plants causes formation of enhanced levels of oxygen free radicals, mediating formation of 8-hydroxyguanine from chloroplast DNA (Floyd et al, 1989). The in vitro reaction of ozone with DNA did not cause formation of 8-hydroxyguanine, however. It was concluded that the interaction of ozone with plant cells and chloroplast-mediated oxygen free radical generations is responsible for the DNA damage.

As reviewed by Elstner et al (1988), a stress feedback chain seems to be reflected onto the photosynthetic electron transport system, inducing a series of changes cooperating in photodynamic damage. These sequences in stress response can be counteracted to a large extent by the antioxidative potential of the chloroplasts in cooperation with peroxisomes. The beginning of this reaction chain is apparently characterized as the onset of photorespiration.

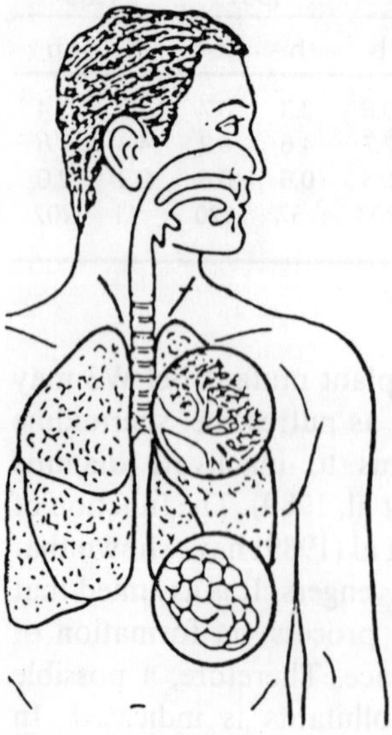

Site of attack	Water solubility	Compounds	Particle size
Eye Larynx Trachea	High	NH_3 HCl HCHO S_2Cl_2 $CH_2{=}CH{-}CHO$	≥10 μm
Bronchi	Medium	SO_2 Cl_2 Br_2 RCOCl $R(NCO)_2$	2–10 μm
Alveoli Capillaries	Low	O_3, O_2 NO_2 $COCl_2$ CdO	≤2 μm

Figure 7. Sites of attack in the respiratory tract in association with water solubility and particle size (Hippeli, 1990).

B. Animals

1. Sites of attack and symptoms

The primary target in higher animals of airborne pollutants is the respiratory system. The site of attack or reaction is less dependent on specific chemical properties than on the solubility of the reactive chemicals in aqueous media and on the diameter and weight of the particles or aerosols. The relationship between the site of reaction, water solubility and particle size is shown in Figure 7.

All lung irritants cause 'chemical' inflammation via the denaturation of proteins. Consecutive reactions depend on the water solubility of the irritants. Highly water-soluble substances cause eye, pharynx and tracheal irritation, and eventually cauterization of the upper epithelial layers.

Less water-soluble substances mainly attack the bronchioli, leading to mucus secretion, throat irritation, bronchoconstriction and spasms, bronchitis and pneumonia. Lipophilic compounds reach the alveolar space, inducing exudative inflammation (Forth et al, 1990).

2. Reaction mechanisms and health effects

Ozone

Ozone at high concentrations causes lung injury (review: Lippmann, 1989). Acute ozone exposure leads to characteristic lesions in the respiratory tract. These include accumulation of macrophages in the proximal alveoli and damage to ciliated cells and type I alveolar epithelial cells which line the alveolar sacs at the blood–air barrier. Subsequent proliferation of type II cells and Clara cells results in population shifts in the epithelial cells lining these regions (HEI, 1990). Changes in cell population are due to local generation of lung-cell specific growth factors (Tanswell, 1989).

Acute exposure to O_3 can stimulate a late nonspecific hyperreactive response in asthmatics with or without an early stimulus of an allergen (Koenig et al, 1987). The direct stimulation of airway epithelial cells results in the release of specific mediators which, in turn, stimulate local nerve endings and result in the contraction of smooth muscles or altered epithelial permeability. This increase in permeability induced by ozone enables inhaled particles to gain access to sensitized cells, finally also causing bronchoconstriction.

Chronic ozone exposure leads to thickening of the epithelium of the alveolar and respiratory bronchioles, and increases in the thickness and volume of the interstitium. In addition, alterations in collagen indicate an association between oxidant injury and the development of chronic lung disease such as pulmonary fibrosis (HEI, 1990). An increased level of xenobiotic-metabolizing activity produced by chronic O_3 exposure was observed (Takahashi and Miura, 1989).

In vitro studies suggested that ozone damage not only occurs as a consequence of the oxidation of membrane constituents, but also results from the inhibition of several intracellular key enzymes, and the depletion of intracellular glutathione (Van Der Zee et al, 1987).

In vitro studies showed that ozonation of α_1-protease inhibitor results in the total loss of inhibitory activity against human neutrophil elastase. This effect is due to the modification of several amino acid residues, resulting in an altered conformation (Johnson 1987). A disturbed balance between proteinase and antiprotease activity is thought to be the reason for the development of emphysema. Johnson (1987) reported that data from early animal studies indicate that exposure to ozone induces emphysematic lesions. But the ozone used in these experiments was generated by irradiating air, which is also known to generate nitrogen oxides. Work with ozone generated by irradiation of pure oxygen did not induce emphysematic abnormalities (Johnson, 1987). Inhaled O_3 may damage the proteinase inhibitor of the lung, but the possible extent of this effect in vivo is difficult to estimate.

NO_x cigarette smoke and peroxides

NO_2 at high levels of exposure is known to cause serious effects, including bronchospasm and pulmonary oedema (Kulle and Clements, 1987). The most common symptoms are coryza, sore throat, and cough without sputum (Rose et al, 1989).

NO_x concentration in urban air is mostly below 1.0 ppm. NO_2 generated indoors by use of gas cooking appliances may reach average levels of 0.025 ppm with peaks as high as 0.2–0.4 ppm (Samet et al, 1987). Mohsenin (1988) reported that indoor concentrations of NO_2 can exceed 2.0 ppm. Nonsmokers exposed to 2.0 ppm NO_2 for 1 h develop an increase in airway reactivity to methacholine aerosol without direct bronchoconstriction (Mohsenin, 1988).

Exposure of animals to NO_2 causes pulmonary oedema, due to increased lung microvasculature permeability caused by endothelial injury (Guidotti, 1978). The mechanism of this NO_2-induced damage is believed to involve free-radical-mediated peroxidative destruction of unsaturated membrane compounds.

Patel and Block (1987) reported on biochemical and metabolic responses to high levels of NO_2. Significant increases in GSSG reductase and glucose-6-phosphate dehydrogenase (G6PDH) activities were measured after exposure of pulmonary artery and aortic endothelial cells to 3 or 5 ppm NO_2 for 24 h. Treatment with 5 ppm NO_2 for 24 h leads to a significant increase in lipid peroxides and lactate dehydrogenase (LDH) release from both cell types. The 5-hydroxytryptamine uptake by endothelial cells was reduced.

In addition, NO_2 may damage the lung directly via its oxidant properties, or indirectly by altering the defence mechanisms of lung cells and their susceptibility to infections by bacterial or viral pathogenes. Rose et al (1989) showed that mice exposed to 5 ppm NO_2 before and after inoculation with murine virus were rendered more sensitive to initial infections by the virus and also to reinfection. These effects were associated with alterations in macrophage function in vivo.

Investigations on the modulation of pulmonary defence mechanisms against viral and bacterial infections by acute exposures of mice to NO_2 (Jakab, 1988) provided the following results. The intrapulmonary killing of *Staphylococcus aureus*, a Gram-positive bacterium, was impaired at 5 ppm (4 h) NO_2. The same effect was also found at 2.5 ppm or less if the NO_2 exposure was superimposed on lungs predisposed to lower resistance via immunosuppression with corticosteroids. An adverse effect of NO_2 occurred at lower concentrations when exposure followed bacterial challenge, suggesting a possible existence of a 'NO_2 high-risk' population, whose altered host status (chronic lung damage, immunosuppression, age) would render them more susceptible to infection. Subjects with chronic obstructive pulmonary disease showed a decrease in forced vital capacity after exposure to only 0.3 ppm NO_2 for 4 h (Morrow and Utell, 1989). Exposure of virus-infected mice to NO_2 had no influence on the infection in the lung but increased virus-associated lung damage (Jakab, 1988). The influence of NO_2 on susceptibilities to respiratory infections has been demonstrated in several species of animals (Fenters et al, 1973; Morrow, 1984; Samet et al, 1987; Jakab, 1980).

There are some uncertainties in animal studies; Kulle and Clements (1987) list the following points. Most workers reported positive effects at 2–3 ppm NO_2 or greater. Ambient levels of NO_2, however, rarely exceed 0.3 ppm. Animal studies exploited mortality as the endpoint of infection, but respiratory infections in humans are rarely lethal. Most animal studies, in addition, have employed bacterial challenge while, generally, human respiratory infections are of viral origin.

Epidemiological studies concerning the association between NO_2 exposure and altered susceptibility to acute respiratory infections have been conducted (Shy et al, 1970; Cohen et al, 1972; Melia et al, 1977; Goldstein et al, 1979; Speizer et al, 1980; Ware et al, 1984). Most results of such studies have been inconsistent and inconclusive (Samet et al, 1987). As specified by Kulle and Clements (1987), some investigators found higher rates of lung disease in children from households with gas stoves and therefore higher ambient NO_2 concentrations. Others found no significant differences in the incidence of respiratory disease of children in homes with gas stoves compared to those in homes with electric stoves. Studies suggesting a causality between indoor and outdoor NO_2 pollution and altered incidence of respiratory diseases, especially in children, are subject to confounding factors, such as exposure to other pollutants or parental cigarette smoking.

Regarding responses of susceptible subpopulations to NO_2, Morrow and Utell (1989) pointed out that the controlled clinical and environmental studies have also produced inconsistent results, 0.3 ppm NO_2 probably being the dose representing the minimal-effect exposure level for susceptible groups.

NO_2 in combination with H_2O_2 inactivates the α_1-proteinase inhibitor (Dooley and Pryor, 1982), a key event for the development of emphysema. The inactivation can be blocked by SOD or mannitol. Therefore, it was suggested that superoxide is formed intermediately:

$$NO_2 + H_2O_2 \longrightarrow HONO + H^+ + O_2^{\bullet -}$$
$$NO_2 + H_2O_2 \longrightarrow NO_2^- + O_2^{\bullet -} + 2H^+$$

NO_2 and H_2O_2 are both components of cigarette smoke; NO_2 may exceed 300 ppm in the smokestream (Forth et al, 1990), and the amount of H_2O_2 formed after a fixed time from the smoke of each cigarette is proportional to its tar content (Nakayama et al, 1989). Cigarette smoking is considered to be the major environmental cause of chronic lung emphysema.

The relationship between cigarette smoking and enhanced risk of cancer is well documented. The frequency of lung cancer (epithelial carcinoma of the bronchial mucous membrane) strongly depends on the extent of cigarette smoking. In addition, components of cigarette smoke are associated with the development of bladder cancer, which is doubled in smokers as compared to nonsmokers. The blood of smokers contains haemoglobin adducts such as 3-aminobiphenyl, 2-naphthylamine, *o*- and *p*-toluidine, 2,4-dimethylamine and 2-ethylaniline. These compounds are metabolized into electrophilic intermediates (hydroxylamines), ultimately leading to DNA adducts in the bladder.

A cooperation between metallic compounds and semiquinones in cigarette tar is probably responsible for single-strand breaks in phages PM2-DNA (Borish et al, 1985). Hydroxyl radicals formed via metal-catalysed decomposition of H_2O_2 may be finally responsible for DNA lesions.

Tar particles of the cigarette smoke cause a characteristic electron spin resonance (ESR) signal, due to internal electron transfer processes between special quinones (Pryor et al, 1983). The mutagenic activity of cigarette smoke can be stimulated by diverse activators. Jackson et al (1987) showed an enhanced rate of phage DNA strand breaks in the presence of asbestos fibres or $FeSO_4$. The increased activity was inhibited by hydroxyl radical scavengers (DMSO, mannitol, benzoate), by metal ion chelators (phenanthroline) and by catalase. These findings suggest an intermediate production of hydroxyl radicals or similar oxidants. H_2O_2 instead of cigarette smoke shows similar effects.

Cigarette smoking inactivates pulmonary NAD^+-dependent 15-hydroxyprostaglandin dehydrogenase, a key enzyme responsible for the biological inactivation of prostaglandins, and 5-lipoxygenase in pulmonary alveolar macrophages, resulting in a significant decrease in leukotriene B_4 (LTB_4) release into bronchoalveolar fluid (Mobley et al, 1987). These interferences with lipid metabolism may play a role in smoking-induced pulmonary diseases. Harats et al (1989) showed that cigarette smoking renders plasma low-density lipoprotein (LDL) more susceptible to subsequent peroxidative modification by cellular elements; this might enhance the atherogenicity of this lipoprotein.

Particles and particle-bound toxins

Diesel soot particles. Diesel exhaust is an airborne pollutant of growing importance. From the environmental standpoint, diesel engines have low hydrocarbon and carbon monoxide emissions. NO_x emissions are comparable to those of gasoline-driven cars equipped with catalysts. Of special importance are the particulate emissions, however, which are 30–70 times higher than those from catalyst-equipped spark ignition engines (McClellan, 1980; Schützle et al, 1981), rising to approximately 1 g km^{-1} (McClellan, 1980). The soot particles are small in size (less than 0.5 μm) (McClellan, 1986, 1987), easily respirable, and have carbon cores with a very large surface area onto which a variety of organic compounds are adsorbed. Fifteen to eighty per cent of the mass consists of extractable organic materials such as polycyclic aromatic compounds, nitroaromatics and quinones (Laresgoiti et al, 1977). The number and structure of these compounds depend on the type of engine and the mode of operation and thus are extremely variable (Laresgoiti et al, 1977). The adsorbed compounds of diesel soot are subject to atmospheric conversions, which make them even more toxic (McCoy et al, 1979; Stärk and Stauff, 1986a,b; Yoshikawa et al, 1985; Pitts et al, 1978; Arey et al, 1986).

Several investigators demonstrated that organic extracts from diesel particles are mutagenic in both bacterial (review: Claxton, 1983) and mammalian cell systems, including sister chromatid exchange, chromosomal aberrations and mutations (Schlipköter et al, 1987; Lewtas, 1982; Mitchell, 1988). McClellan (1987) reviewed the role of organic chemicals associated with diesel soot as causative factors of the carcinogenic response of rodents. Bond et al (1990) demonstrated the concentration and time-dependent formation of DNA adducts in lungs of rats exposed to diesel exhaust. Products from the reaction between diesel fuel and NO_2 seem to be highly cytotoxic and mutagenic (Handa et al, 1983; Henderson et al, 1981). Regarding genotoxic effects of diesel soot extracts, an activation of oncogenes is discussed (Rosenkranz, 1987).

The redox balance of the lung cells can be shifted by inhalation of diesel soot particles (Heinrich et al, 1986). In animal experiments these particles may provoke inflammatory responses accompanied by thickening of the alveolar septa connected with alveolar hypoplasia. In parallel, a decrease of lung clearance was observed. Unfiltered diesel soot may provoke adenocarcinoma in the lung.

As summarized by Schlipköter et al (1987), animal studies concerned with cancer showed that particles from diesel emissions cause lung cancer in at least one kind of animal under conditions which are comparable to certain sites of exposure of humans.

In general, the carcinogenic activity of polycyclic hydrocarbons obtained in diesel emissions has been correlated with their ability to form free radicals (Southorn and Powis, 1988). The generation of free radicals via electron transfer processes in diesel soot particles similar to cigarette smoke could likewise be demonstrated by characteristic ESR signals (Ross et al, 1982).

The cytotoxicity of diesel soot extracts for mammalian cell cultures is diminished in the presence of serum proteins and substances containing free SH-groups (glutathione, cysteine, mercaptoethanol). This protection is enhanced by addition of mitochondrial liver and lung preparations and cofactors such as $NADP^+$, $MgCl_2$ and G6PDH (Li, 1981).

In the presence of cysteine and H_2O_2, diesel soot particles cause lipid peroxidation and pigment bleaching, prevented by catalase and radical scavengers (α-tocopherol, propylgallate or diazobicyclooctane) (Vogl and Elstner, 1989). These findings suggest that cellular antioxidants are able to decrease soot toxicity, which acts via oxygen activation.

Hippeli and Elstner (1989) reported on synergistic effects of aqueous diesel soot suspension and sulphite (the aqueous solution of SO_2), which were suppressed by SOD. The cooperative mechanism observed in the presence of diesel soot particles (DP) and HSO_3^- indicates monovalent oxygen reduction and subsequent lipid peroxidation in the presence of an unsaturated fatty acid. In this process, HSO_3^- is supposed to function both as electron donor and radical-propagating agent:

$$\text{a.} \quad HSO_3^- + O_2 \xrightarrow{DP} HSO_3^{\bullet} + O_2^{\bullet -}$$

$$\text{b.} \quad HSO_3^{\bullet} + O_2 + H_2O \longrightarrow SO_4^{2-} + O_2^{\bullet -} + 3H^+$$

$$\text{c.} \quad O_2^{\bullet -} + DP \longrightarrow DP^{\bullet} + O_2$$

d. $$DP^{\bullet} + HSO_3^{-} \longrightarrow DP^{-} + HSO_3^{\bullet}$$
e. $$DP^{-} + O_2 \longrightarrow DP^{\bullet} + O_2^{\bullet -}$$
f. $$2DP^{\bullet} \longrightarrow DP + DP^{-}$$

g. $$DP^{\bullet} + O_2^{\bullet -} + H^{+} \xrightarrow{SOD} DPH + O_2$$

h. $$O_2^{\bullet -} + O_2^{\bullet -} + 2H^{+} \xrightarrow{SOD} H_2O_2 + O_2$$

As activating principles of DP in reaction (a), both nitroaromatics and naphthoquinones have to be considered, since both classes of compounds have been shown to undergo redox cycling, thus driving oxidative destructions in the presence of appropriate electron donor molecules. Reactions b–e may operate as propagators of the synergistic radical chain reaction observed in the presence of both DP and sulphite. Superoxide plays an important role as mediator between the DP redox factors and different SO_x oxidation states. Disproportionation of $DP^{\bullet}$ (f) and SOD-catalysed dismutations (g, h) represent chain-terminating events, where reaction g would be in agreement with the function of SOD as a superoxide–semiquinone oxidoreductase (Cadenas, 1989).

Since both SO_2 and DP are present in significant concentrations in urban air pollution (smog), the indicated reactions may contribute to certain respiratory disorders discussed in the context of severe air pollution.

In addition, vital functions of human polymorphonuclear neutrophils (PMN) are influenced by aqueous suspensions of DPs in combination with sulphite in vitro. Phagocytosis is increased, and in contrast the generation of activated oxygen during the respiratory burst appears to be decreased (Hippeli and Elstner, 1990). These effects are due to changes of the cell membrane potential. Martin et al (1988) observed a partial depolarization of the alveolar macrophage membrane, closely associated with inhibition of the respiratory burst. Low doses of hydroperoxides could decrease superoxide generation of alveolar macrophages due to a very fast-running depolarization of both plasma and mitochondrial membranes (Forman and Kim, 1989). Scott and Rabito (1988) suggest that superoxide is involved in hydroperoxide-induced change of membrane potential.

$O_2^{\bullet -}$, the main radical generated during the interaction of DPs and sulphite, has been shown to be also involved during influenza infections (Oda et al, 1989). Immune defence against infectious lung diseases in mice is decreased after exposure to diesel exhaust (Lewis et al, 1989). Thus, superoxide-generating systems like DP–sulphite cooperating in the alveolar space may modify immune responses in vivo.

Asbestos fibres. Asbestos fibres are airborne particles with high toxicity, especially indoors. The term asbestos collectively describes inorganic fibres derived from naturally occurring silicate minerals. The various types of asbestos fibres differ in their chemical composition and morphology. Therefore, the biological effects should be considered individually for each fibre type.

Prolonged inhalation of asbestos fibres of a certain size (length $>5\,\mu m$, diameter $<3\,\mu m$) produces fibrosis of the lung (asbestosis) as well as malignant tissue transformations (Selikoff and Lee, 1978; Mossman et al, 1983). Timbrell (1989) reviewed the significance of fibre size in fibre-related lung disease. Symptoms of corresponding diseases are cough and chronic bronchitis up to acute dyspnoea. There seems to exist a significant correlation between asbestos exposition and bronchial or pleural tumours (mesotheliomas), whereby crocidolite proved especially responsible for these rare tumours. Mossman and Landesman (1983) suggested that asbestos fibres enhance the cellular production of deleterious superoxide radicals, whereas other reactive oxygen species are apparently not involved. Incubation of asbestos fibres with H_2O_2 generated spin-trapped ESR signals, strongly associated with hydroxyl radicals and superoxide (Weitzman and Graceffa, 1984). From the influence of certain iron chelators on the formation of these signals, the authors concluded that asbestos fibres were able to generate $^{\bullet}OH$ and $O_2^{\bullet -}$ from H_2O_2 via iron catalysis. Pulmonary epithelial cell injury is mediated by H_2O_2 release from asbestos-activated polymorphonuclear leukocytes (PMNs) (Kamp et al, 1989). Asbestos alone has less cytotoxic effect on epithelial cells in leukocyte-free media; asbestos in combination with PMNs causes significant damage which is dependent on dose. As summarized by Mossman et al (1990), several studies support the concept of a cause and effect relation between activated oxygen species and the development of asbestosis. Therefore, asbestos fibres may act as immobilized catalysts for Fenton-type or Haber–Weiss reactions (Figure 8).

Several damaging activities of asbestos fibres may, furthermore, be potentiated by other air pollutants. Cigarette smoke enhances the frequency of asbestos-induced single-strand breaks of DNA (Jackson et al, 1987). Ozone enhances pulmonary retention of inhaled asbestos fibres (Pinkerton et al, 1989), and sulphite stimulates the production of $^{\bullet}OH$-type oxidants of crocidolite (Elstner et al, 1988). The formation of hydroxyl radicals may be involved in activation of polycyclic aromatic hydrocarbons, which may be adsorbed on the surface of asbestos or stem from other sources in the environment (Eberhardt et al, 1985).

Other particulate components. Besides asbestos, other particulate components of air pollution have been shown to activate oxygen. A series of silicates such as minusile, caoline, talcum, tremolite or diatomite form $^{\bullet}OH$ according to Fenton chemistry (Kennedy et al, 1989). Peroxides formed through adhesion of macrophages or exogenous peroxides as components in dust (Stärk and Stauff, 1986a,b) may be reductively activated by transition metals present in silicates or by gallium trioxide (Ga_2O_3). Induction of bronchial pulmonary inflammations has been observed, and particle clearance in the lung seems to be inhibited (Wolff et al, 1989). Generally, one can assume that opsonized bacteria in the presence of ingested particles increase superoxide formation. This holds for quartz as well as for particles in cigarette smoke and different types of soot. By the combination of particulate and gaseous toxins, the activity and number of leukocytes in lung tissues are changed. It is also worth noting that there are interspecies differences in the response to inhaled particles and gases (Warheit, 1989). Quantitative differences between humans and experimental animals are known to exist due to deposition and mucociliary clearance of inhaled substances. But morphological aspects of the respiratory tract and lung defence mechanisms are qualitatively similar among different species.

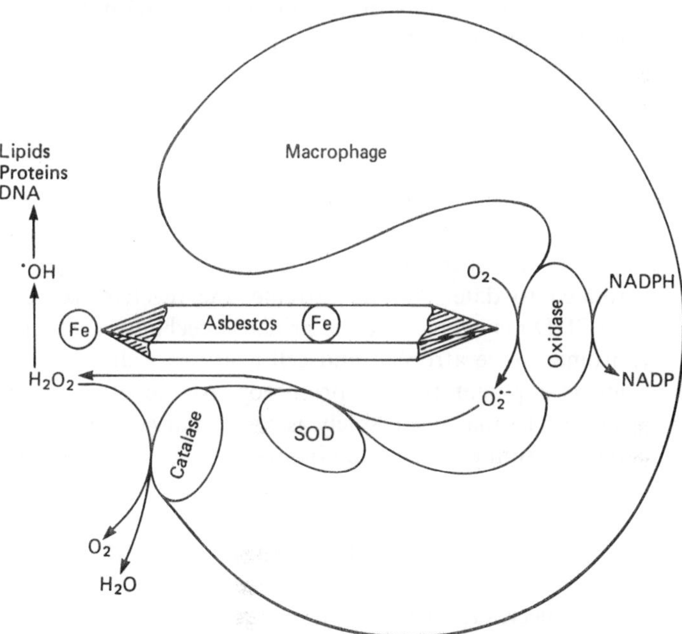

Figure 8. Cooperative effects between asbestos fibres and macrophages (Elstner et al, 1986).

Changes in the metabolism of arachidonic acid have been demonstrated by exposing animals to diverse dusts, aerosols and smoke mixtures (Ishinishi et al, 1986): as documented by König et al (1985), arachidonic acid metabolism seems to play a central role in initiation of bronchial obstructions. This may be of importance, since diseases of the respiratory tract seem to be influenced by genetic and environmental factors which, in turn, are composed of gaseous and particulate toxins.

The carcinogenicity of TCDD has been demonstrated in rats, mice, and monkeys at levels as low as $25\,\mathrm{ng\,kg^{-1}\,day^{-1}}$, where the carcinogenicity is specific to the chlorinated chemical and not the 'core' ring compound (Lilienfeld and Gallo, 1989). A series of studies (review: Stohs, 1990) showed that an oxidative stress is induced in tissues of TCDD-treated animals, including increases of (a) in vivo and in vitro hepatic and extrahepatic lipid peroxidation, (b) hepatic and macrophage DNA damage and (c) urinary excretion of malondialdehyde, and decreases of GSH, $NADPH_2$ and non-protein SH-groups in the liver. TCDD causes porphyria in experimental animals and increases in releasable iron associated with hepatic microsomes. In addition, iron is essential for TCDD-induced lipid peroxidation and DNA fragmentation. The need for iron indicates an involvement of superoxide, hydrogen peroxide and hydroxyl radical in TCDD-induced oxidative tissue damage. Furthermore, observed lipid peroxidation of hepatic microsomal, mitochondrial, and plasma membranes correlates inversely with membrane fluidity. These membrane alterations may be due to calcium influx into hepatocytes as demonstrated with rats. TCDD-induced DNA fragmentation was believed to be due to calcium-dependent activity of endonucleases. Therefore, a potential role of reactive oxygen species in tumour promotion by TCDD is discussed. 'The precise role of the oxidative stress in the expression of the toxic manifestations of TCDD has not been clearly determined to date. At one extreme, the reactive oxygen species produced by TCDD may constitute the primary mechanism associated with wasting syndrome, tissue atrophy, and cell death as well as the ability of TCDD to act as a potent tumour promotor. At the other extreme, the reactive oxygen species may occur solely as a secondary or tertiary effect and may ultimately contribute to the late phases of toxicity' (Stohs, 1990).

V. CONCLUSIONS AND REGULATORY DECISIONS

Reactive oxygen species are not only direct partners in reactions between atmospheric components but also play an important role during manifestation of damage in cells based on atmospheric pollutants. Most of the observed pathological changes in corresponding organs or cells of plants or animals include reactive oxygen species as initiators of symptoms. We differentiate between primary effects with activated oxygen as a damaging factor, and secondary effects mediated through (a) inflammatory processes in animals or (b) limitations in metabolism and thus photodynamic reactions in plants.

Photochemical primary reactions in the atmosphere are preconditions for the formation of photooxidants, especially the very reactive $^{\bullet}OH$ radical. This radical initiates most atmospheric and pollutant-driven conversions. The only significant chemical reaction producing ozone in the atmosphere is the one between molecular and atomic oxygen, where atomic oxygen at altitudes below 20 km is formed mainly by photodissociation of NO_2. During this reaction $NO^{\bullet}$ is formed which quickly reacts with ozone with reformation of NO_2. Thus, accumulation of ozone only occurs if $NO^{\bullet}$ is consumed by another series of reactions, predominantly due to peroxyl radicals. These are formed during reactions between hydrocarbons and the hydroxyl radical which, in turn, is formed by photodissociation of ozone or several carbonyl compounds.

An efficient decrease of photooxidants in the atmosphere seems possible through a consequent decrease of NO_x emission and also VOC emission. In contrast to VOC emissions, NO_x in the atmosphere is mainly due to anthropogenic activities, especially the exhausts of combustion engines and power plants. Thus it seems to be mandatory to decrease the emissions by these sources. A special problem may be seen in the emissions from heavy trucks and aeroplanes, since to date no specific means and corresponding technologies for decreases from these sources have been developed. Alternative structures in transportation have to be envisaged such as transfer of heavy loads from road to rail and decreases in short-distance flights and military air movements. Transportation-based emissions of VOCs would likewise be diminished by the above-mentioned changes.

SO_2 emissions due to changes in industrial structures have been strongly decreased in Western industrialized countries during the last ten years. Investigations on the cost-efficiency decreases of emissions showed that a decrease in SO_2 emission of 70% in Europe is technically possible with minimal financial input.

A special problem resides in particular emissions such as soot particles. Soot particles are surface-active and are carriers of diverse adsorbed toxic substances, thus rendering these particles a very complex toxic system. Health impacts, especially as far as disturbances of the respiratory tract are concerned, have to be investigated with increased emphasis. This holds especially for certain 'high-risk' groups such as children, elderly people or allergic patients. Cooperative effects between particles and gaseous emissions have to be investigated: effects of particles and combinations of emissions have not been examined thoroughly so far.

In a status symposium of PEF (Projekt Europäisches Forschungszentrum für Maßnahmen zur Luftreinhaltung, Karlsruhe), the project carrier comes to the conclusion that all political instruments utilized so far are not sufficient to introduce satisfactory environmental changes rapidly. Legislative political instruments have to be supported by economical instruments, where utilizations of environmental factors have their price. New elements in legislations recommended by the PEF project group contain:

1. taxes on each kg of emitted pollutants, irrespective of the source of emission; and
2. elaboration of threshold emission values, which especially should allow for the avoidance of health risk.

Legislative conclusions in this respect can only be obtained by intensive discussions between basic research groups, industry and political decision makers.

REFERENCES

ABEF (Arbeitsmaterialien des Bundesamtes für Ernährung und Forstwirtschaft) (1989) *Emissionen von Ammoniak – Quellen – Verbleib – Wirkungen – Schützmaßrahmen*. Bundesamt Für Ernährung und Forstwirtschaft, 6000 Frankfurt/M.

Aben JMM, Janssen-Jurkuvicova M & Adema EH (1990) Effects of low-level ozone exposure under ambient conditions on photosynthesis and stomatal control of Vicia faba L. *Plant Cell Environ.* **13:** 463–469.

Adams RM, Rosenzweig C, Peart RM et al (1990) Global climate change and US agriculture. *Nature* **345:** 219–224.

Apostol I, Heinstein PF & Low PS (1989) Rapid stimulation of an oxidative burst during elicitation of cultured plant cells. *Plant Physiol.* **90:** 109–116.

Arey J, Zielinska B, Atkinson R, Winer AM, Ramdahl T & Pitts JN Jr (1986) The formation of nitro-PAH from the gas-phase reactions of fluoranthene and pyrene with the OH radical in the presence of NO_x. *Atmospheric Environ.* **20(12):** 2339–2345.

Becker KH, Brockman KJ & Bechara J (1990) Production of hydrogen peroxide in forest air by reaction of ozone with terpenes. *Nature* **346:** 256–258.

Bieniek D (1985) Dioxin – Was ist das? Woher kommt es? In Klemm C & Haury HJ (eds) Dioxin – durch die Hintertür in die Umwelt, pp 5–9. Munich: Mensch + Umwelt.

Blake DR & Rowland FS (1986) World-wide increase in tropospheric methane, 1978–1983. *J. Atmospheric Chem.* **4:** 43–62.

Blumthaler M & Ambach W (1990) Indication of increasing solar ultraviolet-B radiation flux in alpine regions. *Science* **248:** 206–208.

Bond JA, Mauderly JL & Wolff RK (1990) Concentration- and time-dependent formation of DNA adducts in lung of rats exposed to diesel exhaust. *Toxicology* **60:** 127–135.

Borish ET, Cosgrove JP, Church DF, Deutsch WA & Pryor WA (1985) Cigarette tar causes single-strand breaks in DNA. *Biochem. Biophys. Res. Commun.* **133:** 780–786.

Bors W, Langebartels C, Michel C & Sandermann H (1989) Polyamines as radical scavengers and protectants against ozone damage. *Phytochemistry* **28(6):** 1589–1595.

Bufalini JJ & Arnts RR (eds) (1981) *Atmospheric Biogenic Hydrocarbons*, Vols 1 and 2. Ann Arbor, Michigan: Ann Arbor Science.

Cadenas E (1989) Biochemistry of oxygen toxicity. *Ann. Rev. Biochem.* **58:** 79–110.

Carp H & Janoff A (1978) Possible mechanisms of emphysema in smokers: in vitro suppression of serum elastase-inhibitory capacity by fresh cigarette smoke and its prevention by antioxidants. *Am. Rev. Respir. Dis.* **118:** 617–621.

Castillo FJ, Miller PR & Greppin H. (1987) Extracellular biochemical markers of photochemical oxidant air pollution damage to Norway spruce. *Experientia* **43:** 111–115.

Clarke BB, Henninger MR & Brennan E (1983) An assessment of potato losses caused by oxidant air pollution in New Jersey. *Phytopathology* **73:** 104–108.

Claxton LD (1983) Characterization of automobile emissions by bacterial mutagenesis bioassay: A review. *Environ. Mutagen.* **5:** 609–631.

Cohen CA, Hudson AR, Clausen JL & Knelson, JH (1972) Respiratory symptoms, spirometry, and oxidant air pollution in nonsmoking adults. *Am. Rev. Respir. Dis.* **105:** 251–261.

Cox RA (1990) Atmospheric chemistry—progress and surprises in the past 5 years. *J. Photochem. Photobiol.* **51:** 29–40.

Darrall NM (1989) The effect of air pollutants on physiological processes in plants. *Plant Cell Environ.* **12:** 1–30.

Di Mascio P, Kaiser S & Sies H (1989) Lycopene as the most efficient biological carotenoid single oxygen quencher. *Arch. Biochem. Biophys.* **274(2):** 532–538.

Dominy PJ & Heath RL (1985) Inhibition of the K^+-stimulated ATPase of the plasmalemma of pinto bean leaves by ozone. *Plant Physiol.* **77:** 43–45.

Dooley MM & Mudd JB (1982) Reaction of ozone with lysozyme under different exposure conditions. *Arch. Biochem. Biophys.* **218(2):** 459–471.

Dooley M & Pryor W (1982) Free radical pathology: Inactivation of human α-1-proteinase inhibitor by products from the reaction of NO_2 with H_2O_2 and the etiology of emphysema. *Biochem. Biophys. Res. Commun.* **106:** 981–987.

Eberhardt MK, Roman-Franco AA & Quiles MR (1985) Asbestos-induced decomposition of hydrogen peroxide. *Environ. Res.* **37:** 287–292.

Ehrlich R & Fenters JP (1973) In *Proceedings of the Third International Clean Air Congress, A11–A13.* Sponsored by the International Union of Air Pollution Prevention Association, VDI-Verlag, Düsseldorf.

Elstner EF (1982) Oxygen activation and oxygen toxicity. *Ann. Rev. Plant Physiol.* **33:** 73–96.

Elstner EF (1987) Metabolism of activated oxygen species. In Davies DD (ed.) *The Biochemistry of Plants*, pp 253–315. Academic Press. London and New York.

Elstner EF (ed) (1990) *Der Sauerstoff—Biochemie, Biologie, Medizin.* Mannheim: BI Wissenschaftsverlag.

Elstner EF & Osswald WF (1991) Air pollution: Involvement of oxygen radicals (A Mini Review). *Free Radicals Res. Comm.* (in press).

Elstner EF, Osswald WF & Youngman RJ (1985) Basic mechanisms of pigment bleaching and loss of structural resistance in spruce (Picea abies) needles: advances in phytomedical diagnostics. *Experientia* **41:** 591–597.

Elstner EF, Wagner GA & Schütz W (1988) Activated oxygen in green plants in relation to stress situations. In Randall DD, Blevin DG & Campell UH (eds) *Current Topics in Plant Biochemistry and Biophysics*, Vol. **7**, pp 159–187. Columbia: University of Missouri.

Fenters JD, Erlich R, Findlay J, Spanglerand J & Tolkacz V (1971) Serologic response in squirrel monkeys exposed to nitrogen dioxide and influenza virus. *Am. Rev. Respir. Dis.* **104:** 448–451.

Floyd RA, West MS, Hogsett WE & Tingey DT (1989) Increased 8-hydroxyguanine content of chloroplast DNA from ozone-treated plants. *Plant Physiol.* **91:** 644–647.

Fond der Chem. Industrie (ed.) (1987) *Umweltbereich Luft*. Karlstraße 21, 6000 Frankfurt am Main 1.

Forman HJ & Kim E (1989) Inhibition by linoleic acid hydroperoxide of alveolar macrophage superoxide production: Effect upon mitochondrial and plasma membrane potentials. *Arch. Biochem. Biophys.* **274(2):** 443–452.

Forth W, Henschler D & Rummel W (eds) (1990) *Allgemeine und spezielle Pharmakologie und Toxikologie*, 5th edn. Mannheim/Zürich/Wien: BI Wissenschaftsverlag.

Fuhrer J (1985) Formation of secondary air pollutants and their occurrence in Europe. *Experientia* **41:** 286–301.

Goldstein BD, Melia RJW, Chinn S, du Florey, CV, Clark D & John HH (1979) The relationship between respiratory illness in primary schoolchildren and the use of gas for cooking: II. Factors affecting nitrogen dioxide levels in the home. *Int. J. Epidemiol;* **8:** 339–345.

Guderian R (ed.) (1985) *Air Pollution by Photochemical Oxidants. Formation, Transport, Control and Effects on Plants.* Berlin, Heidelberg, New York, Tokyo: Springer-Verlag.

Guderian R, Klumpp G & Klumpp A (1988) In Öztürk MA (ed.) *Plants and Pollutants in Developed and Developing Countries*, pp 231–268. Int. Symp. Izmir, Turkey. Bornova, Izmir: Ege University.

Guidotti TL (1978) The higher oxides of nitrogen: inhalation toxicology. *Environ. Res.* **15:** 443–472.

Guthrie FE & Perry JJ (eds) (1980) *Environmental Toxicology*. New York: Elsevier.

Hafner L, Endler W, Wendering R & Weese G (1989) Ultrastructural studies on the needles of damaged Scotch pine trees from forest dieback areas in West Berlin. *Aquilo Ser. Bot.* **27:** 7–14.

Handa T, Yamauchi T, Ohnishi M, Hisamatsu Y & Ishii T (1983) Detection and average content levels of carcinogenic and mutagenic compounds from the particulates on diesel and gasoline engine mufflers. *Environ. Int.* **9:** 335–341.

Harats D, Ben-Naim M, Dabach Y, Hollander G, Stein O & Stein Y (1989) Cigarette smoking renders LDL susceptible to peroxidative modification and enhanced metabolism by macrophages. *Atherosclerosis* **79:** 245–252.

Heath RL (1980) Initial events in injury to plants by air pollutants. *Annu. Rev. Plant Physiol.* **31:** 395–431.

HEI (1987) *Automotive methanol vapors and human health: An evaluation of existing scientific information and issues for future research.* Health Effects Institute, 215 First Street, Cambridge, MA 02142.

HEI (1990) *Request for applications and 1990 research agenda.* Health Effects Institute, 215 First Street, Cambridge, MA 02142.

Heinrich U, Muhle H, Takenaka S et al (1986) Chronic effects on the respiratory tract of hamsters, mice and rats after long-term inhalation of high concentrations of filtered and unfiltered diesel engine emissions. *J. Appl. Toxicol.* **6:** 383–395.

Henderson TR, Li AP, Royer RE & Clark CR (1981) Increased cytotoxicity and mutagenicity of diesel fuel after reaction with NO_2. *Environ. Mutagen.* **3:** 211–220.

Henriksen T, Dahlback A, Larsen SHH & Moan J (1990) Ultraviolet-radiation and skin cancer. Effect of an ozone layer depletion. *Photochem. Photobiol.* **51(5):** 579–582.

Hewitt CN, Kok GL & Fall R (1990) Hydroperoxides in plants exposed to ozone mediate air pollution damage to alkene emitters. *Nature* **344:** 56–58.
Hippeli SC (1990) Biochemische Modellreaktionen zur Dieselruß – Toxikologie – Kooperative Effekte mit Sulfit. Dissertation an der Technischen Universität München.
Hippeli SC & Elstner EF (1989) Diesel soot—catalyzed production of reactive oxygen species: Cooperative effects with bisulfite. *Z. Naturforsch.* **44c:** 514–523.
Hippeli SC & Elstner EF (1990) Influence of diesel soot particles and sulfite on functions of polymorphonuclear neutrophils. *Free Radicals Res. Commun.* **2:** 29–38.
Holle G (1892) Vergiftete Nadelhölzer, *Die Gartenlaube* **48:** 795.
Ishinishi N, Kuwabara N, Nagase S, Suzuki T, Ishiwata S & Kohno T (1986) Long-term inhalation studies on effects of exhaust from heavy and light duty diesel engines on F344 rats. In Ishinishi N, Koizumi H, McClellan RO & Stober W (eds) *Carcinogenicity and Mutagenicity of Diesel Engine Exhaust*, pp 329–348. Amsterdam: Elsevier.
Jackson JH, Schraufstädter IU, Hyslop PA *et al* (1987) Role of oxidants in DNA damage—hydroxyl radical mediates the synergistic DNA damaging effects of asbestos and cigarette smoke. *J. Clin. Invest.* **80:** 1090–1095.
Jakab GJ (1980) NO_2-induced susceptibility to acute respiratory illness: A perspective. *Bull. NY Acad. Med.* **56:** 847–855.
Jakab GJ (1988) *Modulation of pulmonary defense mechanisms against viral and bacterial infections by acute exposures to nitrogen dioxide*, Res. Rep. No. 20, Health Effects Institute, 215 First Street, Cambridge, MA 02142.
Janoff A (1985) Elastases and emphysema current assessment of the protease–antiprotease hypothesis. *Am. Rev. Respir. Dis.* **132:** 417–433.
Johnson DA (1987) *Effects of ozone and nitrogen dioxide on human lung proteinase inhibitors*, Res. Rep. No. 24, Health Effects Institute, 215 First Street, Cambridge, MA 02142.
Kamp DW, Dunne M, Weitzman SA & Dunn MM (1989) The interaction of asbestos and neutrophils injures cultured human pulmonary epithelial cells: Role of hydrogen peroxide. *J. Lab. Clin. Med.* **114(5):** 604–612.
Keen NT & Taylor OC (1975) Ozone injury in soybeans. *Plant Physiol.* **55:** 731–733.
Kelly TJ, Daum PH & Schwartz SE (1985) Measurements of peroxides in cloudwater and rain. *J. Geophys. Res.* **90(D5):** 7861–7871.
Kennedy TP, Dodson R, Rao NV et al (1989) Dust causing pneumoconiosis generate $^{\bullet}OH$ and produce hemolysis by acting as Fenton catalysts. *Arch. Biochem. Biophys.* **269(1):** 359–364.
Kerr JA, Calvert JG & Demerjian KL (1976) Free radical reactions in the production of photochemical smog. In Pryor WA (ed.) *Free Radicals in Biology*, pp 159–179. London and New York: Academic Press.
King CM (1988) *Metabolism and biological effect of nitropyrene and related compounds*. Res. Rep. No. 16, Health Effects Institute, 215 First Street, Cambridge, MA 02142.
Klepper L (1990) Comparison between NO_x evolution mechanisms of wild-type and nr_1 mutant soybean leaves. *Plant Physiol.* **93:** 26–32.
Knight KL & Mudd JB (1984) The reactions of ozone with glyceraldehyde-3-phosphate dehydrogenase. *Arch. Biochem. Biophys.* **229(1):** 259–269.
Knox JP & Dodge AD (1985) Review article number 7: Singlet oxygen and plants. *Phytochemistry* **24(5):** 889–896.

König JQ, Covert DS, Morgan MS, Horike M, Marshall SG & Pierson WE (1985) The effect of 0.12 ppm ozone or 0.12 ppm nitrogen dioxide on pulmonary function and asthmatic adolescents. *Am. Rev. Respir. Dis.* **132:** 648–651.

König JQ, Pierson WE, Marshall SG, Covert DS, Morgan MS & Belle GV (1987) *The effects of ozone and nitrogen dioxide on lung function in healthy and asthmatic adolescents.* Res. Rep. No. 14, Health Effects Institute, 215 First Street, Cambridge, MA 02142.

Koziol MJ & Whatley FR (eds) (1984) *Gaseous Air Pollutants and Plant Metabolism.* London: Butterworths.

Kulle TJ & Clements ML (1987) *Susceptibility to virus infection with exposure to nitrogen dioxide.* Res. Rep. No. 15, Health Effects Institute, 215 First Street, Cambridge, MA 02142.

Laresgoiti A, Loos AC & Springer GS (1977) Particulate and smoke emission from light duty diesel engine. *Current Res.* **11(10):** 973–978.

Lashof DA & Ahuja DR (1990) Relative contributions of greenhouse gas emissions to global warming. *Nature* **344:** 529–531.

Laurence JA & Weinstein LH (1981) Effects of air pollutants on plant productivity. *Annu. Rev. Phytopathol.* **19:** 257–271.

Lelieveld J & Crutzen PJ (1990) Influences of cloud photochemical processes on tropospheric ozone. *Nature* **343(6255):** 227–233.

Lewis TR, Green FHY, Moorman WJ, Burg JR & Lynch DW (1989) A chronic inhalation toxicity study of diesel engine emissions and coal dust, alone and combined. *J. Am. Coll. Toxicol.* **8(2):** 345–373.

Lewtas J (1982) Mutagenic activity of diesel emissions. In Lewtas J (ed.) *Toxicological Effects of Emissions from Diesel Engines*, pp 62–82. New York: Elsevier.

Li AP (1981) Antagonistic effects of animal sera, lung and liver cytosols and sulfhydryl compounds on the cytotoxicity of diesel exhaust particle extracts. *Toxicol Appl. Pharmacol.* **57:** 55–62.

Lilienfeld DE & Gallo MA (1989) 2,4-D, 2,4,5-T, and 2,3,7,8-TCDD: An overview. *Epidemiol. Rev.* **11:** 28–58.

Lippmann M (1989) Health effects of ozone: A critical review. *JAPCA* **39:** 672–695.

LIS (1989) *Experimentelle Untersuchungen der LIS zur Aufklärung möglicher Ursachen der neuartigen Waldschäden.* Landesanstalt für Immissionsschutz Nordrhein-Westfalen, Wallneyer Str. 6, 4300 Essen 1.

Magel E & Ziegler H (1986) Einfluß von Ozon and saurem Nebel auf die Struktur der stomatären Wachspfropfen in den Nadeln von Picea abies (L.) Karst. *Forstw. Cbl.* **105(4):** 234–238.

Malanchuk JL & Nilsson J (eds) (1989) *The role of nitrogen in the acidification of soils and surface waters.* Nordic Council of Ministers.

Martin MA, Nauseef WM & Clark RA (1988) Depolarization blunts the oxidative burst of human neutrophils. Parallel effects of monoclonal antibodies. Depolarizing buffers and glycolytic inhibitors. *J. Immunol.* **140:** 3928–3935.

Masuch G, Kettrup A, Mallant RKAM & Slanina J (1986) Effects of H_2O_2-containing acidic fog on young trees. *Int. J. Environ. Anal. Chem.* **27:** 183–213.

McClellan RO (ed.) (1980) *Diesel exhaust emissions. Toxicology program. Status report.* Inhalation Toxicology Research Institute, Lovelace Biomedical and Environmental Research Institute, PO Box 5890, Albuquerque, New Mexico 87115.

McClellan RO (1986) Health effects of diesel exhaust: a case study in risk assessment. *Am. Ind. Hyg. Assoc.* **47:** 1–13.

McClellan RO (1987) Health effects of exposure to diesel exhaust particles. *Annu. Rev. Pharmacol. Toxicol.* **27:** 279–300.

McCord JM & Fridovich I (1969) Superoxide dismutase. An enzymic function for erythrocuprein (hemocuprein). *J. Biol. Chem.* **244:** 6049–6055.

McCoy EC, Hyman J & Rosenkranz HS (1979) Conversion of environmental pollutant to mutagens by visible light. *Biochem. Biophys. Res. Commun.* **89(2):** 729–734.

McManus MS, Altman LC, König JQ et al (1989) Human nasal epithelium: characterization and effects of in vitro exposure to sulfur dioxide. *Exp. Lung Res.* **15**: 849–865.

Mehlhorn H, Seufert G, Schmidt A & Kunert KJ (1986) Effect of SO_2 and O_3 on production of antioxidants in conifers. *Plant Physiol.* **82:** 336–338.

Melia RJW, du Florey CV, Altman DS & Swan AV (1977) Association between gas cooking and respiratory disease in children. *Br. Med. J.* **2:** 149–152.

Mengel K, Breininger MT & Lutz HJ (1990) Effect of simulated acidic fog on carbohydrate leaching, CO_2 assimilation and development of damage symptoms in young spruce trees (Picea abies L. Karst). *Environ. Exp. Bot.* **30:** 165–173.

Metz N (1989) Dieselpartikel-Emissionen in der Bundesrepublik Deutschland. *Ergomed* **13:** 32–35.

Mitchell CE (1988) Damage and repair of mouse lung DNA induced by 1-nitropyrene. *Toxicol. Lett.* **42:** 159–166.

Mobley A, Tanizawa H, Iwanaga T, Tai CL & Tai H (1987) Selective inhibition of 5-lipoxygenase pathway in rat pulmonary alveolar macrophages by cigarette smoking. *Biochim Biophys. Acta* **918:** 115–119.

Mohnen VA (1988) The mountain cloud chemistry program. In *Proceedings of the US/FRG research symposium: Effects of atmospheric pollutants on the spruce-fir forest of the eastern United States and the Federal Republic of Germany*, October 19–23, pp 27–60. Burlington, Vermont. Broomall, PA: Northeastern Forest Experiment Station.

Mohsenin V (1988) Airway responses to 2.0 ppm nitrogen dioxide in normal subjects. *Arch. Environ. Health* **43:** 242–246.

Mohsenin V & Gee BL (1989) Oxidation of α1-protease inhibitor: role of lipid peroxidation products. *J. Appl. Physiol.* **65(5):** 2211–2215.

Morrow PE (1984) Toxicological data on NO_x: An overview. *J. Toxicol. Environ. Health* **13:** 205–227.

Morrow PE & Utell MJ (1989) *Responses of susceptible subpopulations to nitrogen dioxide.* Res. Rep. No. 23, Health Effects Institute, 215 First Street, Cambridge, MA 02142.

Mossman BT & Landesman (1983) Importance of oxygen free-radicals in absestos-induced injury to airway epithelial cells. *Chest* **83:** 503–515.

Mossman BT, Light W & Wei E (1983) Asbestos: Mechanisms of toxicity and carcinogenicity in the respiratory tract. *Annu. Rev. Pharmacol. Toxicol.* **23:** 595–615.

Mossman BT, Bignon J, Seaton A & Lee JBL (1990) Asbestos: Scientific developments and implications for public policy. *Science* **247:** 294–301.

Mudd JB & Kozlowski TT (eds) (1975) *Responses of Plants to Air Pollution.* New York, San Francisco, London: Academic Press.

Nakayama T, Church DF & Pryor WA (1989) Quantitative analysis of the hydrogen peroxide formed in aqueous cigarette tar extracts. *Free Radical Biol. Med.* **7**: 9–15.

NAPAP (1990) *1989 Annual Report to the President and Congress*. The National Acid Precipitation Assessment Program, 722 Jackson Place, NW, Washington DC 20503.

Neftel A, Jacob P & Klockow D (1984) Measurements of hydrogen peroxide in polar ice samples. *Nature* **311(5981):** 43–45.

Novakov T (1982) In Schryer DR (ed.) *Heterogenous atmospheric chemistry*, Washington DC: American Geophysical Union.

Oda T, Akaike T, Hamamoto T, Suzuki F, Hirano T & Maeda H (1989) Oxygen radicals in influenza-induced pathogenesis and treatment with pyran polymer-conjugated SOD. *Science* **244:** 974–976.

Osswald WF & Elstner EF (1986) Fichtenerkrankungen in den Hochlagen der Bayerischen Mittelgebirge. *Ber. Deutsch. Bot. Ges.* **99:** 313–339.

Osswald WF, Höll W & Elstner EF (1986) Ergosterol as a biochemical indicator of fungal infection in spruce and fir needles from different sources. *Z. Naturforsch.* **41c:** 542–546.

Osswald WF, Senger H & Schrammel P (1990) Vergleich der Ascorbinsäure-, Glutathion-, Zucker- und Jonengehalte gesunder und geschädigter Fichten des Standortes Laugna nordwestlich von Augsburg. *Forstw. Cbl.* **109:** 181–189.

Patel JM & Block ER (1987) *Biochemical and metabolic response to nitrogen dioxide-induced endothelian injury*. Res. Rep. No. 9, Health Effects Institute, 215 First Street, Cambridge, MA 02142.

Peiser GD & Yang SF (1978) Chlorophyll destruction in the presence of bisulfite and linolenic acid hydroperoxide. *Phytochemistry* **17:** 79–84.

Peters JL, Castillo FJ & Heath RL (1989) Alteration of extracellular enzymes in pintobean leaves upon exposure to air pollutants, ozone and sulfur dioxide. *Plant Physiol.* **89:** 159–164.

Peters RE & Mudd JB (1982) Inhibition by ozone of the acylation of glycerol 3-phosphate in mitochondria and microsomes from rat lung. *Arch. Biochem. Biophys.* **216(1):** 34–41.

Pinkerton KE, Brody AR, Miller FJ & Crapo JD (1989) Exposure to low levels of ozone results in enhanced pulmonary retension of inhaled asbestos fibers. *Am. Rev. Respir. Dis.* **140:** 1075–1081.

Pitts JN Jr, Van Cauwenberghe KA, Grosjean D et al (1978) Atmospheric reactions of polycyclic aromatic hydrocarbons: Facile formation of mutagenic nitro derivates. *Science* **202:** 515–519.

Poborski PS (1988) Pollutant penetration through the cuticle. In Schulte-Hostede S, Darrall NM, Blank LW & Wellburn AR (eds) *Air Pollution and Plant Metabolism*, pp 19–35. London and New York: Elsevier Applied Science.

Prather MJ & Watson RT (1990) Stratospheric ozone depletion and future levels of atmospheric chlorine and bromine. *Nature* **344:** 729–734.

Prinn R, Cunnold D, Rasmussen R et al (1987) Atmospheric trends in methylchloroform and the global average for the hydroxyl radical. *Science* **238:** 945–950.

Projekt Europäisches Forschungszentrum für Maßnahmen zur Luftreinhaltung (1990) *6. Statuskolloquium des PEF vom 6. bis 8. März 1990 im Kernforschungszentrum Karlsruhe—Zusammenfassung der Projektleitung*. Kernforschungszentrum Karlsruhe.

Projekt Europäisches Forschungszentrum für Maßnahmen zur Luftreinhaltung, (1990) *6. Statuskolloquium des PEF vom 6. bis 8. März 1990 im Kernforschungszentrum Karlsruhe*, Band 2. Kernforschungszentrum Karlsruhe.

Projektgruppe Bayern zur Erforschung der Wirkung von Umweltschadstoffen (ed.) (1988) *Verteilung und Wirkung von Photooxidantien im Alpenraum*, Gesellschaft für Strahlen- und Umweltforschung München (GSF-Bericht 17/88).
Pryor WA (ed.) (1976) *Free Radicals in Biology*, Vol. 2. New York, San Francisco, London: Academic Press.
Pryor WA & Lightsey JW (1981) Mechanisms of nitrogen dioxide reactions: Initiation of lipid peroxidation and the production of nitrous acid. *Science* **214:** 435–437.
Pryor WA, Hales BJ, Premovic PI & Church DF (1983) The radicals in cigarette tar: their nature and suggested physiologic implications. *Science* **220:** 425–427.
Rice-Evans C (ed.) (1987) *Free Radicals, Oxidant Stress and Drug Action*. London: The Richelieu Press.
Robinson DC & Wellburn AR (1983) Light-induced changes in the quenching of 9-aminoacridine fluorescence by photosynthetic membranes due to atmospheric pollutants and their products. *Environ. Pollution* **32:** 109–120.
Rose RM, Pinkston P & Skornik WA (1989) *Altered susceptibility to viral respiratory infection during short-term exposure to nitrogen dioxide*. Res. Rep. No. 24, Health Effects Institute, 215 First Street, Cambridge, MA 02142.
Rosenkranz HS (1987) Diesel emissions revisited: Is the carcinogenicity due to a geotoxic mechanism? *Mutat. Res.* **182:** 1–4.
Ross MM, Chedekel MR, Risby TH, Lestz SS & Yasbin RE (1982) Electron paramagnetic resonance spectrometry of diesel particulate matter. *Environ. Int.* **7:** 325–329.
Russell AG, St Pierre D & Milford JB (1980) Ozone control and methanol fuel use. *Science* **247:** 201–205.
Sakaki T, Kondo N & Sugahara K (1983) Breakdown of photosynthetic pigments and lipids in spinach leaves with ozone fumigation: Role of active oxygens. *Physiol. Plant.* **59:** 28–34.
Samet JM, Marbury MC & Spengler JD (1987) Health effects and sources of indoor air pollution: Part I. *Am. Rev. Respir. Dis.* **136:** 1486–1508.
Sandermann H Jr, Langebartels C & Heller W (1990) Ozonstreß bei Pflanzen. Frühe und 'Memory'-Effekte von Ozon bei Nadelbäumen. *UWSF-Z. Umweltchem. Ökotox.* **2(1):** 14–15.
Schlipköter HW, Brockhaus A, Einbrodt H et al (1987) *Zur Frage der krebserzeugenden Wirkung von Dieselmotorabgasen* Gutachten der Arbeitsgruppe 'Immissionswirkung auf den Menschen', im Auftrag des Ministeriums für Umwelt, Raumordnung und Landwirtschaft des Landes Nordrhein-Westfalen.
Schmailzl UV (1990) Klimagefahr durch den Flugverkehr. *Ökologische Konzepte* **31:** 10–15.
Schryer DR (ed.) (1982) *Heterogenous Atmospheric Chemistry*. Washington DC: American Geophysical Union.
Schützle D, Skewes LM, Fisher GE, Levine SP & Gorse RA Jr (1981) Determination of sulfates in diesel particulates. *Anal. Chem.* **53:** 837–840.
Schulte-Hostede S, Darall NM, Blank LW & Wellburn AR (eds) (1988) *Air Pollution and Plant Metabolism*. London and New York: Elsevier Applied Science.
Scott JA & Rabito CA (1988) Oxygen radicals and plasma membrane potential. *Free Radicals Biol. Med.* **5:** 237–246.
Selikoff IJ & Lee DHK (eds) (1978) *Asbestos and Disease*. New York: Academic Press.

Shy CM, Creason JP, Pearlman MF, McClain KE, Benson FB & Young MM (1970) The Chattanooga school children study: Effects of community exposure to nitrogen dioxide: II. Incidence of acute respiratory illness. *J. Air Pollut. Control Assoc.* **20:** 582–586.

Sies H (ed.) (1985) *Oxidative Stress*. London: Academic Press.

Smith WH (1981) *Air Pollution and Forests*. New York, Heidelberg, Berlin: Springer-Verlag.

Southorn PA & Powis G (1988) Free radicals in medicine. II. Involvement in human disease. *Mayo Clin. Proc.* **63:** 390–408.

Speizer FE, Ferris B, Bishop YMM & Spengler J (1980) In Lee SD (ed.) *Nitrogen Oxides and their Effects on Health.* pp 343–359. Ann Arbor, MI: Ann Arbor Science Publishers.

Stärk G & Stauff J (1986a) Anreicherung hochmolekularer Peroxide auf Fichtennadeln in Reinluftgebieten. *Staub—Reinhalt. Luft* **46(9):** 396–400.

Stärk G & Stauff J (1986b) Nachweis von aromatischen Peroxiden in Schwebstaubextrakten. *Staub—Reinhalt. Luft* **46:** 289–291.

Stohs SJ (1990) Oxidative stress induced by 2,3,7,8-tetrachlorodibenzo-p-dioxin (TCDD). *Free Radical Biol. Med.* **9:** 79–90.

Takahashi Y & Miura T (1989) Effects of nitrogen dioxide and ozone in combination on xenobiotic metabolizing activities of rat lungs. *Toxicology*. **56:** 253–262.

Tanaka K & Sugahara K (1980) Role of superoxide dismutase in defense against SO_2 toxicity and an increase in superoxide dismutase activity with SO_2 fumigation. *Plant Cell Physiol.* **21(4):** 601–611.

Tanswell AK (1989) *Detection of paracrine factors in oxidant lung injury*. Res. Rep. No. 22, Health Effects Institute, 215 First Street, Cambridge, MA 02142.

Timbrell V (1989) Review of the significance of fibre size in fibre-related lung disease: A centrifuge cell for preparing accurate microscope-evaluation specimens from slurries used in innoculation studies. *Ann. Occup. Hyg*. **33(4):** 483–505.

Traub WA, Johnson DG & Chance KV (1990) Stratospheric hydroperoxyl measurements. *Science* **247:** 446–449.

Unsworth MH & Ormrod DP (eds) (1982) *Effects of Gaseous Air Pollution in Agriculture and Horticulture*. London: Butterworths.

Vaida V, Solomon S, Richard EC, Rühl E & Jefferson A (1989) Photoisomerization of OClO: a possible mechanism for polar ozone depletion. *Nature* **342(6248):** 405–408.

Valenzo DP (1987) Photomodification of biological membranes with emphasis on singlet oxygen mechanisms. *Photochem. Photobiol.* **46(1):** 147–160.

Van Der Zee J, Dubbelman TMAR, Raap TK & Van Steveninck J (1987) Toxic effects of ozone on murine L929 fibroblasts. *Biochem. J.* **242a**, 707–712.

Vogl G & Elstner EF (1989) Diesel soot particles catalyze the production of oxy-radicals. *Toxicol. Lett.* **47:** 17–23.

Ware JH, Dockery DW, Spiro A, Speizer FE & Ferris BG (1984) Passive smoking, gas cooking and respiratory health of children, living in six cities. *Am. Rev. Respir. Dis.* **129:** 366–374.

Warheit DB (1989) Interspecies comparisons of lung responses to inhaled particles and gases. *Toxicology* **20**(1): 1–29.

Weitzman SA & Graceffa P (1984) Asbestos catalyzes hydroxyl and superoxide radical generation from hydrogen peroxide. *Arch. Biochem. Biophys.* **228:** 373–376.

Winer AM, Atkinson R & Pitts JN Jr (1984) Gaseous nitrate radical: possible nighttime atmospheric sink for biogenic organic compounds. *Science* **224:** 156–159.

Wolff RK, Barr EB, Bond JA et al (1988) *Factors affecting possible carcinogenicity of inhaled nitropyrene aerosols.* Res. Rep. No. 19, Health Effects Institute, 215 First Street, Cambridge, MA 02142.

Wolff RK, Griffith WC, Henderson RF et al (1989) Effects of repeated inhalation exposures to 1-nitropyrene, benzo[a]pyrene Ga_2O_3 particles, and SO_2 alone and in combinations on particle clearance, bronchoalveolar lavage fluid composition and histopathology. *J. Toxicol. Environ. Health* **27:** 123–138.

Woods AF (1899) The destruction of Chlorophyll by oxidizing enzymes. *Centralblatt Bakt., Paras. Infekt.* **22:** 745–754.

Yoshikawa T, Ruhr LP, Flory W, Giamalva D, Church DF & Pryor WA (1985) Toxicity of polycyclic aromatic hydrocarbons. I. Effect of phenanthrene, pyrene and their ozonized products on blood chemistry in rats. *Toxicol. Appl. Pharmacol.* **79:** 218–226.

nighttime atmosphere: sink for biogenic organic compounds. Science 224: 156–159.

Wolff RK, Barr EB, Bond JA et al (1988) Factors affecting possible carcinogenicity of inhaled nitropyrene aerosols. Res. Rep. No. 19. Health Effects Institute, 215 First Street, Cambridge, MA 02142.

Wolff RK, Griffith WC, Henderson RF et al (1989) Effects of repeated inhalation exposures to 1-nitropyrene, benzo[a]pyrene, Ga_2O_3 particles, and SO_2 alone and in combinations on particle clearance, bronchoalveolar lavage fluid composition and histopathology. J. Toxicol. Environ. Health 27: 123–138.

Wood, AT (1989) The destruction of Chlorophyll by oxidizing enzymes. Chemistry of Rosse Pathol. 12: 745–754.

Yoshikawa T, Ruhr LP, Flory W, Giamalva D, Church DF & Pryor WA (1985) Toxicity of polycyclic aromatic hydrocarbons. I. Effect of phenanthrene, pyrene and their ozonized products on blood chemistry in rats. Toxicol. Appl. Pharmacol. 79: 218–226.

2

UVA (320–380 nm) Radiation as an Oxidative Stress

R. M. TYRRELL.
ISREC, Ch. des Boveresses, CH-1066 Epalinges, Switzerland

I. Introduction 57
II. UVA radiation and the generation of active oxygen intermediates 58
III. UVA radiation and constitutive cellular defence pathways 66
IV. Induction of gene expression by UVA radiation 72
V. Summary 78

I. INTRODUCTION

For biological purposes, the ultraviolet component of sunlight incident on the earth's surface is usually divided into two wavelength regions which are termed UVA (320–380 nm, near-UV) and UVB (290–320 nm, mid-UV). The UVB region overlaps with the tail-end of DNA absorption and is generally believed to be largely responsible for the most damaging effects of sunlight, including skin cancer. However, wavelengths in the UVA region, which penetrate deeply into skin (Bruls et al, 1984), are also potentially carcinogenic and can cause a wide variety of biological effects (Parrish et al, 1978; Jagger, 1985). Human exposure to UVA radiation is actually increasing due, at least in part, to the longer exposure times permitted by the more widespread use of primarily UVB-absorbing sunscreens. Certain individuals also expose themselves to extremely high acute levels of UVA radiation using modern tanning equipment. Along with the expanding volume of work

Oxidative Stress: Oxidants and Antioxidants
ISBN 0-12-642762-3

in this area has come the realization that UVA radiation actually constitutes a very significant oxidative stress. There is considerable overlap between the types of cellular modifications induced by UVA and those generated by oxidative pathways. Furthermore, it now appears that cellular defence against UVA damage has more in common with antioxidant pathways than it does with the multiple repair pathways that act to eliminate DNA damage inflicted by UVC (190–290 nm) and UVB radiations.

The purpose of the first part of this chapter is to gather together lines of evidence in different biological systems which are consistent with the notion that absorption of UVA radiation does indeed lead to cellular oxidation. Since the UVA literature has not previously been summarized from such a viewpoint, some of the work described dates back to the early 1970s when studies of this wavelength region began to interest an increasing number of research groups. The involvement of oxygen in many of the biological effects of UVA radiation is summarized in Table 1 and described in some detail in the following section. The second part of the chapter (Sections III and IV) is concerned with the antioxidant pathways that now appear to be involved in protection of both prokaryotic and eukaryotic cells against the deleterious effects of UVA radiation.

II. UVA RADIATION AND THE GENERATION OF ACTIVE OXYGEN INTERMEDIATES

A. In vitro

Many cellular compounds can absorb UVA radiation, often leading to the generation of active oxygen intermediates. An important example is the photochemical degradation of tryptophan to yield hydrogen peroxide and *N*-formyl kynurenin (McCormick et al, 1976). There is also an indication that hydrogen peroxide may be produced by irradiation of cysteine (McCormick et al, 1982). UVB irradiation of lens crystallin will lead to photooxidation of tryptophan, and these tryptophan derivatives (including *N*-formyl kynurenin) will act as photosensitizers to generate both hydrogen peroxide and superoxide (e.g. Andley and Clark, 1989). These tryptophan oxidation products absorb both in the UVA and UVB regions and accumulate with age so that UVA radiation is believed to be an important factor in human cataract formation (Pitts et al, 1986).

The generation of even low levels of hydrogen peroxide in biological systems may be important since, in vivo, it is believed that naturally occurring iron complexes will react with hydrogen peroxide to generate the highly reactive hydroxyl radical in a superoxide-driven Fenton reaction (Gutteridge, 1985; Imlay et al, 1988). The reaction is driven by the continual reduction of ferric iron to the ferrous state by superoxide anion so that generation of the latter will also contribute to the intracellular formation of hydroxyl radical. Indeed, both hydrogen peroxide and superoxide anion are generated by UVA irradiation of NADH and NADPH (Czochralska et al, 1984; Cunningham et al, 1985).

In addition to the intracellular generation of superoxide anion and hydrogen peroxide through the interaction of UVA radiation with cellular components, cells contain many additional chromophores which may generate singlet oxygen via a type II photodynamic process. In this case, the excited triplet state sensitizer will react with oxygen in the ground state to give ground state sensitizer and singlet oxygen. The most common groups of compounds of this type are the flavins, quinones and porphyrins.

Table 1. Biological actions of UVA radiation that depend upon the presence of oxygen.

DNA damage	Single-strand breaks, DNA–protein crosslinks, inactivation of transforming DNA
Membrane alterations	Lipid peroxidation Permeability Arachidonate release
Inactivation of enzymes	DNA repair enzymes Antioxidant enzymes
Inactivation of microorganisms	Bacteria, fungi, yeast
Inactivation of mammalian cells	Rodent, human
Skin damage	Erythema: immediate and delayed pigment darkening

B. DNA damage

Wavelengths in the UVA range cause various types of DNA damage, including cyclobutane-type pyrimidine dimers, strand breaks and DNA–protein crosslinks (review; Tyrrell, 1984). The ratio of the latter two classes of lesion to pyrimidine dimers increases markedly at longer wavelengths. Dimer formation by radiation at 365 nm is unaffected by reducing the oxygen tension so that active oxygen species are unlikely to be involved (RM Tyrrell, unpublished observation). In contrast, the fluence-dependent induction of DNA strand breaks in the DNA of *E. coli* by radiation at 365 nm is completely prevented by irradiation under anaerobic conditions (Tyrrell et al, 1974). Strand breaks are induced at approximately twice the in vivo rate (aerobic) in extracted bacteriophage T4 DNA and the induction is almost completely prevented by the presence of 2-aminoethylisothiouronium bromide-HBr (AET), a free radical scavenger. These results suggest that if an endogenous photosensitizer is involved in generating the active oxygen intermediates which lead to strand breaks, then it must be tightly bound to the DNA and survive extraction. Alternatively, the in vivo result may reflect the simultaneous oxygen-dependent inhibition of single-strand break rejoining (see Section II.E). The prevention of UVA-induced DNA strand breakage in human fibroblasts by exogenously added catalase (Roza et al, 1985) is evidence that H_2O_2 is generated by UVA irradiation of the buffer medium. Since there was no parallel effect on survival, such damage appears not to contribute to cell death. In more recent experiments, Peak et al (1987a,b) have shown that strand break induction by irradiation of human cells at 365 nm is enhanced in the presence of deuterium oxide (which modifies singlet oxygen lifetimes), possibly implicating singlet oxygen. In contrast, they conclude from other experiments that hydroxyl radical is involved since agents which are known to quench this species (acetate, formate, azide and mannitol) protect against strand break induction by treatment of isolated *Bacillus subtilis* DNA with radiation at 365 nm (Peak and Peak, 1990). However, it is possible that quite different mechanisms are responsible for the induction under in vivo and in vitro conditions.

Gantt et al (1979) have shown that DNA–protein crosslinking in both human and mouse cells by a broad-band radiation source (380–490 nm) is a strictly oxygen-dependent process. Although the wavelength dependence for the oxygen effect has not been determined, the induction of this DNA lesion in human cells by radiation at 405 nm is reduced by irradiation under hypoxic conditions and enhanced in the presence of deuterium oxide (Peak et al, 1985).

C. Inactivation of transforming DNA

The inactivation of bacterial-transforming DNA can be monitored by following the activity of selectable genetic markers after introduction of the damaged DNA into the host cell. In this way it was shown that the UVA inactivation of both *B. subtilis*- and *Haemophilus influenzae*-transforming DNA is dependent upon the presence of oxygen (Cabrera-Juarez, 1964; Peak et al, 1973). Curiously, the oxygen dependence of inactivation of *H. influenzae* DNA was lost after additional purification but was maintained even after stringent purification of *B. subtilis* DNA. The simplest conclusion from the latter observation is that the oxygen dependence is mediated via a photochemically active chromophore which is more or less tightly bound to the DNA. Further evidence for the involvement of oxygen intermediates in the inactivation came from observations that histidine, AET, 1,4 diazocyclo(2.2.2)octane (DABCO) and glycerol all protect against this UVA effect (Peak et al, 1973, 1981; Cabrera-Juarez et al, 1976; Peak and Peak, 1975, 1980). Except for glycerol, all these compounds are known to scavenge both free radicals and singlet oxygen. However, the findings with glycerol suggest that the hydroxyl radical may be involved.

D. Membranes

The lipid membrane is readily susceptible to attack by active oxygen intermediates. There are many reports that UV radiation can lead to peroxidation of membrane lipids (e.g. Desai et al, 1964; Roshchupkin et al, 1975; Putvinsky et al, 1979; Azizova et al, 1980). In vitro studies with lecithin microvesicles have shown UV-induced changes in the microviscosity of membrane bilayers (Dearden et al, 1981) which is correlated with the degree of unsaturation of fatty acid chains (Dearden et al, 1985). Both UVC radiation and the UVA component of sunlight have been shown to cause lipid peroxidation in the liposomal membrane (Mandal and Chatterjee, 1980). Haem proteins such as cytochrome *c* and catalase are known to catalyse lipid peroxidation and peroxidative breakdown of membranes (e.g. Brown and Wüthrich, 1977; Goni et al, 1985; Szebini and Tollin, 1988). A recent study has shown a dose-dependent linear increase in lipid peroxidation of liposomal membranes by UVA radiation (Bose et al, 1989) which could largely be inhibited by butylated hydroxytoluene, a nonspecific scavenger of lipid free radicals. Since both sodium azide and histidine led to a 40–50% inhibition of peroxidation, the authors suggest that singlet oxygen is involved in the initiation of the reaction.

UVA irradiation of liposomes can lead to lipid peroxidation in the absence of photosensitizer molecules (Bose et al, 1989) so that singlet oxygen may arise through direct stimulation of molecular oxygen. However, biological membranes are rich in endogenous photosensitizer molecules such as those involved in electron transport and these may contribute to the peroxidation of lipids observed in biological systems (Jagger, 1985). Membrane damage has long been implicated in UVA lethality in bacteria (Hollaender, 1943) and almost certainly contributes to the sensitivity of UVA-treated populations plated on minimal medium, a phenomenon which is strongly dependent on oxygen (Moss and Smith, 1981). UVA sensitivity has been related to the levels of unsaturated fat in membranes (Klamen and Tuveson, 1982; Chamberlain and Moss, 1987). Furthermore, the presence of deuterium oxide enhances the levels of membrane damage, UVA sensitivity and lipid peroxidation (Chamberlain and Moss, 1987), suggesting a role for singlet oxygen in all three processes. Leakage experiments have also been used to assess UVA-induced membrane damage in yeast and again changes in permeability correlate well with lethality and are strongly oxygen dependent (Ito and Ito, 1983). UVA radiation of cultured fibroblasts of both human and murine origin leads to the release of arachidonate metabolites from the membrane in a fluence-dependent fashion (Hanson and De Leo, 1989). The release is dependent upon the presence of both oxygen and calcium ions and may be related to the induction of cutaneous erythema which is also oxygen-dependent (Section II.H).

E. Inactivation of enzymes

Studies early this century showed that certain enzymes, including amylase, catalase, pepsin and tyrosinase, are inactivated in an oxygen-dependent process by UVA radiation (Agulhon, 1912). It is likely that most enzymes which contain aromatic amino acids will be sensitive to some extent to UVA radiation, since there will be low but significant absorption in this region. Since UVA will also cause DNA damage and generate active oxygen species, it is of relevance to question whether DNA repair enzymes or enzymes involved in scavenging oxygen intermediates are damaged. Indeed, there is indirect evidence that the *recA* gene-dependent repair of DNA strand breaks in *E. coli* is inhibited by UVA (365 nm) radiation and that the inhibition is strongly dependent upon the presence of oxygen (Tyrrell, 1976). The activity of yeast photolyase (as measured by the rate of cyclobutane-type pyrimidine dimer removal) in crude extracts is also destroyed in an oxygen-dependent fashion by radiation at 365 nm (Tyrrell, 1976). The destruction of repair capacity by radiation in the UVA range almost certainly influences the ability of bacterial populations to survive such treatment.

Among the antioxidant enzymes, both peroxidases and catalase are haem-containing proteins which are UVA chromophores and may generate active species. Catalase exhibits an absorption maximum in the UVA range with a peak close to 400 nm. In addition to inactivation by sunlight (Mitchell and Anderson, 1965), more recent studies have shown that it is rapidly destroyed by UVA radiation (Kramer and Ames, 1987). In fact, indirect evidence suggests that endogenous catalase may even act as a photosensitizer since an *E. coli* strain carrying the *kat G* plasmid and which overproduces catalase (Loewen et al, 1983) is actually more sensitive to UVA than the allelic strain not carrying the plasmid, although the catalase-overproducing strain is more resistant to H_2O_2 (Eisenstark and Perrot, 1987). Similar studies using overexpression of a multicopy clone of *kat G* (catalase HP-1 (Triggs-Raine and Loewen, 1987)) in *Salmonella typhimurium* also demonstrated an enhanced sensitivity to UVA killing (Kramer and Ames, 1987). Furthermore, *S. typhimurium oxyR1* mutants which constitutively overexpress several proteins, including both catalase and alkyl hydroperoxide reductase (a flavin-containing protein), are considerably more sensitive to UVA killing than is the wild type (Kramer and Ames, 1987). Double mutants which overexpressed all *oxyR* proteins except catalase or alkyl hydroperoxide reductase were less UVA-sensitive than the *oxyR* strain. Taken together, this evidence strongly suggests that these enzymes may not only be inactivated by UVA but can also act as endogenous sensitizers.

F. Inactivation of microorganisms

Among the earliest studies that demonstrated that oxygen intermediates were involved in the biological effects of UVA radiation are those which defined a clear oxygen dependence for lethal effects. Among bacterial species in which this has been observed are *E. coli* (Webb and Lorenz, 1970), *S. typhimurium* (Eisenstark, 1971), *B. subtilis* (Eisenstark, 1971), *Sarcina lutea* (Denniston et al, 1972) and, more recently, *Proprionibacterium acnes* (Kjeldstad, 1984). This oxygen dependence extends to recombination-deficient mutants of several bacterial strains (Eisenstark, 1971) and to *E. coli* strains deficient in the excision repair of DNA damage (Webb and Lorenz, 1970; Tyrrell, 1976; Webb and Brown, 1976). The effects are generally very large and have been observed both with broad-band UVA and radiation at monochromatic wavelengths; for action spectra, see Webb and Brown (1976). Strong oxygen effects for inactivation by broad-band UVA irradiation have also been observed in yeast (*Saccharomyces cerevisiae* (Ito and Ito, 1983)) and slime mould (*Dictyostelium discoideum*) (Graetzer, 1987)).

Attempts to modulate UVA survival of microorganisms using agents which modify the activity of active oxygen intermediates have given somewhat ambiguous or negative results. For example, glycerol protects *E. coli* populations from the lethal action of radiation at 334 nm and 365 nm but sensitizes quite strongly to longer wavelengths (405 and 454 nm (Peak et al, 1982)). Apparently, glycerol has quite different modes of action, the relative importance of which vary with wavelength. Nevertheless, these experiments do indicate that some potentially lethal radical species which are quenched by glycerol may be generated in the UVA region. UVA inactivation of *Neurospora crassa* conidia (Thomas et al, 1981) is unchanged by the presence of deuterium oxide (which enhances singlet oxygen lifetime) or sodium azide (which quenches singlet oxygen), suggesting that singlet oxygen is not an important intermediate for UVA-induced lethality in this species. In contrast, deuterium oxide does lead to a significant enhancement of the inactivation of the bacterium *P. acnes* by broad-band UVA (Kjeldstad, 1984).

The role of carotenoids (which quench singlet oxygen) in protecting cells against visible light damage mediated by endogenous photosensitizing chromophores has been well documented (Krinsky, 1976). The extent to which carotenoids may protect cells against UVA damage is less clear. Although carotenoids present in *N. crassa* appear to afford only limited protection against these wavelengths (Krinsky, 1976), studies by Webb and coworkers (Webb, 1977) have shown that the small amount of protection afforded by carotenoids against killing of the bacterium *S. lutea* by radiation at monochromatic wavelengths occurs almost exclusively in the UVA region. Much larger carotenoid protection is observed with broad-band radiation sources, suggesting that an interaction between two or more wavelengths is involved. In more recent work, Tuveson et al (1988) have irradiated strains of *E. coli* in which carotenoid genes cloned from *Erwinia herbicola* (a phytopathogenic bacteria) were expressed (Perry et al, 1986). In these experiments, the carotenoids did appear to protect the bacteria against high fluences of UVA radiation. Although indirect, these experiments suggest that potentially lethal singlet oxygen is being generated by UVA irradiation of *E. coli*.

The degree to which singlet oxygen may be generated in different cell types clearly depends on the types of endogenous chromophores present. Most evidence points to porphyrin involvement, e.g. chlorophyll (in plants) and cytochrome *c* in the smut fungus *Ustilago violacea* (Kovacs et al, 1987; Will et al, 1987). Indirect evidence based on action spectra has indicated a role for porphyrin in the UVA inactivation of the bacterium *P. acnes* (Kjeldstad and Johnsson, 1986). However, the strongest evidence for porphyrin involvement in bacterial inactivation derives from studies by Tuveson, Peak and coworkers (Tuveson and Sammartano, 1986; Peak et al, 1987a) with a mutant of *E. coli* deficient in L-aminolevulinate synthase. Such cells are unusually resistant to UVA radiation but resistance returns to normal when cells are grown in L-aminolevulinic acid prior to irradiation. In parallel with the decreased resistance to UVA radiation associated with supplementation of the mutant, membrane permeability (as assessed by rubidium leakage) was also enhanced by radiation at 405 nm and to a lesser extent at 334 nm (Peak et al, 1987a). These results strongly suggest that porphyrins are important UVA sensitizers and that the membrane is a likely target. However, Tuveson and Sammartano (1986) point out the possibility that if other components of the respiratory chain are coordinately regulated, a similar result could be obtained from a deficiency in flavin containing proteins. Nevertheless, the importance of porphyrins was further indicated in a separate study by Sammartano and Tuveson (1987) in which they showed that populations of *E. coli* which overproduce cytochromes were hypersensitive to broad-band UVA radiation.

G. Inactivation of mammalian cells

The inactivation of mammalian cells by broad-spectrum UVA irradiation was first reported by Wang et al (1974) but the effects were entirely due to toxic photoproducts, presumably including hydrogen peroxide, generated in the culture medium in which the cells were irradiated. A later study (Danpure and Tyrrell, 1976) showed that populations of either cultured human cells (Hela) or Chinese hamster cells (V-79) could be inactivated in a salts buffer by near-monochromatic radiation at a wavelength of 365 nm. The inactivation was strongly dependent upon the presence of oxygen, suggesting that, as for bacteria, active oxygen intermediates were involved in the biological effect. Further evidence for the involvement of such intermediates was provided by the observation that glutathione plays a major role in protecting cultured human skin cells against the lethal action of UVA radiation (see Section III.B). As in bacteria, attempts to use hydroxyl radical scavengers to test the role of this radical in the inactivation of skin fibroblasts gave complex results (Tyrrell and Pidoux, 1989). Low concentrations of glycerol and mannitol led to a small protection against UVA (365 nm) radiation but higher concentrations of these alcohols or dimethylsulphoxide (at any concentration tested), led to a sensitization. Irradiation in the presence of aminotriazole, which inhibits catalase and strongly sensitizes cells to hydrogen peroxide treatment, had little effect on survival, suggesting that hydrogen peroxide is not involved. In contrast, irradiation in the presence of sodium azide (a quencher of singlet oxygen) or deuterium oxide (which enhances singlet oxygen lifetimes) led to significant levels of protection or sensitization respectively. Taken together, these results strongly suggest that the generation of singlet oxygen is involved in the UVA inactivation of cultured human fibroblasts.

H. Skin

The delayed erythema response associated with acute sunburn can be produced by both the UVB and UVA components of sunlight (Parrish et al, 1978). However, several lines of evidence suggest that the mechanism of induction of the response differs between the two regions, not least of which is the oxygen dependence of UVA-induced delayed erythema. Initial studies (Tegner et al, 1983) showed that pressing a UVA-transparent acrylic plate against the skin surface during irradiation prevented both delayed erythema and melanogenesis, presumably by the mechanical reduction of oxygen tension in the irradiated skin area. Auletta et al (1986) have taken these studies further by cutting off blood flow using a cuff applied to the upper arm in order to achieve varying degrees of hypoxia in the forearm. The minimal erythemal dose of UVA was increased by a factor of 2.7 under conditions of low oxygen tension. The inhibition of erythema was totally prevented by the transcutaneous diffusion of 100% oxygen. In addition, both immediate and delayed pigment darkening induced by UVA have been shown to depend on the presence of molecular oxygen (Pathak et al, 1962; Tegner et al, 1983; Auletta et al, 1986; Rorsman and Tegner, 1988). These observations are strongly indicative that active oxygen intermediates participate in the acute response of skin to UVA radiation. There is also evidence that the induction of sunburn cells by UVB radiation is mediated via oxygen intermediates (Young et al, 1988) but no studies have yet focused on the UVA range.

Free radical intermediates have also been implicated in UV-induced carcinogenesis [see Black (1987) for an overview] but such studies have not been undertaken specifically with UVA radiation. These investigations were originally stimulated by the isolation of the putative carcinogen, cholesterol epoxide, from UV-irradiated skin (Black and Lo, 1971). Although no causal relationship has been established, this finding is clear evidence that UV irradiation of skin leads to free radical-induced peroxidations. In addition, in animals that were fed an antioxidant supplement containing butylated hydroxytoluene, tocopherol, ascorbic acid and glutathione, a reduction in UV-induced cholesterol epoxide formation was observed as well as a suppression of the numbers and severity of UV-induced squamous cell carcinomas (Black, 1987). The role of active oxygen intermediates, possibly singlet oxygen, in UV-induced carcinogenesis has also been demonstrated by studies in which β-carotene and other carotenoid pigments have been shown to suppress UV-induced carcinogenesis (Epstein, 1977; Mathews-Roth, 1982; Mathews-Roth and Krinsky, 1985, 1987).

III. UVA RADIATION AND CONSTITUTIVE CELLULAR DEFENCE PATHWAYS

A. DNA repair

UVA radiation induces a variety of lesions in DNA, some of which are formed only in the presence of oxygen (Section II.B). The earliest studies of repair of UVA damage focused on bacterial systems which were already being characterized for their involvement in repair of UVC radiation-induced damage. Cyclobutane-type pyrimidine dimers which are induced under both aerobic and anaerobic conditions throughout the UV range are susceptible to photoenzymatic repair, excision repair and the various pathways of recombination repair (Webb, 1977). Recovery of UVA-induced lethal damage as a result of photoenzymatic repair was observed only in strains that were highly deficient in other repair systems (Brown and Webb, 1972; Tyrrell, 1973). Strains deficient in the incision step of excision repair were rather more sensitive in stationary phase (Webb et al, 1976) but exhibited similar sensitivities in exponential phase (Tyrrell and Webb, 1973), and this is believed, at least in part, to reflect simultaneous inhibition of excision repair by radiation at 365 nm.

Although both *recA* and *polA* strains of *E. coli* are sensitive to radiation throughout the UV range, the actual pathways involved in repairing DNA damage induced by UVC or UVA damage are probably different. The *recA* gene is involved in both constitutive and inducible pathways for removal of pyrimidine dimers induced by UVC radiation (Friedberg, 1985). Recombinationless strains of *S. typhimurium*, *B. subtilis* and *E. coli* are sensitive to broad-spectrum UVA radiation in the presence of oxygen (Eisenstark, 1970, 1971). Although this may partly involve defective removal of pyrimidine dimers, an important contribution to sensitivity may also arise from a reduced ability to repair the DNA strand breaks formed under aerobic conditions. A proportion of the strand breaks induced by ionizing radiation are repaired slowly by a full–medium-dependent pathway involving the *recA* gene (Town et al, 1973) which now appears to be primarily a mechanism for repair of double strand breaks (Sargentini and Smith, 1986). Strand breaks induced by radiation at 365 nm also appear to be repaired by this pathway as well as an additional fast *recA*-dependent pathway, which occurs in buffer at a low temperature during the irradiation period (Miguel and Tyrrell, 1986) and which has not been further characterized. The *polA* gene product is not only essential for the fast repair of ionizing radiation-induced strand breaks, which occurs in buffer at low temperature (Town et al, 1973; Boye et al, 1974), but is also involved in a similar rejoining process which acts on strand breaks induced by radiation at 365 nm (Ley et al, 1978; Miguel and Tyrrell, 1986). Thus, it appears that strand break repair pathways originally characterized for their role in removing damage induced by ionizing radiation play an important role in the repair of UVA-induced strand breaks.

Several recent advances have been made in investigations of the mechanisms by which bacteria can repair free radical-mediated DNA damage, and these are described in detail in Chapter 5. To date, the pathway involving exonuclease III has been most closely linked to removal of UVA damage since *E. coli xthA* mutants (lacking exoIII) are sensitive to hydrogen peroxide (Demple et al, 1983) and broad-spectrum UVA radiation (Sammartano and Tuveson, 1983) but not UVC radiation. An important property of the enzyme would appear to be its recognition of apurinic and apyrimidinic sites in DNA. In addition to catalysing the endonucleolytic hydrolysis of 3′-terminal phosphomonoesters and the release of 5′-mononucleotides from 3′ ends of damaged sites (Rogers and Weiss, 1980), the enzyme has 3′ to 5′ exonuclease activity which can trim 3′ termini to permit repair replication by DNA polymerase I (Demple et al, 1986). Thus the *xthA* and *polA* genes may be involved in a coordinated constitutive pathway for removal of UVA-induced DNA damage.

Little is known about the repair of UVA-induced DNA damage in mammalian cells. Excision repair has the potential to remove pyrimidine dimers and other large adducts from the DNA but excision deficient human fibroblast cell lines (derived from xeroderma pigmentosum patients) only exhibit slight sensitivities to UVA radiation relative to cell lines from normal individuals (Smith and Patterson, 1982; Keyse et al, 1983; Tyrrell, unpublished results). As in bacteria, this may reflect both a less important role for pyrimidine dimers in the lethal action of UVA radiation and the simultaneous inhibition of repair synthesis by wavelengths in this range (Parsons and Hayward, 1985). Most UVA-induced strand breaks appear to be repaired rapidly (within 15–30 min) with kinetics similar to those seen for ionizing radiation-induced strand breaks (Bradley et al, 1978; Erickson et al, 1980; Bredberg, 1981; Roza et al, 1985). However, studies with human lymphocytes (Holmberg et al, 1985) have indicated that a slower pathway may also exist. DNA–protein crosslinks induced by certain chemical agents can be repaired via the nucleotide excision repair pathway (Fornace and Seres, 1982). The majority of crosslinks induced by a sunlamp filtered to remove wavelengths <315 nm are removed from two out of three normal human skin fibroblast lines within 24 h (Lai et al, 1987) but the pathway(s) involved have not yet been determined. Based on the observation that crosslinks appear to increase during incubation after UVA radiation, Rosenstein et al (1989) have suggested that crosslinks may actually arise as a consequence of cellular repair of UVA-induced damage.

Cells cultured from patients with actinic reticuloid (a common idiopathic photodermatosis) show a persistent inhibition of RNA synthesis and cytopathic changes when irradiated with broad-band UVA radiation and this may involve a cellular defect which leads to inefficient quenching of free radicals (Botcherby et al, 1984). This is supported by the finding that at an irradiation temperature of 25°C, the increased sensitivity of an actinic reticuloid cell line to monochromatic radiation at 365 nm is prevented when

Trolox-C, a water-soluble analogue of vitamin E, is present in the preirradiation growth medium or the postirradiation plating medium (Kralli and Moss, 1987). However, no human cell types have yet been identified which have a specific defect in repair of UVA radiation-induced DNA damage.

B. Antioxidant molecules

Various cellular compounds, including ascorbate, uric acid, carotenoids and sulphydryls, are potent free radical scavengers (for overview, see Black (1987)). Carotenoids may play a role in protection against UVA damage in certain microorganisms (see Section II.F) and various exogenous carotenoids have been shown to inhibit UV-induced epidermal damage (Mathews-Roth, 1986) and tumour formation (Section II.H).

Glutathione (L-γ-glutamyl-L-cysteinyl-glycine) is the predominant low molecular weight thiol in many prokaryotic and eukaryotic cells and is believed to play an important role in protection of cells against oxidative damage (review: Meister and Anderson, 1983). However, mutants of *E. coli* which lack glutathione as a result of a defect in the *gshA* gene show normal resistance to hydrogen peroxide, cumene hydroperoxide, heat and ionizing radiation (Greenberg and Demple, 1986) and are also not especially sensitive to monochromatic UVA (365 nm) radiation (Tyrrell, unpublished results). In contrast, there is considerable evidence that glutathione depletion causes a sensitization of cultured eukaryotic cells to various types of oxidant stress and heat treatment (e.g. Dethmers and Meister, 1981; Arrick et al, 1982; Biaglow et al, 1983; Freeman et al, 1985; Lach et al, 1986; Clark et al, 1986). Many of these studies have employed buthionine-*S*,*R*-sulphoximine, which is currently the most specific drug available for depleting cellular glutathione levels and acts by inhibition of γ-glutamyl cysteine synthetase (Griffith et al, 1979). Using this drug, we have shown that glutathione depletion of cultured human skin fibroblasts leads to a marked sensitization to UVA (334 nm, 365 nm) and near-visible (405 nm) radiation as well as to radiation in the UVB (302 nm, 313 nm) range (Tyrrell and Pidoux, 1986, 1988). There is a direct correlation between the levels of sensitization and cellular glutathione content. In preliminary experiments using cells cultured from a patient with 5-oxoloprolinurea (25% normal glutathione levels), the sensitization was not as great as expected but such studies need to be extended to a wider range of deficient cell types. Indeed, most of the studies which rely on drug inhibition to delineate a role for glutathione should be substantiated by appropriate experiments using cell lines deficient in glutathione as a result of a genetic defect. With this reservation, we propose that glutathione plays a major role in protecting cultured cells against the lethal effects of both UVA and UVB radiation, and, at least for UVA radiation, the protection appears to be of far greater significance than that provided by DNA excision repair. Although the glutathione effect suggests the involvement of active intermediates in lethal effects, it is not clear at what level the glutathione acts. Sensitization to short-wavelength UVB (302 nm) radiation can be largely reversed by addition of cysteamine, suggesting that the protection afforded by glutathione may be related to free radical quenching (Tyrrell and Pidoux, 1988). However, at longer wavelengths (313 nm, 365 nm) no cysteamine protection is seen except at high fluences suggesting that glutathione may play a more specific role. One possibility is that reduced glutathione levels may influence the activity of glutathione peroxidase for which this thiol is the unique hydrogen donor.

Additional evidence that glutathione is a photoprotective agent in skin cells is derived from experiments which have demonstrated that glutathione levels in both dermis and epidermis are transiently depleted by UVA treatment (Wheeler et al, 1986; Connor and Wheeler, 1987) with a return to normal levels within a few hours. Thus, it is of interest that certain lines of cells cultured from the inherited disorder ataxia telangiectasia (AT) show impaired glutathione synthesis which is only evident after depletion of the thiol (Meredith and Dodson, 1987). The limited synthesis appears to be due to a defect in cysteine transport. However, impaired glutathione resynthesis does not appear to be a general characteristic of all AT cell lines (Dean et al, 1988; Meredith, 1988). Most AT cell lines are sensitive to ionizing radiation (Taylor et al, 1975) and oxidants, including hydrogen peroxide (Shiloh et al, 1983), but only certain lines are slightly sensitive to UVB radiation (Keyse et al, 1985) and there have been no reports of sensitivity to UVA radiation.

Depletion of glutathione also reduces the fluence threshold for the UVA induction of the stress protein, haem oxygenase (Keyse and Tyrrell, 1987; see Section IV.B), suggesting that the response is mediated by UVA-generated active intermediates which are quenched by glutathione. We have also observed that glutathione depletion not only markedly enhances the UVA and hydrogen peroxide induction of haem oxygenase mRNA levels (as expected from the protein studies) but also enhances constitutive levels several-fold in the absence of any additional stress treatment (Lautier and Tyrrell, submitted). Thus, it appears that haem oxygenase expression is not only influenced by exposure of cells to oxidant stress but also by the redox state of the cell.

C. Antioxidant enzymes

If the intracelluar generation of hydrogen peroxide by UVA radiation (Section II.A) is relevant to cell lethality then catalase would be expected to play a protective role. The study of such a functional role for catalase has been facilitated by the isolation of *E. coli* mutants totally deficient in catalase activity and the genetic mapping of relevant loci (Loewen, 1984; Loewen and Triggs, 1984). Mutants defective at the *katF* locus (and lacking the catalase designated as hydroperoxidase II, HP-II) were sensitive to both UVA and H_2O_2 treatment (Sammartano et al, 1986; Eisenstark and Perrot, 1987). However, the expression of HP-II also requires a functional *katE* gene, and mutants at neither this locus nor the *katG* locus (essential for hydroperoxidase I, HP-I, activity) are particularly sensitive to UVA radiation (Sammartano et al, 1986; Eisenstark and Perrot, 1987). These observations have been clarified by a recent study (Sak et al, 1989) which has indicated that the *katF* gene is a transacting regulator of both the exonuclease III (exoIII, product of *xthA*) and the HP-II enzymes. Since *xthA* mutants are also sensitive to UVA radiation (Sammartano and Tuveson, 1983), it appears most likely that it is exoIII, an enzyme involved in repair of apurinic and apyrimidic sites, rather than catalase, which contributes to cell survival. Furthermore, the *nur* mutation which causes sensitivity to both broad-band (Tuveson and Jonas, 1979) and monochromatic UVA radiation (Tuveson et al, 1982; Peak et al, 1983) was later recognized as a mutant allele of the *katF* locus (Sammartano et al, 1986) and probably also sensitizes to UVA by virtue of its control of exoIII production. In view of the apparently minor role played by endogenous catalase, it is somewhat surprising that exogenous catalase added to the plating medium is able to protect strongly against both UVA lethality and mutagenesis in several bacterial strains (Sammartano and Tuveson, 1984). This result suggests that a sustained reaction, such as peroxidation of lipid membrane components, is initiated as a result of H_2O_2 generation by UVA radiation. Evidently, high levels of exogenous catalase but not endogenous catalase are able to prevent such a reaction.

Superoxide anion is also produced by UVA radiation (Section II.A) and the role of this active species in inactivation has been examined using bacterial mutants deficient in superoxide dismutase (SOD). *E. coli* genes have been identified that code for both the inducible Mn-SOD (*sodA*; Touati (1983)) and Fe-SOD (*sodB*; Sakamoto and Touati (1984)). Only the *sodA sodB* double mutant is sensitive to paraquat (Carlioz and Touati, 1986) which generates superoxide anion intracellularly. Eisenstark (1989) has reported that strains carrying a *sodA* plasmid (Bloch and Ausubel, 1986; Carlioz and Touati, 1986) but not *sodB* plasmid (Sakamoto and Touati, 1984) confer some additional resistance to UVA. However, he only observed a slight sensitivity of *sodA*, *sodB* or *sodA sodB* mutants to UVA radiation under aerobic conditions. In our own studies (unpublished) even the *sodA sodB* double mutant did not show enhanced sensitivity to monochromatic radiation at a wavelength of 365 nm.

The results with bacterial mutants suggest that neither endogenous catalase nor SOD play a major role in protecting cells against the lethal effects of UVA irradiation. A similar picture appears to be true for human cells (Tyrrell and Pidoux, 1989). The addition of aminotriazole, a catalase inhibitor, to populations of cultured human skin fibroblasts sensitizes them strongly to the lethal effects of hydrogen peroxide but does not result in a significant sensitization to radiation at 365 nm. The addition of exogenous catalase to buffer in which cells were being irradiated with broad spectrum UVA did not alter the survival of human fibroblast populations although it strongly inhibited the induction of strand breaks (Section II.B), an indication that hydrogen peroxide generated in the irradiation medium can penetrate cells to cause damage but does not contribute to lethality (Roza et al, 1985). Diethyldithiocarbamate, which chelates intracellular copper and inhibits SOD, sensitizes cells to hydrogen peroxide (Mello-Filho et al, 1984) but not UVA radiation (Tyrrell and Pidoux, 1989). Similarly, a SOD mimetic agent (copper diisopropylsalicylate) is without effect.

Although these in vitro results suggest that catalase and SOD do not contribute significantly to protection of UVA-damaged cells, it should be noted that a small inhibition of catalase and glutathione peroxidase but not SOD activity has been observed after an acute UVA treatment of mouse skin (Fuchs et al, 1989). However, the physiological significance of such changes is unknown.

IV. INDUCTION OF GENE EXPRESSION BY UVA RADIATION

Microorganisms

Many genes have now been recognized in various microorganisms whose expression is regulated by oxygen; for recent examples see Cotter and Gunsalus (1989), Hodge et al (1989) Marykwas and Fox (1989), Jayaraman et al (1988), Zagorec *et al* (1988), Ditta et al (1987) and Zitomer et al (1987). In addition, it is now clear that oxidizing agents (in particular hydrogen peroxide) and agents which generate superoxide anion can induce a wide variety of DNA repair and antioxidant enzymes in bacteria (see Chapters 5 and 6). Treatment of bacteria with hydrogen peroxide leads to the enhanced synthesis of at least 30 proteins (Christman et al, 1985; Morgan et al, 1986; Eisenstark, 1989), nine of which are under the control of the *oxyR* gene (Christman et al, 1985). The *oxyR* gene (which is described in detail in Chapter 6) is a 34.4-kDa protein which shows homology to a family of other regulatory proteins and appears to be negatively autoregulated (Christman et al, 1989). Among the proteins induced by H_2O_2 and under *oxyR* control are catalase, glutathione reductase and alkyl hydroperoxide reductase.

The inactivation of *E. coli* by hydrogen peroxide is a complex process which has been divided into mode 1 killing, which requires active metabolism, and has a maximum near 1–3 mM H_2O_2, and mode 2 killing, which occurs in the absence of metabolism at higher concentrations of H_2O_2 (Imlay and Linn, 1986). In addition, exposure of *E. coli* to low concentrations of hydrogen peroxide induces resistance to a subsequent exposure at higher concentrations (Demple and Halbrook, 1983; Tyrrell, 1985; Imlay and Linn, 1987). Imlay and Linn (1986) have concluded from survival studies with *katEkatG*, *gshB* and *sodAsodB* mutants that the *oxyR*-dependent protection induced by hydrogen peroxide is mainly due to enhancement of catalase activity.

Broad-spectrum UVA radiation has also been shown to stimulate an inducible protective mechanism associated with new protein synthesis in *E. coli* (Peters and Jagger, 1981) and later studies demonstrated a crossover of induced resistance to UVA and hydrogen peroxide (Tyrrell, 1985; Sammartano and Tuveson, 1985). However, under different conditions, UVA radiation actually sensitizes bacterial cells to several types of physical and chemical treatment (Tyrrell, 1978), including hydrogen peroxide (Eisenstark, 1989; Kramer and Ames, 1987), even where the interval between treatments is adequate for inducible protection to develop. Nevertheless, an inducible response involving the *oxyR* locus does appear to be important in protection against UVA killing since *oxyR* mutants are hypersensitive to UVA killing (Kramer and Ames, 1987). Catalase has been shown to be induced by monochromatic UVA (334 nm, 365 nm) and near-visible (405 nm) radiation (Tyrrell, unpublished) but not by broad-spectrum UVA radiation (Kramer et al, 1988). UVA does induce a large set of proteins (at least 57; Kramer et al. (1988)) in *S. typhimurium*, including the *oxyR* regulated alkyl hydroperoxide reductase. That the latter enzyme may be of particular importance is indicated by the enhanced sensitivity of Δ*ahp* mutants lacking this enzyme. On the other hand, the inducible catalase, HP-1, does not appear to be important. Moreover, when catalase and alkyl hydroperoxide reductase are overproduced, as in *oxyR1* mutants which constitutively express the regulon, there is a pronounced hypersensitivity to UVA which may result from a photosensitization by these two enzymes (Kramer and Ames, 1987; see Section II.E).

The nucleotide guanosine 5′-diphosphate-3′-diphosphate (ppGpp) is synthesized in response to the UVA-mediated cross-linking of 4-thiouridine in tRNA, the photosensitivity of which is the basis of the UVA-induced growth delays observed in bacteria (Jagger, 1985). Since *E. coli nuv* mutants, which lack 4-thiouridine, are more resistant than wild type to UVA inactivation

(Tsai and Jagger, 1981; Peak et al, 1983), it was proposed that the molecule might act as a photosensitizer. However, *S. typhimurium nuv* mutants irradiated under somewhat different conditions (low fluence rate, irradiation in growth medium) are actually more sensitive to UVA radiation treatment (Kramer et al, 1988). The sensitivity of those strains together with that of *relA* mutants defective in the synthesis of ppGpp (Gallant, 1979) led Kramer et al (1988) to propose that 4-thiouridine acts as a sensor for UVA radiation and that the stimulation of the synthesis of cellular defence proteins occurs as a result of the accumulation of ApppGpp (adenosine 5′,5″-triphosphoguanosine-3″-diphosphate, the adenylated form of ppGpp) whose levels increase over a 100-fold in UVA-irradiated wild-type *S. typhimurium*.

In addition to the *oxyR*-dependent stress response, a whole set of proteins is induced in bacteria by superoxide-generating redox-cycling agents such as menadione and paraquat (see Chapter 5). Certain of these are not induced by H_2O_2 and the signal for induction may be the superoxide anion. Among these proteins are the DNA repair enzyme, endonuclease IV (Chan and Weiss, 1987), and the Mn-SOD (Greenberg and Demple, 1989), which is also induced by increasing oxygen tension (Gregory and Fridovich, 1973). Although UVA radiation may also generate superoxide anion intracellularly, there is no evidence to date that SOD is involved in protecting bacteria against UVA damage.

B. Mammalian cells

In view of the central importance of the UVC-inducible SOS repair response observed in bacteria (Witkin, 1976), several laboratories have searched for an analogous response in mammalian cells and sought to identify proteins induced by UVC (predominantly 254 nm) radiation. Among the proteins increased after UVC and/or UVB treatment of cultured cells are metallothionine (Angel et al, 1986), collagenase (Angel et al, 1986), plasminogen activator (Rotem et al, 1987), the product of the *c-fos* gene (Büscher et al, 1988), interleukin 1 (Kupper et al, 1987) and β-polymerase (Fornace et al, 1989). Although considerable progress has been made in elucidating the signal transduction pathways involved in these inductions (Büscher et al, 1988), their functional significance in terms of cellular defence has not been clarified. However, more recently the UV-stimulated synthesis of a constitutive damage-specific DNA-binding protein has been observed in monkey and human cells (Hirschfield et al, 1990). Since this binding activity is absent in cells derived from xeroderma pigmentosum (group E) patients, the protein may have a function in recognition and repair of DNA damage.

Certain of the RNA transcripts induced by UVC in cultured Chinese hamster cell lines overlap with those induced by hydrogen peroxide (Fornace et al, 1988). Identification of the proteins involved would be a useful contribution towards understanding cellular defence against oxidant stress. A recent report (Morichetti et al, 1989) describes a significant induction of cytochrome P450 and catalase activity in the yeast *S. cerevisiae* after UVC and X-irradiation treatment and the authors propose that the response is related to cellular protection. However, the high levels of stimulation of catalase and/or SOD activities by oxidants that occur in bacteria are not seen in higher eukaryotes. Continuous exposure to paraquat was reported to increase Cu/Zn-SOD (but not Mn-SOD) by a maximum of only 50% in Chinese hamster fibroblasts (Nicotera et al, 1985) although both Cu and Mn-SODs were stimulated to higher levels (two to three times over control) in human Hela cells exposed to paraquat (Krall et al, 1988). Mn-SOD activity is enhanced significantly by both ionizing radiation (Oberley et al, 1987) and tumour necrosis factor (TNF; Wong and Goeddel (1988)), both of which induce oxidative damage (e.g. Zimmerman et al, 1989), and a recent study using cells in which Mn-SOD is overexpressed has clearly shown that this enzyme is essential for protection against the cytotoxicity of TNF (Wong et al, 1989). However, neither constitutive catalase nor SOD protect significantly against UVA-induced cytotoxicity and there is no evidence that treatment of cells with UVA radiation will lead to enhanced levels of these proteins.

There has been considerable interest in the overlap between the heat shock response and the stress response induced by other agents, including oxidants (Lindquist, 1986). One approach has been to study the crossresistance induced by different agents in terms of cell survival. Treatment of Chinese hamster ovary (CHO) fibroblasts with either mild heat or H_2O_2 induces a transient but substantial increase in resistance to a subsequent H_2O_2 challenge with a maximum at 24–36 h (Spitz et al, 1987). Conversely, H_2O_2 treatment induced a small degree of resistance to a 43°C heat challenge, suggesting that this acquired resistance arises by a different mechanism. In an independent study with CHO cells, treatment of cell populations with either low doses of H_2O_2 or xanthine–xanthine oxidase enhanced cellular resistance to both H_2O_2 and ionizing radiation (Laval, 1988). On the other hand, studies with *Drosophila melanogaster* cell lines have indicated that exposure of populations to low levels of H_2O_2 or mild heat, conditions which enhance heat shock gene expression, actually increase sensitivity to a subsequent H_2O_2 challenge. Thus, although it is clear that there is a relationship between the responses induced by heat and oxidants in cells, its nature is little understood.

Protection against oxidant damage may occur on at least three levels:

(1) removal of damage by repair enzymes; (2) prevention of damage by scavenging of active intermediates by antioxidant compounds or enzymes; and (3) prevention of the formation of active intermediates by removal of catalytic molecules, etc. The multiple biological mechanisms by which the level of free iron is kept as low as possible appear to fall in the latter category. Not only free iron but also the iron contained in haem and haem-containing proteins can readily become available to participate in the Fenton reaction in which the active hydroxyl radical is generated from peroxides. Thus, the levels of cellular and circulating haem pools may have a major influence on the generation of active oxygen species under conditions of oxidant stress. There is evidence that haptoglobin and haemopexin (respectively haemoglobin- and haem-transporting proteins) may play an important role in protecting extracellular fluids against radical reactions (Gutteridge, 1987; Gutteridge and Smith, 1988). The cellular uptake and detoxification of iron are rapidly controlled by the levels of transferrin receptor and ferritin respectively (Klausner and Harford, 1989). Thus, the rapid increase in the iron-sequestering protein, ferritin, that occurs in the presence of excess iron may be considered an important inducible cellular defence against iron toxicity.

The cytotoxicity of agents, including UVA radiation, that generate active oxygen intermediates may also be influenced by the levels of intracellular iron and haem. Such a possibility has been given further credibility by our recent finding that treatment of human skin fibroblasts with low fluences of UVA radiation or low concentrations of oxidizing agents such as hydrogen peroxide stimulates the synthesis of high levels of haem oxygenase (Keyse and Tyrrell, 1987, 1989), the principal mammalian enzyme involved in haem catabolism. The phenomenon is not restricted to human fibroblasts and occurs in a variety of cultured human cell types, including epidermal keratinocytes and lymphoblastoid cells, and in cells cultured from a variety of mammals, including monkey, rat, mouse and opossum (Applegate et al, 1991). Since, under the appropriate conditions, haem oxygenase mRNA levels can be enhanced by more than two orders of magnitude within 1–2 h, this stimulation will lead to a dramatic decrease in intracellular haem levels. These observations appear to provide evidence for a major inducible pathway involved in protection against oxidant stress. More detailed studies have shown that the enhancement of gene expression occurs at the level of transcription (Keyse et al, 1990). The presence of iron chelators can prevent the induction of the enzyme (Keyse and Tyrrell, 1990), suggesting that stimulation of gene expression may be mediated by the generation of hydroxyl radicals via the Fenton reaction. Although little is currently known concerning the signal transduction pathways which may lead to the stimulation of gene expression by UVA radiation, recent studies have shown that

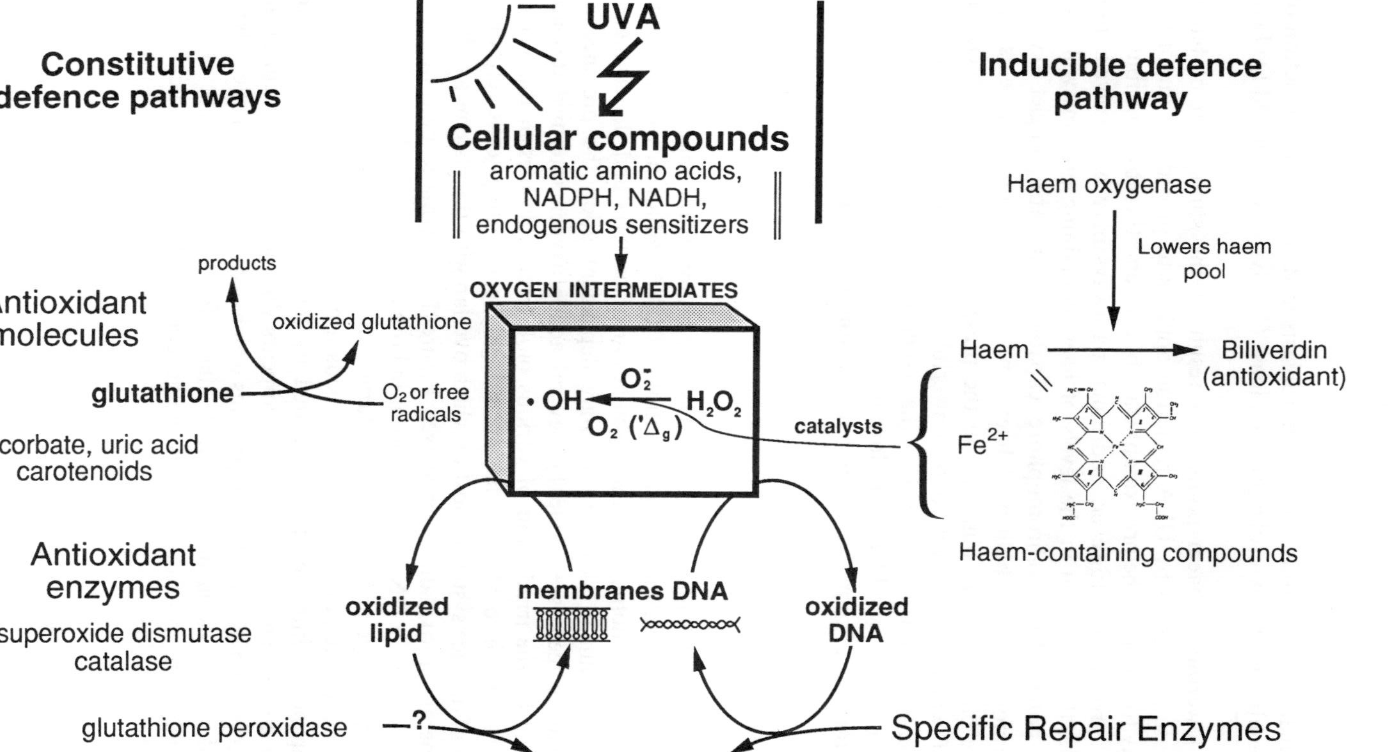

Figure 1. Possible antioxidant defence against UVA radiation damage to mammalian cells.

UVA treatment of cultured mouse fibroblasts leads to a rapid induction of both membrane and cytosolic protein kinase C activity (Matsui and De Leo, 1990).

The lowering of haem pools as a result of the stimulation of haem oxygenase expression by UVA radiation appears to provide an important pathway for altering the prooxidant state of cells and thereby contributes to cellular protection. However, additional pathways may be involved. For example, the reduction in the levels of haem-containing proteins will also remove potential UVA chromophores capable of generating singlet oxygen. Furthermore, the catabolism of haem and haem-containing proteins will produce biliverdin and bilirubin (in the presence of biliverdin reductase), both of which are known to be powerful antioxidants (see Chapter 9). However, the levels of antioxidant that could be generated in skin cells by such a pathway would appear low when compared with the millimolar concentrations of glutathione present constitutively in cells.

V. SUMMARY

Multiple lines of evidence implicate UVA radiation as an important oxidative stress. Many of the biological actions of UVA radiation, which include such diverse phenomena as the inactivation of bacteria and the erythema response in human skin, are strongly dependent upon the presence of molecular oxygen (see Table 1). Several active oxygen intermediates are generated by the irradiation of various biological molecules in this wavelength range. Since UVA radiation is highly penetrating, cells rely on a battery of defence systems, most of which overlap with the systems available for general antioxidant defence (see Figure 1). Enzymatic pathways for repair of oxidative DNA damage occur in both eukaryotes and prokaryotes. However, damage to DNA or other targets may also be completely avoided by the immediate intervention of antioxidant molecules and enzymes which mop up active intermediates generated by the interaction of UVA with cellular components. Finally, it now appears that defence may occur at an even earlier step by induction of proteins which decrease the level of catalytic molecules (such as haem and iron) available for the enhancement of such potentially dangerous reactions as the generation of hydroxyl radical from hydrogen peroxide.

ACKNOWLEDGEMENTS

Research from this laboratory which is mentioned in this review is supported by grants from the Swiss National Science Foundation (3.186.088) and the Swiss League Against Cancer. The author also wishes to express sincere thanks to Drs A Eisenstark, J Jagger, SM Keyse, RD Ley, SH Moss, MJ Peak and KC Smith who were kind enough to read the manuscript and provide helpful comments.

REFERENCES

Agulhon H (1912) *Ann. Inst. Pasteur Paris* **26:** 38–47.

Andley UP & Clark BA (1989) *Photochem. Photobiol.* **50:** 97–105.

Angel P, Pöting A, Mallick U, Rahmsdorf HJ, Schorpp M & Herrlich P (1986) *Mol. Cell Biol.* **6:** 1760–1766.

Applegate LA, Luscher P & Tyrell RM (1991) *Cancer Res.* **51:** 974–978.

Arrick BA, Nathan CF, Griffith OW & Cohn ZA (1982) *J. Biol Chem.* **257:** 1231–1237.

Auletta M, Gange RW, Tan OT & Matzinger E (1986) *J. Invest. Dermatol.* **86:** 649–652.

Azizova AO Islomov AI, Roshchupkin DI, Predvoditelev DA, Remizov AN & Vladimirov YA (1980) *Biophysics* **24:** 407–414.

Biaglow JE, Clark EP, Epp ER, Morse-Guadio M, Varnes ME & Mitchel JB (1983) *Int. J. Radiat. Biol.* **5:** 489–495.

Black HS (1987) *Photochem. Photobiol.* **46:** 213–221.

Black HS & Lo WB (1971) *Nature* **234:** 306–308.

Bloch CA & Ausubel FM (1986) *J. Bacteriol.* **168:** 745–798.

Bose B, Agarwal S & Chatterjee SN (1989) *Radiat. Environ. Biophys.* **28:** 59–65.

Botcherby PK, Magnus IA, Marimo B & Gianneli F (1984) *Photochem. Photobiol.* **39:** 641–649.

Boye E, Johansen I & Brustad T (1974) *J. Bacteriol.* **119:** 522–526.

Bradley MO, Erickson LC & Kohn KW (1978) *Biochim. Biophys. Acta* **520:** 11–20.

Bredberg A (1981) *J. Invest. Dermatol.* **76:** 449–451.

Brown LR & Wüthrich K (1977) *Biochim. Biophys. Acta* **464:** 356–369.

Brown MS & Webb RB (1972) *Mutat. Res.* **15:** 348–352.

Bruls WAG, Sloper H, Van der Leun JC & Berrens L (1984) *Photochem. Photobiol.* **40:** 485–494.

Büscher M, Rahmsdorf HJ, Liftin M, Karin M & Herrlich P (1988) *Oncogene* **3:** 301–311.

Cabrera-Juarez E (1964) *J. Bacteriol.* **177:** 960–964.

Cabrera-Juarez E, Setlow JK, Sevenson PA & Peak MJ (1976) *Photochem. Photobiol.* **23:** 309–314.

Carlioz A & Touati D (1986) *EMBO J.* **5:** 623–630.

Chamberlain J & Moss SH (1987) *Photochem. Photobiol.* **45:** 625–630.

Chan E & Weiss B (1987) *Proc. Natl. Acad. Sci. USA* **84:** 3189–3193.

Christman MF, Morgan RW, Jacobson FS, & Ames BN (1985) *Cell* **41:** 753–762.

Christman MF, Storz G & Ames BN (1989) *Proc. Natl. Acad. Sci. USA* **86:** 3484–3488.

Clark EP, Epp ER, Morse-Gaudio M & Bigalow, JE (1986) *Radiat. Res.* **108:** 238–250.

Connor MJ & Wheeler LA (1987) *Photochem. Photobiol.* **46:** 239–245.

Cotter PA & Gunsalus RP (1989) *J. Bacteriol.* **171:** 3817–3823.
Cunningham ML, Johnson JS, Giovanazzi SM & Peak MJ (1985) *Photochem. Photobiol.* **42:** 125–128.
Czochralska B, Barrosz W & Shugar D (1984) *Biochim. Biophys. Acta* **801:** 403–409.
Danpure HJ & Tyrrell RM (1976) *Photochem. Photobiol.* **23:** 171–177.
Dean SW, Sykes HR, Cole J, Jaspers, NGJ, Linssen P & Verkerk A (1988) *Cancer Res.* **48:** 5374–5376.
Dearden SJ, Hunter TF & Philp J (1981) *Chem. Phys Lett.* **81:** 606–609.
Dearden SJ, Hunter TF & Philp J (1985) *Photochem. Photobiol.* **41:** 213–215.
Demple B & Halbrook J (1983) *Nature* **304:** 466–468.
Demple B, Halbrook J & Linn S (1983) *J. Bacteriol.* **153:** 1079–1082.
Demple B, Johnson A & Fung D (1986) *Proc. Natl. Acad. Sci. USA* **83:** 7731–7735.
Denniston KJ, Webb RB & Brown MS (1972) *Abst. Am. Soc. Microbiol.* 184.
Desai ID, Sawant PL & Tappel AL (1964) *Biochim. Biophys. Acta* **86:** 277–285.
Dethmers JK & Meister A (1981) *Proc. Natl. Acad. Sci. USA* **78:** 7492–7496.
Ditta G, Virts E, Palomares A & Kim CH (1987) *J. Bacteriol.* **169:** 3217–3223.
Eisenstark A (1970) *Mutat. Res.* **10:** 1–6.
Eisenstark A (1971) *Adv. Genet.* **12:** 167–198.
Eisenstark A (1989) *Adv. Genet.* **26:** 99–147.
Eisenstark A & Perrot G (1987) *Mol. Gen. Genet.* **207:** 68–72.
Epstein JH (1977) *Photochem. Photobiol.* **25:** 211–213.
Erickson LC, Bradley MO & Kohn KW (1980) *Biochim. Biophys. Acta* **610:** 105–115.
Fornace AJ & Seres DS (1982) *Mutat. Res.* **94:** 277–284.
Fornace AJ, Alamo I & Hollander MC (1988) *Proc. Natl. Acad. Sci USA* **85:** 8800–8804.
Fornace AJ, Zmudzka B, Hollander MC & Wilson SH (1989) *Mol. Cell. Biol.* **9:** 851–853.
Freeman MF, Malcolm AW & Meredith MJ (1985) *Cancer Res.* **45:** 6313.
Friedberg EC (1985) *DNA Repair*. New York: W.H. Freeman & Co.
Fuchs J, Huflejt MW, Rothfuss LM, Wilson DS, Garcanco G & Packer L (1989) *Photochem. Photobiol.* **50:** 739–744.
Gallant JA (1979) *Annu. Rev. Genet.* **13:** 393–415.
Gantt R, Stephens GM, Stephens EV, Baek AE & Sanford KK (1979) *Biochim. Biophys Acta* **565:** 231–240.
Goni FM, Ondarroa M, Azpiazu I & Macarulla JM (1985) *Biochim. Biophys. Acta* **835:** 549–556.
Graetzer R (1987) *Photochem. Photobiol.* **45:** 425–428.
Greenberg JT & Demple B (1986) *J. Bacteriol.* **168:** 1026–1029.
Greenberg JT & Demple B (1989) *J. Bacteriol.* **171:** 3933–3939.
Gregory EM & Fridovich I (1973) *J. Bacteriol.* **114:** 543–548.
Griffith DW, Anderson MW & Meister A (1979) *J. Biol. Chem.* **254:** 1205–1210.
Gutteridge JMC (1985) *FEBS Lett.* **185:** 19–23.
Gutteridge JMC (1987) *Biochim. Biophys. Acta* **917:** 219–223.
Gutteridge JMC & Smith, A (1988) *Biochem. J.* **256:** 861–865.
Hanson DL & De Leo VA (1989) *Photochem. Photobiol.* **49:** 423–430.
Hirschfield S, Levine AS, Ozato K & Protic M (1990) *Mol. Cell. Biol.* **10:** 2041–2048.
Hodge MR, Kim G, Singh K & Comsky MG (1989) *Mol. Cell. Biol.* **9:** 1958–1964.
Hollaender A (1943) *J. Bacteriol.* **46:** 531–541.
Holmberg M, Almassy Zs, Lagerberg M & Niejahr B (1985) *Photochem. Photobiol.* **41:** 437–444.

Imlay JA & Linn S (1986) *J. Bacteriol.* **166:** 519–527.
Imlay JA & Linn S (1987) *J. Bacteriol.* **169:** 2967–2976.
Imlay JA, Chir SM & Linn S (1988) *Science* **240:** 640–642.
Ito A & Ito T (1983) *Photochem. Photobiol* **37:** 395–401.
Jagger J (1985) *Solar UV Actions on Living Cells.* New York: Praeger.
Jayaraman PS, Gaston KL, Dole JA & Busby SJ (1988) *Mol. Microbiol.* **2:** 527–530.
Keyse SM & Tyrrell RM (1987) *J. Biol. Chem.* **262:** 14821–14825.
Keyse SM & Tyrrell RM (1989) *Proc. Natl. Acad. Sci. USA* **86:** 99–103.
Keyse SM & Tyrrell RM (1990) *Carcinogenesis* **11:** 787–791.
Keyse SM, Moss SH & Davies DJG (1983) *Photochem. Photobiol.* **37:** 307–312.
Keyse SM, McAleer MA, Davies DJG & Moss SH (1985) *Int. J. Radiat. Biol.* **48:** 975–985.
Keyse SM, Applegate LA, Tromvoukis Y & Tyrrell RM (1990) *Mol. Cell Biol.* **10:** 4967–4969.
Kjeldstad B (1984) *Z. Naturforsch.* **39c:** 300–302.
Kjeldstad B & Johnsson A (1986) *Photochem. Photobiol.* **43:** 67–70.
Klamen D & Tuveson RW (1982) *Photochem. Photobiol* **35:** 167–173.
Klausner RD & Harford JB (1989) *Science* **246:** 870–872.
Kovacs AM, Rossing W, Will III OH & Roddat M (1987) *Photochem. Photobiol.* **45:** 493–499.
Krall J, Bagley AC, Mellenbach GT, Hallewell RA & Lynch RE (1988) *J. Biol. Chem.* **263:** 1910–1914.
Kralli A & Moss SH (1987) *Br. J. Dermatol.* **116:** 761–772.
Kramer GF & Ames BN (1987) *J. Bacteriol.* **169:** 2259–2266.
Kramer GF, Baker JC & Ames BN (1988) *J. Bacteriol.* **170:** 2344–2351.
Krinsky NI (1976) In Gray TGR & Postgate JR (eds) *The Survival of Vegetative Microbes*, pp 209–239. Cambridge University Press.
Kupper TS, Chua AO, Flood P, McGuire J & Gubler V (1987) *J. Clin. Invest.* **80:** 430–436.
Lach LW, Hagen TM & Jones DP (1986) *Proc. Natl. Acad. Sci USA* **83:** 4641–4645.
Lai LW, Ducore JM & Rosenstein BS (1987) *Photochem. Photobiol.* **46:** 143–146.
Laval F (1988) *Mutat. Res.* **201:** 73–79.
Ley RD, Sedita BA & Boye E (1978) *Photochem. Photobiol.* **27:** 323–327.
Lindquist S (1986) *Annu. Rev. Biochem.* **55:** 1151–1191.
Loewen PC (1984) *J. Bacteriol.* **157:** 622–626.
Loewen PC & Triggs BL (1984) *J. Bacteriol.* **160:** 668–675.
Loewen PC, Triggs BL, Klassen GR & Weiner, JH (1983) *Can. J. Biochem. Gen. Biol.* **61:** 1315–1321.
Love JD, Vivino AA & Minton KW (1986) *J. Cell. Physiol.* **126:** 60–68.
Mandal TK & Chatterjee SN (1980) *Radiat. Res.* **83:** 290–303.
Marykwas DL & Fox TD (1989) *Mol. Cell. Biol.* **9:** 484–491.
Mathews-Roth MM (1982) *Oncology* **38:** 33–37.
Mathews-Roth MM (1986) *Photochem. Photobiol.* **43:** 91–93.
Mathews-Roth MM & Krinsky NI (1985) *Photochem. Photobiol.* **42:** 35–38.
Mathews-Roth MM & Krinsky NI (1987) *Photochem. Photobiol.* **46:** 507–509.
Matsui MS & De Leo VA (1990) *Carcinogenesis* **11:** 229–234.
McCormick JP, Fisher JR, Pachlatko JP & Eisenstark A (1976) *Science* **198:** 468–469.
McCormick JP, Klita S, Terry J, Schrodt M & Eisenstark A (1982) *Photochem. Photobiol.* **36:** 367–369.

Meister A & Anderson ME (1983) *Annu. Rev. Biochem.* **52:** 711–760.
Mello-Filho AC, Hoffmann ME & Menighini R (1984) *Biochem. J.* **218:** 273–275.
Meredith MJ (1988) *Cancer Res.* **48:** 5374–5376.
Meredith MJ & Dodson MC (1987) *Cancer Res.* **47:** 4576–4581.
Miguel AG & Tyrrell RM (1986) *Biophys J.* **49:** 485–491.
Mitchell RL & Anderson IC (1965) *Science* **150:** 74.
Morgan RW, Christman MF, Jacobson F, Storz G & Ames BN (1986) *Proc. Natl. Acad Sci. USA* **83:** 8059–8063.
Morichetti E, Cundari E, Del Carratore R & Bronzetti G (1989) *Yeast* **5:** 141–148.
Moss SH & Smith KC (1981) *Photochem. Photobiol.* **33:** 203–210.
Nicotera TM, Block AW, Gibas Z & Sandberg AA (1985) *Mutat. Res.* **151:** 263–268.
Oberley LW, St Clair DK, Autor AP & Oberley TD (1987) *Arch. Biochem. Biophys.* **254:** 69–80.
Parrish JA, Rox Anderson R, Urbach F & Pitts D (1978) *UVA, Biological Effects of Ultraviolet Radiation with Emphasis on Human Responses to Longwave Ultraviolet.* New York: John Wiley and Sons.
Parsons PG & Hayward IP (1985) *Photochem. Photobiol.* **42:** 287–293.
Pathak MA, Riley FC & Fitzpatrick TB (1962) *J. Invest. Dermatol.* **39:** 435–443.
Peak JG, Foote CS & Peak MJ (1981) *Photochem. Photobiol.* **34:** 45–49.
Peak JG, Peak MJ & Foote CS (1982) *Photochem. Photobiol.* **36:** 413–417.
Peak JG, Peak MJ & Tuveson RW (1983) *Photochem. Photobiol.* **38:** 541–543.
Peak MJ & Peak JG (1975) *Photochem. Photobiol.* **22:** 147–148.
Peak MJ & Peak JG (1980) *Radiat. Res.* **83:** 553–558.
Peak MJ & Peak JG (1990) *Photochem. Photobiol.* **51** 649–652.
Peak MJ, Peak JG & Webb RB (1973) *Mutat. Res.* **20:** 137–141.
Peak MJ, Peak JG & Nerad L (1983) *Photochem. Photobiol.* **37:** 169–172.
Peak MJ, Peak JG & Jones CA (1985) *Photochem. Photobiol.* **42:** 141–146.
Peak MJ, Johnsson JS, Tuveson RW & Peak JG (1987a) *Photochem. Photobiol.* **45:** 473–478.
Peak MJ, Peak JG & Carnes BA (1987b) *Photochem. Photobiol.* **45:** 381–387.
Perry KL, Simonitch TA, Harrison-Lavoie KJ & Liv S-T (1986) *J. Bacteriol.* **168:** 607–612.
Peters J & Jagger J (1981) *Nature (Lond.)* **289:** 194–195.
Pitts DG, Cameron LL, Lone JG et al (1986) In Waxler M & Hitchins VM (eds) *Optical Radiation and Visual Health*, pp 5–42. Boca Raton, Florida: CRC Press.
Putvinsky AV, Sokolov AV, Roshchupkin DI & Vladimirov YA (1979) *FEBS Lett.* **106:** 53–55.
Rogers SG & Weiss B (1980) *Methods Enzymol.* **65:** 201–211.
Rorsman H, & Tegner E (1988) *Photodermatology* **5:** 30–38.
Rosenstein BS, Lai L, Ducore JM & Rosenstein RB (1989) *Mutat. Res.* **217:** 219–226.
Roshchupkin DI, Pelenitsyn AB, Potapenko YY, Talitsky VV & Vladimirov YA (1975) *Photochem. Photobiol* **21:** 63–69.
Rotem N, Axelrod JH & Miskin R (1987) *Mol. Cell. Biol.* **7:** 622–631.
Roza L, Van der Schans GP & Lohman PHM (1985) *Mutat. Res.* **146;** 89–98.
Sak BD, Eisenstark A & Touati D (1989) *Proc. Natl. Acad. Sci. USA* **86:** 3271–3275.
Sakamoto H & Touati D (1984) *J. Bacteriol.* **159:** 418–420.
Sammartano LJ & Tuveson RW (1983) *J. Bacteriol.* **156:** 904–906.
Sammartano LJ & Tuveson RW (1984) *Photochem. Photobiol.* **40:** 607–612.
Sammartano LJ & Tuveson RW (1985) *Photochem. Photobiol.* **41:** 367–370.
Sammartano LJ & Tuveson RW (1987) *J. Bacteriol.* **169:** 5304–5307.

Sammartano LJ, Tuveson RW & Davenport R (1986) *J. Bacteriol.* **168:** 13–21.
Sargentini, NJ & Smith KC (1986) *Radiat. Res.* **105:** 180–185.
Shiloh Y, Tabor E & Becker Y (1983) *Carcinogenesis* **4:** 1317–1322.
Smith PJ & Patterson MC (1982) *Photochem. Photobiol.* **36:** 333–343.
Spitz DR, Dewey WC & Li GC (1987) *J. Cell. Physiol.* **131:** 364–373.
Szebini J & Tollin G (1988) *Photochem. Photobiol.* **47:** 475–479.
Taylor AMR, Harnden DG, Arlett CF et al (1975) *Nature (Lond.)* **258:** 427–429.
Tegner E, Rosman H & Rosengren E (1983) *Acta Derm. Venereol. (Stockholm)* **63:** 21–25.
Thomas SA, Sargent ML & Tuveson RW (1981) *Photochem. Photobiol.* **33:** 349–354.
Touati D (1983) *J. Bacteriol.* **155:** 1078–1087.
Town CD, Smith KC & Kaplan HS (1973) *Radiat. Res.* **55:** 334–345.
Triggs-Raine BL & Loewen PC (1987) *Gene* **52:** 121–128.
Tsai S-C & Jagger J (1981) *Photochem. Photobiol.* **33:** 825–834.
Tuveson RW & Jonas RB (1979) *Photochem. Photobiol.* **30:** 667–676.
Tuveson RW & Sammartano LJ (1986) *Photochem. Photobiol.* **43:** 621–626.
Tuveson RW, Peak JG & Peak MJ (1982) *Photochem. Photobiol.* **37:** 109–112.
Tuveson RW, Larsson RA & Kagan J (1988) *J. Bacteriol.* **170:** 4675–4680.
Tyrrell RM (1973) *Photochem. Photobiol.* **17:** 69–73.
Tyrrell RM (1976) *Photochem. Photobiol.* **23:** 13–20.
Tyrrell RM (1978) *Photochem. Photobiol. Rev.* **3:** 35–113.
Tyrrell RM (1984) In Hurst A & Nasim A (eds) *Repairable Lesions in Microorganisms,* pp 85–124. London: Academic Press.
Tyrrell RM (1985) *Mutat. Res.* **145:** 129–136.
Tyrrell RM & Pidoux M (1986) *Photochem. Photobiol.* **44:** 561–564.
Tyrrell RM & Pidoux M (1988) *Photochem. Photobiol.* **47:** 405–412.
Tyrrell RM & Pidoux M (1989) *Photochem. Photobiol.* **49:** 407–412.
Tyrrell RM & Webb RB (1973) *Mutat. Res.* **19:** 361–364.
Tyrrell RM, Ley RD & Webb RB (1974) *Photochem. Photobiol.* **20:** 395–398.
Wang RJ, Stoien JD & Landa F (1974) *Nature (Lond.)* **247:** 43–45.
Webb RB (1977) *Photochem. Photobiol. Rev.* **2:** 169–261.
Webb RB & Brown MS (1976) *Photochem. Photobiol.* **24:** 425–432.
Webb RB & Lorenz JR (1970) *Photochem. Photobiol.* **12:** 283–289.
Webb RB, Brown MS & Tyrrell RM (1976) *Mutat. Res.* **37:** 163–172.
Wheeler LA, Aswad A, Connor MJ & Lowe N (1986) *J. Invest. Dermatol.* **87:** 658–662.
Will III OH, Jankowski P, Jorre K et al (1987) *Photochem. Photobiol.* **45:** 609–615.
Witkin EM (1976) *Bacteriol. Rev.* **40:** 869–907.
Wong GHW & Goeddel DV (1988) *Science* **242:** 941–944.
Wong GHW, Elwell JH, Oberley LW & Goeddel DV (1989) *Cell* **58:** 923–931.
Youn JI, Gange RW, Maytum D & Parrish JA (1988) *Photodermatology* **5:** 252–256.
Zagorec M, Buhler JM, Treich I, Keng T, Guarente L & Labbe-Bois R (1988) *J. Biol. Chem.* **263:** 9718–9724.
Zimmerman RJ, Chan A & Leadon SA (1989) *Cancer Res.* **49:** 1644–1648.
Zitomer RS, Sellers JW, McCarter DW, Hastings GA, Wick P & Lowry CV (1987) *Mol. Cell. Biol.* **7:** 2212–2220.

Sargent [illegible] ML (1969) Kenaga [illegible] 183.
Sutton Y, Taber [illegible] Becker Y (1983) [illegible] 1327–1332.
Smith TJ & Paterson MC (1984) Photochem. Photobiol. 39: 33–43.
Spitz DR, [illegible] WC & [illegible] (1987) J. Cell. Physiol. 131: 364–373.
[illegible] & [illegible] (1988) Photochem. Photobiol. 47: 425–429.
Taylor AMR, [illegible] CR, Arlett CF et al (1975) Nature (Lond.) 258: 427–429.
[illegible] (1983) [illegible] J. [illegible] 6: 21–28.
Thomas SA, Sargent ML & [illegible] RW (1981) Photochem. Photobiol. 33: 349–354.
[illegible] P (1983) J. Bacteriol. 155: 1075–1087.
[illegible] Kaplan HS (1962) Bacteriol. Rev. 26: 134–145.
[illegible] (1987) Cell 51: [illegible]
[illegible] (1991) Photochem. Photobiol. [illegible]
[illegible] & Jones RB (1979) Photochem. Photobiol. [illegible]
[illegible] (1986) Photochem. Photobiol. 43: 221–228.
Tuveson RW, Peak JG & Peak MJ (1983) Photochem. Photobiol. 37: 109–112.
Tuveson RW, Larson RA & Kagan J (1988) J. Bacteriol. 170: 4675–4680.
Tyrrell RM (1973) Photochem. Photobiol. 17: 69–73.
Tyrrell RM (1976) Photochem. Photobiol. 23: 13–20.
Tyrrell RM (1978) Photochem. Photobiol. Rev. 3: 35–113.
Tyrrell RM (1984) In Hurst A & Nasim A (eds) Repairable Lesions in Microorganisms, pp 85–124. London: Academic Press.
Tyrrell RM (1985) Mutat. Res. 145: 129–136.
Tyrrell RM & Pidoux M (1986) Photochem. Photobiol. 44: 561–564.
Tyrrell RM & Pidoux M (1988) Photochem. Photobiol. 47: 405–412.
Tyrrell RM & Pidoux M (1989) Photochem. Photobiol. 49: 407–412.
Tyrrell RM & Webb RB (1973) Mutat. Res. 19: 361–364.
Tyrrell RM, Ley RD & Webb RB (1974) Photochem. Photobiol. 20: 395–398.
Wang [illegible], [illegible] (1974) Nature (Lond.) 247: 43–45.
Webb RB (1977) Photochem. Photobiol. Rev. 2: 169–261.
Webb RB & Brown MS (1976) Photochem. Photobiol. 24: 425–432.
Webb RB & [illegible] (1979) Photochem. Photobiol. 12: 483–489.
Webb RB, Brown MS & Tyrrell RM (1978) Mutat. Res. 59: 1–13.
[illegible] LA, [illegible] & [illegible] N (1990) J. Invest. Dermatol. 87: 634–636.
[illegible] (1987) Photochem. Photobiol. 45: 609–615.
[illegible] RM (1986) Microbiol. Rev. 49: [illegible]
[illegible] & [illegible] DV (1988) Science 242: 941–944.
Wong [illegible], [illegible] & Goeddel DV (1989) Cell 58: 923–931.
[illegible] (1988) Photochem. Photobiol. [illegible]
[illegible] (1988) [illegible] 2: [illegible]
Zimmerman RJ, Chan A & Leadon SA (1989) Cancer Res. 49: 1644–1648.
Zmudzka [illegible], [illegible] (1988) Proc. Natl. Acad. Sci. USA 85: [illegible]

3

Non-Radical Mechanisms for the Oxidation of Glutathione

D. M. ZIEGLER

The Clayton Foundation Biochemical Institute, Department of Chemistry and Biochemistry, The University of Texas at Austin, Austin, Texas 78712, USA

I. Introduction 85
II. Non-radical mechanisms for the oxidation of GSH 87
III. Extracellular oxidation of labile glutathione conjugates 94
IV. Concluding remarks 95

I. INTRODUCTION

Glutathione (GSH) is often the first line of defence against tissue injury from administered or metabolically generated electrophiles. The critical role of this tripeptide in protecting tissues from reactive xenobiotic metabolites has stimulated considerable interest in the underlying biochemical mechanisms. Alkylation of electrophilic metabolites or reduction of reactive oxygen species are considered the major protective mechanisms of GSH. (See the following recent reviews for details: Anders (1988), Cotgreave et al (1988), Sies (1988) and Smith et al (1983).) Alkylation, followed by excretion of the conjugate, is a major route for detoxification of metabolically generated soft electrophiles, but when the rate of GSH loss exceeds the rate of GSH biosynthesis, the tissue becomes susceptible to irreversible injury from alkylation of cellular macromolecules by metabolically activated xenobiotics. A detailed description of biochemical mechanisms and of changes

Oxidative Stress: Oxidants and Antioxidants
ISBN 0-12-642762-3

in essential cellular processes resulting from loss of GSH by conjugation is beyond the scope of this chapter. The reader is referred to excellent recent reviews on biochemical mechanisms for alkylation (Anders, 1988), transport of conjugates (Sies et al, 1989) and metabolic effects produced by GSH loss (Reed and Olafsdotter, 1989; Cotgreave et al, 1988).

Drug-dependent oxidation of GSH to GSSG is also a major pathway for loss of this tripeptide induced by some classes of xenobiotics. Drugs bearing functional groups that can transfer one or two electrons to molecular oxygen via enzyme-catalysed reactions can generate damaging quantities of reactive oxygen species in tissues containing reductases capable of sustaining the reaction. The mechanisms of several such reactions and the metabolic consequences are known in considerable detail; see recent reviews by Sies (1988), Cotgreave et al (1988), Smith et al (1983) and Ziegler (1988a). While the superoxide anion is usually the precursor and often an essential intermediate in the metabolic generation of oxy radicals, its direct measurement in whole cells or crude tissue preparations is complicated by the presence of endogenous superoxide dismutases. However, its dismutation product, H_2O_2, can be determined by spectrophotometric or fluorometric methods. Several methods for determining rates of drug-dependent H_2O_2 formation catalysed by tissue homogenates and/or subcellular fractions are available (Hildebrandt et al, 1978; Hyslop and Sklar, 1984). Methods for measuring rates of H_2O_2 formation in live animals or intact tissues have also been described (Chance et al, 1979; Sies, 1981), but the velocities of drug-dependent oxy radical reactions in intact tissues are most often estimated from changes in the rate of GSSG efflux. The glutathione peroxidase catalysed reduction of peroxide formed directly, or by dismutation of O_2^-, yields GSSG. Several reports have shown that when GSH oxidation exceeds the catalytic capacity of GSH reductases, GSSG is actively excreted (Akerboom et al, 1982; Eklöw et al, 1984; Adams et al, 1983). In the liver, GSSG is selectively transported in bile (Sies et al, 1978; Burk et al, 1978; Akerboom et al, 1982) and the biliary efflux of GSSG is one of the best models for measuring drug-dependent oxy radical generation when the underlying biochemical mechanisms are known.

However, not all drugs that stimulate the excretion of GSSG in intact tissue generate oxy radicals, and some radical-independent pathways are described in the following sections. The discussion will be limited to pathways unambiguously supported by both in vitro and in vivo studies.

In addition, evidence supporting a hydroperoxide-independent mechanism for the apparent efflux of GSSG from perfused tissues or cultured cells treated with drugs bearing *N*-methylamine groups will be reviewed. While many of these drugs stimulate the reduction of oxygen to superoxide anions in reactions catalysed by isolated microsomes or by purified microsomal monooxygenases, attempts to demonstrate these reactions in intact tissues have not been successful. The reasons for the discrepancies are not known but some mechanisms for drug-dependent generation of oxy radicals in vitro may not occur at significant rates in vivo.

II. NON-RADICAL MECHANISMS FOR THE OXIDATION OF GSH

A. By diamide

The oxidation of GSH to GSSG by diamide (reaction 1) is perhaps the best known and most widely used method to oxidize GSH by a non-radical mechanism.

$$(CH_3)_2N\overset{\overset{\displaystyle O}{\|}}{C}{-}N{=}N{-}\overset{\overset{\displaystyle O}{\|}}{C}N(CH_3)_2 + 2GSH \longrightarrow GSSG + (CH_3)_2N\overset{\overset{\displaystyle O}{\|}}{C}{-}NHNH{-}\overset{\overset{\displaystyle O}{\|}}{C}N(CH_3)_2 \qquad (1)$$

This agent was specifically developed by Kosower et al (1969) to oxidize GSH to GSSG in intact tissues without intervention or spurious generation of radical intermediates often accompanying the oxidation of thiols by other reagents (Kosower and Kosower, 1974). The reaction appears selective for thiols and is quite rapid even at 2–5°C. The use of diamide for the oxidation of GSH is so well known that further discussion here would be of only marginal interest.

B. By organic nitrates

Metabolic reduction of nitroglycerine was first described by Needleman and Harkey (1971), who reported that nitrite was formed from nitroglycerine perfused into rat liver. While the nature of the reductant was not defined, the fall in liver GSH suggested that this tripeptide was probably involved.

The subsequent studies of Habig et al (1975) and Keen et al (1976) demonstrated that the glutathione *S*-transferases accounted for essentially all of the organic nitrate reductase activity in liver. Most of the glutathione *S*-transferase isoforms catalyse formation of glutathione sulphenyl nitrite from GSH and alkyl nitrates as illustrated in reaction 2.

$$R{-}ONO_2 + GSH \longrightarrow ROH + GS{-}NO_2 \tag{2}$$

$$GS{-}NO_2 + GSH \longrightarrow GSSG + NO_2^- + H^+ \tag{3}$$

The sulphenyl nitrite reacts rapidly and non-enzymatically with another equivalent of GSH, generating nitrite and GSSG (reaction 3).

A more recent report (Hill et al, 1989) demonstrated that infusion of nitroglycerine or isosorbide dinitrate into rat liver stimulates biliary efflux of GSSG. The reduction of the organic nitrates was quite rapid and the efflux of nitrite and GSSG correlated with the substrate specificities of the rat liver glutathione *S*-transferases. For instance, the turnover with nitroglycerine is faster than with isosorbide dinitrate (Keen et al, 1976) and at equivalent concentrations the efflux of GSSG was considerably greater with the former, which supports the conclusion that oxidation of GSH to GSSG by alkyl nitrates occurs as illustrated in reactions 2 and 3. Although not tested, oxygen is apparently not required for the oxidation of GSH by alkyl nitrates, and because glutathione *S*-transferases are the principal (and perhaps the only) enzymes required for the metabolism of these alkyl nitrates, intermediate radicals are not involved.

Oxidation of GSH by xenobiotics bearing other functional groups readily reduced by thiols is also possible but none has been examined in detail. A systematic investigation of other types of reductive reactions potentially catalysed by glutathione *S*-transferases may be a productive area for further investigation.

C. By xenobiotics bearing a thiol function

1. *Reactions*

The oxidation of GSH to GSSG by the agents discussed above is essentially stoichiometric with diamide or alkyl nitrates as indicated in reactions 1–3. In contrast to these, the xenobiotics dealt with in this section appear to act catalytically, and relatively small amounts can lead to the oxidation of several hundred molar equivalents of GSH. Although structurally diverse, all compounds in this category contain a thione or thiol fuction that can be oxygenated by microsomal oxygenases to sulphenic acids (reaction 4). Sulphenic acids are extremely reactive and will form mixed disulphides with any available thiol, including GSH, as shown in reaction 5. Most mixed disulphides will exchange with another equivalent of GSH, regenerating the parent xenobiotic (reaction 6).

$$\mathrm{R{-}SH + NADPH + H^+ + O_2 \longrightarrow RSOH + NADP^+ + H_2O} \qquad (4)$$

$$\mathrm{RSOH + GSH \longrightarrow RS{-}SG + H_2O} \qquad (5)$$

$$\mathrm{RS{-}SG + GSH \rightleftharpoons RSH + GSSG} \qquad (6)$$

$$\mathrm{NADPH + 2GSH + H^+ + O_2 \longrightarrow NADP^+ + GSSG + 2H_2O} \qquad (7)$$

The sum of reactions 4–6 suggests that xenobiotics oxidized to sulphenic acids can catalyse the oxidation of GSH at the expense of NADPH and oxygen (reaction 7) and the velocity will depend largely on the rate of the slowest reaction. Reactions 5 and 6 are probably never rate-limiting in vivo since the non-enzymatic reaction of sulphenic acids with thiols is extremely fast and the exchange (reaction 6) is not only reasonably fast above pH 7.4, but is also catalysed by thioltransferases (Askelöf et al, 1974). The oxidation of GSH initiated by xenobiotics of this type, therefore, depends primarily on the kinetic parameters for the oxidation of the xenobiotic to the sulphenic acid (reaction 4). The oxidation of most xenobiotics bearing a thione or thiol function is catalysed primarily by microsomal flavin-containing monooxygenases (FMOs), and some of the properties of these enzymes are summarized in the following section.

2. *Properties of multisubstrate flavin-containing monooxygenases*

The major steps in the catalytic cycle of these enzymes, deduced primarily from studies on the purified porcine liver flavoprotein (Poulsen and Ziegler, 1979; Beaty and Ballou, 1981a,b), are illustrated in Figure 1. Flavoproteins of this type are present within the cell predominantly in the 4a-hydroperoxyflavin form, and in the first step of the cycle electrophilic attack by the terminal hydroperoxide oxygen on the xenobiotic nucleophile (S) followed by heterolytic cleavage of the enzyme-bound hydroperoxide yield the oxygenated product (SO) and the flavin pseudobase. The oxygenated xenobiotic immediately diffuses from the site and both S and SO are present on the enzyme only in the first step. Steps 2–5 simply regenerate the oxygenating form of the enzyme from NADPH, oxygen and H^+. Oxygen activation in the absence of the oxygenatable substrate is unique to this class of flavoproteins and distinguishes them from all other oxidases or monooxygenases bearing flavin, haem or other redox-active prosthetic groups. The enzyme is apparently present in this, the activated oxygen form, and any soft nucleophile that can contact the enzyme-bound oxidant will be oxidized. Because only a single point of contact between enzyme and substrate is essential for activity, the specificity of FMOs is exceptionally broad. (See a recent review, Ziegler (1988b), for evidence supporting this point.)

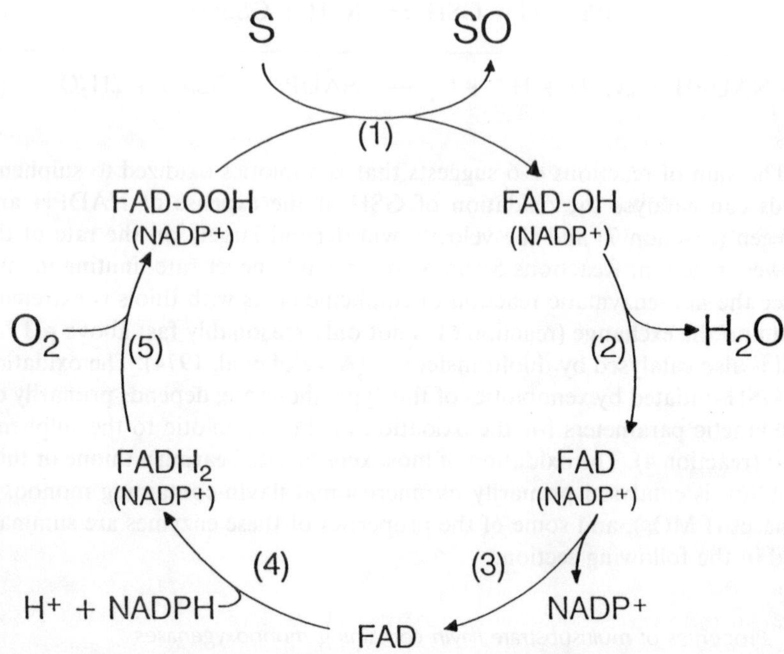

Figure 1. Major steps in the catalytic cycle of multisubstrate flavin-containing monooxygenases. (Figure taken from Ziegler (1991). Reproduced by permission.)

While the basic steps shown in Figure 1 are common to all flavoproteins classified as multisubstrate FMOs, recent studies (Williams et al, 1984a; Tynes et al, 1985; Ozols, 1989) have shown that isoforms with distinct but overlapping substrate specificities are present in some species. For instance, pulmonary FMO, the major microsomal protein in lung tissue from pregnant rabbits (Williams et al, 1984b), does not catalyse the oxidation of chloropromazine, imipramine or thioridazine, which are among the better substrates for hepatic FMO. This is not due to the lack of *N*-oxygenase activity of the pulmonary enzyme, because both the lung and liver flavoproteins catalyse *N*-oxidation of trifluoroperazine and *N*,*N*-dimethylaniline. A recent study (Nagata et al, 1990) suggests that some, but not all, of the differences in substrate specificities between hepatic and pulmonary enzymes may be due to a somewhat narrower substrate channel in the lung enzyme. A nucleophilic heteroatom that projects no more than 5 Å from a bulky group larger than 8 Å in its longest dimension apparently cannot contact the 4a-hydroperoxyflavin in pulmonary FMO whereas nucleophilic centres attached to less bulky groups or on the end of slender side chains more than 6–7 Å in length are readily oxidized.

The more restricted access of large molecules to the 4a-hydroperoxyflavin in the rabbit lung enzyme may explain differences in pulmonary toxicity of phenylthiourea and 1,3-diphenylthiourea in this species. It has been known for some time that thiourea and many of its mono *N*-substituted derivatives produce pulmonary oedema and pleural effusion in susceptible animals (Dieke et al, 1947), whereas symmetrical *N,N′*-diarylthioureas are not pulmonary toxins for rabbits (Smith and Williams, 1961) and are considerably less toxic for rats (Dieke and Richter, 1945). The differences in pulmonary toxicities of mono- and diaryl-substituted thioureas correlate with the substrate specificities of lung and liver FMO. Rabbit lung FMO, unlike liver FMO, does not catalyse *S*-oxygenation of 1,3-diphenylthiourea, whereas thiourea, phenylthiourea and naphthylthiourea are substrates. The substrate specificity of rat lung FMO(s) for these substrates is not known, but the immunochemical measurements of Tynes and Philpot (1987) suggest that, unlike those of rabbits, both 'hepatic' and 'pulmonary' forms of the FMO are expressed in lung tissue from rats. Because the oxidation of thioureas to their sulphenic acids (reaction 4) is the critical step leading to the uncontrolled thiocarbamide-dependent oxidation of GSH to GSSG (reactions 4–6), it would appear that differences in the type and amounts of each FMO expressed in lung tissue may account for species differences in pulmonary toxicities of different thioureas.

The tissue distribution and substrate specificities of FMOs in humans is, at present, unknown. A form similar to the porcine enzyme has been detected in adult (Gold and Ziegler, 1973) and fetal (Rane, 1973) human liver, and is undoubtedly present in other tissues. However, the extreme lability of this flavoprotein in human tissues (Gold and Ziegler, 1973) has hampered in vitro measurements of this enzyme (or enzymes). Recent studies of Smith and his associates in London (Al-Waiz et al, 1987a,b) have described a genetic polymorphism in humans that is undoubtedly due to a defect in one or more of the FMOs. Affected individuals have a decreased capacity for *N*-oxidation of trimethylamine and nicotine (Ayesh et al, 1988). These have been the only two compounds studied in affected individuals, but because of the exceptionally broad specificities of these enzymes, it is quite likely that the metabolism of many different soft nucleophiles, including thiocarbamides and other xenobiotics bearing thiol or thione functions, may be altered in humans bearing this defect.

3. Xenobiotics oxidized to sulphenic acids

In addition to thiocarbamides, other functional groups bearing sulphur that are oxidized to products with the properties of sulphenic acids via FMOs include alkylthiols, arylthiols, mercaptoimidazoles and mercaptopyrimidines. The sulphenic acid derivatives (formamidine sulphenic acids) of thiocarbamides have been isolated and identified as the first oxidation products of phenylthiourea and ethylenethiourea in reactions catalysed by purified preparations of the porcine liver FMO (Poulsen et al, 1979). Although the formation of sulphenic acids has been unambiguously proven only for these two compounds, the properties of the initial oxidation products of aminothiols and mercaptoaromatic heterocyclic amines (Table 1) are also consistent with their respective sulphenic acids. They, like thiocarbamides, are substrates for FMOs and are oxidized to products reduced by GSH (reactions 5 and 6). The drug-dependent oxidation of GSH has been demonstrated in vitro for all of the compounds listed in Table 1.

Table 1. Kinetic constants for the oxidation of sulphur xenobiotics catalysed by the porcine liver flavin-containing monooxygenase.

Xenobiotic	$k_{cat}/K_m \times 10^{-3}$
Thiocarbamides[a]	
Thiourea	1500
Ethylenethiourea	1030
Phenylthiourea	6400
Thiocarbanibide	5300
Mercaptoimidazoles[a]	
Methimazole	5300
2-Mercaptobenzimidazole	2900
Mercaptopyrimidines[a]	
Propylthiouracil	36
Aminothiols[b]	
N,N-Dimethylcysteamine	290
Piperazinylethanethiol	60
Guanidinoethanethiol	50

Kinetic constants based on data from:
[a] Poulsen et al (1979)
[b] Ziegler et al (1983).

However, overall oxidation of GSH to GSSG in the intact tissues can occur only when the velocity of GSH oxidation by the metabolically generated sulphenic acid exceeds the catalytic capacity of cellular GSSG reductive enzymes. This appears to be the case for this type of xenobiotic because the report by Krieter et al (1984) demonstrated that biliary efflux of GSSG increased three- to four-fold in rat liver perfused with micromolar concentrations of thiourea, phenylthiourea and methimazole. The thiol function appears essential, since the oxygen analogues of these sulphur xenobiotics had little or no effect on the rate of GSSG efflux. The oxidation of the xenobiotic to a sulphenic acid appears essential because other sulphur substrates for the FMO that do not yield this intermediate did not affect release of GSSG. The same report also showed that the biliary efflux of GSSG increased dramatically and remained elevated for several hours in live rats treated with less than 200 μmol methimazole (kg body weight)$^{-1}$.

The oxidation of GSH and other cellular thiols in animals treated with radioprotective aminothiols (Mondovi et al, 1962) probably also occurs by the same mechanism. Although the sulphenic acids of aminothiols cannot be isolated, the aminothiol disulphides (the first stable products detected in reactions catalysed by the purified monooxygenase; Ziegler, 1988b) undoubtedly arise from the reaction of a second equivalent of the aminothiol with the enzymatically generated sulphenic acid (reaction 8).

$$\rangle NCH_2CH_2SOH + \rangle NCH_2CH_2SH \longrightarrow \rangle NCH_2CH_2S{-}SCH_2CH_2N\langle + H_2O \qquad (8)$$

Either the disulphide or the sulphenic acid can oxidize GSH to GSSG and the metabolic generation of these products is responsible for the extensive oxidation of tissue thiols in animals treated with radioprotective doses of aminothiols (Eldjarn and Pihl, 1956; Mondovi et al, 1962). More recent studies by Lauterburg and Mitchell (1981) have shown that the biliary excretion of GSSG increases almost 300-fold over the basal rate in rats treated with 5 mmol cysteamine (kg body weight)$^{-1}$. These reports suggest that the aminothiols, like thiocarbamides and the mercaptoaromatic heterocyclic amines, can initiate and sustain a futile cycle that oxidizes GSH to GSSG by a non-radical mechanism.

Although never demonstrated experimentally with intact tissues, studies with the purified FMO indicate that a number of other functional groups bearing sulphur are also oxidized to thiol-reactive products. Thioacids, dithioacids and dithiocarbamates are all oxidized to reactive intermediates via FMO (Taylor and Ziegler, 1987). While thioacids or dithioacids are rarely encountered, a dithiocarbamate disulphide (disulfiram) is used clinically. This disulphide can be reduced by GSH to *N,N*-diethyldithiocarbamate and the latter is oxidized to a thiol-reactive intermediate at the same rate as thiourea by the FMO. If the reduction of disulfiram by GSH is sufficiently rapid in vivo, this drug may, like thiocarbamides, be capable of sustaining a futile cycle oxidizing GSH to GSSG.

III. EXTRACELLULAR OXIDATION OF LABILE GLUTATHIONE CONJUGATES

The generation of reduced oxygen species during the oxidation of drugs catalysed by isolated hepatic microsomes or by purified P450-dependent monooxygenases has been documented (Hildebrandt and Roots, 1975; Powis and Jansson, 1979; Jones et al, 1978). The possible partial uncoupling of the P450-dependent monooxygenases by drug substrates would provide an attractive general mechanism for the loss of hepatic GSH in animals treated with drugs that are preferentially oxidized by these microsomal monooxygenases. However, attempts to demonstrate drug-dependent formation of H_2O_2 by this system in intact cells produced unexpected results. For instance, Sies et al (1978) reported that liver perfused with aminopyrine stimulated the efflux of GSSG equally well in this organ from normal and selenium-deficient animals. This clearly demonstrated that the increased concentration of aminopyrine-dependent GSSG in the perfusate was not due to drug-dependent generation of hydrogen peroxide, since the selenium-deficient animals lack glutathione peroxidase. Furthermore, Jones et al (1981) indicated that only drugs that produce formaldehyde stimulate the release of GSSG from isolated hepatocytes which was presumably formed within the hepatocytes.

A subsequent report by Krieter et al (1985) presented data supporting a different mechanism for these anomalous results. When bile from perfused liver or from conscious rats was collected in tubes containing metaphosphoric acid (to prevent oxidation of GSH to GSSG), only aminopyrine-dependent GSH efflux was detected and the drug-induced release of GSH could be duplicated by infusing formaldehyde. These data suggest that formaldehyde formed by *N*-demethylation (catalysed by P450-dependent monooxygenases) combines reversibly with GSH (reaction 9) and when the rate of formaldehyde generation exceeds the catalytic capacity of formaldehyde dehydrogenase the conjugate is excreted into bile.

$$CH_2O + GSH \rightleftharpoons GS{-}CH_2OH \qquad (9)$$

The reaction of formaldehyde with GSH is rapid and freely reversible. In the hepatocytes the reaction lies far to the right, but at the much lower concentration of GSH in bile the conjugate dissociates to GSH and formaldehyde. The autooxidation of GSH to GSSG renders the dissociation of the conjugate essentially irreversible and is probably responsible for the increased concentration of GSSG in extracellular fluids of hepatic tissue exposed to aminopyrine. Although this model is consistent with all of the available data, the possible excretion of a labile glutathione–hydroxymethylaminopyrine conjugate cannot be ruled out. In either case the apparent stimulation of GSH efflux into bile in animals treated with drugs oxidized by P450-dependent monooxygenases is not due to the generation of oxy radicals. The reasons for the discrepancies between the in vivo and in vitro measurements is not known but some subtle modification of the microsomal enzymes during cell rupture may be responsible for the uncoupling so characteristic of P450 monooxygenases in isolated microsomes.

IV. CONCLUDING REMARKS

Because of sensitive and specific methods available for measuring GSSG (Tietze, 1969), the oxidation of GSH to its disulphide is a convenient model for determining drug-dependent generation of oxygen radicals in intact tissues. There is no question that drugs capable of generating oxy radicals in vivo (at rates sufficient to produce significant tissue damage) will also stimulate efflux of GSSG from liver and most other tissues. On the other hand, some drug-dependent mechanisms for the oxidation of GSH to GSSG do not involve oxy radicals. For example, the non-enzymatic oxidation of GSH to GSSG by diamide or by the glutathione *S*-transferase-catalysed oxidation of this tripeptide by alkyl-nitrates is rapid and apparently oxygen-independent. The oxidation of GSH by xenobiotics bearing thiol or thione functional groups also does not involve non-enzyme-bound reactive oxygen intermediates. These organic sulphur drugs are oxidized by microsomal monooxygenases to sulphenic acids that are reduced non-enzymatically to the parent drug by GSH. While these reactions are perhaps the best-known examples of drug-dependent non-radical mechanisms for the oxidation of GSH, others undoubtedly exist. Determining whether the oxidation of GSH induced by a specific xenobiotic is a radical or non-radical process is often a challenging task. This is especially true for compounds such as the azo esters (Kosower et al, 1969), which can oxidize GSH by radical as well as non-radical mechanisms in vitro. The tools required to determine which process predominates in vivo are available, and judicious use of appropriate controls can often define the molecular mechanisms in intact tissues.

REFERENCES

Adams JD, Lauterburg BH & Mitchell JR (1983) *J. Pharmacol. Exp. Ther.* **227:** 749–754.
Akerboom PM, Bilzer M & Sies H (1982) *J. Biol. Chem.* **257:** 4248–4252.
Al-Waiz M, Ayesh R, Mitchell SC, Idle JR & Smith RL (1987a) *Lancet* **i:** 634–635.
Al-Waiz M, Ayesh R, Mitchell SC, Idle JR & Smith RL (1987b) *Clin. Pharmacol. Ther.* **42:** 588–594.
Anders M (1988) In Arias JM, Jakoby WB, Popper H, Schachter D & Shafritz DA (eds) *The Liver: Biology and Pathobiology*, 2nd edn, pp 389–400. New York: Raven Press.
Askelöf P, Axelsson K, Eriksson S & Mannervik KB (1974). *FEBS Lett.* **38:** 263–267.
Ayesh R, Al-Waiz M, Crothers MJ et al (1988) *Br. J. Clin. Pharmacol.* **25:** 664P–665P.
Beaty NS & Ballou DP (1981a) *J. Biol. Chem.* **256:** 4611–4618.
Beaty NS & Ballou DP (1981b) *J. Biol. Chem.* **256:** 4619–4625.
Burk RF, Nishiki K, Lawrence RA & Chance B (1978) *J. Biol. Chem.* **253:** 43–46.
Chance B, Sies H & Boveris A (1979) *Physiol. Rev.* **58:** 527–605.
Cotgreave IA, Moldeus P. & Orrenius S (1988) *Annu. Rev. Pharmacol. Toxicol.* **28:** 189–212.
Dieke SH & Richter CP (1945) *J. Pharmacol. Exp. Ther.* **85:** 195–202.
Dieke SH, Allen GS & Richter CP (1947) *J. Pharmacol. Exp. Ther.* **90:** 260–270.
Eklöw L, Moldeus P & Orrenius S (1984) *Eur. J. Biochem.* **138:** 459–463.
Eldjarn L & Pihl A (1956) *J. Biol. Chem.* **223:** 341–352.
Gold M & Ziegler DM (1973) *Xenobiotica* **3:** 179–189.
Habig WH, Keen JH & Jakoby WB (1975) *Biochem. Biophys. Res. Commun.* **64:** 501–506.
Hildebrandt AG & Roots I (1975) *Arch. Biochem. Biophys.* **171:** 385–397.
Hildebrandt AG, Roots I, Tjoe M & Heinemeyer G (1978) *Methods Enzymol.* **52:** 385–397.
Hill KE, Ziegler DM, Konz K-H, Haap M, Hunt RW & Burk RF (1989) *Biochem. Pharmacol.* **38:** 3807–3810.
Hyslop PA & Sklar LA (1984) *Anal. Biochem.* **141:** 280–286.
Jones DP, Thor H, Andersson B & Orrenius S (1978) *J. Biol. Chem.* **253:** 6031–6037.
Jones DP, Eklöw L, Thor H & Orrenius S (1981) *Arch. Biochem. Biophys.* **210:** 505–516.
Keen JH, Habig WH & Jakoby WB (1976) *J. Biol. Chem.* **251:** 6153–6188.
Kosower NS & Kosower EM (1974) In Flohe L, Benöhr H, Sies H, Waller D & Wendel A (eds) *Glutathione*, pp 276–287. Stuttgart: Thieme Verlag.
Kosower NS, Kosower EM, Wertheim B & Correa W (1969) *Biochem. Biophys. Res. Commun.* **37:** 593–596.
Krieter PA, Ziegler DM, Hill KE & Burk RF (1984) *Mol. Pharmacol.* **26:** 122–127.
Krieter PA, Ziegler DM, Hill KE & Burk RF (1985) *Biochem. Pharmacol.* **34:** 955–960.
Lauterburg B & Mitchell JR (1981) *Hepatology* **1:** 523.
Mondovi B, Tentori L, De Marco, C & Cavallini D (1962) *Int. J. Radiat. Biol.* **4:** 371–378.
Nagata T, Williams DE & Ziegler DM (1990) *Chem. Res. Toxicol.* **3:** 372–376.
Needleman P & Harkey AB (1971) *Mol. Pharmacol.* **1:** 77–86.

Ozols J (1989) *Biochem. Biophys. Res. Commun.* **163:** 49–55.
Poulsen LL & Ziegler DM (1979) *J. Biol. Chem.* **254:** 6449–6455.
Poulsen LL, Hyslop RM & Ziegler DM (1979) *Arch. Biochem. Biophys.* **198:** 78–88.
Powis G & Jansson I (1979) *Pharmacol. Ther.* **7:** 297–311.
Rane A (1973) *Clin. Pharmacol. Ther.* **15:** 32–38.
Reed DJ & Olafsdottr K (1989) In Taniguchi N, Higashi T, Sakamoto Y & Meister A (eds) *Glutathione Centennial: Molecular Perspectives and Clinical Implications*, pp 35–55. San Diego: Acadmic Press Inc.
Sies H (1981) *Methods Enzymol.* **77:** 15–20.
Sies H (1988) *ISI Atlas of Science: Biochemistry* **1:** 109–114.
Sies H, Bartoli GM, Burk RF & Waydhas C (1978a) *Eur. J. Biochem.* **89:** 113–118.
Sies H, Wahlländer A & Waydhas C (1978b) In Sies H & Wendel A (eds) *Functions of Glutathione in Liver and Kidney*, pp 120–126. Heidelberg, New York: Springer Verlag.
Sies H, Akerboom T & Ishikawa T (1989) In Taniguchi N, Higashi T, Sakamoto Y & Meister A (eds) *Glutathione Centennial: Molecular Perspectives and Clinical Implications*, pp 357–368. San Diego: Academic Press Inc.
Smith CV, Hughes H, Lauterburg BH & Mitchell JR (1983) In Larsson A, Orrenius S, Holmgren A & Mannervik B (eds) *Functions of Glutathione; Biochemical, Physiological, Toxicological and Clinical Aspects*, pp 125–137. New York: Raven Press Ltd.
Smith RL & Williams RT (1961) *J. Med. Pharm. Chem.* **4:** 97–107.
Taylor KL & Ziegler DM (1987) *Biochem. Pharmacol.* **36:** 141–146.
Tietze F (1969) *Anal. Biochem.* **27:** 502–522.
Tynes RE & Philpot RM (1987) *Mol. Pharmacol.* **31:** 569–574.
Tynes RE, Sabourin PJ & Hodgson E (1985) *Biochem. Biophys. Res. Commun.* **126:** 1069–1075.
Williams DE, Ziegler DM, Hordin DJ, Hale SE & Masters BSS (1984a) *Biochem. Biophys. Res. Commun.* **125:** 116–122.
Williams DE, Hale SE, Muerhoff AS & Masters BSS (1984b) *Mol. Pharmacol.* **28:** 381–390.
Ziegler DM (1988a) In Simic MG, Taylor KA, Ward JF & von Sonntag C (eds) *Oxygen Radicals in Biology and Medicine*, pp 729–744. New York and London: Plenum Press.
Ziegler DM (1988b) *Drug. Metab. Rev.* **19:** 1–32.
Ziegler DM (1991) In Hlavica P, Damani LA & Gorrod JW (eds) *Proceedings of the 4th International Symposium on the Biological Oxidation of Nitrogen in Organic Molecules*. London: Chapman and Hall.
Ziegler DM, Poulsen LL & Richardson RB (1983) In Nygaard OF & Simic MG (eds) *Radioprotectors and Anticarcinogens*, pp 191–202. New York: Academic Press, Inc.

4

Formation of 8-Hydroxydeoxyguanosine in DNA by Oxygen Radicals and its Biological Significance

HIROSHI KASAI AND SUSUMU NISHIMURA
Biology Division, National Cancer Center Research Institute, Tsukiji 5-1-1, Chuo-ku, Tokyo 104, Japan

I. Introduction 99
II. Discovery of 8-OH-dG formation 100
III. Formation of 8-OH-dG in vitro 101
IV. Formation of 8-OH-dG in vivo 105
V. Misreading of 8-OH-dG in DNA replication 108
VI. Repair of 8-OH-dG residue in DNA 109
VII. Discussion 111

I. INTRODUCTION

Oxygen radicals are thought to be involved in the aetiology of various chronic diseases, including cancer (Ames, 1983). When cells are irradiated, hydroxyl radicals formed by radiolysis of water induce oxidative DNA damage. Since DNA damage induced by oxygen radicals occurs concomitantly with carcinogenesis, it is considered one of the causative events in radiation carcinogenesis (Schulte-Frohlinde and von Sonntag, 1985). In addition to ionizing radiation, there are many oxygen radical-forming chemical agents in the environment. They can be detected by virtue of their mutagenicity, using *Salmonella* strains such as TA102 and TA104 that were

Oxidative Stress: Oxidants and Antioxidants
ISBN 0-12-642762-3

developed specifically for detection of oxidative DNA damage (Levin et al, 1982). Oxygen radicals are also produced in vivo during cellular metabolism of oxygen. The ubiquitous presence of superoxide dismutase (SOD) and catalase in cells is evidence that oxygen radicals are formed continuously under physiological conditions. Studies of DNA damage induced by oxygen radicals will be important in understanding mechanisms of mutation and carcinogenesis which arise spontaneously as well as those induced by chemical carcinogens or radiation.

During a study of DNA damage induced by heated carbohydrates, a model for cooked foods, a new DNA modification was discovered; namely 8-hydroxydeoxyguanosine (8-OH-dG) (Kasai et al, 1984a; Kasai and Nishimura, 1983, 1984a). We found that oxygen radicals are involved in this hydroxylation of deoxyguanosine residues in DNA. 8-Hydroxydeoxyguanosine had not been isolated previously or chemically synthesized. In this chapter we describe the generation of 8-OH-dG in DNA by various oxygen radical-forming agents in vitro and in vivo, its possible relation to mutagenesis and carcinogenesis, and its repair mechanism.

II. DISCOVERY OF 8-OH-dG FORMATION

Many environmental mutagens and carcinogens are known to react with components of DNA with or without metabolic activation, and to form adducts with adenine, guanine, cytosine and thymine, deoxyribose and phosphate (Singer and Grunberger, 1983). Among these targets, guanine is known to be the most reactive. Based on this information we developed a chemical method to detect mutagens/carcinogens in the environment by analysing the guanine adduct after in vitro reaction of fluorescent guanine derivative with a compound to be analysed (Kasai et al, 1984a).

When adducts were produced by reaction of 2′,3′-isopropylideneguanosine (IPG) with heated glucose in 1983, we found a new type of guanine modification, i.e. formation of 8-hydroxy-IPG (Figure 1) (Kasai et al, 1984a). Heated glucose, which is known to have mutagenic activity with *Salmonella* test, was used as a model for cooked foods. Although various oxidation products of DNA components, particularly those induced by ionizing radiation, had been identified at that time, the hydroxylation of guanine at the C-8 position [formation of 8-hydroxyguanine (8-OH-Gua)] had not been reported. Since then, we have been involved in studies of the mechanism of 8-OH-Gua formation, its relation to mutagenesis and carcinogenesis and the elucidation of mechanisms for its repair.

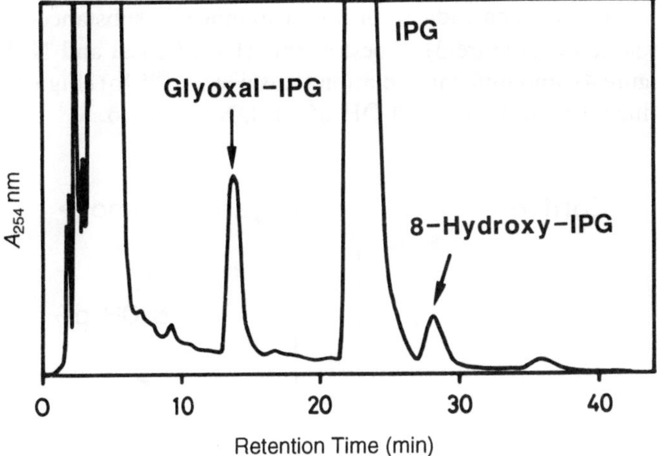

Figure 1. Fractionation of the reaction products of heated glucose and isopropylidene-guanosine (IPG) by HPLC.

III. FORMATION OF 8-OH-dG IN VITRO

The following evidence suggested a mechanism for the formation of 8-OH-dG: (1) 8-methylguanosine is formed by the reaction between guanosine and methyl radical ($\dot{C}H_3$)(Maeda et al, 1974); (2) the carbon atom on the aromatic ring is hydroxylated by the Fenton reaction using Fe^{2+}, EDTA, ascorbic acid and O_2 (Udenfriend et al, 1954). We first observed formation of 8-OH-dG from dG in high yield by Fenton-type $^{\bullet}OH$-generating systems such as that described by Udenfriend (Kasai and Nishimura, 1983) and a catechol–Fe^{3+}–H_2O_2 system (Kasai and Nishimura, 1984b) (Figure 2). In

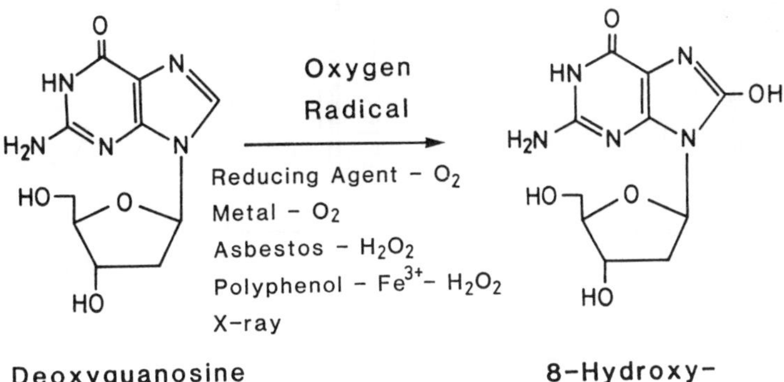

Figure 2. Formation of 8-OH-dG from dG.

addition, various oxygen radical-forming carcinogenic substances, such as cigarette smoke tar (Figure 3), asbestos plus H_2O_2 (Kasai and Nishimura, 1984c) (Figure 4), and ionizing radiation (Kasai et al, 1984b) (Figure 5) were able to induce the formation of 8-OH-dG in DNA in vitro.

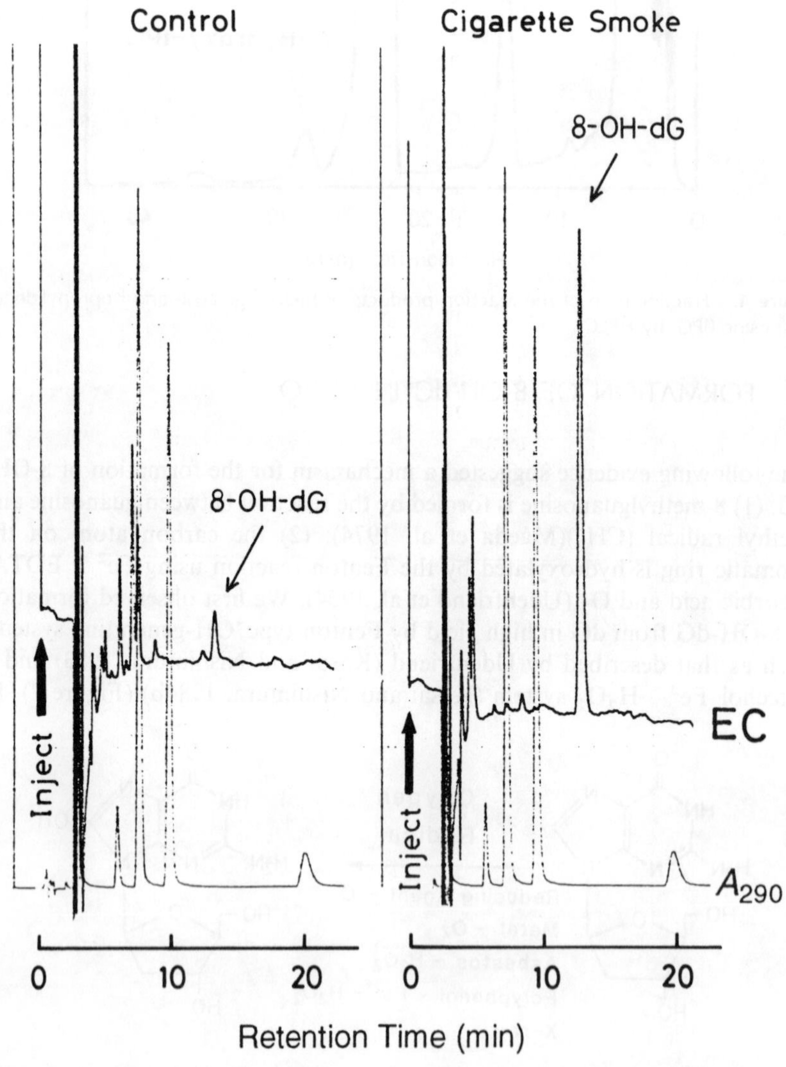

Figure 3. Formation of 8-OH-dG in DNA in vitro by cigarette smoke tar.

When calf thymus DNA dissolved in an aqueous solution was X-irradiated with 20–60 krad, 1–2% of the total dG residues in DNA were converted to 8-OH-dG (Kasai et al, 1984a,b) (Figure 5). The yield of 8-OH-dG was dose-dependent. Formation of 8-OH-dG in DNA by γ-irradiation was reported later by Dizdaroglu (1985). Paul et al (1988) showed that, in addition to imidazolidine products derived from deoxycytidine, 8-OH-dG was a principal product when oligodeoxy-nucleotide d(TpApCpG) was X-irradiated. 8-OH-dG was also a major product when d(GpC) but not when the sequence isomer d(CpG) was irradiated in aqueous solution (Paul et al, 1987). The formation of 8-OH-dG was observed under N_2- or O_2-saturated conditions.

When calf thymus DNA is incubated with H_2O_2, 8-OH-dG is formed (Figure 4). By adding absestos fibres to the incubation mixture containing DNA and H_2O_2 formation of this product was increased (Kasai and Nishimura, 1984c) (Figure 4). This system may be a good model for asbestos-induced carcinogenesis in lung, since asbestos fibres are known to activate lung macrophages after inhalation and H_2O_2 is released from macrophages during phagocytosis. The iron present in asbestos fibre may be involved in generation of $^{\bullet}OH$ from H_2O_2, thus stimulating the formation of 8-OH-dG in DNA of lung tissue. Blakely et al (1990) showed that 8-OH-dG was a major product in H_2O_2-treated DNA, based on GC-MS analysis.

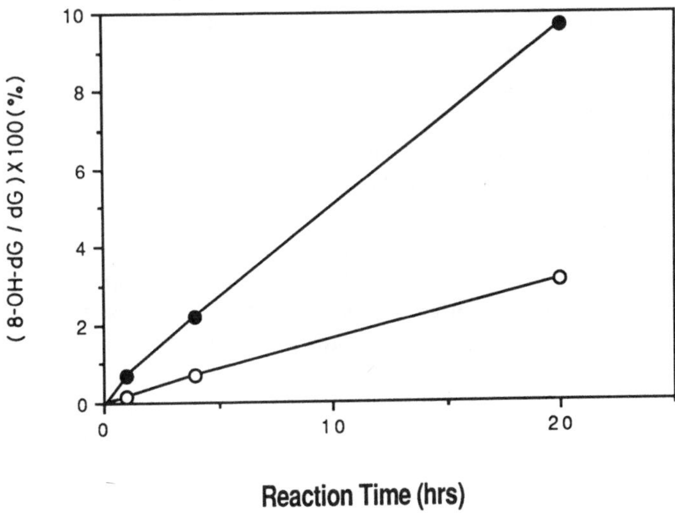

Figure 4. Formation of 8-OH-dG in DNA in vitro by asbestos and H_2O_2. ○, H_2O_2; ●, asbestos and H_2O_2.

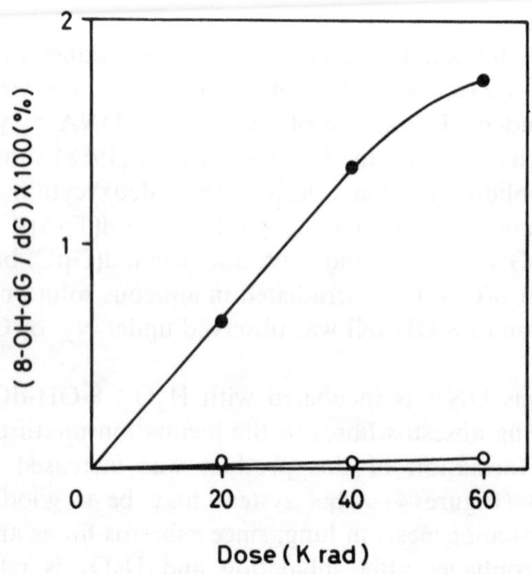

Figure 5. Formation of 8-OH-dG in DNA plotted as a function of X-ray dose. ●, No EtOH; ○, with EtOH.

We also observed the formation of 8-OH-dG in DNA in vitro by autoxidized unsaturated fatty acids (Kasai and Nishimura, 1988). It has been reported that unsaturated fatty acids stimulate rat mammary carcinogenesis and that their autoxidized products are mutagenic. Therefore, it is possible that DNA damage, such as 8-OH-dG or strand breaks induced by lipid peroxides, may be involved in mutagenesis and carcinogenesis.

The formation of 8-OH-dG in DNA was also found after reaction of DNA in vitro with mutagens or carcinogens such as 4-hydroxyaminoquinoline (Kohda et al, 1986), fecapentaene (Shioya et al, 1989), diethylstilboestrol (Rosier and Van Peteghem, 1989), nickel compounds (Kasprzak and Hernandez, 1989) or betel quid ingredients (Nair et al, 1987) used for tobacco chewing. The latter is believed to be a causative agent of oral cavity cancer.

From heated starch, two oxygen radical-forming agents were purified by assay of 8-OH-dG formation, and their structures were identified as methylreductic acid and hydroxymethylreductic acid (Kasai et al, 1989a). They showed weak mutagenic activity to *Salmonella* strains and induced sister chromatid exchange (SCE) in cultured human cells. Therefore, a simple assay to determine formation of 8-OH-dG in DNA or from dG in vitro should be useful in detecting oxygen radical-forming mutagens and/or carcinogens in the environment.

Ames and his coworkers reported that the formation of 8-OH-dG from dG by an oxygen radical-forming model system (Cu^{2+}, ascorbic acid) in vitro is inhibited by carnosine, homocarnosine and anserine present in muscle and brain (Kohen et al, 1988). We also observed inhibition by catechin and quercetin of the formation of 8-OH-dG by lipid peroxide (Kasai and Nishimura, 1988). Thus, in vitro assay of inhibition of 8-OH-dG formation is also useful for the detection of natural antioxidants to prevent oxidative DNA damage.

8-Hydroxydeoxyadenosine, an analogue of 8-OH-dG derived from the deoxyadenosine residue of DNA, has also been identified in DNA irradiated in vitro (Bonicel et al, 1980). However, the mechanism of its formation seems to be somewhat different from that of 8-OH-dG. In general, the former is only produced by ionizing radiation (unpublished results) while the latter is produced by various oxygen radical-forming agents in addition to ionizing radiation. Formation of 8-OH-dG by methylreductic acid, for example, was inhibited by SOD, catalase, EDTA and low molecular weight radical scavengers, such as mannitol and benzoic acid. Therefore, the formation of 8-OH-dG in DNA may be attributable to hydroxyl radical (•OH) which is ultimately produced by a Fenton-type Haber–Weiss reaction in the presence of trace amounts of metal ions.

IV. FORMATION OF 8-OH-dG IN VIVO

It is important to study the in vivo formation of 8-OH-dG in cellular DNA as it relates to mutagenesis/carcinogenesis. A simple and sensitive method to detect 8-OH-dG by high performance liquid chromatography-electrochemical detector (HPLC-ECD) system, developed by Floyd et al (1986b), made it possible to analyse the formation of 8-OH-dG in vivo.

Initially we examined the formation of 8-OH-dG in DNA of HeLa cells by X-irradiation and in DNA of liver by whole body irradiation of mice with γ-rays (Kasai et al, 1986). In these experiments, we observed 8-OH-dG formation with a yield of 0.8–3.2 per 10^7 dG per krad. This yield of 8-OH-dG is roughly 1000 times lower than that of in vitro 8-OH-dG formation. The yield of 8-OH-dG was approximately the same order of magnitude as that of thymine glycol (Leedon and Hanawalt, 1983) (Table 1).

Formation of 8-OH-dG in cellular DNA was observed when bacterial cells were treated with H_2O_2 (Kasai et al, 1986) or when cultured mammalian cells were treated with 4-nitroquinoline-*N*-oxide (4-NQO) (Kohda et al, 1986). Floyd et al (1986a) reported formation of 8-OH-dG in human granulocyte DNA after these cells were incubated with the tumour promoter 12-*O*-tetradecanoylphorbol-13-acetate (TPA).

Table 1. Comparison of yield of 8-OH-dG and thymidine glycol (TG) in cellular DNA induced by ionizing radiation.

8-OH-dG per 10^7 dG per krad			TG per 10^7 T per krad
Mouse liver 0.13 krad min^{-1}	Mouse liver 14 krad min^{-1}	HeLa cell 0.4 krad min^{-1}	BS-C-1 1.5 krad min^{-1}
0.8	3.2	2.0	3.4

To study the correlation between oxidative DNA damage and carcinogenesis, we examined 8-OH-dG formation in various organ DNA after oxygen radical-forming carcinogens were administered to rats. When the renal carcinogen potassium bromate ($KBrO_3$) was orally administered to rats at a dose of 400 mg kg^{-1}, a significant increase in 8-OH-dG was observed 24 h after administration in the DNA of the target organ, kidney, but not in the liver (Figure 6) (Kasai et al, 1987). In contrast, the non-carcinogenic oxidants NaClO and $NaClO_2$ did not stimulate 8-OH-dG formation in kidney DNA.

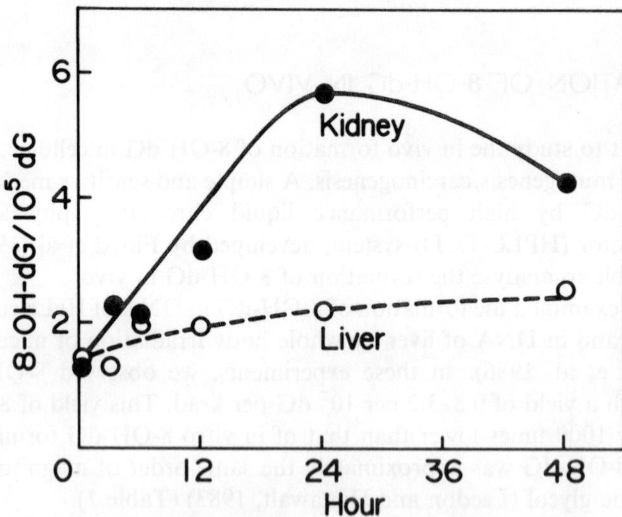

Figure 6. Formation of 8-OH-dG in rat organ DNA after oral administration of $KBrO_3$.

In collaboration with Drs Rao and Reddy, 8-OH-dG levels in liver DNA were analysed after the peroxisome proliferator, ciprofibrate, was orally administered to rats (Kasai et al, 1989b). This compound induces hepatocellular carcinoma in rats and mice, and it is known that the release of H_2O_2 from liver peroxisomes increases after administration to rats. After chronic administration of ciprofibrate in the diet at a concentration 0.025% for ten months, a significant increase of 8-OH-dG was observed in DNA of target organ liver. This result clearly demonstrated, for the first time, that persistent proliferation of peroxisomes leads to specific oxidative DNA damage. Kurokawa and his coworkers also recently reported that, after oral administration of high doses of peroxisome proliferators (1–7.5 g kg^{-1} day^{-1}) to rats for eight days, there were significant increases of 8-OH-dG in DNA of liver (Takagi et al, 1989).

Formation of 8-OH-dG in cellular DNA can be used as a marker for monitoring cellular oxidative DNA damage in evaluating the carcinogenic potential of various oxygen radical-forming agents. It may also be possible to predict the target organ response by measuring the 8-OH-dG level in DNA from various tissues.

An interesting observation concerning induction of oxidative DNA damage in vivo by oxygen radicals is the increased level of 8-OH-dG in rat liver mitochondrial DNA as compared with nuclear DNA (16-fold increase) (Richter et al, 1988). In this case, oxygen radicals endogenously produced during metabolism of oxygen in respiratory processes, which proceed predominantly in mitochondria, probably induce more oxidative DNA damage in mitochondrial DNA.

Fiala et al (1989) reported the formation of 8-hydroxyguanine (8-OH-Gua) from DNA and RNA of liver after the hepatocarcinogen 2-nitropropane was administered to rats. The increase of 8-OH-Gua in RNA (11-fold) was much more pronounced compared to that in DNA (3.6-fold). This difference may be due to rapid removal of 8-OH-Gua in DNA by a repair enzyme or more reactions of cytoplasmic RNA with oxygen radicals produced by the metabolism of 2-nitropropane. The non-carcinogen, 1-nitropropane, caused no increase of 8-OH-Gua in DNA and RNA. Therefore, determination of 8-hydroxy-guanosine (8-OH-G) in RNA could be useful for evaluating cellular oxidative damage induced by oxygen radical-forming agents.

As already mentioned, 8-OH-dG is produced in DNA in vitro by reaction with cigarette smoke condensate. The 8-OH-dG level in human peripheral leukocyte DNA was analysed before and 10 min after smoking two cigarettes. The mean levels of 8-OH-dG in leukocyte DNA increased significantly from 3.3 ± 0.8 to 5.1 ± 2.5 per 10^6 dG after smoking (Kiyosawa et al, 1990). These results indicate that cigarette smoking induces oxidative DNA damage in peripheral blood cells in a relatively short time.

V. MISREADING OF 8-OH-dG IN DNA REPLICATION

When bacterial or mammalian cells are irradiated with X-rays or γ-rays, various types of mutation are found (Glickman et al, 1980; Grosovsky et al, 1988). These induce transitions (GC→AT, AT→GC), transversions (GC→CG, GC→TA, AT→TA, AT→CG), frameshifts and deletion and translocation of large DNA fragments. Although several types of DNA damage induced in vivo by ionizing radiation have been identified, i.e. strand breaks, and formation of thymine glycol, hydroxymethyluracil and 8-OH-Gua, the relationship between particular DNA damage and a particular mutation is not clearly understood. Difficulty in performing these studies is due to the fact that many types of DNA damage are produced simultaneously in DNA by ionizing radiation and other oxygen radical-forming agents. This is in sharp contrast to the situation of alkylating carcinogen and UV radiation. In these cases, DNA modifications such as O^6-methyldeoxyguanosine (O^6-Me-dG) (Goth and Rajewsky, 1974) and O^4-methyldeoxythymidine (O^4-Me-dT) (Singer, 1986), and pyrimidine–pyrimidine dimers (Brash, 1988), are known to play crucial roles in carcinogenesis.

An oligodeoxynucleotide which contains 8-OH-dG at a specific position was chemically synthesized in a stepwise manner and used as a single-stranded DNA template for DNA synthesis in vitro catalysed by *E. coli* DNA polymerase Klenow fragment, and detected by a dideoxy chain-termination sequencing method. In these experiments, bands corresponding to incorporation of dideoxy-A,G,T, in addition to dideoxy-C, were observed at the positions of 8-OH-dG residues in the template (Kuchino et al, 1987). Misincorporation was also observed at neighbouring positions.

Shibutani et al (1991) reported that dCTP and dATP are incorporated preferentially opposite 8-OH-dG residues in the template DNA during DNA synthesis in vitro by *E. coli* polymerase I (Klenow fragment) and by mammalian polymerases α and β. Only when dATP was incorporated opposite 8-OH-dG was rapid chain extension observed. From these results, GC→TA transversions, induced by 8-OH-dG are predicted.

With respect to misreading of 8-OH-dG-containing DNA by DNA polymerases, it is interesting to note that 8-OH-GTP is incorporated into RNA instead of UTP on poly[d(A–T)·d(A–T)] template during RNA synthesis by *E. coli* RNA polymerase (Bruskov and Kuklina, 1988). The incorporation of 8-OH-GTP instead of UTP is stimulated when Mg^{2+} ions are replaced by Mn^{2+} ions. Thus, it is likely that 8-OH-Gua does base-pair with adenine. 8-OH-dG may adopt the *syn*-conformation (Uesugi and Ikehara, 1977) and the C-8 hydroxyl group tautomerizes to the C-8 keto form in aqueous solution (Culp et al, 1989). In fact, X-ray crystallographic analysis had already shown that a derivative of 8-OH-dG, 9-ethyl-8-hydroxyguanine, is in 6,8-diketo form (Kasai et al, 1987). Ab initio molecular orbital calculations on 8-OH-Gua also showed that its stable tautomeric form is the 6,8-diketo form (Aida and Nishimura, 1987). Thus 8-OH-dG can also be called 8-oxo-7,8-dihydrodeoxyguanosine, abbreviated as 8-oxo-dG. In fact, Kouchakdjian et al (1990) recently showed by NMR analysis that the oligodeoxynucleotide containing 8-OH-dG makes base pairs with A in the opposite strand in its *syn*-conformation, although 8-OH-dG in an oligodeoxynucleotide can make a normal base pair with C (Oda et al, 1987). Probably, 8-OH-Gua in the C-8-keto form can base-pair with adenine, as shown in Figure 7.

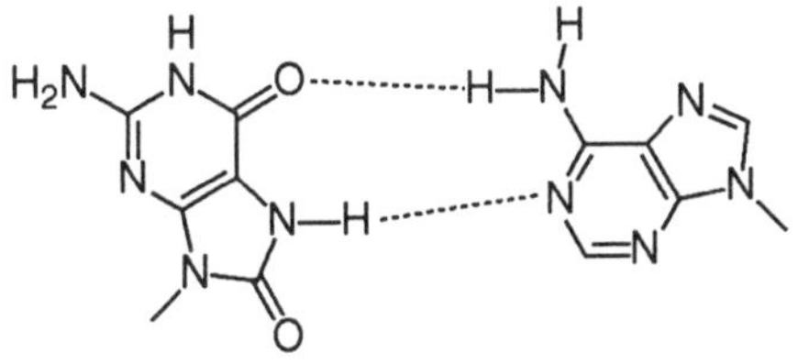

8-OH-Gua (syn, keto) Ade

Figure 7. Possible base pairing between 8-OH-Gua and adenine.

VI. REPAIR OF 8-OH-dG RESIDUE IN DNA

We observed previously that 8-OH-dG residues produced in liver DNA of mice by γ-irradiation decreased over time. The 8-OH-dG content in liver DNA decreased to half of the initial level 90 min after irradiation (Figure 8) (Kasai et al, 1986). This result suggested that liver cells have a repair mechanism for 8-OH-dG. We have purified a repair enzyme for 8-OH-dG from *E. coli.* For this purpose, chemically synthesized double stranded 46-mer DNA having 8-OH-dG at a specific position was used as substrate. The synthetic DNA with ^{32}P-endlabel was incubated with the enzyme fraction and strand cleavage was detected by polyacrylamide gel electrophoresis. We were able to detect an activity that cleaves the DNA strand at the position of 8-OH-dG. After several purification steps by ammonium sulphate precipitation, DEAE-cellulose and Sephadex G-100 column chromatography, the endonuclease was purified without contamination with other DNase. This endonuclease cleaves both phosphodiester bonds, 3′ and 5′ to the 8-OH-dG residue, leaving phosphate on the deoxyribose moiety of neighbouring bases

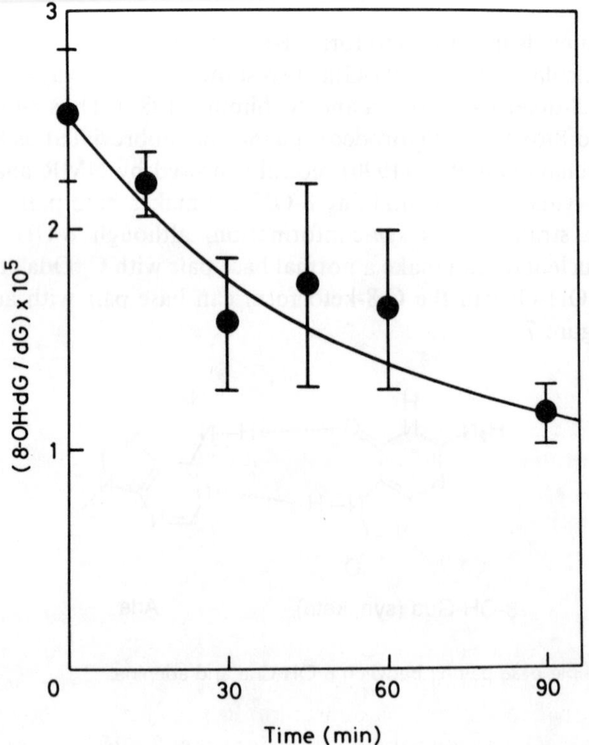

Figure 8. Repair of 8-OH-dG residues in liver DNA after whole body irradiation of mice with γ-rays.

as shown in Figure 9. It cleaved double-stranded DNA but not single-stranded DNA containing 8-OH-dG (Chung et al, 1991). We studied the specificity of this endonuclease using several double-stranded DNAs which contained either mismatches in sequence or other modified bases.

T G-OH C
5'-------- --------3'
P P

Figure 9. Cleavage position of DNA strand by *E. coli* 8-OH-Gua endonuclease.

Strand cleavage was not observed in the position of mismatches such as G·A and C·T. Other modified bases, such as 8-hydroxyadenine, O^6-methylguanine and N^7-methylguanine, were not recognized by this endonuclease. Ring-opened-guanine (Fapy) in DNA was only very weakly cleaved by this enzyme. We found also that the 8-OH-Gua-specific endonculease cleaved an apurinic/apyrimidinic (AP) site in DNA. *E. coli* endonuclease III, which is involved in repair of thymine glycol (Demple and Linn, 1980), another type of oxidative DNA damage, also cleaved the DNA strand at the position of 8-OH-dG. However, *E. coli* endonuclease III cleaved only at the 3′ side of the 8-OH-dG residue. Therefore, the action of 8-OH-dG endonuclease is different from that of *E. coli* endonuclease III. The properties of 8-OH-Gua endonuclease are also different from those of *E. coli* endonuclease IV and exonuclease III. 8-OH-Gua endonuclease is affected by high ionic strength and is easily heat-inactivated. It is fully active in the absence of Mg^{2+}. It should also be mentioned that 8-OH-Gua was detected as the reaction product (unpublished results). Based on the above characteristics, two possible mechanisms for repair of 8-OH-dG are proposed: (1) 8-OH-Gua is first removed, and the AP site is cleaved at the two positions (the repair enzyme has both glycosylase activity and AP endonuclease activity); (2) the repair enzyme directly cleaves at the two positions and 8-OH-dG is released.

We also found 8-OH-Gua endonuclease activity in cultured human cells (unpublished results). In this respect, Ames and his coworkers (Shigenaga et al, 1989; Fraga et al, 1990) showed the presence of 8-OH-dG and 8-OH-Gua in mouse, rat and human urine. The quantity of 8-OH-dG released in urine was proportional to the oxygen consumption per gram of body weight of these mammals. These data supported the idea that 8-OH-dG is produced in cellular DNA by endogenously generated oxygen radicals during metabolism of oxygen. The presence of the 8-OH-Gua-specific endonuclease and detection of 8-OH-Gua in animal urine imply that 8-OH-dG is produced in vivo by oxygen radicals and that cells have acquired a mechanism to repair it. Therefore, 8-OH-dG should be a biologically significant DNA modification.

VII. DISCUSSION

Oxygen radicals produce many DNA modifications in addition to 8-OH-dG. Past efforts were concentrated on modifications of the pyrimidine residue, such as thymine glycol, and on DNA strand scission. More recently, studies on modifications of purine residues, particularly formation of 8-OH-dG, have become popular because of the ease with which 8-OH-dG can be detected by HPLC-ECD and its potential mutagenicity as demonstrated on in vitro DNA synthesis. In addition to DNA strand scission and formation of thymine glycol, 5-hydroxymethyluracil, and 8-OH-Gua, many other modified nucleosides are generated by oxygen radical-forming agents in vitro. It is possible that some of these modified bases are also produced in vivo. An important question is how much 8-OH-Gua in DNA is involved in mutagenesis and/or carcinogenesis, as compared with other types of oxidative DNA damage. Each scientist tends to work only on a particular type of DNA modification. It will be important in the future to estimate the relative contribution of each modification to overall mutagenesis by oxygen radicals. It should also be emphasized that many chemical carcinogens generate oxygen radicals in addition to forming adducts. In these cases, the extent to which DNA damage by oxygen radicals is relevant to carcinogenic potency has not been established.

It is likely that the formation of 8-OH-Gua is proportional to that of other DNA modifications produced by oxygen radicals, regardless of organisms, type of oxygen radical-forming agent and conditions of exposure. Therefore, measurement of 8-OH-Gua in DNA or urine should be very useful in estimating oxygen radical-induced damage and is facilitated by the ease and sensitivity by which this lesion can be detected. The amount of 8-OH-Gua present in DNA (1 per 10^5 guanine) is extremely high as compared with DNA adducts formed by alkylating agents and other carcinogens (1 per 10^7 to 1 per 10^8 for O^6-methylguanine, acetylaminofluorine or aflatoxin). 8-OH-Gua residues in DNA causes exclusive misreading during DNA synthesis *in vitro* (G→T transversion, especially in the case of polymerase α) (Shibutani et al, 1991). In vivo mutations caused by 8-OH-Gua in *E. coli* were also found to produce G→T transversions (Wood et al, 1990; Moriya et al, 1991). The mutation frequency in vivo is not as high as suggested by the in vitro experiments but is similar to the mutation frequency of O^6-methyl-Gua. It is tempting to speculate that the situation of DNA replication in vivo is somewhat different from that in vitro in terms of misreading 8-OH-Gua. Additional protein factor(s) involved in DNA replication in vivo may prevent the misreading. Another factor which could influence mutagenesis by 8-OH-Gua in vivo involves the repair system for 8-OH-Gua, which functions very efficiently in vivo. Even if the mutation frequency of 8-OH-Gua in vivo is not high, one cannot ignore the contribution of 8-OH-Gua to mutagenesis, since the total number of lesions in DNA is over 100-fold greater than that of other DNA adducts.

The isolation of 8-OH-Gua-specific endonuclease from *E. coli* (Chung et al, 1991) and its identification as the enzyme characterized as Fapy glycosylase (O'Connor and Laval, 1989; unpublished results) provides further leads to studies of the mechanism of mutations induced by 8-OH-Gua. For example, it will be important to investigate whether 8-OH-Gua-specific endonuclease is induced in *E. coli* under oxidative stress. The levels of 8-OH-Gua-specific endonuclease in *E. coli* and *Salmonella* mutants sensitive or resistant to exposure to oxygen radicals would also be of interest.

An enzymatic activity similar to *E. coli* 8-OH-Gua-specific endonuclease has recently been purified from mammalian cells (unpublished results). In this connection, one must determine if the mammalian enzyme possesses both glycosylase and AP endonuclease activity in the same molecule or whether removal of 8-OH-Gua is carried out by two separate enzymes. Measurement of the level of 8-OH-Gua-specific repair enzyme in leukocytes or lymphocytes of the peripheral blood of each individual is also of interest. If such enzymatic activity varies among different individuals, it may be relevant to their susceptibility to cancer.

ACKNOWLEDGEMENT

We are very much indebted to Dr Arthur P. Grollman for his valuable suggestions and critical reading of the manuscript.

REFERENCES

Aida M & Nishimura S (1987) An ab initio molecular orbital study on the characteristics of 8-hydroxyguanine. *Mutat. Res.* **192:** 83–89.

Ames BN (1983) Dietary carcinogens and anticarcinogens, oxygen radicals and degenerative diseases. *Science* **221:** 1256–1264.

Blakely WF, Fuciarelli AF, Wegher BJ & Dizdaroglu M (1990) Hydrogen peroxide-induced base damage in deoxyribonucleic acid. *Radiat. Res.* **121:** 338–343.

Bonicel A, Mariaggi N, Hughes E & Teoule R (1980) *In vitro* γ irradiation of DNA: identification of radioinduced chemical modifications of the adenine moiety. *Radiat. Res.* **83:** 19–26.

Brash DE (1988) UV mutagenic photoproducts in *Escherichia coli* and human cells: a molecular genetics perspective on human skin cancer. *Photochem. Photobiol.* **48:** 59–66.

Bruskov VI & Kuklina OV (1988) Manifestation by 8-hydroxy-GTP of the substrate of UTP in the reaction of polynucleotide synthesis by *Escherichia coli* RNA polymerase on a poly [d(AT)·d(AT)] template. *Mol. Biol. (Mosk.)* **22:** 726–730.

Chung MH, Kasai H, Jones DS et al (1990) An endonuclease activity of *Escherichia coli* that specifically removes 8-hydroxyguanine residues from DNA. *Mutat. Res* **254:** 1–12.

Culp SJ, Cho BP, Kadlubar FF & Evans FE (1989) Structural and conformational analyses of 8-hydroxy-2′-deoxyguanosine. *Chem. Res. Toxicol.* **2:** 416–422.

Demple B & Linn S (1980) DNA N-glycosylase and UV repair. *Nature (Lond.)* **287:** 203–208.

Dizdaroglu M (1985) Formation of an 8-hydroxyguanine moiety in deoxyribonucleic acid on γ-irradiation in aqueous solution. *Biochemistry* **24:** 4476–4481.

Fiala ES, Conaway CC & Mathis JE (1989) Oxidative DNA and RNA damage in the liver of sprague-dawley rats treated with the hepatocarcinogen 2-nitropropane. *Cancer Res.* **49:** 5518–5522.

Floyd RA, Watson JJ, Harris J, West M & Wong PK (1986a) Formation of 8-hydroxydeoxyguanosine, hydroxyl free radical adduct of DNA in granulocytes exposed to the tumor promoter, tetradeconyl phorbolacetate. *Biochem. Biophys. Res. Commun.* **137:** 841–846.

Floyd RA, Watson JJ, Wong PK, Altmiller DH & Rickard RC (1986b) Hydroxyl free radical adduct of deoxyguanosine: Sensitive detection and mechanisms of formation. *Free Radicals Res. Commun.* **1:** 163–172.

Fraga CG, Shigenaga MK, Park JW, Degan P & Ames BN (1990) Oxidative damage to DNA during aging: 8-Hydroxy-2′-deoxyguanosine in rat organ DNA and urine. *Proc. Natl. Acad. Sci. USA* **87:** 4533–4537.

Glickman BW, Rietveld K & Aaron CS (1980) γ-Ray induced mutational spectrum in the lac I gene of *Escherichia coli*: Comparison of induced and spontaneous spectra at the molecular level. *Mutat. Res.* **69:** 1–2.

Goth R & Rajewsky MF (1974) Persistence of O6-ethylguanine in rat brain DNA: correlation with nervous system-specific carcinogenesis by ethylnitrosourea. *Proc. Natl. Acad. Sci. USA* **71:** 639–643.

Grosovsky AJ, de Boer JG & Glickman BW (1988) Base substitutions, frameshifts, and small deletions constitute ionizing radiation-induced point mutations in mammalian cells. *Proc. Natl. Acad. Sci. USA* **85:** 185–188.

Hoebee B, Brouwer J, Putte P vd, Loman H & Retèl J (1988) ^{60}Co-γ rays induce predominantly C/G to G/C transversions in double-stranded M13 DNA. *Nucleic Acids Res.* **16:** 8147–8156.

Hoebee B, Ende MA vd, Brower J, Putte P vd, Loman H & Retèl J (1989) Mutation induced by ^{60}Co gamma-irradiation in double-stranded M13 bacteriophage DNA in nitrous-oxide saturated solutions are characterized by high specificity. *Int. J. Radiat. Biol.* **56:** 401–411.

Kasai H & Nishimura S (1983) Hydroxylation of the C-8 position of deoxyguanosine by reducing agents in the presence of oxygen. *Nucleic Acids Symp. Ser.* No. 12, 165–167.

Kasai H & Nishimura S (1984a) Hydroxylation of deoxyguanosine at the C-8 position by ascorbic acid and other reducing agents. *Nucleic Acids Res.* **12:** 2137–2145.

Kasai H & Nishimura S (1984b) Hydroxylation of deoxyguanosine at the C-8 position by polyphenols and aminophenols in the presence of hydrogen peroxide and ferric ion. *Gann* **75:** 565–566.

Kasai H & Nishimura S (1984c) DNA damage induced by asbestos in the presence of hydrogen peroxide. *Gann* **75:** 841–844.

Kasai H & Nishimura S (1988) Formation of 8-hydroxydeoxyguanosine in DNA by auto-oxidized unsaturated fatty acids. In Hayaishi O, Niki E, Kondo M and Yoshikawa T (eds) *Medical, Biochemical and Chemical Aspects of Free Radicals*, pp 1021–1023. Elsevier Science Publishers.

Kasai H, Hayami H, Yamaizumi Z, Saito H & Nishimura S (1984a) Detection and identification of mutagens and carcinogens as their adducts with guanosine derivatives. *Nucleic Acids Res.* **12:** 2127–2136.

Kasai H, Tanooka H & Nishimura S (1984b) Formation of 8-hydroxyguanine residues in DNA by X-irradiation. *Gann* **75:** 1037–1039.

Kasai H, Crain PF, Kuchino Y, Nishimura S, Ootsuyama A & Tanooka H (1986) Formation of 8-hydroxyguanine moiety in cellular DNA by agents producing oxygen radicals and evidence for its repair. *Carcinogenesis* **7:** 1849–1851.

Kasai H, Nishimura S, Kurokawa Y & Hayashi (1987a) Oral administration of the renal carcinogen, potassium bromate, specifically produces 8-hydroxydeoxyguanosine in rat target organ DNA. *Carcinogenesis* **8:** 1959–1961.

Kasai H, Nishimura S, Toriumi Y, Itai T & Iitaka Y (1987b) The crystal structure of 9-ethyl-8-hydroxyguanine. *Bull. Chem. Soc. Jpn* **60:** 3799–3800.

Kasai H, Nakayama M, Toda N, Yamaizumi Z, Oikawa J & Nishimura S (1989a) Methylreductic acid and hydroxymethylreductic acid: oxygen radical forming agents in heated starch. *Mutat. Res.* **214**: 159–164.

Kasai H, Okada Y, Nishimura S, Rao MS & Reddy JK (1989b) Formation of 8-hydroxydeoxyguanosine in liver DNA of rats following long-term exposure to a peroxisome proliferator. *Cancer Res.* **49:** 2603–2605.

Kasprzak KS & Hernandez L (1989) Enhancement of hydroxylation and deglycosylation of 2′-deoxyguanosine by carcinogenic nickel compounds. *Cancer Res.* **49:** 5964–5968.

Kiyosawa H, Suko M, Okudaira H et al (1990) Cigarette smoking induces formation of 8-hydroxydeoxyguanosine, one of the oxidative DNA damages in human peripheral leukocytes. *Free Radical Res. Comm.* **11:** 23–27.

Kohda K, Tada M, Kasai H, Nishimura S & Kawazoe Y (1986) Formation of 8-hydroxyguanine residues in cellular DNA exposed to the carcinogen 4-nitroquinoline 1-oxide. *Biochem. Biophys. Res. Commun.* **139:** 626–632.

Kohen R, Yamamoto Y, Cundy KC & Ames BN (1988) Antioxidant activity of carnosine, homocarnosine, and anserine present in muscle and brain. *Proc. Natl. Acad. Sci. USA* **85:** 3175–3179.

Kouchakdjian M, Bodepudi V, Shibutani S et al (1991) NMR structural studies of ionizing radiation adduct 8-oxo-7-hydro-deoxyguanosine (8-oxo-7H-dG) opposite deoxyadenosine in a DNA duplex. 8-oxo-7H-dG(syn) dA(anti) alignment at lesion site. *Biochemistry*. **30:** 1403–1412.

Kuchino Y, Mori F, Kasai H et al (1987) Misreading of DNA templates containing 8-hydroxydeoxyguanosine at the modified base and at adjacent residues. *Nature (Lond.)* **327:** 77–79.

Leedon SA & Hanawalt PC (1983) Monoclonal antibody to DNA containing thymine glycol. *Mutat. Res.* **112:** 191–200.

Levin DE, Hollstein M, Christman MF, Schwire EA & Ames BN (1982) A new *Salmonella* tester strain (TA102) with A·T base pairs at the site of mutation detects oxidative mutagens. *Proc. Natl. Acad. Sci. USA* **79:** 7445–7449.

Maeda M, Nushi K & Kawazoe Y (1974) Studies on chemical alterations of nucleic acids and their components VII: C-alkylation of purine bases through free radical process catalyzed by ferrous ion. *Tetrahedron* **30:** 2677–2682.

Moriya et al (1991) Site-specific mutagenesis using a gapped duplex vector: a study of translesion synthesis past 8-oxodeoxyguanosine in *E. coli*. *Mutation Res.* (In press).

Nair UJ, Floyd RA, Nair J, Bussachini V, Friesen M & Bartsch H (1987) Formation of reactive oxygen species and of 8-hydroxydeoxyguanosine in DNA in vitro with betel quid ingredients. *Chem. Biol. Interact.* **63:** 157–169.

O'Connor TR & Laval J (1989) Physical association of the 2,6-diamino-4-hydroxy-5N-formamidopyrimidine-DNA glycosylase of *Escherichia coli* and an activity nicking DNA at apurinic/apyrimidinic sites. *Proc. Natl. Acad. Sci. USA* **86:** 5222–5226.

Oda K, Uesugi S, Ikehara M et al (1987) Analysis of DNA structure containing 8-hydroxyguanine. Annual Meeting of Japanese Biochemical Society, Abstract 59, p 911.

Paul CR, Arakali AV, Wallace JC, McReynolds J & Box HC (1987) Radiation chemistry of 2′-deoxycytidylyl-(3′-5′)-2′-deoxyguanosine and its sequence isomer in N_2 and O_2-saturated solutions. *Radiat. Res.* **112:** 464–477.

Paul CR, Wallace JC, Alderfer JL & Box HC (1988) Radiation chemistry of d(TpApCpG) in oxygenated solution. *Int. J. Radiat. Biol.* **54:** 403–415.

Richter C, Park J-W & Ames BN (1988) Normal oxidative damage to mitochondrial and nuclear DNA is extensive. *Proc. Natl. Acad. Sci. USA* **85:** 6465–6467.

Rosier JA & Van Peteghem CH (1989) Peroxidative *in vitro* metabolism of diethylstilbestrol induces formation of 8-hydroxy-2′-deoxyguanosine. *Carcinogenesis* **10:** 405–406.

Schulte-Frohlinde D & von Sonntag C (1985) Radiolysis of DNA and model systems in the presence of oxygen. In Sies H (ed.) *Oxidative Stress*, pp 11–40. London: Academic Press.

Shibutani S, Takeshita M & Grollman AP (1990) Insertion of specific bases during DNA synthesis past the oxidation-damaged base 8-oxodG. *Nature.* **349:** 431–434.

Shigenaga MK, Gimeno CJ & Ames BN (1989) Urinary 8-hydroxy-2′-deoxyguanosine as a biological marker of in vivo oxidative DNA damage. *Proc. Natl. Acad. Sci. USA* **86:** 9697–9701.

Shioya M, Wakabayashi K, Yamashita K, Nagao M & Sugimura T (1989) Formation of 8-hydroxydeoxyguanosine in DNA treated with fecapentaene-12 and -14. *Mutat. Res.* **225:** 91–94.

Singer B (1986) O-Alkylpyrimidines in mutagenesis and carcinogenesis: Occurrence and significance. *Cancer Res.* **46:** 4879–4885.

Singer B & Grunberger D (1983) *Molecular Biology of Mutagens and Carcinogens.* New York: Plenum Press.

Takagi A, Sai K, Umemura T, Hasegawa R, Kurokawa Y & Kasai H (1989) Production of 8-hydroxydeoxyguanosine in rat liver DNA by oral administration of peroxisome proliferators. *Igaku no Ayumi* **149:** 65–66.

Udenfriend S, Clark CT, Axelrod J & Brodie BB (1954) Ascorbic acid in aromatic hydroxylation. I. A model system for aromatic hydroxylation. *J. Biol. Chem.* **208:** 731–739.

Uesugi S & Ikehara M (1977) Carbon-13 magnetic resonance spectra of 8-substituted purine nucleosides. Characteristic shifts for the syn conformation. *J. Am. Chem. Soc.* **99:** 3250–3253.

Umemura T, Sai K, Takagi A, Hasegawa R & Kurokawa Y (1990) Formation of 8-hydroxydeoxyguanosine (8-OH-dG) in rat kidney DNA after intraperitoneal administration of ferric nitrilotriacetate (Fe-NTA). *Carcinogenesis* **11:** 345–347.

Wood ML, Dizdaroglu M, Gajewski E & Essigmann JM (1990) Mechanistic studies of ionizing radiation and oxidative mutagenesis: Genetic effects of a single 8-hydroxyguanine (7-hydro-8-oxoguanine) residue inserted at a unique site in a viral genome. *Biochemistry* **29:** 7024–7032.

DEFENCE AGAINST OXIDANTS

5

Repair Systems for Radical-damaged DNA

BRUCE DEMPLE and JOSHUA D. LEVIN
Laboratory of Toxicology, Harvard School of Public Health, 665 Huntington Avenue, Boston, Mass. 02115, USA

I. Introduction 119
II. Radical-induced DNA damage and its biological consequences 120
III. Repair enzymes for oxy radical lesions in DNA 127
IV. Issues in DNA repair: survival, mutagenesis, and cellular responses 145

I. INTRODUCTION

As should be clear from other chapters in this volume, the sources of reactive oxygen species relevant to biology are widespread, and include both intra- and extracellular mechanisms. The key species for damaging DNA and other macromolecules are probably more limited, however. These include principally the hydroxyl radical $^{\bullet}OH$ and the solvated electron $e^{-}(aq)$ (von Sonntag, 1987), although some evidence suggests that localized formation of radicals mediated by DNA-bound transition metals could also produce damage (Imlay and Linn, 1988; Sagripanti and Kraemer, 1989). Additionally, certain antibiotics (e.g. bleomycin (BLM); see below) damage DNA as the apparently exclusive target.

Oxidative Stress: Oxidants and Antioxidants
ISBN 0-12-642762-3

II. RADICAL-INDUCED DNA DAMAGE AND ITS BIOLOGICAL CONSEQUENCES

Much information on the damage formed by radical-dependent reactions has come from analysis of DNA treated with ionizing radiation (reviewed extensively by Hutchinson (1985) and von Sonntag (1987)). New methods, such as mass spectroscopy with selected ion monitoring (Dizdaroglu, 1984; Dizdaroglu and Gajewski, 1990), have confirmed and extended older studies that employed paper chromatography and HPLC. The bottom line of this work is that every component of DNA is subject to attack, leading to a daunting variety of individual lesions (perhaps as many as 100 distinct forms). This might have suggested the need for as many different DNA repair enzymes, but as we shall see the number of activities is smaller because each enzyme acts on a variety of DNA lesions.

A. Damaged bases

Most of the lesions identified after X-irradiation or other radical-forming treatments are modified or fragmented bases and include derivatives of all four DNA bases (a few examples are given in Figure 1, but this is a far from exhaustive list).

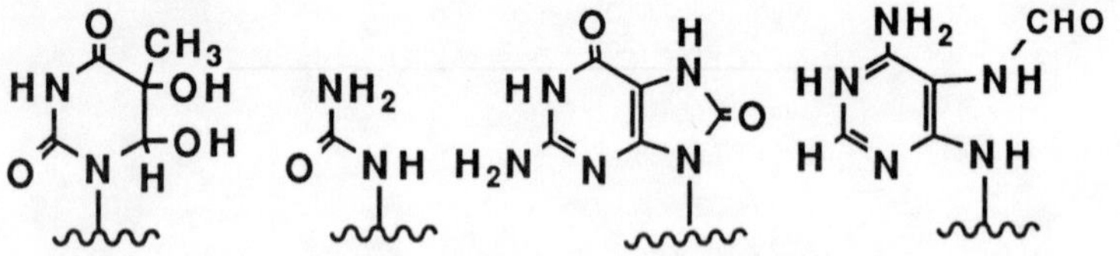

Figure 1. DNA base damage. **1**, Thymine glycol (TG); **2**, urea; **3**, hydroxyguanine (8-OH-G); **4**, formamidopyrimidine (A-FAPy). The wavy line at the bottom of the bases represents the sugar-phosphate backbone of DNA.

Many of these products arise from secondary reactions after the initial radical formation, but it should be evident that most of the base lesions fall into families: hydroxylated bases (**1** and **3** in Figure 1), ring-opened (ruptured) forms (**4** in Figure 1), contracted forms and base fragments (**2** in Figure 1). The relative amounts of the different lesions are affected by the addition of oxygen to base radicals to form base-peroxyl radicals, especially for the pyrimidines (von Sonntag, 1987). Far less work has been done with DNA damaged by chemical oxidants or near-UV (UVB) light, although what information there is points to the formation of the same types of lesions (Demple and Linn, 1982a; Cadet et al, 1986).

Because of the array of oxidative lesions, the biochemical properties of only a few individual types of damage have been examined in any detail in DNA. Thymine glycol (**1** in Figure 1; *cis*-5,6-dihydroxy-5,6-dihydrothymine, here abbreviated *cis*-TG) has been especially well studied, perhaps because simple methods have been developed for its identification and synthesis (Beer et al, 1966; Cadet and Teoule, 1973; Demple and Linn, 1980). TG can exist as any of several stereoisomers, but the *cis* form (probably 5*R*,6*S*) predominates in DNA irradiated with low doses of γ-rays (Frenkel et al, 1981; Basu et al, 1989).

TG residues in synthetic oligonucleotides or single-stranded phage DNA were formed by group-specific reactions with OsO_4 or $KMnO_4$ and tested as substrates for DNA polymerase I of *Escherichia coli*. Four groups (Rouet and Essigmann, 1985; Ide et al, 1985; Hayes and LeClerc, 1986; Clark and Beardsley, 1986) independently agreed that *cis*-TG constitutes a replication block in which the last nucleotide is inserted at (or is turned over opposite) the damaged base. The residue incorporated opposite TG was >90% dAMP, which may be properly hydrogen-bonded to the adduct (Clark et al, 1987), although a molecular modelling study predicts *cis*-TG to be displaced slightly compared to normal DNA thymine (Basu et al, 1989). TG might therefore be termed a replication-blocking rather than a miscoding lesion. Consistent with this interpretation, single-stranded phage DNA treated with OsO_4 and transfected into *E. coli* accumulated mutations at C rather than T residues, which suggests that oxidized cytosines are more potent mutational targets than oxidized thymines (Hayes et al, 1988). Moreover, a uniquely positioned TG in single-stranded viral DNA yielded mutations targeted to the modified site (*cis*-TG : A to C : G transitions) at a frequency of only 0.3% of the input DNA, while the same DNA in the duplex form did not undergo detectable mutagenesis (Basu et al, 1989). Thus, TG certainly can provide a weakly mutagenic target, but its contribution may be overshadowed by those of other oxidative lesions.

In contrast to TG, alkaline degradation of oxidized thymines to urea residues (**2** in Figure 1) gave rise to sites that not only blocked DNA polymerase I, but also prevented the detectable incorporation of any dNMP opposite the lesion (Ide et al, 1985). A similar non-templating effect was caused by 2,6-diamino-4-hydroxy-*N*5-methylformamidopyrimidine (an analogue of the adenine derivative **4** in Figure 1; these are abbreviated as G-FAPy and A-FAPy, respectively) formed by alkali treatment of a template containing *N*7-methylguanine residues (O'Connor et al, 1988). These base-fragmentation types of damage are therefore more potent blocks to DNA synthesis than are TGs. It seems possible that some incorporation of dNMP residues occurs opposite these ruptured bases, but at rates much slower than subsequent dNMP removal during proofreading.

A more subtle type of free radical damage is 5-hydroxymethyluracil (formed by hydroxylation of the exocyclic methyl group of thymine), which is formed in significant quantities by γ-rays, or through the radiolytic decay of [6-^{3}H]thymine in DNA (Teebor et al, 1984). Whether this lesion has any important biological consequences is open to question. For example, certain bacteriophages (e.g. phage SP01, which attacks *B. subtilis*) have a complete replacement of thymine by 5-hydroxymethyluracil (Kallen and Marmur, 1962), without obvious deleterious effects on the viral life cycle or stability. On the other hand, inclusion of 5-hydroxymethyl-2′-deoxyuridine in the culture media of *E. coli* and *S. typhimurium* incited λ-prophage induction (often taken as an assay for the SOS response) and mutagenesis, respectively (Shirname-More et al, 1987). However, these latter effects might be explained by induced imbalances in the dNTP pools, without requiring that the hydroxylated base is itself a target for mutagenesis in DNA.

Exposure of DNA in aqueous solution to γ-rays (Dizdaroglu, 1985) or other conditions that generate active oxygen (e.g. UV light plus H_2O_2 (Floyd et al, 1988), or methylene blue plus visible light (Schneider et al, 1990)) yields 8-hydroxyguanine (8-OH-Gua; drawn in Figure 1 as the 8-oxo form) as a major product. DNA containing 8-OH-Gua has also been examined as a replication substrate. One study indicated that this lesion directs the insertion of any of the four dNMPs (Kuchino et al, 1987), a type of ambiguity not usually seen during in vitro polymerase reactions. Such promiscuity is not easily explained, since lesions that contain no obvious information (e.g. base-free sites) are actually read by DNA polymerases in a stricter way (see below). Some ‘misreading’ of the next nucleotide, just beyond the 8-OH-Gua site, was also recorded (Kuchino et al, 1987). In other laboratories, the misinsertion of only dTMP was detected opposite the adducted site in vitro (see Note added in proof). Moreover, transfection into *E. coli* of a single-stranded viral genome with a uniquely positioned 8-OH-Gua showed an overwhelming predominance of targeted G to T transversion mutations (Wood et al, 1990). This latter result indicates that 8-OH-Gua acts like many other individual lesions to provoke a specific response by the DNA replication apparatus. The results of Kuchino et al (1987) might be explained by intermediates remaining from the synthesis of the 8-OH-Gua containing DNA, and this is a pitfall of the methods often used to generate site-specific lesions. See also Chapter 4.

B. Damage to deoxyribose

Free-radical damage to DNA sugars has been known for many years (e.g. deoxyribonic acid in irradiated DNA (Rhaese and Freese, 1968; Dizdaroglu et al, 1977)), but has generally received less direct attention than the modified bases. In a way, this is ironic, because the strand breaks and alkali-labile sites that have so often been used to measure biological damage by ionizing radiation (von Sonntag, 1987) or H_2O_2 (Demple et al, 1986) arise from attack on the deoxyribose moieties of DNA. The known lesions may be broadly categorized as oxidized base-free (AP) sites, and as fragmented deoxyribose. To the former category belong 2-deoxyribonolactone (the ring form of deoxyribonic acid) and 2-deoxypentos-4-ulose (**5** and **6** in Figure 2, respectively). These are both found in X-irradiated DNA (von Sonntag, 1987), while **5** is also formed specifically by the antitumour agent neocarzinostatin (NCS; Kappen and Goldberg (1989)) and **6** by BLM (Rabow et al, 1986) and by NCS under some conditions (Saito et al, 1989). NCS also induces DNA strand breaks, but these are usually bracketed by a normal nucleotide 3′-hydroxyl on one side and a nucleotide 5′-aldehyde on the other (**7** in Figure 2; Kappen and Goldberg (1983)).

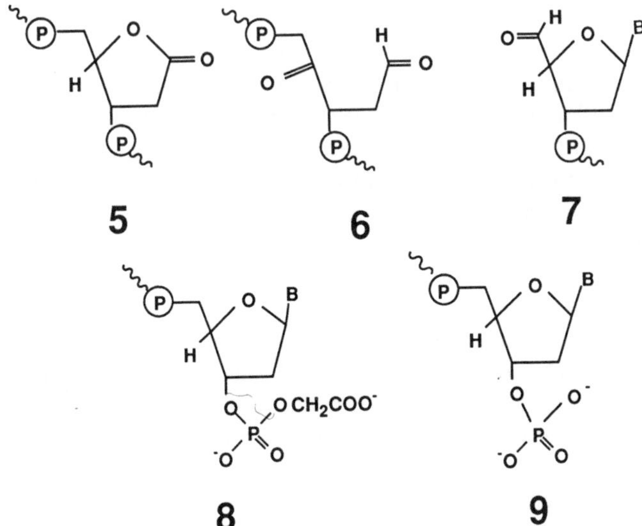

Figure 2. DNA sugar damage **5**, Deoxyribonolactone; **6**, 2-deoxypentos-4-ulose; **7**, nucleotide-5′-aldehyde; **8**, 3′-phosphoglycolate ester; **9**, 3′-phosphate ester. The circled phosphorus residues represent phosphodiester groups, the wavy lines the continuation of the DNA chain.

Fragmented deoxyribose forms include 3′-phosphoglycolate esters and 3′-phosphomonoesters (respectively **8** and **9** in Figure 2). The 3′-phosphoglycolate esters were originally identified as products of the antitumour drug BLM (Giloni et al, 1981) and later as γ-ray products (Henner et al, 1983a). Very unstable products may also be formed: NCS generates 3′-formylphosphate termini in DNA, which hydrolyse spontaneously to create 3′-phosphates (Chin et al, 1987).

Unmodified AP sites generated chemically by heat/acid or DNA alkylation, or by enzymatic removal of altered bases by DNA glycosylases, have been investigated extensively and are mutagenic in some circumstances (Loeb and Preston, 1986). When used as substrates for in vitro DNA synthesis, these ‘conventional’ AP sites direct the insertion predominantly of dAMP by a variety of DNA polymerases (Sagher and Strauss, 1983; Kunkel et al, 1983). Synthetic AP-like sites (tetrahydrofuran, propandiol or ethylene glycol derivatives) produced by total chemical synthesis also lead to preferential incorporation of dAMP during in vitro DNA synthesis (Takeshita et al, 1987), which indicates that DNA polymerases themselves could dictate insertion preferences.

The possible biological consequences of the radical-produced oxidized AP sites and deoxyribose fragments are beginning to receive more attention. Both BLM and NCS are potent mutagens, while each produces only a subset of the lesions formed by X-rays. One intriguing observation has been that the AP-like sites in a fragment of *lacI* DNA treated with NCS are resistant to some AP endonucleases and correlate with mutational hotspots in that gene (Povirk and Goldberg, 1985). Recent analysis indicates that these sites are actually positioned in a DNA strand nearly opposite a frank break in the complementary strand, i.e. closely opposed lesions, and that this juxtaposition may account for the resistance to repair endonucleases, and thus for mutagenesis (Povirk et al, 1988). BLM also produces closely opposed lesions, yielding double-strand breaks.

The deoxyribose fragments produced in DNA by free-radical attack block the action of DNA polymerase and DNA ligase in vitro (Henner et al, 1983b; Demple et al, 1986). This feature accounts for some of the biological effects of these lesions in mutants that lack particular DNA repair activities (see below), leading to cellular sensitivity to H_2O_2, BLM or γ-rays (Demple et al, 1983; Cunningham et al, 1986; D.S. Ramotar et al, unpublished data). The responsible lesions include both 3′-phosphodiesters (such as 3′-phosphoglycolate (Giloni et al, 1981; Henner et al, 1983a; Demple et al, 1986)) and 3′-phosphomonoesters (3′-phosphates (Henner et al, 1983a)). The ratio of mono- to diester products is affected by various factors, including the agent in question and the presence or absence of oxygen, but has been addressed virtually only in in vitro systems. Consequently, the roles of such fragmented lesions as possible mutation targets or specific inducers of cellular responses remain undefined.

C. Intra-DNA cyclic adducts

Some unusual crosslinked structures are formed in DNA by ionizing radiation treatment. Some of these involve covalent bonds between a DNA strand and its complementary strand or another molecule (see next sections), while others arise from the formation of new bonds between elements within a single DNA strand. Examples of the latter type are the 8,5′-cyclo-2′-deoxypurines (**10** in Figure 3).

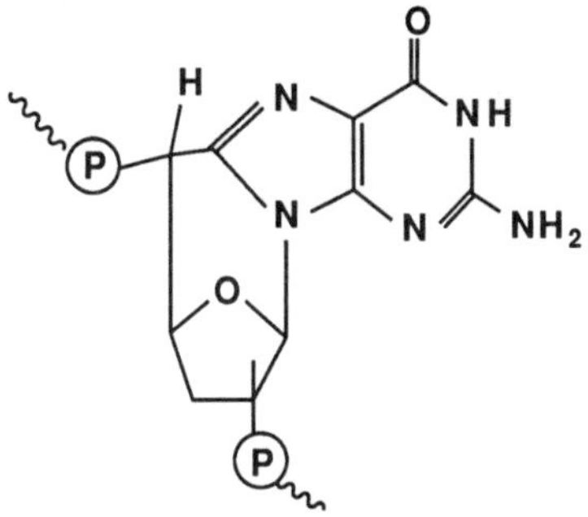

10

Figure 3. 8,5′-Cyclodeoxyguanosine.

Studies with AMP as a model system were the first to demonstrate such a product (8,5′-cyclo-AMP) of γ-irradiation (Keck, 1968; Raleigh et al, 1976). Later work with dG and dGMP (Berger and Cadet, 1983) and with DNA in vitro (Dizdaroglu, 1986) and in vivo (Dizdaroglu et al, 1987) added 8,5′-cyclo-dG to the list. By using gas chromatography–mass spectrometry (GC-MS) for separation and other techniques for analysis (e.g. NMR), both the 5′*S* and 5′*R* diastereoisomers of 8,5′-cyclo-2′-dG were demonstrated (Dizdaroglu, 1986; Dizdaroglu et al, 1987). The investigations with both AMP and DNA showed a greater production of the cyclic damage in the absence of O_2, suggesting that a key radical intermediate is diverted by reaction with molecular oxygen. In the absence of oxygen, these cyclic lesions are formed in significant amounts, with efficiencies (measured as 'G-values') of 1/20 to 1/10 that for single-strand breaks (von Sonntag, 1987; Dirksen et al, 1988). The formation of lesions such as 8,5′-cyclo-dG by agents other than ionizing radiation has not been reported, but the assertion that they result from •OH radical attack (Raleigh et al, 1976; Dizdaroglu, 1986) predicts that the cyclo adducts should be found in DNA exposed to H_2O_2, for example.

D. DNA interstrand crosslinks

Ionizing radiation generates crosslinks between the duplex strands of DNA molecules treated in aqueous solution (Washino and Schnabel, 1982), and also between noncomplementary single strands of poly(dT) homopolymers (Karam et al, 1986). The chemical structures of these crosslinks have not been defined, however, so the mechanism of their formation remains unknown. Two possibilities predominate. Pyrimidine base radicals dimerize easily (Al-Sheikhly and von Sonntag, 1983), and this could certainly occur between base-paired strands or separated poly(dT) molecules. Alternatively, the crosslinks could be formed as secondary reactions, perhaps from the abasic sites formed by free radical attack, a type of reaction that seems to occur slowly with apurinic sites formed in DNA by heating (Goffin and Verly, 1983).

E. Protein–DNA crosslinks

The production of radical species in macromolecules potentiates the formation of covalent bonds between molecules that otherwise touch only non-covalently. In the case of DNA, ionizing radiation (and probably other radical-inducing treatments) produces crosslinks to proteins (Oleinick et al, 1986). Little biochemical detail is as yet available on the types and structures of these adducts. One important, if not surprising, observation is that isolated chromatin irradiated in vitro contains protein–DNA crosslinks that arise predominantly from the core histones H2A, H2B, H3 and H4 (Mee and Adelstein, 1981). The nature of the covalent bonds was not determined in those experiments. It seems likely that crosslinks between DNA and other, rarer proteins are likely to be formed, perhaps with gene regulatory proteins that bind avidly to specific sites. Crosslinks are certainly also formed between DNA and proteins that have little or no affinity for DNA (e.g. BSA), but require relatively high protein/DNA ratios (Minsky and Braun, 1977). The ability of the aldehyde function of AP sites (in the ring-opened form) to form Schiff's base links to protein lysine residues could also mediate protein–DNA crosslinks in secondary reactions.

A biochemically well-characterized example of a radical-induced protein–DNA crosslink comes from model studies of radiation-treated amino acids and DNA bases and involves the 5-methyl group of thymine and the 5-carbon of tyrosine, as determined by GC-MS (Margolis et al, 1988). The formation of both this species and of the histone–DNA crosslinks cited above depended largely on the production of hydroxyl radical, while scavenging of other radicals (e.g. superoxide or solvated electrons) had relatively small effects (Mee and Adelstein, 1981; Margolis et al, 1988). Application of GC-MS analysis to irradiated chromatin revealed protein–DNA crosslinks involving thymine bonded to glycine, alanine, valine, leucine, isoleucine and threonine, although the detailed structure of the links was not determined (Gajewski et al, 1988).

Several important features of radiation-induced protein–DNA crosslinks remain to be elucidated. What proteins become linked to DNA? Are there specifically reactive sites or sequences on the DNA itself, such as actively transcribed regions? How does DNA structure affect crosslinking? What cellular mechanisms remove or otherwise process such damages? What are the consequences when protein–DNA crosslinks go unrepaired?

III. REPAIR ENZYMES FOR OXY RADICAL LESIONS IN DNA

As mentioned above, an emerging theme for enzymes that repair free radical damage to DNA is that they generally have broad substrate specificities. Many of the enzyme systems described below were discovered either using substrates damaged with agents that make many different lesions, or initially with defined substrates and later shown to act on multiple types of damage.

There are also non-enzymatic reactions that might be termed DNA repair. An example is the reversal of DNA-based radicals by glutathione (Raleigh et al, 1987). A radical centred at the 1′-carbon of deoxyribose can revert to a normal nucleotide by the addition of an H atom donated by homolytic cleavage of the S—H bond of reduced glutathione. Two glutathione radicals can then combine to form oxidized glutathione. Since it lacks stereospecificity, this reaction may also produce a rearranged nucleotide that has the α configuration at 1′-C (in contrast to the normal β), with unknown biological consequences. Enzymes mediate much more specific reactions.

A. Repair of damaged bases

1. *DNA glycosylases*

The first of these enzymes was identified only in 1974 (Lindahl, 1974), but at present at least nine distinct types are known that initiate the repair of damage caused by hydrolytic deamination, alkylation, free radicals or replication-induced mismatches (Sancar and Sancar, 1988; Wiebauer and Jiricny, 1989; Au et al, 1989). In contrast to classical nucleases, these enzymes hydrolyse the base–sugar (N–1′C) glycosylic bond to release the unusual or damaged base from DNA, and thereby produce an AP site. Some of the DNA glycosylases also have AP-cleaving activities (see below). A comprehensive and useful summary of DNA glycosylases and AP endonucleases that act on oxidative damage appeared recently (Wallace, 1988).

TG glycosylase

The first example of a mechanistically well-defined DNA repair enzyme specific for oxidative damage in DNA was endonuclease III of *E. coli* (Demple and Linn, 1980). This enzyme was identified initially by its ability to act on DNA that contained extensive damage produced by 254-nm UV light (Radman, 1976; Gates and Linn, 1977a). Although the predominant lesions formed by such treatment are pyrimidine cyclobutane dimers, TG is also produced (at ~1/50 the rate for dimers; Hariharan and Cerutti (1977)). In fact, endonuclease III could cleave DNA treated with OsO_4, which produces TG but not pyrimidine dimers (Gates and Linn, 1977a). Endonuclease III was also found to incise DNA containing AP sites (Gates and Linn, 1977a), which suggested that the activity might actually cleave DNA first as a DNA glycosylase to remove the damaged base, then as an AP endonuclease at the resultant abasic sites. This two-step incision mechanism was confirmed by the demonstration that endonuclease III is a DNA glycosylase for TG (and for 5,6-dihydrothymine), and that TG release precedes strand scission (Demple and Linn, 1980) (see Figure 4).

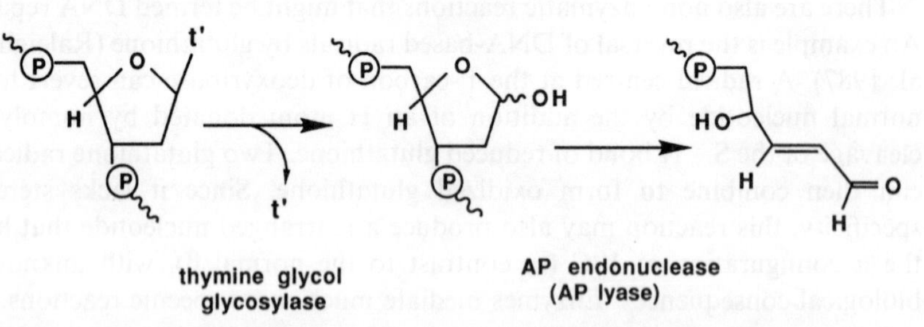

t′ P O H P t′ OH HO H O H

thymine glycol glycosylase

AP endonuclease (AP lyase)

Figure 4. Model for the endonuclease III incision mechanism. t′ represents a damaged pyrimidine base such as TG. Two steps yield a chain break: DNA glycosylase activity removes t′, then the AP lyase cleaves the AP site by β-elimination.

The suggestion then was that endonuclease III is a bifunctional DNA glycosylase–AP endonuclease (Demple and Linn, 1980), although the highly purified enzyme preparation used then was not homogeneous. This assertion was confirmed following additional purification efforts, which also revealed that the TG glycosylase acts on an array of other oxidized, reduced, fractured and ring-contracted forms of thymine, including urea, methyltartronylurea and 5-hydroxy-5-methylhydantoin (Breimer and Lindahl, 1984, 1985; Katcher and Wallace, 1983). Products of the radiolytic decay of [methyl-^{3}H]thymine in DNA (dihydrothymine?) are also attacked by endonuclease III (Demple and Linn, 1980) and its counterpart from cultured mouse cells (Nes, 1981) (see below). Later experiments indicated that endonuclease III also removes damaged cytosine bases from UV-irradiated poly(dG-dC) (Weiss and Duker, 1986; Doetsch et al, 1986). Although the damaged cytosines were not further characterized, one might guess them to be cytosine hydrates. Thus, as mentioned above, TG glycosylase (endonuclease III) is an enzyme that acts on a broad range of pyrimidine lesions. What these lesions all have in common is the loss of aromaticity (base stacking?) and planar configuration compared to normal DNA pyrimidines.

The cleavage of AP sites by a DNA glycosylase was unusual enough, but the product of this reaction was stranger still. Indirect analysis with DNA polymerase I of *E. coli* (Warner et al, 1980) showed that endonuclease III acting on AP sites yielded poor 3′-primers for DNA synthesis, in contrast to prototypical AP endonucleases such as exonuclease III and endonuclease IV, which were known to incise AP sites on the 5′-side. The same feeble priming capacity was found for the products of another bifunctional DNA glycosylase–AP endonuclease (Demple and Linn, 1980), the phage T4 UV endonuclease (Warner et al, 1980). High-resolution studies had already indicated that the analogous UV endonuclease of *Micrococcus luteus* produced unusual 3′-termini that were labile in alkali (Grossman et al, 1978). (Note that the T4 and *M. luteus* UV endonucleases are completely unrelated to the multienzyme complexes that act on pyrimidine dimers in most cells, as typified by the UvrABC enzyme of *E. coli* (Sancar and Sancar, 1988).) The authors of these and related studies concluded that the AP-cleaving activities of endonuclease III and the UV endonucleases hydrolyse AP sites on the 3′-side, and that the resulting 3′-deoxyribose residue is both poorly utilized and poorly removed by DNA polymerase I of *E. coli*.

The actual AP cleavage mechanism of endonuclease III and T4 UV endonuclease is not a hydrolysis. These enzymes break AP sites via β-elimination reactions, producing 3′-terminal 2,3-didehydro-2,3-dideoxyribose residues (Bailly and Verly, 1987; Kim and Linn, 1988; Manoharan et al, 1988). Thus, these are not hydrolytic enzymes but β-lyases, and the β-eliminations they catalyse can also be catalysed by small compounds such as primary amines (Lindahl and Andersson, 1972). Moreover, although the β-elimination cleavage reactions can be readily detected in vitro (Kow and Wallace, 1987), the extent of such cleavage in vivo has not been established.

TG glycosylases like endonuclease III have now been identified in many species and in mitochondria. These enzymes have sometimes been called 'redoxyendonucleases' (Gossett et al, 1988; Doetsch et al, 1986), but this wording confuses the enzyme with the mechanism of DNA damage, and there is no indication that these enzymes carry out redox chemistry. Analogous names were used for less well-characterized activities that act on X-ray damage (Paterson and Setlow, 1972; Bacchetti and Benne, 1975; Strniste and Wallace, 1975). The preferred terms would be TG glycosylase, endonuclease III, yeast TG glycosylase or endonuclease, etc.

The properties of partly purified TG glycosylases from yeast (Gossett et al, 1988; A.W. Johnson and B. Demple, unpublished data) and cultured mouse cells (Nes, 1980a; Kim and Linn, 1989) seem to be similar to endonuclease III, although none of the eukaryotic enzymes is yet purified sufficiently for us to be certain of its substrate specificity. The calf thymus activity was reported to cleave AP sites to yield a small gap bracketed by 3′- and 5′-phosphates (Doetsch et al, 1986). Since the preparation used was only partly purified, the participation of a second activity could not be ruled out. However, recent work with a nearly homogeneous preparation of the calf thymus enzyme has indicated the same cleavage specificity (P. Doetsch, personal communication).

E. coli has not stopped yielding surprises, either. Two additional glycosylases that remove ring-saturated or ruptured pyrimidines and cleave AP sites have been identified in extracts of endonuclease III-deficient strains (Wallace, 1988). These activities, termed endonucleases VIII and IX, can be distinguished by differences in substrate preference compared to each other and to endonuclease III (Wallace, 1988). The biological significance of these activities remains to be established, and it is possible that endonuclease VIII or IX could turn out to act on lesions not yet tested.

The biological role of endonuclease III has been partly established. Progress on this question depended on the isolation (by a mass-screening approach) of the endonuclease III structural gene *nth*, and its subsequent targeted disruption in *E. coli* (Cunningham and Weiss, 1985). It came as somewhat of a surprise that *nth::kan* mutants did not display any demonstrable phenotypic hypersensitivity (to H_2O_2, X-rays, UV light, etc.; Cunningham and Weiss (1985)). Perhaps the newly identified endonucleases mentioned above act as backup activities for the major toxic lesions usually handled by endonuclease III. The *nth::kan* strains did exhibit one tantalizing characteristic, however: namely, a 25-fold elevated frequency of spontaneous (i.e. not experimentally induced) reversion of the *argE3* ochre mutation (Cunningham and Weiss, 1985). A similar increase was found for certain other nonsense mutations, and virtually all of these cases turned out to involve nonsense suppressors (Weiss et al, 1988). A variety of oxidative mutagens (peroxides, streptonigrin) failed to increase the spontaneous reversion rates more than for wild types (Weiss et al, 1988), although it is still possible that the observed effects are due to production of some oxidative damage for which endonuclease III is required during repair. In this regard, it would be of interest to establish whether other loci exhibit *nth*-dependent mutability, and to what extent the observed mutation increase depends on aerobic growth.

The availability of the cloned *nth* gene has allowed investigation of the nature of the protein isolated from overproducing strains. Unexpectedly, endonuclease III is an iron–sulphur protein containing a single 4Fe–4S cluster (Cunningham et al, 1989). The role of this metal centre is unknown. The Fe–S cluster does not readily undergo redox reactions (Cunningham et al, 1989), and redox chemistry is not obviously required for either a DNA glycosylase or a β-lyase.

FAPy glycosylase

As mentioned above, purines with ruptured imidazole rings (FAPy residues) constitute strong blockades to DNA synthesis. Such residues not only arise through radiation damage, but are also produced slowly by base-catalysed hydrolysis of *N*7-alkyated guanines (Haines et al, 1962). A DNA glycosylase that released ring-opened guanines from a *N*7-[^{3}H]methylated, alkali-treated DNA polymer was identified initially in *E. coli* extracts (Chetsanga and Lindahl, 1979), but was partly purified. The same enzyme apparently also removes the FAPy adducts of aflatoxin B_1 to guanine-*N*7 in DNA (Chetsanga and Frenette, 1983). A nearly homogeneous preparation of this enzyme (isolated as a side fraction during the purification of endonuclease III) released the 'natural' FAPy damage in γ-irradiated poly(dA) paired with poly(dT) (Breimer, 1984). The report a few years ago of a 'FAPy cyclase' activity that closes the imidazole ring (Chetsanga and Grigorian, 1985) has not been verified.

The *E. coli* gene encoding the FAPy glycosylase, *fpg*, was isolated by a mass-screening procedure (Boiteux et al, 1987), thus allowing the Fpg protein to be overproduced for detailed analysis. There is a preliminary indication that the enzyme contains one atom of Zn per molecule, although the state of the metal in the enzyme was not defined (Boiteux et al, 1990). The purified Fpg enzyme (M_r 30 200) had an apparent K_m of 7 nM for FAPy residues (derived from *N*7-methylguanine; Boiteux et al. (1990)), which indicates that the methylated residues are an appropriate model for free radical-derived damage (although kinetic measurements should also be made for radical-induced FAPy damage). The glycosylase has an associated AP β-lyase activity similar to that of endonuclease III, but which may also produce 3′-phosphates (Bailley et al, 1989). The observation (Demple and

Linn, 1980) seems to be holding up that class I (β-lyase) AP endonuclease activities are always associated with DNA glycosylases, while class II (hydrolytic) enzymes never are.

No physiological role has yet been demonstrated for the FAPy glycosylase. Directed recombination was used to generate a mutant (*fpg-1* : : Kn^r) that lacks the activity (Boiteux and Huisman, 1989). The resulting strain has normal sensitivity to a variety of DNA-damaging agents (UV light, H_2O_2, γ-rays, mitomycin C, methylmethanesulphonate, nitrosoguanidine) (Boiteux and Huisman, 1989). Perhaps FAPy residues are not the dominant killing lesions produced by these agents in vivo, or there might be an alternative repair mechanism for them. Possible effects on more subtle phenotypes (e.g. spontaneous mutation rates) have not yet been examined (but see Note added in proof).

Hypoxanthine glycosylase

The deaminated form of adenine, hypoxanthine (the nucleoside is inosine), is removed by a DNA glycosylase present in small amounts in crude extracts of *E. coli* (Karran and Lindahl, 1978; Oeda et al, 1978). The bovine hypoxanthine glycosylase was purified extensively and shown to release hypoxanthine produced by hydrolytic deamination (Karran and Lindahl, 1980). Like other DNA glycosylases, the bacterial and mammalian enzymes are small proteins ($M_r \sim 30\,000$) that have no known cofactors (Karran and Lindahl, 1978, 1980; Myrnes et al, 1982). The variation of hypoxanthine glycosylase levels during the mammalian cell cycle has been examined (Dehazya and Sirover, 1986).

Hypoxanthine is not a significant DNA product of ionizing radiation (Hutchinson, 1985; von Sonntag, 1987). However, it is possible that the enzymes described above have among the many other purine lesions, substrate(s) upon which they act more readily.

2. Nucleases

Three nuclease (phosphodiester-hydrolysing) activities have now been identified that act on oxidative damage to DNA bases. One is the well-defined UvrABC system of *E. coli*, which acts on many types of lesions, a second is endonuclease V of *E. coli*, and the third is a newly discovered nuclease activity that attacks 8-hydroxyguanines in DNA. The latter seems to be the first example of a conventional (?) nuclease that is apparently specific for one DNA lesion.

The UvrABC complex

The multiprotein Uvr system of *E. coli* is noteworthy for the variety of lesions that it attacks in DNA (Sancar and Sancar, 1988). The lesions that

UvrABC is most well known to repair are bulky ones, such as pyrimidine cyclobutane dimers or psoralen crosslinks. However, recent work indicates that the complex in vitro can act on both TG and AP sites, in a manner that depends, as for the bulky lesions, on all three proteins (Lin and Sancar, 1989; Kow et al, 1990). Despite this observation, mutants that lack the enzyme (*uvr*$^-$) are not sensitive to ionizing radiation (Howard-Flanders and Boyce, 1966) or H_2O_2 (Imlay and Linn, 1988). The difficulty in isolating mutants (Saporito et al, 1989) that simultaneously lack exonuclease III (*xth*$^-$), endonuclease IV (*nfo*$^-$), and UvrABC (*uvrA*$^-$) was taken to indicate that the Uvr system acts as a backup for situations where other enzyme systems are saturated. However, such triple mutants were isolated when additional mutations were also present: either in the *ung* gene, eliminating uracil–DNA glycosylase, and presumably a good deal of the 'normal' AP site production, or in the *sulA* gene, eliminating a damage-inducible inhibitor of cell division (Foster, 1990). The nonviability of *xth*$^-$ *nfo*$^-$ *uvrA*$^-$ strains might thus also be due to indirect effects of DNA damage, e.g. disrupted regulation of cell division.

Endonuclease V

E. coli endonuclease V incises heavily UV-irradiated DNA, double-stranded DNA treated with OSO_4 or containing apurinic sites, as well as undamaged single-stranded DNA (Gates and Linn, 1977b). The pyrimidine cyclobutane dimers produced by 254-nm UV light are apparently not the sites of endonuclease action. A different method to generate the damage (360-nm light in the presence of the photosensitizer acetophenone) failed to invite incision; TG residues were the presumptive targets (Gates and Linn, 1977b). This enzyme might therefore operate in the repair of free radical damage, but whether a common element in damaged DNA activates the endonuclease remains unclear.

The nature of the mechanism and substrate specificity of endonuclease V was investigated extensively by Demple and Linn (1982b). Endonuclease V degrades uracil-containing DNA with an efficiency that correlates with the degree of replacement of thymine by uracil, although a single G : U base pair in the 10-kbp PM2 phage DNA was not actively cleaved (Demple and Linn, 1982b). The DNA of phage PBS2, in which all the thymine is replaced by uracil, is the best overall substrate for the enzyme (Gates and Linn, 1977b). The only common structural theme noted among this diversity (AP sites, TG residues, multiple A : U base pairs and single-stranded DNA) was that all the structures are expected to disrupt base stacking (Demple and Linn, 1982b). It seems possible that endonuclease V recognizes some overall structural feature, such as might be caused by unstacking of bases, but this question has not yet been further examined.

Endonuclease V has an unexpectedly complex mode of DNA incision.

The enzyme makes both single- and double-strand breaks in a nearly constant ratio ($\sim 8:1$) in the various duplex substrates, but probably also conducts some 'nibbling' from the exposed termini (Demple and Linn, 1982b). Endonuclease V digests DNA molecules processively, and exhibits cooperativity with respect to the endonuclease concentration (Demple and Linn, 1982b). These various features indicate that this enzyme could have a role other than incision at damage sites; it is even possible that endonuclease V does not function in DNA repair, but rather, say, in recombination (cleavage of Holliday junctions?). This question will of course be addressed most directly by the isolation and characterization of enzyme-deficient mutants, but to our knowledge such experiments are not currently under way.

8-OH-Gua nuclease

Recent interest in the biological effects of 8-OH-Gua has prompted investigation of possible repair of this damage. Whole animal studies indicated that the lesion is actively removed from the liver DNA of X-irradiated mice (Kasai et al, 1986). Analysis of bacterial crude extracts has recently revealed an activity that attacks 8-OH-Gua in DNA. In one study, a synthetic duplex oligodeoxynucleotide containing a single 8-OH-Gua residue was used to monitor the presence of a nuclease activity during fractionation of cell-free extracts of *E. coli* (S. Nishimura, personal communication; see also Chapter 4). The partly purified activity cleaved the artificial substrate at the position of 8-OH-Gua, but produced incisions thought to be bracketed by both 3′- and 5′-phosphates. It is unclear whether this unusual structure is the product of more than one enzyme in the impure preparation, and the possible presence of a DNA glycosylase for 8-OH-Gua was not ruled out. However, the purified Fpg glycosylase does not remove 8-OH-Gua from X-irradiated DNA (M. Dizdaroglu and J. Laval, personal communication).

An independent set of experiments suggested that 8-OH-Gua may not be removed by a DNA glycosylase. Analysis of multiple lesions (by GC-MS) in X-irradiated DNA incubated with crude extracts of *E. coli* showed that 8-OH-Gua disappears rapidly from the DNA and does not appear as the free base after ethanol precipitation (B. Demple and M. Dizdaroglu, unpublished data). However, acid hydrolysis of the ethanol-soluble fraction did release free 8-OH-Gua in quantities similar to the amounts removed from DNA.

The further purification of the responsible enzyme(s) is an obvious goal, but even more important will be the eventual isolation of enzyme-deficient strains, which will be essential for establishing the biological function of this novel enzyme (see Note added in proof). To judge from the whole animal experiment cited above (Kasai et al, 1986), there are probably eukaryotic counterparts to the *E. coli* activity.

B. Repair of sugar damage

1. *Bacterial enzymes*

As with most types of DNA repair, the removal of oxidatively damaged sugars has been characterized best in the bacterium *E. coli.* This organism contains a number of activities that attack AP sites or sugar fragments in DNA, and the biological roles of the main enzymes have been established experimentally.

Exonuclease III

Exonuclease III was one of the first bacterial nucleases to be characterized extensively. The enzyme was discovered in side fractions during the purification of *E. coli* DNA polymerase I, by virtue of its ability to activate DNA templates for use by the polymerase (Richardson et al, 1964). This in vitro activating ability is directly related to the enzyme's in vivo function in repair of oxidative damage. Exonuclease III is also familiar to molecular biologists, who have exploited its exonuclease function in genetic engineering (Sambrook et al, 1989).

The activation of primer templates for the DNA polymerase by exonuclease III was correlated with the 3′-phosphatase activity of the enzyme (Richardson and Kornberg, 1964). The purified protein also had 3′→5′ exonuclease activity specific for duplex DNA (Richardson et al, 1964). Although it seemed possible that the 3′-phosphatase of exonuclease III might remove such blocking groups formed in vivo, its physiological role remained unclear in the absence of enzyme-deficient mutants. When such mutants (xth^-) were finally isolated in a mass-screening approach (Milcarek and Weiss, 1972), their properties were initially not very revealing: apart from a slightly increased sensitivity to alkylating agents, the xth^- strains did not reveal any consistent feature. The modest sensitivity to alkylating agents was rationalized when it was later shown that exonuclease III is also the major AP endonuclease of *E. coli* (Weiss, 1976; Gossard and Verly, 1978).

The key observations suggesting a role for exonuclease III in repairing oxidative damage are fairly recent: exonuclease III activates DNA synthesis in DNA damaged in vitro by BLM (Niwa and Moses, 1981), and exonuclease III-deficient strains are exquisitely sensitive to H_2O_2 (Demple et al, 1983). The known 3′→5′ exonuclease and AP endonuclease functions of the enzyme could not easily explain the latter phenotype. The 3′-phosphatase and 3′-phosphoglycolate diesterase (Niwa and Moses, 1981; Henner et al, 1983b) activities of exonuclease III seemed more likely to function specifically in repair of oxidative DNA damage (Demple et al, 1983).

This hypothesis was strongly supported when the nature of the H_2O_2 lesions formed in vivo in *E. coli* was examined (Demple et al, 1986). Exonuclease III-deficient bacteria accumulated DNA single-strand breaks at a ~20-fold greater rate than wild-type cells upon exposure to H_2O_2. The chromosomal DNA was isolated from peroxide-treated cells and examined for its ability to support the activity of DNA polymerase I, as a measure of the presence of undamaged 3′-hydroxyl groups. The result was clear: the DNA from H_2O_2-treated wild-type cells supported abundant DNA synthesis, while that from *xth*$^-$ mutants did not, despite the presence of a greater number of strand breaks. More importantly, treatment of the isolated DNA with purified exonuclease III prior to the polymerase reaction had no effect on the DNA of wild-type origin, but gave up to ten-fold activation of priming for H_2O_2-damaged DNA from *xth*$^-$ mutants. Two other enzymes were examined for their effects in this 'primer activation' reaction. Endonuclease IV, which has 3′-phosphatase and 3′-repair activity (see below), but not 3′→5′ exonuclease activity, gave activation nearly equal to exonuclease III. The 3′-phosphatase activity of phage T4 polynucleotide kinase had little effect. Thus, the critical blocking groups are not 3′-phosphates, but 3′-phosphodiesters of some kind. These results and their interpretation are analogous to those obtained for BLM damage in vitro (Niwa and Moses, 1981).

These experiments show that one important role for exonuclease III in repair of oxidative damage is the removal of lesions that block 3′-termini and prevent DNA repair synthesis; see Figure 5.

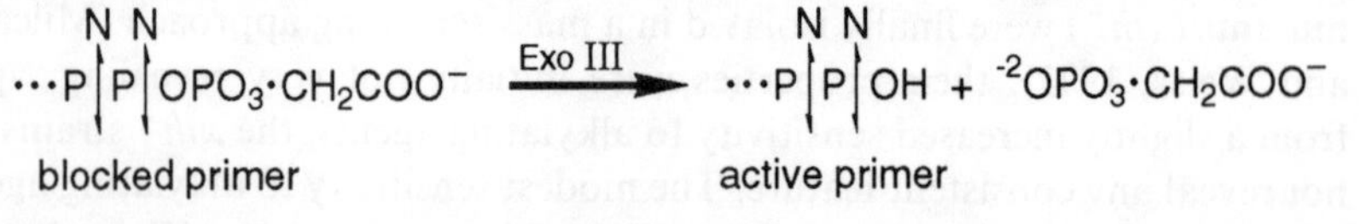

$$\cdots P \overset{N}{|} P \overset{N}{|} OPO_3{\cdot}CH_2COO^- \xrightarrow{\text{Exo III}} \cdots P \overset{N}{|} P \overset{N}{|} OH + {}^{-2}OPO_3{\cdot}CH_2COO^-$$

blocked primer — active primer

Figure 5. Activation by exonuclease III of 3′-blocking groups produced by oxidative attack on DNA.

A convenient model substrate for this reaction is poly(dA-dT) terminated by 3′-[^{32}P]phosphoglycolaldehyde (3′-PGA) residues, which is used to assay directly the release of the labelled deoxyribose fragments by exonuclease III and other 3′-repair enzymes such as endonuclease IV (Demple et al, 1986; Levin et al, 1988; Johnson and Demple, 1988a).

Agents such as γ-rays break DNA via $^{\bullet}OH$ radicals to generate 3′-phosphoglycolate esters (Henner et al, 1983a). Such lesions are likely targets of exonuclease III in vivo, although the structure(s) of the DNA lesions that accumulate in *xth*$^-$ mutants still needs to be established. Work is under way in our laboratory to make this determination. Whatever this 3′-blocking damage is, it does not seem to be dealt with efficiently in vivo by endonuclease IV. Even though the latter enzyme can give full primer activation in vitro (Demple et al, 1986; Johnson and Demple, 1988b), overproduction of endonuclease IV from a multicopy plasmid does not cure the H_2O_2 sensitivity of *xth*$^-$ cells (Ramotar et al, 1991; R.P. Cunningham, personal communication).

The 3′-repair function of exonuclease III is not the only activity proposed to explain the H_2O_2-sensitive phenotype of *xth*$^-$ strains. Purified exonuclease III cleaves DNA containing urea residues (Kow and Wallace, 1985), presumably due to the AP endonuclease activity. Although this feature might explain some aspects of exonuclease III biology, it does not correlate with the excess of unrepaired strand breaks that accumulates in exonuclease-deficient cells treated with H_2O_2 (Demple et al, 1986). If the urea-cleavage activity were predominant in vivo, the opposite effect would be expected, namely a paucity of strand breaks in *xth*$^-$ mutants compared to wild-type cells.

Two additional effects of mutations in *xth* merit mention. First, *xth*$^-$ mutants have a strongly increased frequency of at least one type of intrachromosomal recombination (Zieg et al, 1978). It has not been reported whether this recombination phenotype is affected by the presence of O_2 during growth. Second, *xth*$^-$ mutations alleviate the elevated spontaneous mutagenesis otherwise seen in strains that lack superoxide dismutase activity (Farr et al, 1986). This latter observation seems counterintuitive, in that exonuclease III is involved in removing oxidative DNA damage. Perhaps the enzyme in this case corrects some lesions that would otherwise be lethal, allowing other lesions to be converted to mutations. This model is analogous to mutagenesis caused by the SOS response, which could be merely a byproduct of a survival mechanism.

Endonuclease IV

Endonuclease IV was identified first as an AP endonuclease in extracts of *E. coli* deficient in exonuclease III (Ljungquist et al, 1976; Ljungquist and Lindahl, 1977). Endonuclease IV could also be distinguished from exonuclease III by the former's resistance to EDTA in enzyme assays (Ljungquist et al, 1976), while the exonuclease has a requirement for divalent metals, usually Mg^{2+} (Rogers and Weiss, 1980).

Endonuclease IV received relatively little attention at first and was thought to be merely a backup AP endonuclease, until the enzyme's connection to oxidative DNA damage was discovered. Firstly, as mentioned above, endonuclease IV efficiently activated primers in oxidized DNA

(Demple et al, 1986). Secondly, endonuclease IV was found to be inducible as part of a multicomponent system that is activated by redox-cycling agents such as paraquat (Chan and Weiss, 1987; Greenberg et al, 1990). Purified endonuclease IV is not only an AP endonuclease, but is also a 3′-phosphatase, but the enzyme lacks the uncontrolled exonuclease activity of exonuclease III (Levin et al, 1988). Atomic absorption spectroscopy of purified endonuclease IV indicates the presence of both Zn and Mn in the enzyme; these metals could play roles in binding and phosphodiester cleavage (J.D. Levin et al, unpublished data).

As for exonuclease III, establishment of the biological role of endonuclease IV required the isolation of mutant strains lacking the enzyme. The structural gene for the endonuclease, called *nfo*, was isolated by a mass-screening method analogous to that used for endonuclease III: assay of extracts from strains bearing a plasmid library for overproduction of EDTA-resistant AP endonuclease (Cunningham et al, 1986). A disrupted allele (*nfo* : : *kan*) was then constructed in vitro and recombined into the bacterial chromosome (Cunningham et al, 1986). The resulting nfo^- strains had significant sensitivities to BLM, *t*-butyl hydroperoxide and mitomycin C, and the nfo^- mutation also increased the sensitivity of xth^- cells to H_2O_2 and methylmethanesulphonate (Cunningham et al, 1986). Moreover, the xth^- nfo^- double mutant was hypersensitive to γ-rays, while each of the single mutants displayed wild-type resistance (Cunningham et al, 1986). These observations are consistent with a role for endonuclease IV in backing up the repair of alkylation-induced AP sites and some lethal γ-ray lesions, but also demonstrate that the enzyme is essential for the efficient repair of certain oxidative lesions.

The identity of the BLM-produced damage repaired by endonuclease IV has not been established unambiguously. However, endonuclease IV does incise alkali-labile sites in DNA treated in vitro with BLM more efficiently than does exonuclease III, while both enzymes attack conventional (hydrolytic) AP sites about equally well (Povirk and Goldberg, 1985; Povirk and Houlgrave, 1988; Hagensee and Moses, 1990). Recent work in our laboratory shows that this feature extends to damage formed in vivo, and damage-containing intermediates can be isolated from nfo^- cells (J.D. Levin and B. Demple, unpublished data). Briefly, chromosomal DNA isolated from BLM-treated nfo^- cells contains sites sensitive to endonuclease IV, as detected by the primer activation assay. The degree of activation was considerably less with DNA from cells expressing the endonuclease at levels comparable to those seen after induction with paraquat. Evidently, certain BLM lesions, presumably oxidized AP sites such as 2-deoxypentos-4-ulose (see above), require endonuclease IV in vivo for their repair. We are

We are developing new post-labelling methods to identify these endonuclease IV-specific DNA lesions produced by BLM.

Other 3′-diesterases

E. coli has at least five enzymes that hydrolyse the 3′-PGA substrate (Bernelot-Moens and Demple, 1989). The two major activities are exonuclease III and endonuclease IV, as described above. Gel filtration chromatography of extracts from a *xth*$^-$ *nfo*$^-$ mutant reveals two peaks of activity, one that would correspond to a globular protein of M_r ~55 000, the other to a globular protein of M_r ~30 000. The latter peak generates two bands in the M_r 25 000–27 000 range in 3′-repair activity gels (Bernelot-Moens and Demple, 1989). Although the identities of these various species have not yet been established, the M_r 55 000 activity might correspond to *E. coli* deoxyribophosphodiesterase (dRpase), an enzyme that removes the 5′-terminal deoxyribose 5-phosphate produced by class II AP endonucleases (Franklin and Lindahl, 1988). This possibility seems the more likely because the major 3′-PGA diesterase of HeLa cells is also associated with dRpase (Chen et al, unpublished data).

2. *Yeast AP endonuclease Apn1*

The major 3′-PGA diesterase of the yeast *Saccharomyces cerevisiae* was purified (Johnson and Demple, 1988a) in order to provide detailed insight into the biochemistry of repair of oxidative DNA damage in a eukaryote. This same enzyme turned out to be both a potent DNA 3′-phosphatase and the major AP endonuclease of yeast (Johnson and Demple, 1988b). In the latter capacity, this enzyme probably corresponds to the AP endonuclease 'E' described by Armel and Wallace (1984). The other AP endonucleases found by the latter workers are probably either derived by proteolysis of the 'E' form (Johnson and Demple, 1988b; Popoff et al, 1990) or are AP lyases. The yeast enzyme as originally purified (in buffers containing EDTA and thiols) exhibited an absolute requirement for added metals that was best satisfied by $CoCl_2$ (Johnson and Demple, 1988a), but this requirement was obviated by a redesigned purification from which EDTA and thiols were omitted (Popoff et al, 1990; Ramotar et al, 1990).

The purified AP endonuclease/3′-diesterase was used to generate highly specific polyclonal antisera, which in turn were employed to screen gene expression libraries to identify the cloned structural gene for the enzyme, designated *APN1* (Popoff et al, 1990). A disrupted form of *APN1* was engineered and used to replace the wild-type gene in haploid yeast. The resulting *apn1 : : URA3* mutants lack >97% of the AP endonuclease and 3′-PGA diesterase found in wild-type cells and demonstrate that the Apn1 enzyme is dispensable under normal laboratory growth conditions. However, the *apn1 : : URA3* mutants did exhibit several significant phenotypes (D.S. Ramotar, unpublished data), including hypersensitivity to some oxidants (H_2O_2, *t*-butyl hydroperoxide) and alkylating agents (methylmethanesulphonate, methyl nitrosoguanidine), and a modestly increased rate of spontaneous mutagenesis. Somewhat surprisingly, the increase in spontaneous mutation was not dependent on aerobic growth; the responsible metabolic mutagens are therefore not likely to be related to oxygen radicals, although this is still formally possible.

The predicted polypeptide sequence of Apn1 is obviously homologous to the endonuclease IV protein, including 55% identity at the amino acid level (Popoff et al, 1990). This homology resides entirely within the first ~280 amino acids of Apn1, which correspond roughly to the complete endonuclease IV sequence. The basic features of substrate recognition and phosphodiester cleavage must therefore be contained within the homologous region, because the prokaryotic and eukaryotic enzymes have virtually identical substrate specificity, apparent K_m values and turnover numbers in vitro (Levin et al, 1988: Johnson and Demple, 1988b). The function of the 'extra' ~80 residues in Apn1 is unclear, although this predicted sequence is rich in basic amino acids: 30% lysine plus arginine, a level equal to the most basic histone proteins (Igo-Kemenes et al, 1982). We have speculated that this presumably basic domain could mediate access of the repair domain to lesions bound in chromatin by competing with histones for binding to DNA (Popoff et al, 1990). Other functions are certainly also possible. In order to address this question, experiments are under way to engineer derivatives of Apn1 and endonuclease IV lacking or containing the C-terminal basic region.

3. *Drosophila AP endonucleases*

AP endonucleases have been described from the fruit fly *Drosophila melanogaster*. The activities described so far are rather different from their counterparts in microbes (see above) and human cells (see below). For example, a partly purified AP endonuclease from embyros is large (M_r ~66 000) and appears to leave 3′-phosphate termini after cleavage (Spiering and Deutsch, 1986), rather than the 3′-hydroxyl groups produced by the other main enzymes. Molecular cloning experiments and analysis of proteins separated by SDS gel electrophoresis also indicate the existence of an AP endonuclease of M_r ~35 000, which was not initially detected in standard enzyme purification experiments (Kelley et al, 1989). In neither case has a role in repair of oxidative damage been demonstrated directly, and the necessary enzyme-deficient strains have not yet been constructed. However, an educated guess would point to AP endonucleases having a role in free radical repair in *Drosophila* as they do in other organisms.

4. Human enzymes

AP endonuclease

Enzymes that attack AP sites generated by the hydrolytic action of DNA glycosylases or by heat and acid have been purified from extracts of rat liver (Cesar and Verly, 1983), calf thymus (Sanderson et al, 1989), human placenta (Linsley et al, 1977; Shaper et al, 1982) and cultured rodent (Nes, 1980b) and human cells (Kane and Linn, 1981; Kühnlein, 1985). Those studies did not address whether the enzymes act on oxidative damage to deoxyribose.

Our work has characterized the mammalian enzymes that hydrolyse the synthetic 3′-PGA substrate described above (Section III.B.1). Two major Mg^{2+}-dependent activities were found in HeLa crude extracts and analysed by ion exchange and gel filtration chromatography (Demple et al, 1988). One 3′-PGA diesterase is the HeLa AP endonuclease described previously (Kane and Linn, 1981). Immunoprecipitation with polyclonal antisera specific for the AP endonuclease (Kane and Linn, 1981), and purification of the crossreacting activity, confirmed it to be a polypeptide of M_r 37 000 with substantial AP endonuclease activity (Demple et al, 1988; Chen et al, unpublished data). In contrast to *E. coli* exonuclease III, endonuclease IV and yeast Apn1 protein, the human AP endonuclease is much more active against AP sites than 3′ sites: the ratio of 3′-PGA diesterase to AP endonuclease is $\sim 1:1$ for the microbial enzymes, and $\sim 1:100$ for the AP endonuclease (Demple et al, 1988; Chen et al, unpublished data).

The N-terminal 25 amino acids of HeLa AP endonuclease (Chen et al, in preparation) closely match the N-terminus of the enzyme from calf thymus (Henner et al, 1987). Antisera generated against the protein we purified have now been used to screen cDNA expression libraries for the structural gene for this enzyme. One promising cloned DNA expresses a protein that can be used to immunopurify antibodies specific for authentic AP endonuclease, and yields AP endonuclease and 3′-PGA diesterase activity in *E. coli* (Chen et al, unpublished data).

Another approach to cloning mammalian genes that encode enzymes of AP site repair depends on probing libraries with AP site-containing DNA

(Lenz et al, 1990). This procedure also yielded positive candidate sequences, although none was shown to specify an AP endonuclease.

3′-PGA diesterase

The second major activity in HeLa cells has not been previously characterized in detail, although reports exist in the literature (Kühnlein, 1985; Thibodeau and Verly, 1980) of mammalian AP endonuclease activities distinct from the main M_r 37 000 enzyme. The HeLa 3′-PGA diesterase is of apparently greater size than the main AP endonuclease (gel filtration profile consistent with a globular protein of M_r ~55 000) and does not crossreact detectably with anti-AP endonuclease antibodies (Chen et al, unpublished data). Furthermore, the extensively purified (but not homogeneous) 3′-PGA diesterase enzyme has a substrate specificity that contrasts with the AP endonuclease: 3′-PGA diesterase/AP endonuclease ratio 1 : 1 (Chen et al, unpublished data). The partly purified diesterase also removes 5′-terminal AP sites, i.e. it is analogous to the dRpase activity partly purified from *E. coli* extracts, which also has a native size that corresponds to a globular protein of M_r 55 000 (Franklin and Lindahl, 1988). Indeed, the bacterial enzyme may also have both 5′-dRpase and 3′-PGA diesterase activity (Bernelot-Moens and Demple, 1989). Although it cannot be ruled out that separate proteins mediate the different reactions, the most parsimonious view is that one enzyme has both 3′-repair and 5′-repair activities. Such a broad substrate specificity would be appropriate for the repair of oxidative lesions, which sometimes include both 3′- and 5′-deoxyribose lesions (von Sonntag, 1987).

5. *Dual-function enzymes: exonuclease III, endonuclease IV, and Apn1*

Endonuclease IV and exonuclease III, along with the yeast Apn1 endonuclease (see below), are unusual enzymes that possess both phosphodiesterase and phosphomonoesterase activity. Since the known enzymatic mechanisms of these two types of reaction are quite different (Cleland, 1990), the mechanism(s) of such dual-function enzymes is of interest. All three AP endonucleases are modest in size, so there is no a priori reason to suppose more than one active site. One therefore needs to entertain the notion of either a novel cleavage mechanism, or an active site of unusual versatility.

One clue to the unusual phosphoester cleavage abilities of the three enzymes comes from the intimate involvement of metals in the activity of each.

Exonuclease III, like many other nucleases, has a nearly absolute requirement for divalent metals (Rogers and Weiss, 1980). For all substrates of exonuclease III, this need can be filled by Mg^{2+}, although Ca^{2+} supports the

AP endonuclease but not the exonuclease activity. It is not clear whether these metals are involved in bond cleavage or, for example, substrate binding. Indeed, divalent metals can be dispensed with entirely when the 3′-PGA diesterase activity of exonuclease III is assayed in activity gels, and addition of Mg^{2+} during incubation of these gels gave no detectable activation of exonuclease III (Bernelot-Moens and Demple, 1989). For the same substrate in solution, the enzyme shows the usual metal requirement (Levin et al, 1988). It is therefore possible that the Mg^{2+} in exonuclease III reactions is involved in DNA binding by the enzyme, and that this requirement is alleviated in the restricted quarters of the gel matrix.

Although endonuclease IV and Apn1 are active in the presence of metal chelators, both enzymes are metalloproteins (J.D. Levin et al, unpublished data). Endonuclease IV prepared in different ways contains both Zn and Mn (2.4 and 0.6 atoms per protein molecule, respectively). These transition metals are required for the enzyme's activity, which accounts for the previously observed sensitivity of endonuclease IV to EDTA (Ljungquist, 1977; Levin et al, 1988). The active Apn1 nuclease is also sensitive to chelators, and the enzyme was initially purified as a metal-requiring apoprotein (Johnson and Demple, 1988a). The pattern of metal activation of the EDTA-treated bacterial and yeast enzymes is quite similar, with $CoCl_2$ giving the strongest reactivation in each case, followed by $MnCl_2$ and to a lesser extent $ZnCl_2$ (Johnson and Demple, 1988a; J.D. Levin et al, unpublished data).

One key difference between the enzymes is that chelator-treated endonuclease IV is unstable (J.D. Levin et al, unpublished data), while Apn1 apoprotein could be purified without its native metal (Johnson and Demple, 1988a). This latter observation indicates that at least a fraction of the removable metal plays more of a structural role in endonuclease IV than in Apn1. Obviously, the establishment of the specific functions of the metals in these enzymes awaits more detailed structural investigation, but it is tempting to speculate that one role might be polarization of phosphate esters for cleavage, in a manner analogous to the metals in the 3′→5′ exonuclease of *E. coli* DNA polymerase I (Freemont et al, 1988).

C. Recombinational repair

A complete system for restoring DNA damaged by free radical attack includes components of the recombinational machinery. This has been demonstrated clearly in *E. coli* and the yeast *S. cerevisiae*. *E. coli recA* mutants are hypersensitive to various radical-generating agents, e.g. X-rays (Howard-Flanders and Theriot, 1966; Kapp and Smith, 1970) and H_2O_2

(Ananthaswamy and Eisenstark, 1977; Carlson and Carpenter, 1980; Demple and Halbrook, 1983). *E. coli recBC* mutants are X-ray-sensitive (Howard-Flanders and Theriot, 1966; Kapp and Smith, 1970), but have normal resistance to H_2O_2 (Imlay and Linn, 1988).

In yeast, many mutations that were identified initially as conferring increased sensitivity to X-rays turn out also to affect general genetic recombination (Haynes and Kunz, 1981). Examples include the *RAD52* locus, which is required for most mitotic and meiotic recombination (Resnick, 1987), and *RAD50*, which plays a key role in meiotic recombination and spore formation (Alani et al, 1990). Strains with the *rad52-1* mutation are also hypersensitive to H_2O_2 (Thacker, 1976).

It should not be surprising that recombinational functions contribute to the survival of DNA damage; indeed, the recombination-deficient mutants mentioned above are variously sensitive to other agents in addition to oxidants, including simple alkylating agents and UV light (Friedberg, 1988). However, certain oxidative lesions, particularly those resulting from X-rays, require special treatment owing to the presence of clustered lesions. The double-strand breaks caused abundantly by X-rays are a case in point. Their repair requires the rejoining of contiguous segments of information that would otherwise be lost or scrambled. This process evidently involves the use of a second copy or a homologue of the damaged chromosome, as evidenced by elegant studies in both *E. coli* and yeast. In the bacterial case, X-ray double-strand break repair requires both the $recA^+$ allele and a duplicate copy of the chromosome (Krasin and Hutchinson, 1977), as is found in rapidly dividing cells. In the absence of those requirements, one double-strand break can be lethal. Analogous requirements have been demonstrated in *S. cerevisiae* (Resnick and Martin, 1976). Presumably, strand transfer activities line up the broken molecules along an intact stretch of the homologue, allowing polymerases and strand-joining activities to fill in and reconnect across the break.

D. Roles for DNA polymerases

The individual functions of DNA polymerases in repair of oxidative damage have been examined in the organisms most tractable for genetic studies: *E. coli* and *S. cerevisiae*. In the bacterium, DNA polymerase I has a general role in excision repair that is easily detected in the face of damage by X-rays (Town et al, 1971) and H_2O_2 (Ananthaswamy and Eisenstark, 1977; Hagenesee and Moses, 1989). The key function here is probably the gap filling required after the removal of 3′ lesions or oxidized AP sites. The isolation of *E. coli* mutants that depend on DNA polymerase I rather than polymerase III for semiconservative replication allowed the demonstration of a modest role for DNA polymerase III in recovery from H_2O_2 damage (Hagenesee and Moses, 1986). It is possible, however, that the recruitment of DNA polymerase I for replication in these mutants forces DNA polymerase III into a role that it does not usually have.

In yeast, the main replicative polymerase has been examined at the biochemical and the molecular biological level. Deletion of the enzyme's structural gene (*POL1*) does not have an obvious effect on the cellular sensitivity to X-rays; other oxidants were not tested (Budd et al, 1989). Two other DNA polymerases have been identified in this organism, and, to date, neither has been assigned a defined role in repair of oxidative damage.

Some elegant biochemical studies have indicated roles for DNA polymerases α, δ and β in the repair of oxidative damage in human cells. Dresler and colleagues (DiGiuseppe and Dresler, 1989) used permeabilized cells to show that DNA repair synthesis after damage by BLM showed only partial blockage by chemical inhibitors of either polymerase β or of α and δ (e.g. dideoxynucleoside triphosphates and butylated dGTP). The two types of polymerase appeared to be involved in the production of short (~4 nucleotides) and long (~20 nucleotides) repair patches, respectively (DiGuiseppe and Dresler, 1989). This partial dependence on polymerase β for repair of BLM damage contrasts with the apparent requirement for DNA polymerase δ after UV damage, also demonstrated with permeabilized cells (Dresler et al, 1988). The key difference may be the size of the repair patch produced in the different cases. Indeed, polymerase β does seem to be involved in repair synthesis after damage by simple alkylating agents (Dresler and Lieberman, 1983), which is also expected to generate small patches of DNA repair synthesis (Kataoka and Sekiguchi, 1982).

IV. ISSUES IN DNA REPAIR: SURVIVAL, MUTAGENESIS, AND CELLULAR RESPONSES

In surveying the repair mechanisms for oxidative DNA damage (Figure 6), we encounter some key issues that arise with respect to repair of all types of DNA damage. A lesion formed in DNA can have any of several possible fates. The damage may be removed, as described above for many examples. If repair is not initiated sufficiently soon after the damage occurs, or if the appropriate repair enzyme does not exist, the damage may be encountered by the DNA replication apparatus. This encounter may cause replication to halt, as in the case of TG or urea, or lead to the incorporation of an 'incorrect' nucleotide, as for 8-OH-Gua (see Section II.A). A further possible outcome is that the unrepaired damage triggers a cellular response, either through its ability to stall DNA synthesis, or via specific sensors that detect the lesion. Examples of these phenomena include respectively, triggering of the SOS response by high-level exposure to H_2O_2 (vanBogelen et al, 1987; Imlay and Linn, 1988), or the induction of the adaptive response to alkylating agents through interaction of Ada protein with O^6-methylguanine in DNA (Sancar and Sancar, 1988).

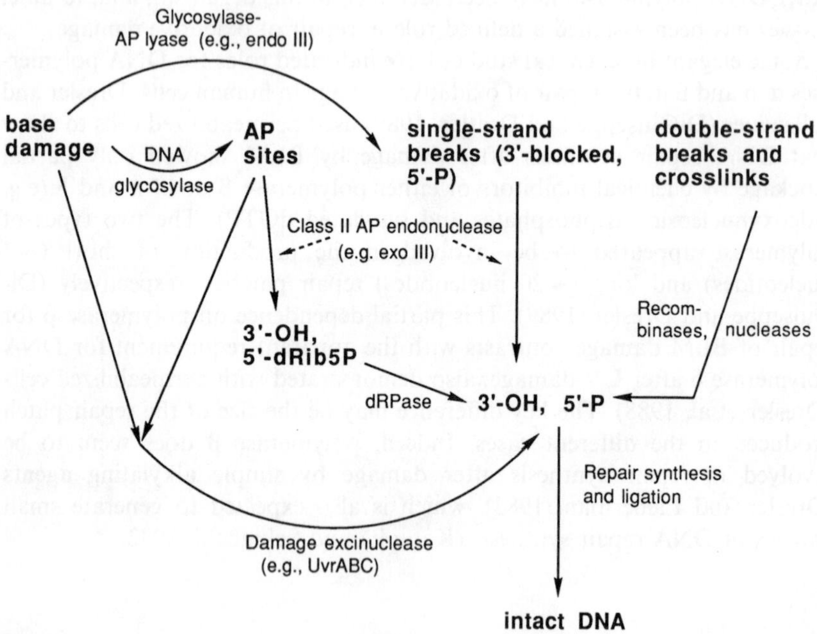

Figure 6. Pathways of repair for oxidative DNA damage. DNA lesions and repair intermediates are shown in bold type; enzymatic processes are shown in plain type. dRib5P, deoxyribose-5-phosphate; ddRib5P, (2,3-didehydro-2,3-dideoxy)ribose-5-phosphate. Endo III and exo III refer to endonuclease III and exonuclease III, respectively.

These various outcomes are not mutually exclusive (see Figure 7). A lesion may halt replication and trigger the SOS response, which leads in turn to the enhanced repair of the lesion. The activation of the SOS system might also induce DNA replication to proceed past the lesion, with the chance of inserting a nucleotide that is fixed as a mutation. The restoration of double-stranded DNA might then also allow repair, which is usually inhibited in single-stranded regions of DNA such as are found in the vicinity of a replication fork.

It is possible to generate other such scenarios, and it should be plain that the processing of any one DNA lesion in an actively growing cell is unpredictable. Moreover, the probability that a given lesion will be processed by one or another pathway also depends on the degree of damage to the cell: how many lesions, how many stalled replication forks, what other effects the damaging agent has (e.g. disruption of energy production to diminish dNTP supplies), etc. The establishment of individual repair mechanisms is certainly important, but such knowledge is only partial until the competing pathways and the possible regulation of the repair activities are also known.

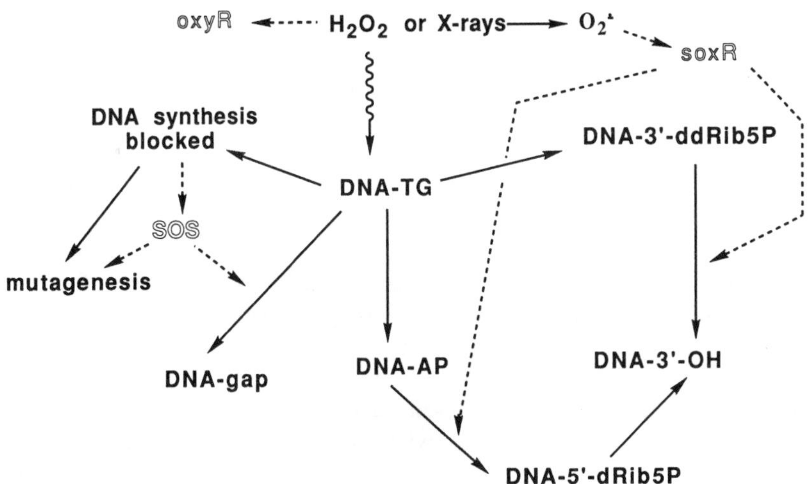

Figure 7. Competing fates for a single lesion (illustrated for thymine glycol, TG). Physical processes and structures are shown in bold type, inducible global regulators in outlined type. The solid arrows represent processes which are known to be direct (see text), the dotted arrows those that are less well understood. The abbreviations are as used in Figure 6.

The above considerations merit special attention with respect to the toxicity of oxygen radical agents. Oxidative damage to DNA produces a complex array of lesions that require repair, and this complexity is paralleled by the growing number of enzymatic repair pathways now defined. The promiscuous reactivity of radicals such as $^{\cdot}OH$ also causes damage to many other macromolecules. For example, oxidative attack on membranes compromises this permeability barrier and collapses membrane potential and concentration gradients, thus disrupting energy production and some transport mechanisms (Farr et al, 1988).

The cellular energy balance is also upset by the redox-cycling agents, which generate superoxide by diverting electrons from NADPH (Kappus and Sies, 1981). These complex effects can be dealt with by equally complex responses on the part of the cell. In *E. coli*, redox-cycling agents induce the synthesis of ~40 proteins in addition to the ~40 switched on by H_2O_2 (Greenberg and Demple, 1989). The way in which most of these inducible proteins are regulated is unknown. However, the redox-inducible proteins include nine that are controlled as a group, whose structural genes constitute the *soxR* regulon (Greenberg et al, 1990; Tsaneva and Weiss, 1990). The *soxR* regulon proteins include an energy-producing component, glucose-6-phosphate dehydrogenase, and a DNA repair component, endonuclease IV. These and the other *soxR* components (e.g. Mn-SOD) operate together to alleviate the multiple effects of this kind of oxidative stress. Another coregulated group of eight proteins responds to H_2O_2 and is controlled by the *oxyR* gene (Christman et al, 1985; Greenberg and Demple, 1989; see also Chapter 6).

Additional oxidative DNA lesions probably await discovery. Not even for all the known lesions has the biological relevance been determined. The means to do this is somewhat circular, in that repair enzyme-deficient mutants often provide the best evidence for the biological effects of a lesion. When a new lesion is discovered or begins to receive new attention, as in the case of 8-OH-Gua, the process begins anew: repair activities are identified and characterized, structural genes are cloned and mutants generated, enzyme-deficient strains are analysed for mutagenesis and killing by certain agents, and the data are correlated with the specific defect in repair. This has been a productive paradigm, but, as stressed above, such work lays out only one pathway in a network of competing reactions.

ACKNOWLEDGEMENTS

Work in the authors' laboratory was supported by grants to Bruce Demple from the National Cancer Institute (CA37831), the National Institute of Environmental Health Sciences (ES03926), the National Institute of General Medical Sciences (GM40000), and the Camille and Henry Dreyfus Foundation. Bruce Demple was a Dreyfus Foundation Teacher-Scholar.

NOTE ADDED IN PROOF

Recent results from several laboratories together indicate the fundamental importance of the repair of 8-hydroxyguanine. Grollman's group now reports that translesion synthesis past 8-hydroxyguanine in a gapped duplex vector yielded G to T transversions as the main detectable mutations both in vitro (Shibutani et al, 1991) and in vivo (Moriya et al, 1991). Moreover, the 8-hydroxyguanine nuclease identified in Nishimura's group is evidently identical with the FAPy glycosylase (Tchou et al, 1991). Finally, the *mutM* gene, cloned and sequenced by J.H. Miller's group, is identical with the *fpg* gene encoding FAPy glycosylase (J.H. Miller, personal communication). The *mutM* mutants have a strongly elevated spontaneous rate of G to T transversions (Cabrera et al, 1988), which suggests that 8-hydroxyguanine is formed at a significant level by endogenous processes and that the Fpg/MutM protein is the prime enzyme for its repair.

Cabrera M, Nghiem Y & Miller JH (1988) *J. Bacteriol* **170:** 5405–5407.
Moriya M, Ou C, Bodepudi V et al (1991) *Mutat. Res.* (in press).
Takeshita M, Shibutani S, Grollman LL et al (1991) *Nature* **349:** 431–434.
Tchou J, Kasai H, Shibutani S et al (1991) *Proc. Natl. Acad. Sci. USA* (in press).

REFERENCES

Alani E, Padmore R & Kleckner N (1990) *Cell* **61:** 419–436.
Al-Sheikhly M & von Sonntag C (1983) *Z. Naturforsch. B: Anorgan. Chem., Org. Chemie* **38B:** 1622–1629.
Ananthaswamy HM & Eisenstark A (1977) *J. Bacteriol.* **130:** 187–191.
Armel PR & Wallace SS (1984) *J. Bacteriol.* **160:** 895–902.
Au KG, Clark S, Miller JH & Modrich P (1989) *Proc. Natl. Acad. Sci USA* **86:** 8877–8881.
Bacchetti S & Benne R (1975) *Biochem. Biophys. Acta* **390:** 285–297.
Bailly V & Verly WG (1987) *Biochem. J.* **242:** 565–572.
Bailly V, Verly WG, O'Connor, TR & Laval J (1989). *Biochem. J.* **262:** 581–589.
Basu AK, Loechler EL, Leadon, SA & Essigmann JM (1989) *Proc. Natl. Acad. Sci. USA* **86:** 7677–7681.
Beer M, Stern S, Carmalt D & Mohlhenrich KH (1966) *Biochemistry* **5:** 2283–2288.
Berger M & Cadet J (1983) *Chem. Lett.* 435–438.
Bernelot-Moens C & Demple B (1989) *Nucleic Acids Res.* **17:** 587–600.
Boiteux S & Huisman O (1989) *Mol. Gen. Genet.* **215:** 300–305.
Boiteux S, O'Connor TR & Laval J (1987) *EMBO J.* **6:** 3177–3183.
Boiteux S, O'Connor TR, Lederer F, Gouyette A & Laval J (1990) *J. Biol. Chem.* **265:** 3916–3922.

Breimer LH (1984) *Nucleic Acids Res.* **12:** 6359–6367.
Breimer LH & Lindahl T (1984) *J. Biol. Chem.* **259:** 5543–5548.
Breimer LH & Lindahl T (1985) *Biochemistry* **24:** 4018–4022.
Budd ME, Wittrup KD, Bailey JE & Campbell JL (1989) *Mol. Cell. Biol.* **9:** 365–376.
Cadet J & Teoule R (1973) *Carbohydr. Res.* **29:** 345–361.
Cadet J, Berger M, Decarroz C et al (1986) *Biochimie* **68:** 813–834.
Carlson J & Carpenter VS (1980) *J. Bacteriol.* **142:** 319–321.
Cesar R & Verly WG (1983) *Eur. J. Biochem.* **129:** 509–517.
Chan E & Weiss B (1987) *Proc. Natl. Acad. Sci. USA* **84:** 3189–3193.
Chetsanga CJ & Frenette GP (1983) *Carcinogenesis* **4:** 997–1000.
Chetsanga CJ & Grigorian C (1985) *Proc. Natl. Acad. Sci. USA* **82:** 633–637.
Chetsanga CJ & Lindahl T (1979) *Nucleic Acids Res.* **6:** 3673–3684.
Chin D-H. Kappen LS & Goldberg IH (1987) *Proc. Natl. Acad. Sci. USA* **84:** 7070–7074.
Christman MF, Morgan RB, Jacobson FS & Ames BN (1985) *Cell* **41:** 753–762.
Clark JM & Beardsley GP (1986) *Nucleic Acids Res.* **14:** 737–749.
Cleland WW (1990) *FASEB J.* **4:** 2899–2905.
Cunningham RP & Weiss B (1985) *Proc. Natl. Acad. Sci. USA* **82:** 474–478.
Cunningham RP, Saporito S, Spitzer SG & Weiss B (1986) *J. Bacteriol.* **168:** 1120–1127.
Cunningham RP, Asahara H, Bank JF et al (1989) *Biochemistry* **28:** 4450–4455.
Dehazya P & Sirover MA (1986) *Cancer Res.* **46:** 3756–3761.
Demple B & Halbrook JH (1983) *Nature (Lond.)* **304:** 466–468.
Demple B & Linn S (1980) *Nature (Lond.)* **287:** 203–208.
Demple B & Linn S (1982a) *Nucleic Acids Res.* **10:** 3781–3789.
Demple B & Linn S (1982b) *J. Biol. Chem.* **227:** 2848–2855.
Demple B, Halbrook JH & Linn S (1983) *J. Bacteriol.* **153:** 1079–1082.
Demple B, Johnson AW & Fung D (1986) *Proc. Natl. Acad. Sci. USA* **83:** 7731–7735.
Demple B, Greenberg JT, Johnson A & Levin JD (1988). In Friedberg EC & Hanawalt PC (eds) *Mechanisms and Consequences of DNA Damage Processing.* UCLA Symposium on Molecular and Cellular Biology, New Series, Vol. 83, pp 151–155. New York: Alan R. Liss.
DiGuiseppe JA & Dresler SL (1989) *Biochemistry* **28:** 9515–9520.
Dirksen M-L, Blakely WF, Holwitt E & Dizdaroglu M (1988) *Int. J. Radiat. Biol.* **54:** 195–204.
Dizdaroglu M (1984) *J. Chromatogr.* **295:** 103–121.
Dizdaroglu M (1985) *Biochemistry* **24:** 4476–4480.
Dizdaroglu M (1986) *Biochem. J.* **238:** 247–254.
Dizdaroglu M & Gajewski E (1990) *Methods Enzymol.* **186:** 530–544.
Dizdaroglu M, Schulte-Frohlinde D & von Sonntag C (1977) *Int. J. Radiat. Biol.* **32:** 481–483.
Dizdaroglu M, Dirksen M-L, Jiang H & Robbins JH. (1987) *Biochem. J.* **241:** 929–932.
Doetsch PW, Helland DE & Haseltine WA (1986) *Biochemistry* **25:** 2212–2220.
Dresler SL & Lieberman MW (1983) *J. Biol. Chem.* **258:** 9990–9994.
Dresler SL, Gowans BJ, Robinson-Hill RM & Hunting DJ (1988) *Biochemistry* **27:** 6379–6383.
Farr SB, D'Ari R & Touati D (1986) *Proc. Natl. Acad. Sci. USA* **83:** 8268–8272.
Farr SB, Touati D & Kogoma T (1988) *J. Bacteriol.* **170:** 1837–1842.

Floyd RA, West MS, Eneff KL, Hogsett WE & Tingey DT (1988) *Arch. Biochem. Biophys.* **262:** 266–272.
Foster PL (1990) *J. Bacteriol.* **172:** 4719–4720.
Franklin W & Lindahl T (1988) *EMBO J.* **7:** 3617–3622.
Freemont PS, Friedman JM, Beese LS, Sanderson MR & Steitz TA (1988) *Proc. Natl. Acad. Sci. USA* **85:** 8924–8928.
Frenkel K, Goldstein MS & Teebor GW (1981) *Biochemistry* **20:** 7566–7571.
Friedberg EC (1988) *Microbiol. Rev.* **52:** 70–102.
Gajewski E, Fuciarelli AF & Dizdaroglu M (1988) *Int. J. Radiat. Biol.* **54:** 445–459.
Gates FT & Linn S (1977a) *J. Biol. Chem.* **252:** 2802–2807.
Gates FT & Linn S (1977b) *J. Biol. Chem.* **252:** 1647–1653.
Giloni L, Takeshita M, Johnson F, Iden C & Grollman AP (1981) *J. Biol. Chem.* **256:** 8606–8615.
Goffin C & Verly WG (1983) *FEBS Lett.* **161:** 140–144.
Gossard F & Verly WG (1978) *Eur. J. Biochem.* **82:** 321–332.
Gossett J, Lee K, Cunningham RP & Doetsch PW (1988) *Biochemistry* **27:** 2629–2634.
Greenberg JT & Demple B (1989) *J. Bacteriol.* **171:** 3933–3939.
Greenberg JT, Monach P, Chou JH, Josephy PD & Demple B (1990) *Proc. Natl. Acad. Sci. USA* **87:** 6181–6185.
Grossman L, Riazuddin S, Haseltine W & Lindan C (1978) *Cold Spring Harbor Symp. Quant. Biol.* **43:** 947–955.
Hagenesee ME & Moses RE (1986) *J. Bacteriol.* **186:** 1059–1065.
Hagenesee ME & Moses RE (1989) *J. Bacteriol.* **171:** 991–995.
Hagenesee ME & Moses RE (1990) *Biochim. Biophys. Acta* **1048:** 19–23.
Haines JA, Reese CB & Lord Todd (1962) *J. Chem. Soc.* 5281–5288.
Hariharan P & Cerutti P (1977) *Biochemistry* **16:** 2791–2795.
Hayes RC & LeClerc JE (1986) *Nucleic Acids Res.* **14:** 1045–1061.
Hayes RC, Petrullo LA, Huang H, Wallace SS & LeClerc JE (1988) *J. Mol. Biol.* **201:** 239–246.
Haynes RH & Kunz B (1981) In Strathern JN, Jones EW & Broach JR (eds) *The Yeast Saccharomyces: Life Cycle and Inheritance*, pp 371–414. Cold Spring Harbor, NY: Cold Spring Harbor Laboratory.
Henner WD, Rodriguez LO, Hecht SM & Haseltine WA (1983a) *J. Biol. Chem.* **258:** 711–713.
Henner WD, Grunberg SM & Haseltine WA (1983b) *J. Biol. Chem.* **258:** 15198–15205.
Henner WD, Kiker NP, Jorgensen TJ & Munck J (1987) *Nucleic Acids Res.* **15:** 5529–5544.
Howard-Flanders P & Boyce RP (1966) *Radiat. Res. Suppl.* **6:** 156–184.
Howard-Flanders P & Theriot L (1966) *Genetics* **53:** 1137–1150.
Hutchinson F (1985) *Prog. Nucleic Acid Res. Mol. Biol.* **32:** 115–154.
Ide H, Kow YM & Wallace SS (1985) *Nucleic Acids Res.* **13:** 8035–8052.
Igo-Kemenes T, Hörz W & Zachau HG (1982) *Annu. Rev. Biochem.* **51:** 89–122.
Imlay JA & Linn S (1988) *Science* **240:** 1302–1309.
Johnson AW & Demple B (1988a) *J. Biol. Chem.* **263:** 18009–18016.
Johnson AW & Demple B (1988b) *J. Biol. Chem.* **263:** 18017–18022.
Kallen RG & Marmur J (1962) *J. Mol. Biol.* **5:** 248–250.
Kane CM & Linn S (1981) *J. Biol. Chem.* **256:** 3405–3414.

Kapp DS & Smith KC (1970) *J. Bacteriol.* **103:** 49–54.
Kappen LS & Goldberg IH (1983) *Biochemistry* **22:** 4872–4878.
Kappen LS & Goldberg IH (1989) *Biochemistry* **28:** 1027–1032.
Kappus H & Sies H (1981) *Experientia* **37:** 1233–1241.
Karam LR, Simic MG & Dizdaroglu M (1986) *Int. J. Radiat. Biol.* **49:** 67–75.
Karran P. & Lindahl T (1978) *J. Biol. Chem.* **253:** 5877–5879.
Karran P. & Lindahl T (1980) *Biochemistry* **19:** 6005–6011.
Kasai H, Crain PF, Kuchino Y, Nishimura S, Ootsuyama A & Tanooka H (1986) *Carcinogenesis* **7:** 1849–1851.
Kataoka H & Sekiguchi M (1982) *J. Biochem.* **92:** 971–975.
Katcher HL & Wallace SS (1983) *Biochemistry* **22:** 4071–4081.
Keck K (1968) *Z. Naturforsch. B. Anorg. Chem. Org. Chem. Biochem. Biophys. Biol.* **23:** 1034–1043.
Kelley MR, Venugopal S, Harless J & Deutsch WA (1989) *Mol. Cell. Biol.* **9:** 965–973.
Kim J & Linn S (1988) *Nucleic Acids Res.* **16:** 1135–1141.
Kim J & Linn S (1989) *J. Biol. Chem.* **264:** 2739–2745.
Kow YW & Wallace SS (1985) *Proc. Natl. Acad. Sci. USA* **82:** 8354–8358.
Kow YW & Wallace SS (1987) *Biochemistry* **26:** 8200–8206.
Kow YW, Wallace SS & van Houten B (1990) *Mutat. Res.* **235:** 147–156.
Krasin F & Hutchinson F (1977) *J. Mol. Biol.* **116:** 81–98.
Kuchino Y, Mori F, Kasai H et al (1987) *Nature (Lond.)* **327:** 77–79.
Kühnlein U (1985) *J. Biol. Chem.* **260:** 14918–14924.
Kunkel TA, Schaaper RM & Loeb LA (1983) *Biochemistry* **22:** 2378–2384.
Lenz J, Okenquist SA, LoSardo JE, Hamilton KK & Doetsch PW (1990) *Proc. Natl. Acad. Sci. USA* **87:** 3396–3400.
Levin JD, Johnson AW & Demple B (1988) *J. Biol. Chem.* **263:** 8066–8071.
Lin J-J & Sancar S (1989) *Biochemistry* **28:** 7979–7984.
Lindahl T (1974) *Proc. Natl. Acad. Sci. USA* **71:** 3649–3654.
Lindahl T & Andersson A (1972) *Biochemistry* **11:** 3618–3623.
Linsley WS, Penhoet EE & Linn S (1977) *J. Biol. Chem.* **252:** 1235–1242.
Ljungquist S (1977) *J. Biol. Chem.* **252:** 2808–2814.
Ljungquist S & Lindahl T (1977) *Nucleic Acids Res.* **4:** 2871–2879.
Ljungquist S, Lindahl T & Howard-Flanders P (1976) *J. Bacteriol.* **126:** 646–653.
Loeb LA & Preston B (1986) *Annu. Rev. Genet.* **20:** 201–230.
Manoharan M, Mazunder A, Ranson SC, Gerlt JA & Bolton PH (1988) *J. Am. Chem. Soc.* **110:** 2690–2691.
Margolis SA, Coxon B, Gajewski E & Dizdaroglu M (1988) *Biochemistry* **27:** 6353–6359.
Mee LK & Adelstein SJ (1981) *Proc. Natl. Acad. Sci. USA* **78:** 2194–2198.
Milcarek C & Weiss B (1972) *J. Mol. Biol.* **68:** 303–318.
Minsky BD & Braun A (1977) *Radiat. Res.* **71:** 505–515.
Myrnes B, Guddal P-H & Krokan H (1982) *Nucleic Acids Res.* **10:** 3693–3701.
Nes I (1980a) *Eur. J. Biochem.* **112:** 161–168.
Nes I (1980b) *Nucleic Acids Res.* **8:** 1575–1589.
Nes IF (1981) *FEBS Lett.* **133:** 217–220.
Niwa O & Moses RE (1981) *Biochemistry* **20:** 238–244.
O'Connor TR, Boiteux S & Laval J (1988) *Nucleic Acids Res.* **16:** 5879–5894.
Oeda K, Shimizu K & Sekiguchi M (1978) *J. Biochem.* **84:** 1165–1169.

Oleinick NL, Chiu S, Friedman LR, Xue L & Ramakrishnan (N (1986) In Simic MG, Grossman L & Upton AC (eds) *Mechanisms of DNA Damage and Repair*, pp 181–192. New York: Plenum Press.
Patterson MC & Setlow RB (1972) *Proc. Natl. Acad. Sci. USA* **69:** 2927–2931.
Popoff SC, Spira AI, Johnson AW & Demple B (1990) *Proc. Natl. Acad. Sci. USA* **87:** 4193–4197.
Povirk LF & Goldberg IH (1985) *Proc. Natl. Acad. Sci. USA* **82:** 3182–3186.
Povirk LF & Houlgrave CW (1988) *Biochemistry* **27:** 3850–3857.
Povirk LF, Houlgrave CW & Han Y-H (1988) *J. Biol. Chem.* **263:** 19263–19266.
Rabow L, Stubbe J, Kozarich JW & Gerlt JA (1986) *J. Am. Chem. Soc.* **108:** 7130–7131.
Radman M (1976) *J. Biol. Chem.* **251:** 1438–1445.
Raleigh JA, Kremers W & Whitehouse R (1976) *Radiat. Res.* **65:** 414–422.
Raleigh JA, Fuciarelli AF & Kulatunga CR (1987) In Cerutti PA, Nygaard OF & Simic MG (eds) *Anticarcinogenesis and Radiation Protection*, pp 33–39. New York: Plenum Press.
Ramotar DS, Popoff SC & Demple B (1991) *Mol. Microbiol.* **5:** 149–155.
Resnick MA (1987) In Moens P (ed.) *Meiosis*, pp 157–212. New York: Academic Press.
Resnick MA & Martin P. (1976) *Mol. Gen. Genet.* **143:** 119–129.
Rhaese H-J & Freese E (1968) *Biochim. Biophys. Acta* **155:** 476–490.
Richardson CC & Kornberg A (1964) *J. Biol. Chem.* **239:** 242–250.
Richardson CC, Lehman IR & Kornberg A (1964) *J. Biol. Chem.* **239:** 251–258.
Rogers S & Weiss B (1980) *Methods Enzymol.* **65:** 201–211.
Rouet P & Essigmann JM (1985) *Cancer Res.* **45:** 6113–6118.
Sagher D & Strauss B (1983) *Biochemistry* **22:** 4518–4526.
Sagripanti J-L & Kraemer KH (1989) *J. Biol. Chem.* **264:** 1729–1734.
Saito I, Kawabata H, Fujiwara T, Sugiyama H & Matsuura T (1989) *J. Am. Chem. Soc.* **111:** 8302–8303.
Sambrook J, Fritsch EF & Maniatis T (1989) *Molecular Cloning—A Laboratory Manual*, 2nd edn. Cold Spring Harbor Laboratory. New York: Cold Spring Harbor.
Sancar A & Sancar G (1988) *Annu. Rev. Biochem.* **57:** 29–67.
Sanderson BJS, Chang C-N, Grollman AP & Henner WD (1989) *Biochemistry* **28:** 3892–3901.
Saporito SM, Gedenk M & Cunningham RP (1989) *J. Bacteriol.* **171:** 2542–2546.
Schaaper RM, Kunkel TA & Loeb LA (1983) *Proc. Natl. Acad. Sci. USA* **80:** 487–491.
Schneider JE, Price S, Maidt L, Gutteridge JMC & Floyd RA (1990) *Nucleic Acids Res.* **18:** 631–635.
Shaper NL, Grafstrom RH & Grossman L (1982) *J. Biol. Chem.* **257:** 13455–13458.
Shirname-More L, Rossman TG, Troll W, Teebor GW & Frenkel K (1987) *Mutat Res.* **178:** 177–186.
Spiering AL & Deutsch WA (1986) *J. Biol. Chem.* **261:** 3222–3228.
Strniste GF & Wallace SS (1975) *Proc. Natl. Acad. Sci. USA* **72:** 1997–2001.
Takeshita M, Chang CN, Johnson F, Will S & Grollman AP (1987) *J. Biol. Chem.* **262:** 10171–10179.
Teebor GW, Frenkel K & Goldstein MS (1981) *Proc. Natl. Acad. Sci. USA* **81:** 318–321.

Thacker J (1976) *Radiat. Res.* **68:** 371–380.
Thibodeau L & Verly WG (1980) *Eur. J. Biochem.* **107:** 555–563.
Town CD, Smith KC & Kaplan HS (1971) *Science* **172:** 851–854.
Tsaneva IR & Weiss B (1990) *J. Bacteriol.* **172:** 4197–4205.
van Bogelen R, Kelley PM & Neidhardt FC (1987) *J. Bacteriol.* **169:** 26–32.
von Sonntag C (1987) *The Chemical Basis of Radiation Biology*. London: Taylor & Francis.
Wallace SS (1988) *Environ. Mol. Mutagen.* **12:** 431–477.
Warner HR, Demple B, Deutsch WA, Kane CM & Linn S (1980) *Proc. Natl. Acad. Sci. USA* **77:** 4602–4606.
Washino K & Schnabel W (1982) *Makromol. Chem.* **183:** 697–709.
Weiss B (1976) *J. Biol. Chem.* **251:** 1896–1901.
Weiss B, Cunningham RP, Chan E & Tsaneva IR (1988) In Friedberg EC & Hanawalt PC (eds) *Mechanisms and Consequences of DNA Damage Processing*, pp 133–142. New York: Alan R. Liss.
Weiss RB & Duker NJ (1986) *Nucleic Acids Res.* **14:** 6621–6631.
Wiebauer K & Jiricny J (1989) *Nature (Lond.)* **339:** 234–236.
Wood ML, Dizdaroglu M, Gajewski E & Essigmann JM (1990) *Biochemistry* **29:** 7024–7032.
Zieg J, Maples VF & Kushner SR (1978) *J. Bacteriol.* **134:** 958–966.

6

The Bacterial Adaptation to Hydrogen Peroxide Stress

LOUIS A. TARTAGLIA, GISELA STORZ, SPENCER B. FARR and BRUCE N. AMES

Division of Biochemistry and Molecular Biology, University of California, Berkeley, CA 94720, USA

I. Introduction 155
II. Bacterial killing and damage by hydrogen peroxide 156
III. Constitutive defences against hydrogen peroxide 157
IV. Inducible defences in response to hydrogen peroxide challenge (the OxyR regulon) 159
V. OxyR-mediated defences against mutagenesis and membrane damage . 162
VI. Mechanism of OxyR regulation 163
VII. Concluding remarks 167

1. INTRODUCTION

The well-developed genetic and recombinant DNA techniques that are available in *Salmonella typhimurium* and *Escherichia coli* have made these organisms excellent model systems for the study of oxidative stress. Bacteria, like all organisms that take advantage of aerobic respiration, must be able to defend against the toxic oxidant species that are generated through the incomplete reduction of molecular oxygen. *E. coli* and *S. typhimurium* must also withstand some of the same exogenous oxidative stresses as do mammalian cells, such as the oxidants generated by stimulated macrophages and radiation. It therefore seems likely that much of what is learned in bacteria

Oxidative Stress: Oxidants and Antioxidants
ISBN 0-12-642762-3

about the deleterious effects of oxidative stresses and the cellular defence mechanisms that protect against them will serve as a guide for investigation in higher systems.

Macromolecular damage by exposure to the oxidant hydrogen peroxide has been widely documented in both bacterial and mammalian systems and it is thought that peroxide, acting either directly or indirectly, is responsible for much of the oxidative stress that occurs in living cells. Exposure to hydrogen peroxide causes chromosomal aberrations and sister chromatid exchanges (Oya et al, 1986; Larramendy et al, 1987), single-strand breaks (Hagensee and Moses, 1986), DNA–protein crosslinks (Lesko et al, 1982), thymine glycol adducts (Demple and Linn, 1982), and other types of specific DNA damage (Chance et al, 1979). Hydrogen peroxide has also been shown to be a mutagen in bacteria (Levin et al, 1982). In addition, membrane damage from hydrogen peroxide exposure has been documented in *E. coli* (Farr et al, 1988) and several proteins are known to be hydrogen peroxide sensitive (Levine et al, 1981; Brot and Weissbach, 1983). Although a complete understanding of the mechanisms of hydrogen peroxide-induced damage and the identity of the critical cellular targets has not yet been realized, a number of important insights in these areas have come from the study of bacteria.

One interesting finding that has come from the investigation of hydrogen peroxide stress in bacteria is that bacteria possess not only constitutive defences against hydrogen peroxide, but also a sophisticated network of inducible defences. Under normal aerobic growing conditions both *E. coli* and *S. typhimurium* express a number of activities whose primary purpose appears to be protection against hydrogen peroxide. However, upon sensing increased levels of hydrogen peroxide in their environment, these bacteria can induce the expression of key defence activities. In recent years it has become clear that mammalian cells also regulate their defences against hydrogen peroxide in order to cope with severe transient oxidative stresses (Niwa et al, 1990; Shibanuma et al, 1990; Keyse and Tyrell, 1989). The analysis of the bacterial response to oxidants has provided important insights into the molecular details of how a cell can sense an oxidative stress and then induce the appropriate response.

II. BACTERIAL KILLING AND DAMAGE BY HYDROGEN PEROXIDE

Exposure of *E. coli* to hydrogen peroxide leads to two kinetically distinct modes of cell killing (Imlay and Linn, 1986, 1987, 1988; Imlay et al, 1988).

Mode 1 killing is observed at low concentrations of hydrogen peroxide where the killing rate increases with hydrogen peroxide concentration to a maximum at about 1 mM, and then drops to a rate that is approximately half maximal at 10 mM. The cell killing that results from mode 1 concentrations of hydrogen peroxide requires active metabolism during exposure. Mode 2 killing requires a relatively high dosage of hydrogen peroxide (>20 mM), and does not require active metabolism. The site (or sites) of the lethal damage that occurs during mode 2 killing has not yet been identified.

The mechanism of mode 1 killing, however, has been studied extensively, since it occurs at more physiologically relevant doses of hydrogen peroxide. Mode 1 killing is a result of DNA damage as evidenced by the extreme sensitivity to mode 1 killing of a number of DNA repair mutant strains. Mutant strains that are defective in recombinational or excision DNA repair are especially sensitive to mode 1 killing, and both mutagenesis and the induction of the SOS DNA repair regulon (indicators of DNA damage) accompany challenges with mode 1 doses of hydrogen peroxide.

The dependence of mode 1 killing on active metabolism during hydrogen peroxide exposure suggests that available reducing equivalents are required to convert hydrogen peroxide to more toxic oxidants. Both in vivo and in vitro DNA damage assays have shown that the generation of these toxic oxidants also depends upon the availability of a free iron species. The identity of the direct DNA-damaging oxidants has not been established decisively; however, a number of lines of evidence suggest that the damaging oxidants are hydroxyl radicals produced by the Fenton reaction and ethanol-resistant ferryl radicals (a hydroxyl radical complexed to Fe^{3+}) that may be generated as intermediates during the Fenton reaction.

III. CONSTITUTIVE DEFENCES AGAINST HYDROGEN PEROXIDE

Normally growing bacteria express low levels of a number of defences that help to protect them against hydrogen peroxide-mediated damage (see Table 1). One class of defence activities acts to detoxify peroxides. Most notable among these activities are catalase and an alkyl hydroperoxide reductase which directly detoxify hydrogen peroxide and organic hydroperoxides. The catalase and peroxidase activities of *S. typhimurium* and *E. coli* and the genes that encode them will be discussed in greater detail below. The second class of hydrogen peroxide defence activities is composed of repair functions. Several enzymatic activities that serve to repair DNA damage caused by exposure to low doses of hydrogen peroxide have been identified in *E. coli*. Exonuclease III, the major apurinic/apyrimidinic endonuclease (AP endo-

Table 1. *E. coli* defences against hydrogen peroxide stress.

Activity		Gene	Reference
Catalase	HPI	*katG*[a]	Loewen et al (1985)
	HPII	*katE*	
Alkyl hydroperoxide reductase	C22,F52	*ahpC,*[a] *ahpF*[a]	Jacobson et al (1989); Storz et al (1989)
Glutathione reductase		*gorA*[a]	Christman et al (1985)
Apurinic/apyrimidinic endonuclease	Exonuclease III	*xthA*	Demple et al (1983)
Endonuclease IV		*nfo*	Cunningham and Weiss (1985)
Endonuclease III		*nth*	Weiss and Cunningham (1985)
DNA polymerase		*polA*	Ananthaswamy and Eisenstark (1977); Demple et al (1983)
Excision nuclease		*uvrA*	Goerlich et al (1989)
		uvrB	
Exonuclease		*recB*	Imlay and Linn (1987)
		recC	
OxyR regulator		*oxyR*[a]	Christman et al (1985)

[a] Regulated by OxyR.

nuclease) activity found in *E. coli*, has been shown to remove replication blocks from the 3′-terminal of oxidized DNA in vitro and is encoded by *xthA* (Demple et al, 1983, 1986). A second AP endonuclease capable of removing 3′ replication blocks is endonuclease IV, which is encoded by *nfo* (Cunningham and Weiss, 1985; Demple et al, 1986). Other DNA repair activities found to be important for a defence against oxidative stress are: endonuclease III encoded by *nth*, DNA polymerase I encoded by *polA*, the excision nuclease activity encoded by *uvrAB*, and the exonuclease activities associated with *recBC* (Weiss and Cunningham, 1985; Ananthaswamy and Eisenstark, 1977; Demple et al, 1983; Imlay and Linn, 1987; Goerlich et al, 1989).

IV. INDUCIBLE DEFENCES IN RESPONSE TO HYDROGEN PEROXIDE CHALLENGE (THE OxyR REGULON)

Since most life is continuously exposed to oxidative stress, it is not surprising that cells constitutively express some basal level of oxidative defence activities. However, organisms are occasionally exposed to unusually high levels of reactive oxygen species. In mammalian tissues this can be a result of shock, trauma, bacterial infection, or induction of tumour necrosis factor (Cross et al, 1987; Wong et al, 1989). In the case of enteric bacteria, sudden exposure to high levels of oxidative stress may be a result of the sudden switch to an aerobic environment when voided from their host, contact with macrophages, or contact with environmental oxidants or radiation. It would therefore seem reasonable that most cells would be equipped with gene regulatory mechanisms that could sense a high oxidative stress potential and consequently induce higher levels of defence activities. Inducible defences in response to oxidative stress have indeed been observed in a number of both prokaryotic and eukaryotic systems. The best documented of these is the adaptation to hydrogen peroxide exposure in *S. typhimurium* and *E. coli* (Demple and Halbrook, 1983; Christman et al, 1985).

When *E. coli* and *S. typhimurium* cells are treated with low doses (60 μM) of hydrogen peroxide, they become resistant to subsequent lethal doses (10 mM) of hydrogen peroxide and also a variety of other oxidants (Demple and Halbrook, 1983; Christman et al, 1985). This adaptive response requires the induction of protein synthesis, since cells treated with low doses of hydrogen peroxide in the presence of chloramphenicol, a protein synthesis inhibitor, are no longer resistant to the subsequent lethal challenges (Christman et al, 1985). Consistent with the requirement for protein synthesis, treatment with low doses of hydrogen peroxide induces the synthesis of approximately 30 proteins as seen on two-dimensional gels (Morgan et al, 1986; VanBogelen et al, 1987).

A subset of the hydrogen peroxide-inducible proteins overlaps with proteins induced by other stresses. One example is the overlapping protein induction between the adaptive response to hydrogen peroxide and the adaptive response to superoxide-generating agents. Although the induced protein synthesis after treatment with these agents overlaps significantly, the hydrogen peroxide and superoxide responses are controlled by different regulators (Greenberg et al, 1990; Tsaneva and Weiss, 1990). It is not known whether this overlapping protein induction is a result of genes being responsive to both regulatory pathways or, alternatively, interconversion between the two types of oxygen species within the cell. It has also been found that cells treated with low doses of hydrogen peroxide become crossresistant to heat shock (Christman et al, 1985). This cross-resistance might be explained by the observation that some of the proteins induced by hydrogen peroxide or superoxide radicals overlap with heat shock proteins (Morgan et al, 1986; VanBogelen et al, 1987; Walkup and Kogoma, 1989; Greenberg and Demple, 1989). The overlapping regulation of these different stresses is not well understood, but it is hoped that further characterization of the overlaps will lead to a better understanding of how cells respond to stress.

Time-course experiments showed that the 30 hydrogen peroxide-inducible proteins can be grouped into two temporal classes; 12 proteins are maximally induced within the first 10 min after treatment with hydrogen peroxide and the other 18 proteins are induced within the subsequent hour (Christman et al, 1985). The expression of nine of the early hydrogen peroxide-inducible proteins is regulated by the *oxyR* gene. *E. coli* and *S. typhimurium* strains carrying deletions of *oxyR* are hypersensitive to hydrogen peroxide and are unable to induce the expression of the nine proteins. Strains containing dominant missense *oxyR* mutant alleles (*oxyR1* in *S. typhimurium* and *oxyR2* in *E. coli*) have also been isolated. These strains are more resistant to hydrogen peroxide than wild-type strains and constitutively overexpress the nine proteins (Christman et al, 1985).

Several of the proteins whose expression is regulated by *oxyR* have been identified (Table 1). All of these activities can be implicated in a defence against oxidative stress. *E. coli* has two isozymes of catalase (Loewen et al, 1985), an enzyme that converts hydrogen peroxide to oxygen and water (Chance, 1947). The expression of one of the catalase activities, denoted HPI (position D69/D71 on two-dimensional gels) and encoded by *katG*, is regulated by *oxyR*. Strains carrying the *oxyR1* mutation have 50-fold higher levels of HPI catalase than the corresponding wild-type strains (Christman et al, 1985). Glutathione reductase (G35) encoded by *gorA* is also elevated six-fold in the *oxyR1* mutant strains (Christman et al, 1985). Glutathione reductase probably plays a role in protecting against oxidation by maintaining a pool of reduced glutathione which in turn can serve to maintain the reduced state of cellular proteins.

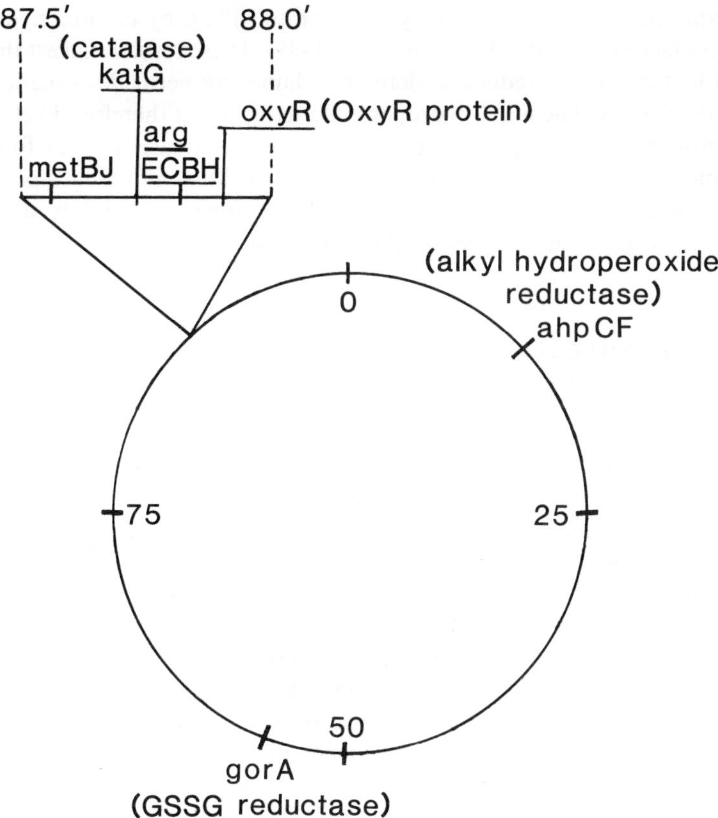

Figure 1. Map positions on the *E. coli* chromosome of the OxyR-regulated *ahpCF*, *katG*, *gorA* and *oxyR* genes.

oxyR1 mutant strains were also found to overproduce a novel alkyl hydroperoxide reductase activity (20-fold) (Christman et al, 1985). This activity was purified to homogeneity and found to be a flavoprotein composed of two identical 57-kDa subunits (F52) encoded by *ahpF* and a 22-kDa protein (C22) encoded by *ahpC* (Jacobson et al, 1989). Recent sequence analysis of the *ahpCF* operon revealed that the *ahpF*-encoded protein (F52) is highly homologous to the *E. coli* thioredoxin reductase protein, particularly in the active-site region, suggesting that the alkyl hydroperoxide reductase is a disulphide oxidoreductase (Tartaglia et al, 1990). The purified enzyme is capable of reducing physiological hydroperoxides such as thymine hydroperoxide and linoleic hydroperoxides to their corresponding alcohols, and mutant strains that lack the alkyl hydroperoxide reductase are extremely sensitive to killing by organic hydroperoxides (Jacobson et al, 1989; Storz et al, 1989). These results suggest that the alkyl hydroperoxide reductase detoxifies damaging peroxides generated by oxidative stress. The alkyl hydroperoxide reductase is therefore likely to be the prokaryotic analogue of the glutathione peroxidase activity found in mammalian cells. It is interesting to note that the known hydrogen peroxide-inducible genes are not cotranscribed within a single operon, but rather are dispersed throughout the bacterial chromosome (Figure 1).

V. OxyR-MEDIATED DEFENCES AGAINST MUTAGENESIS AND MEMBRANE DAMAGE

Both *E. coli* and *S. typhimurium* strains carrying deletions of *oxyR* have significantly increased frequencies of mutagenesis (Table 2; Storz et al, 1987; Greenberg and Demple, 1988). The increased mutation frequency is most pronounced under aerobic conditions. When the mutations generated in the $oxyR^+$ and the *oxyR* deletion strains were characterized, it was found that the type of mutation whose frequency was most elevated was T : A to A : T transversions, the type of mutation most frequently caused by oxidative mutagens. These mutations may in part be generated as a consequence of SOS induction, a response induced by several types of DNA damage (Goerlich et al, 1989). The high frequency of mutagenesis in *oxyR* deletion strains was suppressed by multicopy plasmids expressing high levels of catalase (*katG*), alkyl hydroperoxide reductase (*ahpCF*) or superoxide dismutase (*sodA*) activity (Storz et al, 1987) and by suppressor mutations that caused increased expression of *katG* and *ahpCF* genes (Greenberg and Demple, 1988). These observations provide evidence that the *oxyR* regulon plays an important role in protecting against oxidative DNA damage that is otherwise converted into mutations.

Studies on the uptake of labelled lactose, guanosine, uracil, or methylglucopyranoside by *E. coli* strains after treatment with hydrogen peroxide have also shown that oxidative membrane damage leads to altered membrane function (Farr et al, 1988). Cells treated with 5 mM hydrogen peroxide showed a rapid loss of both proton motive force-dependent (lactose, guanosine) and -independent (methylglucopyranoside) transport within 5 min after treatment with hydrogen peroxide. If the cells were first pretreated with lower doses (35 μM) of hydrogen peroxide, transport was recovered rapidly. This adaptation was not seen in strains with mutations in *oxyR* or *katG*, suggesting that increased expression of hydrogen peroxide-scavenging activities is required to protect cells from membrane damage by oxidative stress.

Table 2. *oxyR* deletion strains have increased frequencies of spontaneous mutagenesis.

Strain	Number of mutants/plate[a]
oxyR$^+$ (wild type)	6
oxyRΔ2 (*oxyR* deletion)	76
oxyR$^+$/pKM101[b]	57
oxyRΔ2/pKM101	3102
oxyRΔ2/pKM101/pACYC184 (vector)	946
oxyRΔ2/pKM101/pAQ5 (*oxyR*)	33
oxyRΔ2/pKM101/pAQ6 (*sodA*)	196
oxyRΔ2/pKM101/pAQ7 (*katG*)	47
oxyRΔ2/pKM101/pAQ8 (*ahp*)	31

[a] The frequency of mutagenesis in the *S. typhimurium oxyR* mutant strains was assayed by the reversion of His− auxotrophy to His+ prototrophy (taken from Storz et al, 1987).
[b] pKM101 encodes *mucA* and *mucB* (analogues of the *E. coli umuC* and *umuD* genes) that make strains more susceptible to mutagenesis by a number of mutagens.

VI. MECHANISM OF OxyR REGULATION

The mechanism by which *oxyR* senses oxidative stress and regulates the hydrogen peroxide-inducible proteins has been investigated at the molecular level. The *oxyR* gene was cloned on the basis of its ability to complement the hydrogen peroxide sensitivity of a strain carrying a deletion of the *oxyR* locus (Christman et al, 1989). Sequence comparisons showed that OxyR is homologous to an expanding family of bacterial regulators, including LysR (*E. coli* regulator of *lysA* involved in lysine biosynthesis), NodD (*Rhizobium* regulator of genes involved in the formation of nodules), IlvY (*E. coli* regulator of genes in the isoleucine–valine biosynthetic pathway), CysB (*E. coli* and *S. typhimurium* regulator of a cysteine regulon), MetR (*S. typhimurium* regulator of *metE* and *metH*) and AmpR (*Enterobacter cloacae* regulator of the *ampC* gene which confers resistance to cephalosporin) (Christman et al, 1989; Bölker and Kahmann, 1989; Tao et al, 1989; Warne et al, 1990). The evolutionary significance of the homology among these proteins is not clear but it may be noteworthy that the fixation of nitrogen in nodules is very sensitive to oxygen, and the amino acids, cysteine, methionine, and isoleucine are all especially susceptible to oxidation. All of the above regulators serve as activators of heterologous genes, and many, including OxyR, have also been found to regulate their own expression negatively.

Dot blots of total cellular RNA isolated from *oxyR*$^+$ 'wild-type' and *oxyR1* 'constitutive' mutant strains probed with the cloned *katG* gene showed that the level of the *katG* message is elevated 50-fold in the *oxyR1* mutant strain (Morgan et al, 1986). Primer extension studies showed the same to be true for the *ahpC* transcript and allowed identification of the start of the *oxyR*-regulated *katG* and *ahpC* transcripts (Tartaglia et al, 1989). Deletion analysis of sequences directly upstream of the *ahpC* promoter defined a region (−67 to −46 relative to the start of transcription) required for *oxyR* activation. A fragment carrying 141 bp of upstream *katG* sequence could confer *oxyR*-dependent hydrogen peroxide induction to a reporter *lacZ* gene. The regions found to be important for oxyR activation upstream of the *ahpC* and *katG* genes were found to be protected from DNase I digestion by purified OxyR protein. Both of these OxyR-binding sites were found to extend into the −35 sigma 70 consensus sequence, which also typically makes contacts with RNA polymerase. These results suggested that OxyR activates the expression of *katG* and *ahpC* by binding to promoter sequences upstream of the hydrogen peroxide-inducible genes and increasing the rate of transcription initiation via direct contacts with RNA polymerase.

In addition to the OxyR-binding sites upstream of the *ahpC* and *katG* genes, four other *oxy*R-binding sites have been identified. One of these is located upstream of the *oxyR* gene itself, allowing OxyR to both negatively regulate its own gene and positively regulate a transcript of unknown function that is transcribed opposite the *oxyR* gene (Christman et al, 1989). OxyR has also been shown to bind upstream of the phage Mu *mom* gene where it acts as a negative regulator (Bölker and Kahmann, 1989), upstream of the *E. coli gorA* gene where it serves to mediate the induction of glutathione reductase in response to peroxide (Storz et al, unpublished results), and upstream of an open reading frame of unknown function running opposite the *ahpC* gene (Tartaglia et al, unpublished results). Intriguingly, the sequences protected by OxyR show very little homology (Tartaglia et al, 1989; Tartaglia et al, unpublished results), which is atypical of prokaryotic promoters that are regulated by a common protein. The mechanism by which OxyR recognizes its dissimilar binding sites is currently being investigated and appears to involve the recognition of a highly degenerate consensus sequence.

The mechanism by which the cell senses hydrogen peroxide and activates

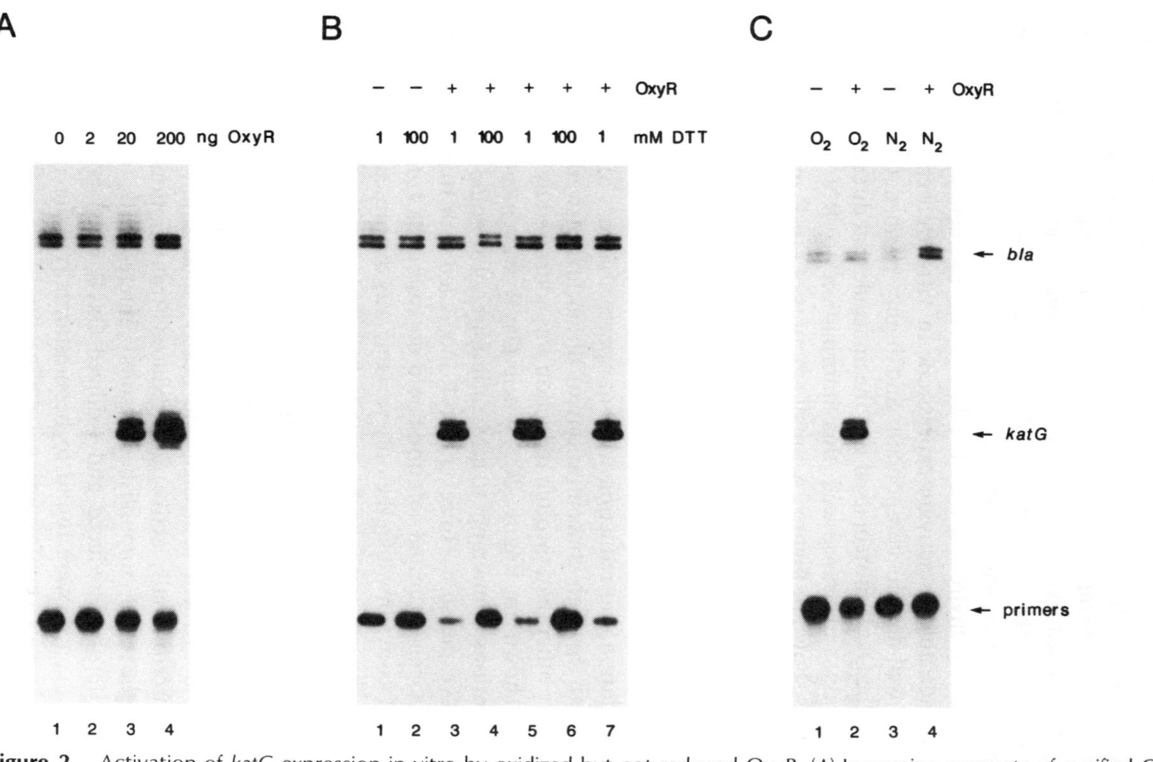

Figure 2. Activation of *katG* expression in vitro by oxidized but not reduced OxyR. (A) Increasing amounts of purified OxyR protein were added to RNA polymerase and a plasmid carrying the *katG* and *bla* genes and the resulting in vitro transcription products were examined by primer extension. (B) OxyR assayed for the ability to activate *katG* transcription in the presence of 1 or 100 mM dithiothreitol. The same sample of OxyR protein was taken through several cycles of oxidation and reduction. (C) OxyR assayed for the ability to activate *katG* transcription in the presence of air- or nitrogen-saturated buffers. Reprinted wth permission from *Science*.

transcription was illuminated by studies on the ability of OxyR-overproducing extracts and purified OxyR to activate the expression of the *oxyR*-regulated *ahpCF* and *katG* genes in vitro (Storz et al, 1990). Immunoblots and immunoprecipitation experiments clearly showed that neither the levels nor the synthesis of the OxyR protein increased after exposure to hydrogen peroxide. These observations suggested that the activation of existing levels of OxyR was responsible for the adaptive response to hydrogen peroxide. It was therefore surprising when in vitro transcription experiments revealed that OxyR-enriched extracts isolated from normally growing cells could strongly stimulate the expression of the *katG* and *ahp* genes. Subsequent experiments, however, showed that the extracts became active upon release from the reducing environment of the *E. coli* cell. Extracts prepared under argon were not able to activate expression. In vitro transcription experiments using well-defined components and highly purified OxyR protein showed that the OxyR protein itself was the molecule that directly sensed the oxidative stress and transduced this information to RNA polymerase (Figure 2). As shown in Figure 2A, small amounts of OxyR protein purified in air-saturated solutions were able to stimulate transcription by RNA polymerase strongly. However, after a short incubation in 100 mM dithiothreitol the OxyR protein was inactivated (Figure 2B). This inactivation was readily reversed, since removal of the dithiothreitol could restore activity to OxyR. The same sample of OxyR protein could be taken through several cycles of oxidation and reduction without appreciable loss of activity. The possibility that the high concentrations of dithiothreitol were inactivating OxyR in a way other than through reduction was addressed by the anaerobic experiments shown in Figure 2C. When OxyR was incubated in air-saturated solutions containing 10 mM dithiothreitol it could still strongly stimulate *katG* expression. However, OxyR incubated with 10 mM dithiothreitol in a nitrogen-saturated solution could not stimulate *katG* expression. These parallel assays in which the only variable was the presence or absence of dissolved oxygen strongly suggested that the activity of OxyR was dependent on its redox state.

Although the ability of OxyR to stimulate transcription is highly regulated, this is not the case with its sequence-specific DNA-binding activity. Experiments with several OxyR-binding sites have shown that OxyR is capable of binding them with high and nearly equal affinity in both the oxidized (transcriptionally active) and reduced (transcriptionally inactive) forms (Tartaglia et al, unpublished results). Interestingly, the DNase I protection patterns of reduced and oxidized OxyR when bound to their recognition sequences were found to be substantially different, suggesting that OxyR undergoes a conformational change during its transition from the reduced to the oxidized form (Figure 3; Storz et al, 1990). These findings indicate that OxyR senses oxidative stress directly by becoming oxidized, and that oxidation brings about a conformational change by which OxyR transduces the oxidative stress signal to RNA polymerase.

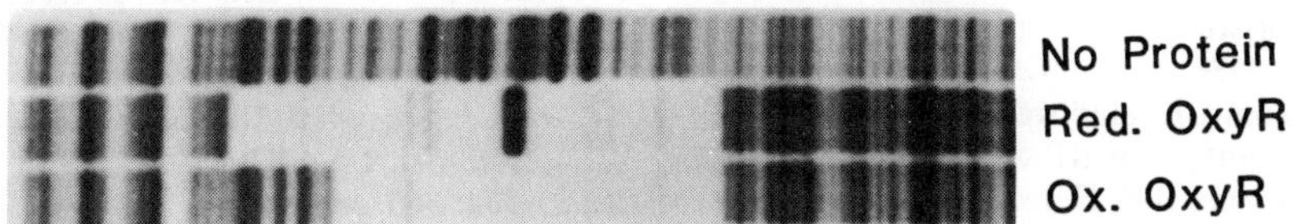

Figure 3. Binding of reduced and oxidized OxyR to the *oxyR* promoter. Samples of purified OxyR were incubated with an *oxyR* promoter fragment in the presence of 1 and 100 mM dithiothreitol and then digested with DNase I as described previously (Tartaglia et al, 1989; Storz et al, 1990).

VII. CONCLUDING REMARKS

E. coli and *S. typhimurium* possess both constitutive and inducible defences against the reactive oxygen species hydrogen peroxide. The inducible defences may be important in defending against transient oxidative stresses such as the oxidative burst of macrophages during a bacterial infection. In support of this hypothesis, *S. typhimurium* mutant strains that are defective in their ability to respond to hydrogen peroxide challenge have been shown to be less virulent in mice (Fields et al, 1986). The regulation of the hydrogen peroxide-inducible proteins is coordinated by the OxyR protein. The OxyR protein can sense increased concentrations of peroxide by being oxidized directly. It is likely that the activities of other regulators of gene expression are modulated by oxidation and reduction. Mammalian tissues must also cope with the oxidative burst of macrophages, and therefore mammalian cells may also contain redox-sensitive regulators which allow them to avoid lethal damage. It has recently been found that the activity of the Fos and Jun proteins (products of the protooncogenes *c-fos* and *c-jun*) is redox-regulated (Abate et al, 1990). As additional sensory and transduction pathways are characterized, it will be interesting to see if oxidation–reduction is a general mechanism of regulation.

REFERENCES

Abate C, Patel L, Rauscher III FJ & Curran T (1990) *Science* **249:** 1157–1161.
Ananthaswamy HN & Eisenstark A (1977) *J. Bacteriol.* **130:** 187–191.

Bölker M & Kahmann R (1989) *EMBO J* **8:** 2403–2410.
Brot N & Weissbach H (1983) *Arch. Biochem. Biophys.* **223:** 271–281.
Chance B (1947) *Acta Chem. Scand.* **1:** 236–267.
Chance B, Sies H & Boveris A (1979) *Physiol. Rev.* **59:** 572–605.
Christman MF, Morgan RW, Jacobson FS & Ames BN (1985) *Cell* **41:** 753–762.
Christman MF, Storz G & Ames BN (1989) *Proc. Natl. Acad. Sci. USA* **86:** 3484–3488.
Cross CE, Halliwell B, Borish ET et al (1987) *Ann. Intern. Med.* **107:** 526–545.
Cunningham RP & Weiss B (1985) *Proc. Natl. Acad. Sci. USA* **82:** 474–478.
Demple B & Halbrook J. (1983) *Nature (Lond.)* **304:** 466–468.
Demple B & Linn S (1982) *Nucleic Acids Res.* **10:** 3781–3789.
Demple B, Halbrook J & Linn S (1983) *J. Bacteriol.* **153:** 1079–1082.
Demple B, Johnson A & Fung D (1986) *Proc. Natl. Acad. Sci. USA* **83:** 7731–7735.
Farr SB, Touati D & Kogoma T (1988) *J. Bacteriol.* **170:** 1837–1842.
Fields PI, Swanson RV, Haidaris CG & Heffron F (1986) *Proc. Natl. Acad. Sci. USA* **83:** 5189–5193.
Goerlich O, Quillardet P & Hofnung M (1989) *J. Bacteriol.* **171:** 6141–6147.
Greenberg JT & Demple B (1988) *EMBO J.* **7:** 2611–2617.
Greenberg JT & Demple B (1989) *J. Bacteriol.* **171:** 3933–3939.
Greenberg JT, Monach P, Chou JH, Josephy PD & Demple B (1990). *Proc. Natl. Acad. Sci. USA* **87:** 6181–6185.
Hagensee ME & Moses RE (1986) *J. Bacteriol.* **168:** 1059–1065.
Imlay JA & Linn S (1986) *J. Bacteriol.* **166:** 519–527.
Imlay JA & Linn S (1987) *J. Bacteriol.* **169:** 2967–2976.
Imlay JA & Linn S (1988) *Science* **240:** 1302–1309.
Imlay JA, Chin SM & Linn S (1988) *Science* **240:** 640–642.
Jacobson FS, Morgan RW, Christman MF & Ames BN (1989) *J. Biol. Chem.* **264:** 1488–1496.
Keyse SM & Tyrell RM (1989) *Proc. Natl. Acad. Sci. USA* **86:** 99–103.
Larramendy M, Mello-Filho, AC, Leme Martins EA & Meneghini R (1987) *Mutat. Res.* **178:** 57–63.
Lesko SA, Drocourt J-L & Yang S-U (1982) *Biochemistry* **21:** 5010–5015.
Levin DE, Hollstein M, Christman MF, Schwiers EA & Ames BN (1982) *Proc. Natl. Acad. Sci. USA* **79:** 7445–7449.
Levine RL, Oliver CN, Fulks RM & Stadtman ER (1981) *Proc. Natl. Acad. Sci. USA* **78:** 2120–2124.
Loewen PC, Switala J & Triggs-Raine BL (1985) *Arch. Biochem. Biophys.* **243:** 144–149.
Morgan RW, Christman MF, Jacobson FS, Storz G & Ames BN (1986) *Proc. Natl. Acad. Sci. USA* **83:** 8059–8063.
Niwa Y, Ishimoto K & Kanoh T (1990) *Blood* **76:** 835–841.
Oya Y, Yamamoto K & Tonomura A (1986) *Mutat. Res.* **172:** 245–253.
Shibanuma M, Kuroki T & Nose K (1990) *Oncogene* **5:** 1025–1032.
Storz G, Christman MF, Sies H & Ames BN (1987) *Proc. Natl. Acad. Sci. USA* **84:** 8917–8921.
Storz G, Jacobson FS, Tartaglia LA, Morgan RW, Silveira LA & Ames BN (1989) *J. Bacteriol.* **171:** 2049–2055.
Storz G, Tartaglia LA & Ames BN (1990) *Science* **248:** 189–194.
Tao K, Makino K, Yonei S, Nakata A & Shinagawa H (1989) *Mol. Gen. Genet.* **218:** 371–376.

Tartaglia LA, Storz G & Ames BN (1989) *J. Mol. Biol.* **210:** 709–719.
Tartaglia LA, Storz G, Brodsky MH, Lai A & Ames BN (1990) *J. Biol. Chem.* **265:** 10535–10540.
Tsaneva IR & Weiss B (1990) *J. Bacteriol.* **172:** 4197–4205.
VanBogelen RA, Kelley PM & Neidhardt FC (1987) *J. Bacteriol.* **169:** 26–32.
Walkup LKB & Kogoma T (1989) *J. Bacteriol.* **171:** 1476–1484.
Warne SR, Varley JM, Boulnois GJ & Norton MG (1990) *J. Gen. Microbiol.* **136:** 455–462.
Weiss B & Cunningham RP (1985) *J. Bacteriol.* **162:** 607–610.
Wong GHW, Elwell JH, Oberley LW & Goeddel DV (1989) *Cell* **58:** 923–931.

7

Glutathione Transferases and Products of Reactive Oxygen

BRIAN KETTERER and BRIAN COLES
Cancer Research Campaign Molecular Toxicology Research Group, Department of Biochemistry, University College and Middlesex School of Medicine, Cleveland Street, London W1P 6DB, UK

I. Introduction 171
II. Glutathione transferase isoenzymes 174
III. GSTs and endogenous peroxides 181
IV. Human glutathione transferases 189
V. Conclusions 190

1. INTRODUCTION

Glutathione transferases (GSTs) (E.C. 2.5.1.18) catalyse the attack of glutathione (GSH) on a number of lipophilic compounds containing electrophilic carbon, oxygen and nitrogen (Ketterer et al, 1988). These electrophiles are frequently a consequence of either the oxidative metabolism of xenobiotics or oxygen-centred radical reactions. The common factor in the generation of electrophiles by these oxidation reactions is the introduction of the electronegative oxygen substituent and the subsequent generation of an adjacent electron-deficient (C, N or O) atom.

The major oxidative enzymes concerned are the mixed function oxygenases, notably cytochromes P450 (Guengerich, 1988), which are involved in the formation of a number of carbon- and nitrogen-centred electrophiles. For example, the common environmental pollutants benzo(a)pyrene (BP)

Oxidative Stress: Oxidants and Antioxidants
ISBN 0-12-642762-3

and aminobiphenyl (ABP) are metabolized to BP-7,8-diol-9,10-oxide (BPDE) and *N*-acetoxy aminobiphenyl respectively, which are carcinogenic electrophiles, while paracetamol, an analgesic drug, is converted to *N*-acetylbenzoquinone imine (NAPQ1), a cytotoxic electrophile. BPDE and NAPQ1 have been shown to be substrates for GSTs (see below) while *N*-acetoxy ABP is an example of an electrophile which, although known to give a GSH conjugate, has yet to be tested as a substrate for GSTs (Hinson and Kadlubar, 1988) (see Figure 1).

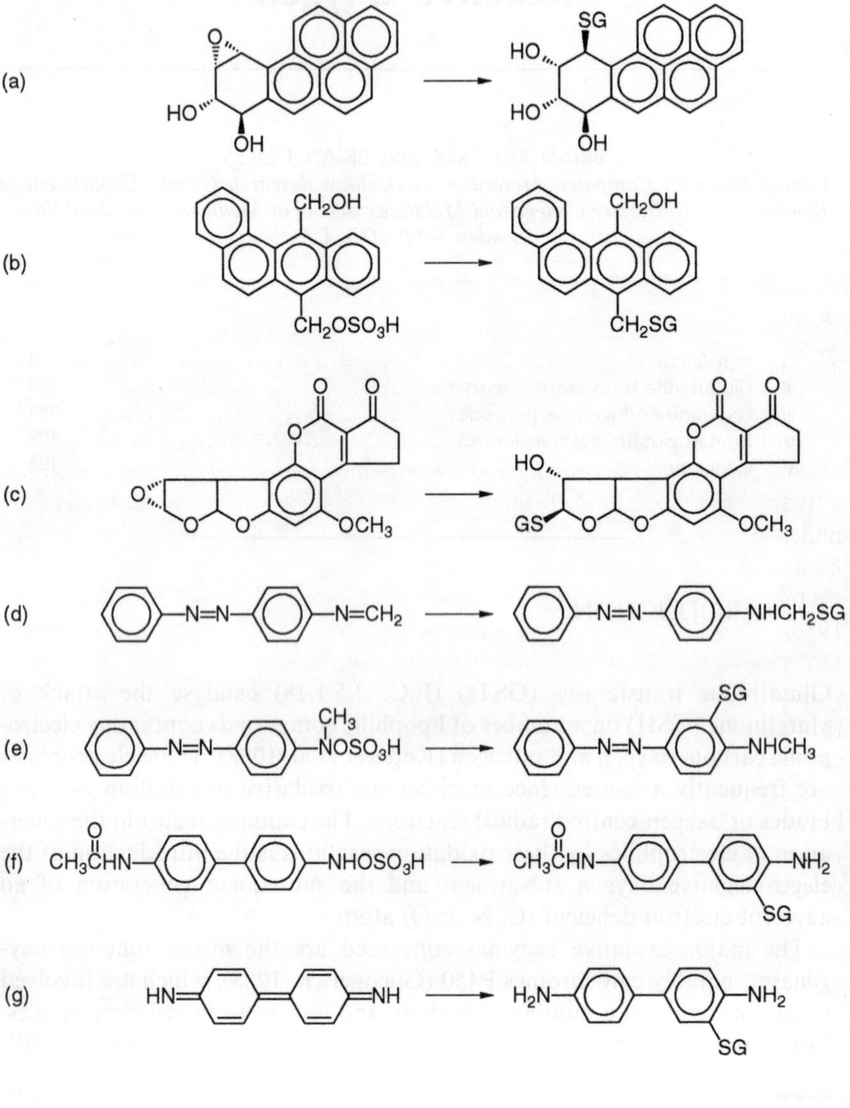

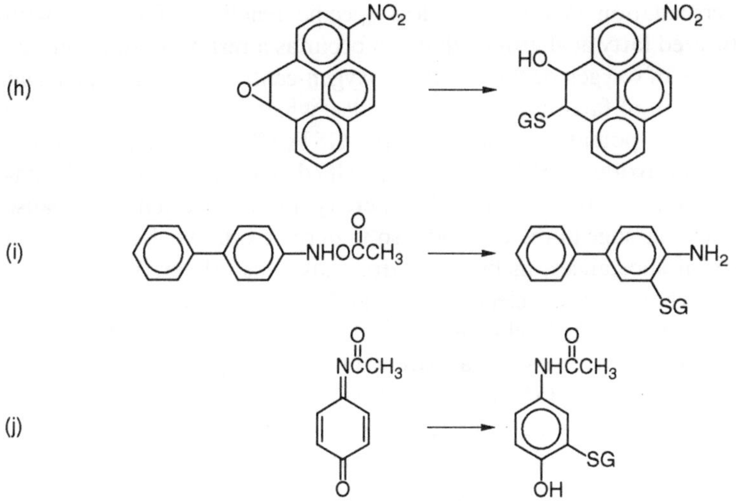

(h)

(i)

(j)

Figure 1. The GSH conjugation of: (a) (+)*anti*-benzo(a)pyrene-7,8-diol-9,10-oxide; (b) 7-sulphonyloxymethyl-12-hydroxymethylbenz(a)anthracene; (c) *exo*-aflatoxin B_1-8,9-oxide; (d) aminoazobenzene methyleneimine; (e) *N*-sulphonyloxy-*N*-methyl-4-aminoazobenzene; (f) *N*-sulphonyloxy-*N*′-acetyl benzidine; (g) benzidine diimine; (h) 1-nitropyrene-4,5-oxide; (i) *N*-acetoxyaminobiphenyl; (j) *N*-acetylbenzoquinone imine.

Another oxidizing enzyme system which may be important in certain extrahepatic tissues is prostaglandin synthetase or cyclooxygenase, which by a hydroperoxide-dependent mechanism can also produce genotoxic electrophiles from xenobiotics; one example is the industrial dye precursor and bladder carcinogen, benzidine, which, in the presence of prostaglandin synthetase, gives benzidine diimine and DNA adducts (Yamazoe et al, 1986). In vitro GSH has been shown to reduce adduct formation with DNA (Wise et al, 1985), but catalysis by GSTs has not been tested as yet (Figure 1g). Prostaglandin synthetase also produces BPDE from BP-7,8-diol (Marnett et al, 1982) and may be responsible for its presence in a number of extrahepatic tissues (Sivarajah et al, 1981).

Cyclooxygenase has the physiological role of catalysing a specific reaction between arachidonate and oxygen to give the hydroperoxy endoperoxide, prostaglandin PGG_2, and its reduction product PGH_2, initiating pathways of prostaglandin biosynthesis and metabolism. Physiologically related enzymes are the lipoxygenases, which give rise to stereospecific reactions of oxygen with arachidonate, forming hydroperoxides and initiating pathways of leukotriene biosynthesis and metabolism. In both the prostaglandin and leukotriene pathways, substrates for GSTs can also be found (Smith, 1989) (Figures 1, 3, 4 and 5).

Lipid peroxidation is the much less specific reaction of oxygen with polyunsaturated fatty acyl groups that can occur as a result of radical attack in the presence of oxygen or the attack of oxygen-centred radicals. The result is the production of fatty acyl hydroperoxides, which, being oxygen-centred electrophiles, are potentially substrates for GSTs (Prohaska and Ganther, 1976). Lipid peroxides which are not detoxified may undergo metal-catalysed conversion to alkoxy and alkyl peroxy radicals which may cause further radical damage in the cell and also undergo oxidative decomposition to a variety of substances, including hydroxyalkenals (Slater, 1984), which are toxic carbon-centred electrophiles and also substrates for GSTs (Danielson et al, 1987). High-energy irradiation of DNA in vitro in the presence of oxygen gives rise to apparent DNA hydroperoxides which are also substrates for GSTs (Meyer et al, 1991) (Figure 4).

Since electrophiles derived from xenobiotics, lipids and nucleic acids may be substrates, the range of structures encountered by GSTs is very large indeed and demands a multiplicity of isoenzymes with differing substrate requirements.

II. GLUTATHIONE TRANSFERASE ISOENZYMES

Most of the exploratory work on GSTs has been done in the rat; therefore, in terms of structure, function and tissue distribution, they provide the standards with which GSTs from other species, including man, are compared. Soluble GSTs are dimeric, 13 subunits having been characterized, with more known to exist. In the nomenclature used in this chapter each subunit is given a number based on the chronological order of its characterization (Jakoby et al, 1984).

A. A supergene family

1. Soluble enzymes

The soluble GSTs fall into four families referred to as alpha, mu, pi (Mannervik, 1985) and theta (Meyer et al, 1991). On the basis of present evidence, including cDNA sequence, gene structure, amino acid sequences, enzymatic activity and immunological crossreactivity, the genetic relationship between all 13 subunits is believed to be as follows: alpha family subunits 1a, 1b, 2, 8 and 10; mu family subunits 3, 4, 6, 9 and 11; pi family subunit 7 only; and theta family subunits 5 and 12. Most of the information on which these designations are based is in the review of Ketterer et al

(1988). More recent information concerning the designation of subunits can be found as follows: 1a and 1b in Hayes et al (1990); 8 in Meyer et al (1989); 10 in Meyer et al (1985); and 5 and 12 in Meyer et al (1991).

Within each multigene family, certain subunits may form heterodimers, GSTs 1–2, 3–4, 3–6, 4–6, 6–9 so far having been identified (Ketterer et al, 1988).

The upstream regulatory regions of these genes are of interest with respect to tissue-specific expression and enzyme induction. Those for subunits 1 and 7 have been mapped and show substantial differences. The subunit 7 gene has sequence motifs associated with the binding of SP1 and AP1 transcription factors near the start site and a sequence resembling a palindrome of an AP1 site further upstream which behaves as a major enhancer and is TPA responsive (Sakai et al, 1988). The 5′-flanking region of subunit 1 does not appear to contain motifs associated with either SP1 or AP1 binding but contains two elements associated with the inducibility of this subunit, namely a xenobiotic-responsive element such as is found in some P450 isoenzymes and an additional element, first recognized as being β-naphthoflavone-responsive and now known as an antioxidant-responsive element (Rushmore et al, 1990). Subunit 7 is of interest because it is induced in hepatic neoplasia; in this respect its TPA-responsive enhancer may be relevant. Subunit 1 is one of several which is inducible by xenobiotics and its inducibility is clearly related to the xenobiotic- and antioxidant-responsive elements which have been observed.

Reverse phase HPLC analysis of the soluble GST fraction obtained from a number of tissues by GSH-agarose affinity chromatography (Ostlund Farrants et al, 1987) reveals large differences in GST content from tissue to tissue. Analyses of the liver, kidney, lung, interstitial cells and spermatogenic tubules of the testis are shown in Figure 2. These analyses do not show members of the theta family, which have a high K_m for GSH and are not retained by the affinity column.

Immunohistochemistry has revealed interesting differences within tissues; it has shown that in the normal liver, subunit 7 is present in all bile duct cells and occasionally in hepatocytes (Tatematsu et al, 1985), and that, although subunits 1, 2, 3 and 4 are present in all hepatocytes, they are more abundant in the area around the central vein than in the periportal region (Tatematsu et al, 1985; Redick et al, 1982). In the brain, GSTs, like GSH, have been detected only in the glial, endothelial and ependymal cells and not at all in the neuronal stroma (Abramowitz et al, 1988).

2. *Enzymatic activity*

Much of our knowledge of the enzymatic properties of GSTs involves

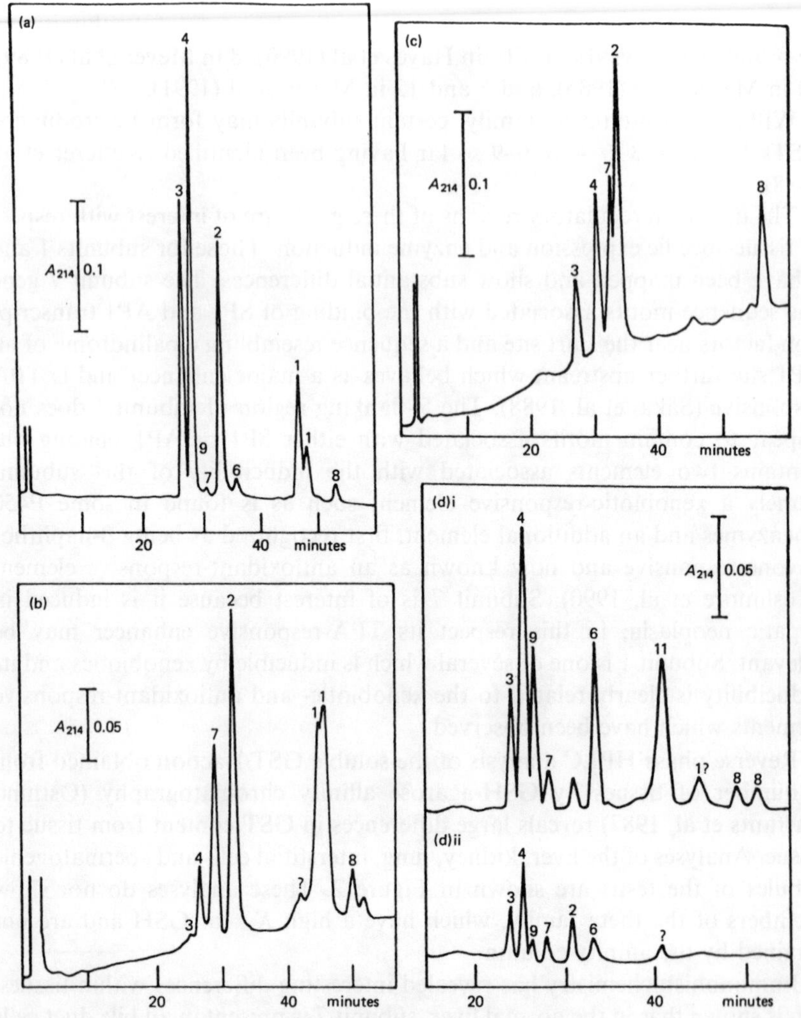

Figure 2. GSH transferase composition of the soluble fraction of: (a) liver; (b) kidney; (c) lung; (d)(i) spermatogenic tubules and (d)(ii) interstitial cells of testis determined by reverse phase HPLC after GSH-agarose affinity chromatography (Ostlund Farrants et al, 1987).

substrates chosen, not because of their biological significance, but because of their stability and therefore ease of handling and also their ability to give a convenient spectrophotometric assay. The most commonly used is 1-chloro-2,4-dinitrobenzene (CDNB), since it is utilized well by all of the known rat subunits with the exception of subunits 5 and 12 (Meyer et al, 1991). Other substrates have been chosen because they are relatively specific for a particular subunit and useful for its detection and quantification. Substrates of physiological significance are usually highly reactive and difficult to synthesize. As a result, knowledge concerning their enzymatic GSH conjugation has been slow to accumulate (Ketterer et al, 1988).

Not all organic electrophiles are substrates for GSTs. Examples are those which are so polarized in their ground state that enzymes can confer no further activation, such as the carbonium ions and nitrenium ions produced by oxidation of certain arylamines and *N*-nitroso compounds. At the other extreme there are Michael acceptors which have such a low activation energy for reaction with GSH that the non-catalytic rate is already rapid at physiological concentrations of GSH (e.g. 5 mM) (Coles et al, 1984/5). Aminobenzene methimine is an example (Ketterer et al, 1982); on the other hand, *N*-acetyl benzoquinone imine not only has a high spontaneous rate of reaction with GSH, but is also a good substrate and, since the K_m for GSH is in the region of 0.1 mM, high rates of catalytic reaction can be maintained under conditions of GSH depletion (Coles et al, 1988).

3. *Membrane-bound enzymes*

Although soluble GSTs have been the most intensively studied, membrane-bound forms also occur. They have been observed in hepatic microsomes and mitochondria and membrane fractions from other cells (Morgenstern and DePierre, 1988), e.g. the leukotriene C_4 synthetase activity of the particulate fraction of rat basophil leukaemia cells (Bach et al, 1986). When isolated from rat liver, microsomes contain both strongly adsorbed soluble GSTs, and at least one intrinsic enzyme, referred to as microsomal GST (Morgenstern and DePierre, 1988).

Microsomal GST has a molecular weight of 17 200 and a primary structure which has no apparent homology with the amino acid sequences of known soluble GST subunits (Morgenstern and DePierre, 1988). This enzyme has appreciable activity towards CDNB and some other model substrates, and low activity towards a carcinogenic electrophile, namely 1-nitroquinoline-*N*-oxide and the BP metabolite BP-4,5-oxide (Morgenstern and DePierre, 1988). Toxicologically its most important function appears to be its catalysis of the first step in the formation of nephrotoxins from hexachlorobutadiene and other halogenated alkenes (Dekant et al, 1988).

In the rat, microsomal GST is found principally in the liver, although it has been detected in extrahepatic tissues (Morgenstern and DePierre, 1988).

B. Examples of xenobiotic electrophiles

1. *Polycyclic aromatic hydrocarbons (PAH)*

A number of polycyclic aromatic hydrocarbon carcinogens have been shown to participate in the GSH conjugation pathway. These include benz(a)anthracene (Boyland and Sims, 1965), 7-methylbenzanthracene, 7,12-dimethylbenzanthracene (DMBA) (Sims, 1967), benzo(a)pyrene (BP) (Waterfall and Sims, 1972) and 3-methylcholanthrene (Sims, 1966). The electrophiles involved are predominantly the k-region epoxides, which are substrates for GSTs and in the case of BP account for about 70% of the biliary metabolites of BP-4,5-oxide in the rat (Smith and Bend, 1979).

The metabolites responsible for the carcinogenic effect of PAH are 'Bay Region' diol epoxides, as mentioned above. They bear a highly reactive benzylic epoxide carbon atom (Jerina and Lehr, 1977), an example being benzo(a)pyrene-7,8-diol-9,10-oxide (BPDE) (Figure 1a). BPDE and several other Bay Region diol epoxides have been shown to be substrates for GSTs (see Table 1). The presence or absence in a tissue of sufficient quantities of isoenzymes for which BPDE is a substrate (see Table 1) is believed to determine the susceptibility of such a tissue to carcinogenesis by BP. For example, the skin and mammary gland, which contain low levels of GSTs, are susceptible to BP carcinogenesis, while the liver, which contains very high levels, is not. Epoxidation of PAH is not the only reaction which gives rise to genotoxic electrophiles. In the case of 7,12-dimethylbenz(a)anthracene, activation to a DNA-binding species occurs by methyl oxidation to 7,12-dihydroxymethylbenz(a)anthracene followed by sulphation. 7,12-Dihydroxymethylbenz(a)anthracene-7-sulphate has been shown to be a good substrate for GSTs, forming *S*-(12-hydroxymethylbenz(a)-anthracene-7-yl)methyl glutathione (Figure 1b, Table 1). Recent evidence indicates that this conjugating activity may be a property of subunit 12 of the theta family (Hiratsuka et al, 1990).

2. *Aflatoxin B_1 (AFB_1)*

A GSH conjugate has been found as a major biliary metabolite of this potent mycotoxin and hepatocarcinogen (Newberne and Butler, 1969; Wogan and Shank, 1970). Activation of AFB_1 gives aflatoxin 8,9-oxide (AFBO), and the attack of GSH which occurs at the 8-position is catalysed by GSTs (Figure 1c, Table 1), low activity being present in all GST classes (F.P. Guengerich, K.D. Raney, D-H. Kim, T. Shunada, D.J. Meyer, B. Ketterer and J.M. Harris, personal communication) (see Table 1). Although GSTs which utilize AFBO are abundant in the rat liver, their activity is not

Table 1. Rat GSH transferases. Activities (μmol $min^{-1} mg^{-1}$) towards substrates of biological importance.

	Enzymes by class								
	alpha			mu				pi	theta
Substrate	1–1	2–2	8–8	3–3	4–4	6–6	11–11	7–7	5–5
1-Chloro-2,4-dinitrobenzene	40.0	38.0	10	50.0	20.0	360	30.0	20.0	—
1,2-Epoxy-3(*p*-nitrophenoxy)propane	0.7	0.9	nd	0.2	0.9	—	—	1.0	170
N-Acetyl-*p*-benzoquinone imine	24.0	48.0	nd	6.0	3.0	nd	nd	60.0	nd
exo-Aflatoxin B_1-8,9-oxide	0.001	0.001	nd	—	—	nd	nd	—	nd
Benzo(a)pyrene-7,8-diol-9,10-oxide	0.1	0.08	nd	0.03	0.33	nd	nd	0.33	nd
1-Nitropyrene-4,5-oxide	0.01	0.03	nd	0.30	0.30	nd	nd	0.02	nd
1-Nitropyrene-9,10-oxide	0.06	0.01	nd	0.40	0.20	nd	nd	0.08	nd
(*R*)-α-Bromoisovalerylurea	0.005	0.05	nd	0.13	0.11	nd	nd	—	nd
(*S*)-α-Bromoisovalerylurea	0.016	0.14	nd	0.055	0.015	nd	nd	—	nd
Ethylene dibromide	0.011	0.117	nd	0.070	0.021	nd	0.003	nd	nd

nd, not detectable.

sufficient to prevent AFB_1 carcinogenesis. However, if powerful inducers of GSTs are fed, such as ethoxyquin and dithiolthiones, AFB_1 carcinogenesis can largely be abolished (Kensler et al, 1986).

3. *N-Methyl-4-aminobenzene (MAB)*

GSH conjugates are important constituents of the biliary metabolites of the arylamine hepatocarcinogen, MAB. The major series of GSH conjugates, which may form as much as 70% of the biliary metabolites in the rat, are derived from aminoazobenzene methimines (Ketterer et al, 1982; Coles et al, 1983a,b). The sequence of reactions involved is methyl oxidation by cytochrome P450 to the *N*-hydroxymethyl-4-aminoazobenzene. This metabolite dehydrates (presumably spontaneously) to the methimine which reacts rapidly with GSH to give *N*-methylene glutathion-*S*-yl-4-aminoazobenzene (Figure 1d). The methimine does not appear to be a substrate for GSTs. The carcinogenic metabolite of MAB is the result of *N*-oxidation by flavin mixed function oxygenase followed by the sulphation of the resulting *N*-hydroxy group. This compound loses sulphate spontaneously to give a nitrenium ion and its carbonium ion tautomers which are highly genotoxic and is an example of a fully polarized electrophile (Figure 1e). It reacts poorly with GSH and is not a substrate for GSTs, presumably because of its highly polarized electrophilic centre and perhaps also some unfavourable structural feature that it has in common with the methimine metabolites (Figure 1d,e). It has been noted above that the more stable and therefore more polarizable *O*-sulphate ester, 7,12-dihydroxymethylbenzanthracene-7 sulphate, is a GST substrate.

4. *Benzidine*

Like other aromatic amines, benzidine is activated to a DNA-binding electrophile by a pathway involving *N*-acetylation, cytochrome P450-dependent *N'*-hydroxylation, and subsequent *N-O*-acetylation (Martin et al, 1982). The GSH conjugate of this electrophile, glutathion-*S*-yl-*N,N'*-diacetylbenzidine, has been detected in bile (Lynn et al, 1984). The synthetic *N*-sulphate reacts with GSH to produce a GSH conjugate (Figure 1f) (B. Coles, F. Kadlubar and F. Beland, unpublished information). Benzidine is also metabolized to an electrophile by ram seminal vesicle and human bladder prostaglandin synthetase-mediated oxidation to benzidinediimine. The diimine reacts with DNA (Yamazoe et al, 1986; Kadlubar et al, 1988) and GSH (Wise et al, 1985), the latter reaction giving a 3-(glutathion-*S*-yl)benzidine conjugate (Figure 1g). GSH has been shown to inhibit DNA binding of the diimine by more than 98%. However, it is not yet known whether any of these electrophiles derived from benzidine are substrates for GSTs.

5. *Nitropolycyclic aromatic hydrocarbons*

The carcinogenicity of nitropolycyclic aromatic hydrocarbons is associated with reductive metabolism of the nitro group to the *N*-hydroxy compound, which is then thought to be activated by esterification in a similar manner to other aromatic amine carcinogens (Beland et al, 1986; Rosenkranz and Mermelstein, 1986; Tokiwa and Ohnishi, 1986). However, since these carcinogens are also polycyclic aromatic hydrocarbons they are susceptible to epoxidation.

The only member of the group in which GSH conjugation has been studied in any detail is 1-nitropyrene, which is the predominant nitrated PAH identified in environmental samples. 1-Nitropyrene is a powerful indirect bacterial mutagen and is tumorigenic in the rat and mouse (Tokiwa and Ohnishi, 1986; Hirose et al, 1984; El-Bayoumy et al, 1984). GSH conjugates form approximately one-third of the biliary metabolites of 1-nitropyrene in the rat (Djuric et al, 1987), these conjugates arising from the k-region oxides. In vitro, both k-region oxides have been shown to be substrates for the GSTs, isoenzymes containing the subunits 3 and 4 being the most effective enzymes (Table 1). Figure 1h shows the GSH conjugation of 1-nitropyrene-4,5-oxide. It has been suggested that the extensive metabolism of 1-nitropyrene to epoxides rather than by nitro reduction is a major contributing factor to its low carcinogenicity. In contrast, the highly carcinogenic 1,6- and 1,8-dinitropyrenes are able to undergo nitro reduction without extensive competitive metabolism to form epoxides (Djuric et al, 1987).

III. GSTs AND ENDOGENOUS PEROXIDES

GSTs participate in the metabolism of endogenous peroxides arising from two sources, one being intermediates in the metabolism of the highly specific biosynthetic pathway for eicosanoid cell mediators and the other being due to the effects of oxidative stress on lipid and nucleic acids.

A. GSTs in eicosanoid metabolism

1. *The leukotriene pathway*

5-Lipoxygenase/leukotriene A_4 synthetase converts arachidonate via (*S*)-

5-hydroperoxy-6-*trans*-8,11,14-*cis*-eicosatetraenoic acid (HPETE) to 5,6-epoxy-7,9-*trans*-11,14-*cis*-eicosatetraenoic acid (leukotriene A_4, LTA_4) (Schewe et al, 1987). LTA_4 is then conjugated with GSH to give LTC_4 (Hammarström, 1983; Tsuchida et al, 1987) (Figure 3a). The GSH conjugation of LTA_4 is catalysed by both membrane-bound and soluble GSTs (Söderström, 1988; Izumi et al, 1988). Among the soluble GSTs, those in the μ class are the best, particularly subunit 6 (see Table 2) (Tsuchida et al, 1987); however, at least one membrane-bound form also occurs (Bach et al, 1986). So far no complete purification of membrane-bound LCT_4 synthetase has been published. The examination of a range of tissues, in the rat, for membrane-bound versus soluble enzymes converting LTA_4 to LTC_4 shows that membrane-bound activity is sometimes, but not always, higher than soluble activity. In the lung and spleen, which have the highest LTC_4 synthetase activity, only membrane-bound activity is detectable; however, in the brain membrane-bound and soluble activity are about equal (Izumi et al, 1988). The membrane-bound form is not microsomal GST. In this pathway GST-catalysed attack on the hydroperoxy group of HPETE, bringing about reduction to HETE, could also occur. Both LTA_4 and HETE are powerful cell mediators (Smith, 1989).

2. *Cyclooxygenase pathway*

Cyclooxygenase converts arachidonate to the hydroperoxyendoperoxide prostaglandin PGG_2 and reduces it to the hydroxyendoperoxide PGH_2 (Kulmacz and Lands, 1987). PGH_2 can then undergo three conversions by GSH-dependent mechanisms. It can be reduced to $PGF_{2\alpha}$ (Figure 4a) or isomerized to PGE_2 or PGD_2 by a GSH-dependent reaction, the mechanism of which is not understood (Christ-Hazelhof and Nugteren, 1979) (Figure 5a,b). GSTs catalyse all the above reactions in vitro (Urade et al, 1985; Meyer and Ketterer, 1987; Ujihara et al, 1988) (see Table 2). The alpha family enzymes are best at both the GSH peroxidase activity, which reduces PGH_2 to $PGF_{2\alpha}$ (Figure 4a), and the isomerase activity, which converts PGH_2 to PGE_2 and PGD_2 (Figure 5). It would appear that the endoperoxide PGH_2 is a relatively poor substrate for the Se-dependent GSH peroxidase, and Se deficiency which induces GST alpha enzymes causes an increased formation of $PGF_{2\alpha}$ in vivo (Hong et al, 1989).

Recently a PGD synthetase from the rat spleen was isolated and shown to be a GST, which, on the basis of limited amino acid sequence data, appeared to be a member of yet another gene family. $PGF_{2\alpha}$, PGD_2 and PGE_2 synthetases unrelated to GSTs have also been isolated. There is, for example, a GSH-independent $PGF_{2\alpha}$ synthetase from bovine lung (Watanabe et al, 1985), a GSH-independent PGD_2 synthetase from rat brain (Shimizu et al,

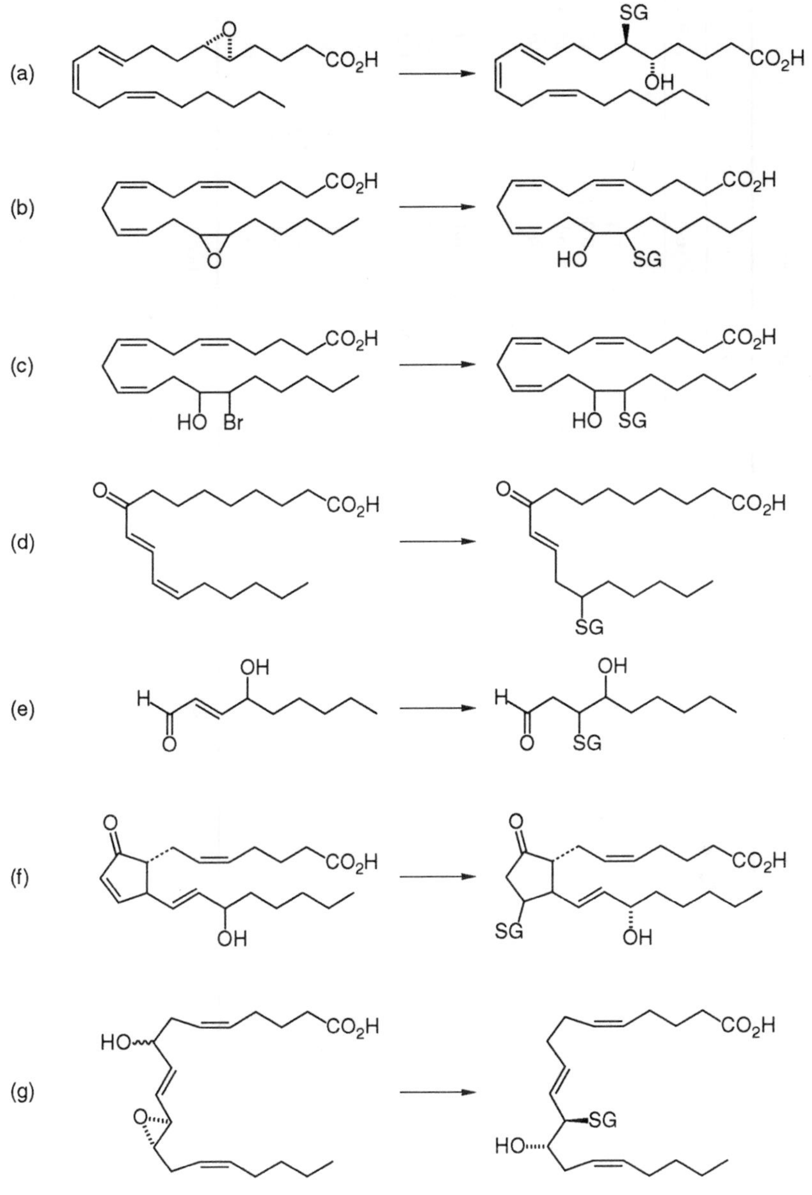

Figure 3. Glutathione conjugation of: (a) leukotriene A_4; (b) 14,15-epoxy-*cis*-eicosatrienoic acid; (c) eicosatrienoic acid-14,15-bromohydrin; (d) 9-oxo-10-*trans*-12-*cis*-octadecadienoic acid; (e) 4-hydroxy-non-2-enal; (f) prostaglandin A_2; (g) hepoxilin A_3.

Table 2. Substrates for rat GSH transferases associated with lipid and DNA peroxides (activities expressed as μmol min^{-1} mg^{-1})

Substrate	GST isoenzyme							
	1–1	2–2	3–3	4–4	5–5	6–6	7–7	8–8
(a) Eicosanoid substrate								
Leukotriene A_4[a]	0.001	0.0006	0.0006	0.028	nd	0.1	0.013	nd
Prostaglandin $H_2 \rightarrow E_2$[b]	1.4	0.420	0.057	0.092	nd	0.063	0.033	nd
Prostaglandin $H_2 \rightarrow D_2$[b]	0.170	0.060	0.016	0.016	nd	nil	0.014	nd
Prostaglandin $H_2 \rightarrow F_{2\alpha}$[b]	1.3	0.390	0.024	0.028	nd	nil	0.014	nd
(b) Other substrates producing fatty acyl GSH conjugates								
Epoxyeicosatrienoic acid[c]	0.003	0.002	0.022	0.001	0.036	nd	nd	nd
Oxolinoleic acid[d]	0.6	0.7	0.1	0.6	nd	nd	0.7	36
Oleic acid bromohydrin[e]	*x* >	*y* =	*y* =	*y*	nd	nd >	*z*	nd
(c) Products of lipid and DNA damage								
Linoleic acid hydroperoxide	3.0	1.6	0.2	0.2	5.3	nd	1.5	nd
4-Hydroxynon-2-enal	2.6	0.7	2.7	6.9	nd	nd	1.4	170
Cholesterol-5,6-oxide	0.0001	nd	nil	nil	nd	nd	nd	nd
Irradiated DNA (DNA hydroperoxide)	nil	nil	0.02	0.03	0.03	nd	0.01	nd

[a] Tsuchida et al (1987); [b] Ujihara et al (1988); [c] Spearman et al (1985); [d] Bull, Lebecq, Meyer, Marnett and Ketterer (unpublished); [e] Taylor, Coles, Meyer and Ketterer (unpublished).

nd, not detectable. *x*, *y* and *z* represent enzymic rates for which only approximate figures are available, i.e. the activity of GST 1–1 > 2–2 = 3–3 = 4–4 > 7–7 in approximate terms.

1979) and labile membrane-associated GSH-dependent PGE_2 synthetases from sheep seminal vesicles (Mooren et al, 1982; Tanaka and Smith, 1985).

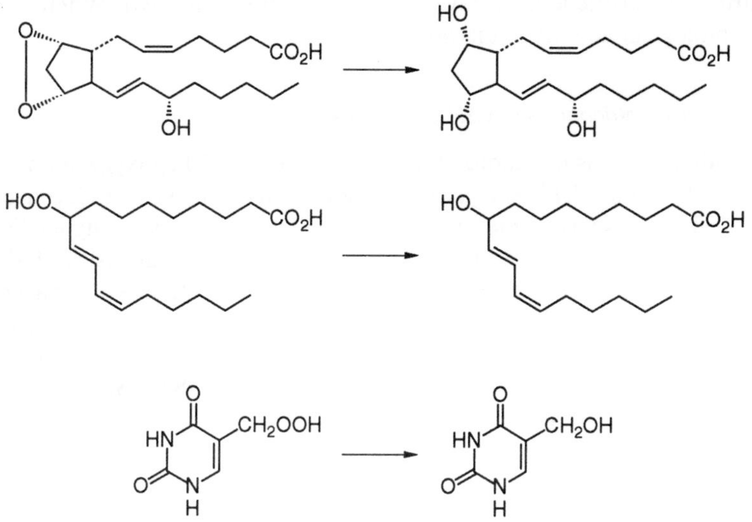

Figure 4. Hydroperoxide reduction of: (a) prostaglandin H_2 to $F_{2\alpha}$; (b) 9-linoleic acid hydroperoxide; (c) 5-hydroperoxymethyluracil.

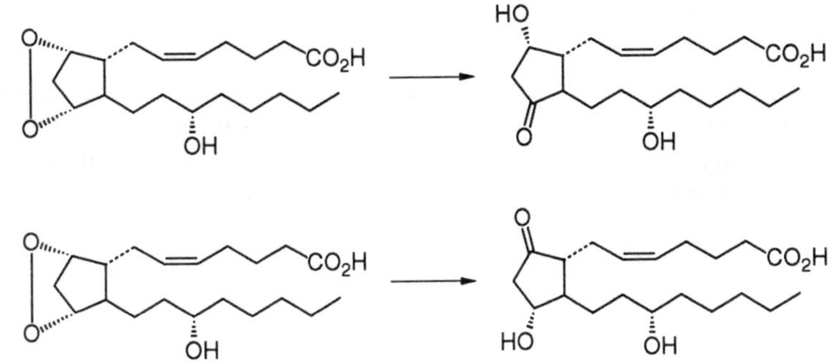

Figure 5. Glutathione-dependent isomerization of: (a) prostaglandin H_2 to D_2; (b) prostaglandin H_2 to E_2.

B. 'Leukotriene-like' compounds

Several types of GSH conjugates derived from fatty acids, which many regard as leukotriene-like, have been demonstrated in vivo. Whether they have biological activity is unknown.

1. *Epoxyeicosatrienoic acid GSH conjugates*

Arachidonic acid is metabolized by a cytochrome P450 epoxygenase to 5,6-, 8,9-, 11,12- and 14,15-epoxyeicosatrienoic acids (EETs) (Capdevila et al, 1982). The EETs have a number of biological activities, including mediation of the in vitro release of insulin and glucagon from pancreatic islets (Falck et al, 1983) and effects on the Na^+/K^+ flux of the kidney tubule (Schwartzman et al, 1985). They are inhibitors of cyclooxygenase and when exposed to epoxide hydrolase give diols which inhibit platelet aggregation (Fitzpatrick et al, 1986). They are also conjugated with GSH by GSTs (Spearman et al, 1985) (see Table 2b), but it is not known whether the resulting eicosatrienoic GSH conjugates have any biological activity. It is noteworthy that biological activity in LTC_4 is stereochemically specific. There is also a degree of stereospecificity in the biosynthesis and pharmacological activity of EETs (Fitzpatrick et al, 1986); but to date none of their GSH conjugates has been tested for biological activity.

2. *Arachidonate halohydrin GSH conjugates*

It is possible that either hypochlorous or hypobromous acids released in the oxidative burst of inflammatory cells could add across the double bond of arachidonic acid to give either a chloro- or bromohydrin. These halohydrins are electrophiles with the potential for nucleophilic displacement of the halide by GSH and its catalysis by GSTs. The result would presumably be hydroxyglutathion-*S*-yl eicosatrienoic acids similar to those described above for EETs (Fig. 3c); however, since the reaction is spontaneous, the stereoselectivity of synthesis reported for the epoxygenase is unlikely. Preliminary work using synthetic oleic acid halohydrins shows the bromohydrin, but not the chlorohydrin, to be a good substrate for GSTs. Comparative but not numerical data are shown in Table 2 (Taylor, Coles, Meyer and Ketterer, personal communication). Once more, the resulting GSH conjugates, while detoxifying halohydrins, may also have biological activity.

3. *GSH conjugates derived from fatty acids containing α,β-unsaturated ketones*

Unsaturated fatty acid hydroperoxides, present in the diet, have been shown to be toxic to colon explants in culture, but so too have their corresponding alcohols (reduction products, which could arise from the action of GSH peroxidase activity in vivo) and α,β-unsaturated ketones, perhaps resulting from subsequent oxidation of the unsaturated alcohols by enzymes such as alcohol dehydrogenase or 15-prostaglandin dehydrogenase (Bull et al, 1988). By analogy with the hydroxyalkenals (Table 2), these α,β-unsaturated ketones are expected to be substrates for GSTs. Oxolinoleic acid (a mixture of 9-oxo-10-*trans*-12-*cis*-octadecadienoic acid (Figure 3d) and 13-oxo-9-*trans*-11-*cis*-octadecadienoic acid) has been synthesized and shown to be a good substrate for GSTs, particularly GST 8–8, with which it gives rates second only to that obtained with 4-hydroxynon-2-enal (Figure 3e) (R. Lebecq, L. Marnett, A. Bull, D.J. Meyer and B. Ketterer, personal communication) (Table 2). This conjugation reaction (Figure 3) may serve primarily to detoxify an α,β-unsaturated ketone which is reactive with protein thiols and results in cytotoxicity. However, it is possible that the GSH conjugate is a mitogenic leukotriene-like compound.

4. *GSH conjugates of prostaglandins containing an α,β-unsaturated ketone*

Prostaglandins E_2 and D_2 undergo dehydration to PGA_2 (Figure 3f) and $\Delta^{12}PGJ_2$, both having a reactive α,β-unsaturated ketone in the cyclopentenone ring. The α,β-unsaturated ketones are cytotoxic, having antiproliferative effects associated with their localization in the nucleus. Conjugation with GSH (see Figure 3f), which is GST catalysed, abolishes toxicity. The isoenzyme specificity of this reaction is not yet known. PGA_2 and $\Delta^{12}PGJ_2$ are probably responsible for the in vitro antiproliferative activity attributed to PGE_2 and PGD_2 in the past (Honn and Marnett, 1985; Atsom et al, 1990).

5. *Hepoxilins*

Hepoxilins are epoxy alcohols formed from arachidonic acids by 12-lipoxygenation followed by intramolecular rearrangement. Two positional isomers have been identified: hepoxilin A_3 (8-hydroxy-11,12-epoxyeicosatrienoic acid) and the 10-hydroxy isomer hepoxilin B_3 (Pace-Asciak et al, 1983). Hepoxilin A_3 can be conjugated by GSTs to give 11-glutathion-*S*-yl hepoxylin A_3 (Pace-Asciak et al, 1990) (Figure 3g). Again the isoenzyme specificity of this reaction is not known. Apart from LTC_4 this is the only lipid–GSH conjugate so far which has been shown incontrovertibly to have biological activity. It has been identified in the rat hippocampus and has a neuromodulatory function in hippocampal CAR neurones; however, its physiological significance has yet to be evaluated.

C. Lipid and DNA peroxides produced during oxidative stress

1. *Inhibition and repair of lipid peroxidation*

GSTs can intervene in lipid peroxidation and reduce fatty acid hydroperoxides by catalysing the nucleophilic attack of GSH on electrophilic oxygen (Prohaska, 1980; Meyer et al, 1985). The specific activity of a number of GST isoenzymes from the rat for linoleic acid hydroperoxides (Figure 4b) is shown in Table 2. Although GSTs reduce free fatty acid hydroperoxides, they cannot reduce esterified fatty acid hydroperoxides. Nevertheless, with the cooperation of phospholipase they are able to inhibit lipid peroxidation in cell membranes (Tan et al, 1984). It is assumed that phospholipase A_2 releases free fatty acid hydroperoxide, which is then detoxified by GSTs, leaving behind in the membrane a lysophosphatide, which is reacylated to bring about repair of the membrane (van Kuijk et al, 1987). Under physiological concentrations of GST 1–1 and GSH, microsomal lipid peroxidation initiated by Fe^{2+} and NADPH is completely abolished (Tan et al, 1984).

If the inhibition of lipid peroxidation is not complete, various products which are toxic or have toxic potential are produced. Metal catalysis gives rise to alkoxy and alkylperoxy radicals which are more stable than carbon-centred radicals and have the capacity to diffuse some distance within the cell before being quenched. Apart from their reactivity as radicals, they can undergo oxidative decomposition to a range of compounds, including hydroxyalkenals such as 4-hydroxynon-2-enal (HNE) (Figure 3e), which has been reported to be cytotoxic and genotoxic (White and Rees, 1984; Marnett et al, 1985) and, in addition, to have pharmacological effects which may be important in cell regulation (Dianzani, 1982; Schauenstein, 1982). HNE is detoxified by GSH conjugation and, as mentioned above, is a good substrate for most GST isoenzymes, particularly GST 8–8, with which it gives the highest rate of catalysis yet observed with GSTs (see Table 2). It is assumed that most of the lipid peroxidation is inhibited; however, should it occur and produce hydroxyalkenals, these should be largely detoxified by GSTs.

Our results suggest that microsomal GSTs have a low capacity for the detoxification of lipid hydroperoxides. The GSTs and Se-dependent GSH peroxidase appear to be more important, but their relative contribution depends on the species and the organ under consideration. The GSTs make a greater contribution to overall GSH peroxidase activity than the Se-dependent enzyme in the human liver compared with the rat liver, and in the rat testis the GSTs contribute more activity than the Se-dependent enzyme (Lawrence and Burk, 1978). Enzymes with GSH peroxidase activity towards phospholipid hydroperoxides have also been reported, but their relative importance is not as yet understood (Ursini et al, 1985; Duan et al, 1988; however, see Chapter 12). It is clear that multiple GSH-dependent mechanisms can be called upon for the protection of membrane lipid from oxidative stress.

2. *Repair of DNA peroxidation*

Knowledge of the nature of the oxy radical damage to DNA in vivo is fragmentary and largely indirect. Thymine glycols, hydroxymethyluracil and 8-hydroxyguanine have all been observed in vivo (Ames and Saul, 1985; Bengtold et al, 1988; Frenkel et al, 1986; Kasai et al, 1990). In vitro studies suggest that other bases may be damaged by oxy radicals (von Sonntag, 1987) and bacteriophage DNA, when damaged by oxy radicals, on transfection into *E. coli* gives mutations which appear to arise from damage to other bases (McBride et al, 1990). We have been interested in oxidative attack on thymine residues which is expected to give 4-hydroxy-5-hydroperoxy-4,5-dihydrothymine, 4-hydroperoxy-5-hydroxy-4,5-dihydrothymine and 5-hydroperoxymethyluracil residues (Figure 4c). These hydroperoxides are expected to be substrates for GSTs. Some evidence that free thymine hydroperoxides are substrates has been presented (Tan et al, 1986) and irradiated DNA which contains hydroperoxides has also proved to be a substrate (Tan et al, 1988) (Table 2). Although GSTs have activity towards irradiated DNA they are not as effective as Se-dependent GSH peroxidase (Meyer et al, 1991). Whichever enzyme operates in the nucleus, thymine hydroperoxides and 5-hydroperoxymethyluracil residues are reduced to the respective glycols and hydroxymethyl derivative and released by a DNA glycosylase. The resulting apyrimidinic DNA should then undergo repair.

IV. HUMAN GLUTATHIONE TRANSFERASES

Human GSTs are now known for all gene families identified in the rat (see Ketterer et al (1988) for gene families alpha, mu and pi; Meyer et al (1991) for gene family theta; and De Jong et al (1988) for human microsomal GST). Homologies are very close, particularly in the mu family (Pemble et al, 1991). As far as can be judged on the basis of the small amount of data available, similarity in structure is usually associated with similarity in enzymatic activity (Robertson and Jernström, 1986; Coles et al, 1988; Cmarik et al, 1990; F.P. Guengerich, K.D. Raney, D.-H. Kim, T. Shunada, D.J. Meyer, B. Ketterer, and J.M. Harris, personal communication). At present almost no information is available relating human GSTs to eicosanoid biosynthesis nor to the production of 'leukotriene-like' compounds.

V. CONCLUSIONS

There is little doubt that GSTs detoxify many carbon-centred electrophiles arising from the oxidative action of cytochrome P450s and play an essential role in protection against carcinogens and drugs which are substrates for GSTs. In vitro evidence suggests that GSTs may share in the detoxification of lipid and DNA hydroperoxides. On consideration of eicosanoid biosynthesis, whereas it is clear that leukotriene A_4 undergoes a catalysed GSH conjugation to give LCT_4, it is not clear to what extent GSTs are involved. GST 6–6 is as active with LTA_4 as any enzyme, but whether or not it is present in cells such as leukocytes which are active in synthesizing leukotrienes is not known as yet. It is an important component in brain and testis, but the extent of the leukotriene biosynthesis in these organs is not known.

The alpha family is active in the formation of $PGF_{2\alpha}$, PGD_2 and PGE_2 from PGH_2. In the rat, the alpha family is abundant in hepatocytes and the proximal convoluted tubule of the kidney, but cyclooxygenase is most abundant in vesicular gland, caecum, gall bladder and lung (Yoshimoto et al, 1986). It is present in the kidney but in the medulla, not the cortex where the proximal convoluted tubules are to be found. It is present in the liver, but at very low levels, and in this organ arachidonate is more likely to be esterified and secreted than converted to PGH_2. The presence of these enzyme activities in apparently inappropriate situations is an enigma which remains to be solved. However, in all cases where GSH conjugates are produced excretion is bound to occur and this may be the function of GSTs with such eicosanoids as PGA_2 and $\Delta^{12}PGJ_2$ and leukotriene-like compounds such as EETs, α,β-unsaturated keto lipids, etc. The fact that leukotriene C_4 is a GSH conjugate ensures a short lifespan since GSH conjugates and related peptides of the mercapturic acid pathway are programmed for excretion (Ishikawa et al, 1990). The physiological role of GSTs in prostaglandin $F_{2\alpha}$, D_2 and E_2 synthetase, despite the relatively high rates of their GST-catalysed formation, remains to be established.

ACKNOWLEDGEMENT

The authors are grateful to the Cancer Research Campaign UK for their generous support.

REFERENCES

Abramowitz M, Homma H, Ishigaki S, Tansey F, Cammer W & Listowsky I (1988) *J. Neurochem.* **50:** 50.

Ames BN & Saul R (1985) In Ramil C, Lambert B & Magnusson J (eds) *Genetic Toxicology of Environmental Chemicals*, p 1. New York: Alan Liss.

Atsom J, Freeman ML, Meredith MJ, Sweetman BJ & Roberts LJ (1990) *Cancer Res.* **50:** 1879.

Bach MK, Brashler JR, Peck RE & Morton DR (1986) *J. Allergy Clin. Immunol.* **74:** 353.

Beland FA, Heflich RH, Howard PC & Fu PP (1986). In Harvey RG (ed.) *Polycyclic Hydrocarbons and Carcinogenesis*, p 371. Washington DC: American Chemical Society.

Bengtold DS, Simic MG, Alessio H & Cutler RG (1988) In Simic MG, Taylor KA, Ward JF & von Sonntag C (eds) *Oxygen Radicals in Biology and Medicine*, p 483. New York: Plenum Press.

Boyland E & Sims P (1965) *Biochem. J.* **97:** 7.

Bull AW, Migro ND & Marnett LJ (1988) *Cancer Res.* **48:** 1771.

Capdevila J, Marnett L, Charcos N, Prough R & Eastabrook R (1982) *Proc. Natl. Acad. Sci. USA* **79:** 767.

Christ-Hazelhof E & Nugteren DH (1979) *Biochim. Biophys. Acta* **572:** 43.

Cmarik JL, Inskeep PB, Meredith MJ, Meyer DJ, Ketterer B & Guengerich FP (1990) *Cancer Res.* **50:** 2747.

Coles B (1984/5) *Drug Metab. Rev.* **15:** 1307.

Coles B, Srai SKS, Ketterer B, Waynforth HB & Kadlubar FF (1983a) *Chem. Biol. Interact.* **43:** 123.

Coles B, Srai SKS, Waynforth B & Ketterer B (1983b) *Chem. Biol. Interact.* **47:** 307.

Coles B, Wilson I, Wardman P, Hinson JA, Nelson SD & Ketterer B (1988) *Arch. Biochem. Biophys.* **264:** 253.

Danielson UH, Esterbauer H & Mannervik B (1987) *Biochem. J.* **247:** 707.

De Jong JL, Morgenstern R, Jörnavall H, DePierre JW & Tu C-P (1988) *J. Biol. Chem.* **263:** 8430.

Dekant W, Lash LH & Anders MW (1988) In Sies H & Ketterer B (eds) *Glutathione Conjugation*, p 415. London: Academic Press.

Dianzani MU (1982) In McBrien DCH & Slater TF (eds) *Free Radicals, Lipid Peroxidation and Cancer*, p 129. London: Academic Press.

Djuric Z, Coles B, Fifer EK, Ketterer B & Beland FA (1987) *Carcinogenesis* **8:** 1781.

Duan Y-J, Komura S, Fiszer-Szafarz B, Szafarz D & Yagi K (1988) *J. Biol. Chem.* **263:** 19003.

El-Bayoumy K, Hecht SS, Sackl T & Stoner GD (1984) *Carcinogenesis* **5:** 1449.

Falck JR, Manna S, Moltz J, Chacos N & Capdevila J (1983) *Biochem. Biophys. Res. Commun.* **114:** 743.

Fitzpatrick FA, Ennis MD, Baze ME, Wynalda MA, McGee JE & Ligett WF (1986) *J. Biol. Chem.* **261:** 15334.

Frenkel K, Chrzan K, Troll W, Teebor GW & Steinberg JJ (1986) *Cancer Res.* **46:** 5533.

Guengerich FP (1988) *Cancer Res.* **48:** 2946.

Hammarström S (1983) *Annu. Rev. Biochem.* **52:** 355.

Hayes JD, Kerr LA, Harrison DJ, Conshaw AD, Ross AG & Neale GE (1990) *Biochem. J.* **268:** 295.

Hinson, JA & Kadlubar FF (1988) In Sies H & Ketterer B (eds) *Glutathione Conjugation*, p 235. London: Academic Press.
Hiratsuka A, Sebata N, Kawashima K et al (1990) *J. Biol. Chem.* **265:** 11973.
Hirose M, Lee M-S, Wang CY & King CM (1984) *Cancer Res.* **44:** 1158.
Hong Y, Li C-H, Burgess JR et al (1989) *J. Biol. Chem.* **264:** 13793.
Honn KV & Marnett LJ (1985) *Biochem. Biophys. Res. Commun.* **129:** 34–40.
Ishikawa T, Müller M, Klünemann C, Schaub T & Keppler O (1990) *J. Biol. Chem.* **265:** 19279–19286.
Izumi T, Honda Z, Ohishi N et al (1988) *Biochim. Biophys. Acta* **959:** 305.
Jakoby WB, Ketterer B & Mannervik B (1984) *Biochem. Pharmacol.* **33:** 2539.
Jerina DM & Lehr RE (1977) In Ullrich V, Roots I, Hildebrandt AG, Estabrook RW & Conney AH (eds) *Microsomes and Drug Oxidation*, p 709. Oxford: Pergamon Press.
Kadlubar FF, Butler MA, Hayes BE, Beland FA & Guengerich FP (1988) In Miners J, Birkett DJ, Drew R & McManus M (eds) *Microsomes and Drug Oxidations*, p 370. London: Taylor and Francis.
Kasai H, Chung M-H, Nishimura S, Kamiya H & Ohitsuka E (1990) *J. Cancer Res. Clin. Oncol.* **66 (supplement II):** 1031.
Kensler TW, Egner PA, Davidson NE, Roebuck BD, Pikul A & Groopman JD (1986) *Cancer Res.* **46:** 3924.
Ketterer B, Srai SKS, Waynforth B, Tullis DL, Evans FE & Kadlubar FF (1982) *Chem. Biol. Interact.* **38:** 287.
Ketterer B, Meyer DJ & Clark AG (1988) In Sies H & Ketterer B (eds) *Glutathione Conjugation*, p 73. London: Academic Press.
Kulmacz R & Lands WEM (1987) In Benedetto C, McDonald-Gibson RG, Nigam S & Slater TF (eds) *Prostaglandins and Related Substances*, p 209. Oxford: IRL Press.
Lawrence RA & Burk RF (1978) *J. Nutr.* **108:** 211.
Lynn RK, Garvie-Gould CT, Milain DF et al (1984) *Toxicol. Appl. Pharmacol.* **42:** 285–300.
Mannervik B (1985) *Adv. Enzymol.* **57:** 357.
Marnett LJ, Panthananickel A & Reed GA (1982) *Drug Metab. Rev.* **13:** 235.
Marnett LJ, Hurd HK, Hollstein MC, Levin DE, Esterbauer H & Ames BN (1985) *Mutat. Res.* **148:** 25.
Martin CN, Beland FA, Roth RW & Kadlubar FF (1982) *Cancer Res.* **42:** 2678.
McBride TJ, Tkeshekashvili UK & Loeb LA (1990) *J. Cancer Res. Clin. Oncol.* **116 (supplement II):** 1154.
Meyer DJ & Ketterer B (1987) In Mantle TJ, Pickett CB & Hayes JD (eds) *Glutathione S-Transferases and Carcinogenesis*, p 57. London: Taylor and Francis.
Meyer DJ, Tan KH, Christodoulides LG & Ketterer B (1985) In Poli G, Cheeseman KH, Dianzani MU & Slater TF (eds) *Free Radicals in Liver Injury*, p 221. Oxford: IRL Press.
Meyer DJ, Lalor E, Coles B et al (1989) *Biochem. J.* **260:** 785.
Meyer DJ, Coles B, Pemble SE, Gilmore KS, Fraser GM & Ketterer B (1991) *Biochem. J.* **274:** (in press).
Mooren P, Buytenlick M & Nugteren DH (1982) *Methods Enzymol.* **86:** 84.
Morgenstern R & DePierre JW (1988) In Sies H & Ketterer B (eds) *Glutathione Conjugation*, p 157. London: Academic Press.
Newberne PM & Butler WH (1969) *Cancer Res.* **29:** 236.

Ostlund Farrants A-K, Meyer DJ, Coles B et al (1987) *Biochem. J.* **245:** 423.
Pace-Asciak CR, Granström E & Sammuelson B (1983) *J. Biol. Chem.* **258:** 6335.
Pace-Asciak CR, Lanenville O, Su W-G et al (1990) *Proc. Natl. Acad. Sci. USA* **87:** 3037.
Pemble SE, Taylor JB & Ketterer B (1986) *Biochem. J.* **240:** 885.
Prohaska JR (1980) *Biochem. Biophys. Res. Commun.* **611:** 87.
Prohaska JR & Ganther HE (1976) *J. Neurochem.* **27:** 1379.
Redick JA, Jakoby WB & Baron J (1982) *J. Biol. Chem.* **257:** 15200.
Robertson IGC & Jernström B (1986) *Carcinogenesis* **7:** 1633.
Rosenkranz HS & Mermelstein RJ (1986) *Environ. Sci. Health* **23:** 221.
Rushmore TH, King RK, Paulson KE & Pickett CB (1990) *Proc. Natl. Acad. Sci. USA* **87:** 3826.
Sakai M, Okuda A & Muramatsu M (1988) *Proc. Natl. Acad. Sci. USA* **85:** 9456.
Schauenstein E (1982) In McBrien DCH & Slater TF (eds) *Free Radicals, Lipid Peroxidation and Cancer*, p 159. London: Academic Press.
Schewe T, Kühn H & Rapoport SM (1987) In Benedetto C, McDonald-Gibson RG, Nigam S & Slater TF (eds) *Prostaglandins and Related Substances*, p 229. Oxford: IRL Press.
Schwartzman M, Ferreri N, Carrol M, Songu-Mize E & McGiff J (1985) *Nature (Lond.)* **314:** 620.
Shimizu T, Yanamoto S & Hayaishi O (1979) *J. Biol. Chem.* **254:** 5222–5228.
Sims P (1966) *Biochem. J.* **98:** 215.
Sims P (1967) *Biochem. J.* **105:** 591.
Sivarajah K, Lasker JM & Eling TE (1981) *Cancer Res.* **41:** 1834.
Slater TF (1984) *Biochem. J.* **222:** 1.
Smith BR & Bend JR (1979) *Toxicol. Appl. Pharmacol.* **49:** 313.
Smith WL (1989) *Biochem. J.* **259:** 315.
Söderström, M, Hammarström, S & Mannervik B (1988) *Biochem. J.* **250:** 713.
Spearman ME, Prough RA, Eastabrook RW et al (1985) *Arch. Biochem. Biophys.* **242:** 225.
Tan KH, Meyer DJ, Belin J & Ketterer B (1984) *Biochem. J.* **220:** 243.
Tan KH, Meyer DJ, Coles B & Ketterer B (1986) *FEBS Lett.* **207:** 231–233.
Tan KH, Meyer DJ, Gillies N & Ketterer B (1988) *Biochem. J.* **254:** 841–845.
Tanaka Y & Smith WL (1985) *Adv. Prostaglandin Thromboxane Leukotriene Res.* **15:** 147.
Tatematsu M, Mera Y, Ito N, Satoh K & Sato K (1985) *Carcinogenesis* **6:** 1621.
Tokiwa H & Ohnishi Y (1986) *CRC Crit. Rev. Toxicol.* **17:** 23.
Tsuchida S, Izumi T, Shimiza T et al (1987) *Eur. J. Biochem.* **170:** 159.
Ujihara M, Tsuchida S, Satoh K, Sato K & Urade Y (1988) *Arch. Biochem. Biophys.* **264:** 428.
Urade Y, Fujimoto M & Hayaishi O (1985) *J. Biol. Chem.* **260:** 12410.
Ursini F, Maiorino M & Gregolin C (1985) *Biochim. Biophys. Acta* **839:** 62.
van Kuijk FJGM, Sevanian A, Handelman GJ & Dratz EA (1987) *TIBS* **12:** 31.
von Sonntag C (1987) *The Chemical Basis of Radiation Biology*. London: Taylor and Francis.
Watanabe K, Yoshida R, Shimizu T & Hayaishi O (1985) *Adv. Prostaglandin Thromboxane Leukotriene Res.* **15:** 151.
Waterfall JF & Sims P (1972) *Biochem. J.* **128:** 265.
White JS & Rees KR (1984) *Chem. Biol. Interact.* **52:** 233.

Wise RW, Zenser TV & Davis B (1985) *Carcinogenesis* **6:** 579.
Wogan GN & Shank RC (1970) In Petts JN & Metcalf RI (eds) *Advances in Environmental Science and Technology*, p 321. New York: John Wiley and Sons Inc.
Yamazoe Y, Roth RW & Kadlubar FF (1986) *Carcinogenesis* **7:** 179.
Yoshimoto T, Magata K, Ehara H, Mizuno K & Yanamoto S (1986) *Biochim. Biophys. Acta* **877:** 141.

8

The Role of NAD(P)H : Quinone Reductase in Protection against the Toxicity of Quinones and Related Agents

HANS J. PROCHASKA*† and PAUL TALALAY*
* *Department of Pharmacology and Molecular Sciences, and*
† *Department of Medicine, The Johns Hopkins University School of Medicine, Baltimore, Maryland 21205, USA*

I. Introduction 195
II. How quinones mediate their toxic effects 197
III. Mechanisms of protection by quinone reductase 199
IV. Value of quinone reductase activity as a marker of a chemoprotected state 205
V. Summary 207

I. INTRODUCTION

Many vertebrate cells express a predominantly cytosolic FAD-containing flavoprotein that promotes the nicotinamide nucleotide-dependent reduction of a wide variety of quinones to hydroquinones. Notable characteristics of this quinone reductase (QR)* are that: (a) NADH and NADPH serve as equally efficient electron donors; (b) the reduction of quinones is mediated via an obligatory two-electron mechanism (i.e. no semiquinone intermediates are detectable); (c) the enzyme is potently inhibited by dicoumarol and other anticoagulants; and (d) QR is induced by an astonishing number of xenobiotics in many cells and tissues.

Oxidative Stress: Oxidants and Antioxidants
ISBN 0-12-642762-3

The molecular, catalytic, and crystallographic characteristics of QR isolated from rodent tissues have been the subject of recent reviews (Lind et al, 1990), including a symposium devoted entirely to this enzyme (Ernster et al, 1987).* Nevertheless, identification of the functions of QR has presented a number of problems. The original suggestions that this enzyme might play a role in electron transport or as a P450 reductase are no longer tenable (Martius, 1959; Ernster et al, 1960; Lind and Ernster, 1974). Because of its potent inhibition by dicoumarol and other anticoagulants (Märki and Martius, 1960; Ernster et al, 1960) and its ability to reduce vitamin K_1 (Fasco and Principe, 1982; Martius et al, 1975), QR has been implicated in the biosynthesis of γ-carboxyglutamate residues of blood coagulation proteins since the hydroquinone form of vitamin K is required to affix CO_2 to glutamate, and oxidation of hydroquinone to quinone occurs during the carboxylation reaction (Stenflo, 1978; Suttie, 1980). Several mechanisms for the cyclic reduction of vitamin K have been described, and although QR can clearly participate in this process in reconstructed systems (Wallin et al, 1978; Wallin and Suttie, 1981), its importance in these reactions in vivo remains uncertain (Wallin et al, 1987).

The possibility that QR plays a major role in protecting cells against the toxicities of quinones and related compounds has attracted increasing attention (Huggins et al, 1964; Benson et al, 1980; Lind et al, 1982). The observations that the toxicities of quinones are abolished by elevated levels of QR and are greatly enhanced by dicoumarol support this view. Furthermore, a wide variety of chemical agents that induce QR also protect cells against the toxicities of quinones and their metabolic precursors (review: Talalay et al, 1987; Talalay, 1989). In this chapter, we review the experimental evidence for the participation of QR in the protection of cells against the toxic effects of quinones and their progenitors (e.g. phenols, polycyclic aromatic hydrocarbons, and diethylstilboestrol) and possibly also certain metals. We also marshal the evidence that the induction of QR by test

---

* Abbreviations and trivial names: QR, quinone reductase, which is officially designated NAD(P)H : (quinone-acceptor) oxidoreductase (EC 1.6.99.2) and is also known as DT-diaphorase, menadione reductase, vitamin K reductase, NMOR [NAD(P)H : menadione oxidoreductase], and QAO (quinone acceptor oxidoreductase); BHA, 2(3)-*tert*-butyl-4-hydroxyanisole; SOD, superoxide dismutase (EC 1.15.1.1); DMBA, 7,12-dimethylbenz(a)anthracene; phase I enzymes, a superfamily of microsomal proteins that functionalize compounds by oxidation or reduction and are also referred to as mixed function oxidases, monooxygenases, or the cytochrome P450 enzymes; phase II enzymes, a series of enzyme families that perform conjugation reactions with endogenous ligands such as glutathione and glucuronic acid. Although QR does not carry out conjugation reactions, we consider it to be a phase II enzyme since it does not introduce functional groups and is induced coordinately with other phase II enzymes.

compounds is an indicator of anticarcinogenic activity and constitutes a useful method for identifying novel anticarcinogens of potential medicinal interest.

II. HOW QUINONES MEDIATE THEIR TOXIC EFFECTS

Quinones manifest toxicity via two pathways (Kappus and Sies, 1981; see Figure 1); the relative importance of these pathways is dependent upon the electrochemical and steric properties of the quinones (Gant et al, 1988; Smith et al, 1985; Thor et al, 1988). First, quinones are potent electrophiles (Michael reaction acceptors) subject to attack by nucleophiles such as the thiol group of reduced glutathione which is abundant in many cells. Since maintenance of reduced glutathione pools plays an important role in preventing the arylation of critical cellular macromolecules, depletion of reduced glutathione can result in damage of DNA. Second, a number of flavoproteins (e.g. cytochrome P450 reductase) catalyse nicotinamide nucleotide-linked one-electron reductions of quinones to semiquinones (Iyanagi and Yamazaki, 1970). Although semiquinones can react with nucleophiles, they also react spontaneously with molecular oxygen to regenerate the parent quinone and to form superoxide ($O_2^{\overline{\cdot}}$). Quinones thus become trapped as mediators of 'redox cycles', resulting in oxidative stress by expenditure of NADPH and molecular oxygen to form superoxide. Superoxide can in turn undergo dismutation to form H_2O_2. Furthermore, $O_2^{\overline{\cdot}}$ and H_2O_2 are subject to metal catalysed conversion to extremely reactive hydroxyl radicals ($HO^{\bullet}$) and singlet oxygen ($^1O_2$) (see Kappus and Sies, 1981). Therefore, the net effect of redox cycling is that: (a) reactive oxygen metabolites are produced (Thor et al, 1982; Wefers and Sies, 1983; Smith et al, 1984, 1985); (b) reduced thiol and nicotinamide nucleotide pools are depleted (Thor et al, 1982; Di Monte et al, 1984a); and (c) loss of noncritical protective thiols renders macromolecules susceptible to arylation reactions (Di Monte et al, 1984b; Thor et al, 1988; Mirabelli et al, 1988, 1989).

These deleterious effects of quinones are not restricted to simple benzoquinones and naphthoquinones. Quinone-containing antineoplastic agents (e.g. anthracyclines) mediate their therapeutic (i.e. cytotoxic) effects at least in part as 'site-specific' redox cyclers based on their high affinity for DNA (Bachur et al, 1978, 1979). Furthermore, quinone metabolites contribute significantly to the toxic and carcinogenic effects of aromatic hydrocarbons. For example, 6,12-, 1,6-, and 3,6-benzo(a)pyrenediones, which represent a substantial fraction of benzo(a)pyrene metabolites (Lind et al, 1978), are exceedingly toxic to Chinese hamster ovary cells when exposed to air

(Lorentzen et al, 1979). These quinones are efficient catalysts for redox cycling (Lorentzen and Ts'o, 1977). The oxidation products of phenol (presumably *ortho*- and *para*-benzoquinones rather than benzene oxide) are now thought to represent the toxic and leukaemogenic metabolites of benzene (Sawahata and Neal, 1983). Similarly, the toxicity of naphthol is the result of further metabolism to naphthoquinones (Thornalley et al, 1984; d'Arcy Doherty and Cohen, 1984; Fluck et al, 1984). Thus, quinones represent a substantial threat to cellular homeostasis since they are ubiquitous in the environment, are major metabolites of aromatic hydrocarbons, and also occur endogenously.

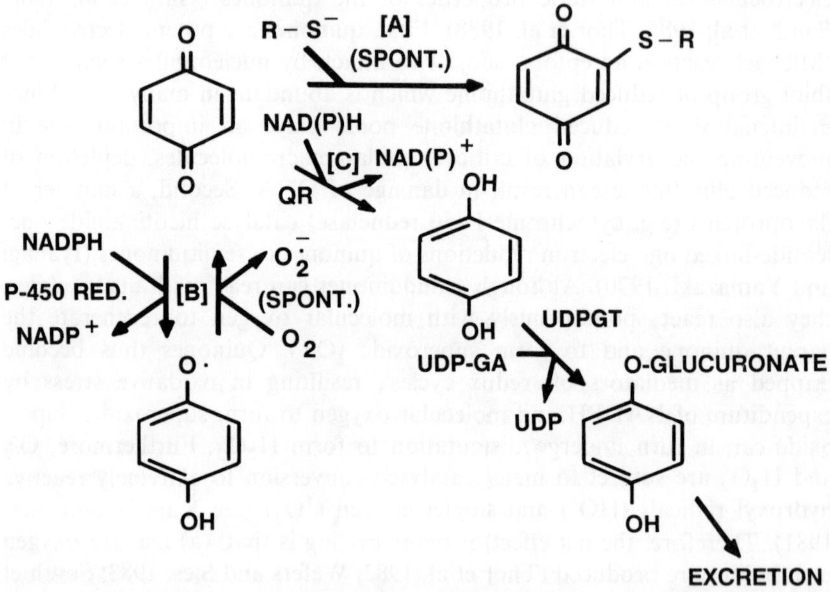

Figure 1. Mechanisms of quinone-mediated toxicity and the role of quinone reductase (QR) in quinone detoxification. Quinones manifest their toxicity in two possible ways, depending on their electrochemical properties. Reaction A: Quinones are potent electrophiles and can be attacked by nucleophiles such as glutathione (R—S$^-$) as well as RNA, DNA, and protein. Reaction B: Quinones also become trapped as catalysts in redox cycles whereby they undergo enzymatically catalysed one-electron reductions to semiquinones (e.g. by cytochrome P450 reductase); semiquinones in turn react spontaneously with molecular oxygen to form the parent quinone and generate superoxide. This results in depletion of reduced nicotinamide nucleotide and glutathione pools, increasing the susceptibility of cells to arylation reactions. Superoxide dismutase stimulates these one-electron redox cycles by depleting $O_2^{\bar{}}$, which favours the formation of quinones available for enzymatic reduction. Reaction C: QR can prevent the toxicity of quinones by catalysing the obligatory two-electron reduction of quinones to hydroquinones, which are generally stable and undergo metabolic inactivation by glucuronidation in the presence of UDP-glucuronic acid (UDP-GA) and UDP-glucuronosyl transferase (UDPGT).

III. MECHANISMS OF PROTECTION BY QUINONE REDUCTASE

A. Detoxification of quinones

QR, in contrast to many other flavoproteins, catalyses the unusual obligatory two-electron reduction of a wide variety of quinones to hydroquinones, which are generally relatively stable and subject to metabolic inactivation by glucuronidation. Major determinants of the stability of hydroquinones are the hydroquinone/semiquinone redox potential relative to that of $O_2/O_2^{\overline{\cdot}}$ and their state of protonation (Öllinger et al, 1990). Thus, protection by QR is thought to result from the detoxification of quinones before they react covalently with macromolecules (arylation) and/or become involved in redox cycles.

Many in vitro models have demonstrated that the deleterious effects of quinones and their metabolic precursors are blocked by QR. Furthermore, the protective effects that QR exerts in these systems are abolished by dicoumarol, a potent inhibitor of QR:

(a) Direct addition of QR greatly enhances the efficiency of glucuronidation of benzo(a)pyrenediones by microsomal fractions (Lind et al, 1978; Lind, 1985a, 1985b).
(b) Menadione-induced consumption of NADPH and O_2 by microsomes is inhibited by QR (Lind et al, 1982).
(c) Menadione-induced $^1O_2$ production (as measured by emission of low-level chemiluminescence) is greatly reduced in the postmitochondrial supernatant fractions obtained from mice fed diets containing 2(3)-*tert*-butyl-4-hydroxyanisole (BHA) (Wefers et al, 1984). BHA is a potent inhibitor of chemically induced neoplasms (Wattenberg, 1983) and an inducer of QR and other phase II enzymes (Benson et al, 1978, 1979, 1980). Although BHA dramatically alters the levels of a number of proteins, as shown by two-dimensional gel electrophoresis of mouse liver (Pearson et al, 1983), the reduction of chemiluminescence by BHA treatment can be reproduced quantitatively by addition of pure QR (Prochaska et al, 1987).
(d) Entrapment of pure QR in erythrocytes inhibits menadione-induced methaemoglobin production (Benatti et al, 1987).
(e) Binding of [^{14}C]phenol to microsomal proteins is potently inhibited by addition of partially purified QR (Smart and Zannoni, 1984).
(f) Redox cycling by the synthetic oestrogen, diethylstilboestrol, is inhibited by QR. Furthermore, the susceptibility of hamster kidney to the neoplastic effects of diethylstilboestrol has been attributed to partici-

pation of this compound in redox cycles, as well as to the specific depression (by 80%) of the levels of renal QR by chronic oestrogen administration (Roy and Liehr, 1988).

Cell culture and in vivo evidence for a protective role of QR generally rests upon experiments demonstrating the enhancement of the toxicity of quinones by dicoumarol. The addition of dicoumarol to menadione-treated hepatocytes results in the depletion of reduced thiol and nicotinamide nucleotide pools as well as disturbances of Ca^{2+} homeostasis and plasma membrane integrity (Thor et al, 1982; Di Monte et al, 1984a,b). The loss of plasma membrane integrity has been associated with the arylation of cytoskeletal proteins (Thor et al, 1988; Mirabelli et al, 1988, 1989). Similarly, dicoumarol increases the toxicity of daunorubicin and several simpler quinones in cardiocytes (Galaris and Rydström, 1983; Galaris et al, 1985) and murine embryo cells (Atallah et al, 1987, 1988). The toxicity of polycyclic aromatic hydrocarbons is decreased by the addition of semipurified preparations of QR to cultures of Chinese hamster ovary cells containing Aroclor-treated rat liver microsomes (Swanson et al, 1986). It is notable that differences in susceptibility to the toxicity of benzene in two mouse strains as well as in bone marrow stromal cell types have been correlated with the presence of higher QR specific activities in the resistant cells (Thomas et al, 1990; Twerdok and Trush, 1990).

Prior treatment of animals with dicoumarol increases the toxic and lethal effects of streptonigrin, 3-methylcholanthrene, and benzo(a)pyrene (Schor et al, 1978a,b; Atallah et al, 1987). Low-level chemiluminescence induced by menadione (reflecting $^1O_2$ production) in livers is increased when dicoumarol is added to the perfusion fluid (Wefers and Sies, 1983). The higher risk of bladder cancer in male in comparison to female rabbits has been related to higher levels of cytochrome P450 reductase and lower levels of QR in the male transitional bladder epithelium (Mohandas et al, 1986). Although much evidence suggests that inducers of QR protect against the toxic, mutagenic, and neoplastic effects of a wide variety of carcinogens (see below), the relative contribution of QR to this protection is difficult to estimate since a large number of other xenobiotic-metabolizing enzymes are also induced by such treatments.

B. Role of superoxide dismutase as a protective adjuvant

It has recently been suggested that superoxide dismutase (SOD) may be an important partner with QR in preventing quinone-mediated toxicity (see Figure 2; Cadenas et al, 1988; Segura-Aguilar and Lind, 1989). During the course of studying the QR-catalysed reduction of 2-epoxybenzoquinone to 2-hydroxyhydroquinone, NAD(P)H and O_2 were noted to be continuously consumed under aerobic conditions (Brunmark et al, 1987). After the added NAD(P)H had been depleted, 2-hydroxyhydroquinone was rapidly converted to 2-hydroxybenzoquinone. Brunmark et al (1987) proposed the following scheme (Figure 2) to explain these results. Reaction A: 2-hydroxyhydroquinone is unstable and is oxidized by O_2 to form 2-hydroxysemiquinone. Reaction B: 2-hydroxysemiquinone undergoes further oxidation by O_2 to form 2-hydroxybenzoquinone. Reaction C: 2-hydroxybenzoquinone is reduced by QR to regenerate 2-hydroxyhydroquinone. These reactions form a QR-driven redox cycle. Hence two superoxide molecules are produced for each equivalent of NAD(P)H consumed. Other quinones (including aminochrome [2,3-dihydroindole-5,6-quinone], which is postulated to represent the toxic metabolites of dopamine implicated in the pathogenesis of Parkinson's disease) have been demonstrated to participate in these QR-catalysed redox cycles (Brunmark et al, 1987; Segura-Aguilar and Lind, 1989; Öllinger et al, 1990).

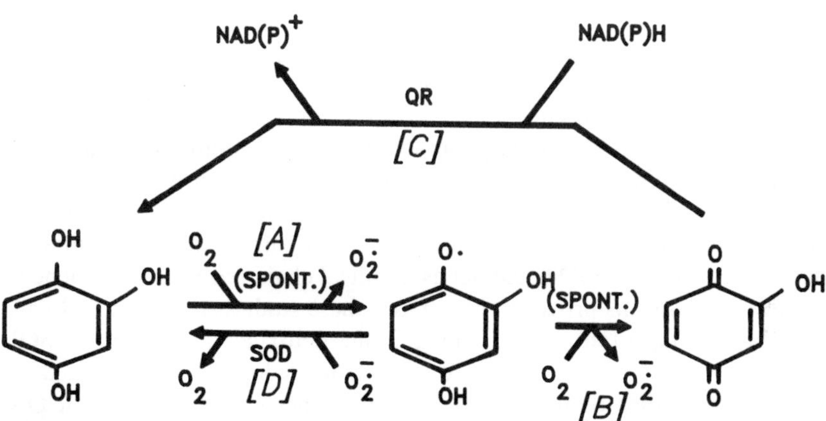

Figure 2. Quinone reductase-driven redox cycling and possible role of superoxide dismutase in quinone detoxification. Some quinones, particularly those with hydroxyl (as in the example above) and thiol substitutions (Öllinger et al, 1990) are susceptible to oxidation by molecular oxygen (reaction A). The resultant semiquinone can also react with O_2 to form superoxide and quinone (reaction B). These quinones are substrates for QR and are reduced to the cognate hydroquinone form (reaction C). Reactions A, B, and C therefore comprise a redox cycle wherein reactive oxygen intermediates are produced at the expense of NAD(P)H. SOD prevents QR-driven redox cycling possibly by acting as a semiquinone:superoxide oxidoreductase providing a catalytically mediated process for reversing reaction A (reaction D). Levels of the semiquinone form would depend on the redox potential of the hydroquinone/semiquinone couple rather than the kinetics of its production and consumption.

Interestingly, Cadenas et al (1988) found that in the presence of SOD, QR could greatly suppress the consumption of O_2 and NAD(P)H promoted by these quinones. Furthermore, addition of SOD to cytochrome P450 reductase (a one-electron quinone reductase) generally *enhanced* redox cycling (Öllinger et al, 1990). The latter observation has been ascribed to depletion of $O_2^{\overline{\cdot}}$ by SOD, which favours the formation of quinone available for enzymatic reduction (Figure 1; pathway B). Two mechanisms have been suggested to account for the SOD-mediated inhibition of QR-driven redox cycling (Cadenas et al, 1988; Öllinger et al, 1990). First, SOD can act as a superoxide:semiquinone oxidoreductase by catalysing the reduction of semiquinone by superoxide, resulting in the formation of hydroquinone and oxygen (Figure 2; reaction D). This provides an enzyme-mediated process for reversing reaction A (Figure 2) described above. The pools of hydroquinone and semiquinone are therefore set at equilibrium; the pool of semiquinone available for further oxidation to the quinone is dependent on the redox potential of the corresponding hydroquinones and semiquinones (in which the hydroquinone form is favoured). This possibility has been questioned by Öllinger et al (1990) based on thermodynamic arguments and the finding that Cu/Zn- and Mn-SODs are both effective inhibitors of hydroquinone autoxidation. Second, SOD may act as a radical chain terminator by consuming $O_2^{\overline{\cdot}}$. Öllinger et al (1990) have suggested that the oxidation of hydroquinones by molecular oxygen occurs slowly (reaction A, Figure 2). However, once produced, $O_2^{\overline{\cdot}}$ can oxidize hydroquinones efficiently to produce semiquinones and H_2O_2 (Figure 3; this reaction would in essence substitute for reaction A in Figure 2); semiquinones can then undergo further oxidation to generate more $O_2^{\overline{\cdot}}$. Thus, superoxide may initiate a chain reaction; and SOD could act as a radical chain terminator. Whichever of these mechanisms proves to be correct, these studies show that SOD and QR may be important partners in quinone detoxification since: (a) SOD generally enhances the rate of one-electron redox cycles mediated by flavoproteins other than QR; (b) some quinones can be trapped in QR-driven two-electron redox cycles; and (c) SOD inhibits QR-mediated redox cycling.

C. Other reactions promoted by quinone reductase

Are there other metabolic functions of QR? QR has been shown to function as an azoreductase. However, the significance of this observation is unclear since methyl red (2-[[4-(dimethylamino)phenyl]azo]benzoic acid) is the only substrate described that will undergo appreciable QR-mediated azo reduction (Huang et al, 1979). De Flora et al (1985) have demonstrated that the reduction of the mutagenicity of Cr(VI) compounds by NAD(P)H-fortified cytosols could be reversed by dicoumarol and other inhibitors of QR. Although Cr(VI) is not detectably reduced by QR, its mutagenicity in the Ames assay is lowered by pure QR and restored by dicoumarol (De Flora et al, 1988). Whether QR plays a role as a reductase for metals with high oxidation states is still unclear.

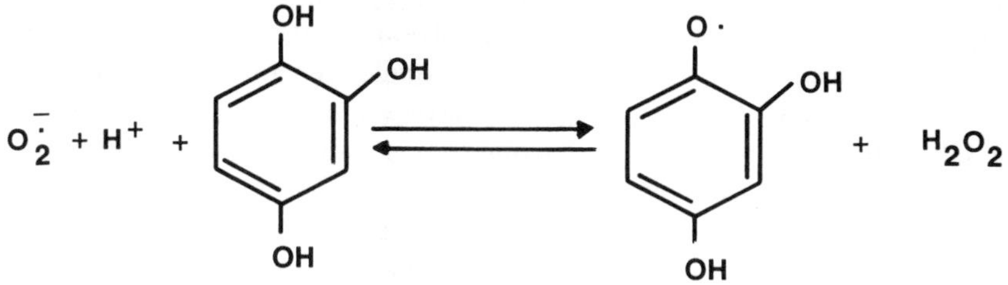

OH OH O· OH

$O_2^{\cdot-}$ + H^+ + ⇌ + H_2O_2

OH OH

Figure 3. Role of superoxide in hydroquinone oxidation. Öllinger et al (1990) have proposed that superoxide, rather than oxygen, is the primary oxidant of hydroquinone in QR-driven redox cycling and acts as a radical chain initiator. They propose that oxidation of hydroquinone by O_2 is slow (reaction A, Figure 2); however, the superoxide formed by this reaction is an efficient oxidant of hydroquinone (reaction shown above). The semiquinone product is easily oxidized by O_2, resulting in the formation of superoxide, which can further oxidize hydroquinone (reaction B, Figure 2). Thus, QR-driven redox cycling results in the net production of one mole of H_2O_2 rather than two moles of superoxide for each mole of NAD(P)H consumed by QR. SOD prevents QR-mediated redox cycling by removing $O_2^{\cdot-}$ before it can oxidize hydroquinone.

There is an intriguing report that the toxicity of glutamate, an excitatory neurotransmitter in neuroblastoma cells, is enhanced by dicoumarol and reduced by prior treatment of these cells with inducers of QR (Murphy et al, 1991). Although the mechanism of the protective role of QR in this system is unclear, it may be related to relief of glutamate-induced oxidative stress. Lastly, we should note that QR possesses nitroreductase activity responsible for the activation of the antineoplastic agent, 5-(aziridin-1-yl)-2,4-dinitrobenzamide, to a cytotoxic nitroso derivative in Walker rat tumour cells (Knox et al, 1988a,b). QR has also been shown to reduce 4-nitroquinoline-*N*-oxide to its hydroxylamino derivative; the mutagenicity of 4-nitroquinoline-*N*-oxide in the Ames assay is lowered by the addition of pure QR and NAD(P)H (De Flora et al, 1988). Thus, although the range of nitroaromatics shown to be substrates is limited, QR may play an important role in the conversion of these compounds to either more or less cytotoxic species.

D. Significance of elevated quinone reductase activity in preneoplastic nodules and tumours

The observation that a number of rodent tumour cells have increased levels of QR and other phase II enzymes (reviewed by Schor, 1987) is of interest because these cells tend to be more resistant to the toxic effects of xenobiotics. Similarly, persistent hepatic nodules, which are presumed to be precancerous lesions, also exhibit increased phase II enzymatic activities while displaying lowered phase I enzyme levels (Roomi et al, 1985; Farber, 1984). This 'resistance phenotype' is observed in nodules generated by a variety of different protocols. Pickett et al (1984) have demonstrated that elevations in QR activity in persistent (preneoplastic) hepatic nodules were due to increased levels of the cognate mRNA for QR. The QR levels of these nodules were still susceptible to induction by planar aromatics. DNA coding for QR in these nodules has subsequently been shown to be hypomethylated, perhaps explaining the persistently elevated QR activities expressed by these nodules (Williams et al, 1986).

As stated above, tumour lines with elevated phase II enzyme levels appear to have enhanced resistance to the toxic effects of xenobiotics. This has been particularly well documented with persistent hepatic nodules (review: Farber, 1987). These nodules have a decreased capacity to activate carcinogens (due primarily to decreased phase I enzyme levels), and display increased resistance to compounds such as CCl_4 and dimethylnitrosamines, and decreased covalent binding of carcinogens to macromolecules. Although it is impossible to ascribe these pleiotropic effects to the activity of QR alone, it does appear that the elevation of QR and other phase II enzymes endows neoplastic tissues with a significant survival advantage. These alterations in phase II enzyme patterns may have profound effects on the ability of chemotherapeutic agents to destroy malignant cells. Such infomation may be a valuable guide for the design of new cytotoxic agents. It is therefore intriguing that the levels of QR are significantly elevated (3.5- to 17-fold) in common human cancers (breast, lung, and colon) over normal tissue (Schlager and Powis, 1990), and raises the possibility that cytotoxic agents activated by QR could be useful chemotherapeutic agents in these cancers (see Knox et al, 1988a,b).

E. Conclusions

Much evidence indicates that QR protects against the toxicities of quinones, their metabolic precursors, and perhaps even certain high-valency metals. Nevertheless, the *relative* contribution of QR to the overall defensive posture of the cell is still not well understood. Agents that protect against carcinogen-induced neoplasms induce QR as well as a battery of other xenobiotic-metabolizing enzymes, including glutathione *S*-transferases, UDP-glucuronosyltransferases, and also enzymes involved in maintenance of reduced glutathione and reduced nicotinamide nucleotide pools (review: Talalay et al, 1987). The enhancement by dicoumarol of quinone-mediated toxicity may not result from the inhibition of QR alone, since dicoumarol has other effects such as the uncoupling of oxidative phosphorylation and the inhibition of cytochrome P450 (Ernster et al, 1960). Furthermore, the role of QR in humans is more difficult to elucidate since its basal activity is much lower (approximately 2–20%) in human than in rodent tissues (Wallin et al, 1987; Martin et al, 1987; Schlager and Powis, 1990), and carbonyl reductase (EC 1.1.1.184) may play a more prominent role in the reduction of quinones in man (Wermuth et al, 1986). However, Schlager and Powis (1990) report that carbonyl reductase represents only 7–23% of the total QR activity in normal human tissues and point out that the livers obtained by Wermuth et al (1986) were removed 6–20 h after death.

Although the importance of QR in protection against xenobiotic-induced toxicity is debatable, studies over the past 30 years have established that elevation of QR by a compound is a clear marker for a chemoprotected state. Moreover, QR activity is elevated in primates in response to administration of the anticarcinogenic enzyme inducer BHA (Wallin et al, 1987). Consequently, we review below the evidence showing that the elevation of QR by chemical agents is a sensitive and reliable method for the identification of compounds that protect against carcinogens and other toxic agents.

IV. VALUE OF QUINONE REDUCTASE ACTIVITY AS A MARKER OF A CHEMOPROTECTED STATE

In 1961, Williams-Ashman and Huggins (1961) reported that QR is induced in a number of tissues by low doses of polycyclic aromatic hydrocarbons under conditions that protect rats against the toxic and lethal effects of 7,12-dimethylbenz(a)anthracene (DMBA) (Huggins and Fukunishi, 1964; Huggins et al, 1964). These workers perceived that other, perhaps nontoxic, classes of protective compounds might be identified by utilizing the induction of QR as a marker for a 'chemoprotected' state. Huggins and Pataki (1965) described a number of nontoxic azo dyes that also protected rats from DMBA-induced toxicity. Furthermore, the rank order of protection was closely correlated with the potency with which these compounds induced hepatic QR. In rats, the most potent of these compounds, Sudan III [1-(4-phenylazophenylazo)-2-naphthol], has been shown to inhibit polycyclic aromatic hydrocarbon-induced mammary cancer (Huggins and Pataki, 1965), leukaemia (Huggins et al, 1978), and chromosomal aberrations in bone marrow (Ito et al, 1982, 1984).

In the 1970s, Wattenberg and colleagues demonstrated that a number of phenols, aromatic isothiocyanates, and coumarins were effective inhibitors against chemically induced neoplasms (Wattenberg, 1978, 1983, 1985; Wattenberg et al, 1986). Protection was observed against a large number of carcinogens and in a variety of target tissues. The mechanism for protection was unclear until is was shown that the above compounds are inducers of phase II enzymes (QR, glutathione transferases, and UDP-glucuronosyltransferases) as well as enzymes involved in the maintenance of reduced glutathione and nicotinamide nucleotide pools (Benson et al, 1978, 1979, 1980; Cha and Bueding, 1979). Subsequently other anticarcinogens, including Sudan III, have been shown to produce these pleiotropic enzyme inductions (review: Talalay et al, 1987).

The close structural correlation between the ability of xenobiotics to induce phase II enzymes and to act as chemoprotective agents suggested to us the possibility that elevation of the activities of these enzymes could provide a means for the identification of novel anticarcinogens. Indeed, 1,2-dithiole-3-thiones were discovered to be anticarcinogens based on their ability to induce phase II enzymes (Ansher et al, 1983, 1986; Wattenberg and Bueding, 1986). Anticarcinogenic diterpenes were isolated from green coffee beans by utilizing the induction of glutathione transferase in mice as a marker for anticarcinogenic activity (Lam et al, 1982; Wattenberg and Lam, 1984). Unfortunately, the systematic screening of compounds for chemoprotective activity in whole animals is prohibitively costly and time-consuming.

In order to simplify the identification of compounds that possess inductive (and by inference anticarcinogenic) activity, we developed a cell culture system. QR is elevated in Hepa 1c1c7 murine hepatoma cells by most compounds that induce QR in vivo (Prochaska et al, 1985; De Long et al, 1986, 1987). This cell culture assay was further simplified by measuring QR directly in hepatocytes grown and induced in 96-well microtitre plates (Prochaska and Santamaria, 1988). Furthermore, by interfacing the optical microtitre plate scanner with a computer, data processing has been greatly simplified, and the assessment of inducer potency can be accomplished in days rather than weeks or months. This system has been used to identify the electrophilic chemical signal responsible for inductive activity, to elucidate the mechanism of QR induction, and to identify the new inducers of potential medicinal interest (Prochaska and Talalay, 1988; Talalay et al,

1988; Talalay, 1989; Spencer et al, 1990). Moreover, the microtitre system appears to be well suited as an assay for monitoring the isolation of naturally occurring anticarcinogens from complex mixtures such as foods.

Epidemiological studies have demonstrated that the consumption of certain vegetables, particularly crucifers, protects against a wide variety of epithelial tumours (Graham et al, 1978; Graham, 1983; Kune et al, 1987; Le Marchand et al, 1989; You et al, 1989). Since these vegetables induce phase II enzymes in vivo (Sparnins et al, 1982; Wattenberg, 1983; Pantuck et al, 1984; Whitty and Bjeldanes, 1987), protect rodents from toxins and carcinogens (Boyd et al, 1982; Wattenberg, 1983; Wattenberg et al, 1986) and contain constituents closely related to recognized anticarcinogens (Wattenberg, 1983; Wattenberg et al, 1986; Talalay et al, 1988; Spencer et al, 1990), we became interested in the possibility that the microtitre assay could be employed for the systematic identification of protective vegetables and their anticarcinogenic components. We (unpublished data) have recently tested organic solvent extracts of a large number of organically grown vegetables for their ability to induce QR in Hepa 1c1c7 murine hepatoma cells. Although there were differences within families, we found that members of the Cruciferae (e.g. broccoli, cabbage, brussels sprouts) and Liliaceae (e.g. onions, garlic) families were the most effective vegetables tested. Members of the Solanaceae family (e.g. potatoes, tomatoes) were remarkable for their low inductive activity. Thus, the microtitre assay identified those vegetables that have been demonstrated to: (a) induce phase II enzymes in vivo; (b) protect rodents from carcinogens; and (c) be associated with reduction of cancer risk in humans.

V. SUMMARY

Although evidence has accumulated that QR can protect against the toxicity of a number of xenobiotics, its overall importance in the detoxification process is unclear. This is especially true in humans. The elevation of QR by anticarcinogenic enzyme inducers is only one of many biochemical alterations evoked by chemoprotective substances, and QR may therefore serve as a marker for the chemoprotected state. Nevertheless, the finding that QR levels are reliably increased by anticarcinogens and are so easily measured in immortalized cell lines may be of considerable interest and importance. Measurement of the induction of this enzyme provides the opportunity for systematic identification of compounds with potential anticarcinogenic activity, and for elucidating the relationship between diet and cancer. It is our hope that the utilization of QR expression as a method to identify novel anticarcinogens will lead to practical methods for reducing the risk of cancer in man.

ACKNOWLEDGEMENTS

Studies from the laboratories of the authors were supported by a Grant from the National Cancer Institute, Department of Health and Human Services (1 PO1-CA-44530), and by Grants from the American Cancer Society (SIG-3 and RP-30). We thank Sharon R. Spencer for searching the quinone reductase literature. We are also grateful to a number of authors for supplying manuscripts in advance of publication.

REFERENCES

Ansher SS, Dolan P & Bueding E (1983) *Hepatology* **3:** 932–935.
Ansher SS, Dolan P & Bueding E (1986) *Food Chem. Toxicol.* **24:** 405–415.
Atallah AS, Landolph JR & Hochstein P (1987) *Chem. Scripta* **27A:** 141–144.
Atallah AS, Landolph JR, Ernster L & Hochstein P (1988) *Biochem. Pharmacol.* **37:** 2451–2459.
Bachur NR, Gordon S & Gee MV (1978) *Cancer Res.* **38:** 1745–1750.
Bachur NR, Gordon S, Gee MV & Kon H (1979) *Proc. Natl. Acad. Sci. USA* **76:** 954–957.
Benatti U, Guida L, De Flora A & Hochstein P (1987) *Chem. Scripta* **27A:** 169–171.
Benson AM, Batzinger RP, Ou S-YL, Bueding E, Cha Y-N & Talalay P (1978) *Cancer Res.* **38:** 4486–4495.
Benson AM, Cha Y-N, Bueding E, Heine HS & Talalay P (1979) *Cancer Res.* **39:** 2971–2977.
Benson AM, Hunkeler MJ & Talalay P (1980) *Proc. Natl. Acad. Sci. USA* **77:** 5216–5220.
Boyd JN, Babish JG & Stoewsand GS (1982) *Food Chem. Toxicol.* **20:** 47–52.
Brunmark A, Cadenas E, Lind C, Segura-Aguilar J & Ernster L (1987) *Free Radical Biol. Med.* **3:** 169–180.
Cadenas E, Mira D, Brunmark A, Lind C, Segura-Aguilar J & Ernster L (1988) *Free Radical Biol. Med.* **5:** 71–79.
Cha Y-N & Bueding E (1979) *Biochem. Pharmacol.* **28:** 1917–1921.
d'Arcy Doherty M & Cohen GM (1984) *Biochem. Pharmacol.* **33:** 3201–3208.
De Flora S, Morelli A, Basso C, Ramano M, Serra D & De Flora A (1985) *Cancer Res.* **45:** 3188–3196.
De Flora S, Bennicelli C, Camoirano A, Serra D & Hochstein P (1988) *Carcinogenesis* **9:** 611–617.
De Long MJ, Prochaska HJ & Talalay P (1986) *Proc. Natl. Acad. Sci. USA* **83:** 787–791.
De Long MJ, Santamaria AB & Talalay P (1987) *Carcinogenesis* **8:** 1549–1553.

Di Monte D, Ross D, Bellomo G, Eklow L & Orrenius S (1984a) *Arch. Biochem. Biophys.* **235:** 334–342.
Di Monte D, Bellomo G, Thor H, Nicotera P & Orrenius S (1984b) *Arch. Biochem. Biophys.* **235:** 343–350.
Ernster L, Ljunggren M & Danielson L (1960) *Biochem. Biophys. Res. Commun.* **2:** 88–92.
Ernster L, Eastabrook RW, Hochstein P & Orrenius S (eds) (1987) *Chem. Scripta* **27A:** 1–207.
Farber E (1984) *Can. J. Biochem. Cell Biol.* **62:** 486–494.
Farber E (1987) *Chem. Scripta* **27A:** 131–133.
Fasco MJ & Principe LM (1982) *Biochem. Biophys. Res. Commun.* **104:** 187–192.
Fluck DS, Rappaport SM, Eastmond DA & Smith MT (1984) *Arch. Biochem. Biophys.* **235:** 351–358.
Galaris D & Rydström J (1983) *Biochem. Biophys. Res. Commun.* **110:** 364–370.
Galaris D, Georgellis A & Rydström J (1985) *Biochem. Pharmacol.* **34:** 989–995.
Gant TW, Ramakrishna Rao DN, Mason RP & Cohen GM (1988) *Chem. Biol. Interact.* **65:** 157–173.
Graham S (1983) *Cancer Res.* **43:** 2409s–2413s.
Graham S, Dayal H, Swanson M, Mittelman A & Wilkinson G (1978) *J. Natl. Cancer Inst.* **61:** 709–714.
Huang M-T, Miura GT & Lu AYH (1979) *J. Biol. Chem.* **254:** 3930–3934.
Huggins C & Fukunishi R (1964) *J. Exp. Med.* **119:** 923–942.
Huggins C & Pataki J (1965) *Proc. Natl. Acad. Sci. USA* **53:** 791–796.
Huggins C, Ford E, Fukunishi R & Jensen EV (1964) *J. Exp. Med.* **119:** 943–954.
Huggins CB, Ueda N & Russo A (1978) *Proc. Natl. Acad. Sci. USA* **75:** 4524–4527.
Ito Y, Maeda S, Fujihara T, Ueda N & Sugiyama T (1982) *J. Natl. Cancer Inst.* **69:** 1343–1346.
Ito Y, Maeda S, Souno K, Ueda N & Sugiyama T (1984) *J. Natl. Cancer Inst.* **73:** 177–183.
Iyanagi T & Yamazaki I (1970) *Biochim. Biophys. Acta* **216:** 282–294.
Kappus H & Sies H (1981) *Experientia* **37:** 1233–1241.
Knox RJ, Boland MP, Friedlos F, Coles B, Southan C & Roberts JJ (1988a) *Biochem. Pharmacol.* **37:** 4671–4677.
Knox RJ, Friedlos F, Jarman M & Roberts JJ (1988b) *Biochem. Pharmacol.* **37:** 4661–4669.
Kune S, Kune GA & Watson LF (1987) *Nutr. Cancer* **9:** 21–42.
Lam LKT, Sparnins VL & Wattenberg LW (1982) *Cancer Res.* **42:** 1193–1198.
Le Marchand L, Yoshizawa CN, Kolonel LN, Hankin JH & Goodman MT (1989) *J. Natl. Cancer Inst.* **81:** 1158–1164.
Lind C (1985a) *Arch. Biochem. Biophys.* **240:** 226–235.
Lind C (1985b) *Biochem. Pharmacol.* **34:** 895–897.
Lind C & Ernster L (1974) *Biochem. Biophys. Res. Commun.* **56:** 392–400.
Lind C, Vadi H & Ernster L (1978) *Arch. Biochem. Biophys.* **190:** 97–108.
Lind C, Hochstein P & Ernster L (1982) *Arch. Biochem. Biophys.* **216:** 178–185.
Lind C, Cadenas E, Hochstein P & Ernster L (1990) *Methods Enzymol.* **186:** 287–301.
Lorentzen RJ & Ts'o POP (1977) *Biochemistry* **16:** 1467–1473.
Lorentzen RJ, Lesko SA, McDonald K & Ts'o POP (1979) *Cancer Res.* **39:** 3194–3198.
Märki F & Martius C (1960) *Biochem. Z.* **333:** 111–135.
Martin LF, Patrick SD & Wallin R (1987) *Cancer Lett.* **36:** 341–347.

Martius C (1959) In Wolstenholme GEW & O'Connor CM (eds) *Ciba Foundation Symposium on Cell Metabolism*, pp 194–200. London: Churchill.
Martius C, Ganser R & Viviani A (1975) *FEBS Lett.* **59:** 13–14.
Mirabelli F, Salis A, Marinoni V et al (1988) *Arch. Biochem. Biophys.* **264:** 261–269.
Mirabelli F, Salis A, Vairetti M, Bellomo G, Thor H & Orrenius S (1989) *Arch. Biochem. Biophys.* **270:** 478–488.
Mohandas J, Chennel AF, Duggin GG, Horwath JS & Tiller DJ (1986) *Carcinogenesis* **7:** 353–356.
Murphy TH, De Long MJ & Coyle JT (1991) *J. Neurochem.* **56:** 990–995.
Öllinger K, Buffinton G, Ernster L & Cadenas E (1990) *Chem. Biol. Interact.* **73:** 53–76.
Pantuck EJ, Pantuck CB, Anderson KE, Wattenberg LW, Conney AH & Kappas A (1984) *Clin. Pharmacol. Ther.* **35:** 161–169.
Pearson WR, Windle JJ, Morrow JF, Benson AM & Talalay P (1983) *J. Biol. Chem.* **258:** 2052–2062.
Pickett CB, Williams JB, Lu AYH & Cameron RG (1984) *Proc. Natl. Acad. Sci. USA* **81:** 5091–5095.
Prochaska HJ & Santamaria AB (1988) *Anal. Biochem.* **169:** 328–336.
Prochaska HJ & Talalay P (1988) *Cancer Res.* **48:** 4776–4782.
Prochaska HJ, De Long MJ & Talalay P (1985) *Proc. Natl. Acad. Sci. USA* **82:** 8232–8236.
Prochaska HJ, Talalay P & Sies H (1987) *J. Biol. Chem.* **262:** 1931–1934.
Roomi MW, Ho RK, Sarma DSR & Farber E (1985) *Cancer Res.* **45:** 564–571.
Roy D & Liehr JG (1988) *J. Biol. Chem.* **263:** 3646–3651.
Salbe AD & Bjeldanes LF (1986) *Food Chem. Toxicol.* **24:** 851–856.
Sawahata T & Neal RA (1983) *Mol. Pharmacol.* **23:** 453–460.
Schlager JJ & Powis G (1990) *Int. J. Cancer* **45:** 403–409.
Schor NA (1987) *Chem. Scripta* **27A:** 135–139.
Schor NA, Boh E & Burke VT (1978a) *Enzyme* **23:** 217–224.
Schor NA, Ogawa K, Lee G & Farber E (1978b) *Cancer Lett.* **5:** 167–171.
Segura-Aguilar J & Lind C (1989) *Chem. Biol. Interact.* **72:** 309–324.
Smart RC & Zannoni VG (1984) *Mol. Pharmacol.* **26:** 105–111.
Smith MT, Thor H & Orrenius S (1984) *Methods Enzymol.* **105:** 505–510.
Smith MT, Evans CG, Thor H & Orrenius S (1985) In Sies H (ed.) *Oxidative Stress*, pp 91–113. London: Academic Press.
Sparnins VL, Vanegas PL & Wattenberg LW (1982) *J. Natl. Cancer Inst.* **68:** 493–496.
Spencer SR, Wilczak CA & Talalay P (1990) *Cancer Res.* **50:** 7871–7875.
Stenflo J (1978) *Adv. Enzymol.* **46:** 1–31.
Suttie JW (1980) *CRC Crit. Rev. Biochem.* **8:** 191–223.
Swanson MS, Haugen DA, Reilly Jr CA & Stamoudis VC (1986) *Toxicol. Appl. Pharmacol.* **84:** 336–345.
Talalay P (1989) *Adv. Enzyme Regul.* **28:** 237–250.
Talalay P, De Long MJ & Prochaska HJ (1987) In Cory JG & Szentivanyi A (eds) *Proceedings of the International Symposium in Cancer Biology and Therapeutics*, pp 197–216. New York: Plenum Press.
Talalay P, De Long MJ & Prochaska HJ (1988) *Proc. Natl. Acad. Sci. USA* **85:** 8261–8265.
Thomas DJ, Sadler A, Subrahmanyam VV et al (1990) *Mol. Pharmacol.* **37:** 255–262.

Thor H, Smith MT, Hartzell P, Bellomo G, Jewell SA & Orrenius S (1982) *J. Biol. Chem.* **257:** 12419–12425.
Thor H, Mirabelli F, Salis, A, Cohen GM, Bellomo G & Orrenius S (1988) *Arch. Biochem. Biophys.* **266:** 397–407.
Thornalley PJ, d'Arcy Doherty M, Smith MT, Bannister JV & Cohen GM (1984) *Chem. Biol. Interact.* **48:** 195–206.
Twerdok LE & Trush MA (1990) *Res. Commun. Chem. Pathol. Pharmacol.* **67:** 375–386.
Wallin R & Suttie JW (1981) *Biochem. J.* **194:** 983–988.
Wallin R, Gebhardt O & Prydz H (1978) *Biochem. J.* **169:** 95–101.
Wallin R, Rannels SR & Martin LF (1987) *Chem. Scripta* **27A:** 193–202.
Wattenberg LW (1978) *Adv. Cancer Res.* **26:** 197–226.
Wattenberg LW (1983) *Cancer Res.* **43:** 2448s–2453s.
Wattenberg LW (1985) *Cancer Res.* **45:** 1–8.
Wattenberg LW & Bueding E (1986) *Carcinogenesis* **7:** 1379–1381.
Wattenberg LW & Lam LKT (1984) In MacMahon B & Sugimura T (eds) *Coffee and Health*, Banbury Report No. 17, pp 137–145. New York: Cold Spring Harbor Laboratory, CSH.
Wattenberg LW, Hanley AB, Barany G, Sparnins VL, Lam LKT & Fenwick GR (1986) In Hayashi Y et al (eds) *Diet, Nutrition and Cancer*, pp 193–203. Utrecht: Japan Sci. Soc. Press, Tokyo/VNU Sci. Press.
Wefers H & Sies H (1983) *Arch. Biochem. Biophys.* **224:** 568–578.
Wefers H, Komai T, Talalay P & Sies H (1984) *FEBS Lett.* **169:** 63–66.
Wermuth B, Platts KL, Siedel A & Oesch F (1986) *Biochem. Pharmacol.* **35:** 1277–1282.
Whitty JP & Bjeldanes LF (1987) *Food Chem. Toxicol.* **25:** 581–587.
Williams JB, Lu AYH, Cameron RG & Pickett CB (1986) *J. Biol. Chem.* **261:** 5524–5528.
Williams-Ashman HG & Huggins C (1961) *Med. Exp.* **4:** 223–226.
You W-C, Blot WJ, Chang Y-S et al (1989) *J. Natl. Cancer Inst.* **81:** 162–164.

9

Endogenous Antioxidant Defences in Human Blood Plasma

ROLAND STOCKER AND BALZ FREI*

*The Heart Research Institute, 145 Missenden Road, Camperdown, Sydney, NSW 2050, Australia and *Department of Nutrition, Harvard School of Public Health, 665 Huntington Avenue, Boston, MA 02115, USA.*

I. Introduction 213
II. Antioxidant defences present in human blood plasma 214
III. Antioxidant defences provided by blood cells 234
IV. Conclusions 237

I. INTRODUCTION

The concept that free radicals play a significant role in the pathogenesis of certain human diseases such as atherosclerosis, cancer, rheumatoid arthritis, drug-associated toxicity, postischaemic reoxygenation injury, and parasitic and viral infections, is becoming increasingly recognized (Sies, 1985; Halliwell and Gutteridge, 1985, 1986; Clark et al, 1986, Peterhans et al, 1988). As a result of this there is a growing interest in mechanisms of antioxidant protection as well as in the development of adequate indices of both oxidative stress and the damage that may result from such stress. Blood plasma functions as transport machinery and as a regulator of osmotic pressure. It is exposed to oxygen and a great variety of compounds derived from cellular metabolism and dietary intake, such as bile pigments, a wide range of lipids and vitamins, and certain drugs. In addition, it is also exposed to oxidants produced mainly from activated phagocytes. Through scaveng-

Oxidative Stress: Oxidants and Antioxidants
ISBN 0-12-642762-3

ing of these oxidants plasma provides protection to the endothelial cell-lined vasculature and lipid-carrying lipoproteins. Oxidative damage to low-density lipoprotein and endothelial cells are events postulated to be of prime importance in the development of fatty streaks, the earliest lesion in atherogenesis (Steinberg et al, 1989). Such reasons, together with its ready availability from human subjects, make blood plasma a suitable model of extracellular fluids and ideal biological material to monitor in vivo free radical reactions.

In this chapter we extend and update previous reports on antioxidant defences present in normal human blood plasma (Halliwell and Gutteridge, 1985, 1986).

II. ANTIOXIDANT DEFENCES PRESENT IN HUMAN BLOOD PLASMA

A. Enzymatic antioxidants

While the intracellular enzymatic antioxidant defence is provided by superoxide dismutase (SOD), catalase, glutathione peroxidase (GSH-Px), glutathione reductase, phospholipid hydroperoxide glutathione peroxidase and glutathione *S*-transferase, these enzymes are either absent or present only at much smaller concentrations in human blood plasma (Table 1). Marklund (1987) has characterized a Cu- and Zn-containing extracellular SOD (EC-SOD) that is distinct from the cellular form. EC-SOD is a glycoprotein secreted mainly from endothelial cells and is heterogeneous with regard to its affinity to heparin-Sepharose. Although the actual amount of EC-SOD in plasma is very small, its circulating concentration increases markedly upon injection of heparin, as a result of release of the high heparin affinity subtype of EC-SOD from endothelial cell surfaces (Karlsson and Marklund, 1987, 1988). Binding of EC-SOD to the cellular surface of endothelial cells but not blood cells (Karlsson and Marklund, 1989) seems to be mediated through heparin sulphates (Adachi and Marklund, 1989), thereby providing these former cells with a protective coat of EC-SOD.

Human blood plasma also contains peroxidase activity (Takahashi and Cohen, 1986; Maddipati et al, 1987; Frei et al, 1988a; Terao et al, 1988). A selenium-dependent GSH-Px, antigenically different from that present in human red cells, has recently been purified and characterized by several independent groups (Takahashi et al, 1987; Broderick et al, 1987; Maddipati and Marnett, 1987) although its biological role remains unclear. The enzyme, like EC-SOD, is glycosylated (Takahashi et al, 1987) but seems to be secreted from hepatic rather than endothelial cells (Avissar et al, 1989). Plasma GSH-Px has an apparent K_m value of between 4.3 mM (Maddipati and Marnett, 1987) and 5.3 mM (Takahashi et al, 1987) for reduced glutathione (GSH), i.e. several orders of magnitude higher than the actual concentrations of GSH present in freshly obtained human plasma (Wendel and Cikryt, 1980). In order to observe GSH-dependent peroxidase activity, the plasma has to be supplemented with exogenous GSH (Maddipati et al, 1987). This is in contrast to reports showing the presence in freshly obtained, unsupplemented human blood plasma of peroxidase-like activity for phosphatidylcholine, fatty acid and cholesterol hydroperoxides, but not triglyceride and cholesterylester hydroperoxides (Frei et al, 1988a), and arachidonate but neither phosphatidylcholine nor phosphatidylethanolamine hydroperoxides (Terao et al, 1988; Terao, 1989a). Further work is required to investigate whether these conflicting reports result from more than one peroxidase activity present in human plasma, instability of the enzyme(s) in isolated plasma, different methods used to quantitate lipid hydroperoxides, or still other reasons.

Table 1. Typical levels of enzymatic antioxidants in human plasma and red blood cells.

	Typical concentration	
Antioxidant enzyme	Plasma ($U\ ml^{-1}$)	Red blood cells ($U\ 10^{-10}$ cells)[a]
Superoxide dismutase	5–20[b]	550–800
Catalase	?	3800–5400
(GSH) peroxidase	0.4[c]	7.8–10.6
GSSG reductase	0.03[c]	2.7–3.7
GSH transferase	0.005[c]	1.5–2.5

[a] Values are calculated from data presented by Lentner (1984) and Beutler (1984), using a haematocrit value of 50% (v/v) and an average of 5×10^9 red cell or 150 mg of haemoglobin per ml of whole blood (Lentner, 1984). As 10^{10} erythrocytes correspond to about 1 ml of packed cells, the values presented in the two columns can be compared directly.

[b] Concentration as determined and given by Karlsson and Marklund (1987).

[c] Concentration as defined and listed by Lentner (1984).

Although it is generally believed that plasma from healthy individuals is poorly endowed with catalase (Halliwell and Gutteridge, 1985, 1986), there are some reports in the literature describing the presence of catalase in plasma or serum (Góth, 1987; Link and Riley, 1988; Frei et al, 1988b). Plasma catalase may protect ascorbic acid from oxidation by H_2O_2, thereby sparing this important antioxidant (Frei et al, 1988b). Whether catalase found in plasma is released from erythrocytes by lysis during blood collection or preparation of plasma, or reflects haemolysis in vivo, or is distinct from the erythrocyte enzyme (Góth, 1987), remains to be investigated.

B. Non-enzymatic proteinaceous antioxidants

1. *Proteins that bind metals or biological iron complexes*

In the presence of transition metals (mainly iron and copper) loosely bound or complexed to biological macromolecules, primary oxygen reduction products can produce secondary and more reactive oxidants, particularly hydroxyl radicals ($^{\bullet}OH$) (Halliwell and Gutteridge, 1985). To prevent such reactions from taking place, human plasma contains various proteins that tightly bind metals and biological iron complexes (Table 2).

Table 2. Plasma proteins that bind metals or biological iron complexes to prevent them from participating in free radical reactions.

Ceruloplasmin ($0.18–0.4\ g\ l^{-1}$)	Ferroxidase activity: $CP\text{-}[Cu(II)]_4 + 4\ Fe(II) \rightarrow CP\text{-}[Cu(I)]_4 + 4Fe(III)$ $CP\text{-}[Cu(I)]_4 + O_2 + 4H^+ \rightarrow CP\text{-}[Cu(II)]_4 + 2H_2O$ Inhibits iron- and copper-dependent lipid peroxidation Reincorporates iron mobilized from the iron-storage protein ferritin back into ferritin Does *not* react directly with peroxyl radicals
Transferrin ($1.8–3.3\ g\ l^{-1}$)	Binds iron and stops or slows its participation in lipid peroxidation and iron-catalysed Haber–Weiss reactions Can take up iron released from ferritin, thereby inhibiting ferritin-dependent lipid peroxidation
Lactoferrin ($0.0002\ g\ l^{-1}$)	Action similar to that of transferrin
Albumin ($50\ g\ l^{-1}$)	Binds copper tightly and iron weakly. The function of this property of albumin is *not* clear Albumin-bound copper is still redox active and therefore may participate in Fenton reactions, giving rise to $^{\bullet}OH$
Haptoglobin ($0.5–3.6\ g\ l^{-1}$)	Binds free haemoglobin (Hb) and met-Hb, thereby preventing these proteins from catalysing lipid peroxidation Transferrin and haemopexin *do not* inhibit Hb-mediated lipid peroxidation
Haemopexin ($0.6–1.0\ g\ l^{-1}$)	High affinity for haem ($K_d < 10^{-13}\ mol\ l^{-1}$). Inhibits haem-mediated lipid peroxidation Transferrin and haptoglobin *do not* inhibit haem-catalysed lipid peroxidation

Table modified from Halliwell and Gutteridge (1986). The values in parentheses indicate the concentrations of the proteins in human plasma (Lentner, 1984). See text for further references.

Ceruloplasmin, the copper-transporting acute-phase protein accounting for practically all the plasma copper, has some ability to scavenge superoxide radicals ($O_2^{\cdot-}$) (Goldstein et al, 1979) while it fails to scavenge peroxyl radicals (Wayner et al, 1987; Samokyszyn et al, 1989). However, of greater biological importance is ceruloplasmin's ferroxidase activity, i.e. its ability to catalyse the oxidation of ferrous to ferric ions (Gutteridge et al, 1980; Gutteridge, 1983; Samokyszyn et al, 1989), thereby inhibiting both iron-stimulated lipid peroxidation and the Fenton reaction. In addition to these ferroxidase-related antioxidant activities, ceruloplasmin also inhibits copper-stimulated lipid peroxidation (Gutteridge et al, 1980; Gutteridge, 1983), probably by its ability to bind copper.

Most of the iron in human plasma is bound to transferrin. There are good experimental data supporting the view that, at least under physiological conditions with only about 20% of the protein's binding capacity occupied, transferrin-bound iron is not available for catalysis of lipid peroxidation (Winterbourn and Sutton, 1984; Aruoma and Halliwell, 1987). A potential source of catalytic iron in human blood plasma, however, is ferritin (Thomas et al, 1985; Thomas and Aust, 1985). This ubiquitous iron-storage protein, usually considered to be an intracelluar protein, can be detected in apparently healthy non-anaemic adults (Jacobs et al, 1975). Mobilization of iron from ferritin for the promotion of $^{\cdot}OH$ production and lipid peroxidation requires a reductant of small size and appropriate reduction potential. One of the potential physiological reductants is $O_2^{\cdot-}$ (Samokyszyn et al, 1989), the primary oxygen reduction product released into the circulation by activated phagocytes. Indeed, isolated human polymorphonuclear leukocytes (PMN) can mobilize iron from human ferritin in vitro (Biemond et al, 1984). However, whether such iron released from ferritin will be available for catalytic activity in biological tissues such as human blood must be questioned. Aust and coworkers recently showed that $O_2^{\cdot-}$ and ferritin-dependent lipid peroxidation is inhibited completely by ceruloplasmin, which reductively reincorporates mobilized iron back into ferritin (Samokyszyn et al, 1989). Furthermore, apotransferrin and apolactoferrin have also been shown to take up iron released from ferritin, thereby greatly inhibiting $O_2^{\cdot-}$ and ferritin-dependent lipid peroxidation (Monteiro and Winterbourn, 1988). Lactoferrin, another iron-binding protein, thought to derive from the secondary granules of PMNs (Baggiolini et al, 1970), is present in small amounts in plasma of healthy subjects (Rosenmund et al, 1988). Like the situation with transferrin, lactoferrin-bound iron is a very poor catalyst of $^{\cdot}OH$ production and lipid peroxidation (Winterbourn, 1983; Baldwin et al, 1984).

These observations suggest that ceruloplasmin functions as a primary protective agent against metal-catalysed oxidation in human plasma (see also Stocks et al, 1974). Partially iron-saturated transferrin and lactoferrin provide additional protection to inhibit ferritin-dependent reactions (Monteiro and Winterbourn, 1988). Like free copper and iron ions, free haemoglobin and haem can potentially participate in radical reactions either directly or through release of iron (Tappel, 1955; Sadrzadeh et al, 1984; Halliwell and Gutteridge, 1986). Although occurring to only a minor extent in health, these biological iron complexes are released into the circulation by intravascular haemolysis accompanying haemorrhages and a number of disease states involving haemoglobinopathies or mechanical stress to erythrocytes. Conditions of oxidative stress may lead to the formation of methaemoglobin, in which ferrihaem and globin dissociate relatively readily (Bunn and Jandl, 1968) followed by release of free haem into the circulation. In the extracellular space, haemoglobin and haem are bound immediately by haptoglobin and haemopexin, respectively. The complexes formed are removed from the circulation by the liver and taken up selectively by hepatic parenchymal cells where haem is degraded to bilirubin. Recent studies from several independent laboratories have shown that both haptoglobin (Sadrzadeh et al, 1984; Gutteridge, 1987) and haemopexin (Vincent et al, 1988; Gutteridge and Smith, 1988) are effective inhibitors of lipid peroxidation catalysed by haemoglobin and haem, respectively. Haemopexin has an exceptionally high affinity for haem ($K_d < 10^{-13}$ mol l^{-1}) (Hrkal et al, 1974) that is several orders of magnitude higher than the haem-binding affinity of human albumin (Morgan et al, 1976). Nether haptoglobin nor transferrin can prevent haem-catalysed lipid peroxidation, while haemopexin and transferrin fail to inhibit haemoglobin-mediated lipid peroxidation (Gutteridge and Smith, 1988). Thus, just as iron-binding proteins inhibit iron-mediated lipid peroxidation, haptoglobin and haemopexin specifically inhibit haemoglobin- and haem-dependent lipid peroxidation, respectively, providing these acute-phase proteins with a unique role in antioxidant defence mechanisms within the circulation and extravascular space.

2. *Protein sulphydryl groups*

Human albumin, the most prominent of all plasma proteins, has specific binding sites for a vast array of ligands, including fatty acids, copper (Laussac and Sarkar, 1984) and tetrapyrroles (Peters, 1985). Its possible function as a 'sacrificial' antioxidant in plasma and the extravascular space has been reviewed recently (Halliwell, 1988). Though albumin does not normally inhibit iron-mediated radical reactions, it provides additional protection against copper- (Gutteridge, 1983; Halliwell, 1988) and haem-catalysed (Vincent et al, 1988; Gutteridge and Smith, 1988) oxidations, in addition to ceruloplasmin and haemopexin, respectively. The highly stable structure of human albumin is provided by 17 disulphide bridges. The single remaining cysteine residue is thought to contribute significantly to the plasma sulphydryls that react with peroxyl radicals (Wayner et al, 1985, 1987). Perhaps more importantly, however, albumin appears to be largely responsible for the protective activity of human serum against the myeloperoxidase-derived oxidant hypchlorous acid (HOCl) (Wasil et al, 1987; Halliwell, 1988).

Whether oxidation of the plasma proteins' thiols should be considered oxidative damage or antioxidant protection is not clear for at least two reasons. Firstly, oxidation of critical thiol groups of plasma proteins other than albumin, e.g. enzymes or antibodies, might lead to partial or total loss of the biological activity of these proteins. Secondly, the thiyl radical that can be formed from protein thiols upon scavenging of a free radical by hydrogen transfer is itself a potential source of reactive oxidants. Thiyl free radicals can abstract bisallylic hydrogens from polyunsaturated fatty acids and thereby initiate lipid peroxidation (Schönreich et al, 1989). Vitamin A (retinol) and ascorbic acid can inhibit lipid peroxidation mediated by thiyl free radicals, because they react about 100 and 10 times faster, respectively, with thiyl radicals than these do with polyunsaturated fatty acids (D'Aquino et al, 1989). DeLange and Glazer (1989) have shown that the protective effects of albumin and other proteins against free radical damage decrease proportionally at higher concentrations, and have suggested that this is because at higher concentrations protein free radicals not only decay through intramolecular reactions but also react with, i.e. oxidatively damage, other molecules.

In summary, the antioxidant defences provided by plasma proteins are related to their ability to prevent metal ion-catalysed free radical reactions from taking place (Halliwell and Gutteridge, 1985, 1986; Gutteridge, 1987; Gutteridge and Smith, 1988). In contrast to intracellular enzymatic antioxidant defence mechanisms that are aimed mainly at the general removal of reactive oxygen species, EC-SOD seems to act more locally at the surface of endothelial cells. An understanding of the biological role of plasma peroxidase activity requires further investigations.

C. Non-enzymatic small molecular antioxidants

Despite the apparent lack of enzymatic defences against reactive oxygen species, human blood plasma possesses highly efficient small molecular antioxidants (Table 3) that protect important target molecules against possible oxidative damage. Oxidative stress in the forms of $O_2^{\cdot-}$, H_2O_2, HOCl, $^{\cdot}OH$, and possibly the ferryl ion (FeO_2^{+}) or perferryl ion ($Fe^{2+}{}^{-}O_2$), can exist within the circulation when blood phagocytes become activated by

Table 3. Non-enzymatic small molecular antioxidants in human plasma.

Antioxidant	Plasma concentration (μM)
Water-soluble	
Glucose	4500
Pyruvate	30–70
Uric acid	160–450
Ascorbic acid	30–150
Bilirubin	5–20
GSH	<2
Lipid-soluble	
α-Tocopherol	15–40
Ubiquinol-10	0.4–1.0
Lycopene	0.5–1.0
β-Carotene	0.3–0.6
Lutein	0.1–0.3
Zeaxanthin	0.1–0.2
α-Carotene	0.05–0.1

certain immunological stimuli (Hamers and Roos, 1985). In addition, peroxyl and thiyl radicals may be formed as a result of hydrogen abstraction from appropriate biological molecules, followed, in the case of the former, by reaction of the carbon-centred radical with oxygen (Pryor, 1976; Schönreich et al, 1989).

1. *Water-soluble antioxidants*

Ascorbic acid, or vitamin C, has been known for a long time to be essential for protection of humans against scurvy. The ascorbic activity of vitamin C lies in its role as an essential cofactor in hydroxylation reactions involved in the biosynthesis of stable crosslinked collagen (Jaffe, 1984). This and other metabolic functions of ascorbic acid are derived from its strong reducing potential. The same property makes the vitamin an excellent antioxidant, capable of scavenging a wide variety of different oxidants. For example, ascorbic acid has been shown to scavenge $O_2^{\bar{\cdot}}$, H_2O_2, $^{\bullet}OH$, HOCl, aqueous peroxyl radicals, and singlet oxygen effectively (Nishikimi, 1975; Bodannes and Chan, 1979; Cabelli and Bielski, 1983; Bendich et al, 1986; Halliwell et al, 1987; Rose, 1987; Kwon and Foote, 1988; Frei et al, 1989a). During its antioxidant action, ascorbic acid undergoes a two-electron oxidation to dehydroascorbic acid with intermediate formation of the relatively unreactive ascorbyl radical (Bielski and Richter, 1975). While dehydroascorbic acid is relatively unstable and hydrolyses readily to diketogulonic acid, it can be reduced back to ascorbic acid in erythrocytes and other blood cells (Section III). Both ascorbic acid and dehydroascorbic acid are biologically active.

Although ascorbic acid is present in the aqueous phase of extracellular fluids and within cells, it can protect the lipids of lipoproteins and biomembranes against peroxidative damage by directly intercepting oxidants generated in the aqueous phase before they can attack lipids. However, while ascorbic acid is able to interact synergistically with membrane-bound α-tocopherol (Section II.C.5) it does not efficiently scavenge lipid radicals within membranes. As discussed below (Section II.C.4), ascorbic acid plays a pivotal role in protecting plasma lipids against peroxidative damage by many different types of oxidants. In particular, ascorbic acid deactivates the extracellular oxidants generated by activated neutrophils (Halliwell et al, 1987; Frei et al, 1988a; Thomas et al, 1988), while protecting the neutrophil membrane and leaving intracellular generation of antimicrobial oxidants intact (Anderson and Lukey, 1987).

Uric acid efficiently scavenges radicals (Matsushita et al, 1963; Ames et al, 1981), and it was proposed in 1981 by Ames that the increase in lifespan during human evolution may be partly explained by the protective action provided by uric acid in human plasma (Ames et al, 1981). Plasma levels of uric acid have increased in human evolution during the descent from prosimians over the past 60 million years, coinciding with a large increase in both lifespan and the loss of the ability to synthesize vitamin C. In addition to directly scavenging oxygen radicals, uric acid at physiological concentrations has been shown to stablilize ascorbic acid in human serum (Sevanian et al, 1985), possibly through formation of stable coordination complexes with iron ions (Davies et al, 1986).

Ames' work on uric acid led to the general hypothesis that end-products of degradative metabolic pathways may play important roles as protective agents, and that, in this context, bilirubin was a likely candidate for such a role (Stocker et al, 1987b). There is some existing and emerging evidence in the literature in support of such a general hypothesis (Wright et al, 1986; Cadenas et al, 1989; DeLange and Glazer, 1989; Christen et al, 1990). However, while the metabolites under investigation in these papers may have a protective function in certain tissues or areas of ongoing inflammation, they are less likely to function in human blood plasma.

In contrast, human albumin-bound bilirubin appears to contribute significantly to the non-enzymatic antioxidant defences in human plasma as judged by in vitro studies (Stocker et al, 1987b; Frei et al, 1988a; Lindeman et al, 1989). Bilirubin, normally regarded merely as a useless catabolic waste product, was first suggested to serve as a natural antioxidant in 1954 (Bernhard et al, 1954). Albumin-bound bilirubin efficiently protects albumin-bound linoleic acid, present at a 100-fold molar excess compared to the pigment, from peroxyl radical-induced oxidation in vitro (Stocker et al, 1987a). Unlike the situation with the pigment present in homogeneous solution, biliverdin is formed almost stoichiometrically as the oxidation product during this reaction (Stocker et al, 1987a,b). Protective activity of albumin-bound bilirubin is observed with mammalian but not avian species of albumin (Stocker et al, 1989). In this context it is interesting to note that in birds, biliverdin appears to be the end-product of haem catabolism (Colleran and O'Carra, 1977). In vitro antioxidant activity of albumin-bound bilirubin has also been reported in haemoprotein-catalysed oxidation of linoleic acid (Kaufmann and Garloff, 1961), photooxidation of albumin (Pedersen et al, 1977), iron-dependent peroxidation of rat liver microsomes (Brass et al, 1989), and protection of human myocytes against xanthine/xanthine oxidase-mediated damage (Wu et al, 1989).

Albumin-bound bilirubin, like uric acid, does not react rapidly with HOCl (Stocker et al, 1989), in contrast to conjugated bilirubin (Stocker and Peterhans, 1989a). The latter, also an efficient reducing substance and a scavenger of peroxyl radicals (Stocker and Ames, 1987), is present only at submicromolar concentrations in plasma of healthy adults and is hence of minor importance. Free and conjugated bilirubin are also efficient scavengers of singlet oxygen ($^1\Delta_gO_2$) (Stevens and Small, 1974; Di Mascio et al, 1989), the non-radical form of oxygen that can be produced in vivo by exposure to light if endogenous or exogenous photosensitizers are present in skin or subcutaneous tissue.

In addition to ascorbic acid, uric acid and bilirubin, normal human plasma contains glucose (4–5 mM) and pyruvate (30–80 μM), metabolites that can react with $^{\bullet}OH$ radicals (Sagone et al, 1983) and H_2O_2, respectively (O'Donnell-Tormey et al, 1985). While these antioxidant reactivities of glucose and pyruvate may be of importance under special circumstances, e.g. cell cultures (O'Donnell-Tormey et al, 1987), their physiological roles remain to be tested. Glucose does not react efficiently with radicals that are less reactive than $^{\bullet}OH$, as its additional inclusion at 5.6 mM in human plasma changed neither the peroxyl radical-induced disappearance of ascorbic acid and other small molecular antioxidants, nor the appearance of lipid hydroperoxides (B. Frei, unpublished). Perhaps more importantly, glucose serves as an energy source for the production of reducing equivalents within blood cells (Section III).

In summary, human plasma is endowed with a number of different water-soluble compounds that rapidly react with various types of oxidants. While these antioxidant properties are most likely of physiological relevance in the case of ascorbic acid (Section II.C.4), additional studies are required to substantiate such function for albumin-bound bilirubin, uric acid, pyruvate, and glucose.

2. *Lipid-soluble antioxidants*

In recent years the general role of α-tocopherol (Burton and Ingold, 1981) and β-carotene (Burton and Ingold, 1984) as lipid-soluble, chain-breaking antioxidants present in human plasma has become well recognized. The experimental work on radical-scavenging properties of these antioxidants has been pioneered by Ingold's group. Using controlled experimental parameters, based on a chain of lipid oxidation initiated at a uniform and reproducible rate by the thermal decomposition of a lipid-soluble azo compound, RN=NR (where $R = (CH_3)_2C(CN)$, reactions 1–5), they initially measured the rate of reaction of tocopherols and lipid peroxyl radicals. For tocopherols, this rate decreased in the order $\alpha > \beta > \gamma > \delta$, in analogy to the biological potencies of these forms of vitamin E (Burton and Ingold, 1981). They subsequently demonstrated that vitamin E is the major (possibly sole) lipid-soluble, chain-breaking antioxidant in plasma (Burton et al, 1982) and red cells (Burton et al, 1983) of normal humans, and even in plasma of individuals suffering severe vitamin E deficiency (Ingold et al, 1987). These findings indirectly support a function of α-tocopherol as the most important physiological lipid-soluble antioxidant in human blood plasma and cells, and indeed are widely quoted in this context. Notwithstanding this, it is important to realize that the experimental design used by Ingold's group is not without potential pitfalls. Thus, their work is carried out under the highly non-physiological oxygen pressure of 760 torr or 100% O_2, and only considers quantitative aspects (see Section II.C.4) and lipid-soluble antioxidants that partition into *n*-octane and chloroform in two-phasic, aqueous alcohol-containing extracts of plasma and red cells, respectively. For example, such extraction procedures are not suitable for the quantitation of membrane-bound bilirubin (R. Stocker, unpublished). The bile pigment is a very potent lipid-soluble antioxidant (Stocker et al, 1987b), that conceivably associates with biological membranes (Leonard et al, 1989).

$$RN{=}NR \longrightarrow N_2 + 2R^\bullet \quad (1)$$

$$R^\bullet + O_2 \longrightarrow ROO^\bullet \quad (2)$$

$$ROO^\bullet + LH \longrightarrow ROOH + L^\bullet \quad (3)$$

$$L^\bullet + O_2 \longrightarrow LOO^\bullet \quad (4)$$

$$LOO^\bullet + LH \longrightarrow LOOH + L^\bullet \quad (5)$$

The peroxyl radical-trapping activity of β-carotene and possibly other carotenoids is dependent on the partial pressure of oxygen applied (Burton and Ingold, 1984). It is less efficient under conditions of air, but becomes a good peroxyl radical trap at the low pO_2 (Burton and Ingold, 1984; Stocker et al, 1987b) that prevails in biological tissues. At the very low pO_2 of 4 torr, β-carotene inhibited adriamycin-enhanced microsomal lipid peroxidation even more efficiently than α-tocopherol, while the antioxidant protection observed with retinol did not show a dependency on oxygen concentration

(Vile and Winterbourn, 1988). Like carotenoids, membrane-bound bilirubin scavenges peroxyl radicals also more efficiently under low pO_2 where it can become as efficient as α-tocopherol (Stocker et al, 1987b). These observations are of potential physiological significance as the oxygen pressure in human arterial and venous blood is around 68–93 and 40–61 torr, respectively (Gibbs et al, 1942; Lentner, 1984), and decreases further within tissues.

Ubiquinol-10 (reduced coenzyme Q_{10}) is also an effective, yet underappreciated, lipid-soluble, chain-breaking antioxidant (Mellors and Tappel, 1966; Frei et al, 1990; Stocker et al, 1991) present in human blood plasma (Table 3). Its antioxidant properties have been reviewed (Beyer et al, 1987; Beyer and Ernster, 1990). Ubiquinol-10 scavenges lipid peroxyl radicals with slightly higher efficiency than α-tocopherol (Mellors and Tappel, 1966; Frei et al, 1990; Stocker et al, 1991), and appears to be able to regenerate membrane-bound α-tocopherol from the α-tocopherol radical (Maguire et al, 1989; Frei et al, 1990b). Coenzyme Q_{10} is distributed among all lipoproteins (Elmberger et al, 1989). In plasma from healthy people about 60–80% of it is in the reduced, i.e. antioxidant active, form (Okamoto et al, 1988; Yamamoto and Niki, 1989; Frei et al, 1990), while only 35% is reduced in patients suffering from adult respiratory distress syndrome (Cross et al, 1990; B. Frei, unpublished), a condition proposed to be associated with oxidative stress (Repine, 1985).

3. *Antioxidant versus prooxidant activity*

By the very nature of their biological function, highly efficient antioxidants are expected to easily undergo autoxidation, and this is especially true for ascorbic acid, the first line of non-enzymatic antioxidant defence in plasma (Section II.C.4). Indeed, ascorbic acid is commonly employed as a prooxidant in peroxidative reactions involving transition metals (e.g. Haase and Dunkley, 1969), a mechanism that is thought to be mediated through the reduction of the metal ions by the vitamin. Transition metals, particularly copper, greatly enhance autoxidation of ascorbic acid which is accompanied by the production of $O_2^{\cdot-}$ and H_2O_2. Hydrogen peroxide can oxidize the reduced metal ion via the Fenton reaction, giving rise to the highly reactive $^{\cdot}OH$ radical. If the transition metal is bound to a biological target molecule a 'site-specific' Fenton reaction can take place (Samuni et al, 1983). Ascorbic acid has also been suggested to reductively decompose *tert*-butylhydroperoxide and possibly lipid hydroperoxides to their corresponding alkoxyl radicals (Baysal et al, 1989) which then can initiate lipid peroxidation. However, the peroxide-induced oxidation was inhibited by the iron chelator desferrioxamine, indicating that the prooxidant activity observed was dependent on the presence of contaminating transition metals. A similar observation was made with ascorbic acid-enhanced lipid peroxidation in photosensitized red cell membranes (Bachowski et al, 1988). Other antioxidants with strong reducing activities, like α-tocopherol (Terao and Matsushita, 1986; Fukuzawa et al, 1988; Yamamoto and Niki, 1988), bilirubin (Girotti, 1978), and GSH (Liebler et al, 1986), have also been reported to act as prooxidants in in vitro systems dependent on transition metals.

Results reported from several independent groups working with different in vitro models of lipid peroxidation indicate that the balance between pro- and antioxidant activity of ascorbic acid may ultimately depend on the α-tocopherol status within the lipid phase (Liebler et al, 1986; Terao and Matsushita, 1986; Wefers and Sies, 1988). While these results indirectly support an interaction between the two vitamins (see Section II.C.5), they contrast with observations made by others (Doba et al, 1985; Niki et al, 1985; Stocker and Peterhans, 1989b). Using purified lipids and test systems essentially free of transition metals, these latter studies showed that ascorbic acid, even at low micromolar concentrations and in the absence of α-tocopherol, had no prooxidative activity whether oxidation was initiated in the lipid or the aqueous phase.

In summary, available evidence suggests that, in the absence of lipid hydroperoxides and transition metals, physiological amounts of ascorbic acid and other known antioxidants do *not* exert prooxidant activity in vitro. Since in healthy human blood plasma the levels of lipid hydroperoxides do not exceed 1–5 nM (Frei et al, 1988b; Yamamoto and Niki, 1989) and there are no detectable free transition metals (Halliwell and Gutteridge, 1985; Evans et al, 1989), it follows that in this fluid antioxidants, particularly ascorbic acid, function as antioxidants only. This conclusion is corroborated by the observation that vitamin C added to freshly isolated human plasma up to millimolar concentrations prolonged the induction period of radical-induced lipid peroxidation in a dose-dependent manner showing no prooxidant activities (Frei et al, 1989a).

4. *Relative importance of individual antioxidants*

Numerous in vitro studies have revealed that the relative importance of individual antioxidants in plasma depends on the nature of the oxidant and the type of target molecule to be protected against oxidative damage. In addition, one has to distinguish between the quantitative importance of an antioxidant (which depends on its concentration in plasma and the number of oxidant molecules scavenged by each molecule of this antioxidant, i.e. its n value), and the qualitative importance of an antioxidant (which depends on both its concentration in plasma and its reaction rate with the oxidant relative to the reaction rate of the oxidant with the target molecule).

Table 4. Contributing antioxidant parameter (CAP) of individual plasma antioxidants to TRAP of human plasma.

Component	*n* Value	CAP[a]	CAP[b]
Uric acid	1.3	58 ± 18	32.5
Protein thiols	0.33	21 ± 10	15.6
Ascorbic acid	1.7/0.8[c]	14 ± 8	6.4
α-Tocopherol	2.0	7 ± 2	7.1
Bilirubin	3.67[d]	nc[e]	7.3
Lipids	1.0[f]	nc	30.3

[a] CAP values, given in percentages of total antioxidant activity, are as reported by Wayner et al (1987), or [b] as determined from the results given in Figure 1 of Frei et al (1988a).

[c] While an *n* value of 1.7 has been used by Wayner et al (1987) for ascorbic acid in diluted plasma samples, we used undiluted samples where the corresponding *n* value is approximately 0.8 (Wayner et al, 1986, 1987).

[d] The value of 3.67 takes into account that about 37% of albumin-bound bilirubin with an *n* value of 1.9 is oxidized to biliverdin, which itself has an *n* value of 4.7 (Stocker et al, 1987a; Stocker and Ames, 1987).

[e] Not considered.

[f] We assume that during the phase of inhibited lipid peroxidation each molecule of oxidized lipid has reacted with one peroxyl radical.

The most comprehensive studies on the relative importance of antioxidants in plasma have used aqueous radicals generated from water-soluble azo compounds as oxidants (see reactions 1 and 2), and lipid peroxidation as the measure of oxidative damage. The group of Ingold and Burton measured the peroxyl radical-trapping capability of human blood plasma and introduced the concept of the TRAP value (Total Radical-trapping Antioxidant Parameter) of plasma as a possible indicator of a person's antioxidant status (Wayner et al, 1985, 1987). TRAP in plasma of healthy individuals is about 820 μM (Wayner et al, 1987), i.e. during the phase of inhibited lipid peroxidation (the so-called 'induction period') 820 μmol of aqueous peroxyl radicals are scavenged per litre of plasma. TRAP correlates with the sum of the individual radical-trapping capabilities of the four plasma antioxidants uric acid, protein thiols, ascorbic acid, and α-tocopherol. The individual radical-trapping capability of an antioxidant in plasma is calculated by multiplying its concentration in plasma (see Table 3) by its *n*-value (Table 4). Thus: TRAP $\propto$ 1.3[uric acid] + 0.33[protein thiols] + 1.7 (or 0.8) [ascorbic acid] + 2.0[α-tocopherol].

The quantitative contributions of the four major plasma antioxidants to TRAP in plasma from healthy individuals are given in Table 4 as determined by Wayner et al (1987) and also measured by Thurnham et al (1987). However, we (Frei et al, 1988a) and others (Lindeman et al, 1989) observed substantial discrepancies between the measured TRAP and the sum of the individual radical-trapping capabilities of the above four antioxidants. We suggested that albumin-bound bilirubin and lipids themselves contribute significantly to the radical 'scavenging' during the phase of inhibited lipid peroxidation (Table 4), i.e. there is considerable peroxidative damage to the plasma lipids during the 'induction period'. TRAP is significantly lowered in patients suffering from rheumatoid arthritis (Thurnham et al, 1987), while it is higher in newborns than in healthy adults, mostly because of the higher plasma levels of ascorbic acid and bilirubin in newborns (Lindeman et al, 1989).

While the TRAP assay gives quantitative information about the relative importance of the major plasma antioxidants in scavenging aqueous peroxyl radicals, no qualitative information can be derived from it. Wayner et al (1987) noted that the sequence of plasma antioxidant consumption is protein thiols > uric acid > α-tocopherol. This indicates that these antioxidants are differentially reactive with aqueous peroxyl radicals, and that the antioxidants in plasma form several lines of antioxidant defence. Stocker et al, (1987a) observed that the antioxidants react with aqueous peroxyl radicals in the order ascorbic acid > albumin-bound bilirubin > uric acid. These studies were extended and it was shown that in plasma the sequence of antioxidant consumption is ascorbic acid = protein thiols > bilirubin > uric acid > α-tocopherol (Frei et al, 1988b, 1989a,b) (Figure 1). Most importantly, these latter studies showed that ascorbic acid is the only endogenous antioxidant in plasma that can fully protect all lipid classes against detectable peroxidative damage. Bilirubin, in a site-specific manner, appears to protect albumin-bound fatty acids against detectable peroxidation completely (Stocker et al, 1987a; Frei et al, 1988a; Section II.C.6). All the other major plasma antioxidants merely lower the rate of peroxyl radical-mediated lipid peroxidation. In whole blood, the sequence of plasma antioxidant consumption is the same as in plasma (Niki et al, 1988), although the complete protection of lipids against detectable peroxidation lasts about twice as long in whole blood as in plasma (B. Frei and B.N. Ames, unpublished), possibly due to regeneration of some plasma ascorbic acid from dehydroascorbic acid by blood cells.

In summary, experiments using aqueous peroxyl radicals as oxidants and lipid peroxidation as a measure of oxidative damage revealed that uric acid quantitatively contributes most to the antioxidant defences in human blood plasma, while qualitatively ascorbic acid appears to be the single most important antioxidant.

When a lipid-soluble, rather than a water-soluble, radical initiator is used as oxidant in plasma, lipid-soluble α-tocopherol belongs to the first line of antioxidant defence, together with ascorbic acid (Frei et al, 1989b). As soon

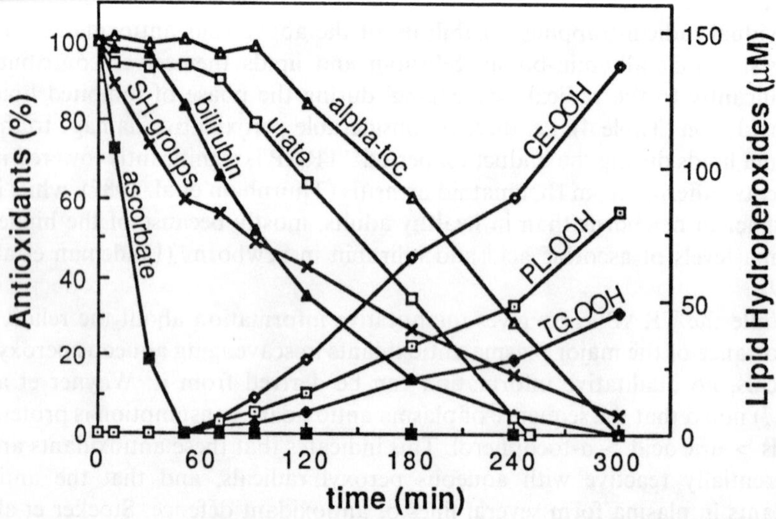

Figure 1. Antioxidant defences and lipid peroxidation in human plasma exposed to the water-soluble radical initiator AAPH. Plasma was incubated at 37°C in the presence of 50 mM AAPH. The levels of the antioxidants ascorbate (initial concentration 72 μM), sulphydryl groups (SH-groups, 425 μM), bilirubin (18 μM), urate (225 μM), and α-tocopherol (alpha-toc, 32 μM) are given as percentages of the initial concentrations. The levels of the lipid hydroperoxides triglyceride hydroperoxides (TG-OOH), cholesterol ester hydroperoxides (CE-OOH), and phospholipid hydroperoxides (PL-OOH) are given in micromolar concentrations (right ordinate). From Frei et al (1988a).

as ascorbic acid has been depleted completely, albumin-bound bilirubin becomes oxidized, indicating that ascorbic acid and albumin-bound bilirubin in the aqueous phase interact with α-tocopherol present in the lipoproteins (see Section II.C.5). Uric acid and protein thiols are not consumed in plasma exposed to lipid-soluble peroxyl radicals, thereby indirectly supporting the view that radicals are produced solely in the lipid phase of plasma. Lipid peroxidation concurs with consumption of α-tocopherol (Frei et al, 1989b). Obviously this vitamin is not as effective in protecting plasma lipids against lipid-soluble peroxyl radicals as is ascorbic acid against aqueous peroxyl radicals. This is not surprising since α-tocopherol prevents *propagation* of lipid peroxidation (once it has been initiated) by scavenging the chain-carrying lipid peroxyl radical ($LOO^\bullet$) in reaction 6, whereas ascorbic acid prevents *initiation* of lipid peroxidation by scavenging peroxyl radicals in the aqueous phase of plasma before they can diffuse into plasma lipids.

$$LOO^\bullet + \alpha\text{-tocopherol} \longrightarrow LOOH + \alpha\text{-tocopheroxyl radical} \qquad (6)$$

Thus far, the relative importance among the lipid-soluble antioxidants of plasma has not been investigated in whole plasma. From Table 3 it is evident that quantitatively α-tocopherol contributes most to the lipid-soluble antioxidant capacity of plasma (Burton et al, 1982) because it is present in concentrations 15 times or more higher than any other lipid-soluble antioxidant.

Studies by Esterbauer and associates (1987, 1989a,b) provided insights into the qualitative importance of lipid-soluble antioxidants in isolated low-density lipoproteins (LDL). Using Cu^{2+} as oxidant they showed that the antioxidants are progessively consumed in the sequence α-tocopherol > γ-tocopherol > phytofluene* = lycopene > β carotene. Lipid peroxidation occurs continuously at a slow rate during the consumption of antioxidants in LDL. As mentioned above, this is expected because scavenging of lipid peroxyl radicals by vitamin E and other lipid-soluble, chain-breaking antioxidants leads to formation of lipid hydroperoxides (reaction 6). When the LDL particle is depleted of its antioxidants, lipid peroxidation enters the propagating phase (Esterbauer et al, 1989a,b). Ascorbic acid or uric acid added to isolated LDL fully protect the lipids and endogenous antioxidants present in LDL (Esterbauer et al, 1989a,b) from Cu^{2+}-induced oxidation. However, there is recent evidence that an LDL's content of α-tocopherol does not predict the ease with which the lipoprotein becomes oxidized (Jessup et al, 1990), indicating the presence of additional important antioxidants associated with the LDL. Indeed, ubiquinol-10, the most common form of ubiquinol present in human plasma (Table 3) is asociated with LDL and is consumed before α-tocopherol, both in plasma exposed to activated neutrophils (Cross et al, 1990) and isolated LDL exposed to Cu^{2+}, aqueous peroxyl radicals and other types of oxidizing conditions (Stocker et al, 1990).

Terao (1989b) investigated the effectiveness of α-tocopherol and carotenoids at trapping peroxyl radicals in organic solution and found the order to be α-tocopherol > astaxanthin = canthaxanthin > β-carotene = zeaxanthin. The reactivity of lipid-soluble antioxidants in organic solvents with the eosinophil-derived oxidant singlet oxygen was shown to decrease in the order lycopene > α-carotene > β-carotene > lutein > bilirubin > α-tocopherol (Di Mascioi et al, 1989). Taking into consideration the concentrations of these antioxidants in plasma, the authors concluded that lycopene and bilirubin are qualitatively about equally important in quenching singlet oxygen in plasma, followed by β-carotene and α-tocopherol.

In summary, α-tocopherol is by far the most concentrated lipid-soluble, chain-breaking antioxidant in plasma. However, ubiquinol-10 appears to form the first line of antioxidant defence in LDL exposed to a variety of

* The compound originally described by Esterbauer et al (1989a) as retinyl stearate was later identified as phytofluene.

different oxidants. Lycopene and bilirubin appear to be qualitatively most important in quenching singlet oxygen. Since by their very nature chain-breaking antioxidants in the lipids scavenge lipid radicals, consumption of lipid-soluble antioxidants is always accompanied by formation of lipid hydroperoxides.

When the relative importance of antioxidants in plasma or serum is investigated in vitro in systems that are based on metal-catalysed lipid peroxidation, the metal-binding proteins of plasma become of primary importance (Barber, 1961; Wills, 1965; Stocks et al, 1974; Section II.B.1). Thus, transferrin, lactoferrin, ceruloplasmin, albumin, haemopexin, and haptoglobin can be expected to form the first line of antioxidant defence in plasma when iron, copper, haem, or haemoglobin are released into it. Uric acid might also be important as plasma antioxidant under some of these conditions, as it strongly chelates iron (Davies et al, 1986), and has been shown to delay Cu^{2+}-induced lipid peroxidation in isolated LDL (Esterbauer et al, 1989a,b).

An important source of oxidants in vivo are activated phagocytes. In plasma challenged with phorbol myristate-activated PMN, the order of antioxidant consumption is ascorbic acid = protein thiols = bilirubin > uric acid, without appreciable consumption of α-tocopherol (Frei et al, 1988a; R. Stocker, unpublished). As is the case for aqueous peroxyl radicals, ascorbic acid is the only endogenous antioxidant that can completely protect all lipids against detectable peroxidative damage if plasma is exposed to oxidants produced by activated PMNs. Plasma ascorbic acid shows the same antiperoxidative efficacy against other, pathologically relevant types of oxidant stress, including $O_2^{\bar{\cdot}}$ and H_2O_2 production by the enzyme xanthine oxidase (B. Frei and B.N. Ames, unpublished), and the oxidants present in the gas phase of cigarette smoke (Frei et al, 1991).

In summary, these findings do *not* support the hypothesis put forward by Halliwell and Gutteridge (1985, 1986) that small molecular antioxidants are of secondary importance, only becoming active after the metal-binding capacity of plasma has been overwhelmed. Rather, the findings show that in plasma, and most likely in other extracellular fluids, ascorbic acid is qualitatively the most important antioxidant and suggest that this vitamin protects lipoproteins against peroxidative damage induced by a number of different, physiologically and pathologically relevant types of oxidant stress. These results support the concept that the redox potential of human blood is dependent on the dehydroascorbic acid/ascorbic acid ratio (Ziegler, 1960), and that this ratio can be used as a sensitive index of oxidative stress (Stocker et al, 1986a,b,c, unpublished).

As mentioned earlier, the relative importance of antioxidants is dependent on the type of oxidative damage measured, and all the studies mentioned thus far used lipid peroxidation as an index of oxidative damage. Inactivation of the main circulating antiprotease, α_1-antiproteinase, is a pathologically relevant type of oxidative damage as it may result in decreased tissue protection against proteolytic attack by elastase, released from activated PMNs (Weiss, 1989). In serum, protection against inactivation by H_2O_2 or the PMN-derived oxidant, HOCl, appears to be largely due to albumin (Wasil et al, 1987). When protein thiol oxidation is regarded as oxidative damage, rather than antioxidant protection (see Section II.B.1), none of the plasma antioxidants is completely protective against aqueous peroxyl radicals, the oxidants generated by activated PMNs, $O_2^{\overline{\cdot}}$ and H_2O_2, or the oxidants present in the gas phase of cigarette smoke (Wayner et al, 1987; Frei et al, 1988a, 1989a,b, 1990, 1991). This could be because of self-propagating autoxidation of the plasma proteins' thiols once their oxidation has been initiated (Frei et al, 1988a).

5. *Interaction between antioxidants*

In addition to directly scavenging radicals, some of the non-enzymatic small molecular antioxidants may also be capable of interacting with each other. A cooperative interaction between water-soluble vitamin C and lipid-soluble tocopherols has long been suggested to exist in vivo (Tappel, 1968), and the mechanism underlying this interaction has been clarified using in vitro test systems (Packer et al, 1979; Barclay et al, 1985; Doba et al, 1985; Niki et al, 1985). It has been shown that ascorbic acid present in the aqueous phase can reduce the tocopheroxyl radical (i.e. the one-electron oxidation product of tocopherols) via a hydrogen transfer reaction, thereby regenerating vitamin E. Highly efficient in vitro interaction with membrane-bound vitamin E has also been observed with conjugated bilirubin (Stocker and Peterhans, 1989b) while other water-soluble antioxidants, such as cysteine (Motoyama et al, 1989), glutathione (Tsuchiya et al, 1985, Stocker and Perterhans, 1989b) have little, and uric acid (Niki et al, 1986), has no synergistic activity with vitamin E, respectively. While these in vitro studies, pioneered by Niki and coworkers, have provided useful insights into the chemical nature of the synergistic interaction between the two vitamins, they do not, of course, demonstrate that such interaction takes place in biological systems.

Working with rat liver microsomes as an experimental biological system, Wefers and Sies (1988) showed that both ascorbic acid and GSH prolong the lag phase of NADPH-induced lipid peroxidation. The authors interpreted their results as evidence for an interaction between the water-soluble antioxidants and endogenous vitamin E, although such interpretation is dependent on the site of radical production by the test system (Niki, 1987). NADPH-dependent radical formation by microsomes is likely to produce radicals in the aqueous phase, as it is mediated by cytochrome P450 (Kuthan and Ullrich, 1982), the active site of which is localized outside the membrane (Heinemann and Ozols, 1983; Kunz et al, 1989). Therefore, the effects seen could conceivably be explained by direct scavenging of aqueous radicals by ascorbic acid and GSH. To achieve selective oxidation of α-tocopherol within biological membranes, Packer and associates incubated rat liver microsomes or mitochondria with peroxidases and hydrophobic substrates. Using such systems they observed that addition of either ascorbic acid (Mehlhorn et al, 1989) or the respiratory chain substrates NADH and succinate (Maguire et al, 1989) prevented the formation of α-tocopheroxyl radical measured by electron spin resonance (ESR). They suggested that reduction of the α-tocopheroxyl radical in mitochondrial and microsomal membranes occurs mainly via transmembrane electron transport systems with ubiquinol as actual reductant. While these findings are provocative and put into doubt the physiological relevance of an interaction between ascorbic acid and vitamin E, some questions remain open. Firstly, in order to detect ESR signals mitochondria and microsomes had to be 'loaded' substantially with additional α-tocopherol, the location of which has not been investigated. Secondly while electron transport chain-dependent reduction of α-tocopheryl radical may represent the main route for regeneration of α-tocopherol in plasma membranes, microsomes, and mitochondria, it obviously cannot be of great importance in the protection of lipoprotein-associated α-tocopherol. In the latter situation, studies with the lipid-soluble radical initiator added to human plasma provide indirect evidence for regeneration of lipoprotein-associated vitamin E by both ascorbic acid and albumin-bound bilirubin (Frei et al, 1989b) or isolated LDL (Sato et al, 1990).

While reports on cooperations of water- and lipid-soluble antioxidants are limited to vitamin E, some interactions between water-soluble antioxidants themselves have been noted. Dehydroascorbic acid is reduced non-enzymatically to ascorbic acid by GSH, especially under slightly acidic conditions (Stocker et al, 1986b), and ascorbic acid can regenerate uric acid from uric acid radical (Maples and Mason, 1988). While levels of GSH in human plasma are very low, the tripeptide represents the major cellular reductant, and regeneration of the reduced form of vitamin C by blood cells may be important in the maintenance of plasma ascorbic acid. In contrast, as discussed above (Section II.C.4), plasma ascorbic acid always seems to be oxidized before uric acid, so that regeneration of uric acid by vitamin C is not likely to be of physiological importance.

In summary, most studies on non-enzymatic interactions between small molecular antioxidants have focused on and support such activities for lipid-soluble vitamin E and water-soluble vitamin C. However, its direct demonstration in biological systems remains to be shown unambiguously. This is particularly so for human plasma, where regeneration of lipoprotein-associated vitamin E by ascorbic acid may be of greater importance than is the case in cellular membranes.

6. *Site-specific antioxidant protection*

Since its introduction in 1983 (Samuni et al, 1983), the perception of site-specific radical production with concomitant site-specific oxidative damage in biological systems has been well supported (Chevion, 1988). According to this mechanism, transition metal-binding sites on macromolecules serve as centres for repeated production of $^{\bullet}OH$ that are generated via the Fenton reaction. It seems plausible to argue that if site-specific free radical damage represents an important mechanism of deleterious processes in biological systems, site-specific antioxidant protection may also be an important mechanism of antioxidant defence. In analogy to site-specific radical production, the latter considers compartmentalization of biological systems, providing local protection to macromolecular structures. Such a mechanism of antioxidant defence therefore differs from that of 'general' antioxidant protection provided by radical scavengers in homogeneous systems.

Perhaps surprisingly, despite the enormous amount of literature available on site-specific radical formation, the concept of site-specific antioxidant protection has received only little attention. Fukuzawa et al (1988) described the inhibitory action of α-tocopherol incorporated into negatively charged micelles against H_2O_2–Fe^{2+}-induced lipid peroxidation as 'site specific', on the basis of separate localization of the vitamin and transition metal. This contrasts with the situation obtained with positively charged micelles, where α-tocopherol acts as prooxidant through regeneration of Fe^{2+} (Fukuzawa et al, 1988).

While this 'site specific' antioxidant action defined by Fukuzawa et al (1988) is clearly different to the one we defined, EC-SOD could be regarded as a site-specific antioxidant within our definition. Much as cell surface-associated SOD can protect certain bacteria against oxidative inactivation by neutrophils (Beaman et al, 1985), extracellular SOD bound to the surface of endothelial cells is likely to protect these cells site-specifically. Similarly, while albumin-bound bilirubin within human plasma fails to protect lipids completely within lipoproteins from radical-mediated oxidation, it very efficiently protects circulating fatty acids bound to albumin against peroxidation (Stocker et al, 1987a; Frei et al, 1988a). In the light of these findings it seems not only possible that bilirubin provides site-specific protection to other proteins to which it avidly binds (e.g. its intrahepatic carrier protein GSH *S*-transferase; Litwack et al (1971), but also that this type of antioxidant defence is extended to other proteins and antioxidant active ligands.

III. ANTIOXIDANT DEFENCES PROVIDED BY BLOOD CELLS

A. Red blood cells

Circulating erythrocytes are exposed to high oxygen tensions as a direct consequence of their main function. As the red cell membrane contains a variety of proteins that are susceptible to oxidative crosslinking and are embedded in phospholipids rich in unsaturated fatty acids, erythrocytes might be expected to be especially vulnerable to oxidative damage. Furthermore, under physiological conditions each day about 3% of the total haemoglobin is converted to its oxidized form, met-haemoglobin (Eder et al, 1949). Autoxidation of haemoglobin results in generation of $O_2^{\cdot-}$ (Misra and Fridovich, 1972; Carrell et al, 1975), and therefore red cells are continuously exposed to intracellular oxidative stress. Despite these potential hazards, erythrocytes are exceptionally resistant to oxidative damage. This is probably best demonstrated by their ability to prevent oxidative damage to various biological targets in vitro and in vivo (Weinberg and Hibbs, 1977; Hand, 1984; Toth et al, 1984; Van Asbeck et al, 1985; Agar et al, 1986; Winterbourn and Stern, 1987). The high degree of resistance of red cells to oxidative damage can be ascribed to their great efficiencies in both energy supply and antioxidant defences, reviewed in Stern (1985) and Hunt and Stocker (1990). In addition to ATP, glucose is the chief energy source for erythrocytes, providing the cells with reducing equivalents in the form of NADH, lactate, NADPH, and GSH. Through interaction with met-haemoglobin reductase, the enzymes of the glutathione cycle, catalase, and ascorbic acid, these reductants form the basis of the cell's antioxidant defence systems.

Total reactive thiols of normal adult human erythrocytes are composed of the sulphydryl groups of haemoglobin (80–85%), GSH (10–15%), and membrane proteins (5%) (Morell et al, 1976). GSH at concentrations approaching 2 mM and representing the largest pool of mobile cellular thiols reacts with thiol-reactive agents more readily than does haemoglobin (Morell et al, 1976). Through its relationship to GSH-Px and glutathione reductase, the redox status of GSH indirectly reflects the steady-state concentrations of intraerythrocytic reactive oxygen species. This intracellular redox system is extended to the red cell membrane proteins and possibly lipids through thiol exchange reactions and interactions between ascorbic acid and vitamin E, respectively. Regulation by erythrocytic GSH of the thiol status of the cell membrane also includes exofacial proteins (Kosower et al, 1982). Thus, when cellular GSH is absent or reduced substantially, membrane protein thiols are oxidized with concomitant formation of intra- and interchain disulphides (Kosower et al, 1982).

Amongst the GSH-regulated, thiol-dependent membrane proteins, the hexose transport system is particularly interesting as it mediates the facilitated diffusion of glucose across red cells and constitutes about 60% of all erythrocytic exofacial thiols. For the enzyme to be catalytically active exofacial thiol groups are required (Batt et al, 1976; Abbot et al, 1986; Reglinski et al, 1988). It is believed that these thiols are not only linked to intracellular GSH (via the transport protein and spectrin) but also interact with various sulphydryl populations of plasma proteins (Reglinski et al, 1988). Therefore, the redox status of erythrocytic GSH appears to be linked to that of plasma thiol groups, which themselves are influenced by other plasma antioxidants, especially ascorbic acid (Section II.C.4).

In addition to thiol exchange reactions, erythrocytic and plasma antioxidant defences may also be interrelated through ascorbic acid. The transport of ascorbic acid into blood cells and other tissues has been reviewed recently by Rose (1988). Vitamin C enters erythrocytes preferentially in its oxidized form, dehydroascorbic acid, through facilitated diffusion. Although dehydroascorbic acid can be a ligand of the hexose transporter under in vitro conditions of no or low sugar concentrations, a saturable, glucose-independent transport system seems to be physiologically relevant (Bianchi and Rose, 1986). Uptake of dehydroascorbic acid is subject to *cis* inhibition and *trans* stimulation by ascorbic acid (Bianchi and Rose, 1986), indicating that the transport system recognizes both oxidized and reduced forms of vitamin C. Once taken up, dehydroascorbic acid is reduced to ascorbic acid by GSH (Hughes and Maton, 1968), most likely non-enzymatically (De Chatelet et al, 1972; Stocker et al, 1986b). Intraerythrocytic concentration of ascorbic acid is around 50 μM (Stocker et al, 1986c; Moser, 1987), i.e. very similar to that in plasma (Table 3). For this reason and because the transporter recognizes both ascorbic acid and dehydroascorbic acid, this system might effectively handle the dual role of bringing dehydroascorbic acid into the cell and releasing ascorbic acid (Orringer and Roer, 1979; Rose, 1988), thereby maintaining plasma vitamin C in the reduced state.

As a consequence of the interrelationship between red cell and plasma antioxidant defence mechanisms, potential cellular responses need to be considered when addressing the effects of conditions of oxidative stress on plasma constituents. While red cells do not have the capacity to synthesize 'stress' proteins, they can nevertheless respond to in vivo oxidative stress by upregulation of glycolysis and the hexose monophosphate shunt (Stern, 1985). Plasma concentrations of lactate and pyruvate mainly control red cell glycolysis, and hence their levels of NADH, in response to oxidative stress (Sullivan and Stern, 1983, 1984), while the need for increased production of NADPH is met by a larger proportion of glucose passing through the hexose monophosphate shunt. The latter results in greater availability of reducing equivalents for the glutathione cycle to remove peroxides and to maintain the cellular thiol status. An increase in cellular GSH can also be achieved by upregulation of its biosynthesis as the result of increased uptake of cystine under oxidative stress (Ohtsuka et al, 1988).

B. White blood cells

In comparison to erythrocytes, much less is known about possible interactions of white blood cell antioxidant defences with those of plasma. In addition, we are also not aware of reports that directly address the possible regeneration of plasma ascorbic acid by leukocytes. This section is therefore limited to uptake and storage of vitamin C by white cells in the light of a potential function of these cells as regenerators of plasma ascorbic acid.

Leukocytes, unlike erythrocytes, accumulate ascorbic acid to concentrations of 1–3 mM against a gradient (Moser, 1987; Washko et al, 1989). Previous work indicates that almost all of the intracellular vitamin C exists in the reduced form and that dehydroascorbic acid is the form of the vitamin that crosses the plasma membrane (Bigley and Stankova, 1974; Rose, 1988). Like the situation with red cells, the dehydroascorbic acid taken up is reduced intracellularly in a non-enzymatic fashion (Stahl et al, 1985), with reducing equivalents supplied by the hexose monophosphate shunt. The activity of this pathway is increased in neutrophils exposed to ascorbic acid or dehydroascorbic acid (DeChatelet et al, 1972). More recent results indicate that human neutrophils have high- and low-affinity transporters for ascorbic acid with K_m values of 2–5 μM and 6–7 mM, respectively (Washko et al, 1989). However, these authors could not exclude the possibility that some of the ascorbic acid became oxidized to dehydroascorbic acid in the medium and/or at the cell membrane prior to its uptake. The authors further showed that almost all of the intracellular ascorbic acid remains localized in the cytosol, is not protein-bound, and is therefore at least potentially available for secretion into the plasma. Like red cells, leukocytes also respond to conditions of oxidative stress by upregulating mechanisms that produce reducing equivalents (review: Hamers and Roos, 1985). Upon activation of neutrophils, extracellular as well as part of the intracellular ascorbic acid is converted to dehydroascorbic acid (Winterbourn and Vissers, 1983), indicating the presence of oxidative stress in both compartments. Activated neutrophils take up vitamin C more rapidly than unstimulated cells (Moser, 1987). Whether this activity of the most prominent type of white cells in human blood is aimed at regenerating extra- or intracellular ascorbic acid, or both, remains to be established.

In summary, there is experimental evidence that supports a link between intraerythrocytic antioxidant defence mechanisms and those present in plasma. In particular, it seems possible that red cells may regenerate plasma ascorbic acid through processes ultimately linked to the cells' large reducing capacities. Their relatively low cellular level of vitamin C when compared with that in leukocytes is more than compensated for by their numbers (8000 red cells for each white cell). The situation is much less clear with white blood cells in this context, although they accumulate and concentrate ascorbic acid and hence could serve as potential additional reservoirs for this important plasma antioxidant. Both erythrocytes and leukocytes respond to conditions of oxidative stress by upregulating production of reducing equivalents. The above processes need to be considered when assessing the physiological significance of antioxidants based on results obtained with isolated human plasma.

IV. CONCLUSIONS

Human blood plasma contains a variety of antioxidant active biomolecules that can potentially protect circulating lipids, proteins and cells as well as the vasculature-lining endothelial cells from oxidative damage. We have reviewed their properties, emphasizing qualitative rather than quantitative aspects. Enzymatic antioxidants are present only in small concentrations in plasma and their physiological function remains unclear. In contrast, proteins whose primary biological function is related to the transport or storage of iron and copper ions or biological iron complexes provide antioxidant protection by sequestering these transition metals in forms incapable of stimulating free radical reactions. In the general absence of metal ion-catalysed reactions, ascorbic acid is qualitatively the single most important small molecular plasma antioxidant. It completely protects all classes of lipids from oxidation under a number of relevant types of oxidant stress while other non-enzymatic antioxidants merely lower the rate of oxidation or act in a more restricted, local environment. There is evidence in the literature that the antioxidant defences in plasma are interrelated with those of blood cells. In particular, the maintenance of reduced plasma ascorbic acid may be linked to the redox and energy status of red cells, indicating the multitude of biochemical events that regulate the status of plasma antioxidants. In the light of the potential importance of plasma antioxidants in protecting humans against diseases associated with oxidative stress, further studies are warranted that thoroughly address such interactions of plasma antioxidants with blood cells and hopefully also include additional cells such as endothelial cells.

ACKNOWLEDGEMENTS

We are very grateful to Dr B.N. Ames for his helpful comments and enthusiasm over the years, and thank Dr N.H. Hunt for carefully reading the manuscript.

REFERENCES

Abbott RE, Schachter D, Batt ER & Flamm M (1986) *Am. J. Physiol.* **250:** C853–C860.

Adachi T & Marklund SL (1989) *J. Biol. Chem.* **264:** 8537–8541.

Agar NS, Sadrzadeh SMH, Hallaway PE & Eaton JW (1986) *J. Clin. Invest.* **77:** 319–321.

Ames BN, Cathcart R, Schwiers E & Hochstein P (1981) *Proc. Natl. Acad. Sci. USA* **78:** 6858–6862.

Anderson R & Lukey PT (1987) *Ann. NY Acad. Sci.* **498:** 229–247.

Aruoma OI & Halliwell B (1987) *Biochem. J.* **241:** 273–278.

Avissar N, Within JC, Allen PZ, Wagner DD, Liegey P & Cohen HJ (1989) *J. Biol. Chem.* **264:** 15850–15855.

Bachowski GJ, Thomas JP & Girotti AW (1988) *Lipids* **23:** 580–586.

Baggiolini M, De Duve C, Masson PL & Heremans JF (1970) *J. Exp. Med.* **131:** 559–570.

Baldwin DA, Jenny ER & Aisen P (1984). *J. Biol. Chem.* **259:** 13391–13394.

Barber AA (1961) *Arch. Biochem. Biophys.* **92:** 38–43.

Barclay LRC, Locke SJ & MacNeil JM (1985) *Can. J. Chem.* **63:** 366–374.

Batt ER, Abbott RE & Schachter D (1976) *J. Biol. Chem.* **251:** 7184–7190.

Baysal E, Sullivan SG & Stern A (1989) *Biochem. Pharmacol.* **21:** 1109–1113.

Beaman BL, Black CM, Doughty F & Beaman L (1985) *Infect. Immun.* **47:** 135–141.

Bendich A, Machlin LJ, Scandurra O, Burton GW & Wayner DDM (1986) *Adv. Free Radicals Biol. Med.* **2:** 419–444.

Bernhard K, Ritzel G & Steiner KU (1954) *Helv. Chim. Acta* **37:** 306–313.

Beutler E (ed.) (1984) *Red Cell Metabolism*, 3rd edn. New York: Grune & Stratton.

Beyer RE & Ernster L (1990) In Lenaz G, Barnabei O, Rabbi A & Battino M (eds) *Highlights in Ubiquinone Research*, pp 191–213. London: Taylor and Francis.

Beyer RE, Nordenbrand K & Ernster L (1987) *Chem. Scripta* **27:** 145–153.

Bianchi J & Rose RC (1986) *Proc. Soc. Exp. Biol. Med.* **181:** 333–337.

Bielski BHJ & Richter HW (1975) *Ann. NY Acad. Sci.* **258:** 231–237.

Biemond P, van Eijk HG, Swaak AJG & Koster JF (1984) *J. Clin. Invest.* **73:** 1576–1579.

Bigley RH & Stankova L (1974) *J. Exp. Med.* **139:** 1084–1092.

Bodannes RS & Chan PC (1979) *FEBS Lett.* **105:** 195–196.

Brass CA, Wrchota EM & Gollan JL (1989) 40th Annual Meeting of the American Association for the study of Liver Diseases (Chicago, IL 28–31 Oct. 1989), abstracts.

Broderick DJ, Deagen JT & Whanger PD (1987) *J. Inorg. Biochem.* **30:** 299–308.

Bunn HF & Jandl JH (1968) *J. Biol. Chem.* **243:** 465–475.

Burton GW & Ingold KU (1981) *J. Am. Chem. Soc.* **103:** 6472–6477.
Burton GW & Ingold KU (1984) *Science* **224:** 569–573.
Burton GW, Joyce A & Ingold KU (1982) *Lancet* **2:** 327–328.
Burton GW, Joyce A & Ingold KU (1983) *Arch. Biochem. Biophys.* **221:** 281–290.
Cabelli DE & Bielski BHJ (1983) *J. Phys. Chem.* **87:** 1809–1812.
Cadenas E, Simic MG & Sies H (1989) *Free Radical Res. Commun.* **6:** 11–17.
Carrell RW, Winterbourn CC & Rachmilewitz EA (1975) *Br. J. Haematol.* **30:** 259–264.
Chevion M (1988) *Free Radicals Biol. Med.* **5:** 27–37.
Christen S, Peterhans E & Stocker R (1990) *Proc. Natl. Acad. Sci. USA* **87:** 2506–2510.
Clark IA, Hunt NH & Cowden WB (1986) *Adv. Parasitol.* **25:** 1–44.
Colleran E & O'Carra P (1977) In Berk PD & Berlin NI (eds) *Chemistry and Physiology of Bile Pigments*, pp 69–80. Washington: GPO.
Cross CE, Forte TM, Stocker R et al (1990) *J. Lab. Clin. Med.* **115:** 396–404.
D'Aquino M, Dunster C & Willson RL (1989) *Biochem. Biophys. Res. Commun.* **161:** 1199–1203.
Davies KJA, Sevanian A, Muakkassah-Kelly SF & Hochstein P (1986) *Biochem. J.* **235:** 747–754.
DeChatelet LR, Cooper MR & McCall CE (1972) *Antimicrob. Agents Chemother.* **1:** 12–16.
DeLange RJ & Glazer AN (1989) *Anal. Biochem.* **177:** 300–306.
De Mascio P, Kaiser S & Sies H (1989) *Arch. Biochem. Biophys.* **274:** 532–538.
Doba T, Burton GW & Ingold KU (1985) *Biochim. Biophys. Acta* **835:** 298–302.
Eder HA, Finch C & McKee RW (1949) *J. Clin. Invest.* **28:** 265–272.
Elmberger PG, Kalen A, Brunk UT & Dallner G (1989) *Lipids* **24:** 919–930.
Esterbauer H, Jürgens G, Quehenberger O & Koller E (1987) *J. Lipid Res.* **28:** 495–509.
Esterbauer H, Striegl G, Puhl H & Rotheneder M (1989a) *Free Radical Res. Commun.* **6:** 67–75.
Esterbauer H, Striegl G, Puhl H et al (1989b) *Ann. NY Acad. Sci.* **570:** 254–267.
Evans PJ, Bomford A & Halliwell B (1989) *Free Radical Res. Commun.* **7:** 55–62.
Frei B, Stocker R & Ames BN (1988a) *Proc. Natl. Acad. Sci. USA* **85:** 9748–9752.
Frei B, Yamamoto Y, Niclas D & Ames BN (1988b) *Anal. Biochem.* **175:** 120–130.
Frei B, England L & Ames BN (1989a) *Proc. Natl. Acad. Sci. USA* **86:** 6377–6381.
Frei B, Stocker R, England L & Ames BN (1989b) In Emerit I & Packer L (eds) *Antioxidants in Therapy and Preventive Medicine*, pp 155–163. London: Plenum Press.
Frei B, Forte T, Ames BN & Cross CE (1991) *Biochem. J.* (in press).
Frei B, Kim MC & Ames BN (1990) *Proc. Natl. Acad. Sci. USA* **87:** 4879–4883.
Fukuzawa K, Kishikawa K, Tadokoro T, Tokumura A, Tsukatani H & Gebicki JM (1988) *Arch. Biochem. Biophys.* **260:** 153–160.
Gibbs EL, Lennox WG, Nims LF & Gibbs FA (1942) *J. Biol. Chem.* **144:** 325–332.
Girotti AW (1978) *J. Biol. Chem.* **253:** 7186–7193.
Goldstein IM, Kaplan HB, Edelson HS & Weissman GR (1979) *J. Biol. Chem.* **254:** 4040–4045.
Góth L (1987) *Clin. Chem.* **33:** 2302–2303,
Gutteridge JMC (1983) *FEBS Lett.* **157:** 37–40.
Gutteridge JMC (1987) *Biochim. Biophys. Acta* **917:** 219–223.
Gutteridge JMC & Smith A (1988) *Biochem. J.* **256:** 861–865.
Gutteridge JMC, Richmond R & Halliwell B (1980) *FEBS Lett.* **112:** 269–272.

Haase G & Dunkley WL (1969) *J. Lipid Res.* **10:** 561–566.
Halliwell B (1988) *Biochem. Pharmacol.* **37:** 569–571.
Halliwell B & Gutteridge JMC (1985) *Free Radicals in Biology and Medicine.* London: Oxford University Press (Clarendon).
Halliwell B & Gutteridge JMC (1986) *Arch. Biochem. Biophys.* **246:** 501–514.
Halliwell B, Wasil M & Grootveld M (1987) *FEBS Lett.* **213:** 15–18.
Hamers MN & Roos D (1985) In Sies H (ed.) *Oxidative Stress*, pp 351–381. New York and London: Academic Press.
Hand WL (1984) *Infect. Immun.* **44:** 465–468.
Heinemann FS & Ozols J (1983) *J. Biol. Chem.* **258:** 4195–4201.
Hrkal Z, Vodrázka Z & Kalousek I (1974) *Eur. J. Biochem.* **43:** 73–78.
Hughes RE & Maton SC (1968) *Br. J. Haematol.* **14:** 247–253.
Hunt NH & Stocker R (1990) *Blood Cells* **16:** 499–530.
Ingold KU, Webb AC, Witter D, Burton GW, Metcalfe TA & Muller DPR (1987) *Arch. Biochem. Biophys.* **259:** 224–225.
Jacobs A, Path FRC & Worwood M (1975) *N. Engl. J. Med.* **292:** 951–956.
Jaffe GM (1984) In Machlin LJ (ed.) *Handbook of Vitamins: Nutritional, Biochemical, and Clinical Aspects*, pp 199–244. New York: Marcel Dekker.
Jessup W, Rankin SM, De Whalley CV, Hoult JRS, Scott J & Leake DS (1990) *Biochem. J.* **265:** 399–405.
Karlsson K & Marklund SL (1987) *Biochem. J.* **242:** 55–59.
Karlsson K & Marklund SL (1988) *J. Clin. Invest.* **82:** 762–766.
Karlsson K & Marklund SL (1989) *Lab. Invest.* **60:** 659–666.
Kaufmann HP & Garloff H (1961) *Fette Seifen Anstrichmittel* **63:** 334–344.
Kosower NS, Zisper Y & Fatlin Z (1982) *Biochim. Biophys. Acta* **691:** 345–352.
Kunz BC, Rehorek M, Hauser H, Winterhalter KH & Richter C (1989) *Biochemistry* **24:** 2889–2895.
Kuthan H & Ullrich V (1982) *Eur. J. Biochem.* **126:** 583–588.
Kwon B-M & Foote CS (1988) *J. Am. Chem. Soc.* **110:** 6582–6583.
Laussac J-P & Sarkar B (1984) *Biochemistry* **23:** 2832–2838.
Lentner C (ed.) (1984) *Geigy Scientific Tables*, Vol. 3, 8th revised and enlarged edition. Basle: Ciba-Geigy.
Leonard M, Noy N & Zakim D (1989) *J. Biol. Chem.* **264:** 5648–5652.
Liebler DC, Kling DS & Reed DJ (1986) *J. Biol. Chem.* **261:** 12114–12119.
Lindeman JHN, Van Zoeren-Grobben D, Schrijver J, Speek AJ, Poorthius BJHM & Berger HM (1989) *Pediatr. Res.* **26:** 20–24.
Link EM & Riley PA (1988) *Biochem. J.* **249:** 391–399.
Litwack G, Ketterer B & Arias IM (1971) *Nature (Lond.)* **234:** 466–467.
Maddipati KR & Marnett LJ (1987) *J. Biol. Chem.* **262:** 17398–17403.
Maddipati KR, Gasparski C & Marnett LJ (1987) *Arch. Biochem. Biophys.* **254:** 9–17.
Maguire JJ, Wilson DS & Packer L (1989) *J. Biol. Chem.* **264:** 21462–21465.
Maples KR & Mason RP (1988) *J. Biol. Chem.* **263:** 1709–1712.
Marklund SL (1987) *Proc. Natl. Acad. Sci. USA* **84:** 6634–6638.
Matsushita S, Ibuki F & Aoki A (1963) *Arch. Biochem. Biophys.* **102:** 446–451.
Mehlhorn RJ, Sumida S & Packer L (1989) *J. Biol. Chem.* **264:** 13448–13452.
Mellors A & Tappel AL (1966) *J. Biol. Chem.* **241:** 4353–4356.
Misra HP & Fridovich I (1972) *J. Biol. Chem.* **247:** 6960–6962.
Monteiro HP & Winterbourn CC (1988) *Biochem. J.* **256:** 923–928.

Morell SA, Ayers, VE, Greenwalt TJ & Hoffman P (1976) *J. Biol. Chem.* **239:** 2696–2705.
Morgan WT, Liem HH, Sutor RP & Müller-Eberhard U (1976) *Biochim. Biophys. Acta* **444:** 435–445.
Moser U (1987) *Ann. NY Acad. Sci.* **498:** 200–215.
Motoyama T, Miki M, Mino M, Takahashi M & Niki E (1989) *Arch. Biochem. Biophys.* **270:** 655–661.
Niki E (1987) *Chem. Phys. Lipids* **44:** 227–253.
Niki E, Kawakami A, Yamamoto Y & Kamiya Y (1985) *Bull. Chem. Soc. Jpn* **58:** 1971–1975.
Niki E, Saito M, Yoshikawa Y, Yamamoto Y & Kamiya Y (1986) *Bull. Chem. Soc. Jpn* **59:** 471–477.
Niki E, Yamamoto Y, Takahashi M et al (1988) *J. Nutr. Sci. Vitaminol.* **34:** 507–512.
Nishikimi M (1975) *Biochem. Biophys. Res. Commun.* **63:** 463–468.
O'Donnell-Tormey J, DeBoer CJ & Nathan CF (1985) *J. Clin. Invest.* **76:** 80–86.
O'Donnell-Tormey J, Nathan CF, Lanks K, DeBoer CJ & De La Harpe J (1987) *J. Exp. Med.* **165:** 500–514.
Ohtsuka Y, Kondo T & Kawakami Y (1988) *Biochem. Biophys. Res. Commun.* **155:** 160–166.
Okamoto T, Fukunaga Y, Ida Y & Kishi T (1988) *J. Chromatogr.* **430:** 11–19.
Orringer EP & Roer MES (1979) *J. Clin. Invest.* **63:** 53–58.
Packer JE, Slater TF & Wilson RL (1979) *Nature (Lond.)* **278:** 737–738.
Pedersen AO, Schønheyder F & Brodersen R (1977) *Eur. J. Biochem.* **72:** 213–221.
Peterhans E, Jungi TW & Stocker R (1988) In Cerutti P, Fridovich I & McCord J (eds) *Oxy-Radicals in Molecular Biology and Pathology*, pp 543–562. New York: Alan R. Liss.
Peters T (1985) *Adv. Protein Chem.* **37:** 161–184.
Pryor WA (1976) In Pryor WA (ed.) *Free Radicals in Biology*, pp 1–49. New York and London: Academic Press.
Reglinski J, Hoey S, Smith WE & Sturrock RD (1988) *J. Biol. Chem.* **263:** 12360–12366.
Repine JE (1985) In Said SI (ed.) *The Pulmonary Circulation and Acute Lung Injury*, pp 249–281. Mount Kisco, NY: Futura Pub.
Rose RC (1987) *Ann. NY Acad. Sci.* **498:** 506–508.
Rose RC (1988) *Biochim. Biophys. Acta* **947:** 335–366.
Rosenmund A, Freidli J, Bebié H & Straub PW (1988) *Acta Haematol.* **80:** 40–48.
Sadrzadeh SMH, Graf E, Panter SS, Hallaway PE & Eaton JW (1984) *J. Biol. Chem.* **259:** 14354–14356.
Sagone AL Jr, Greenwald J, Kraut EH, Bianchine J & Singh D (1983) *J. Lab. Clin. Med.* **101:** 97–104.
Samokyszyn VM, Miller DM, Reif DW & Aust SD (1989) *J. Biol. Chem.* **264:** 21–26.
Samuni A, Aronovitch J, Godinger D, Chevion M & Czapski G (1983) *Eur. J. Biochem.* **137:** 119–124.
Sato K, Niki E & Shimasaki H (1990) *Arch. Biochem. Biophys.* **279:** 402–405.
Schönreich C, Asmus K-D, Dillinger U & v. Bruchhausen F (1989) *Biochem. Biophys. Res. Commun.* **161:** 113–120.
Sevanian A, Davies KJA & Hochstein P (1985) *J. Free Radicals Biol. Med.* **1:** 117–124.
Sies H (ed.) (1985) *Oxidative Stress*. New York and London: Academic Press.
Stahl RL, Liebes LF & Silber R (1985) *Biochim. Biophys. Acta* **839:** 119–121.

Steinberg D, Parthasarathy S, Carew TE, Khoo JC & Witztum JL (1989) *N. Engl. J. Med.* **320:** 915–924.
Stern A (1985) In Sies H (ed.) *Oxidative Stress*, pp 331–349. New York and London: Academic Press.
Stevens B & Small RD (1974) *Photochem. Photobiol.* **23:** 33–36.
Stocker R & Ames BN (1987) *Proc. Natl. Acad. Sci. USA* **84:** 8130–8134.
Stocker R & Peterhans E (1989a) *Free Radical Res. Commun.* **6:** 57–66.
Stocker R & Peterhans E (1989b) *Biochim. Biophys. Acta* **1002:** 238–244.
Stocker R, Hunt NH & Weidemann MJ (1986a) *Biochem. Biophys. Res. Commun.* **134:** 152–158.
Stocker R, Hunt NH & Weidemann MJ (1986b) *Biochim. Biophys. Acta* **881:** 391–397.
Stocker R, Hunt NH & Weidemann MJ (1986c) *Biochim. Biophys. Acta* **876:** 294–299.
Stocker R, Glazer AN & Ames BN (1987a) *Proc. Natl. Acad. Sci. USA* **84:** 5918–5922.
Stocker R, Yamamoto Y, McDonagh AF, Glazer AN & Ames BN (1987b) *Science* **235:** 1043–1046.
Stocker R, Lai A, Peterhans E & Ames BN (1989) In Hayaishi O, Niki E, Kondo M & Yoshikawa T (eds) *Medical, Biochemical and Chemical Aspects of Free Radicals*, pp 465–468. New York and Amsterdam: Elsevier/North Holland.
Stocker R, Bowry VB & Frei B (1991) *Proc. Natl. Acad. Sci. USA* **88:** 1646–1650.
Stocks J, Gutteridge JMC, Sharp RJ & Dormandy TL (1974) *Clin. Sci. Mol. Med.* **47:** 223–233.
Sullivan SG & Stern A (1983) *Biochem. Pharmacol.* **32:** 2891–2902.
Sullivan SG & Stern A (1984) *Biochem. Pharmacol.* **33:** 1417–1421.
Takahashi K & Cohen HJ (1986) *Blood* **68:** 640–645.
Takahashi K, Avissar N, Within J & Cohen H (1987) *Arch. Biochem. Biophys.* **256:** 677–686.
Tappel AL (1955) *J. Biol. Chem.* **217:** 721–733.
Tappel AL (1968) *Geriatrics* **23:** 97–105.
Terao J (1989a) *Biochim. Biophys. Acta* **1003:** 221–224.
Terao J (1989b) *Lipids* **24:** 659–661.
Terao J & Matsushita S (1986) *Lipids* **21:** 255–260.
Terao J, Shibata SS & Matsushita S (1988) *Anal. Biochem.* **169:** 415–423.
Thomas CE & Aust SD (1985) *J. Free Radicals Biol. Med.* **1:** 293–300.
Thomas CE, Morehouse LA & Aust SD (1985) *J. Biol. Chem.* **260:** 3275–3280.
Thomas EL, Learn DB, Jefferson MM & Weatherred W (1988) *J. Biol. Chem.* **263:** 2178–2186.
Thurnham DI, Situnayake RD, Koottathep S, McConkey B & Davis M (1987) In Rice-Evans C (ed.) *Free Radicals, Oxidant Stress and Drug Action*, pp 169–192. London: Richelieu Press.
Toth KM, Clifford DP, Berger EM, White CW & Repine JE (1984) *J. Clin. Invest.* **74:** 292–295.
Tsuchiya J, Yamada T, Niki E & Kamiya Y (1985) *Bull. Chem. Soc. Jpn* **58:** 326–330.
Van Asbeck BS, Hoidal J, Vercellotti GM, Schwartz BA, Moldow CF & Jacob HS (1985) *Science* **227:** 756–759.
Vile GF & Winterbourn CC (1988) *FEBS Lett.* **238:** 353–356.
Vincent SH, Grady RW, Shaklai N, Snider JM & Müller-Eberhard U (1988) *Arch. Biochem. Biophys.* **265:** 539–550.

Washko P, Rotrosen D & Levine M (1989) *J. Biol. Chem.* **264:** 18996–19002.
Wasil M, Halliwell B, Hutchinson DCS & Baum H (1987) *Biochem. J.* **243:** 219–223.
Wayner DDM, Burton GW, Ingold KU & Locke S (1985) *FEBS Lett.* **187:** 33–37.
Wayner DDM, Burton GW & Ingold KU (1986) *Biochim. Biophys. Acta* **884:** 119–123.
Wayner DDM, Burton GW, Ingold KU, Barclay LRC & Locke SJ (1987) *Biochim. Biophys. Acta* **924:** 408–419.
Wefers H & Sies H (1988) *Eur. J. Biochem.* **174:** 353–357.
Weinberg JB & Hibbs JB Jr (1977) *Nature (Lond.)* **269:** 245–247.
Weiss SI (1989) *N. Engl. J. Med.* **320:** 365–376.
Wendel A & Cikryt P (1980) *FEBS Lett.* **120:** 209–211.
Wills ED (1965) *Biochim. Biophys. Acta* **98:** 238–251.
Winterbourn CC (1983) *Biochem. J.* **210:** 15–19.
Winterbourn CC & Stern A (1987) *J. Clin. Invest.* **80:** 1486–1491.
Winterbourn CC & Sutton HC (1984) *Arch. Biochem. Biophys.* **235:** 116–126.
Winterbourn CC & Vissers MCM (1983) *Biochim. Biophys. Acta* **763:** 175–179.
Wright CE, Tallan WH, Lin YY & Gaull GE (1986) *Annu. Rev. Biochem.* **55:** 427–453.
Wu T-W, Au J-X & Mickle DAG (1989) Scientific Conference on Myocardial Cell Viability, 10–12 July, 1989 (Santa Fe, Mexico), abstracts.
Yamamoto K & Niki E (1988) *Biochim. Biophys. Acta* **958:** 19–23.
Yamamoto Y & Niki E (1989) *Biochem. Biophys. Res. Commun.* **165:** 988–993.
Ziegler E (1960) In *Messung und Bedeutung des Redoxpotentials im Blut in vivo und in vitro*, supplement 10, Arzneimittel-Forschung. Aulendorf: Editio Cantor.

10

Protective and Adverse Biological Actions of Phenolic Antioxidants

REGINE KAHL
Abteilung Klinische Pharmakologie, Universität Göttingen, Robert-Koch-Str 40, D 3400 Göttingen, Germany

I. Introduction 245
II. Protective biological actions of phenolic antioxidants 246
III. Adverse biological actions of phenolic antioxidants 259
IV. Relation between adverse effects of phenolic antioxidants and their oxidative metabolism and prooxidative activity 263
V. Conclusion 267

I. INTRODUCTION

The term 'antioxidant' has been used in a broad sense, referring to agents capable of interfering with processes involved in oxidative stress. In a more restricted sense, the term denotes a chain-breaking action during the autoxidation of lipids. Phenolic compounds are prototypic chain-breaking antioxidants and have technologically been used for this purpose. In addition to the tocopherols (not considered in this chapter), a number of *tert*-butylphenol-derived monophenols and gallic acid-derived polyphenols are applied in food preservation and for suppression of lipid peroxidation in biological material (Figure 1). The intake of these synthetic antioxidants has been estimated from the maximum allowed levels in food and dietary recall data on food consumption. From such estimates, it has been concluded that in the early 1970s less than 0.4 mg butylated hydroxyanisole (BHA) and other

Oxidative Stress: Oxidants and Antioxidants
ISBN 0-12-642762-3

permitted synthetic phenolic antioxidants, butylated hydroxytoluene (BHT) and propyl gallate, per kg body weight was consumed daily by the Canadian population (Kirkpatrick and Lauer, 1986). A daily intake of 0.12–0.35 mg BHA per kg body weight has been calculated for the US population (National Academy of Sciences, 1979). Similarly, the Scientific Committee for Food of the Commission of the European Communities (1987) concludes that total intake of antioxidants may be 14 mg per person and day based on theoretical calculations but is probably only about 1 mg per person and day based on statistics on production and sale of antioxidants in the EEC. In samples of human adipose tissue, 0.01–0.03 ppm BHA and 0.07–0.19 ppm BHT were detected (Conacher et al, 1986), indicating that BHT may be accumulated more extensively than BHA. A second class of phenolic antioxidants in food is provided by the large number of polyphenolic plant flavonoids. A prototype is quercetin (Figure 1). These agents were used in folk medicine long before their chain-breaking potential was known (Havsteen, 1983; Middleton, 1984). The average Western diet contains about 1 g per day of plant flavonoids (Middleton, 1984).

In addition to lipid hydroperoxyl radicals, other radical species, including the hydroxyl radical, can be trapped by phenolic antioxidants; a number of them interact with the superoxide anion radical. The metal-complexing properties shared with other phenolic compounds contribute to their antioxidative potential by removing the catalysts required for the formation of highly reactive oxidants. The relationships between these basic chemical properties and the biological actions are not fully understood. Phenolic antioxidants exert complex effects on enzymes, including inhibition (e.g. lipoxygenase, cytochromes P450), activation (e.g. prostacyclin synthase) and induction (e.g. glutathione transferase, quinone reductase). It is tempting to speculate that the common mechanism of action is interference with the regulation of enzyme activity via endogenous oxidants. However, this hypothesis has still to be proven in detail.

II. PROTECTIVE BIOLOGICAL ACTIONS OF PHENOLIC ANTIOXIDANTS

Flavonoid antioxidants as well as *tert*-butylphenol derivatives and other synthetic phenols and polyphenols have been subjected to biochemical and pharmacological experiments in order to evaluate their therapeutic potential, especially for use in inflammatory disease. The following provides a discussion of selected examples of protection observed in animals or potentially applicable to diseased animals or humans.

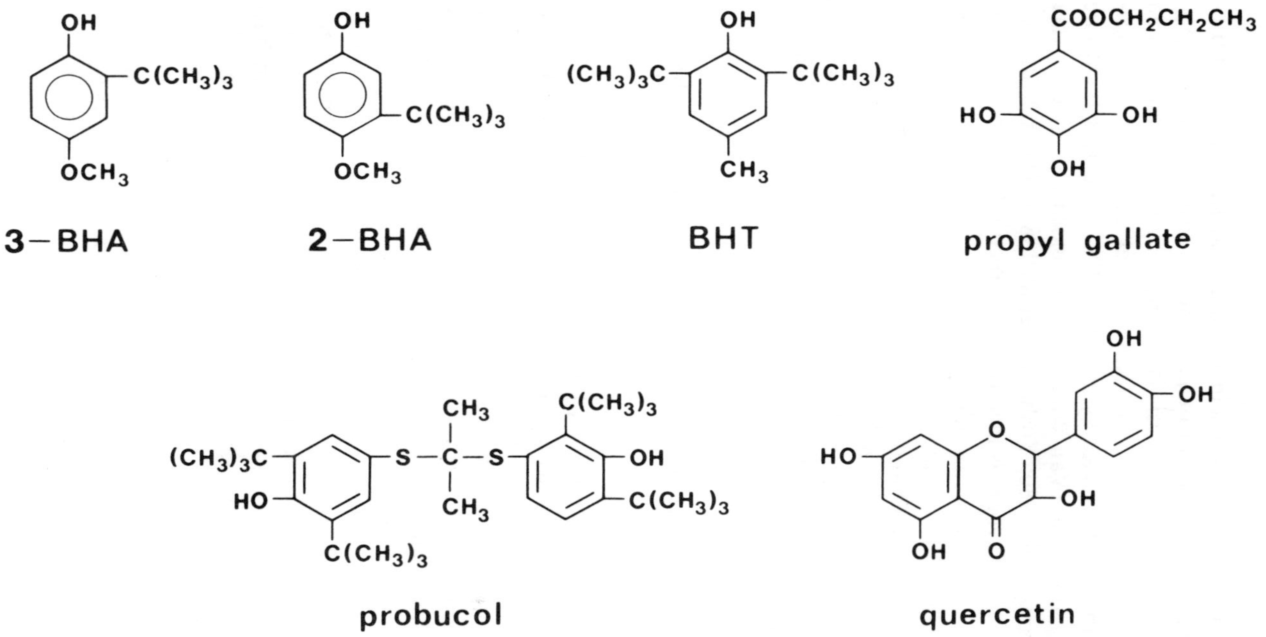

3–BHA

2–BHA

BHT

propyl gallate

probucol

quercetin

Figure 1. Chemical structure of prototypic phenolic antioxidants.

A. Antiinflammatory action

Interference with a variety of processes involved in mediator release contributes to the antiinflammatory action of phenolic antioxidants. The lipoxygenase pathway of eicosanoid synthesis is inhibited by propyl gallate (Van Wauwe and Goossens, 1983), BHA (Nakadate et al, 1984), the *tert*-butylphenol derivatives E-5110 (Katayama et al, 1987) and R-830 (Moore and Swingle, 1982), quercetin (Van Wauwe and Goossens, 1983) and other flavonoids (Baumann et al, 1980; Yoshimoto et al, 1983). The unhydroxylated parent compound, flavone, is inactive; the structural requirements for lipoxygenase inhibition have been studied (Landolfi et al, 1984; Wheeler and Berry, 1986). Prostaglandin formation is less consistently inhibited and may even be enhanced (Van Wauwe and Goossens, 1983; Gryglewski et al, 1987; Kuehl et al, 1977; Aishita et al, 1983).

Phenolic antioxidants interfere with transmembrane signal transduction and Ca^{2+}-regulated events (Figure 2). Flavonoids inhibit the Ca^{2+}-dependent ATPase activity of skeletal muscle sarcoplasmic reticulum and of the plasma membrane (Fewtrell and Gomperts, 1977; Sokolove et al, 1986). The flavonoid quercetin acts as a calmodulin antagonist (Nishino et al, 1984). Calmodulin is involved in the regulation of cyclic nucleotide phosphodiesterase activity, and inhibition of phosphodiesterase activity and an increase in cAMP levels is caused by flavonoids (Graziani and Chayoth, 1979; Ferrell et al, 1979; Landolfi et al, 1984). The cAMP-induced fall in free cytosolic Ca^{2+} concentration may, in turn, suppress eicosanoid synthesis, in superposition to a direct effect of the antioxidants on arachidonic acid metabolism. The antioxidants can, however, also act on the phosphoinositide pathway. In platelets, the rise of cytosolic Ca^{2+} concentration due to agonists at this pathway was blocked by BHT and nordihydroguaiaretic acid (NDGA) (Alexandre et al, 1986). For some flavonoid antioxidants, including quercetin, an inhibitory action on protein kinase C (PKC) activated by the phosphoinositide pathway-related endogenous activator diacylglycerol (Ferriola et al, 1989) and by tumour promoters (Gschwendt et al, 1983) has been demonstrated. The structural requirements for PKC inhibition have been studied in a series of flavonoids (Ferriola et al, 1989). Whether PKC inhibition is a general property of phenolic antioxidants is not yet apparent. PKC is involved in the regulation of phospholipase A_2 activity; antioxidant inhibition of PKC may therefore also contribute to the inhibition of eicosanoid synthesis. Histamine liberation is dependent on an increase in cytosolic Ca^{2+}; a number of flavonoids inhibit histamine release from rat mast cells and human basophils (Fewtrell and Gomperts, 1977; Hope et al, 1983). In brain synaptosomes, the depolarization-dependent Ca^{2+} uptake is inhibited by BHT, BHA, propyl gallate and NDGA, and it is hypothesized

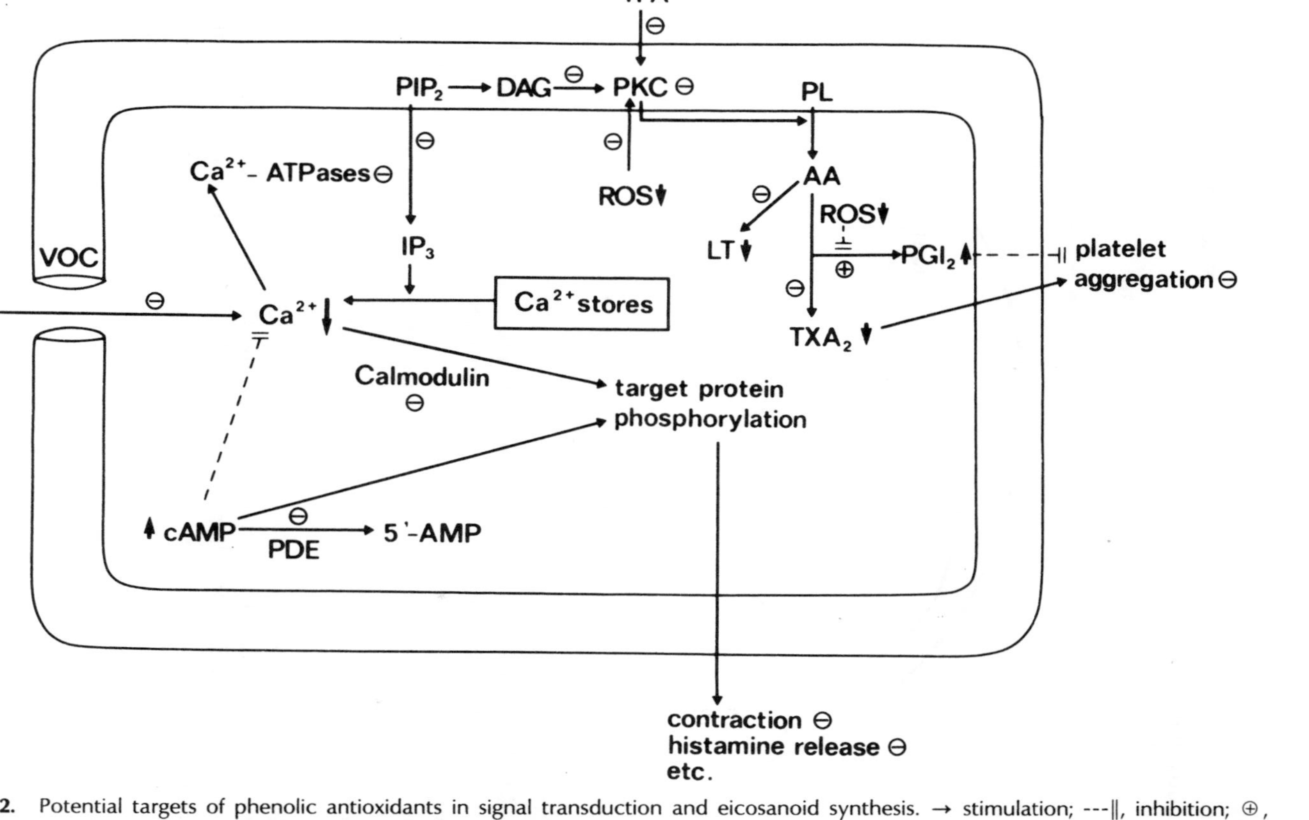

Figure 2. Potential targets of phenolic antioxidants in signal transduction and eicosanoid synthesis. → stimulation; ---‖, inhibition; ⊕, activity increased by antioxidant; ⊖, activity decreased by antioxidant; ↑, concentration increased by antioxidant; ↓, concentration decreased by antioxidant; AA, arachidonic acid; cAMP, cyclic adenosine monophosphate; DAG, diacylglycerol; IP_3, inositol 1,4,5-trisphosphate; LT, leukotrienes; PDE, phosphodiesterase; PGI_2, prostacyclin; PIP_2, phosphatidylinositol 4,5-bisphosphate; PKC, protein kinase C; PL, phospholipids; ROS, reactive oxygen species; TPA, 12-*O*-tetradecanoylphorbol-13-acetate; TXA_2, thromboxane; VOC, voltage-operated Ca^{2+} channel.

that coupling of depolarization with the opening of voltage-gated calcium channels is effected by a reactive oxygen species which can be removed by antioxidants (Zoccarato et al, 1987).

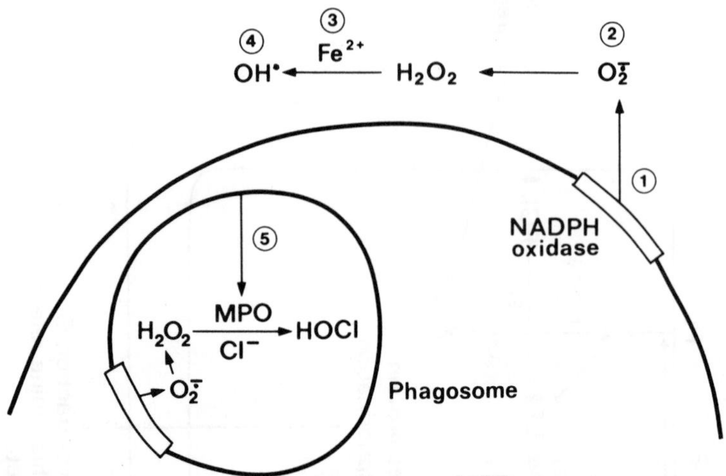

Figure 3. Potential targets of phenolic antioxidants in the respiratory burst. 1, Interference with the NADPH oxidase in the plasma (or phagosome) membrane; 2, scavenging of the superoxide anion radical; 3, binding of catalytic metal; 4, scavenging of the hydroxyl radical; 5, inhibition of the release and/or activity of myeloperoxidase (MPO).

Modulation of immune response by phenolic antioxidants comprises effects on both cellular and humoral defence mechanisms (Ball et al, 1986). Suppression of interleukin 1 release from monocytes by E-5110 (Shirota et al, 1989) and from murine peritoneal macrophages by probucol (Ku et al, 1990) has been observed. The respiratory burst of neutrophils can be inhibited by antioxidants such as E-5110 (Katayama et al, 1987), quercetin and other flavonoids (Schwarz et al, 1984; Pagonis et al, 1986) or propyl gallate (Müller-Peddinghaus and Wurl, 1987). This effect is not specific for phenolic antioxidants but is shared by a variety of antiinflammatory compounds (Müller-Peddinghaus and Wurl, 1987). It is not clear whether the decrease of $O_2^{\cdot-}$ concentration is mainly due to inhibition of phagocyte function or to $O_2^{\cdot-}$ scavenging. Inhibition of neutrophil NADPH oxidase by quercetin has been reported (Tauber et al, 1983). On the other hand, lowering of $O_2^{\cdot-}$ concentration in model systems such as the xanthine oxidase system has been observed with quercetin, cyanidanol, some other flavonoids, propyl gallate and BHA (Robak and Gryglewski, 1988; Blasig et al, 1988; Kahl et al, 1989). Certain flavonoids inhibit the release and activity of myeloperoxidase in neutrophils, thus preventing the formation of hypochlorous acid (T'Hart et al, 1990). Protection against $^{\cdot}OH$, to which part of the damaging effects of the respiratory burst in host tissue has been attributed, may also be provided by phenolic antioxidants, either by metal-binding properties (Blasig et al, 1988; Afanas'ev et al, 1989) or by $^{\bullet}$OH-scavenging properties ascribed to antioxidants, e.g. silibin (Mira et al, 1987), propyl gallate (Weinke et al, 1987) and the *tert*-butylphenol derivatives ONO-3144 and MK-447 (Cheeseman and Forni, 1988). Figure 3 depicts potential mechanisms. Aruoma and Halliwell (1988) have proposed that protection against $^{\bullet}$OH by direct reaction as well as by iron chelation may be a more general feature of antiinflammatory action not confined to genuine antioxidants.

B. Inhibition of platelet aggregation

Flavonoids inhibit platelet aggregation (Beretz et al, 1982). Imbalance between thromboxane and prostacyclin synthesis may be involved in this effect: thromboxane synthesis in platelets is suppressed by phenolic antioxidants, e.g. by quercetin and some other flavonoids (Landolfi et al, 1984; Corvazier and Maclouf, 1985) or ONO-3144 (Aishita et al, 1983). In contrast, prostacyclin synthase appears to be resistant to or even dependent on antioxidants. An increase of PGI_2 formation by propyl gallate has been shown in model systems such as ram seminal vesicle microsomes. This has been interpreted to be due to the scavenging of an oxidant species destructive to prostacyclin synthase formed during the cyclooxygenase reaction and/or to the prevention of the formation of 15-HPETE, which is also destructive to the enzyme (Beetens et al, 1981). Protection of both prostacyclin synthesis by removal of peroxides and of endothelium-derived relaxing factor (EDRF) by removal of superoxide anion has been proposed to be brought about by quercetin or myricetin tightly bound to platelets and thus forming a carpet over the vascular endothelium (Gryglewski et al, 1987). Antioxidant effects in platelet aggregation are, however, probably not only directed to arachidonic acid metabolism: Alexandre et al (1986) have demonstrated that the protective effect with BHT and NDGA also occurs in aspirin-pretreated platelets, together with inhibition of agonist-induced rise of cytosolic Ca^{2+} concentration. The antioxidant effects on second messenger concentrations described above have in part been detected in platelets. Thus, PGI_2-dependent elevation of cAMP concentration in platelets will be potentiated by antioxidants (Landolfi et al, 1984). Increased cAMP levels on the one hand and inositol trisphosphate-induced Ca^{2+} release from intracellular stores and diacylglycercol-induced PKC activity on the other will contribute to antioxidant inhibition of platelet aggregation.

C. Inhibition of smooth muscle contraction

Quercetin inhibits acetylcholine release in the guinea-pig ileum, and this has been claimed to be the reason for the antidiarrhoeal effect of flavonoid-containing folk medicine (Lutterodt, 1989). In the guinea-pig ileum, trachea and pulmonary artery, stimulation by various agonists can be abolished by certain flavonoids (Fanning et al, 1983; Abdalla et al, 1989). Among food-preserving antioxidants, inhibition of the contractility of smooth muscle has been described for BHT (Gad et al, 1979), BHA (Franchi-Micheli et al, 1980; Stewart et al, 1981) and gallates (Posati et al, 1970). In rat ileum longitudinal muscle, BHA exerts a nifedipine-like action, since it is inactive in the presence of a calcium ionophore (Sgaragli et al, 1989). Bioflavonoids lower vasotonus and have previously been alleged to be a vasoprotective vitamin, vitamin P. This assignment has been abandoned but is still discussed in recent literature (Roger, 1988). A beneficial action of flavonoid-containing drugs on venous hypertension and thus an improvement of microcirculation has been claimed to occur in patients (Belcaro et al, 1989).

D. Prevention of atherosclerosis

The beneficial effect of the *tert*-butylphenol derivative probucol in atherosclerosis (Buckley et al, 1989) directly links radical scavenging to therapeutic use. It has been established during the last decade that modification of low-density lipoprotein (LDL) by lipid peroxidation leads to its increased uptake into macrophages via the scavenger receptor and to subsequent transformation of the macrophages into foam cells. Thus, the atherogenicity of LDL is considerably enhanced (Steinberg et al, 1989). Probucol is an inhibitor of lipid peroxidation in various systems (Minakami et al, 1989) and prevents LDL lipid peroxidation (Parthasarathy et al, 1986). This is paralleled by the conversion of probucol into a diphenoquinone via expulsion of the dithiol group connecting the two phenol rings (Barnhart et al, 1989).

The chain-breaking potency of probucol is not extraordinarily high as compared to other phenolic antioxidants. It is conceivable that other antioxidants can also fulfil the function of probucol in preventing LDL oxidation. BHT (Zhang et al, 1989) and some flavonoids (DeWhalley et al, 1990) have been demonstrated to prevent LDL oxidation. Probucol has, however, the advantage of being an established drug for the treatment of hypercholesterolaemia. Moreover, it may possess favourable pharmacokinetic properties in that it is transported in the plasma primarily with LDL itself (Steinberg et al, 1989). Clinical trials with other chain-breaking antioxidants are reportedly being performed at present (Halliwell, 1989).

E. Anticarcinogenic properties

Protection against chemical carcinogenesis is probably the most challenging experimental finding obtained with phenolic antioxidants. Table 1 is a compilation of such effects. Protection has most often been localized in the initiation phase. Postinitiation protection does, however, also occur as evidenced by a number of genuine post-treatment experiments (McCormick et al, 1984; Moore et al, 1984; Hirose et al, 1988; Mizumoto et al, 1989). Figure 4 depicts the main events in the overall process and indicates sites of antioxidant interference.

Initiation is assumed to occur by a mutational event. Antioxidants are potent antimutagenic agents and inhibitors of DNA binding of carcinogenic compounds. The ultimate DNA-binding and mutagenic agent is in most cases metabolically derived from an inert precarcinogen. A specific suitability of phenolic antioxidants for a protective shift in the activation–inactivation balance in carcinogen metabolism can be derived from two properties: (1) inhibition of cytochrome P450 monooxygenase activities required for metabolic activation of precarcinogens; and (2) selective induction of detoxification enzymes such as glutathione transferase, glucuronyl transferase, epoxide hydrolase and quinone reductase (see Chapter 8). Interference with activation–inactivation balance, DNA binding and mutagenicity has been reviewed for food antioxidants (Kahl, 1984). Bioflavonoids also inhibit DNA binding and mutagenic activity of chemicals (Das et al, 1987; Francis et al, 1989).

Induction of glutathione transferase has been a favourite candidate for explaining the anticarcinogenic activity of phenolic antioxidants. Wattenberg (1982) has provided evidence for a close correlation between the induction of the enzyme and antioxidant protection and has pointed out that this correlation holds not only for antioxidants but also for a variety of other compounds. However, in the majority of studies on DNA binding and mutation, protection was achieved by addition of the antioxidant in vitro, thus creating a situation in which enzyme induction cannot play a role. Most carcinogenicity studies applying antioxidants do not allow for discrimination between enzyme induction and monooxygenase inhibition, because the antioxidant has been present concomitantly with the carcinogen for too long a period to exclude an inducer action, or the time interval between antioxidant administration and carcinogen administration has been too short to allow for complete disappearance of the antioxidant, precluding the disproof of a direct inhibitory effect. The fact that the endogenous antioxidants, vitamin C and vitamin E, are also anticarcinogens but do not induce glutathione transferase (Kahl, 1986) further points to antioxidant action other than enzyme induction.

Table 1. Antitumourigenic action of phenolic antioxidants.

Carcinogen	Tissue	Antioxidant	Reference
2-Acetylaminofluorene	Rat liver	BHT, propyl gallate	Ulland et al (1973), McCay et al (1980), Maeura et al (1984)
Aflatoxin B_1	Rat liver	BHA, BHT	Williams et al (1986)
5-Amino-3-(2-(5-nitro-2-furyl)vinyl)-1,2,4-oxadiazole	Various mouse tissues	BHA	Dunsford et al (1984)
Azaserine	Rat pancreas foci	BHA	Roebuck et al (1984)
	Hamster pancreas foci (pretreated)	BHT	Thornton et al (1989)
	Hamster liver foci (post-treated)		
Azoxymethane	Mouse intestine	BHT	Weisburger et al (1977)
Benzo(a)pyrene	Mouse lung	BHA	Wattenberg (1972a), Witschi and Doherty (1984)
	Mouse forestomach	BHA, BHT	Wattenberg (1972a)
	Mouse skin	Quercetin	Slaga et al (1978)
		Rutin, morin	Van Duuren et al (1971)
		Quercetin, myricetin, tannic acid	Mukhtar et al (1988)
Benzo(a)pyrene-7,8-diol	Mouse lung, forestomach	BHA	Wattenberg et al (1979)
	Lymphoid tissue		
Benzo(a)pyrene-7,8-diol-9,10-epoxide	Mouse skin	Ellagic acid	Chang et al (1985)
		Quercetin, tannic acid	Khan et al (1988)
	Mouse lung	Quercetin, robinetin, ellagic acid	Chang et al (1985)
N-Butyl-*N*-(4-hydroxybutyl)-nitrosamine	Rat liver foci	BHA, BHT	Imaida et al (1983), Thamavit et al (1989)
Ciprofibrate	Rat liver	BHA	Rao et al (1984)

Dibenz(a)anthracene	Mouse lung	BHA	Wattenberg (1973)
Diethylnitrosamine	Mouse lung	BHA	Wattenberg (1972b)
	Mouse forestomach	BHT	Clapp et al (1978)
	Rat liver foci	BHA, BHT	Tsuda et al (1984), Moore et al (1984)
	Hamster liver foci	BHA	Moore et al (1987a)
Diethylnitrosamine/ phenobarbital (*promotion*)	Rat liver	Propyl gallate, quercetin	Denda et al (1989)
4-Dimethylaminoazobenzene	Rat liver	BHT	Frankfurt et al (1967)
7,12-Dimethylbenz(a)anthracene	Mouse lung	BHA	Wattenberg (1972a)
	Mouse forestomach	BHA	Wattenberg (1972a)
	Mouse skin	BHA, BHT	Slaga and Bracken (1977), Slaga et al (1982)
		Quercetin, myricetin, tannic acid	Mukhtar et al (1988)
	Various mouse tissues	BHA	Rao (1982)
	Rat mamma	BHA, BHT, TBHQ	Wattenberg (1972a), McCay et al (1980)
		Propyl gallate, quercetin	Cohen et al (1984), McCormick et al (1984), Hirose et al (1988), Verma et al (1988)
	Mouse skin	BHA, BHT	Slaga et al (1982)
7,12-Dimethylbenz(a)-anthracene/12-*O*-tetradeca-noylphorbol-13-acetate (*promotion*)	Mouse skin	Quercetin, nordihydro-guaiaretic acid, morin	Kato et al (1983), Nakadate et al (1982), Nakadate et al (1984)
7,12-Dimethylbenz(a)-anthracene/teleocidin (*promotion*)	Mouse skin	Quercetin	Fujiki et al (1986)
1,2-Dimethylhydrazine	Mouse intestine	BHT	Clapp et al (1979)
	Rat intestine	BHT	Jones et al (1984), Shirai et al (1985)

Table 1. (*Continued*).

Carcinogen	Tissue	Antioxidant	Reference
Dimethylnitrosamine	Mouse lung	BHA	Chung et al (1986)
2,2′-Dioxo-*N*-nitroso-dipropylamine	Hamster liver foci Hamster pancreas foci	BHA	Moore et al (1987b)
N-Ethyl-*N*-hydroxyethyl-nitrosamine	Rat liver, rat liver foci	BHA BHA	Tsuda et al (1984) Thamavit et al (1989)
7-Hydroxymethyl-12-methyl-benz(a)anthracene	Mouse lung	BHA	Wattenberg (1973)
N-bis-(2-Hydroxypropyl)-nitrosamine	Rat liver foci	BHT	Thamavit et al (1989)
Isoniazid	Mouse lung	BHA, BHT	Maru and Bhide (1982)
Methylazoxymethanol acetate	Mouse intestine	BHA	Wattenberg and Sparnins (1979), Reddy and Maeura (1984)
3-Methylcholanthrene	Mouse skin	Quercetin, myricetin, tannic acid	Mukhtar et al (1988)
3′-Methyl-4-dimethyl-aminoazobenzene	Rat liver	BHT	Daoud and Griffin (1980)
N-Methyl-*N*′-nitro-*N*-nitrosoguanidine	Rat stomach	BHT	Tatsuta et al (1983)
N-Methyl-*N*-nitrosourea	Rat mamma	Quercetin	Verma et al (1988)
	Mouse skin	Quercetin, myricetin, tannic acid	Mukhtar et al (1988)
4-Nitroquinoline-*N*-oxide	Mouse lung	BHA	Wattenberg (1972b)
N-Nitrosobis-(2-oxopropyl)amine	Hamster pancreas	BHA	Mizumoto et al (1989)
Uracil mustard	Mouse lung	BHA	Wattenberg (1973)
Urethane	Mouse lung	BHA	Wattenberg (1973), Witschi (1981), Malkinson and Beer (1984), Witschi and Doherty (1984)

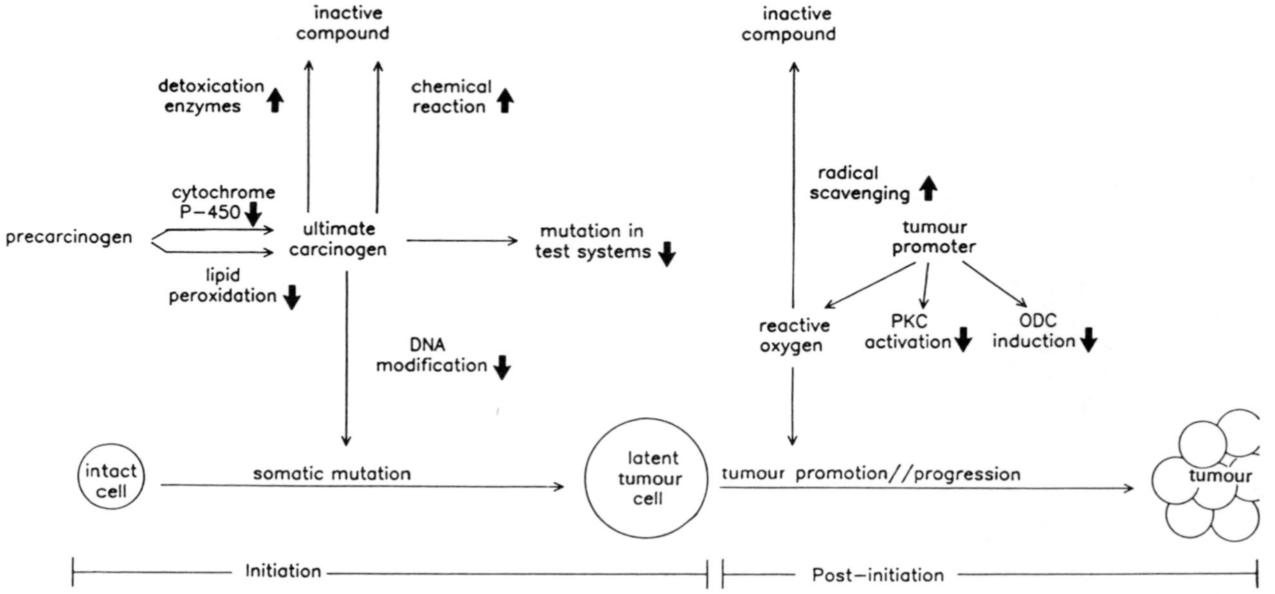

Figure 4. Potential targets of the phenolic antioxidants in two-stage carcinogenesis. ↑, Stimulation by antioxidant; ↓, suppression by antioxidant.

Destruction of radicals involved in the tumourigenic process may partially explain the anticarcinogenic action of phenolic antioxidants. Various targets of a scavenger action are conceivable:

1. Antioxidants may remove radical metabolites contributing to the DNA-modifying action of chemical carcinogens. Mikuni et al (1987) have demonstrated that BHT is capable of scavenging the radical formed from *N*-methyl-*N′*-nitro-*N*-nitrosoguanidine (MNNG) and hydrogen peroxide. The authors assume that a relationship exists between this scavenger effect and the protection by BHT against MNNG-induced gastric cancer. Limited evidence suggests that direct interaction of phenolic antioxidants with reactive species may occur which does not necessarily involve radical scavenging. Various plant phenols remove benzo(a)pyrene 7,8-diol-9,10-epoxide from aqueous solution in the absence of a metabolizing system and exert an antimutagenic and anticarcinogenic effect against the diol epoxide (Wood et al, 1982; Chang et al, 1985; Khan et al, 1988).

2. Lipid peroxidation is involved in the formation of ultimate carcinogenic species (Dix and Marnett, 1983; Gower, 1988). Antioxidants may prevent the formation of reactive species produced in this way.

3. In the rat mammary tumour model (McCay et al, 1980), a dependency of antioxidant protection on the lipid composition of the diet has been described. In some model systems, inhibition of lipid peroxidation parallels inhibition of carcinogenesis (Perera et al, 1985; Black, 1987). Galeotti et al (1986) have suggested that oxy radical-induced structural and functional alterations are in part responsible for the properties of tumour cell membranes. It is conceivable that inhibition of membrane peroxidation contributes to the anticarcinogenic action of antioxidants.

4. The most advanced hypothesis involving radical scavenging relates to tumour promotion. The role of reactive oxygen species in tumour promotion (Troll and Wiesner, 1985; Kensler and Taffe, 1986) and the effects of antioxidants on mouse skin tumour promotion (Perchellet and Perchellet, 1989) have been reviewed; see also Chapter 11. The protective effect may be due to the removal of oxygen radicals involved in the DNA rearrangement and subsequent alteration of gene expression occurring during tumour promotion. Antioxidants also suppress the induction of ornithine decarboxylase by the tumour promoter 12-*O*-tetradecanoylphorbol-13-acetate (TPA) (Nakadate et al, 1984; Kozumbo et al, 1983; Kato et al, 1983). Ornithine decarboxylase induction appears to be linked to arachidonic acid metabolism, and its inhibition has been related to the suppressing effect of phenols on lipoxygenase activity (Kato et al, 1983; Nakadate et al, 1984). Quercetin suppresses inhibition of cell–cell communication by TPA and another tumour promoter, DDT (Warngard et al, 1987). Propyl gallate, quercetin

and BHA inhibit the progression of cell transformation caused by TPA in JB-6 mouse epidermal cells (Nakamura et al, 1985). As discussed above, a variety of phenolic antioxidants suppress PKC activity. PKC is the receptor for TPA and related tumour promoters, and thus inhibition of PKC may be one of the mechanisms which allow for the antipromotional activity of antioxidants (Gschwendt et al, 1984).

III. ADVERSE BIOLOGICAL ACTIONS OF PHENOLIC ANTIOXIDANTS

The interest in extending the protective potential of phenolic antioxidants to the chemoprevention of human cancer is lessened by the perception that some of the most widely studied phenolic compounds are not devoid of adverse reactions when used at high dosage. BHT has become a model lung toxicant used as a tool in animal mimicks of respiratory distress and interstitial pulmonary fibrosis (Witschi et al, 1989). In the rat, the preponderant sign of subacute BHT toxicity is haemorrhage; mice and guinea-pigs will also develop coagulation disorders with BHT, and in hamsters and quails, but not in rabbits and dogs, asymptomatic hypoprothrombinaemia occurs (Takahashi and Hiraga, 1978). The main biochemical target of this adverse effect has been found to be vitamin K epoxide reduction, since the vitamin K-dependent synthesis of clotting factors II, VII, IX and X is decreased (Takahashi, 1987). BHT and BHA induce hypertrophy in the liver of various animal species, including rats, mice, dogs, pigs and monkeys. This is accompanied by the typical pattern of enzyme induction described above (Kahl, 1984). In rats, oral administration of a high dose of BHT leads to centrilobular necrosis accompanied by initial glutathione depletion (Nakagawa et al, 1984). Kidney injury by BHT (Meyer et al, 1978; Nakagawa and Tayama, 1988), has also been described.

With BHA, the finding which attracted most concern was the formation of papillomas and squamous-cell carcinomas in the forestomach of rat, hamster and, less unequivocally, mouse due to long-term feeding with high doses (Ito et al, 1986). No comparable changes were observed in the stomach of animal species not possessing a forestomach, such as dog (Ikeda et al, 1986; Tobe et al, 1986) and monkey (Iverson et al, 1986). The superficial hyperplasia and many of the papillomas regressed within a few months but basal cell hyperplasia and dysplasia persisted after withdrawal of BHA (Masui et al, 1987; Hirose et al, 1990).

The significance for human health of forestomach carcinogenesis in rodents by BHA has been questioned (Wester and Kroes, 1988; Grice, 1988):

1. The main target tissue of BHA is forestomach, while tumourigenic alterations at other sites are rarely reported (Sato et al, 1987; Amo et al, 1990). Squamous epithelium similar to that of the rodent forestomach is present in human oesophagus, pharynx, larynx and oral cavity. In rodents, BHA did not lead to hyperplastic changes in the oesophagus. Parakeratosis and proliferative changes have been claimed to occur in the oesophagus and the oesophageal part of the stomach in pigs (Würtzen and Olsen, 1986) but this finding has been questioned. It is concluded that a carcinogenic action of BHA in the oesophagus obviously does not exist.

2. The doses required for obtaining carcinomas have been extremely high as compared to current food levels. Thus, feeding BHA at 2% in the diet (corresponding to about 1000 mg (kg body weight)$^{-1}$) for nearly the whole lifetime resulted in malignancies in rats. BHA has been tested in a variety of mutagenicity tests, and negative results have been obtained in most studies (Kahl, 1984; Williams, 1986; Hageman et al, 1988). However, two recent clastogenicity studies were positive with certain BHA metabolites (Phillips et al, 1989; Matsuoka et al, 1990; see below). Due to the lack of mutagenicity, a threshold for the carcinogenic action is assumed by most agencies, and it is assumed that food levels of BHA will not induce cancer.

The mechanisms of BHA-induced forestomach tumourigenesis is most probably the expression of latent initiation via a tumour-promoting potential of the antioxidant. BHA and BHT meet many of the criteria previously established for tumour promoters in accepted model systems such as mouse skin promotion by TPA and rat liver promotion by phenobarbitone:

(a) Suppression of metabolic cell–cell cooperation by BHT (Trosko et al, 1982) and by BHA (Williams, 1986; Masui et al, 1988).
(b) Potentiation of cell transformation in vitro by BHT (Djurhuus and Lillehaug, 1982; Potenberg et al, 1986) and BHA (Potenberg et al, 1986). Table 2 shows potentiation and inhibition of cell transformation in the same model system.
(c) Induction of ornithine decarboxylase in rat liver by BHT (Kitchin and Brown, 1987). In mouse skin, however, induction of ornithine decarboxylase by BHA, BHT, *tert*-butylhydroquinone (TBHQ) and prostaglandin did not occur (Kozumbo et al, 1983).

Numerous protumourigenic effects of synthetic food antioxidants have been described (Table 3). They occur in a large variety of tissues and are directed towards a large variety of chemical carcinogens. Some authors have attributed the enhancement of tumours to an increase in lifespan by the antioxidant which allowed for detection of tumours that would also have been produced in control animals if they had only survived long enough (e.g. Mirvish et al, 1983). Antioxidants sometimes induce a change in target organ preference. For instance, BHA protects against the development of liver foci due to nitrosamines but enhances bladder and forestomach carcinogenesis by the same class of agents (Tables 1 and 3).

Table 2. Potentiation and inhibition of morphological transformation of cultured Syrian hamster embryo cells by BHT and BHA.

First test phase (2 days) initiation phase	Second test phase (5 days) promotion phase	Transformation frequency (%)
DMSO	DMSO	0
Benzo(a)pyrene 0.1 μg ml^{-1}	DMSO	0.27
Benzo(a)pyrene 0.1 μg ml^{-1}	BHT 100 μM	0.77
Benzo(a)pyrene 0.1 μg ml^{-1}	BHA 100 μM	0.49
Benzo(a)pyrene 1 μg ml^{-1}	DMSO	0.64
Benzo(a)pyrene 1 μg ml^{-1} + BHT 100 μM	DMSO	0.15
Benzo(a)pyrene 1 μg ml^{-1} + BHA 100 μM	DMSO	0.17

Sequential exposure to the initiator, benzo(a)pyrene (BP), and the antioxidants leads to an increase in transformation frequency. In contrast, inhibition of transformation is obtained by concomitant exposure to BP and the antioxidants (Potenberg et al, 1986).

The antioxidant most often used in tumour promotion studies is BHT. It is likely that the carcinogenic effect of this antioxidant in rodent liver (Olsen et al, 1986; Inai et al, 1988) is also due to its activity as a tumour promoter, since genotoxic activity of BHT has in most tests not been detected (Kahl, 1984). For the gallic acid ester antioxidants and for the BHA metabolite TBHQ, studies meeting current standards of toxicological testing are rare (Van der Heijden et al, 1986; Van Esch, 1986). TBHQ has protumourigenic activity (Tamano et al, 1987); moreover, it induces forestomach hyperplasia, though less efficiently than BHA (Altmann et al, 1986). A genotoxic potential has not been excluded (Van Esch, 1986). For propyl gallate, a recent carcinogenicity study has not revealed a carcinogenic potential (Abdo et al, 1983).

With the bioflavonoid antioxidants, the topic of greatest concern is their genotoxic potential (MacGregor, 1984). The flavonoid most often tested in bacterial and mammalian mutagenicity tests (MacGregor, 1984) and for DNA-breaking activity (Alvi et al, 1986; Rahman et al, 1989) is quercetin. Most carcinogenicity studies, including a recent one using up to 5% quercetin in the diet (Ito et al, 1989), have concluded with a negative result; however, an increase in tumour incidence has also been reported (MacGregor, 1984). The tumour-promoting potential of quercetin and other flavonoid antioxidants has obviously not been tested.

Table 3. Enhancement of chemical tumourigenesis by phenolic antioxidants.

Carcinogen	Tissue	Antioxidant	Reference
2-Acetylaminofluorene	Rat liver	BHT	Peraino et al (1977)
	Rat liver foci	BHT	Maeura and Williams (1984)
	Rat bladder	BHT	Maeura et al (1984)
Azaserine	Rat pancreas foci	BHT	Thornton et al (1989)
Azoxymethane	Rat colon	BHT	Weisburger et al (1977)
Benzo(a)pyrene	Mouse lung	BHT	Witschi and Morse (1983)
N-Bis-(2-hydroxypropyl)-nitrosamine	Rat bladder, thyroid	BHT	Thamavit et al (1989)
N-Butyl-*N*-(4-hydroxybutyl)-nitrosamine	Rat bladder	BHA, BHT, TBHQ	Sakata et al (1984), Tsuda et al (1984), Fukishima et al (1987b), Tamano et al (1987)
	Rat bladder, thyroid	BHT	Thamavit et al (1989)
N,N-Dibutylnitrosamine	Rat liver	BHA, BHT	Funato et al (1987), Imaida et al (1988)
	Rat forestomach	BHA	Fukushima et al (1987a)
	Rat oesophagus	BHT	Fukushima et al (1987a)
Diethylnitrosamine	Mouse lung	BHT	Clapp et al (1974)
Dimethylbenz(a)anthracene	Mouse skin	BHT-OOH	Taffe and Kensler (1988)
1,2-Dimethylhydrazine	Mouse colon	BHT	Lindenschmidt et al (1986)
	Rat colon	BHT	Shirai et al (1985)
	Rat gastrointestinal tract	BHT	Lindenschmidt et al (1987)
Dimethylnitrosamine	Mouse lung	BHT	Witschi and Morse (1983)
N-Ethyl-*N*-hydroxyethyl-nitrosamine	Rat bladder, thyroid	BHT	Thamavit et al (1989)
3-Methylcholanthrene	Mouse lung	BHT	Witschi and Morse (1983)
N-Methyl-*N*′-nitro-*N*-nitrosoguanidine	Rat forestomach	BHA, BHT plus NaCl	Shirai et al (1984), Newberne et al (1986)
N-Methyl-*N*-nitrosourea	Rat forestomach	BHA	Imaida et al (1984)
	Rat bladder	BHA, BHT	Imaida et al (1984)
	Rat thyroid	BHA	Imaida et al (1984)
N-Nitrosopyrrolizidine	Mouse lung	BHA	Chung et al (1986)
Urethane	Mouse lung	BHT	Witschi et al (1977), Witschi and Morse (1983), Malkinson and Beer (1983)
	Mouse lung	BHT-BuOH	Thompson et al (1989b)

IV. RELATION BETWEEN ADVERSE EFFECTS OF PHENOLIC ANTIOXIDANTS AND THEIR OXIDATIVE METABOLISM AND PROOXIDATIVE ACTIVITY

By its very mechanism of action, a phenolic antioxidant is converted into a phenoxyl radical. The fate of BHA- and/or BHT-derived phenoxyl radicals has been studied in model systems using well-defined substrates such as a specific linoleic acid hydroperoxide (Kaneko and Matsuo, 1985) or *tert*-butylhydroperoxide (Kurechi and Kato, 1983). In biological systems, phenoxyl radicals can be the basis of a cascade of prooxidative events. The phenoxyl radical from BHT and BHA can be formed enzymatically via π-electron oxidation of the benzyl ring. Peroxidatic oxidation of BHA leading to BHA dimer(s) and covalently binding products has been observed in sheep seminal vesicle microsomes and with horseradish peroxidase or prostaglandin H synthase (Rahimtula, 1983; Thompson et al, 1989a). Valoti et al (1989) observed that in a horseradish peroxidase system the BHT phenoxyl radical was much more stable than the BHA phenoxyl radical. The rat intestinal peroxidase for which BHA is a substrate (Guarna et al, 1983) has recently been purified and characterized (Valoti et al, 1988).

In the case of BHT, the cyclohexa-2,5-dienone products arising from π-electron oxidation have been studied extensively, and it has been shown that the hydroperoxide product of this pathway, BHT-OOH, and the quinone methide probably arising from disproportionation of the phenoxyl radical are active metabolites of BHT. Moreover, a BHT metabolite oxidized in one of the *tert*-butyl moieties, BHT-BuOH, also appears to be an active metabolite (see Chapter 22).

In addition to the formation of reactive metabolites by oxidative pathways, reactive oxygen species can also be produced, e.g. by autoxidation of a diphenol or polyphenol concomitant with univalent reduction of molecular oxygen, followed by dismutation of the superoxide formed and subsequent formation of hydroxyl radicals in a Fenton-type reaction. Passi et al (1987) have demonstrated that di- and triphenolic compounds are more cytotoxic than monophenols because they produce much larger quantities of reactive oxygen metabolites in the extracellular space. In contrast, monophenols require an intracellular action in order to gain cytotoxicity. Flavonoid antioxidants are polyphenols. DNA damage by quercetin has been ascribed to H_2O_2 formed during autoxidation of the flavonoid (Ochiai et al, 1984). Laughton et al (1989) have studied the production of hydroxyl radicals and the enhancement of bleomycin-induced DNA damage in the presence of the polyphenols quercetin, myricetin and gossypol. They demonstrated that superoxide and H_2O_2 were intermediates and provided evidence that the reduction of Fe^{3+}-EDTA by the phenols was involved in the process.

Rahman et al (1989) have studied the mechanism of DNA breakage by quercetin plus Cu(II) in a metal-catalysed autoxidation process. They propose that Cu(II) oxidizes the phenol and reoxidation of Cu(I) in turn leads to the formation of superoxide. In a study on flavonoid-induced suppression of mitochondrial respiration, four out of 15 compounds unexpectedly led to a cyanide-insensitive increase in oxygen consumption which was ascribed to autoxidation (Hodnick et al, 1986).

The monophenol BHA is not readily autoxidizable but may become so after metabolic introduction of further hydroxyl substituents. Addition of BHA to rat liver microsomes leads to extra production of superoxide, hydrogen peroxide and hydroxyl radicals, as well as to increased consumption of NADPH (Cummings and Prough, 1983; Rössing et al, 1985; Cummings et al, 1985; Weinke et al, 1987; Thompson and Trush, 1988b; Kahl et al, 1989). The superoxide burst induced by BHA in liver microsomes requires metabolic activation to the hydroquinone metabolite, *tert*-butylhydroquinone (TBHQ) (Kahl et al, 1989). TBHQ is converted to the semiquinone, which further autoxidizes to the corresponding quinone, TBQ, with the concomitant production of superoxide (Bergmann et al, 1989). In microsomes, protein adducts appear to be formed extensively from TBQ and/or other metabolites of BHA (Rahimtula, 1983; Armstrong and Wattenberg, 1985; DeStafney et al, 1986). In microsomes, TBQ is much more active as a prooxidant than TBHQ (Cummings and Prough, 1983; Kahl et al, 1989) (Figure 5). Therefore, the quinone can be assumed to be the ultimate active compound formed from BHA in microsomes with respect to oxygen activation. Redox cycling between the quinone and the semiquinone radical occurs via reducing enzyme activity, as suggested by the detection of the semiquinone radical from TBQ in microsomes (Bergmann et al, 1989).

It is not clear at present whether the carcinogenic and tumour-promoting action of BHA can be ascribed to its metabolism to the redox-active TBHQ. H_2O_2 formed from the BHA metabolite TBHQ was involved in the clastogenic action of BHA (Phillips et al, 1989). In another study, however, further oxidative metabolism of TBHQ was required for clastogenic activity (Matsuoka et al, 1990). TBQ exhibits initiating activity at much lower concentration than BHA in a cell transformation assay (Sakai et al, 1990). Increased production of O_2^{-} by the addition of BHA, TBHQ and TBQ to rat forestomach preparations has been described (Kahl et al, 1989); however, the formation of TBHQ from BHA could not be shown (Hirose et al, 1987; Kahl et al, 1989). TBHQ was less potent than BHA in the induction of forestomach hyperplasia in the rat (Altmann et al, 1986). It has been suggested that another diphenol metabolite of BHA, the *O*-hydroxy compound (catechol) (Armstrong and Wattenberg, 1985), may be even more readily oxidizable than TBHQ (DeStafney et al, 1986). However, experimen-

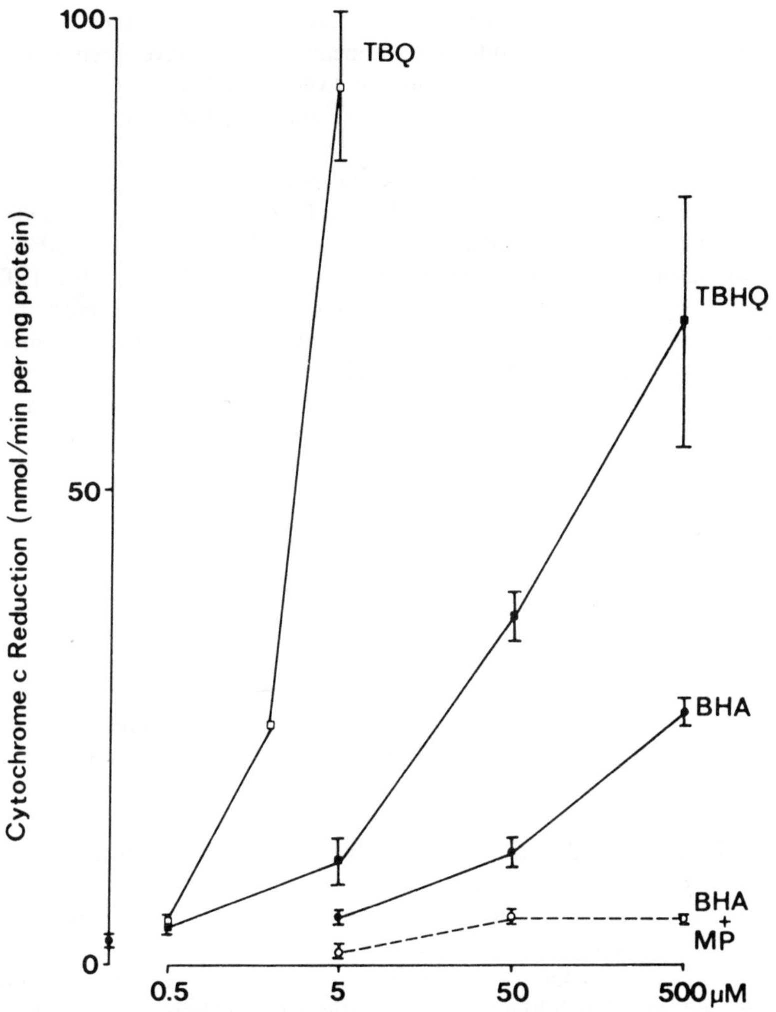

Figure 5. Stimulation of SOD-inhibitable cytochrome *c* reduction in rat liver microsomes by BHA, TBHQ and TBQ. Ordinate: control value. The increase in $O_2^{\cdot-}$ formation by BHA is prevented by the monooxygenase inhibitor metyrapone (MP, 100 μM), indicating that metabolic activation of BHA is required. BHA metabolites TBHQ and TBQ are much more active in producing $O_2^{\cdot-}$ than the parent compound, BHA (Kahl et al, 1989).

tal data as to a role of the catechol metabolite in BHA toxicity appear to be lacking.

Quinoid metabolites are also formed from BHT (see Chapter 12). The tumour promoting activity of BHT-OOH appears to depend on the formation of the quinone methide (Guyton et al, 1991). Whether the dibutyl quinone or the quinone methide are subject to redox cycling, thus leading to a burst of superoxide production, appears not to have been studied. However, increased H_2O_2 formation in liver microsomes (Rössing et al, 1985) and especially in lung microsomes (Kahl et al, 1987) can be obtained with BHT.

In animals, synergism between BHA and BHT has caused aggravation of the pulmonary toxicity of BHT by BHA (Thompson and Trush, 1988a). BHA dramatically stimulated the peroxidatic oxidation of BHT, leading to the formation of BHT quinone methide (Thompson and Trush, 1989). The authors suggest that the observed synergism proceeds via the mechanism of action previously demonstrated in the *tert*-butylhydroperoxide model system (Kurechi and Kato, 1983): BHA is recovered from the BHA phenoxyl radical via reduction by BHT, which in turn is itself converted into the BHT phenoxyl radical and subsequently into the quinone methide:

$$2\text{BHA}^{\bullet} + \text{BHT} \longrightarrow 2\text{BHA} + \text{BHT quinone methide}$$

Redox cycling of the BHA metabolite TBHQ has been claimed to be involved in the synergistic action of BHA on BHT pneumotoxicity. This assumption was supported by the finding that preincubation of NADPH–cytochrome P450 reductase with TBHQ increased the stimulating action of BHA on the formation of BHT quinone methide in a horseradish peroxidase system. It was concluded that TBHQ provides the H_2O_2 required for the peroxidase action on BHA providing the BHA phenoxyl radical (Thompson and Trush, 1988b). No synergism between TBHQ itself and BHT was detected, indicating that the mere presence of H_2O_2 is insufficient to elicit increased quinone methide production (Thompson and Trush, 1988a). Instead, diversion of NADPH from monooxygenation via TBHQ-dependent NADPH oxidation and concomitant support of the peroxidase function of cytochrome P450 will lead to the conversion of BHA to its phenoxyl radical and will thus initiate the synergistic action (Thompson and Trush, 1988b).

Excess production of H_2O_2 in microsomes has been observed with a number of phenolic antioxidants other than BHA and BHT, including the gallates (Rössing et al, 1985), cyanidanol (Ritter et al, 1985) and quercetin (Sousa and Marletta, 1985). For antioxidants capable of scavenging superoxide, the excess H_2O_2 may arise from their reducing action, e.g. propyl gallate (Deeble et al, 1988). However, no evidence for prooxidative effects or hydroxyl radical formation from propyl gallate has been obtained (Weinke et al, 1987; Kahl et al, 1989). It may be worthwhile to examine whether gallates possess a favourable benefit–risk quotient in biological systems.

V. CONCLUSION

Two groups of phenolic antioxidants, the synthetic food preservatives derived from *tert*-butylphenol or gallic acid and the polyphenolic plant flavonoids, were examined for their biological activity. Diverse biochemical and pharmacological actions are known, both beneficial and potentially dangerous to the organism. Examples of protective effects are antiinflammatory activity, inhibition of platelet aggregation, spasmolytic properties, prevention of LDL oxidation and antimutagenic and anticarcinogenic action. Mechanisms underlying protection can be identified in signal transduction pathways and in the regulation of enzyme activities involved in carcinogen metabolism. An obvious link between radical-scavenging properties and protection appears to exist in the antipromotional action of the antioxidants. Among the potentially dangerous antioxidant effects, the carcinogenic activity observed with high doses of BHA and BHT and the mutagenic potential of flavonoids have attracted concern. BHA and BHT are considered nonmutagenic, and a threshold for their carcinogenic action may exist. These two antioxidants also exhibit tumour-promoting activity. A relationship appears to exist between some of the adverse actions and the formation of oxidized metabolites and/or the production of reactive oxygen via redox cycling of the phenolic compounds. The broad spectrum of biological activity of the radical-scavenging phenolic antioxidants indicates that regulation of biological processes via endogenous oxidants is important. However, many of the targets of such endogenous oxidants have still to be identified.

REFERENCES

Abdalla S, Zarga MA, Afifi F, al Khalil S, Mahasneh A & Sabri S (1989) *J. Pharm. Pharmacol.* **41:** 138–141.

Abdo KM, Huff JE, Haseman JK et al (1983) *J. Am. Coll. Toxicol.* **2:** 425.

Afanas'ev IB, Dorozhko AI, Brodskii AV, Kostyuk VA & Potapovitch AI (1989) *Biochem. Pharmacol.* **11:** 1763–1769.

Aishita H, Morimura T, Obata T et al (1983) *Arch. Int. Pharmacodyn. Ther.* **261:** 316–327.

Alexandre A, Doni MG, Padoin E & Deana R (1986) *Biochem. Biophys. Res. Commun.* **139:** 509–514.

Altmann H-J, Grunow W, Mohr U, Richter-Reichhelm HB & Wester PW (1986) *Food. Chem. Toxicol.* **24:** 1183–1188.

Alvi NK, Rizvi RY & Hadi SM (1986) *Biosci. Rep.* **6:** 861–868.

Amo H, Kubota H, Lu J & Matsuyama M (1990) *Carcinogenesis* **11:** 151–154.

Armstrong KE & Wattenberg LW (1985) *Cancer Res.* **45:** 1507–1510.
Aruoma OI & Halliwell B (1988) *Xenobiotica* **18:** 459–470.
Ball SS, Weindruch R & Walford RL (1986) In Johnson JE Jr, Walford R, Harman D & Miquel J (eds) *Free Radicals, Aging, and Degenerative Diseases*, pp 427–456. New York: Alan R. Liss, Inc.
Barnhart RL, Busch SJ & Jackson RL (1989) *J. Lipid Res.* **30:** 1703–1710.
Baumann J, von Bruchhausen F & Wurm G (1980) *Prostaglandins* **20:** 627–639.
Beetens JR, Claeys M & Herman AG (1981) *Biochem. Pharmacol.* **30:** 2811–2815.
Belcaro G, Rulo A & Candiani C (1989) *Vasa* **18:** 146–151.
Beretz A, Cazenave JP & Anton R (1982) *Agents Actions* **12:** 382–387.
Bergmann B, Dohrmann J & Kahl R (1989) *Naunyn-Schmiedeberg's Arch. Pharmacol.* **339 (supplement):** R16.
Black HS (1987) *Fed. Proc.* **46:** 1901–1905.
Blasig IE, Loewe H & Ebert B (1988) *Biomed. Biochim. Acta* **47:** 252–255.
Buckley MMT, Goa KL, Price AH & Brogden RN (1989) *Drugs* **37:** 761–800.
Chang RL, Huang MT, Wood AW et al (1985) *Carcinogenesis* **6:** 1127–1133.
Cheeseman KH & Forni LG (1988) *Biochem. Pharmacol.* **37:** 4225–4233.
Chung F-L, Wang M, Carmella SG & Hecht SS (1986) *Cancer Res.* **46:** 165–168.
Clapp NK, Tyndall RL, Cumming RB & Otten JA (1974) *Food Cosmet. Toxicol.* **12:** 367–371.
Clapp NK, Tyndall RL, Satterfield LC, Klima WC & Bowles ND (1978) *J. Natl. Cancer Inst.* **61:** 177–180.
Clapp NK, Bowles ND, Satterfield LC & Klima WC (1979) *J. Natl. Cancer Inst.* **63:** 1081–1087.
Cohen LA, Polansky M, Furuya K, Reddy M, Berke B & Weisburger JH (1984) *J. Natl. Cancer Inst.* **72:** 165–173.
Conacher HBS, Iverson F, Lau P-Y & Page BD (1986) *Food Chem. Toxicol.* **24:** 1159–1162.
Corvazier E & Maclouf J (1985) *Biochim. Biophys. Acta* **835:** 315–321.
Cummings SW & Prough RA (1983) *J. Biol. Chem.* **258:** 12315–12319.
Cummings SW, Ansari GAS, Guengerich FP, Crouch LS & Prough RA (1985) *Cancer Res.* **45:** 5617–5624.
Daoud AH & Griffin AC (1980) *Cancer Lett.* **9:** 299–304.
Das M, Khan WA, Asokan P, Bickers DR & Mukhtar H (1987) *Cancer Res.* **47:** 767–773.
Deeble DJ, Parsons BJ, Phillips GO, Schuchmann HP & von Sonntag C (1988) *Int. J. Radiat. Biol.* **54:** 179–193.
Denda A, Ura H, Tsujiuchi T et al (1989) *Carcinogenesis* **10:** 1929–1935.
DeStafney CM, Prabhu UDG, Sparnins VL & Wattenberg LW (1986) *Food Chem. Toxicol.* **24:** 1149–1157.
De Whalley CV, Rankin SM, Hoult JRS, Jessup W & Leake DS (1990) *Biochem. Pharmacol.* **39:** 1743–1750.
Dix TA & Marnett LJ (1983) *Science* **221:** 77–79.
Djurhuus R & Lillehaug JR (1982) *Bull. Environ. Contam. Toxicol.* **29:** 115–120.
Dunsford HA, Dolan PM, Seed JL & Bueding E (1984) *J. Natl. Cancer Inst.* **73:** 161–168.
Fanning MJ, Macander P, Drzewiecki G & Middleton Jr E (1983) *Int. Arch. Allergy Appl. Immunol.* **71:** 371–373.
Ferrell JE, Chang Sing PGD, Loew G, King R, Mansour JM & Mansour TE (1979) *Mol. Pharmacol.* **16:** 556–568.
Ferriola PC, Cody V & Middleton Jr E (1989) *Biochem. Pharmacol.* **38:** 1617–1624.

Fewtrell CMS & Gomperts BD (1977) *Nature (Lond.)* **265:** 635–636.
Franchi-Micheli S, Della Corte L, Puliti R et al (1980) *Boll. Soc. Ital. Sper.* **56:** 2521–2524.
Francis AR, Shetty TK & Bhattacharya RK (1989) *Mutat. Res.* **222:** 393–401.
Frankfurt OS, Sepchina LP & Bunto TV (1967) *Bull. Exp. Biol. Med.* **8:** 86–88.
Fujiki H, Horiuchi T, Yamashita K et al (1986) *Prog. Clin. Biol. Res.* **213:** 429–440.
Fukushima S, Sakata T, Tagawa Y, Shibata MA, Hirose M & Ito N (1987a) *Cancer Res.* **47:** 2113–2116.
Fukushima S, Ogiso T, Kurata Y, Hirose M & Ito N (1987b) *Cancer Lett.* **34:** 83–90.
Funato T, Yokota J, Sakamoto H et al (1987) *Gann* **78:** 689–694.
Gad SC, Leslie SW & Acosta D (1979) *Toxicol. Appl. Pharmacol.* **49:** 45–52.
Galeotti T, Borello S, Minotti G & Masotti L (1986) *Ann. NY Acad. Sci.* **488:** 468–480.
Gower JD (1988) *Free Radicals Biol. Med.* **5:** 95–111.
Graziani Y & Chayoth R (1979) *Biochem. Pharmacol.* **28:** 397–403.
Grice HC (1988) *Food Chem. Toxicol.* **26:** 717–723.
Gryglewski RJ, Korbut R, Robak J & Swies J (1987) *Biochem. Pharmacol.* **36:** 317–322.
Gschwendt M, Horn F, Kittstein W & Marks F (1983) *Biochem. Biophys. Res. Commun.* **117:** 444–447.
Gschwendt M, Kittstein W & Marks F (1984) *Cancer Lett.* **22:** 219–225.
Guarna A, Della Corte L, Giovannini MG, De Sarlo F & Sgaragli G (1983) *Drug Metabol. Dispos.* **11:** 581–584.
Guyton KZ, Bhan P, Kuppusamy P, Zweier JL, Trush MA & Kensler TW (1991) *Proc. Natl. Acad. Sci. USA* **88:** 946–950.
Hageman GJ, Verhagen H & Kleinjans JCS (1988) *Mutat. Res.* **208:** 207–211.
Halliwell B (1989) *Br. J. Exp. Pathol.* **70:** 737–757.
Havsteen B (1983) *Biochem. Pharmacol.* **32:** 1141–1148.
Hirose M, Asamoto M, Hagiwara A et al (1987) *Toxicology* **45:** 13–24.
Hirose M, Masuda A, Fukushima S & Ito N (1988) *Carcinogenesis* **9:** 101–104.
Hirose M, Masuda A, Hasegawa R, Wada S & Ito N (1990) *Carcinogenesis* **11:** 239–244.
Hodnick WF, Kung FS, Roettger WJ, Bohmont CW & Pardini RS (1986) *Biochem. Pharmacol.* **35:** 2345–2357.
Hope WC, Welton AF, Fiedler-Nagy C, Batula-Bernardo C & Coffey JW (1983) *Biochem. Pharmacol.* **32:** 367–371.
Ikeda GJ, Stewart JE, Sapienza PP et al (1986) *Food Chem. Toxicol.* **24:** 1201–1221.
Imaida K, Fukushima S, Shirai T, Ohtani M, Nakanishi K & Ito N (1983) *Carcinogenesis* **4:** 895–899.
Imaida K, Fukushima S, Shirai T, Masui T, Ogiso T & Ito N (1984) *Gann* **75:** 769–775.
Imaida K, Fukushima S, Inoue K, Masui T, Hirose M & Ito N (1988) *Cancer Lett.* **43:** 167–172.
Inai K, Kobuke T, Nambu S et al (1988) *Gann* **79:** 49–58.
Ito N, Hirose M, Fukushima S, Tsuda H, Shirai T & Tatematsu M (1986) *Food. Chem. Toxicol.* **24:** 1971–1982.
Ito N, Hagiwara A, Tamano S et al (1989) *Gann* **80:** 317–325.
Iverson F, Truelove J, Nera E, Lok E, Clayson DB & Wong J (1986) *Food Chem. Toxicol.* **24:** 1197–1200.
Jones FE, Komorowski RA & Condon RE (1984) *J. Surg. Oncol.* **25:** 54–60.
Kahl R (1984) *Toxicology* **33:** 185–228.

Kahl R (1986) *J. Environ. Sci. Health* **C4:** 47–92.
Kahl R, Weimann A, Weinke S & Hildebrandt AG (1987) *Arch. Toxicol.* **60:** 158–162.
Kahl R, Weinke S & Kappus H (1989) *Toxicology* **59:** 179–194.
Kaneko T & Matsuo M (1985) *Chem. Pharm. Bull. (Tokyo)* **33:** 1899–1905.
Kappus H & Sies H (1981) *Experientia* **37:** 1233–1241.
Katayama K, Shirota H, Kobayashi S, Terato K, Ikuta H & Yamatsu I (1987) *Agents Actions* **21:** 269–271.
Kato R, Nakadate T, Yamamoto S & Sugimura T (1983) *Carcinogenesis* **4:** 1301–1305.
Kensler TW & Taffe BG (1986) *Free Radicals Biol. Med.* **2:** 347–387.
Khan WA, Wang ZY, Athar M, Bickers DR & Mukhtar H (1988) *Cancer Lett.* **42:** 7–12.
Kirkpatrick DC & Lauer BH (1986) *Food Chem. Toxicol.* **24:** 1035–1037.
Kitchin KT & Brown JL (1987) *Food Chem. Toxicol.* **25:** 603–607.
Kozumbo WJ, Seed JL & Kensler TW (1983) *Cancer Res.* **43:** 2555–2559.
Ku G, Doherty NS, Schmidt LF, Jackson RL & Dinerstein RJ (1990) *FASEB J.* **4:** 1645–1653.
Kuehl Jr FA, Humes JL, Egan RW, Ham EA, Beveridge GC & Van Arman CG (1977) *Nature (Lond.)* **265:** 170–173.
Kurechi T & Kato T (1983) *Chem. Pharm. Bull. (Tokyo)* **31:** 1772–1776.
Landolfi R, Mower RL & Steiner M (1984) *Biochem. Pharmacol.* **33:** 1525–1530.
Laughton MJ, Halliwell B, Evans PJ & Hoult JRS (1989) *Biochem. Pharmacol.* **38:** 2859–2865.
Lindenschmidt RC, Tryka AF, Goad ME & Witschi H (1986) *Toxicology* **38:** 151–160.
Lindenschmidt RC, Tryka AF & Witschi H (1987) *Fundam. Appl. Toxicol.* **8:** 474–481.
Lutterodt GD (1989) *J. Ethnopharmacol.* **25:** 235–247.
MacGregor JT (1984) *Adv. Exp. Med. Biol.* **177:** 497–526.
Maeura Y & Williams GM (1984) *Food Chem. Toxicol.* **22:** 191–198.
Maeura Y, Weisburger JH & Williams GM (1984) *Cancer Res.* **44:** 1604–1610.
Malkinson AM & Beer DS (1983) *J. Natl. Cancer Inst.* **70:** 981–986.
Malkinson AM & Beer DS (1984) *J. Natl. Cancer Inst.* **73:** 925–933.
Maru GB & Bhide SV (1982) *Cancer Lett.* **17:** 75–80.
Masui T, Asamoto M, Hirose M, Fukushima S & Ito N (1987) *Cancer Res.* **47:** 5171–5174.
Masui T, Fukushima S, Katoh F, Yamasaki H & Ito N (1988) *Carcinogenesis* **9:** 1143–1146.
Matsuoka A, Matsui M, Miyata N, Sofuni T & Ishidate Jr M (1990) *Mutat. Res.* **241:** 125–132.
McCay PB, King M, Rikans L & Pitha JV (1980) *J. Environ. Pathol. Toxicol.* **3:** 451–465.
McCormick DL, Major N & Moon RC (1984) *Cancer Res.* **44:** 2858–2863.
Meyer O, Blom L & Olsen P (1978) *Arch. Toxicol.* **1:** 355–358.
Middleton Jr E (1984) *Trends Pharmacol. Sci.* **5:** 335–338.
Mikuni T, Tatsuta M & Kamachi M (1987) *J. Natl. Cancer Inst.* **79:** 281–283.
Minakami H, Sotomatsu A, Ohma C & Nakano M (1989) *Arzneimittelforsch./Drug Res.* **39:** 1090–1091.
Mira ML, Azevedo MS & Manso C (1987) *Free Radical Res. Commun.* **4:** 125–129.

Mirvish SS, Salmasi S, Cohen SM, Patil K & Mahboubi E (1983) *J. Natl. Cancer Inst.* **71:** 81–85.

Mizumoto K, Ito S, Kitazawa S, Tsutsumi M, Denda A & Konishi Y (1989) *Carcinogenesis* **10:** 1491–1494.

Moore GGI & Swingle KF (1982) *Agents Actions* **12:** 674–683.

Moore MA, Tsuda H, Ogiso T, Mera Y & Ito N (1984) *Cancer Lett.* **25:** 145–151.

Moore MA, Thamavit W & Ito N (1987a) *J. Natl. Cancer Inst.* **78:** 295–301.

Moore MA, Tsuda H, Thamavit W, Masui T & Ito N (1987b) *J. Natl. Cancer Inst.* **78:** 289–293.

Müller-Peddinghaus R & Wurl M (1987) *Biochem. Pharmacol.* **36:** 1125–1132.

Mukhtar H, Das M, Khan WA, Wang ZY, Bik DP & Bickers DR (1988) *Cancer Res.* **48:** 2361–2365.

Nakadate T, Yamamoto S, Iseki H et al (1982) *Gann* **73:** 841–843.

Nakadate T, Yamamoto S, Aizu E & Kato R (1984) *Gann* **75:** 214–222.

Nakagawa Y & Tayama K (1988) *Arch. Toxicol.* **61:** 359–365.

Nakagawa Y, Tayama K, Nakao T & Hiraga K (1984) *Biochem. Pharmacol.* **33:** 2669–2674.

Nakamura Y, Colburn NH & Gindhart TD (1985) *Carcinogenesis* **6:** 229–235.

National Academy of Sciences (1979) *1977 Survey of Industry on the Use of Food Additives*, Part 2, p 963; Part 3, pp 127, 296. Washington, DC: NAS.

Newberne PM, Charnley G, Adams K et al (1986) *Food Chem. Toxicol.* **24:** 1111–1119.

Nishino H, Naito E, Iwashima A et al (1984) *Gann* **74:** 311–316.

Ochiai M, Nagao M, Wakabayashi K & Sugimura T (1984) *Mutat. Res.* **129:** 19–24.

Olsen P, Meyer O, Bille N & Würtzen G (1986) *Food Chem. Toxicol.* **24:** 1–12.

Pagonis C, Tauber AI, Pavlotsky N & Simons ER (1986) *Biochem. Pharmacol.* **35:** 237–245.

Parthasarathy S, Young SG, Witztum JL, Pittman RC & Steinberg D (1986) *J. Clin. Invest.* **77:** 641–644.

Passi S, Picardo M & Nazzaro-Porro M (1987) *Biochem. J.* **245:** 537–542.

Peraino C, Fry RJM, Staffeldt E & Christopher JP (1977) *Food Cosmet. Toxicol.* **15:** 93–96.

Perchellet JP & Perchellet EM (1989) *Free Radicals Biol. Med.* **7:** 377–408.

Perera MIR, Demetris AJ, Katyal SL & Shinozuka H (1985) *Cancer Res.* **45:** 2533–2538.

Phillips BJ, Carroll PA, Tee AC & Anderson D (1989) *Mutat. Res.* **214:** 105–114.

Posati LP, Fox KK & Pallansch MJ (1970) *J. Agric. Food Chem.* **18:** 632–635.

Potenberg J, Schiffmann D, Kahl R, Hildebrandt AG & Henschler D (1986) *Cancer Lett.* **33:** 189–198.

Rahimtula A (1983) *Chem. Biol. Interact.* **45:** 125–135.

Rahman A, Shahabuddin, Hadi SM, Parish JH & Ainley K (1989) *Carcinogenesis* **10:** 1833–1839.

Rao AR (1982) *Int. J. Cancer* **30:** 121–124.

Rao MS, Lalwani ND, Watanabe TK & Reddy JK (1984) *Cancer Res.* **44:** 1072–1076.

Reddy BS & Maeura Y (1984) *J. Natl. Cancer Inst.* **72:** 1181–1187.

Ritter J, Kahl R & Hildebrandt AG (1985) *Res. Commun. Chem. Pathol. Pharmacol.* **47:** 48–58.

Robak J & Gryglewski RJ (1988) *Biochem. Pharmacol.* **37:** 837–841.

Roebuck BD, MacMillan DL, Bush DM & Kensler TW (1984) *J. Natl. Cancer Inst.* **72:** 1405–1410.

Roger CR (1988) *Experientia* **44:** 725–804.

Rössing D, Kahl R & Hildebrandt AG (1985) *Toxicology* **34:** 67–77.

Sakai A, Miyata N & Takahashi A (1990) *Carcinogenesis* **11:** 1985–1988.

Sakata T, Shirai T, Fukushima S, Hasegawa R & Ito N (1984) *Gann* **75:** 950–956.

Sato H, Takahashi M, Furukawa F et al (1987) *Cancer Lett.* **38:** 49–56.

Schwarz M, Peres G, Kunz W, Furstenberger G, Kittstein W & Marks F (1984) *Carcinogenesis* **5:** 1663–1670.

Scientific Committee for Food of the Commission of the European Communities (1987) *Report on Antioxidants*, p. 6. Brussels: Commission of the European Communities.

Sgaragli GP, Valoti M, Palmi M & Mantovani P (1989) *Pharmacol. Res.* **21:** 649–650.

Shirai T, Fukushima S, Ohshima M, Masuda A & Ito N (1984) *J. Natl. Cancer Inst.* **72:** 1189–1198.

Shirai T, Ikawa E, Hirose M, Thamavit W & Ito N (1985) *Carcinogenesis* **6:** 637–639.

Shirota H, Goto M, Hashida R, Yamatsu I & Katayama K (1989) *Agents Actions* **27:** 322–324.

Slaga TJ & Bracken WM (1977) *Cancer Res.* **37:** 1631–1635.

Slaga TJ, Bracken WM, Viaje A, Berry DL, Fischer SM & Miller DR (1978) *J. Natl. Cancer Inst.* **61:** 451–455.

Slaga TJ, Fischer SM, Weeks CE, Klein-Szanto AJP & Reiners J (1982) *J. Cell. Biochem.* **18:** 207–227.

Sokolove PM, Albuquerque EX, Kauffman FC, Spande TF & Daly JW (1986) *FEBS Lett.* **203:** 121–126.

Sousa RL & Marletta MA (1985) *Arch. Biochem. Biophys.* **240:** 345–357.

Steinberg D, Parthasarathy S, Carew TE, Khoo JC & Witztum JL (1989) *N. Engl. J. Med.* **320:** 915–924.

Stewart RM, Weir EK, Montgomery MR & Niewoehner DE (1981) *Respir. Physiol.* **45:** 333–342.

Taffe BG & Kensler TW (1988) *Res. Commun. Chem. Pathol. Pharmacol.* **61:** 291–303.

Takahashi O (1987) *Food Chem. Toxicol.* **25:** 219–224.

Takahashi O & Hiraga K (1978) *Food Cosmet. Toxicol.* **18:** 229–235.

Tamano S, Fukushima S, Shirai T, Hirose M & Ito N (1987) *Cancer Lett.* **35:** 39–46.

Tatsuta M, Mikuni T & Taniguchi H (1983) *Int. J. Cancer* **32:** 253–254.

Tauber AI, Fay JR & Marletta MA (1983) *Fed. Proc. Fed. Am. Soc. Exp. Biol.* **42:** 2063.

Thamavit W, Fukushima S, Kurata Y, Asamoto M & Ito N (1989) *Cancer Lett.* **45:** 93–101.

T'Hart BA, Ip Vai Ching TRAM, Van Dijk H & Labadie R (1990) *Chem. Biol. Interact.* **73:** 323–335.

Thompson DC & Trush MA (1988a) *Toxicol. Appl. Pharmacol.* **96:** 115–121.

Thompson DC & Trush MA (1988b) *Toxicol. Appl. Pharmacol.* **96:** 122–131.

Thompson DC & Trush MA (1989) *Chem. Biol. Interact.* **72:** 157–173.

Thompson DC, Cha YN & Trush MA (1989a) *J. Biol. Chem.* **264:** 3957–3965.

Thompson JA, Schullek KM, Fernandez CA & Malkinson AM (1989b) *Carcinogenesis* **10:** 773–775.

Thornton M, Moore MA & Ito N (1989) *Carcinogenesis* **10:** 407–410.

Tobe M, Furuya T, Kawasaki Y et al (1986) *Food Chem. Toxicol.* **24:** 1223–1228.

Troll W & Wiesner R (1985) *Annu. Rev. Pharmacol. Toxicol.* **25:** 509–528.
Trosko JE, Yotti LP, Warren ST, Tsushimoto G & Chang C-C (1982) *Carcinogenesis* **7:** 565–585.
Tsuda H, Fukushima S, Imaida K, Sakata T & Ito N (1984) *Acta Pharmacol. Toxicol.* **55 (suppl 2):** 125–143.
Ulland BM, Weisburger JH, Yamamoto RS & Weisburger EK (1973) *Food Cosmet. Toxicol.* **11:** 199–207.
Valoti M, Della Corte L, Tipton KF & Sgaragli G (1988) *Biochem. J.* **250:** 501–507.
Valoti M, Sipe Jr HJ, Sgaragli G & Mason RP (1989) *Arch. Biochem. Biophys.* **269:** 423–432.
Van der Heijden CA, Janssen PJCM & Strik JJTWA (1986) *Food Chem. Toxicol.* **24:** 1067–1070.
Van Duuren BL, Blazej T, Goldschmidt BM, Katz C, Melchionne S & Sivak A (1971) *J. Natl. Cancer Inst.* **46:** 1039–1044.
Van Esch GJ (1986) *Food Chem. Toxicol.* **24:** 1063–1065.
Van Wauwe J & Goossens J (1983) *Prostaglandins* **26:** 725–730.
Verma AK, Johnson JA, Gould MN & Tanner MA (1988) *Cancer Res.* **48:** 5754–5758.
Warngard L, Flodstrom S, Ljungquist S & Ahlborg UG (1987) *Carcinogensis* **8:** 1201–1205.
Wattenberg LW (1972a) *J. Natl. Cancer Inst.* **48:** 1425–1430.
Wattenberg LW (1972b) *Fed. Proc.* **31:** 633.
Wattenberg LW (1973) *J. Natl. Cancer Inst.* **50:** 1541–1544.
Wattenberg LW (1982) In Arnott MS, van Eys J & Wang YM (eds) *Molecular Interrelations of Nutrition and Cancer*, pp 43–56. New York: Raven Press.
Wattenberg LW & Sparnins VL (1979) *J. Natl. Cancer Inst.* **63:** 219–222.
Wattenberg LW, Jerina DM, Lam LKT & Yagi H (1979) *J. Natl. Cancer Inst.* **62:** 1103–1106.
Weinke S, Kahl R & Kappus H (1987) *Toxicol. Lett.* **35:** 247–251.
Weisburger EK, Evarts RP & Wenk ML (1977) *Food Cosmet. Toxicol.* **15:** 139–141.
Wester PW & Kroes R (1988) *Toxicol. Pathol.* **16:** 165–171.
Wheeler EL & Berry DL (1986) *Carcinogenesis* **7:** 33–36.
Williams GM (1986) *Food Chem. Toxicol.* **24:** 1163–1166.
Williams GM, Tanaka T & Maeura Y (1986) *Carcinogenesis* **7:** 1043–1050.
Witschi HP (1981) *Toxicology* **21:** 95–104.
Witschi HP & Doherty DG (1984) *Fundam. Appl. Toxicol.* **4:** 795–801.
Witschi H & Morse CC (1983) *J. Natl. Cancer Inst.* **71:** 859–866.
Witschi H, Williamson D & Lock S (1977) *J. Natl. Cancer Inst.* **58:** 301–305.
Witschi H, Malkinson AM & Thompson JA (1989) *Pharmacol. Ther.* **42:** 89–113.
Wood AW, Huang MT, Chang RL et al (1982) *Proc. Natl. Acad. Sci. USA* **79:** 5513–5517.
Würtzen G & Olsen P (1986) *Food Chem. Toxicol.* **24:** 1121–1125.
Yoshimoto T, Furukawa M, Yamamoto S, Horie T & Watanabe-Kohno S (1983) *Biochem. Biophys. Res. Commun.* **116:** 612–618.
Zhang HF, Davis WB, Chen XS, Whisler RL & Cornwell DG (1989) *J. Lipid. Res.* **30:** 141–148.
Zoccarato F, Pandolfo M, Deana R & Alexandre A (1987) *Biochem. Biophys. Res. Commun.* **146:** 603–610.

PROCESSES AND CELL RESPONSES

11

Role of Free Radicals in Carcinogen Activation

MICHAEL A. TRUSH and THOMAS W. KENSLER
Department of Environmental Health Sciences, Johns Hopkins School of Hygiene and Public Health, 615 N. Wolfe Street, Baltimore, Maryland 21205, USA

I. Introduction 277
II. The multistage nature of chemical carcinogenesis 278
III. Free radical-mediated activation of procarcinogens 280
IV. Oxidation of complete carcinogens to free radicals 292
V. Free radical metabolites of tumour promoters and progressors 304
VI. Summary 310

I. INTRODUCTION

Over two centuries ago Hill (1761) and Pott (1775) suggested that discrete chemical agents were involved in the aetiology of cancer. Nearly a century ago Clunet observed that neoplasia could be produced by irradiation in experimental animals, and physicians noted radiation cancers in scientists who utilized radioactive sources (see Shimkin, 1977). Concordantly, it has long been recognized that oxygen has a primary role in radiation toxicity and that radicals and peroxides derived from molecular oxygen participate in radiation damage to cells. However, the possibility that free radicals might also be implicated in chemical carcinogenesis has been more difficult to establish. Nonetheless, nearly 50 years ago Kensler et al (1942) suggested that the proximate carcinogens formed from azo dyes such as butter yellow were probably their free radical forms. In the intervening decades consider-

Oxidative Stress: Oxidants and Antioxidants
ISBN 0-12-642762-3

able evidence has accumulated to demonstrate that many different classes of carcinogens can either be metabolized to free radical derivatives or can undergo metabolic activation through interactions with free radicals, particularly those derived from molecular oxygen. It is the purpose of this chapter to highlight some of the key elements of the mechanisms by which free radicals mediate the activation of carcinogens and to discuss the roles of these reactive intermediates in the carcinogenic process.

II. THE MULTISTAGE NATURE OF CHEMICAL CARCINOGENESIS

Experimental and epidemiological studies have served to define many causative agents in carcinogenesis and have established that the induction of cancer by chemicals involves stages of cellular evolution from normal, through preneoplastic and premalignant cells, to highly malignant neoplasia. In several model systems the stages of initiation, promotion and progression can be operationally defined through the use of different chemicals. Initiation is a phenomenon of gene alteration such as might be effected by carcinogen interaction with DNA and subsequent activation of proto-oncogenes. A key component in the understanding of the initial events of carcinogenesis was the recognition by the Millers that many chemical carcinogens are not chemically reactive per se, but must undergo metabolic activation to form an electrophilic reactant (Miller, 1970; Miller and Miller, 1985). These reactive species can interact with nucleophilic groups in DNA to induce point mutations and other genetic lesions. The importance of metabolic activation in carcinogenesis is highlighted by the fact that target organ specificities and even species susceptibilities can be determined through the presence or absence of metabolic activation pathways. The metabolism of chemicals to a proximate carcinogen often involves an initial two-electron oxidation to a hydroxylated or epoxidated product and is typically catalysed by the cytochrome P450 system. However, it is becoming increasingly apparent that one-electron oxidations or reductions of procarcinogens yielding a radical intermediate having an odd or spin-unpaired electron in its outer orbital may also play critical roles in the activation of carcinogens to DNA-damaging species. If this DNA damage is not repaired, a round of cell replication serves to fix the lesions in DNA and completes the process of initiation. However, the initiated cell as such is a dormant tumour cell in which the potential neoplastic phenotype remains unexpressed in the absence of subsequent exposure to tumour promoters.

Tumour promotion is a stage in carcinogenesis which encompasses events involved in the development of a benign neoplasia from initiated but preneoplastic tissue. This stage has been characterized by the use of chemical agents which are not by themselves carcinogenic, but which can modulate phenotypic expression in both initiated and non-initiated cells. In some circumstances, tumour promoters may modulate gene expression to result in the proliferation rather than differentiation of initiated cells. Coordinately, selection and clonal expansion of initiated cells may be facilitated when tumour promoters accelerate terminal differentiation of surrounding non-initiated cells or when they are particularly toxic to them. Although tumour promoters typically do not produce damage to DNA and are considered to be epigenetic in action, their effects are often ultimately on the genome. The molecular mechanisms by which tumour promoters affect gene expression and cell proliferation are incompletely resolved, but clearly involve the signal transduction cascades employed by growth factors and other regulators of cellular homeostasis. One component of the action of tumour promoters on tissues involves the elaboration of free radicals. While detailed reviews of the involvement of free radicals in tumour promotion can be found elsewhere (Cerutti, 1985; Kensler and Taffe, 1986), the experimental observations can be summarized briefly. The bulk of the evidence supporting a role for free radicals in these processes comes indirectly from studies demonstrating inhibition of tumour promotion and/or progression by superoxide dismutase (SOD)-mimicking agents, the synthetic antioxidants butylated hydroxyanisole (BHA) and butylated hydroxytoluene (BHT), or the naturally occurring antioxidants α-tocopherol and glutathione (reviews: Kensler and Trush, 1985; Perchellet and Perchellet, 1989; Kensler and Guyton, 1991). Further, reactive oxygen-generating systems are able to mimic some of the biochemical actions of tumour promoters, such as enhancement of transformation in vitro and altered gene expression (Nakamura et al, 1988; Crawford et al, 1989). Tumour promoters are also known to stimulate the endogenous production of reactive oxygen species in several cell types, including inflammatory cells and keratinocytes (DeChatelet et al, 1976; Fischer et al, 1986; Robertson et al, 1990). Tumour promoters also provoke a rapid and sustained decrease in cellular antioxidant defences, including SOD, catalase, and glutathione peroxidase activities (Solanki et al, 1981), which in turn may enhance the prooxidant status of cells. The most direct evidence to date that free radicals are involved in the later stages of carcinogenesis are the observations that free radical-generating compounds are effective tumour promoters and progressors (Slaga et al, 1981). The metabolism of these types of compounds is reviewed in Section IV.B.

An additional component of carcinogenesis is the progression of a relatively benign lesion to a highly malignant, rapidly growing neoplasm. While repetitive treatment of initiated mouse epidermis with phorbol esters is not particularly effective in the conversion of papillomas to carcinomas, subsequent treatment with initiating agents or peroxides enhances this progression to malignancy (Hennings et al, 1983; O'Connell et al, 1986). This final stage may involve a second, discrete, inheritable event that involves either further direct chemical interaction with DNA or the transposition of genetic material. The net result will be the irreversible transformation of one or more cells in the benign tumour into autonomously growing cancer cells.

III. FREE RADICAL-MEDIATED ACTIVATION OF PROCARCINOGENS

A. Enzymatic and non-enzymatic activation of carcinogens and other xenobiotics

One of the basic premises underlying mechanistic toxicology is the demonstration that many xenobiotics, including most mutagens and carcinogens, are biotransformed from inert chemicals to highly reactive metabolites capable of interacting with cellular constituents (Gillete et al, 1974; Miller and Miller, 1985) (Figure 1). It is generally acknowledged that ultimate carcinogenic metabolites are electrophilic, with the initiation of chemical carcinogenesis closely linked to the formation of a specific carcinogen–DNA adduct(s). Subsequently, we have come to appreciate that reactive metabolite-dependent interactions with other biomolecules and the ensuing alteration of cellular activities may also be critical to both the promotion and the progression stages of neoplastic development. While it has been possible to characterize the chemical identity of many reactive metabolites and to examine their interaction with potential target molecules, in many instances the identity of the critical target molecule(s) and the sequence of ensuing changes in cellular function are presently lacking.

The metabolic activation of carcinogens is highly dependent upon the presence of appropriate activating enzymes. The best-characterized enzyme system identified in the bioactivation of procarcinogens is the mixed function oxidase system (MFO). This system consists of two flavoproteins, NADPH–cytochrome P450 (cytochrome *c*) reductase and NADH–cytochrome b_5 reductase, and a terminal oxidase, cytochrome P450 (Lu and West, 1978). Within this system, hydroxylation and epoxidation reactions are catalysed exclusively by one of the many cytochrome P450s found in the endoplasmic reticulum. At the molecular level, it appears that these reactions are mediated by an oxidant bound to the haem iron (White and Coon, 1980). Depending upon the particular substrate molecule, one-electron oxidations or reductions yielding radical intermediates can also be catalysed by either cytochrome P450 or NADPH–cytochrome P450 reductase. Carbon tetrachloride is a classic example of a chemical metabolized to a radical intermediate by cytochrome P450, whereas paraquat, a herbicide, exemplifies a chemical oxidized by NADPH–cytochrome P450 reductase to a radical intermediate (Mason, 1979). Radicals derived from chemicals such as paraquat, nitrofurantoin and adriamycin are capable of activating molecular oxygen by univalent reduction (Trush et al, 1982d). The continued oxidation–reduction, or redox cycling, of such agents results in the cellular accumulation of active oxygen metabolites, resulting in an oxidative stress state. The MFO system also generates superoxide (O_2^-), hydrogen peroxide (H_2O_2), singlet oxygen ($^1O_2$) and a hydroxyl-like radical which can participate in xenobiotic activation mechanisms (White and Coon, 1980). For example, benzo(a)pyrene (BP), a carcinogenic polycyclic aromatic hydrocarbon, is metabolized to dihydrodiols by H_2O_2 in the presence of hepatic microsomes prepared from phenobarbitone-pretreated rats (Reneeberg et al, 1981). Likewise, Johansson and Ingelman-Sundberg (1983) demonstrated that SOD inhibits the conversion of benzene, a leukaemogenic agent, to phenol by a reconstituted cytochrome P450 system. In this system, the metabolism of benzene to phenol and the covalent binding of metabolites to protein were also inhibited by hydroxyl radical scavengers. It was hypothesized that the interaction of benzene with cytochrome P450-generated hydroxyl radicals resulted in the formation of a hydroxycyclohexadienyl radical.

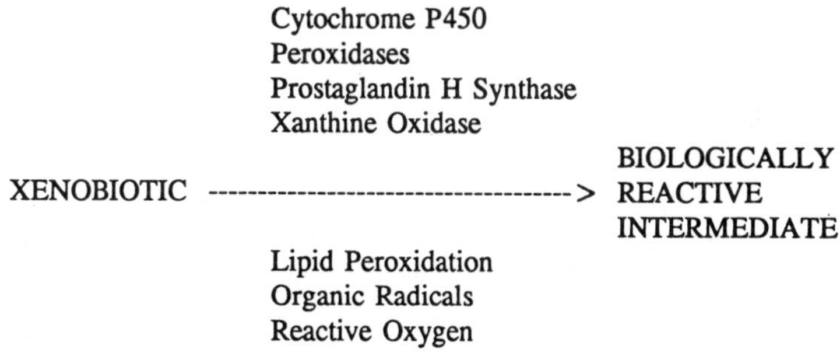

Figure 1. Various processes by which xenobiotics can be converted to biologically reactive intermediates. These reactive intermediates may function as tumour initiators, promoters, or progressors as well as mediators of other toxic manifestations.

The antibiotic bleomycin is an example of a chemical which is activated by microsomal-derived superoxide to a DNA-damaging intermediate (Trush et al, 1982a). Studies with purified NADPH–cytochrome *c* reductase and intact microsomes have shown that SOD inhibits bleomycin-mediated cleavage of DNA deoxyribose (Kappus and Mahmutoglu, 1986). A proposed role for O_2^- is the reduction of bleomycin-bound ferric iron to the more reactive ferrous form. Isolated lung nuclei can also catalyse the activation of bleomycin by a SOD-inhibitable reaction (Trush et al, 1983). While this activation mechanism is relevant to bleomycin-induced lung toxicity, it may also be critical to the transforming and genotoxic actions of bleomycin (Borek and Troll, 1983; Vig and Lewis, 1978). The risk of induction of secondary tumours by antineoplastic agents such as bleomycin is of public health concern.

Prostaglandin H synthase (PHS) is another membrane-associated enzyme that has been demonstrated to participate in the metabolic activation of xenobiotics, including a number of carcinogens (Marnett, 1981; Reed, 1988; Eling et al, 1990). The PHS system consists of a cyclooxygenase and a hydroperoxidase component. Mechanistic studies have demonstrated that xenobiotic cooxidations are facilitated by the hydroperoxidase component through both enzyme-free and enzyme-bound oxidants. Reactive oxygen has been proposed as an intermediate in the catalytic actions of this system. Although the identity of the free oxidant is unknown, it has been suggested that its behaviour is similar in many respects to that of the hydroxyl radical (Kuehl et al, 1980). Using low-level chemiluminescence technology, Sies et al (1984) have also provided evidence for the generation of $^1O_2$ during the conversion of PGG_2 to PGH_2 by PHS. Support for an oxidant-mediated reaction in PHS-catalysed cooxygenation reactions stems from the ability of antioxidants such as BHA and BHT to inhibit the cooxygenation reaction. However, BHA and BHT are in themselves cooxidation substrates for this enzyme (Thompson et al, 1989b), clouding this conjecture. Recently Hughes et al (1989) have demonstrated that lipoxygenase, another enzyme involved in cellular arachidonic acid metabolism, mediated the cooxidation of benzo[a]pyrene-7,8-dihydrodiol (BP-7,8-diol), the proximate carcinogenic metabolite of BP, via an enzyme-associated peroxyl radical. The activation of both BP and BP-7,8-diol by PHS has been extensively studied using various microsomal preparations as well as intact cells. Because PHS activity is found in target organs for chemical carcinogenesis, such as bladder, lung and skin, PHS-dependent cooxidation may play an important role in carcinogen activation. However, the overall contribution of the PHS system to the in vivo metabolism or activation of xenobiotics has yet to be firmly established. In this regard, the best evidence to date is for the nitrofuran FANFT, a bladder carcinogen which is activated by PHS to a reactive intermediate in intact animals (Zenser et al, 1984).

Like the hydroperoxidase component of PHS, a number of other cell- or tissue-specific peroxidases could participate in the metabolic activation of procarcinogens (O'Brien, 1988), although, unlike PHS, they do not form their own hydroperoxide substrate. These peroxidases include lactoperoxidase, uterine peroxidase, thyroid peroxidase and the myeloperoxidase of polymorphonuclear leukocytes (PMNs). Activation via peroxidases may account for the cell- and/or tissue-specific toxicities often observed with xenobiotics. Likewise, the activation of xenobiotics by PMNs via myeloperoxidase-dependent mechanisms is one of several possible ways in which inflammatory states could contribute to the carcinogenic process (Eastmond and Smith, 1990; Trush and Kensler, 1991).

While membrane-associated enzymes clearly play a role in the activation of exogenous chemicals, cytoplasmic enzymes, such as xanthine oxidase, may also be involved in xenobiotic processing and activation. For example, xanthine oxidase facilitates the activation of α-methyldopa and 2-hydroxyoestrogen to intermediates which covalently bind to proteins in a reaction inhibited by SOD (Dybing et al, 1976; Nelson et al, 1976). Enhanced bleomycin-induced DNA damage is also observed in the presence of xanthine oxidase. Additionally, xanthine oxidase catalyses the reduction of a number of compounds to radical intermediates (Boyd et al, 1979; Josephy et al, 1981; Pan and Bachur, 1980). Although little xanthine oxidase is normally found in cells, hypoxic conditions result in the increased conversion of xanthine dehydrogenase to xanthine oxidase (McCord, 1985). Reiners et al (1987) have also demonstrated a three-fold increase in xanthine oxidase activity related to cell hyperplasia and terminal differentiation of keratinocytes. This increase in xanthine activity requires several days to reach maximum and may be due to de novo synthesis of xanthine dehydrogenase as well as the conversion of both newly synthesized and pre-existing dehydrogenase to the oxidase form (Pence and Reiners, 1987). Given that the skin is a target organ for many carcinogens, such a reaction may be pertinent to xenobiotic processing and the oxidative state of the skin (Fischer et al, 1989).

In addition to enzyme-mediated chemical activation, several chemical-initiated reactions of toxicological significance have also been shown to contribute to xenobiotic activation. For example, Hewer et al (1979) demonstrated the formation of various dihydrodiols, including BP-7,8-diol, from BP by an ascorbic acid–ferrous iron–EDTA system. Ferric iron addition has also been shown to enhance the formation of BP-7,8-diol from BP by rat liver microsomes. The presence of iron is well known to enhance lipid peroxidation. In fact, Dix and Marnett (1983) have demonstrated that peroxidative epoxidation of BP-7,8-diol to diolepoxide occurs during the lipid peroxidation of rat liver microsomes initiated by either a NADPH, ADP–Fe^{3+} and EDTA–Fe^{2+} system or an ascorbate, ADP–Fe^{3+} system.

Marshall and O'Brien (1984) have also demonstrated that addition of blue asbestos to rat liver microsomes resulted in microsomal lipid peroxidation and the activation and covalent binding of 2-fluorenylacetamide (2-FAA). Both lipid peroxidation and 2-FAA activation were inhibited by the addition of EDTA, suggesting that iron is involved. Indeed, previous studies have established that asbestos-associated iron can undergo redox cycling, resulting in the generation of reactive oxygen metabolites. Asbestos has also been shown to facilitate the metabolism of BP through a reactive oxygen-mediated reaction (Graceffa and Weitzman, 1987). Such observations provide one oxidative-based mechanism which could contribute to the pulmonary carcinogenic actions of asbestos (Mossman et al, 1990).

As with asbestos, concern has been expressed regarding the possible relationship between ozone (O_3) exposure and pulmonary carcinogenesis (Witschi, 1988). It is well known that the interaction of O_3 with biological systems results in the initiation of a number of oxidative processes, including lipid peroxidation, the breakdown of lipids through ozonolysis and the generation of H_2O_2 (Mudd and Freeman, 1977). Ozone is also known to interact with and activate a number of xenobiotics (Burleson et al, 1979; Caulfield et al, 1979). Recently, Kozumbo et al (1989) have demonstrated that O_3 can directly activate the tobacco smoke constituent 1-naphthylamine (1-NA) to a product that elicits DNA damage in human lung cells as monitored by alkaline elution. Catalase partially inhibited this clastogenic activity of 1-NA, suggesting that the activation of 1-NA by O_3 is mediated in part by H_2O_2. Lungs isolated from ozone-exposed rats have also been shown to have altered BP metabolism as demonstrated by increased BP-quinone formation (Bassett et al, 1988). In intact organisms another important aspect of ozone action is its ability to induce pulmonary inflammation (Koren et al, 1989). PMNs recruited into the lung following ozone exposure can metabolically activate BP-7,8-diol to a DNA-damaging species (Esterline et al, 1989).

B. Enhanced chemical activation via radical-mediated reactions: a basis for chemical–chemical interactions

It is well established that the simultaneous or sequential exposure of biological systems to one chemical can enhance or promote the toxicological response induced by another agent. The exposure to tumour promoters following initiation by a carcinogen is an excellent example of this concept. Recently, a number of examples have been described in which oxygen-derived or organic free radicals have been shown to enhance the metabolic activation of another chemical (Figure 2). Such a mechanism provides a novel basis for chemical–chemical interactions. For example, McManus and

Davies (1980) demonstrated that O_2^- generated by the enzyme-catalysed redox cycling of paraquat enhanced the activation of L-dopa to a covalent-binding species. As expected, SOD inhibited this process. In a similar fashion, superoxide-generating compounds, including mitomycin C, nitrofurantoin and paraquat, have been shown to stimulate the iron-dependent activation of bleomycin A_2 to a DNA-damaging intermediate (Trush et al, 1982b). Such a chemical–chemical interaction provides a basis for the enhancing effect of mitomycin C on both the chemotherapeutic effect and the pulmonary toxic actions of bleomycin.

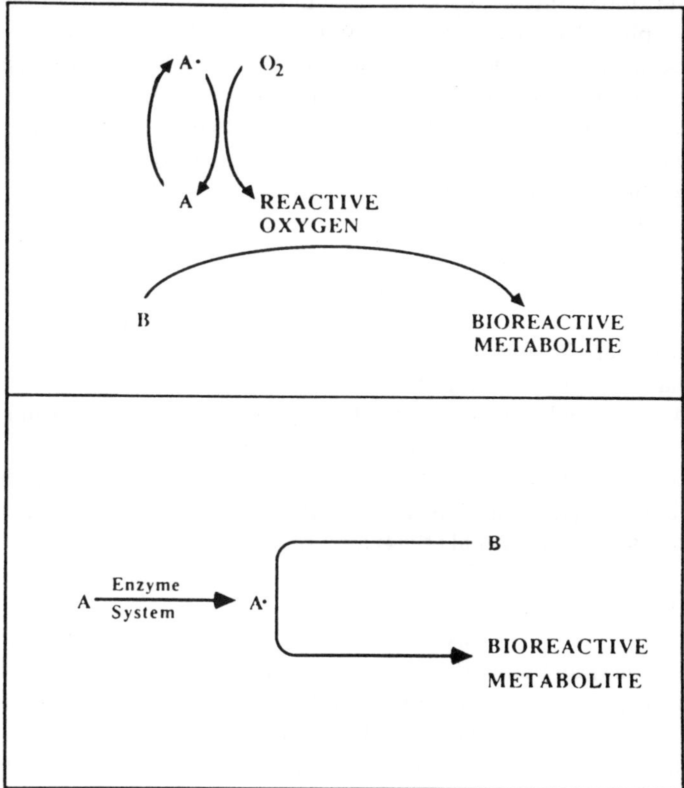

Figure 2. Diagrammatic representation of two general mechanisms whereby one chemical can enhance the metabolic activation of a second chemical. Reactive oxygen species generated by the enzyme-catalysed redox cycling of chemical A interacts with chemical B, resulting in the activation of B to an intermediate which can alter biomolecules (top), or the radical intermediate of A directly interacts with B, resulting in the enhanced activation of B (bottom).

Organic free radicals, including peroxyl and phenoxyl radicals, may also act to enhance the biotransformation of secondary chemicals (Marnett, 1987; Trush and Thompson, 1988). Lovstad (1979), for example, observed that a chlorpromazine radical enhanced the peroxidase-catalysed oxidation of catecholamines. Subrahmanyam and O'Brien (1985) have demonstrated that phenol stimulated the peroxidase-catalysed oxidation of *O-O*-biphenol, resulting in an intermediate that bound to DNA. Eastmond et al (1987) have subsequently shown that phenol also enhances the peroxidase-catalysed activation of hydroquinone. Phenol and hydroquinone are both metabolites of benzene and both are capable of forming radical intermediates (Irons and Sawahata, 1985). Interestingly, the simultaneous, but not separate, administration of phenol and hydroquinone to mice resulted in myelotoxicity similar to that seen with benzene (Eastmond et al, 1987). This observation suggests that such chemical–chemical interactions could be relevant in vivo. Similarly, a series of studies by Thompson et al (1989a) and Thompson and Trush (1989) has recently shown that a number of peroxidase-elicited phenoxyl radicals can enhance the activation of BHT to an electrophilic intermediate, BHT-quinone methide. Consistent with these in vitro observations, BHA was shown to enhance the pulmonary oedema resulting from the administration of subthreshold toxic doses of BHT (Thompson and Trush, 1988). Such a radical-mediated interaction provides an alternative mechanism for the formation of BHT-quinone methide in target tissues other than by cytochrome P450 (Thompson et al, 1987; Malkinson et al, 1989). As discussed in Section V.C, BHT-quinone methide has been recently implicated in the tumour-promoting actions of BHT.

As previously mentioned, the epoxidation of BP-7,8-diol at the 9,10 double bond results in its activation to an ultimate mutagenic and carcinogenic diolepoxide. Reed et al (1984) have demonstrated that a PHS-catalysed peroxyl radical of phenylbutazone, an antiinflammatory agent, facilitated the epoxidation of BP-7,8-diol as monitored by the generation of tetraols, which are stable hydrolysis products of diolepoxides. Previous work by Marnett (1987) and coworkers has established the involvement of peroxyl radicals as epoxidizing agents for BP-7,8-diol. In a similar fashion, Reed et al (1988) have also shown that the activation of cyclopento(c,d)pyrene (CPP), an environmental and occupational contaminant, was activated to a mutagenic derivative via a peroxyl radical-mediated reaction. Of particular interest here is that CPP is the first environmentally relevant carcinogenic polycyclic aromatic hydrocarbon found to be directly activated by a peroxyl radical system without undergoing prior epoxidation by cytochrome P450 to a diol derivative.

Activation of CPP to a bacterial mutagenic species was also observed during (bi)sulphite autoxidation (Reed, 1987). It has previously been shown that the metal-catalysed oxidation of (bi)sulphite (SO_3^{2-}), the form of sulphur dioxide found in aqueous solutions near neutral pH, yields the sulphur trioxide radical anion ($^{\bullet}SO_3^-$) (Reed et al, 1986). The reaction of $^{\bullet}SO_3^-$ with molecular oxygen results in a peroxyl free radical ($^-O_3SOO^{\bullet}$) which exhibits epoxidizing activity. Reed et al (1986) and Reed (1987) have previously shown that BP-7,8-diol is activated to diolepoxide during the autoxidation of (bi)sulphite. The conversion of SO_3^{2-} to $^{\bullet}SO_3^-$ can also be catalysed by a peroxidase with the $^{\bullet}SO_3^-$ reacting either with O_2 to yield $^-O_3SOO^{\bullet}$ or with BP-7,8-diol to yield a sulphonate adduct(s) (Curtis et al, 1988). More recently, Green and Reed (1990) have shown that the sulphonate adduct resulting from the nucleophilic addition of sulphite to the 9,10-oxirane ring of anti-BP-diolepoxide facilitates covalent binding to calf thymus DNA as well as enhanced bacterial mutagenicity. These observations are significant in light of the demonstration that sulphur dioxide was cocarcinogenic for the development of BP-induced pulmonary carcinomas.

One of the consequences of an oxidative stress state in cells is the generation of thiyl radicals ($GS^{\bullet}$) resulting from the oxidation of cellular glutatione (GSH). Peroxidases can also catalyse the oxidation of GSH to $GS^{\bullet}$ (Turkall and Tsan, 1982; Ross et al, 1985; Schreiber et al, 1989). Over the last several years, a number of interesting reactions of $GS^{\bullet}$ have been described which are of relevance to mechanisms of chemical activation. This concept has recently been reviewed by O'Brien (1988). As shown by Wefers and Sies (1983), thiol oxidation can also result in oxygen activation to O_2^- and $^1O_2$. Likewise, $GS^{\bullet}$ has been shown to support the oxygen-dependent cooxidation of styrene to styrene oxide (Stock et al, 1986). The active epoxidizing species is believed to be a thiyl peroxyl radical ($GSOO^{\bullet}$). Styrene oxide is a mutagenic and carcinogenic metabolite of styrene (Sugiura and Goto, 1981; Lijinsky, 1986).

C. Activation of xenobiotics by inflammatory cells via oxidant-dependent mechanisms

Although various flavoproteins and haem-proteins confer upon cells in most tissues the capability to generate oxidants, inflammatory cells normally exhibit this activity to the greatest extent. These cells include PMNs, monocytes and a variety of macrophages. The generation of superoxide by inflammatory cells is attributed to the protein kinase C-mediated activation of a plasma membrane NADPH oxidase. Utilization of superoxide-derived H_2O_2 by myeloperoxidase (MPO) results in the generation of hypohalous acids, such as HOCl, and an entity with singlet oxygen-like reactivity

(Klebanoff, 1988). HOCl can interact with amines, such as taurine, resulting in the generation of a more long-lived chloramine oxidant (Weiss et al, 1982). In contrast to PMNs and monocytes, alveolar macrophages and probably other tissue macrophages lack MPO. Therefore, alveolar macrophages lack the ability to activate xenobiotics through MPO-dependent reactions. Macrophages, by contrast, exhibit the ability to generate a variety of nitrogen-derived oxidants (Marletta, 1988). This class of oxidants has been shown to interact with various amines, resulting in the generation of carcinogenic *N*-nitrosamines (Miwa et al, 1987). Thus, inflammatory cells exhibit an array of biochemical processes which generate a spectrum of oxidants capable of contributing to xenobiotic biotransformation and activation (Figure 3 and Table 1).

Table 1. Some xenobiotics which have been shown to be metabolically activated by PMN-derived oxidants or MPO.

Acetominophen	Oestradiol
Aminofluorene	Eugenol
Arylamines	Hydroquinone
Benzidine	*N*-Methylaminoazobenzene
Benzo[a]pyrene-7,8-diol	1-Naphthol
Bleomycin	Phenetidine
Diethylstilboesterol	Phenol

The activation of the redox metabolism of inflammatory cells creates an oxidative stress state in organs, leading to tissue damage and pathology. Over the last several years, a number of in vitro and in vivo models have been developed which demonstrate this concept. Included among the disease states that have been implicated in chronic inflammatory states is the induction of cancer. Pertinent to this issue are the observations that phagocyte-generated oxidants have been shown to: (1) cause various lesions in DNA; (2) be mutagenic to bacteria and malignant cells; (3) alter oncogene methylation in cells; (4) cause the malignant transformation of cells in culture; (5) potentiate the carcinogenicity of several xenobiotics in animals; and (6) activate procarcinogens in genotoxic intermediates in vitro and in vivo. Each of these effects has been reviewed in more detail elsewhere (Weitzman and Gordon, 1990; Trush and Kensler, 1991).

With regard to the activation of xenobiotics by inflammatory cells, such a mechanism of activation could contribute to each of the stages of carcinogenesis, depending upon the chemical involved. Such an activation mechanism would be particularly pertinent to organs in which inflammatory cells are normally found, such as the bone marrow, or in which they accumulate

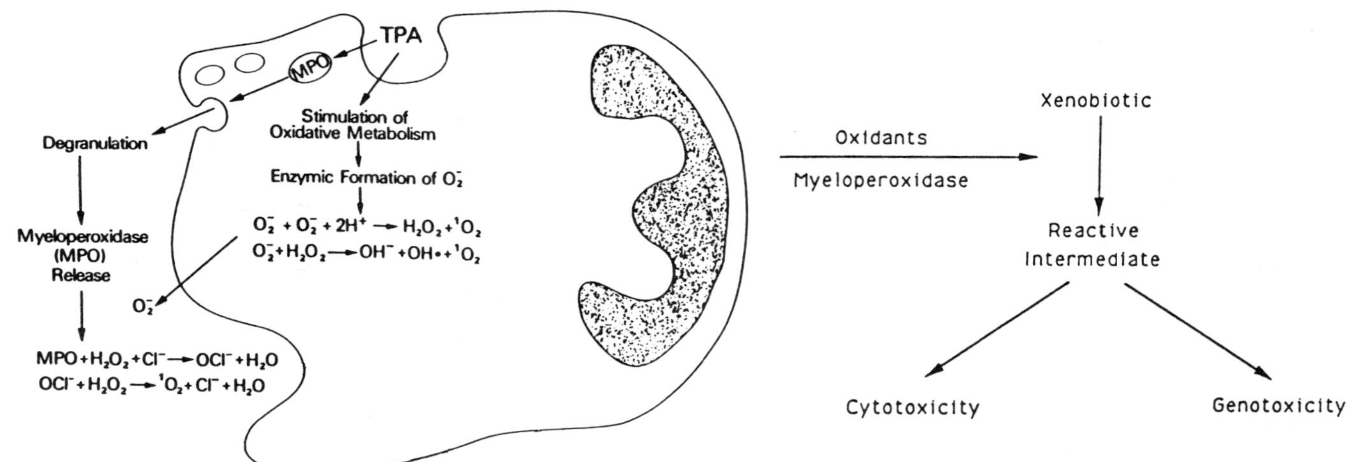

Figure 3. General mechanism whereby the interaction of a xenobiotic with PMN-derived oxidants or myeloperoxidase could result in the generation of reactive intermediates, resulting in cytotoxicity and/or genotoxicity (see Table 1).

in response to exposure to biological and chemical agents. A number of studies have previously examined the metabolism and activation of chemicals by cytochrome P450 in immunological and inflammatory cells, particularly lymphocytes and macrophages (Gram et al, 1986). Over the last several years, however, substantial evidence has been put forth demonstrating that the interaction of xenobiotics with either phagocyte-generated oxidants, O_2^-, $^{\bullet}OH$, HOCl, or MPO results in their activation to reactive intermediates (Eastmond and Smith, 1990; Trush and Kensler, 1991). Some of the first studies along these lines demonstrated the oxidation of oestradiol by the peroxidase found in PMNs and eosinophils. Klebanoff (1977) demonstrated by autoradiography that oestradiol can be activated to a covalent-binding species by human PMNs undergoing phagocytosis. Monocytes exhibited this activity to a lesser degree than PMNs, while lymphocytes were inactive. Additional studies demonstrated that the activation of oestradiol by PMNs was predominantly MPO-dependent and probably involved the conversion of oestradiol to phenoxyl radicals. Further, Geuskens et al (1977) have demonstrated with ultrastructural autoradiography the binding of oestradiol to eosinophils and neutrophils in tissue fragments, including the bone marrow, and in PMNs in a cholangioma produced by administration of 3′-methyl-4,4′-dimethylaminobenzene. In this study, most of the binding was observed in the peroxidase-containing granules. Other xenobiotics that can generate phenoxyl radicals, such as eugenol, naphthol and phenol, have also been shown to be activated by either PMNs or a MPO–H_2O_2 system to covalent-binding species (Eastmond et al, 1987; Thompson et al, 1989b; Eastmond and Smith, 1990).

Recent evidence has been put forth which indicates that the presence of MPO in bone marrow myeloid cells could be an underlying determinant of chemically induced toxicity to bone marrow in general as well as to specific cell types within this organ. For example, a number of metabolites of benzene have been shown to be activated by peroxidases, including MPO and PHS, and include catechol, hydroquinone and phenol (Eastmond et al, 1987; Sadler et al, 1988). Much of the recent focus of benzene toxicity has centred on hydroquinone, both in terms of factors which regulate its bioactivation by peroxidases (Smith et al, 1989; Schlosser et al, 1990) and its detoxification (Twerdok and Trush, 1990). It has been shown that hydroquinone is toxic to bone marrow stromal cells and as such dysregulates both myelopoiesis and lymphopoiesis (King et al, 1989; Thomas et al, 1989). The stromal element of the bone marrow provides an important regulatory microenvironment and consists of macrophages and fibroblasts. Thomas et al (1989) have shown that hydroquinone is preferentially toxic to stromal macrophages rather than to a fibroblast cell line. More recently, Thomas et al (1990) have demonstrated that stromal macrophages have a greater ability to bioactivate hydroquinone relative to a fibroblastoid cell line. In a macrophage homogenate, the bioactivation of hydroquinone to a covalent-binding species is H_2O_2-dependent.

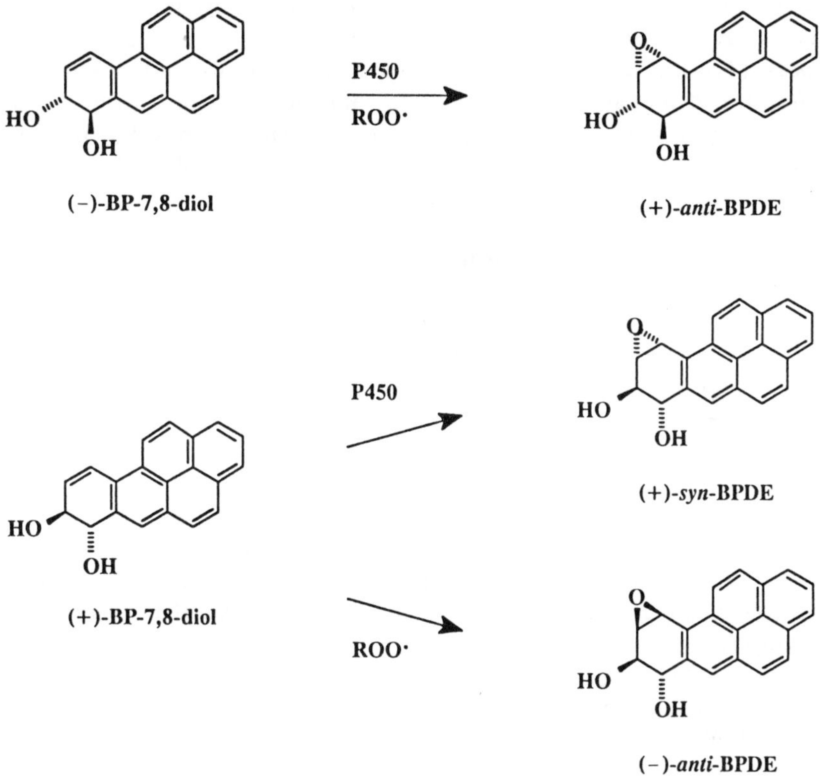

(−)-BP-7,8-diol → P450 / ROO· → (+)-*anti*-BPDE

(+)-BP-7,8-diol → P450 → (+)-*syn*-BPDE

(+)-BP-7,8-diol → ROO· → (−)-*anti*-BPDE

Figure 4. Stereochemistry of BP-7,8-diol epoxidation by peroxyl radicals and cytochrome P450.

Like benzene, polycyclic aromatic hydrocarbons, including BP, elicit bone marrow toxic effects, including the induction of leukaemogenesis (Nebert et al, 1980). DBA/2 mice are more susceptible to the bone marrow toxic effects of BP than are C57B1/6 mice. Correspondingly, Twerdok and Trush (1988a) have demonstrated that bone marrow-derived neutrophils from mice can activate BP-7,8-diol to a DNA-binding intermediate. Further, they have observed that bone marrow preparations from DBA/2 mice, which have more MPO than do preparations from C57B1/6 mice, can activate BP-7,8-diol to a greater extent (Twerdok and Trush, 1988b). Human PMNs can also activate BP-7,8-diol to a chemiluminescent dioxetane intermediate and an intermediate which covalently binds to DNA (Trush et al, 1985). In addition, this chemical–cell interaction elicits mutagenicity in bacteria and clastogenicity in V79-cells (Trush et al, 1985; Seed et al, 1990). Structure–activity studies have provided evidence that these genotoxic effects are due to the activation of BP-7,8-diol at the 9,10 double bond. Likewise, cell biology studies are consistent with the involvement of MPO in this activation process (Trush et al, 1985, 1991; Esterline et al, 1989). Recently it has been demonstrated that either TPA-stimulated human PMNs or a MPO–H_2O_2 system activates BP-7,8-diol to predominantly *anti*-diolepoxides, particularly (+)-*anti*-diolepoxide (Trush et al, 1990). (+)-*anti*-BP-diolepoxide is the ultimate carcinogenic metabolite of BP. As depicted in Figure 4, the stereospecific analysis of BP-7,8-diol metabolism was originally developed by Marnett (1987) as a monitor for peroxyl radical-dependent activation.

As an extension of the aforementioned in vitro studies, Kensler et al (1987) have provided evidence that the activation of BP-7,8-diol by PMNs can occur in vivo as well. In this study dual exposures to the phorbol ester TPA were used first to elicit the recruitment of PMNs into the mouse epidermis and subsequently to activate their redox metabolism. Under these conditions, increased binding of BP-7,8-diol to epidermal DNA was observed, whereas single exposures to TPA were without effect. Recently, using a similar protocol, Marnett (1990) has demonstrated enhanced formation of *anti*-diolepoxides using a ^{32}P-postlabelling assay capable of separating *anti*- and *syn*-derived diolepoxide adducts. Earlier studies by Marnett and coworkers have shown both in vitro and in vivo that there is an oxidative entity in naive keratinocytes that epoxidizes BP-7,8-diol to predominantly the *anti* enantiomer, with *anti*/*syn* ratios greater than 4 (Eling et al, 1986; Pruess-Schwartz et al, 1989). While the addition of antioxidants, such as BHA, prevents this epoxidation of BP-7,8-diol, these studies have not provided evidence which supports either PHS or MPO as the source of peroxidase involved.

PAH–DNA adducts in leukocytes have also been used as biomonitors of exposure (Harris et al, 1985). Recently, Rothman et al (1990) have observed transient PAH–DNA adducts in leukocytes, presumably PMNs, of individuals fed a charcoal-broiled beef diet for seven days. This observation may be reflective of the peroxidase-mediated activation of PAHs in bone marrow myeloid cells, since in DBA/2 mice given BP orally the bone marrow is the primary target organ.

IV. OXIDATION OF COMPLETE CARCINOGENS TO FREE RADICALS

The previous section has focused on the mechanisms by which chemicals may be metabolically activated by free radicals. However, a large number of chemical carcinogens have also been shown to undergo metabolism to free radical derivatives. In this section the one-electron metabolisms of several carcinogens of particular relevance to human health are reviewed. However, in many cases, the primary roles of free radical over electrophilic metabolites in the aetiology of cancer have yet to be established. Other classes of carcinogens, most notably the peroxisome proliferators, are thought to exert their effect through an indirect action by enhancing endogenous formation of reactive oxygen species and consequent oxidative stress. Although this is beyond the scope of this chapter, the reader is referred to recent reviews on this topic (Reddy and Lalwai, 1983; Cattley et al, 1990).

A. Benzene

Because of its extensive use in the rubber, paint, plastics and petroleum industries, benzene is one of the most widely distributed environmental pollutants (Snyder et al, 1980). Chronic exposure of humans, primarily via inhalation, is associated with the development of haematopoietic dyscrasias, especially aplastic anaemia and acute myelogenous leukaemia. Under severe chemical conditions benzene can be transformed into three distinct, highly reactive free radicals, One-electron oxidation, the removal of an electron from the π-electrons, results in the formation of the benzene cation radical. The addition of an electron, the one-electron reduction of benzene, results in the formation of the benzene cation radical. The third free radical is formed by the homolytic cleavage of one of the C–H bonds to form a hydrogen atom and the phenyl radical. Although these free radical metabolites are probably not biologically relevant, metabolism by the microsomal MFO system appears to play a pre-eminent role in the toxic action of benzene. A number of metabolites of benzene have been characterized, including phenol, hydroquinone, catechol, 1,2,4-benzenetriol, muconic acid, benzene dihydrodiol, and the sulphate and glucuronic acid conjugates of the phenol-derived metabolites (Irons and Sawahata, 1985). Rickert et al (1979) observed that, following inhalation exposure of rats to benzene, appreciable quantities of phenol, catechol and hydroquinone were found in bone marrow. Catechol and hydroquinone persisted longer than either phenol or benzene at this site. In vivo studies have also demonstrated covalent binding of benzene-derived metabolites to both liver and bone marrow nucleic acids (Gill and Ahmed, 1981).

Andrews et al (1979) demonstrated that microsomes isolated from bone marrow could catalyse the generation of phenol from benzene, and while hydroquinone and catechol were also detected, their formation was not dependent on the presence of NADPH. Using an in situ perfused bone preparation, Irons et al (1980) found benzene-derived phenol, hydroquinone and catechol in the perfusate and additionally noted significant covalent binding in the marrow. Greenlee et al (1981) have observed covalent binding in bone marrow 24 h after intravenous injection of radiolabelled catechol and hydroquinone. Tunek et al (1980) were able to block the cytochrome P450-catalysed covalent binding of either benzene or phenol with SOD and suggested that O_2^- mediated the oxidation of hydroquinone to an electrophilic reactant, possibly benzosemiquinone and/or *p*-benzoquinone. The inhibition of covalent binding by SOD was accompanied by the accumulation of hydroquinone. Using much higher concentrations of phenol, Sawahata and Neal (1983) were not able to inhibit covalent binding mediated by rat liver microsomes with SOD. They did, however, observe the formation of both hydroquinone and catechol. Johansson and Ingelman-Sundberg (1983) were able to use SOD to inhibit the conversion of benzene to phenol by a reconstituted cytochrome P450 system. Hydroquinone, catechol and other metabolites were also detected in this system. Conversion of benzene to phenol and covalent binding of benzene metabolites to protein were also inhibited by hydroxyl radical scavengers. It was hypothesized that the interaction of benzene with $^{\bullet}OH$ generated by cytochrome P450 resulted in the formation of a hydroxycyclohexadienyl radical which was capable both of covalent binding and serving as a precursor to phenol. In addition to microsomes, mitochondria may also be a source of reactive oxygen in the metabolic activation of benzene. In light of recent reports that mitochondrial DNA might be an important target of chemical carcinogens (Bandy and Davison, 1990), it is interesting to note that high levels of covalent binding of benzene were observed in mitochondria (Gill and Ahmed, 1981).

While the aforementioned studies suggest that reactive oxygen metabolites may be involved in the primary activation of benzene and the secondary activation of benzene-derived metabolites to electrophiles, an important characteristic of polyphenolic compounds such as hydroquinone is their ability to autoxidize and generate O_2^-. In fact, Greenlee et al. (1981) observed significant autoxidation of both hydroquinone and 1,2,4-benzenetriol and concomitant SOD-inhibitable oxidation of adrenaline to adrenochrome. Adrenochrome formation mediated by 1,2,4-benzenetriol, a product of catechol oxidation, was significantly enhanced by incubation with the 3000 *g* supernatant from bone marrow. Quinone reductase and horseradish peroxidase also influence the covalent binding of reactive metabolites of benzene (Smart and Zannoni, 1984). Morimoto et al (1983) demonstrated that metabolic activation of benzene, catechol, hydroquinone and phenol by the 'S-9' fraction of rat liver caused an increased induction of sister chromatid exchanges in cultured human lymphocytes. Catechol also produced a significant enhancement in sister chromatid exchanges in the absence of 'S-9'. 1,2,4-Benzenetriol produces alkali-labile sites in DNA (Kawanishi et al, 1989). This DNA damage, which was Cu(II)-dependent, could be inhibited by SOD, catalase, or methional. Electron paramagnetic resonance (EPR) and spin-trapping studies suggested that the DNA-damaging species produced by the autoxidation of 1,2,4-benzenetriol was not $^{\bullet}OH$. The PHS-catalysed oxidation of hydroquinone also generates a metabolite that causes DNA nicking (Schlosser et al, 1990). The biological relevance of these findings is supported by the observations that chromosomal aberrations have been noted in humans exposed to benzene (Forni et al, 1971) and inhalation exposure of mice results in sister chromatid exchanges (Tice et al, 1980).

B. Polycyclic aromatic hydrocarbons

Polycyclic aromatic hydrocarbons (PAHs) represent an important and diverse class of environmental pollutants. Major sources of PAHs include commercial and industrial power generators, emissions from transportation systems and cigarette smoke. BP is one of the most extensively studied PAHs and has been particularly useful in dissecting the relationship between metabolic activation and chemical carcinogenesis. At least 40 different metabolites of BP have been characterized (Gelboin, 1980) and work in a number of laboratories has demonstrated that the diastereomeric BP-7,8-diol-9,10-epoxide is an ultimate carcinogenic metabolite. This metabolite arises via the conversion of BP to BP-7,8-epoxide and to BP-7,8-diol catalysed sequentially by cytochrome P450, epoxide hydrolase and cytochrome P450. As discussed earlier, peroxidases also convert BP-7,8-diol, but not BP, to diolepoxides. The major DNA adduct formed in vivo is a deoxyguanosine adduct that arises from the *trans* opening of the *anti*-dihydrodiol epoxide by the 2-exocyclic amino group of guanine in DNA.

While the cytochrome P450-catalysed epoxidation of BP results in the metabolic activation of BP to genotoxic metabolites, this is not the only pathway which generates reactive BP intermediates. One-electron oxidation at the C-6 position of BP results in the formation of a radical cation which can undergo nucleophilic attack by water to yield the labile 6-hydroxy-BP derivative which in turn rearranges to the 6-oxy radical. Interaction of molecular oxygen with this radical yields, 1,6-, 3,6- and 6,12-quinone metabolites of BP (Lorentzen et al, 1975). The radical cation, the 6-oxy-BP radical and the semiquinone radical intermediates are electrophiles capable of covalently binding to DNA (Rogan et al, 1979; Kodama et al, 1977). Rogan et al (1988) have described two DNA adducts derived by either the electrochemical oxidation or horseradish peroxidase-catalysed reaction of BP with DNA. The predominant adduct derived from the one-electron oxidation of BP contains BP bound at C-6 to the N-7 atom of guanine, whereas the second contains a bond between C-6 of BP and C-8 of deoxyguanosine. In vivo analysis of BP–DNA adducts in mouse skin by ^{32}P-postlabelling suggests that these radical cation-derived DNA adducts are very minor species compared to the monooxygenase-derived adducts. However, the BP–N^7-guanine adduct is rapidly lost from DNA by depurination and is possibly subject to substantial underestimation in vivo. Thus, the contribution of one-electron oxidation to the initiation of BP-mediated carcinogenesis remains to be established.

Keller and Jefcoate (1983) have also demonstrated that the quinone-containing metabolites can reversibly bind to DNA which may serve as an effective mechanism to sequester these metabolites at critical sites. Further, enzyme-catalysed oxidation–reduction of these quinones results in the formation of semiquinone radical intermediates, which, as previously mentioned, are capable of covalent binding to DNA. On the other hand, in the presence of molecular oxygen, the redox cycling of these quinones is accompanied initially by the generation of O_2^- and subsequently H_2O_2 and $^{\bullet}OH$ (Lesko et al, 1975; Lorentzen and Ts'o, 1977). The above series of reactions results in SOD-inhibitable DNA strand scission. In accord with this observation, Lorentzen et al (1979) have shown that the toxicity of BP-quinones in Syrian hamster embryo fibroblasts is oxygen-dependent. More recently, Leadon et al (1988) have reported that treatment of human mammary epithelial cells with BP produces oxidative DNA damage in the form of thymine glycols. Exposure of cells to BP in the presence of SOD reduced the levels of thymine glycols and increased cell survival but had no effect on levels of BP-diolepoxide-derived DNA adducts. Since DNA strand breakage and transformation of initiated cells can be enhanced by superoxide-dependent reactions, it is quite possible that the enzyme-catalysed redox cycling of BP-quinone metabolites may contribute to the complete carcinogenicity of BP and, in particular, its tumour-promoting aspects. This view can also be inferred from the observation that while BP-7,8-diol-9,10-epoxide is a potent initiator in the murine multistage model, it lacks complete carcinogen activity in mouse skin (Slaga and Fischer, 1983). However, the tumour-promoting activities of BP-quinones have not been directly examined. The overall metabolism of BP to radical intermediates is depicted in Figure 5.

C. Quinone-containing carcinogens

The anthracycline antibiotic antitumour drugs adriamycin and daunomycin

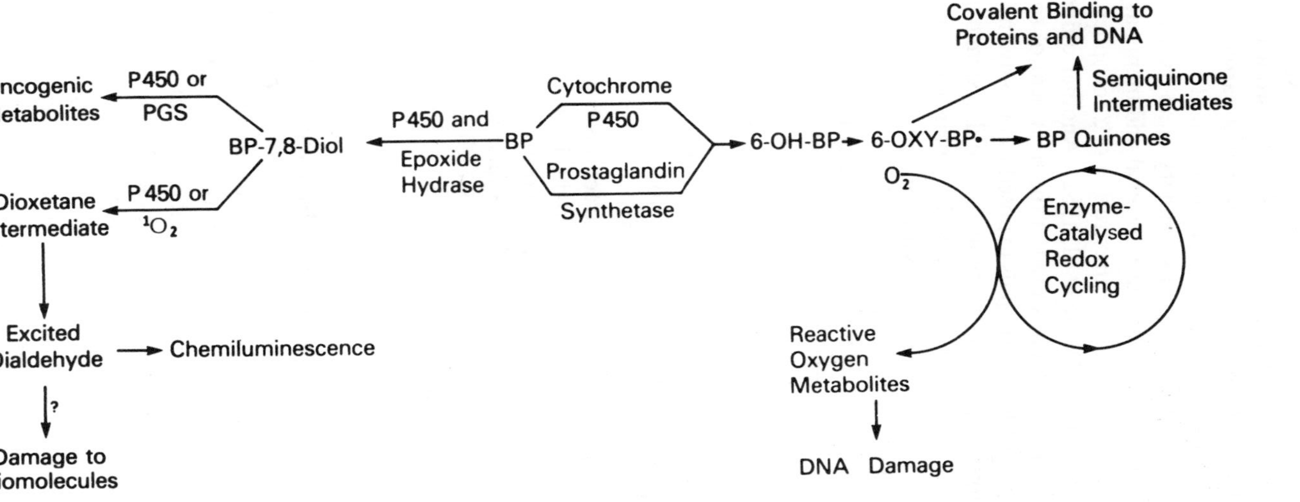

Figure 5. Sequential metabolism of benzo(a)pyrene (BP) to radical intermediates.

are widely active chemotherapeutic agents against several human tumours, including lung, breast, and leukaemias. Several problems have emerged with the clinical use of anthracyclines, including cardiotoxicity, frequent emergence and selective dominance of drug-resistant cells within tumours, and the potential for development of secondary tumours. Rats given a single intravenous dose of daunomycin develop renal and mammary tumours, the latter of which is inhibited by coadministration of α-tocopherol (Sternberg et al, 1972; Solcia et al, 1978; Wang et al, 1982). Another quinone-containing antibiotic, mitomycin C, which is used in the treatment of solid tumours, is also carcinogenic to rodents. Animals exposed to mitomycin C develop local and distant tumours following subcutaneous and systemic administration, respectively (IARC, 1976). Elevations in the mutagenic activity in the urine have been observed in oncology nurses and pharmacists as well as in patients following handling or treatment with these drugs (Falck et al, 1979; Nguyen et al, 1982). However, no evidence to date directly links these agents with the induction of human neoplasia. While the mechanisms of carcinogenicity for these agents have not been well studied, substantial investigation has been undertaken regarding their mechanisms of cytotoxicity and drug resistance. Overall, mechanistic studies indicate that the anthracycline drugs can intercalate into DNA, form reactive metabolites which bind to cellular macromolecules, induce lipid peroxidation, inactivate a variety of enzymes, and induce topoisomerase II-dependent DNA strand breaks (review: Sinha and Mimnaugh, 1990). Some of these cytotoxic actions appear to be mediated by free radicals. Additionally, altered antioxidant defence mechanisms (such as elevated glutathione peroxidase, glutathione transferase, and SOD activities) may play a role in the development of resistance to these drugs by tumour cells. The general chemical mechanisms by which quinoid compounds exert their biological effects have been elegantly reviewed by Brunmark and Cadenas (1989).

As quinones, both adriamycin and mitomycin C can undergo one-electron reduction to the semiquinone and two-electron reduction to the hydroquinone. These reactions can be catalysed by a number of enzymes, including NADPH–cytochrome P450 reductase, cytochrome b_5 reductase, NADH dehydrogenase and xanthine oxidase. Under aerobic conditions the semiquinone can react with molecular oxygen to form O_2^-, H_2O_2 and then $^{\bullet}OH$. Bachur (1979) has proposed that 'site-specific free radical' generation may be important to the cytotoxic actions of these drugs, and reactive oxygen-dependent lipid peroxidation has been implicated in the respective target organ toxicities of adriamycin and mitomycin C (Mimnaugh et al, 1983). Nagasawa and Little (1983) have demonstrated that SOD suppressed the cytotoxic effect of mitomycin C in Fanconi's anaemia fibroblasts. Mitomycin C will also induce transformation of C3H10T$\frac{1}{2}$ cells. Concomitant incubation with SOD enhances the survival of these cells, but suppresses mitomycin C-induced transformation (Little et al, 1983).

Studies by Lown et al (1976, 1977) have demonstrated that chemical reduction of adriamycin, daunomycin, or mitomycin C leads to DNA nicking. This damage is oxygen-dependent and can be inhibited by SOD, catalase, and sodium benzoate. Likewise, the enzymatic reduction of adriamycin to a semiquinone free radical by purified NADPH–cytochrome P450 reductase produces DNA cleavage by a mechanism dependent on molecular oxygen (Berlin and Haseltine, 1981), suggesting the involvement of reactive oxygen in these actions. However, both the generation of reactive oxygen species and the formation of alkylating species may be important to the carcinogenic actions of anthracyclines or mitomycin C. Under reduced oxygen tension, the metabolism of both adriamycin and mitomycin C results in the formation of bioalkylating species. Administration of radiolabelled adriamycin to rats results in covalent binding to DNA, RNA and protein (Sinha and Sik, 1980). Both the one-electron (C-7 free radical) and the two-electron (quinone methide) reduction products have been proposed as the species involved in these covalent interactions. Tomasz and Lipman (1981) demonstrated that the enzymatic reduction of mitomycin C to the hydroquinone is accompanied by the spontaneous loss of methanol, enhancing the displacement of the C-1 aziridine leaving group by nucleophiles and resulting in an alkylating species.

The synthetic oestrogen diethylstilboestrol (DES) can be readily metabolized to a quinone derivative. DES has been widely used in human medicine and in agriculture and has been associated with breast cancer and transplacentally induced cancers of the female genital tract. Although the mechanisms of action remain controversial, quinone metabolites of DES as well as of catechol oestrogens have been postulated to play central roles in the induction of hamster kidney tumours (Liehr and Roy, 1990). In this model tumourigenicity can be clearly dissociated from oestrogenic potency (Liehr et al, 1986). DES can be oxidized in vitro to diethylstilboestrol-4′,4″-quinone by horseradish or rat uterine peroxidase and H_2O_2, or by the peroxidatic activity of cytochrome P450 and organic hydroperoxide. Recently, Roy and Liehr (1989) have detected DES quinone in the livers and kidneys of hamsters injected with DES. DES quinone has been shown to form DNA adducts both in vitro and in vivo (Gladeck and Liehr, 1989). Tissue levels of DES quinone were reduced by pretreatment with vitamin C and α-naphthoflavone, substances known to inhibit DES-induced renal carcinogenesis. α-Naphthoflavone blocks the cytochrome P450-dependent formation of DES quinone, while vitamin C enhances the non-enzymatic reduction of the quinone back to DES. As shown in Figure 6, this reduction can also be facilitated by NADP(H):quinone reductase; pretreatment with inducers of this reductase such as BHA or 1,2-dithiole-3-thione also protects against DES renal tumourigenesis. However, there is no correlation between DES quinone concentrations in vivo and target organ specificity, suggesting that the oxidation of DES and the subsequent genotoxicity of DES quinone may be a necessary but not sufficient component of tumour development.

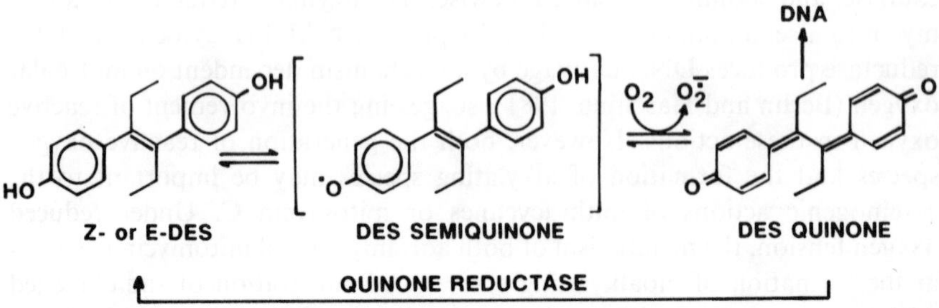

Figure 6. Redox cycling of diethylstilboestrol.

D. Streptozotocin

Streptozotocin (2-deoxy-2-(3-methyl-3-nitrosoureido)-D-glucopyranose) (STZ) is an antibiotic isolated from *Streptomyces achromogenes* and exhibits antibacterial, tumouricidal, diabetogenic, and carcinogenic properties. Tumours are produced in several organs in experimental animals, particularly the liver, kidney, and pancreas (IARC, 1978). Although a weak pancreatic carcinogen, STZ is very toxic to the insulin-producing β-cells. The glucose moiety of STZ facilitates uptake by the pancreas and confers specificity for β-cell toxicity. Coadministration of nicotinamide with STZ protects against the diabetogenic action but greatly increases the incidence of pancreatic islet cell tumours (Rakieten et al, 1971). Following scission of STZ to a *N*-nitrosoureido moiety, either methylcarbonium ions or methyl radicals are formed and produce crosslinking of DNA and strand breakage. This DNA damage appears to activate poly(ADP-ribose) synthetase, which is involved in DNA repair. Activation of this enzyme is thought to be responsible for the depletion of NAD observed in STZ-treated islet cells. Nicotinamide, an inhibitor of this synthetase, blocks the depletion of NAD levels in STZ-treated tissues and, as mentioned, blocks the diabetogenic action of STZ (Schein et al, 1967). Presumably, the inhibition of DNA repair by nicotinamide facilitates the oncogenic process by increasing the incidence of initiated cells. Although an overall oxidant stress mechanism appears to be involved in the action of STZ, it is not clear what role methyl radicals or oxygen-derived free radicals play in this process. Free radical scavenging agents, including SOD, biomimetic SOD, the hydroxyl radical scavenger dimethylurea, and α-tocopherol, protect against the diabetogenic effect of STZ (review: Loven and Oberley, 1983). However, SOD does not protect against the DNA-damaging actions of STZ in islet cells, suggesting, perhaps, a role for methyl radicals (Uchigata et al, 1982).

E. Aromatic amines and nitro compounds

Aromatic amines are an important class of chemicals found in the workplace, environment and food supply. Several aromatic amines, such as benzidine, 4-aminobiphenyl and 2-naphthylamine, are carcinogenic in humans. Aromatic amines produce tumours in the liver and several extrahepatic tissues. Although aromatic amines undergo facile *N*-hydroxylation by hepatic cytochromes P450, PHS may be particularly important in the activation of these compounds in extrahepatic tissues, such as the urinary bladder epithelium. Aromatic amines are metabolized by PHS peroxidase via one-electron oxidation to nitrogen-centred free radicals. Benzidine, for example, is an excellent cosubstrate for PHS; however, acetylation reduces the ease of oxidation by peroxidases (Josephy et al, 1983). Thus, acetylbenzidine is a much poorer substrate than is benzidine for PHS cooxidation. The fact that PHS does not utilize acetylated aromatic amines may in part explain the observation that slow-acetylator individuals are at higher risk for developing bladder cancer.

The one-electron oxidation of aromatic amines has been extensively described (Eling et al, 1990; Floyd, 1982; Josephy, 1986; O'Brien, 1988). As shown in Figure 7, benzidine can undergo one-electron oxidation by peroxidase to a cation radical that can subsequently undergo a second one-electron oxidation to the benzidine diimine. The one-electron oxidation products have been detected by EPR and the diimine characterized by other spectroscopic techniques. Benzidine is known to undergo PHS-catalysed metabolic activation to mutagenic intermediates that bind covalently to cellular macromolecules (Kadlubar et al, 1982). Benzidine diimine has been shown to form an *N*-(deoxyguanosin-8-yl)-benzidine adduct in DNA through a reaction mechanism that involves a simple deprotonation of the cationic diimine, formation of an electrophilic nitrenium ion, and covalent binding to guanine in DNA (Yamazoe et al, 1986). The formation of nitrenium ions by decomposition of protonated *N*-hydroxy arylamines or their *O*-esters as well as from free radical intermediates has been implicated in the binding of many different aromatic amines to DNA to form C-8 deoxyguanosine adducts (Beland and Kadlubar, 1985).

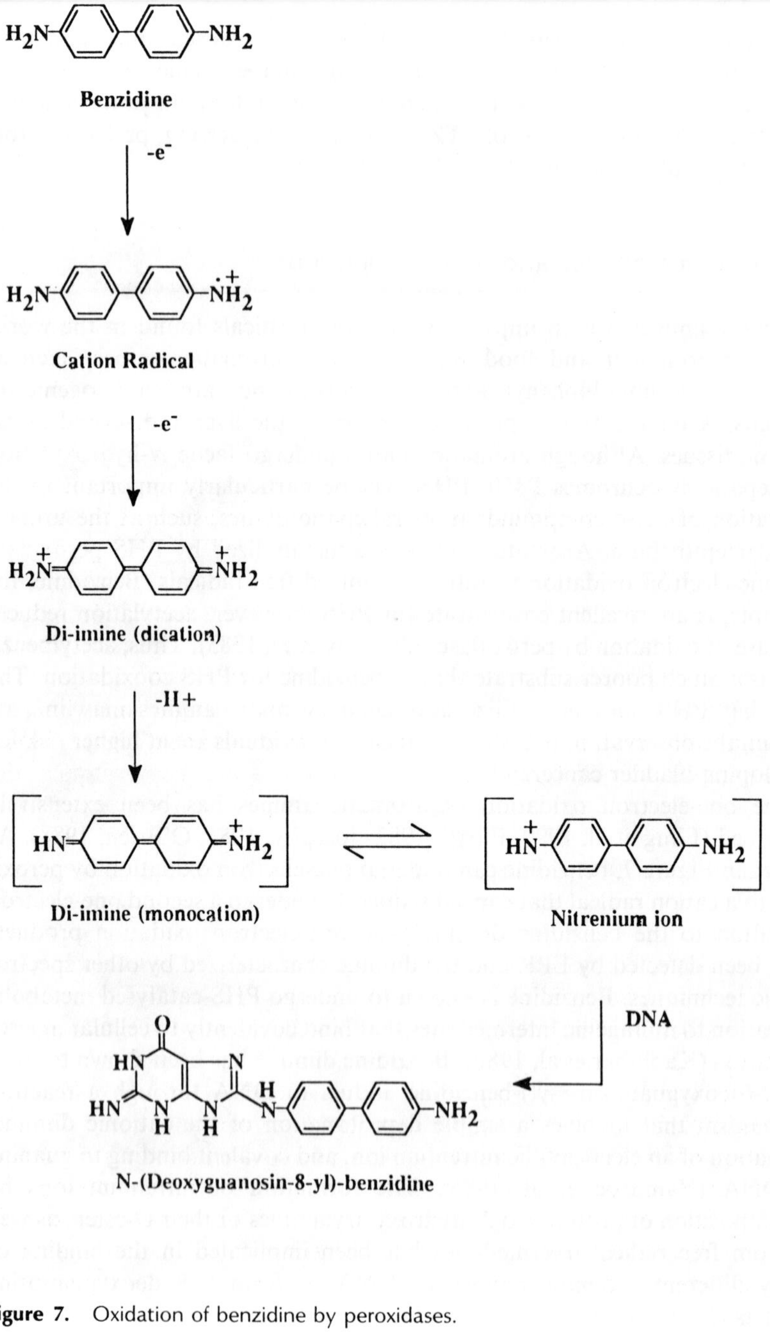

Benzidine

$-e^-$

Cation Radical

$-e^-$

Di-imine (dication)

$-H^+$

Di-imine (monocation)

Nitrenium ion

DNA

N-(Deoxyguanosin-8-yl)-benzidine

Figure 7. Oxidation of benzidine by peroxidases.

N-hydroxamic acid metabolites are common to the metabolic pathways of several arylamides and arylamines. For example, *N*-hydroxy-*N*-(2-fluorenyl)-acetamide (*N*-OH-2-FAA) is a proximate metabolite of 2-fluorenylacetamide (2-FAA). However, hydroxamic acids require further bioactivation before chemical reactions with macromolecules occur. The carcinogenic *N*-fluorenylhydroxamic acids undergo one-electron oxidation to nitroxyl free radicals catalysed by peroxidase–hydroperoxide systems. Two molecules of the nitroxyl radical then dismutate to equimolar yields of 2-nitrosofluorene and an ester of the *N*-fluorenylhydroxamic acid, *N*-acetoxy-*N*-(2-fluorenyl)-acetamide (*N*-AcO-2-FAA) (Bartsch et al, 1971). The electrophilicity of *N*-AcO-2-FAA, yielding adducts with DNA, and the direct mutagenicity of 2-nitrosofluorene have established one-electron oxidation as a potentially important mechanism of arylamine activation. However, *N*-2-substituted fluorenylhydroxamic acids can also be converted directly to 2-nitrosofluorene by haloperoxidase–H_2O_2–Br^- systems. This oxidation has been shown with lactoperoxidase- and eosinophil peroxidase-rich tissues such as uterus, the peroxidase of the mammary gland and MPO of neutrophils (Malejka-Giganti et al, 1986; Ritter and Malejka-Giganti, 1989) and may be very important biologically. Although the nitroso functional group is known to react readily with protein sulphydryls to form protein adducts, it is possible that further metabolism is required for reaction with nucleic acids. 2-Nitrosofluorene also forms covalent adducts with unsaturated membrane lipids (Floyd et al, 1978). Interestingly, a nitrosofluorene–methyl oleate adduct is mutagenic, possibly due to its ability to redox cycle (Floyd, 1981). Nakayama et al (1983) and Stier et al (1980) have demonstrated the generation of reactive oxygen by aromatic amines, but the relevance of this phenomenon to carcinogenic action is unclear.

Another functional moiety commonly found on a number of drugs and environmental chemicals is the nitro group (Bryan, 1978). There has been considerable interest in nitroaromatic compounds, such as nitroimidazoles and nitrofurans, as radiosensitizers and as cytotoxic agents selective for hypoxic cells (Fowler et al, 1977). In addition, nitropolycyclic aromatic hydrocarbons and nitrobenzene are prevalent in the environment; and, in this regard, nitropyrene is of particular interest because of its presence in diesel soot and the carbon-black of xerographic toners (Lofroth et al, 1980; Wei and Shu, 1983). Many nitro-containing compounds have been shown to be mutagenic in the Ames *Salmonella typhimurium* reversion assay and thus are considered as potential human carcinogens (Cohen, 1978; Mermelstein et al, 1981). Administration of 4-nitroquinoline 1-oxide (4-NQO) to rodents results in the induction of pulmonary tumours and leukaemia (Nishizuka et al, 1964). The cytotoxic, mutagenic and carcinogenic properties of these compounds have been attributed to the metabolic reduction of the nitro group, initially to an anion radical and then, through various intermediates, to the reactive hydroxylamine or amine (Swaminathan and Lowes, 1978; Biaglow, 1981). These reactive metabolites are formed under reduced oxygen tension, since the nitro anion radical is also capable of reacting with molecular oxygen, yielding O_2^- and H_2O_2 (Biaglow et al, 1977; Mason and Holtzman, 1975; Sasame and Boyd, 1979). Moreover, as pointed out by Biaglow (1981), hydroxylamine compounds can react with oxygen or H_2O_2 through a peroxidase-dependent process to produce the hydroxylamine radical and O_2^-. Biaglow has also proposed that the nitro anion radical itself reacts with O_2^- to generate a nitroso intermediate which serves as a precursor to the hydroxylamine radical. The data of Peterson et al (1982), demonstrating inhibition of the nitroreduction of nitrofurantoin by SOD, support such an interaction between the nitro anion radical and O_2^-. In addition to the metabolism of nitro compounds to the anion radical by NADPH–cytochrome P450 reductase and xanthine oxidase (McManus et al, 1982; Howard and Beland, 1982), Biaglow et al (1977) observed that 4-NQO mediated an oxidation of ascorbate yielding 4-NQO anion radical and $^{\bullet}OH$. A combination of SOD and catalase blocks the consumption of oxygen which accompanies this chemical–chemical interaction.

The reduction of nitroheterocyclic compounds by mammalian cells results in both DNA damage, as indicated by single-strand scissions (Sugimura et al, 1968), and covalent binding to DNA and other macromolecules (Howard et al, 1983). Ishizawa and Endo (1967) observed that 4-hydroxylamine 1-oxide, a carcinogenic metabolite of 4-NQO, inactivated bacteriophage T_4 in the presence of oxygen with the target molecule being DNA. Interestingly, in the presence of oxygen, a marked production of H_2O_2 was observed from 4-hydroxylamine 1-oxide, but not 4-NQO or 4-aminoquinoline 1-oxide (Hozumi, 1969). An additional component of the cellular interactions of 4-NQO is its acylation by seryl-AMP, yielding a product which covalently binds to macromolecules (Tada and Tada, 1972). While suggestive, the involvement of reactive oxygen in DNA damage mediated by nitro-containing compounds has not been well characterized. However, the redox cycling of nitrofurantoin (Trush et al, 1982d) and 4-NQO (unpublished observation) does augment SOD-inhibitable microsomal lipid peroxidation.

V. FREE RADICAL METABOLITES OF TUMOUR PROMOTERS AND PROGRESSORS

Many tumour promoters, particularly those typified by the phorbol esters, are thought to function through receptor-mediated pathways. Metabolic activation of these tumour promoters is not required, although metabolic inactivation may occur (Müller et al, 1990). However, some classes of tumour promoters and progressors may require cellular activation to reactive intermediates, particularly free radicals, as do many complete carcinogens. This section will highlight some of the mechanisms by which tumour promoters and progressors can be metabolized to free radical derivatives.

A. Anthrones

Anthralin (1,8-dihydroxy-9-anthrone) is used for the topical treatment of psoriasis, a proliferative skin disease, and is also a tumour promoter in mouse skin (Bock and Burns, 1963). Anthralin is a strong reductant that is readily oxidized by air, light or metal ions. The initial oxidation products of anthralin are the 1,8-dihydroxy-9-dianthrone-10-yl radical (10-anthranyl) and the hydroperoxyl radical, which can then react with basic groups to form O_2^- (Ashton et al, 1983). The 10-anthranyl radical undergoes further oxidation to anthraquinone and anthralin dimer, which is the major metabolite formed in skin. However, neither of these end-products of anthralin oxidation has tumour-promoting activity (Segal et al, 1971). Fuchs and Packer (1989) have directly detected anthralin radicals in mouse skin and their formation is inhibited by α-tocopherol. Both the oxygen free radicals and the 10-anthranyl radical may be involved in the biological actions of anthralin. This view is also supported by the observations of DiGiovanni et al (1988) that the actions of anthralin and a related anthrone tumour promoter, chrysorobin, can be inhibited by a biomimetic SOD. Further, these investigators noted a correlation between tumour-promoting activity, ability to induce epidermal ornithine carboxylase (ODC) activity, and ability to undergo base-catalysed oxidation among a series of C_{10}-acyl analogues of anthralin.

B. Organic hydroperoxides and peroxides

A number of free radical-generating peroxides and hydroperoxides used in the chemical and pharmaceutical industries are skin irritants and tumour promoters. Benzoyl peroxide, lauroyl peroxide, decanoyl peroxide, and dicumyl peroxide are quite active in the initiation–promotion model of the mouse epidermis, while *tert*-butylhydroperoxide, cumeme hydroperoxide, and methyl ethyl ketone peroxide can be described as weak promoters. Promoter activity of these peroxides is ascribed to their ability to generate free radical derivatives, although they are generally quite stable to unimolecular homolysis at body temperature. Several of the hydroperoxides have been observed to undergo accelerated decomposition in the presence of haematin to form radical species that can be detected by spin trapping and EPR spectroscopy (Kalyanaraman et al, 1983). Free radical products of the organic hydroperoxides *tert*-butylhydroperoxide and cumene hydroperoxide have recently been demonstrated and characterized during incubation with either isolated murine (Taffe et al, 1987) or human (Athar et al, 1989) keratinocytes. Methyl radicals were the primary species observed. These tumour promoters require a cell-mediated activation to form free radical products. This activation may be catalysed by a variety of haemoproteins, including peroxidases, as well as by non-enzymatic metal-mediated reactions between endogenous reductants and hydroperoxides. In this setting both peroxyl ($ROO^{\bullet}$) and alkoxyl ($RO^{\bullet}$) radicals appear to be formed from the hydroperoxide, although neither species has been trapped under physiological conditions:

$$ROOH + \text{Protein}(Fe^{3+}) \longrightarrow ROO^{\bullet} + \text{Protein}\ (Fe^{2+})$$

$$ROO^{\bullet} + \text{Protein}(Fe^{2+}) \longrightarrow RO^{\bullet} + \text{Protein}(Fe^{3+})$$

The formation of the methyl radical ($CH_3^{\bullet}$) on incubation with either *tert*-butyl- or cumene hydroperoxide can be explained through the instability of the alkoxyl radicals. These tertiary alkoxyl radicals undergo β-scission to form the corresponding ketone and the methyl radical:

$$R\text{—}\underset{\underset{R}{|}}{\overset{\overset{CH_3}{|}}{C}}\text{—}O^{\bullet} \longrightarrow CH_3^{\bullet} + R\text{—}\overset{\overset{O}{\|}}{C}\text{—}R$$

Methylation reactions are well described in chemical carcinogenesis. Methylating carcinogens, particularly the mono- and disubstituted hydrazines, have been shown to generate alkyl radicals on metabolic activation (Augusto et al, 1985). Studies which assess the binding to cellular macromolecules on exposure to dimethylhydrazine have demonstrated methylation of DNA, RNA, and protein, with the highest specific activity reportedly found in the covalently modified cellular RNA (Boffa et al, 1982). Amino acid residues that have been reported to be methylated are the basic amino acids, lysine, arginine, and histidine (Paik et al, 1984). One functional effect of methylation of these basic amino acid residues in histones may be to alter their binding to DNA, which may in turn contribute to alteration of gene expression. Whether the methyl radicals formed from these organic hydroperoxide tumour promoters act in a similar manner is unknown. Methyl radicals also readily react with molecular oxygen to form peroxyl radicals; however, again, the role of peroxyl radicals in tumour promotion is not well resolved (Marnett, 1987).

Organic peroxide tumour promoters, such as benzoyl peroxide (BzPO), also form free radicals in target cells. BzPO is used extensively in the plastics and rubber industries as a catalyst in polymerization reactions and by the pharmaceutical industry in over-the-counter acne preparations. Possible concerns over the commerical use of this peroxide arise from the reports that BzPO is an effective tumour promoter and enhancer of malignant progression in mouse skin (Slaga et al, 1981; O'Connell et al, 1986). However, BzPO is neither an initiator nor a complete carcinogen in this multistage carcinogenesis model. Exposure to BzPO has been shown to result in the transformation of the JB6 mouse epidermal cell line (Gindhart et al, 1985) as well as in the induction of ornithine decarboxylase activity in cultured epithelial cells (Saladino et al, 1985) and mouse epidermis in vivo (Binder et al, 1989), all actions associated with many classes of tumour promoters. Although not mutagenic (Ishidate et al, 1980), BzPO has been demonstrated to cause DNA strand breaks in several types of cells, including keratinocytes, bronchial epithelial cells, and leukocytes (Hartley et al, 1985; Saladino et al, 1985; Birnboim, 1986). The toxicological activities of this compound in mouse skin have been ascribed to its free radical-generating capabilities, in part from the observation that tumour promotion by BzPO is inhibited by the antioxidants BHA and BHT (Slaga et al, 1981). Recently, Swauger et al (1990) have characterized the metabolism of BzPO in keratinocytes, as summarized in Figure 8. EPR studies indicate that both alkoxyl (benzoyloxyl) and alkyl (phenyl) radicals are formed. The principal stable metabolites formed from BzPO in cells, benzoic acid and CO_2, are not likely to be involved in the toxicological manifestations of BzPO. In particular, the DNA-damaging actions of BzPO appear to be mediated by benzoyloxyl radicals. The extent of DNA strand breakage in cells is modulated by changes in cellular thiol status, and damage can be blocked by free radical scavengers such as the spin trap phenyl-*tert*-butylnitrone.

C. Butylated hydroxytoluene

BHT, a widely used food antioxidant, is paradoxically both an effective inhibitor of tumour promotion in mouse skin and a tumour promoter itself in a variety of animal models for multistage carcinogenesis. Inasmuch as inhibitors of cytochrome P450 block tumour promotion by BHT in the lung, the activity of BHT as a tumour promoter may relate to its metabolism to

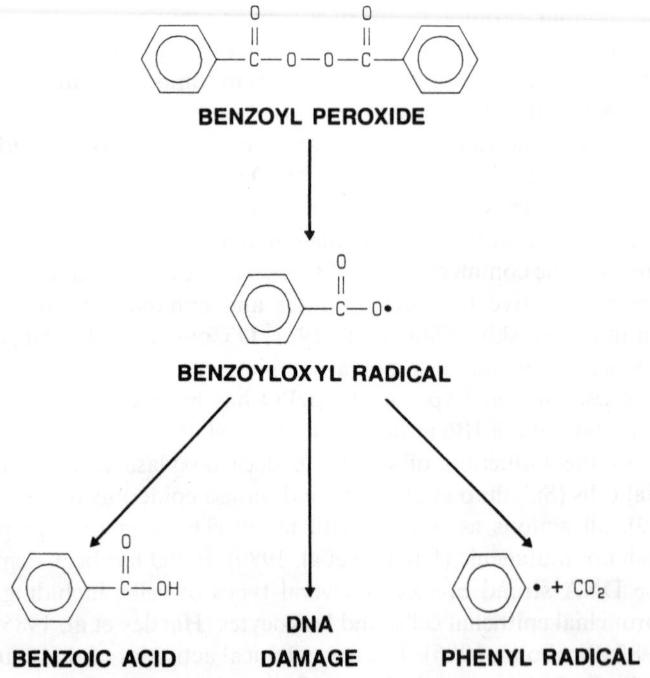

Figure 8. Pathway for the activation of benzoyl peroxide to free radical and DNA-damaging species in cells.

a reactive species. Additionally, Thompson et al (1989c) have recently observed that a P450-derived metabolite of BHT (BHT-BuOH) is a more potent tumour promoter in the mouse lung than is BHT. This finding provides the first direct evidence for a role of metabolism in the action of tumour promoters in internal organs. There appear to be two main routes for the oxidative metabolism of BHT. One pathway involves the hydroxylation of an alkyl substituent and the other oxidation of the π-electron system. The major route of metabolism involves the sequential oxidation of the 4-methyl group to the 4-benzyl alcohol, then to the aldehyde, and finally to the carboxylic acid, which in turn is conjugated and excreted primarily as a glucuronide. The π-electron oxidative pathway leads to products with a cyclohexadienone structure which presumably arise through the formation of a BHT phenoxyl radical intermediate. One metabolite of this pathway of particular interest is 2,6-di-*tert*-butyl-4-hydroperoxyl-4-methyl-2,5-cyclohexadienone (BHTOOH). BHTOOH appears to be formed following the addition of molecular oxygen to the BHT phenoxyl radical at the 4 position of the ring, followed by one-electron reduction. BHTOOH has been isolated in vitro from rat liver microsomal incubations containing BHT (Shaw and Chen, 1972), and from faeces of rats and mice following administration of an intraperitoneal dose of BHT (Matsuo et al, 1984). Interestingly, BHTOOH is an effective tumour promoter in mouse skin, although BHT itself is inactive in this tissue (Taffe and Kensler, 1988).

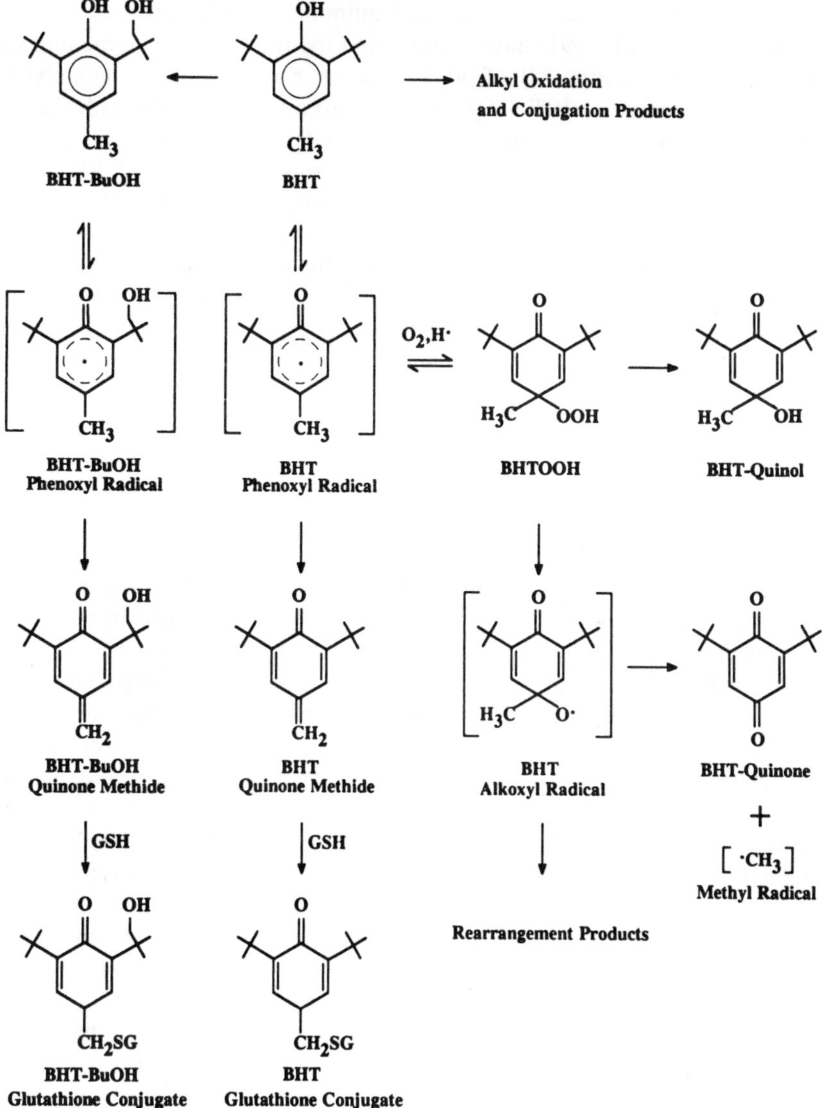

Figure 9. Scheme of the reactive intermediates known to be generated from BHT.

BHTOOH is extensively metabolized by keratinocytes, with BHT quinol representing the major stable metabolite (Taffe et al, 1989). However, as shown in Figure 9, a number of free radicals, including phenoxyl, alkoxyl, and methyl radicals, are also produced and may be of particular relevance to the tumour-promoting activity of BHTOOH. The BHT phenoxyl radical, which is readily detectable by EPR in keratinocytes, can undergo a disproportionation reaction to form the BHT-quinone methide, a reactive electrophile. Guyton et al (1991) have investigated the role of quinone methides in the promoting activity of BHTOOH by comparing the promoting activities of 4-trideuteromethyl BHTOOH and 4-*tert*-butyl BHTOOH with that of BHTOOH. These compounds have a reduced ability or inability, respectively, to form a quinone methide. The deuterated analogue of BHTOOH was several-fold less active as a tumour promoter than BHTOOH, while the 4-*tert*-butyl analogue was inactive. Bolton et al (1990) have noted that the promoting metabolite of BHT found in the lung, BHT-BuOH, also readily forms a quinone methide upon incubation with either liver or lung microsomes. Thus, it appears that an electrophilic metabolite formed through a free radical intermediate may mediate the tumour-promoting activities of BHT and BHTOOH.

VI. SUMMARY

Substantial experimental evidence has been put forth supporting a role for biologically and chemically derived oxidants and free radicals in the initiation, promotion and progression of cancer. In order for most chemicals to contribute to the carcinogenic process, they must first undergo metabolic activation to a biologically reactive intermediate. The metabolic activation of a particular xenobiotic may involve either an initial two-electron oxidation to a hydroxylated or epoxidated product or a one-electron oxidation or reduction to a radical intermediate. For some chemicals, this activation process may involve both types of reactions. As discussed extensively in this chapter, the metabolic activation of chemicals can occur through direct enzymatic catalysis as well as by oxidative entities generated by enzymes. Similarly, free radicals generated as a result of oxidation of biomolecules such as lipids or glutathione can also facilitate the bioactivation of xenobiotics. Further, there are a number of examples whereby a radical intermediate or an oxidant generated as a result of the activation of one xenobiotic can result in the enhanced metabolic activation of a second xenobiotic. Such reactions provide a novel molecular mechanism for chemical–chemical interactions. It is expected that the further understanding of the role(s) of free radicals in the activation of biologically reactive intermediates as well as the characterization of the resultant critical modifications will provide a basis for both predicting toxicological risk to organisms and developing protective strategies.

ACKNOWLEDGEMENTS

We gratefully acknowledge the research support of the National Institutes of Health (CA 44530, ES 03760 and ES 05131) and the American Cancer Society (SIG-3). T.W.K. is recipient of Research Career Development Award CA 01230.

REFERENCES

Andrews LS, Sasame HA & Gillette JR (1979) *Life Sci.* **25:** 567–572.
Ashton RE, Andre P, Lowe N & Whitefield M (1983) *Semin. Dermatol.* **2:** 287–303.
Athar M, Mukhtar H, Bickers DR, Khan IU & Kalyanaraman B (1989) *Carcinogenesis* **10:** 1499–1503.
Augusto O, Du Plessis LR & Weingrill CLV (1985) *Biochem. Biophys. Res. Commun.* **126:** 853–858.
Bachur NR (1979) *Cancer Treat. Rep.* **63:** 817–820.
Bandy B & Davison AJ (1990) *Free Radicals Biol. Med.* **8:** 523–539.
Bartsch H, Traut M & Hecker E (1971) *Biochim. Biophys. Acta* **237:** 556–566.
Bassett DJP, Bowen-Kelley E & Seed JL (1988) *Res. Commun. Chem. Pathol. Pharmacol.* **60:** 291–307.
Beland FA & Kadlubar FF (1985) *Environ. Health Perspect.* **62:** 19–30.
Berlin V & Haseltine WA (1981) *J. Biol. Chem.* **256:** 4747–4756.
Biaglow JE (1981) *Radiat. Res.* **86:** 212–242.
Biaglow JE, Jacobson BE & Nygaard OF (1977) *Cancer Res.* **37:** 3306–3313.
Binder RL, Volpenhein ME & Motz AA (1989) *Carcinogenesis* **10:** 2351–2357.
Birnboim HC (1986) *Carcinogenesis* **7:** 1511–1517.
Bock FG & Burns R (1963) *J. Natl. Cancer Inst.* **30:** 393–397.
Bodell WJ, Devanesan PD, Rogan EG & Cavalieri EL (1989) *Chem. Res. Toxicol.* **2:** 312–315.
Boffa LC, Gruss RJ & Allfrey VG (1982) *Cancer Res.* **42:** 382–388.
Bolton JL, Sevestre H, Ibe BO & Thompson JA (1990) *Chem. Res. Toxicol.* **3:** 65–70.
Borek C & Troll W (1983) *Proc. Natl. Acad. Sci. USA* **80:** 1304–1307.
Boyd MR, Stiko AW & Sasame HA (1979) *Biochem. Pharmacol.* **28:** 601–606.
Brunmark A & Cadenas E (1989) *Free Radicals Biol. Med.* **7:** 435–477.
Bryan GT (1978) In Bryan GT (ed.) *Carcinogenesis—A Comprehensive Survey: Nitrofurans*, Vol. 4, pp 1–11. New York: Raven Press.
Burleson GR, Caulfield MJ & Pollard M (1979) *Cancer Res.* **39:** 2149–2154.
Cattley RC, Marsman DS & Popp JA (1990) In Mendelsohn ML & Albertini RS (eds) *Mutation and the Environment. Part D: Carcinogenesis*, pp 123–132. New York: Wiley–Liss.

Caulfield MJ, Burleson GR & Pollard M (1979) *Cancer Res.* **39:** 2155–2159.
Cerutti PA (1985) *Science* **227:** 375–381.
Cohen SM (1978) In Bryan GT (ed.) *Carcinogenesis—A Comprehensive Survey: Nitrofurans*, Vol. 4, pp 171–231. New York: Raven Press.
Crawford D, Zbinden I, Amstad P & Cerutti P (1988) *Oncogene* **3:** 27–32.
Curtis JF, Hughes MF, Mason RP & Eling TE (1988) *Carcinogenesis* **9:** 2015–2021.
DeChatelet LR, Shirley PS, Johnston RB Jr (1976) *Blood* **47:** 545–554.
DiGiovanni J, Kruszewski FH, Coombs MM, Bhatt TS & Pezeshk A (1988) *Carcinogenesis* **9:** 1437–1443.
Dix TA & Marnett LJ (1983) *Science* **221:** 77–79.
Dybing E, Nelson SD, Mitchell JR, Sasame HA & Gillette JR (1976) *Mol. Pharmacol.* **12:** 911–920.
Eastmond DA & Smith MT (1990) In Packer L & Glazer AN (eds) *Methods in Enzymology*, Vol. 186, pp 579–585. New York: Academic Press.
Eastmond DA, Smith MT, Irons RD (1987) *Toxicol. Appl. Pharmacol.* **91:** 85–95.
Eling TE, Curtis J, Battista JR & Marnett LJ (1986) *Carcinogenesis* **7:** 1957–1963.
Eling TE, Thompson DC, Foureman GL, Curtis JF & Hughes MR (1990) *Annu. Rev. Pharmacol. Toxicol.* **30:** 1–45.
Esterline RL, Bassett DJP & Trush MA (1989) *Toxicol. Appl. Pharmacol.* **99:** 229–239.
Falck K, Grohn P, Sorsa M, Vainio H, Heinonen E & Hosti LR (1979) *Lancet* **1:** 1250–1251.
Fischer SM, Baldwin JK & Adams LM (1986) *Carcinogenesis* **7:** 915–918.
Fischer SM, Kokolus WJ, Baldwin JL & Patrick KE (1989) *Toxicol. In Vitro* **3:** 293–298.
Floyd RA (1981) *Radiat. Res.* **86:** 243–263.
Floyd RA (1982) In Floyd RA (ed.) *Free Radicals and Cancer*, pp 361–396. New York: Marcel Dekker.
Floyd RA, Soong LM, Stuart MA & Reigh DL (1978) *Arch. Biochem. Biophys.* **185:** 450–457.
Forni AM, Pacifico E & Limonata A (1971) *Arch. Environ. Health* **22:** 373–378.
Fowler JF, Adams GE & Denekamp J (1977) *Cancer Treat. Rev.* **3:** 227–256.
Fuchs J & Packer L (1989) *J. Invest. Dermatol.* **92:** 677–682.
Gelboin HV (1980) *Physiol. Rev.* **60:** 1107–1166.
Geuskens M, Burglen MJ & Uriel J (1977) *Virchows Arch. (Cell Pathol.)* **24:** 67–78.
Gill DP & Ahmed AE (1981) *Biochem. Pharmacol.* **50:** 1127–1131.
Gillette JR, Mitchell JR & Brodie BB (1974) *Annu. Rev. Pharmacol.* **4:** 271–288.
Gindhart TD, Srinivas L & Colburn NH (1985) *Carcinogenesis* **6:** 309–311.
Gladeck A & Liehr JG (1989) *J. Biol. Chem.* **264:** 16847–16852.
Graceffa Y & Weitzman SA (1987) *Arch. Biochem. Biophys.* **257:** 481–484.
Gram TE, Okine LK & Gram RA (1986) *Annu. Rev. Pharmacol. Toxicol.* **26:** 259–291.
Green JL & Reed GA (1990) *Chem. Res. Toxicol.* **3:** 59–64.
Greenlee WF, Sun JD & Bus JS (1981) *Toxicol. Appl. Pharmacol.* **59:** 187–195.
Guyton KZ, Phan P, Kuppusumy P et al. (1991) *Proc. Natl. Acad. Sci. USA* **88:** 946–950.
Harris CC, Vahakangas K, Newman MJ et al (1985) *Proc. Natl. Acad. Sci. USA* **82:** 6672–6676.
Hartley JA, Gibson NW, Zwelling LA & Yuspa SH (1985) *Cancer Res.* **45:** 4864–4870.

Hennings H, Shores R, Wenk ML, Spangler EF, Tarone R & Yuspa SH (1983) *Nature (Lond.)* **304:** 67–69.
Hewer A, Ribeiro O, Walsh C, Grover P & Sims P (1979) *Chem. Biol. Interact.* **26:** 147–154.
Hill, J (1761) In *Cautions Against the Immoderate Use of Snuff*, pp 30–31. London: Baldwin and Jackson.
Howard PC & Beland FA (1982) *Biochem. Biophys. Res. Commun.* **104:** 727–732.
Howard PC, Heflich RH, Evans FE & Beland FA (1983) *Cancer Res.* **43:** 2052–2058.
Hozumi M (1969) *Gann* **60:** 83–90.
Hughes MF, Chamulerat W, Mason RP & Eling TE (1989) *Carcinogenesis* **10:** 2075–2080.
IARC (1976) *Monographs on the Evaluation of Carcinogenic Risk of Chemicals to Man* **10:** 171–179.
IARC (1978) *Monographs on the Evaluation of Carcinogenic Risk of Chemicals to Man* **17:** 337–349.
Irons RD & Sawahata T (1985) In Anders MW (ed.) *Bioactivation of Foreign Compounds*, pp 259–281. Orlando: Academic Press.
Irons RD, Dent JG, Baker TS & Rickert DE (1980) *Chem. Biol. Interact.* **30:** 241–245.
Ishidate M, Sofuni T & Yoshikawa K (1980) *Mutagens Toxicol.* **3:** 80–82.
Ishizawa M & Endo H (1967) *Biochem. Pharmacol.* **16:** 637–646.
Johansson I & Ingelman-Sundberg M (1983) *J. Biol. Chem.* **258:** 7311–7316.
Josephy PD (1986) *Fed. Proc.* **45:** 2465–2470.
Josephy PD, Palcic, B & Skarsgard LD (1981) *Biochem. Pharmacol.* **30:** 849–853.
Josephy PD, Eling TE & Mason RP (1983) *Mol. Pharmacol.* **23:** 766–770.
Kadlubar FF, Frederick CB, Weis CC & Zenser TV (1982) *Biochem. Biophys. Res. Commun.* **108:** 253–258.
Kalyanaraman B, Mottley C & Mason RP (1983) *J. Biol. Chem.* **258:** 3855–3858.
Kappus H & Mahmutoglu I (1986) In Kocsis JJ, Jollow DJ, Witmer CM, Nelson JO & Snyder R (eds) *Biological Reactive Intermediates III: Mechanisms of Action in Animal Models and Human Disease*, pp 273–280. New York: Plenum Press.
Kawanishi S, Inoue S & Kawanishi M (1989) *Cancer Res.* **49:** 164–168.
Keller GM & Jefcoate CR (1983) *Mol. Pharmacol.* **23:** 735–742.
Kensler CJ, Dexter SO & Rhoades CP (1942) *Cancer Res.* **2:** 1–10.
Kensler TW & Guyton KZ (1991) In Bloom A & Spatz L (eds) *Biological Consequences of Oxidative Damage: Implications for Cancer and Cardiovascular Disease*, New York: Oxford University Press (in press).
Kensler TW & Taffe BG (1986) *Adv. Free Radicals Biol. Med.* **2:** 347–387.
Kensler TW & Trush MA (1985) In Oberley LW (ed.) *Superoxide Dismutase*, Vol. III, pp 191–236. Boca Raton: CRC Press.
Kensler TW, Egner PA, Moore KG, Taffe BG, Twerdok LE & Trush MA (1987) *Toxicol. Appl. Pharmacol.* **90:** 337–346.
King AF, Landreth KS & Wierda D (1989) *J. Pharmacol. Exp. Ther.* **250:** 582–590.
Klebanoff SJ (1977) *J. Exp. Med.* **145:** 983–996.
Klebanoff SJ (1988) In Gallin JI, Goldstein IM & Snyderman R (eds) *Inflammation: Basic Principles and Clinical Correlates*, pp 391–444. New York: Raven Press.
Kodama M, Ioki Y & Nagata C (1977) *Gann* **68:** 253–254.
Koren HS, Devlin RB, Graham DE et al (1989) *Am. Rev. Respir. Dis.* **139:** 407–415.
Kozumbo WJ, Agarwal S & Koren HS (1989) *Proc. Am. Assoc. Cancer Res.* **30:** 151.

Kuehl FA Jr, Humes JL, Ham EA, Egan RW & Dougherty HW (1980) In Samuelson B, Ramwell PW & Paleotti R (eds) *Prostaglandin and Thromboxane Research*, pp 77–84. New York: Raven Press.

Leadon SA, Stampfer MR & Bartley J (1988) *Proc. Natl. Acad. Sci. USA* **85:** 4365–4368.

Lesko S, Caspary W, Lorentzen R & Ts'o POP (1975) *Biochemistry* **14:** 3978–3984.

Liehr JG & Roy D (1990) *Free Radicals Biol. Med.* **8:** 415–423.

Liehr JG, Stancel GM, Chorich LP, Bousfield GR & Ulubelen AA (1986) *Chem. Biol. Interact.* **59:** 173–184.

Lijinsky W (1986) *J. Natl. Cancer Inst.* **77:** 471–476.

Little JB, Kennedy AR & Nagasawa H (1983) In Nygaard OF and Simic MG (eds) *Radioprotectors and Anticarcinogens*, pp 487–493. New York: Academic Press.

Lofroth G, Hefner E, Alfheim I & Holler M (1980) *Science* **209:** 1037–1039.

Lorentzen RJ & Ts'o POP (1977) *Biochemistry* **16:** 1467–1473.

Lorentzen RJ, Caspary WJ, Lesko SA & Ts'o POP (1975) *Biochemistry* **14:** 3970–3977.

Lorentzen RJ, Lesko SA, McDonald K & Ts'o POP (1979) *Cancer Res.* **39:** 3194–3198.

Loven DP & Oberley LW (1985) In Oberley LW (ed.) *Superoxide Dismutase*, Vol. III, pp 151–189. Boca Raton: CRC Press.

Lovstad RA (1979) *Gen. Pharmacol.* **10:** 437–440.

Lown JW, Begleiter A, Johnson D & Morgan R (1976) *Can. J. Biochem.* **54:** 110–119.

Lown JW, Simms SS, Majumdar KC & Change RY (1977) *Biochem. Biophys. Res. Commun.* **76:** 705–710.

Lu YHA & West SB (1978) *Pharmacol. Ther. A.* **2:** 337–358.

Malejka-Giganti D, Ritter CL, Decker RW & Suilman JM (1986) *Cancer Res.* **46:** 6200–6206.

Malkinson AM, Thaete LG, Blumenthal EJ & Thompson JA (1989) *Toxicol. Appl. Pharmacol.* **101:** 196–204.

Marletta MA (1988) *Chem. Res. Toxicol.* **1:** 249–257.

Marnett LJ (1981) *Life Sci.* **29:** 531–546.

Marnett LJ (1987) *Carcinogenesis* **8:** 1365–1373.

Marnett LJ (1991) In Sipes IG (ed.) *Biological Reactive Intermediates IV. Molecular and Cellular Effects and their Impact on Human Health.* New York: Plenum Press. In press.

Marshall W & O'Brien PJ (1984) In Thaler-Dao, H, dePaulet A & Paoletti R (eds) *Isocanoids and Cancer*, pp 49–61. New York: Raven Press.

Mason RP (1979) In Hodgson E, Bend JR & Philpot RM (eds) *Reviews in Biochemical Toxicology*, Vol. 1, pp 151–200. New York: Elsevier/North Holland.

Mason RP & Holtzman JL (1975) *Biochem. Biophys. Res. Commun.* **67:** 1267–1274.

Matsuo M, Mihara K, Okauno M, Ohkawa H & Miyamoto J (1984) *Food Chem. Toxicol.* **22:** 342–354.

McCord JM (1985) *N. Engl. J. Med.* **312:** 159–163.

McManus ME & Davies DS (1980) *Xenobiotica* **10:** 745–752.

McManus ME, Lang, MA, Stuart K & Strong J (1982) *Biochem. Pharmacol.* **31:** 547–552.

Mermelstein R, Kirazides DK, Butler M, McCoy EC & Rosenkranz HS (1981) *Mutat. Res.* **89:** 187–196.

Miller EC & Miller JA (1985) In Anders MW (ed.) *Bioactivation of Foreign Compounds*, pp 3–28. Orlando: Academic Press.

Miller JA (1970) *Cancer Res.* **30:** 559–576.
Mimnaugh EG, Gram TE & Trush MA (1983) *J. Pharmacol. Exp. Ther.* **266:** 806–816.
Miwa M, Stuehr DJ, Marletta MA, Wishnok JS & Tannenbaum SR (1987) *Carcinogenesis* **8:** 955–958.
Morimoto K, Wolf S & Koizumi A (1983) *Mutat. Res.* **119:** 355–360.
Mossman BT, Bignon J, Corm M, Senton A & Gee JBL (1990) *Science* **247:** 294–300.
Mudd JB & Freeman BA (1977) In Lee SD (ed.) *Biochemical Effects of Environmental Pollutants*, pp 97–133. Ann Arbor, MI: Ann Arbor Science.
Müller G, Hergenhahn M, Roeser H, Tremp GL, Schmidt R & Hecker E (1990) *Carcinogenesis* **11:** 1127–1132.
Nagasawa H & Little JB (1983) *Carcinogenesis* **4:** 795–798.
Nakamura Y, Gindhart TD, Winterstein D, Tomita I, Seed JL & Colburn NH (1988) *Carcinogenesis* **9:** 203–207.
Nakayama T, Kimura T, Kodama M & Nagata C (1983) *Carcinogenesis* **4:** 765–769.
Nebert DW, Jensen NM, Levitt RC & Felton JS (1980) *Clin. Toxicol.* **16:** 99–122.
Nelson SD, Mitchell JR, Dybing E & Sasame HA (1976) *Biochem. Biophys. Res. Commun.* **70:** 1157–1165.
Nguyen TV, Theiss JS & Matney TS (1982) *Cancer Res.* **42:** 4792–4796.
Nishizuka Y, Nakakuki K & Sakakura T (1964) *Gann* **55:** 495–508.
O'Brien PJ (1988) *Free Radicals Biol. Med.* **4:** 169–183.
O'Connell JF, Klein-Szanto AJP, DiGiovanni DM, Freis JW & Slaga TJ (1986) *Cancer Res.* **46:** 2863–2865.
Paik WK, Dimaria P, Kim, S, Magee PN & Lotlikar PD (1984) *Cancer Lett.* **23:** 9–17.
Pan S-S & Bachur NR (1980) *Mol. Pharmacol.* **17:** 95–99.
Pence BC & Reiners JJ Jr (1987) *Cancer Res.* **47:** 6388–6392.
Perchellet JP & Perchellet EM (1989) *Free Radicals Biol. Med.* **7:** 377–408.
Peterson FJ, Combs GF, Holtzman JL & Mason RP (1982) *Toxicol. Appl. Pharmacol.* **65:** 162–169.
Pott P (1775) In *Chirurgical Observations Relative to the Cataract, the Polypus of the Nose, the Cancer of the Scrotum, the Different Kinds of Ruptures and the Mortification of the Toes and Feet*, pp 63–68. London: Hawes, Clarke and Collins.
Pruess-Schwartz D, Nimesheim A & Marnett LJ (1989) *Cancer Res.* **49:** 1732–1737.
Rakieten N, Gordon BS, Beaty A, Cooney DA, Davis RD & Schein PS (1971) *Proc. Soc. Exp. Biol. Med.* **137:** 280–283.
Reddy JL & Lalwai ND (1983) *CRC Crit. Rev. Toxicol.* **12:** 1–58.
Reed GA (1987) *Carcinogenesis* **8:** 1145–1148.
Reed GA (1988) *Environ. Carcinogen. Rev.* **C6:** 223–259.
Reed GA, Brooks EA & Eling TE (1984) *J. Biol. Chem.* **259:** 5591–5595.
Reed GA, Curtis JF, Mottley C, Eling TE & Mason RP (1986) *Proc. Natl. Acad. Sci. USA* **83:** 7499–7502.
Reed GA, Layton ME & Ryan MJ (1988) *Carcinogenesis* **9:** 2291–2295.
Reiners JJ Jr, Pence BC, Barcus MCS & Cantu AR (1987) *Cancer Res.* **47:** 1775–1779.
Reneeberg R, Capdevila J, Chacos N, Estabrook RW & Prough RA (1981) *Biochem. Pharmacol.* **30:** 843–848.
Rickert DE, Baker TS, Bus JS, Barrow CS & Irons RD (1979) *Toxicol. Appl. Pharmacol.* **49:** 417–423.
Ritter CL & Malejka-Giganti D (1989) *Chem. Res. Toxicol.* **2:** 325–333.

Robertson FM, Beavis AJ, Oberszyw TM et al (1990) *Cancer Research* **50:** 6062–6067.
Rogan EG, Katomski PA, Roth RW & Cavalieri EL (1979) *J. Biol. Chem.* **254:** 7055–7059.
Rogan EG, Cavalieri EL, Tibbels SR et al (1988). *J. Am. Chem. Soc.* **110:** 4023–4029.
Ross D, Mehlhorn RJ & Moldeus P (1985) *J. Biol. Chem.* **260:** 15028–15032.
Rothman N, Poirier MC, Baser ME et al (1990) *Carcinogenesis* **11:** 1241–1243.
Roy D & Liehr JG (1989) *Carcinogenesis* **10:** 1241–1245.
Sadler A, Subrahmanyam VV & Ross D (1988) *Toxicol. Appl. Pharmacol.* **93:** 62–71.
Saladino AJ, Willey JC, Lechner JF, Grafstrom RC, LaVeck M & Harris CC (1985) *Cancer Res.* **45:** 2522–2526.
Sasame HA & Boyd MR (1979) *Life Sci.* **24:** 1091–1096.
Sawahata T & Neal RA (1983) *Mol. Pharmacol.* **23:** 453–460.
Schein PS, Cooney DA & Vernon ML (1967) *Cancer Res.* **27:** 2324–2332.
Schlosser MJ, Shurina RD & Kalf GF (1990) *Chem. Res. Toxicol.* **3:** 333–339.
Schrieber J, Foureman GL, Hughes MF, Mason RP & Eling TE (1989) *J. Biol. Chem.* **264:** 7936–7943.
Seed JL, Kensler TW, Elia M & Trush MA (1990) *Res. Commun. Chem. Pathol. Pharmacol.* **67:** 349–360.
Segal A, Katz C & Van Duuren BL (1971) *J. Med. Chem.* **14:** 1152–1154.
Shaw Y-S & Chen C (1972) *Biochem. J.* **128:** 1285–1291.
Shimkin MB (1977) In *Contrary to Nature*, pp 95, 247. Washington DC: DHEW.
Sies H, Wefers H & Cadenas E (1984) In Thaler-Dao, H, de Paulet A & Paoletti R (eds) *Icosanoids and Cancer*, pp 11–19. New York: Raven Press.
Sinha BK & Mimnaugh EG (1990) *Free Radicals Biol. Med.* **8:** 567–581.
Sinha BK & Sik RH (1980) *Biochem. Pharmacol.* **29:** 1867–1868.
Slaga TJ & Fischer SM (1983) *Prog. Exp. Tumor Res.* **26:** 85–109.
Slaga TJ, Klein-Szanto AJP, Triplett LL & Yotti LP (1981) *Science* **213:** 1023–1025.
Smart RC & Zannoni VG (1984) *Mol. Pharmacol.* **26:** 105–111.
Smith MT, Yager JW, Steinmetz KL & Eastmond DA (1989) *Environ. Health Perspect.* **82:** 23–29.
Snyder R, Longacre SL, Whitmer CM, Kocsis JJ, Andrews LS & Lee EW (1980) In Hodgson E, Bend JR & Philpot RM (eds) *Reviews in Biochemical Toxicology*, Vol. 3, pp 123–153. New York: Elsevier/North Holland.
Solanki V, Rana RS & Slaga TJ (1981) *Carcinogenesis* **2:** 1141–1146.
Solcia E, Ballerini L, Bellini D, Sala L & Bertazolli C (1978) *Cancer Res.* **38:** 1444–1446.
Sternberg SS, Philips F & Cronin AP (1972) *Cancer Res.* **32:** 1029–1036.
Stier A, Clauss R, Lucke A & Reitz I (1980) *Xenobiotica* **10:** 661–673.
Stock BH, Bend JR & Eling TE (1986) *J. Biol. Chem.* **261:** 5959–5964.
Subrahmanyam VV & O'Brien PJ (1985) *Xenobiotica* **15:** 873–885.
Sugimura T, Otake H & Matsushima T (1968) *Nature (Lond.)* **218:** 392.
Sugiura K & Goto M (1981) *Chem. Biol. Interact.* **35:** 71–91.
Swaminathan S & Lowes GM (1978) In Bryan GT (ed.) *Carcinogenesis—A Comprehensive Survey: Nitrofurans*, Vol. 4, pp 59–97. New York: Raven Press.
Swauger JE, Dolan PM & Kensler TW (1990) In Mendelsohn ML & Albertini RJ (eds) *Mutation and the Environment. Part D: Carcinogenesis*, pp 143–152. New York: Wiley-Liss.
Tada M & Tada M (1972) *Biochem. Biophys. Res. Commun.* **46:** 1025–1032.
Taffe BG & Kensler TW (1988) *Res. Commun. Pathol. Pharmacol.* **61:** 291–303.

Taffe BG, Takahashi N, Kensler TW & Mason RP (1987) *J. Biol. Chem.* **262:** 12143–12149.
Taffe BG, Zweier JL, Pannell LK & Kensler TW (1989) *Carcinogenesis* **10:** 1261–1268.
Thomas DJ, Reasor MJ & Wierda D (1989) *Toxicol. Appl. Pharmacol.* **97:** 440–453.
Thomas, DJ, Sadler A, Subrahmanyam VV et al (1990) *Mol. Pharmacol.* **37:** 255–262.
Thompson DC & Trush MA (1988) *Toxicol. Appl. Pharmacol.* **96:** 115–121.
Thompson DC & Trush MA (1989) *Chem. Biol. Interact.* **72:** 157–173.
Thompson DC, Cha YN & Trush MA (1989a) *J. Biol. Chem.* **264:** 3957–3965.
Thompson DC, Constantin-Teodosiu D, Norbeck K, Svensson B & Moldeus P (1989b) *Chem. Res. Toxicol.* **2:** 186–192.
Thompson JA, Malkinson AM, Wand MD et al (1987) *Drug Metab. Dispos.* **5:** 833–840.
Thompson JA, Schullek KM, Fernandez CA & Malkinson AM (1989) *Carcinogenesis* **10:** 773–775.
Tice RR, Costa DL & Drew RT (1980) *Proc. Natl. Acad. Sci. USA* **77:** 2148–2152.
Tomasz M & Lipman R (1981) *Biochemistry* **20:** 5056–5061.
Trush MA & Thompson DC (1988) In Simic M, Taylor KA, Ward JF & von Sonntag C (eds) *Oxygen Radicals in Biology and Medicine*, pp 739–744. New York: Plenum Pub. Corp.
Trush MA & Kensler TW (1991) *Free Radicals Biol. Med.* (in press).
Trush MA, Mimnaugh EG, Ginsburg E & Gram TE (1982a) *J. Pharmacol. Exp. Ther.* **221:** 152–158.
Trush MA, Mimnaugh EG, Ginsburg E & Gram TE (1982b) *J. Pharmacol. Exp. Ther.* **221:** 159–165.
Trush MA, Mimnaugh EG, Ginsburg E & Gram TE (1982c) *Biochem. Pharmacol.* **31:** 805–814.
Trush MA, Mimnaugh EG & Gram TE (1982d) *Biochem. Pharmacol.* **21:** 3335–3346.
Trush M, Kennedy KA, Sinha BK & Mimnaugh EG (1983) *The Pharmacologist* **25:** 225.
Trush MA, Seed JL & Kensler TW (1985) *Proc. Natl. Acad. Sci. USA* **82:** 5194–5198.
Trush MA, Twerdok LE & Esterline RL (1990) *Xenobiotica* **20:** 925–932.
Trush MA, Esterline RL, Mallet WG, Mosebrook PR & Twerdok LE (1991) In Sipes IG (ed.) *Biological Reactive Intermediates IV. Molecular and Cellular Effects and Their Impact on Human Health*, pp 399–401. New York: Plenum Press.
Tunek A, Platt KL, Przybylski M & Oesch F (1980) *Chem. Biol. Interact.* **33:** 1–17.
Turkall RM & Tsan M-F (1982) *Res. J. Reticuloendothelial Soc.* **31:** 353–360.
Twerdok LE & Trush MA (1988a) *Chem. Biol. Interact.* **65:** 261–273.
Twerdok LE & Trush MA (1988b) In Simic MG, Taylor KA, Ward JF & von Sonntag C (eds) *Oxygen Radicals in Biology and Medicine*, pp 255–259. New York: Plenum Pub. Corp.
Twerdok LE & Trush MA (1990) *Res. Commun. Chem. Pathol. Pharmacol.* **67:** 375–386.
Uchigata Y, Yamamoto H, Kawamura A & Okamoto H (1982) *J. Biol. Chem.* **257:** 6084–6088.
Vig BV & Lewis R (1978) *Mutat. Res.* **55:** 121–145.
Wang YM, Howell SK, Kimball JC, Tsai CC, Sato J & Gleiser CA (1982) In Arnolt MS, van Eys J & Wang EM (eds) *Molecular Interactions of Nutrition and Cancer*, pp 369–374. New York: Raven Press.

Wefers H & Sies H (1983) *Eur. J. Biochem.* **137:** 29–36.
Wei ET & Shu HP (1983) *Am. J. Public Health* **73:** 1085–1088.
Weiss SJ, Klein R, Slivka A & Wie M (1982) *J. Clin. Invest.* **70:** 598–607.
Weitzman SA & Gordon LI (1990) *Blood* **76:** 655–663.
White RE & Coon MJ (1980) *Annu. Rev. Biochem.* **49:** 315–356.
Witschi H (1988) *Toxicology* **48:** 1–20.
Yamazoe Y, Roth RW & Kadlubar FF (1986) *Carcinogenesis* **7:** 179–182.
Zenser TV, Mattammal MP, Cohen SM, Palmier MO, Wise RW & Davis BB (1984) In Thaler-Dao H, de Paulet A & Paoletti R (eds) *Icosanoids and Cancer*, pp 71–78. New York: Raven Press.

12

Membrane Hydroperoxides

FULVIO URSINI*, MATILDE MAIORINO* and ALEX SEVANIAN[†]
* *Department of Biological Chemistry, University of Padova, Via F. Trieste 75, I-35100 Padova, Italy*
[†] *Institute for Toxicology, University of Southern California, Los Angeles, California, USA*

I. Introduction ... 319
II. Hydroperoxide formation in membranes ... 320
III. Phospholipase A_2-mediated elimination of phospholipid hydroperoxides .. 324
IV. Phospholipid hydroperoxide glutathione peroxidase ... 327
V. Membrane hydroperoxides: perspectives on possible physiological effects ... 329

I. INTRODUCTION

Polyunsaturated fatty acids (PUFA) are among the most susceptible molecules to oxidative degradation present in living tissues. Moreover, most tissues are exposed to oxygen and contain an array of catalysts that participate in biological oxidation reactions. The best known of these are transition metals, particularly iron complexes and haemoproteins which activate oxygen and catalyse both metabolic oxygenation reactions (Estabrook and Werringloer, 1977) and nonspecific peroxidation and oxidative degradation of organic molecules (Chan and Newby, 1980; Vladimirov et al, 1980). Tissue oxygen tensions are known to be sufficient to catalyse the peroxidation of lipids, and the kinetics of PUFA peroxidation appear to be minimally affected at concentrations as low as 20 mmHg (Salaris and Babbs,

Oxidative Stress: Oxidants and Antioxidants
ISBN 0-12-642762-3

1989; Jones, 1985). Therefore, an oxidative challenge is intrinsically linked to aerobic life and is particularly facile in systems rich in PUFAs. The reaction between ground state oxygen (triplet) and ground state oxidizable substrates (singlet) that is permitted by oxygen activation is not only prominent in bioenergetic reactions but is also mechanistically intrinsic to the oxygenation of compounds bearing singlet multiplicity, namely PUFAs. The long-recognized tendency for peroxidation of lipids based on the number of double bonds per molecule (Witting, 1965) also holds for phospholipids containing unsaturated fatty acyl groups (Witting, 1965; Sevanian and Hochstein, 1985; Mowri et al, 1984). The kinetics of peroxidation for phospholipids in bulk phase versus those in membrane arrangements are similar, with deviations largely accounted for by physical constraints of lipids and by the impact of radical termination reactions which can affect product profiles (Wu et al, 1982; Chatterjee and Agarwal, 1988; Yamamoto et al, 1984; Barclay et al, 1984). Hence, the susceptibility to peroxidation of membrane lipids is largely influenced by their degree of unsaturation but contributions by factors such as arrangement or order (Barclay et al, 1984; Coolbear and Keough, 1983) cannot be ignored.

The intrinsic propensity for peroxidation in biological membranes may be viewed in terms of a homeostatic equilibrium where the conservation of a structure facing thermodynamically driven degradation must be actively conserved by specialized protective and repair systems. Although the chemistry and consequences of lipid peroxidation have been extensively reviewed (Vladimirov et al, 1980; Sevanian and Hochstein, 1985; Coolbear and Keough, 1983; Kappus, 1985), less is known about the biological effects of lipid hydroperoxides in membranes. Aside from their association with pathological events, the participation of these hydroperoxides in biochemical and physiological processes has only recently been considered. The aim of this chapter is to concentrate on the effects of hydroperoxides on membranes in terms of protection and repair systems and the effects that these hydroperoxides have on membrane structure and function. In this vein we also offer some perspectives on the possible physiological role of membrane hydroperoxides.

II. HYDROPEROXIDE FORMATION IN MEMBRANES

Hydroperoxides are considered as the major and primary products of the peroxidation of membrane lipids. Lipid peroxidation (i.e. formation of lipid hydroperoxides) is defined as a propagative process involving peroxyl radical-dependent chain reactions among unsaturated fatty acyl moieties.

$$\text{LOO}^{\bullet} + \text{L'H} \longrightarrow \text{LOOH} + \text{L'}^{\bullet} \quad (1)$$

This definition does not ignore the role of initiation reactions:

$$\text{LH} \longrightarrow \text{L}^{\bullet} + e^{-} + \text{H}^{+} \quad (2)$$

$$\text{L}^{\bullet} + \text{O}_2 \longrightarrow \text{LOO}^{\bullet} \quad (3)$$

but we propose that, in biological systems, such as during microsomal lipid peroxidation, hydroperoxide-dependent lipid peroxidation comprises the bulk of lipid peroxidation and the kinetic shift from peroxide-independent (initiation-type) to peroxide-dependent (propagation-type) reactions, accounting for the lag phase and progressive acceleration of lipid peroxidation (Maiorino et al, 1989):

$$\text{LOOH} + \text{M}^{(n)+} \longrightarrow \text{LO}^{\bullet} + \text{OH}^{-} + \text{M}^{(n+1)+} \quad (4)$$

$$\text{LO}^{\bullet} + \text{L'H} \longrightarrow \text{LOH} + \text{L'}^{\bullet} \quad (5)$$

The oxidative decomposition of hydroperoxides, giving rise to hydroperoxyl radicals, is kinetically less favoured than the reductive reaction above (Aust and Svingen, 1982):

$$\text{LOOH} + \text{M}^{(n+1)} \longrightarrow \text{LOO}^{\bullet} + \text{M}^{(n)+} \quad (6)$$

During the early stages of peroxidation, where termination reactions are limited due to low concentrations of radicals, formation of lipid peroxyl radicals (reaction 3) is the predominant kinetic reaction (Uri, 1961). In membranes bearing high concentrations of hydrogen donors (e.g. vitamin E and PUFA), peroxidation reaction probably proceeds to the formation of hydroperoxides (reaction 1). This definition must also recognize the role of 'antiperoxidation' factors which inhibit hydroperoxide formation by scavenging or preventing the formation of free radicals involved in initiation and propagation reactions. Examples of such antiperoxidants include peroxidases operating on initiating reactions, and scavengers participating in and abating propagation reactions (Burton and Ingold, 1984). A further consideration of these reactions will be presented later.

Hydroperoxides can be generated in membrane phospholipids directly by the action of singlet oxygen (Foote, 1976) or indirectly by iron–oxo complexes (Aust and Svingen, 1982; Ursini et al, 1989) via hypervalent iron intermediates. The latter are widely believed to be responsible for in vivo peroxidation as facilitated by a variety of iron-reducing agents, including

NADPH (cytochrome P450-linked lipid peroxidation), ascorbate, and O_2^- (Aust and Svingen, 1982; Ernster et al, 1982). The oxidation of haem iron to hypervalent states is known to produce a very powerful oxidant (i.e. Fe^{4+} OH). Such hypervalent states have been described for haemoglobin and myoglobin upon reaction with H_2O_2 or lipid hydroperoxides (Galaris et al, 1990), and these oxo-haem oxidants readily initiate lipid peroxidation (Kaschitz and Hatefi, 1975). Ascorbate and O_2^- appear to induce peroxidation via reduction of ferric iron, which then becomes able to bind and activate oxygen, generating the primary initiating species (Ernster et al, 1982). On the other hand, ferrous iron is also implicated in heterolytic decomposition of peroxides, leading to alkoxyl radical (Aust and Svingen, 1982), which in turn can start peroxidative chain reactions (secondary initiations or chain branching). Reactions such as these are driven by an iron redox cycle, supplying oxidizing equivalents to membrane phospholipids, and are suggested as criteria for initiating lipid peroxidation (Ursini et al, 1989). Redox-cycling reactions involving iron are supported also by other reductants such as ascorbate, glutathione, and redox-labile compounds, through O_2^- generation and by iron reduction.

The tendency for hydroperoxidation is directly related to the PUFA content of membrane phospholipids. However, the rate and extent of hydroperoxide generation is not determined simply by kinetic constraints, and, excluding the effects of antioxidants, a physical compartmentalization of phospholipid species can influence the rates and degree of membrane lipid peroxidation. For example, initiation rates of lipid peroxidation for phospholipids in lamellar arrangements are considerably slower than those in micellar arrangements or in organic solvents (Barclay et al, 1984, 1985; Winterle and Mill, 1984). The ordered arrangement of methylenic groups in unsaturated membrane bilayers was suggested by Mead et al (1980) to yield non-autocatalytic kinetics, in contrast to the kinetics observed in bulk phase with formation of intramolecular rearrangement products of hydroperoxides. Recently, Volkov et al (personal communication) applied a two-component model for lipid peroxidation in membranes based on the lateral distribution of lipids. The phase-transition properties of membrane phospholipids were proposed to be influenced by ratios of saturated to unsaturated phospholipid species. Membranes in a liquid-crystalline state thereby experience either an influx of saturated phospholipids into unsaturated clusters at lower phase-transition temperatures or an influx of unsaturated phospholipids into saturated clusters at higher phase-transition temperatures. Membranes undergoing lipid peroxidation among unsaturated components are accordingly expected to accumulate clusters of peroxidized lipids with higher phase-transition temperatures. Computer modelling studies revealed that unsaturated phospholipids are drawn into these clusters as part of the phase-compensating phenomenon. A process such as this may ‘feed’ the peroxyl radical-propagating reactions within the clusters of peroxidized lipid, competing for termination reactions and promoting lipid peroxidation. Radical chain reaction lengths could therefore be increased above those theoretically expected in homogeneous systems.

The basic role for phenolic antioxidants in membranes, primarily accounted for by vitamin E (vitE-OH), is the breaking of peroxidative chain reactions (Niki, 1987). This takes place by reactions between these antioxidants and peroxyl radicals:

$$LOO^{\bullet} + \text{vitE-OH} \longrightarrow LOOH + \text{vitE-O}^{\bullet} \qquad (7)$$

This reaction prevents propagation since: (1) the rate constant for interaction of $LOO^{\bullet}$ with vitE-OH (Burton et al, 1983) is several orders of magnitude greater than the rate constants for interaction with PUFA (Howard and Ingold, 1967); and (2) vitE-$O^{\bullet}$ does not favourably participate in propagation reactions and can be easily reduced back to vitE-OH by ascorbate (Niki, 1987; Scarpa et al, 1984) or possibly by enzymatic systems. This reaction depicts the peculiar features necessary for an antioxidant to be biologically active, that is, a kinetic advantage for the reaction with the free radical where PUFAs are spared, and the formation of a relatively stable free radical. Reaction of vitE-OH with $LO^{\bullet}$ is also quite fast but does not show a significant kinetic advantage over the reaction of $LO^{\bullet}$ with a variety of biological targets such as PUFAs, as deduced from measured rate constants in homogeneous solution (Erben-Russ et al, 1987). In this respect vitE-OH is a good antioxidant when lipid peroxidation proceeds via $LOO^{\bullet}$, whilst the antioxidant activity is negligible when reactions are initiated by $LO^{\bullet}$ (Maiorino et al, 1989).

Although vitE-OH inhibits peroxidative chain propagation, its activity leads to the formation of LOOH (reaction 7). Since LOOH (1) are peroxidation products, and (2) can initiate new peroxidative chains via their decomposition, an effective mechanism of protection requires elimination of these LOOH. Two systems have been described which can accomplish this function. The first is the hydrolysis of LOOH from membrane phospholipids followed by reduction through glutathione peroxidase, and the second is a direct reduction by phospholipid hydroperoxide glutathione peroxidase (Scheme 1). Although measurements of LOOH concentrations in cell membranes have not been made directly, approximations of intracellular levels can be made from studies on the activation of prostaglandin synthetase (Lands et al, 1984), since prostaglandin synthetase activation requires hydroperoxides over a concentration range of 10^{-8} to 10^{-7} M. It is surmised that such hydroperoxide levels might be achieved and maintained through a balance of prooxidative versus antioxidative reactions. This (peroxide threshold) can be maintained by a combination of peroxidase activities which display maximal kinetics over the concentration range indicated above. However, it must be established whether cellular levels of LOOH can fluctuate independently of the major pools from which they are derived—namely membrane phospholipid PUFAs.

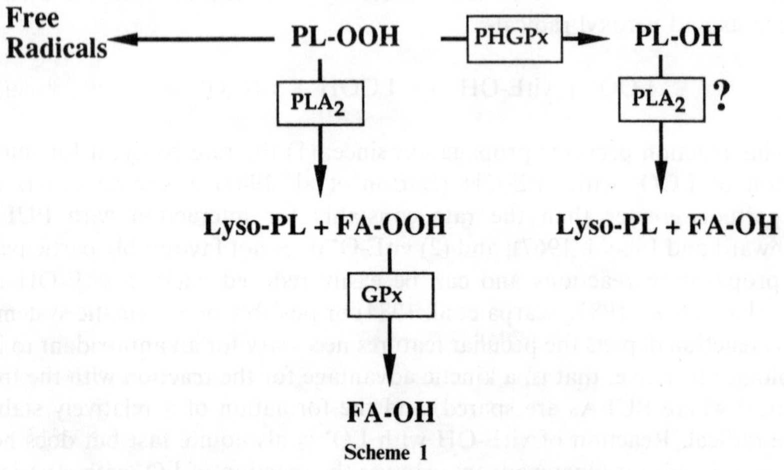

Scheme 1

III. PHOSPHOLIPASE A_2-MEDIATED ELIMINATION OF PHOSPOLIPID HYDROPEROXIDES

Phospholipase A_2 (PLA_2) activity is found in every cell type and plays a crucial role in the metabolism and turnover of membrane phospholipids (Van Den Bosch, 1980). In addition to this basic housekeeping function, PLA_2 serves to supply fatty acid substrates for eicosanoid biosynthesis under the influence of specific stimuli (Axelrod et al, 1988; Slivka and Insel, 1987; Nolan et al, 1981). A variety of agonists have been shown to regulate PLA_2 activity in addition to specific proteins which either activate or inhibit the enzyme. Indirect effects have been described through either structural alterations in membrane phospholipids or by fluctuations in peroxide levels. In the latter case it has been shown (Cao et al, 1987) that vitE-OH inhibits PLA_2 activity by purportedly inhibiting lipid peroxidation. In this context it has been recently reported (Robison et al, 1991) that nordihydroguaiaretic acid (NDGA) inhibits PLA_2 activity in cell culture through reduction in lipid peroxide levels.

The apparent substrate specificity of PLA_2 for hydroperoxy fatty acids may also account for its ability to protect cell membranes from oxidative damage. PLA_2 is uniquely suited for this function since the bulk of membrane PUFAs are located in the sn-2 position of phospholipids, and LOOH formed among these PUFAs would also be expected substrates for the enzyme. Recent studies have shown that the enzyme's order of specificity towards oxidized phospholipids is extremely high and under these circumstances there is also a hydrolytic preference for PUFA compared to the hydrolytic activity seen in unoxidized membranes (Sevanian and Kim, 1987). Peroxidized phospholipids are preferentially hydrolysed by PLA_2 and substantial amounts of unsaturated free fatty acids (particularly arachidonic acid) are frequently isolated from oxidized membrane preparations (Sevanian and Kim, 1987). The suppression of PLA_2 activity by NDGA was found to be accompanied by a marked reduction in arachidonic acid release and eicosanoid biosynthesis in cell cultures (Robison et al, 1991). The molecular basis for PLA_2 hydrolytic specificity remains uncertain; however, an alteration of specific biophysical parameters of the membranes is offered as a possible explanation (Sevanian et al, 1988). In general the enzyme appears to respond to disturbances in membrane organization by catalysing transacylation reactions which tailor the fatty acid composition of phospholipids and thus maintain membrane structure and fluidity characteristics appropriate to physicochemical and environmental constraints (Dawson, 1982; Cossins and Sinensky, 1984; Majewska et al, 1978).

The enhanced activity of PLA_2 in model and isolated cellular membranes subjected to free radical reactions offers an explanation for analogous events described in intact tissues experiencing oxidative stress. Detection of high levels of free fatty acids in ischaemic brain (Chan et al, 1984) and heart (Bentham et al, 1987) is attributed to the activation of membrane phospholipases. Selective hydrolysis of oxidized fatty acids is proposed to serve two important protective functions for the cell. The first is removal of the polar and often reactive hydroperoxides, aldehydes, epoxides and ketones that are generated during lipid peroxidation. The presence of these products in the hydrophobic matrix of the membrane creates disturbances in structure evidenced by decreased fluidity (Galeotti et al, 1984; Dobretsov et al, 1977), increased phospholipid flip-flop and fusion (Coolbear and Keough, 1983; Kornberg and McConnell, 1971) and phase separations which impart packing imperfections, creating phospholipid microdomains (Coolbear and Keough, 1983; Kornberg and McConnell, 1971; Borsukov et al, 1980; Wratten et al, 1989). PLA_2 has been shown to display a high specific activity towards these defective zones consistent with its hydrolytic preference for phospholipids in multiphasic arrangements (Sevanian, 1988).

The second step of antioxidant protection involves the metabolism of lipid hydroperoxides by cytosolic glutathione peroxidase (GSH-Px) (Flohé, 1982). This cytosolic enzyme displays negligible activity against membrane hydroperoxides. PLA_2 is proposed to facilitate the activity of GSH-Px by releasing fatty acid hydroperoxides from membrane phospholipids that are then promptly reduced (Van Kujik et al, 1987). Acting together, PLA_2 and GSH-Px appear to fulfil one part of the general membrane antioxidant activity.

Elimination of damaged fatty acids by hydrolysis and reduction of lipid hydroperoxides represents only half of a membrane repair process. This process is completed by reacylation of the lysophospholipids produced by the hydrolytic action of PLA_2. The reacylation phase is energy-dependent, requiring fatty acyl CoA synthesis. Recent studies have shown that the acyltransferase step may be rate-limiting, since under certain conditions phospholipase activity exceeds reacylation, leading to lysophospholipid accumulation (Zimmerman and Keys, 1988). Prooxidant conditions appear to be one situation where such an imbalance could prevail (Lubin et al, 1972). Zimmerman and Keys (1988), using rod outer segment membranes, have shown that deacylation and reacylation reactions are stimulated under oxidizing conditions with no apparent selectivity for fatty acyl CoA species reinserted into the sn-2 position of phospholipids. Fatty acid incorporation into phosphatidylethanolamine is stimulated when human erythrocytes are exposed to *tert*-butyl hydroperoxide (Dise and Goodman, 1986). Possible mechanisms for the stimulated reacylation may include: membrane structural changes following oxidation, increased lysophospholipid concentrations, increased free fatty acid levels and accumulation of oxidized lipid products.

Zaleska and Wilson (1989) have pointed out that peroxidizing conditions may, in some cases, inhibit rather than stimulate acylation of phospholipids and they attributed the inhibitory effect to fatty acid hydroperoxides. Lipid hydroperoxides perturb cell functions at much lower concentrations than *tert*-butyl hydroperoxide. Enhanced acylation is observed during the early stages of membrane lipid peroxidation, preceding the accumulation of oxidized lipid products. Progressive oxidation is, however, accompanied by an apparent inhibition of reacylation (Dise et al, 1987) but the mechanism for this inhibition remains unknown. Accordingly, the ability to repair oxidatively damaged membranes may be influenced by the degree of lipid peroxidation where peroxidation does not exceed the capacity of the cell to reduce and eliminate fatty acid hydroperoxides. This capacity appears not to be determined by PLA_2, since the enzyme displays enormous catalytic capacity following various stimuli or phospholipid perturbation (Shier, 1979). Thus, enzyme unit activities can vary from a fraction of a unit for homogeneous bilayers composed of single phospholipid species to hundreds of units when phospholipids are present as mixed micelles or detergent dispersions (Barlow et al, 1988).

IV. PHOSPHOLIPID HYDROPEROXIDE GLUTATHIONE PEROXIDASE

Phospholipid hydroperoxide glutathione peroxidase (PHGSH-Px) (Ursini, 1988; Maiorino et al, 1990) is a monomeric protein with a molecular mass of approximately 20 kDA that contains a selenocysteine, apparently involved in its catalytic reaction (Ursini et al, 1985). This enzyme, which reduces lipid hydroperoxides in biological membranes and is the second selenoenzyme thus far identified and characterized in mammals, was first identified on the basis of its antiperoxidative activity. For this reason PHGSH-Px was initially named peroxidation inhibiting protein (PIP) (Ursini et al, 1982). PHGSH-Px inhibits microsomal lipid peroxidation by reducing lipid hydroperoxides. The reduction of lipid hydroperoxides by this enzyme was demonstrated by thin-layer chromatography, high-performance liquid chromatography and mass spectroscopic analysis of the reaction products (Ursini et al, 1984, 1985; Daolio et al, 1983). Furthermore, by reduction to hydroxy derivatives the hydroperoxides, which are formed by interaction of lipid hydroperoxyl radicals with vitamin E, are effectively eliminated by PHGSH-Px. In this way PHGSH-Px also spares vitamin E and accounts for the synergistic antiperoxidant effect of selenium and vitamin E (Maiorino et al, 1989).

The peroxidase activity on intact phospholipid hydroperoxides was demonstrated in lipid dispersions, liposomes, Triton-solubilized mixed micelles (Ursini et al, 1982, 1985), cellular membranes (Maiorino et al, 1985) and red blood cell ghosts (Thomas et al, 1990). Triton dispersion of phospholipids was most effective in supporting maximal rates of enzyme activity.

Phosphatidylcholine hydroperoxide is the typical substrate used to measure PHGSH-Px activity, although all the other phospholipid classes tested as well as cumene hydroperoxide, *tert*-butyl hydroperoxide and hydrogen peroxide are also reduced. Recently, it was demonstrated that PHGSH-Px reduces cholesterol hydroperoxides in Triton-dispersed micelles, red blood cell ghosts, unilamellar liposomes and oxidized human LDLs (Thomas et al, 1990; Maiorino et al, 1991). Interestingly, PHGSH-Px reduces all lipid hydroperoxides in isolated LDL, including those of esterified cholesterol, and this might be an important clue to evaluating the atherogenic potential of oxidatively modified lipoproteins. In fact the cytotoxicity of oxidatively modified LDL on different cells might be related to the expression in these cells of this enzyme.

The peroxidase activity of PHGSH-Px is usually measured spectrophotometrically, where the specific oxidation of GSH in the presence of GSSG reductase and phosphatidylcholine hydroperoxides is monitored by NADPH oxidation (Maiorino et al, 1990a; Ursini et al, 1982). Using this test and taking advantage of the specificity for phospholipid hydroperoxides, on which neither 'classical' GSH-Px, nor Se-independent GSH-Px, are active, PHGSH-Px activity was measured in different tissue samples (Zhang et al, 1989). PHGSH-Px was identified in almost all rat, mouse, pig, dog and beef tissues, a notable exception being red blood cells.

PHGSH-Px was described as a cytosolic enzyme active on membranes. This was deduced from the fact that: (1) the enzyme was identified and purified from the soluble compartment of the cells (liver, heart, brain); (2) the purified enzyme is active on peroxidized membranes; and (3) subcellular membranes (microsomes, mitochondrial outer membrane) contain PHGSH-Px activity that can be partially released in high-ionic-strength media. However, it has been recently observed that membrane fraction of rat testes homogenate contains high PHGSH-Px activity accounted for by a protein that in Western blot analysis appears identical to liver and heart cytosolic PHGSH-Px. This enzyme is only partially solubilized under high-ionic-strength or detergent treatment. The identification of this membrane form of the enzyme apparently complicates its mode of action in terms of transfer of the enzyme from the soluble compartment and anchorage to the membrane, or suggests the presence of multiple forms with more or less affinity to the membrane.

Sequence analysis of PHGSH-Px indicates that sequences with some analogy with 'classical' GSH-Px are present in PHGSH-Px, although sequences which are completely different are also found (Schuckelt et al, 1991). Furthermore, polyclonal antibodies against pig heart PHGSH-Px that recognize PHGSH-Px from all rat and pig tissues do not recognize any forms of GSH-Px from rat, beef and pig tissues. A selenium-deficient diet causes a much faster depletion of GSH-Px activity than PHGSH-Px (Weitzel et al, 1989). In fact, in mice fed a diet containing less than 12 ppb Se, GSH-Px decreases to undetectable levels within 130 days, whilst the residual activity of PHGSH-Px remains at levels 30–70% of controls in different organs even after 250 days.

The two selenoenzymes are, therefore, different but related, at least in the terms of a selenocysteine at the active site, peroxidase activity and the kinetic mechanism of reaction (Ursini and Bindoli, 1987). In both enzymes the selenocysteine is oxidized by the peroxide and the oxidized form of Se is then reduced by two GSH molecules, giving rise to the native form of the enzyme and GSSG. This is a rather peculiar aspect of peroxidase reactions catalysed by selenoenzymes; that is, by this mechanism free radicals or oxidized forms of GSH higher than GSSG are not generated. This is not the case for haem-containing peroxidases or the Se-independent GSH peroxidases. GSH-Px and PHGSH-Px seem to reduce hydroperoxides in different cellular milieu, GSH-Px in the water phase and PHGSH-Px in the membrane phase, possibly due to a different lipophilicity of the substrate binding and catalytic sites. This is suggested also by the different effect of detergents on either enzyme (Maiorino et al, 1986). A ping-pong reaction mechanism very similar to that described for selenium-dependent peroxidases has been reported for the seleno compound Ebselen. Furthermore, the kinetic analysis of the reaction catalysed by this 'artifical enzyme' showed that phospholipids are among the best substrates, indicating that this compound is a mimetic more of PHGSH-Px than of GSH-Px (Maiorino et al, 1988).

Although membrane hydroperoxides are resistant to GSH-Px, this enzyme does reduce fatty acid hydroperoxides released by a phospholipase. This mechanism is possibly relevant physiologically and likely coexists with the direct reduction by PHGSH-Px. It is noteworthy, however, that the antiperoxidative action exerted by GSH-Px plus phospholipase is not complete, since cholesterol hydroperoxides are metabolized only by PHGSH-Px (Thomas et al, 1990). The in vivo observation that hydroxy derivatives of phospholipid fatty acids are major lipid oxidation products (Hughes et al, 1983) supports the idea that the first line of defence of membranes against hydroperoxide-dependent peroxidation is the direct reduction of these lipid hydroperoxides.

V. MEMBRANE HYDROPEROXIDES: PERSPECTIVES ON POSSIBLE PHYSIOLOGICAL EFFECTS

In the previous sections the chemical mechanisms for membrane lipid hydroperoxide generation have been summarized. Key reactions (reactions 1–6) have been identified as committed steps of the general phenomenon of lipid peroxidation, a free radical oxidative chain leading to membrane damage. Different membrane functions are compromised by the modifications of both membrane composition and physical status linked to the formation of hydroperoxy derivatives of lipids and their decomposition products (Vladimirov et al, 1980). The study of the relationships between functional cellular derangement caused by free radical-dependent lipid peroxidation and the mechanism of lipid peroxidation in experimental and spontaneous diseases accounts for the rapidly growing research field of 'free radical pathology'. A complete clarification of the steps between free radical reactions (taking place in microseconds) and pathophysiological consequences (taking place in variable times from minutes to months) would be among the most exciting achievements of modern medicine. The aim of this section is not to review the pathologies where free radicals and lipid peroxidation are thought to be involved but to discuss briefly the appealing possibility that membrane hydroperoxides could be involved in some physiological functions.

The rationale for this hypothesis is that free radicals and lipid hydroperoxide-generating systems could have a physiological function that is achieved through the effect of lipid hydroperoxides on membrane functions such as modification of structure (related to physical parameters), ion permeability and enzyme activities. For such a mechanism to be physiologically relevant the first constraint is that lipid hydroperoxides have to be generated in a specific manner with respect to space and time, without any chain reactions leading to uncontrolled peroxidation. The other constraint is that a counteracting system must operate to abate the physiological signals originating from lipid hydroperoxides and hence modulate their effect by metabolizing them. To this end we postulate a lipoxygenase–peroxidase system where the concentration of hydroperoxides in the membrane could be fine-tuned. Furthermore, this peroxidase activity would provide the terminal switch-off event.

It is known that enzymatic peroxidation is catalysed by different lipoxygenases active on PUFAs (basically arachidonic acid) (Pace-Asciak and Asotra, 1989). However, the products of lipoxygenase reactions, which are relatively water-soluble, can diffuse from the site of generation to other cellular compartments or other cells. For this reason the products of hydroperoxidation of free fatty acids do not fit well the prerequisite of our hypothesis where, instead, a lipoxygenase acting on intact phospholipids is envisaged. Such a lipoxygenase would generate hydroperoxy phospholipids which could exert biophysical and biochemical effects on the membrane analogous to the actions of oxygenated fatty acids (viz. eicosanoids) when they are released into the cytosol. Although such an enzymatic system has yet to be described, there is suggestive evidence for such reactions from several studies (Jung et al, 1985; Funk and Powell, 1985; Rapoport et al, 1979). Notable examples include: (1) the aorta, which generates hydroxy derivatives of phospholipids, apparently by sequential hydroperoxidation and reduction reactions (Funk and Powell, 1985); and (2) reticulocyte lipoxygenase that appears to be active on intact phospholipids (Rapoport et al, 1979). This enzyme is expressed during erythrocyte differentiation and is instrumental in the destruction of mitochondria and other intracellular membranes. The activity of this enzyme in reticulocytes represents an interesting example of a physiological effect, i.e. the terminal differentiation of red blood cells, which is achieved by a physiologically controlled peroxidation of cell membranes. An important consideration concerning this peroxidation-dependent destruction of cellular membranes is that the peroxidation process does not account, per se, for the complete disappearance of membranes, since subsequent reactions must follow which facilitate elimination of hydroperoxide-bearing membranes. The plausible enzymatic step after membrane hydroperoxidation is the hydrolysis of the affected lipids by a specific phospholipase (see previous section). Once activated, PLA_2 is capable of hydrolysing unoxidized phospholipids as well. The formation of the initial hydroperoxide species could merely be a signal marking these membrane loci for destruction. The enzymatic reduction of membrane hydroperoxides would be the counteracting stop signal.

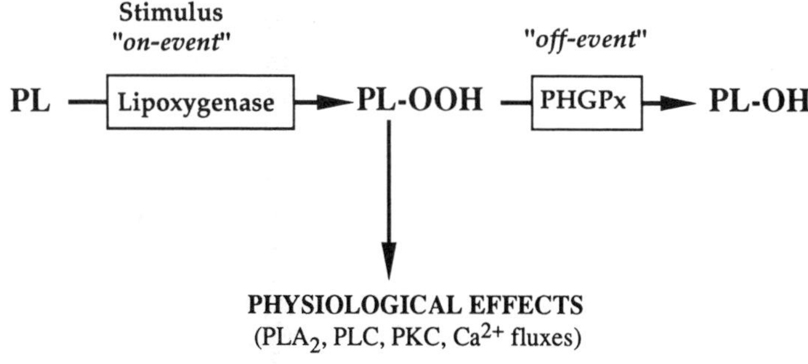

Scheme 2

Based on present knowledge, it is tempting to speculate that such a mechanism could apply not only to reticulocytes but also to other differentiating tissues. Apart from the peroxidative destruction of membranes, a controlled phospholipid peroxidation could be involved in other regulatory mechanisms. Lipid peroxides in membranes have an ionophoric effect on calcium (Levitsky et al, 1988) and this should not only be considered as a pathological event. Furthermore, the 'peroxide tone' (Lands et al, 1984) of membranes can control the activity of key enzymes such as phospholipases (Sevanian, 1988; Gamache et al, 1988), cyclooxygenase (Hughes et al, 1983), and possibly protein kinase C (Bulkley, 1983). We have recently observed that calcium dependency of protein kinase C is shifted to the left by phosphatidylserine hydroperoxides, suggesting their possible activating role (Maiorino et al, unpublished). In this context we note the observation by U.F. Shade (personal communication) that phospholipids extracted from endotoxin-activated macrophages (cell line RAW-264.7) contain 13-hydroxyoctadecanoic acid. It was found that phosphatidylserine, phosphatidylinositol and phosphatidylethanolamine from stimulated cells contained substantially higher amounts of this hydroxy acid (2–4-fold) than unstimulated cells, while phosphatidylcholine was not altered.

The control of cellular proliferation may be another major physiological effect of membrane hydroperoxides (Cornwell and Morisaki, 1984), possibly through the activity of the aforementioned enzymes.

Future research should clarify this potentially important area, and if new lipoxygenases or new lipoxygenase activities are discovered a scenario could be depicted where hydroperoxidation of membrane lipids acts as a 'switch-on' and PHGPx as a 'switch-off' signal for specific physiological events (Scheme 2).

REFERENCES

Aust SD & Svingen BA (1982) The role of iron in enzymatic lipid peroxidation. In Pryor WA (ed.) *Free Radicals in Biology* Vol. V, pp. 1–28. New York: Academic Press.

Axelrod J, Burch RM & Jelsema CL (1988) Receptor mediated activation of phospholipase A2 via GTP binding proteins: arachidonic acid and its metabolites as second messengers. *Trends Neurosci.* **11:** 117–123.

Barclay LRC, Locke SJ, MacNeil JM, Van Kessel J, Burton GW & Ingold KU (1984) Autooxidation of micelles and model membranes: Quantitative kinetic measurements can be made by using either water soluble or lipid soluble initiators with water soluble or lipid soluble chain breaking antioxidants. *J. Am. Chem. Soc.* **106:** 2479–2481.

Barclay LRC, Locke SJ, MacNeil JM & Van Kessel J (1985) Quantitative studies of the autoxidation of linoleate monomers sequestered in phosphatidylcholine bilayers. Absolute rate constants in bilayers *Can. J. Chem.* **63:** 2633–2638.

Barlow PN, Lister MD, Sigler PB & Dennis EA (1988) Probing the role of substrate conformation in phospholipase A2 action on aggregated phospholipids using constrained phosphatidylcholine analogues. *J. Biol. Chem.* **263:** 12954–12958.

Bentham JM, Higgins AJ & Woodward B (1987) The effects of ischemia, lysophosphatidylcholine and palmitoylcarnitine on rat heart phospholipase A2 activity. *Basic Res. Cardiol.* **82:** 127–135.

Borsukov LI, Victorov AV, Vasilenko IA, Evstigneeva RP & Bergelson LD (1980) Investigation of the inside–outside distribution, intermembrane exchange and transbilayer movement of phospholipids of sonicated vesicles by shift reagent NMR. *Biochim. Biophys. Acta* **598:** 153–168.

Bulkley GB (1983) The role of oxygen free radicals in human disease processes. *Surgery* **94:** 407–411.

Burton GW & Ingold KU (1980) β-Carotene: an unusual type of lipid antioxidant. *Science* **224:** 569–575.

Burton GW, Cheeseman KH, Doba T, Ingold KU & Slater TF (1983). Vitamin E as an antioxidant in vitro and in vivo. In *Biology of Vitamin E*, CIBA Foundation Symposium 101, pp 4–14, London: Pitman.

Cao YZOK, Choy PC & Chan AC (1987) Regulation by vitamin E of phosphatidylcholine metabolism in rat heart. *Biochem. J.* **247:** 135–140.

Chan HW-S & Newby VK (1980) Haemoprotein and transition metal ion-catalyzed oxidation of linoleic acid. *Biochim. Biophys. Acta* **617:** 353–362.

Chan PH, Fishman RA, Schmidley JW & Chen SF (1984) Release of polyunsaturated fatty acids from phospholipids and alteration of brain membrane integrity by oxygen-derived free radicals. *J. Neurosci. Res.* **12:** 595–605.
Chatterjee SN & Agarwal S (1988) Liposomes as membrane model for study of lipid peroxidation. *Free Radicals Biol. Med.* **4:** 51–72.
Coolbear KP & Keough KMW (1983) Lipid oxidation and gel to liquid-crystalline transition temperatures of synthetic polyunsaturated mixed-acid phosphatidylcholines. *Biochim. Biophys. Acta* **732:** 531–540.
Cornwell DG & Morisaki N (1984) Fatty acid paradoxes in the control of cell proliferation: Prostaglandins, lipid peroxides and cooxidation reactions. In Pryor WA (ed.) *Free Radicals in Biology*, Vol. VI, pp. 95–148. New York: Academic Press.
Cossins AR & Sinensky M. (1984) Adaptation of membranes to temperature, pressure and exogenous lipids. In Shinitzky M. (ed.) *Physiology of Membrane Fluidity*, pp 7–21. Boca Raton, FL: CRC Press, Inc.
Daolio S, Traldi P, Ursini F, Maiorino M & Gregolin C (1983) Evidence of the peroxidase activity of the peroxidation inhibiting protein on dilynoleyl phosphatidyl choline hydroperoxides as obtained in direct electron impact conditions. *Biomed. Mass Spectrosc.* **10:** 499–504.
Dawson RMC (1982) Phospholipid structure as a modulator of intracellular turnover. *J. Am. Oil Chem. Soc.* **59:** 401–406.
Dise CA & Goodman DPB (1986) *t*-Butyl hydroperoxide alters fatty acid incorporation into erythrocyte membrane phospholipid. *Biochim. Biophys. Acta* **859:** 69–78.
Dise GA, Clark JM, Lambersten CJ & Goodman DBP (1987) Hyperbaric hyperoxia reversibly inhibits erythrocyte phospholipid fatty acid turnover. *J. Appl. Physiol.* **62:** 533–538.
Dobretsov GE, Borschevskaya TA, Petrov VA & Vladimirov YA (1977) The increase of phospholipid bilayer rigidity after lipid peroxidation. *FEBS Lett.* **84:** 125–128.
Erben-Russ M, Michel C, Bors W & Saran M (1987) Absolute rate constants of alkoxyl radical reactions in aqueous solution. *J. Phys. Chem.* **91:** 2362–2365.
Ernster L, Nordenbrand K & Orrenius S (1982) Microsomal lipid peroxidation: Mechanisms and some biomedical implications. In Yagi K (ed.) *Lipid Peroxides in Biology and Medicine*, pp 55–79. New York: Academic Press.
Estabrook RW & Werringloer J (1977) Cytochrome P-450: Its role in oxygen activation for drug metabolism. In Jerina DM (ed.) *ACS Symp. Ser. Drug Metabolism Concepts*, Vol. 44, pp 1–54. Washington DC: American Chemical Society.
Flohé L (1982) Glutathione peroxidase brought into focus. In Pryor WA (ed.) *Free Radicals in Biology*, Vol. V, pp 223–254. New York: Academic Press.
Foote CS (1976) Photosensitized oxidation and singlet oxygen: Consequences in biological systems. In Pryor WA (ed.) *Free Radicals in Biology*, Vol. II, pp 85–133. New York: Academic Press.
Funk CD & Powell WS (1985) Release of prostaglandins and monohydroxy and trihydroxy metabolites of linolenic and arachidonic acids by adult and fetal aortae and ductus arteriosus *J. Biol. Chem.* **260:** 7481–7488.
Galaris D, Buffington G, Hochstein P & Cadenas E (1990) Role of ferryl myoglobin in lipid peroxidation and its reduction to met- or oxymyoglobin by glutathione, quinones, quinone thioether derivatives and ascorbate. In Vigo-Pelfrey C (ed.) *Membrane Lipid Oxidation*, Vol. I, pp 269–283. Boca Raton, FL: CRC Reviews.

Galeotti T, Borrello S, Palombini G et al (1984) Lipid peroxidation and fluidity of plasma membranes from Morris hepatoma 3924A. *FEBS Lett.* **169:** 169–173.

Gamache DA, Fawzy AA & Franson RC (1988) Preferential hydrolysis of peroxidized phospholipids by lysosomal phospholipase C. *Biochim. Biophys. Acta* **958:** 116–123.

Howard JA & Ingold KU (1967) Absolute rate constants for hydrocarbon autoxidation. VI. Alkyl aromatic and olefinic hydrocarbons. *Can. J. Chem.* **45:** 793–797.

Hughes H, Smith CV, Horning EC & Mitchel JR (1983) High performance liquid chromatography and gas chromatography-mass spectrometry determination of specific lipid peroxidation products in vivo. *Anal. Biochem.* **130:** 431–436.

Jones DP (1985) The role of oxygen concentration in oxidative stress: Hypoxic and hyperoxic models. In Sies H (ed.) *Oxidative Stress*, pp 151–195. London: Academic Press.

Jung G, Yang DC & Nokao A (1985) Oxygenation of phosphatidylcholine in human polymorphonuclear leukocyte 15- lipoxygenase. *Biochem. Biophys. Res. Commun.* **130:** 559–561.

Kappus H (1985) Lipid peroxidation: Mechanisms, analysis, enzymology and biological relevance. In Sies H (ed.) *Oxidative Stress*, pp 273–310. London: Academic Press.

Kaschnitz RM & Hatefi Y (1975) Lipid oxidation in biological membranes: Electron transfer proteins as initiators of lipid autoxidation. *Arch. Biochem. Biophys.* **171**: 292–304.

Kornberg RD & McConnell HM (1971) Inside–outside transitions of phospholipids in vesicle membranes. *Biochemistry* **10:** 1111–1120.

Lands WEM, Kulmacz RJ & Marshall PJ (1984) Lipid peroxide actions in the regulation of prostaglandin biosynthesis. In Pryor WA (ed.) *Free Radicals in Biology* Vol. VI, pp 39–61. New York: Academic Press.

Levitsky DO, Lebedev AV, Kuzmin AV & Brovkopvich VM (1988) Role of lipid peroxidation in increasing calcium permeability of model and natural membranes. In L'Abbate A & Ursini F (eds) *The Role of Oxygen Radicals in Cardiovascular Diseases*, pp 127–142. Dordrecht: Kluwer.

Lubin BH, Shohet SB & Nathan DG (1972) Changes in fatty acid metabolism after erythrocyte peroxidation: stimulation of a membrane repair process *J. Clin. Invest.* **51:** 338–344.

Maiorino M, Roveri A, Ursini F & Gregolin C (1985) Enzymatic determination of membrane lipid peroxidation. *J. Free Radicals Biol. Med.* **1:** 203–209.

Maiorino M, Roveri A, Gregolin C & Ursini F (1986) Different effect of Triton X 100, deoxycholate and fatty acids on the kinetics of glutathione peroxidase and phospholipid hydroperoxide glutathione peroxidase. *Arch. Biochem. Biophys.* **251:** 600–605.

Maiorino M, Roveri A, Coassin M & Ursini F (1988) Kinetic mechanism and substrate specificity of glutathione peroxidase activity of Ebselen (Pz 51). *Biochem. Pharmacol.* **37:** 2267–2271.

Maiorino M, Coassin M, Roveri A & Ursini F (1989) Microsomal lipid peroxidation: effect of vitamin E and its functional interactions with phospholipid hydroperoxide glutathione peroxidase, *Lipids* **24:** 721–726.

Maiorino M, Gregolin C & Ursini F (1990) Phospholipid hydroperoxide glutathione peroxidase. *Methods Enzymol.* **186:** 448–457.

Maiorino M, Thomas JP, Girotti AW & Ursini F (1991) Reactivity of phospholipid hydroperoxide glutathione peroxidase with cholesterol hydroperoxides. *Free*

Radicals Res. Commun. **12:** 131–135.

Majewska MC, Strosznajder J & Lazarewicz JW (1978) Effects of ischemic anoxia and barbiturate anesthesia on free radical oxidation of mitochondrial phospholipid. *Brain Res.* **158:** 423–434.

Mead JF, Stein RA, Wu GS, Sevanian A & Gan-Elepano M (1980) Peroxidation of lipids in model systems and in biomembranes. In Simic M & Kaerl M (eds) *Autoxidation in Foods and Biological Systems*, pp 413–425. New York: Plenum Press.

Mowri H, Nojima S & Inoue K (1984) Effect of lipid composition of liposomes on their sensitivity to peroxidation. *J. Biochem.* **95:** 551–558.

Niki E (1987) Antioxidants in relation to lipid peroxidation. *Chem. Phys. Lipids* **44:** 227–253.

Nolan RD, Dusting GJ & Martin TJ (1981) Phospholipase inhibition and the mechanism of angiotensin-induced prostacyclin release from rat mesenteric vasculature. *Biochem. Pharmacol.* **30:** 2121–2125.

O'Brian CA, Ward NE, Weinstein B, Bull AW & Marnett LJ (1988) Activation of rat brain protein kinase C by lipid oxidation products. *Biochem. Biophys. Res. Commun.* **155:** 1374–1380.

Pace-Asciak CR & Asotra S (1989) Biosynthesis, catabolism and biological properties of HPETEs, hydroperoxide derivatives of arachidonic acid. *Free Radicals Biol. Med.* **7:** 409–433.

Rapoport SM, Schewe T, Weisner et al (1979) The lipoxygenase of reticulocytes. *Eur. J. Biochem.* **96:** 545–561.

Robison TW, Sevanian A & Forman HJ (1990) Inhibition of arachidonic acid release by nordihydroguaiaretic acid and its antioxidant action in alveolar macrophages and Chinese hamster lung fibroblast. *Toxicol Appl. Pharmacol.* **105**.

Salaris SC & Babbs CF (1989) Effect of oxygen concentration on the formation of malonaldehyde-like material in a model of tissue ischemia and reoxygenation. *Free. Radicals Biol. Med.* **7:** 603–610.

Scarpa M, Rigo A, Maiorino M, Ursini F & Gregolin C (1984) Formation of alpha tocopherol radical and recycling of alpha tocopherol by ascorbate during peroxidation of phosphatidyl choline liposomes. An electron paramagnetic resonance study. *Biochim. Biophys. Acta* **801:** 215–219.

Schuckelt R, Brigelius-Flohé L, Maiorino M et al (1991) *Free Radicals Res. Commun.* (in press).

Sevanian A (1988) Lipid peroxidation, membrane damage, and phospholipase A2 action. In Chow CK (ed.) *Cellular Antioxidant Defense Mechanisms*, pp 78–95. Boca Raton, FL: CRC Press, Inc.

Sevanian A & Hochstein P (1985) Mechanisms and consequences of lipid peroxidation in biological systems. *Annu. Rev. Nutr.* **5:** 365–390.

Sevanian A & Kim E. (1987) Phospholipase A2 dependent release of fatty acids from peroxidized membranes. *J. Free Radicals Biol. Med.* **1**: 263–271.

Sevanian A, Wratten ML, McLeod LL & Kim E (1988) Lipid peroxidation and phospholipase A2 activity in liposomes composed of unsaturated phospholipids: A structural basis for enzyme activation. *Biochim. Biophys. Acta* **961:** 316–327.

Shier WT (1979) Activation of high levels of endogenous phospholipase A_2 in cultured cells. *Proc. Natl. Acad. Sci. USA* **76:** 195–202.

Slivka SR & Insel PA (1987) Alpha-1-adrenergic receptor-mediated phosphoinositide hydrolysis and prostaglandin E2 formation in Madin-Darby canine kidney cells. *J. Biol. Chem.* **262:** 4200–4207.

Thomas JP, Maiorino M, Ursini F & Girotti AW (1990) Protective action of phospholipid hydroperoxide glutathione peroxidase against membrane-damaging lipid peroxidation: in situ reduction of phospholipid and cholesterol hydroperoxides. *J. Biol. Chem.* **265:** 454–461.

Uri N (1961) Physico-chemical aspects of autoxidation. In Lundberg WO (ed.) *Autoxidation and Antioxidants*, pp 55–106. New York: Wiley Interscience.

Ursini F (1988) The phospholipid hydroperoxide glutathione peroxidase: A peroxidation inhibiting enzyme. In Chow CK (ed.) *Cellular Antioxidant Defense Mechanisms*, pp 69–76. Roca Baton, FL: CRC Press.

Ursini F & Bindoli A (1987) The role of selenium peroxidases in the protection against oxidative damage of membranes. *Chem. Phys. Lipids* **44:** 255–276.

Ursini F, Maiorino M, Valente M, Ferri L & Gregolin C (1982) Purification from pig liver of a protein which protects liposomes and biomembranes from peroxidative degradation and exhibits glutathione peroxidase activity on phosphatidyl-choline liposomes. *Biochim. Biophys. Acta* **710:** 197–211.

Ursini F, Bonaldo L, Maiorino M & Gregolin C (1984) High performance liquid chromatography of hydroperoxy derivatives of stearoyl linoleoyl phosphatidyl choline and of their enzymatic reduction products. *J. Chromatogr.* **270:** 301–308.

Ursini F, Maiorino M & Gregolin C (1985) The selenoenzyme phospholipid hydroperoxide glutathione peroxidase. *Biochim. Biophys. Acta* **839:** 62–70.

Ursini F, Maiorino M, Hochstein P & Ernster L (1989) Microsomal lipid peroxidation: Mechanism of initiation. *Free Radicals Biol. Med.* **6:** 31–36.

Van Den Bosch H (1980). Intracellular phospholipases A. *Biochim. Biophys. Acta* **604:** 191–246.

Van Kujik FJGM, Sevanian A, Handelman GJ & Dratz EA (1987) A new role for phospholipase A2: protection of membranes from lipid peroxidation damage. *Trends Biochem. Sci.* **12:** 31–34.

Vladimirov, Yu A, Olenev VI, Suslova TB & Cheremisina ZP (1980) Lipid peroxidation in mitochondrial membranes. *Adv. Lipid Res.* **17:** 173–249.

Weitzel F, Ursini F & Wendel A (1989) Dependence of mouse liver phospholipid hydroperoxide glutathione peroxidase on dietary selenium. In: Wendel A (ed.) *Selenium in Biology and Medicine*, pp 29–32. Berlin: Springer Verlag.

Winterle JS & Mill T (1984) Free radical dynamics in organized lipid bilayers. *J. Am. Chem. Soc.* **102:** 6336–6339.

Witting LA (1965) Lipid peroxidation in vivo. *J. Am. Oil Chem. Soc.* **42:** 908–913.

Wratten ML, Gratton E, Van de Ven M & Sevanian A (1989) DFH lifetime in vesicles containing phospholipid hydroperoxides. *Biochim. Biophys. Res. Commun.* **164:** 169–175.

Wu GS, Stein RA & Mead JF (1982) Autoxidation of phosphatidylcholine liposomes. *Lipids* **17:** 403–413.

Yamamoto Y, Niki E, Kamiya Y & Shimasaki H (1984) Oxidation of lipids: Oxidation of phosphatidylcholines in homogeneous solution and in water dispersion. *Biochim. Biophys. Acta* **795:** 332–340.

Zaleska MM & Wilson DF (1989) Lipid hydroperoxides inhibit reacylation of phospholipids in neuronal membranes. *J. Neurochem.* **52:** 255–260.

Zhang L, Maiorino M, Roveri A & Ursini F (1989) Phospholipid hydroperoxide glutathione peroxidase: Specific activity in tissues of rats of different age and comparison with other glutathione peroxidases. *Biochim. Biophys. Acta* **1006:** 140–143.

Zimmerman WF & Keys S (1988) Acylation and deacylation of phospholipids in isolated bovine rod outer segments. *Exp. Eye Res.* **47:** 247–260.

13

Biological Activities of 4-Hydroxyalkenals

HELMWARD ZOLLNER, RUDOLF JÖRG SCHAUR and HERMANN ESTERBAUER
Institute of Biochemistry, University of Graz, Schubertstrasse 1, A-8010 Graz, Austria

Dedicated to Professor Dr G. Zigeuner on the occasion of his 70th birthday

I. Introduction 337
II. Metabolism of 4-hydroxyalkenals 340
III. Cytotoxic effects 345
IV. Effects on growth capability of mammalian cells 350
V. Genotoxic effects 355

I. INTRODUCTION

Lipid peroxidation in biological systems is always accompanied by the generation of a complex pattern of aldehydes (reviews: Esterbauer, 1982, 1985; Esterbauer et al, 1988, 1990a). These aldehydes are mainly formed by β-cleavage reactions of alkoxy radicals arising from hydroperoxides of polyunsaturated fatty acids (PUFAs) bound in phospholipids, glycolipids, triglycerides and cholesteryl esters. The principal PUFAs in mammalian tissue are linoleic acid (18 : 2), arachidonic acid (20 : 4) and docosahexaenoic acid (22 : 6), which give in a lipid peroxidation process two (18 : 2), six (20 : 4) and ten (22 : 6) positional isomeric hydroperoxides. Upon scission of the two carbon–carbon bonds adjacent to the OOH group, each of these hydroperoxides can give an aldehyde with a methyl terminus and an aldehyde still linked by the ester bond to the parent lipid molecule. Clearly, the latter group of aldehydes is much more complex, since the three major PUFAs giving 18 different positional isomeric hydroperoxides will result in

Oxidative Stress: Oxidants and Antioxidants
ISBN 0-12-642762-3

several dozens of lipid–aldehyde species. Hitherto no method has been developed which would allow the separation and analysis of such a complex mixture of substances. Furthermore, the lack of reference substances makes it difficult to detect such aldehydes in peroxidized samples and impedes the investigation of possible biological activities. In peroxidized rat liver microsomes and in the liver of haloalkane-intoxicated rats the overall content of such phospholipid-bound aldehydes has been determined (reviews: Comporti, 1985, 1989) and this index correlated well with the damage. However, no attempts were made to ascertain the structure of these aldehydes. That such phospholipids with aldehyde functions might play an important biological role is suggested by the finding that an oxidized derivative of phosphatidylcholine with a short-chain aldehyde at the sn-2 position (1-alkyl-2-(5-oxovaleroyl)-sn-glycero-3-phosphocholine) is a substrate (K_M = 11.3 μM) for the platelet-activating factor acetylhydrolase from human plasma (Stremler et al, 1989). It was also found that a C-12 trienal carboxylic acid (12-oxo-dodeca-5,8,10-trienoic acid) is a major product in arachidonic acid-stimulated porcine leukocytes (Glasgow et al, 1986).

The possible number of aldehydes resulting from the methyl terminus of the parent fatty acid is much smaller since a particular polyunsaturated fatty acid will always give the same aldehydes independently of the type of lipid, e.g. phospholipid, triglyceride, in which it is present. Nevertheless, this class of aldehydes is also rather complex and their complete analysis is still at the early stage (review: Esterbauer and Zollner, 1989; Esterbauer et al, 1990a). In true biological samples such as tissues, plasma, cells, and subcellular fractions, a number of n-alkanals (propanal, butanal, pentanal, hexanal, heptanal, octanal, nonanal, decanal, undecanal, dodecanal), 2-alkenals (acrolein, pentenal, hexenal, heptenal, octenal, nonenal), 2,4-alkadienals (heptadienal, octadienal, decadienal), 4-hydroxy-2,5-undecadienal, 4,5-dihydroxydecenal, 5-hydroxyoctanal, 4-hydroxy-2-alkenals (4-hydroxyhexenal, 4-hydroxyoctenal, 4-hydroxynonenal) and malonaldehyde have been identified, although in some instances only tentatively, by HPLC retention times. Not all of these aldehydes are simple β-cleavage products, but some of them may also result from consecutive reactions or other more complex processes occurring during lipid peroxidation. A large number of additional aldehydes were determined in mildly autoxidized polyunsaturated fatty acids (Esterbauer et al, 1990a; Grosch, 1987).

Three principal mechanisms for free radical-induced cell injury and tissue damage were discussed (review: Esterbauer, 1985): (1) the damaging reaction is caused by the free radicals themselves through their direct attack on sensitive and essential targets; (2) the damaging reaction is the destruction of the lipid bilayer promoted by free radical-induced lipid peroxidation processes—such a destruction of the membrane or certain domains in the membrane could lead to an impairment of enzymes and dysfunction of membrane permeabilities; (3) the ultimate damaging event is a reaction of one or more of the toxic non-radical lipid degradation products, e.g. reactive aldehydes, with critical and essential targets. The latter hypothesis is supported by various findings (review: Slater, 1984; Cerutti, 1985; Comporti, 1985, 1989; Kappus, 1985; Kagan, 1988), which suggest that the damage is causally associated with lipid peroxidation and that at least some of the damage seen in free radical reactions can be reproduced by reactive aldehydic lipid peroxidation products, e.g. hydroxyalkenals, malonaldehyde (MDA), and others. From the quantitative analysis performed so far in native or peroxidized biological samples it appears that the major aldehydes are always MDA, hexanal, and 4-hydroxynonenal (HNE) (Esterbauer et al, 1990a).

Basal, physiological levels of HNE as determined by various authors are given in Table 1. Under conditions of oxidative stress these values can be significantly increased (Esterbauer and Zollner, 1989; Esterbauer et al, 1990a).

Table 1. Physiological levels of HNE as determined by various authors.

Organ	Content	Reference
Rat liver	2.82 ± 0.53 nmol (g wet weight)$^{-1}$	Yoshino et al (1986), Tomita et al (1987)
Rat liver	0.48 ± 0.17 nmol (g wet weight)$^{-1}$	Celotto (1985), Esterbauer and Zollner (1989)
Rat liver	4.60 ± 0.9 pmol (mg protein)$^{-1}$	Norsten-Höög (1989)
Rat plasma	0.86 ± 0.20 nmol ml^{-1}	Yoshino et al (1986), Tomita et al (1987)
Rat retina	0.64 ± 0.64 nmol (g wet weight)$^{-1}$	van Kuijk (1988)
Rat hepatocytes	1.30 ± 0.50 nmol per 10^6 cells	Poli et al (1985)
Rat hepatocytes	Present, not quantified	Norsten-Höög (1989)
Human plasma	0.68 ± 0.42 nmol ml^{-1}	Selley et al (1989)
Human monocytes	0.04 ± 0.01 nmol per 10^6 cell	Selley et al (1989)
Human LDL	0.14 ± 0.17 nmol (mg LDL)$^{-1}$	Esterbauer et al (1987)

We will focus in this chapter on the biological activities of 4-hydroxyalkenals, emphasizing the present knowledge on their metabolism, cytotoxic properties, inhibitory effects on cell proliferation and genotoxicity. The chemotactic properties of these compounds were recently reviewed by Curzio (1988), as were methods for their determination (Esterbauer and Zollner, 1989; Esterbauer and Cheeseman, 1990). Furthermore, several special review articles already exist which deal with some selected aspects of the biological effects of 4-hydroxyalkenals and other aldehydes (Schauenstein et al, 1977; Esterbauer, 1982, 1985; Comporti, 1985; Esterbauer et al, 1988; Witz, 1989; Schaur et al, 1991).

At first glance the reader of this chapter may obtain the impression that most of the investigations come only from one laboratory, i.e. the authors' laboratory. This is in fact the case for most of the early studies between the early 1960s and 1970s. But after the discovery that HNE is a major cytotoxic product formed by peroxidizing microsomes (Benedetti et al, 1980), the interest in this class of compounds rapidly broadened and a large number of laboratories throughout the world have made valuable contributions which show that such compounds do in fact occur in mammalian tissue and exhibit a wide range of biological activities. The interest in this class of compounds has been further augmented by the discovery (Segall et al, 1985) that certain plant alkaloids, e.g. senecionine, are metabolized in the liver to 4-hydroxyhexenal and that this aldehyde is probably responsible for the cytotoxic and genotoxic effects of these alkaloids. HNE and other 4-hydroxyalkenals are among the very few lipid peroxidation products which have high biological activities and can be prepared in pure form and larger quantities by chemical synthesis. The easy availability of a biologically active compound is an indispensable prerequisite of detailed biochemical and biological tests, and this had in our opinion great importance for studies of hydroxyalkenals. Thus with HNE the situation is presently very similar to what it had previously been for MDA, another major product of lipid peroxidation which can be obtained by chemical synthesis. We have repeatedly (review: Schaur et al, 1991) pointed out that in our opinion HNE and other 4-hydroxyalkenals should, however, be considered rather as model substances for reactive lipid peroxidation products than as the only candidates with important biological properties.

II. METABOLISM OF 4-HYDROXYALKENALS

Investigations from various laboratories revealed that 4-hydroxyalkenals can be oxidized enzymatically by NAD(P)-dependent aldehyde dehydrogenases to the corresponding carboxylic acid R—CHOH—CH═CH—COOH, reduced by NADH-dependent alcohol dehydrogenase to the alcohol R—CHOH—CH═CH—CH_2OH, and conjugated with glutathione by glutathione transferases to the HNE–glutathione conjugate R—CHOH—CH(SG)—CH_2—CHO (Figure 1).

Aldehyde dehydrogenases accepting HNE or 4-hydroxyhexenal as substrates were found in mitochondria, microsomes and cytosol of rat liver (Mitchel and Petersen, 1987, 1989), and mouse liver (Antonenkov et al,

1987; Algar and Holmes, 1989a; Lame and Segall, 1986; Hakki and Nodes, 1979) and in mouse stomach (Algar and Holmes, 1989b). It seems that the enzymes are always present in several isoforms differing in chromatographic behaviour, kinetic constants, specific activity and substrate specificity.

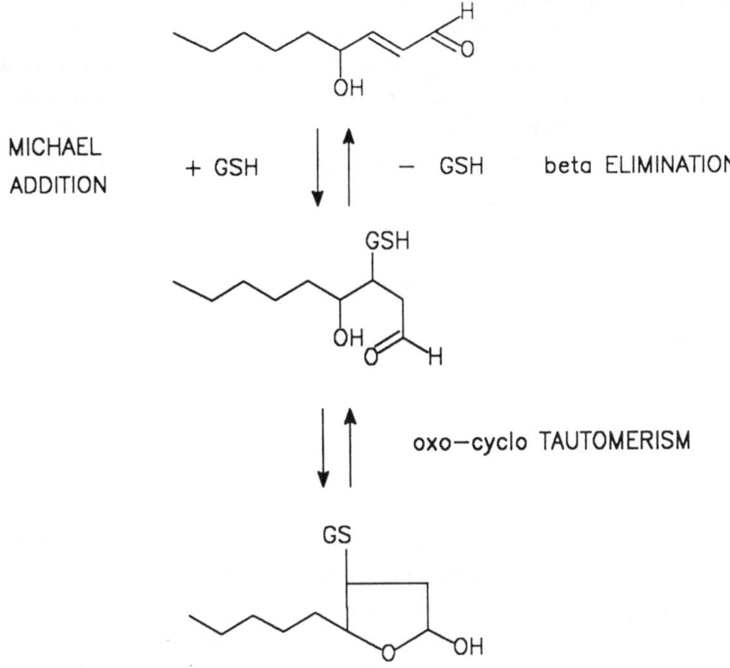

Figure 1. Reaction of HNE with glutathione.

For instance, four HNE-oxidizing aldehyde dehydrogenases were found in rat liver; two of them with K_M values of about 2×10^{-5} M and 6×10^{-5} M were in the mitochondria (Mitchel and Petersen, 1987, 1989), one in the cytosol, K_M 1.8×10^{-5} M, and one in the microsomes, K_M 7×10^{-5} M (Antonenkov et al, 1987). The microsomal aldehyde dehydrogenase oxidized, as well as HNE, a number of other aldehydic lipid peroxidation products and the enzyme could be induced by treatment of rats with clofibrate. This hypolipidaemic drug induces in the rat liver proliferation of peroxisomes, and associated with this is an increased production of reactive oxygen radicals and an increased lipid peroxidation (review: Rao and Reddy, 1987). It is therefore conceivable that the induction of the microsomal aldehyde dehydrogenase is part of a defence system and is of importance for the elimination of toxic aldehydes arising from lipid peroxidation. The inducibility of the microsomal enzymes may also be the reason that in another study (Lame and Segall, 1986) no aldehyde dehydrogenase was found in rat liver microsomes.

Rat liver cytosol supplemented with NADH caused a rapid consumption of HNE (Esterbauer et al, 1985). From the 1:1 stoichiometry of the reaction (NADH consumed/HNE consumed = 1.08–0.91:1) and its inhibition by pyrazol and several other findings, it was concluded that HNE was reduced enzymatically by liver alcohol dehydrogenase to the corresponding alcohol. The K_M value for HNE determined in the whole cytosol was 100 μM. No attempt was made to further characterize or purify this HNE-reducing enzyme activity.

Like other 2-alkenals, HNE is a strong alkylating electrophilic agent which can spontaneously react with nucleophiles HX (e.g. glutathione, cysteine, SH proteins) in a Michael-type addition reaction:

$$\mathrm{R{-}CHOH{-}CH{=}CH{-}CHO + HX \longrightarrow R{-}CHOH{-}CHX{-}CH_2{-}CHO}$$

(Esterbauer et al, 1975, 1976; Schauenstein and Esterbauer, 1979; Esterbauer, 1982) (Figure 1).

The group of Mannervik (Alin et al, 1985; Jensson et al, 1986; Danielson et al, 1987; Mannervik and Danielson, 1988) has found that the spontaneous conjugation reaction of HNE with glutathione can be considerably (up to 300–600 times) accelerated by glutathione transferases. Among the rat liver isoenzymes the glutathione transferase 8–8 has the highest specific activity of 170 units (mg protein)$^{-1}$. With artificial substrates, the specific activities for glutathione transferase 8–8 are in the range of 0.1–20 units (mg protein)$^{-1}$.

Rather large species differences in HNE-metabolizing glutathione transferases seem to exist, since it was found that in human liver only the isoenzyme μ shows a high activity with HNE, and in mouse liver all glutathione transferases show only a weak activity with HNE (Jensson et al, 1986). Also probably important for the detoxification process in vivo is the fact that the glutathione transferase activity is inhibited by low concentrations of the HNE–glutathione conjugate. It has been shown, for instance, that the glutathione transferase 4–4-catalysed conjugation of glutathione with 1-chloro-2,4-dinitrobenzene is inhibited by 50% in the presence of 5 μM HNE–glutathione conjugate. The neutral form of glutathione transferase from rat heart also accepts HNE as substrate with a K_M of 50 μM and is also inhibited by the HNE–glutathione conjugate, K_i = 85 μM (Ishikawa et al, 1986). The catalytic efficiency (k_{cat}/K_M) of glutathione transferase 8–8 increases with increasing chain length, i.e. lipophilicity of the 4-hydroxyalkenals; for 4-hydroxyhexenal, HNE and 4-hydroxyundecenal the k_{cat}/K_M values were, for example, 560, 7850 and 16 100 nM^{-1} s^{-1}.

The high catalytic efficiency also indicates that these enzymes have an important physiological role in the detoxification of such aldehydes. In evolution, such enzymes with a high efficiency and a rather broad substrate specificity may have played an important role in protecting aerobic cells against a wide range of toxic aldehydes either present in the environment or generated intracellularly by peroxidizing membrane lipids. The biological importance of HNE-metabolizing glutathione transferases is further supported by the finding of Ishikawa et al (1986) in rat heart perfused with 4-hydroxynonenal. It was found in these studies that HNE is very rapidly conjugated to glutathione, the half-life of HNE in the rat heart being less than 4 s, and that the HNE–glutathione conjugate is transported out of the cells by a carrier-mediated process. This elimination may be important in diminishing the inhibitory effect of the HNE–glutathione conjugate on the intracellular glutathione transferases and glutathione reductases (Ishikawa et al, 1986).

The three aldehyde-metabolizing enzyme activities present in rat liver, i.e. aldehyde dehydrogenase, alcohol dehydrogenase and glutathione transferases, and the many other possible ways in which HNE reacts spontaneously with cellular nucleophiles make it clear that the overall metabolism of HNE and cells is rather complex and depends on many factors, such as cell type and the particular concentration of HNE, HNE–S conjugates, glutathione, NADH and NADP.

Isolated rat hepatocytes rapidly metabolized HNE (Esterbauer et al, 1985). For instance, when 100 μM HNE was added to hepatocyte suspension with 10^6 cells ml^{-1}, 95% of the aldehyde disappeared within 4 min (Table 2). The maximum rate of HNE consumption was about 200 nmol per min per 10^6 cells.

Table 2. Metabolism of HNE by rat hepatocytes.

	nmol per 10^6 cells	
	20 s	240 s
HNE metabolized	40	95
HNE–glutathione conjugate	25	24
4-Hydroxynonenoic acid	6	20
1,4-Dihydroxynonene	5	3
Protein-bound HNE	—	3
Unidentified products	4	45

Tritium-labelled HNE (100 μM) was added to a suspension of rat hepatocytes (1×10^6 cells ml^{-1}) and the time-course of the disappearance of HNE and the formation of metabolites was followed over 240 s. Given are the values found after 20 and 240 s of incubation (Schaur et al, 1991), and the amount of radioactivity bound to the cell protein.

We had originally assumed (Esterbauer et al, 1985) that the main metabolite is the 1,4-dihydroxynonen, because rat liver cytosol supplemented with NADH could reproduce the rapid HNE consumption (see above). In continuation of this work (review: Schaur et al, 1991) and using tritium-labelled HNE, it is now clear that more than one metabolite is formed by hepatocytes and that their distribution depends on the experimental conditions, e.g. incubation time and initial HNE concentration (Table 2). The metabolites identified so far are the HNE–glutathione conjugate, the 4-hydroxynonenoic acid and 1,4-dihydroxynonen. In the initial phase (20 s) of the HNE-metabolizing process, the three metabolites comprised about 90%, but at later phases (240 s) only 50%. This indicates that the primary metabolites are further converted into other not yet identified products. The incubation system contained radioactive water and this suggests that the metabolite 4-hydroxynonenoic acid was further oxidized in the mitochondria to carbon dioxide and water. Only a rather small percentage (3% with 100 nmol HNE per 10^6 cells) of HNE was bound to the liver cell proteins, most likely through covalent binding to SH groups. The metabolism of HNE was also studied in Morris hepatoma MH_1C_1 cells. Based on equal amounts of cell protein, these cells have nearly the same capacity to metabolize HNE as rat hepatocytes (50 versus 60 nmol min^{-1} (mg protein)$^{-1}$). The cytosol of these cells is low in alcohol dehydrogenase activity and it was assumed that MH_1C_1 cells metabolized HNE mainly to the HNE–glutathione conjugate and to the 4-hydroxynonenoic acid (Ferro et al, 1988). The kinetics of the disappearance of HNE were also measured with Ehrlich ascites tumour cells (Schauenstein, 1982), human fibroblasts (Kaneko et al, 1987) and human platelets (Hurst et al, 1987). Recently it was shown that diethylnitrosamine-induced nodules and hepatoma contained higher activities of HNE-metabolizing cytosolic and microsomal aldehyde dehydrogenases than normal liver (Canuto et al, 1989). Only a few investigations were made to elucidate the metabolism and the detoxification of 4-hydroxyalkenals in whole animals. When tritium-labelled 4-hydroxyhexenal was injected into the portal vein of rats, about 70% of the radioactivity was excreted in the urine within 8 h. The major fraction (75%) of the excreted substances was identified as the C-3 mercapturic acid conjugate of 4-hydroxyhexenal (Winter et al, 1987).

Glutathione conjugates formed in the liver are transferred through the canalicular membrane into the bile and either degraded by the enzymes of the luminal membrane of the biliary epithelium and the bile canalicular membrane or metabolized by the small intestinal epithelium to the corresponding cysteine conjugates. Cysteine conjugates formed in the small intestine may then be translocated to the kidney where they are *N*-acetylated to the mercapturic acid (Dekant et al, 1988).

So far no attempts have been made to demonstrate whether HNE-derived mercapturic acids are normal constituents of the urine of mammals. This is certainly a difficult task, but nevertheless more attention should be given to this subject in the future, since it could offer the possibility of obtaining a good index for in vivo lipid peroxidation.

III. CYTOTOXIC EFFECTS

Mammalian cells treated with millimolar concentrations of 4-hydroxyalkenals become severely damaged and die within one hour or less. Such acute toxic effects of high concentrations of hydroxyalkenals were frequently observed, but only a few studies were done to examine the dose and time dependence of such lethal effects. The toxicity was measured either by counting the number of cells stainable with trypan blue or by measuring the lactate dehydrogenase release. The dose dependence of lethal effects of HNE was determined for cultured human fibroblast (Kaneko et al, 1987; Poot et al, 1987), endothelial cells (Kaneko et al, 1988), Chinese hamster ovary cells (Brambilla et al, 1986), Ehrlich ascites tumour cells (Hauptlorenz et al, 1985), and primary cultures of rat hepatocytes (Griffin and Segall, 1986; Eckl and Esterbauer, 1989; Esterbauer et al, 1990b) and peritoneal Ehrlich ascites tumour cells (Khoschsorur et al, 1981; Schauenstein, 1982). Despite the many variations in the experimental conditions chosen (presence or absence of fetal calf serum, incubation time, method of determining cell death, cell density), we have compiled the data scattered in the literature into Table 3, which lists the approximate HNE concentration leading to 50% cell death within 1.5–20 h. Large differences exist in the toxic effects between different cell types, and HNE concentrations producing lethal effects upon prolonged incubation are much lower than those producing cell death within 3 h.

From the investigation of Kaneko et al (1987) it can be deduced that a single dose of 5 μM HNE added to the culture medium is sufficient to kill growth-inhibited fibroblasts within 20 h. In the logarithmic growth phase, the cells are somewhat less sensitive (LD_{50} = 18 μM). This is different from the report by Poot et al (1987), who observed that doses of 40 μM HNE added to the culture medium of fibroblasts had no effects on cell survival examined by microscopic examination and protein content in normal fibroblast after one week of culturing. Glutathione synthesis-deficient fibroblasts did not survive such a treatment with 40 μM HNE. Rat hepatocytes in primary cultures died within 20 h when the culture medium was supplemented with 100 μM HNE. This has been observed independently by Griffin and Segall (1986), Eckl and Esterbauer (1989) and Esterbauer et al (1990b).

Cultured hepatocytes normally have a polyhedral shape and one to two prominent nuclei, and all cells are well attached to the surface of the dishes. Cultures treated with 100 μM 4-hydroxynonenal for 3 h showed a large number of spherical cells with an intense granular structure and pycnotic nuclei (Eckl and Esterbauer, 1989). These cells also appeared to have lost contact with the surface of the dishes. The cells did not recover from this damage; on the contrary, the damage continued to spread and 24 h later all the cells were dead. This is also in agreement with the increased lactate dehydrogenase release in cultures treated with 120 μM HNE (Griffin and Segall, 1986).

On the other hand, if freshly isolated rat hepatocytes (2×10^6 cells ml^{-1}) (Esterbauer et al, 1985) and also Morris' hepatoma cells MH_1C_1 (1×10^6 cells ml^{-1}) (Ferro et al, 1988) were treated with 100 μM HNE, no increase in the number of dead cells occurred within 1 h. Such a concentration of HNE was very efficiently metabolized (review: Schaur et al, 1991) to the corresponding alcohol, carboxylic acid and glutathione conjugate, and the initial HNE concentration was in fact decreased by 99% to concentrations below the detection limit of about 1 μM. In hepatocytes a transient decrease of the intracellular glutathione by about 30% occurred within 5 min, but thereafter the glutathione concentration slowly increased again. From early studies with rat liver slices and 4-hydroxypentenal, a homologue to HNE, it can be assumed that 100 μM HNE did not decrease protein thiols and had no effect on cell respiration (review: Schauenstein et al, 1977). The acute lethal HNE dose leading to the death of hepatocytes within 30 min is in the range of about 0.5–1 mM. Such concentrations lead to a depletion of glutathione within 5 min and subsequently to a burst in low-level chemiluminescence and alkane production (Cadenas et al, 1983). This suggests that lipid peroxidation was associated with HNE-mediated cell damage and cell death. Whether this temporal relationship is casual or causal is not known.

Haenen et al. (1987) have shown that HNE diminishes the glutathione-dependent protection in rat liver microsomes and induces free radical reaction and lipid peroxidation.

A strong difference between short-term and long-term toxicity also exists in Ehrlich ascites tumour cells. A dose of 50 μM HNE in the culture medium led to 30% trypan blue cells in 6 h (Hauptlorenz et al, 1985); the damage continued and at 36 h 95% of the cells were dead. On the other hand, Ehrlich ascites tumour cells harvested from the peritoneum of mice and incubated for 30 min with 160 μM HNE were not damaged at all and showed full growth capability if implanted into mice again (Schauenstein, 1982) (see Table 5).

Possible explanations for the large differences in HNE toxicity (Table 3)

Table 3. Lethal effects of HNE on various mammalian cells.

	LD_{50} (μM)	Reference
1. Human diploid fibroblasts in growth-arrested phase after 20 h culturing in DME + 10% FCS supplemented with HNE	5	Kaneko et al (1987)
2. As 1, but in logarithmic growth phase	18	Kaneko et al (1987)
3. As 1, growth-arrested phase, after one week culturing in DME + 8% FCS supplemented with HNE	>40	Poot et al (1988a)
4. Glutathione synthetase-deficient fibroblasts, GM3878 incubated as in 3	<20	Poot et al (1988b)
5. Human umbilical vein endothelial cells after 3 h treatment with HNE in Earles' solution	25	Kaneko et al (1988)
6. Chinese hamster ovary cells after 90 min incubation in serum-free medium containing HNE	100	Brambilla et al (1986)
7. Morris' hepatoma cells MH_1C_1 after 60 min incubation in Ham's F10 + 10% serum + HNE	>100	Ferro et al (1988)
8. Isolated hepatocytes (2×10^6 cells ml^{-1}) after 30 min incubation in balanced salt solution + glucose + HNE	1000	Esterbauer et al (1985), Grasse et al (1985)
9. Primary cultures of rat hepatocytes after 20 h culturing in full medium supplemented with HNE	100	Eckl and Esterbauer (1989), Esterbauer et al (1990b)
10. Ehrlich ascites tumour cells after 120 min incubation in PBS + HNE	1000	Khoschsorur et al (1981), Schauenstein (1982)
11. Cultured Ehrlich ascites tumour cells after 6 h culturing in MEM + 8% serum + HNE	100	Hauptlorenz et al (1985)
12. Erythrocytes	450	Benedetti et al (1980)

are that cultured cells are more sensitive than freshly isolated cells or that HNE is more toxic in the presence of fetal calf serum, which was used in all (except endothelial cells) experiments with cultured cells. It could also be that low HNE concentrations cause some kind of permanent damage which programmes cells to die at a later stage of about 24 h, but the Ehrlich ascites tumour cell experiments are clearly not consistent with this, since peritoneal Ehrlich ascites tumour cells (freshly harvested from the peritoneum) treated with 160 μM HNE and then reimplanted were not irreversibly damaged and showed full growth capability, but cultured Ehrlich ascites tumour cells treated with only 100 μM HNE died. A comparison of the experimental protocols on which the toxicity data of Table 3 are based reveals that the largest methodological variations lie in the cell number exposed to a certain HNE concentration or, in other words, in the ratio of HNE to cells. Based on the reported number of cells plated into culture dishes and cultivation time, we made a rough calculation of the cell numbers present in a particular toxicity experiment. The result of this estimate is that in all studies where HNE showed high toxicity the cell density was in the range of about 10^5 cells ml^{-1} or less ($<10^5$ cells cm^{-2}, 0.5 ml culture medium cm^{-2}), whereas in experiments where HNE showed a lower toxicity, the cell density was about 1×10^6 cells ml^{-1} or more. In the case of cultured endothelial cells or cultured Ehrlich ascites tumour cells, the toxic concentration of 25 or 100 μM corresponds in fact to 825 or 2000 nmol HNE per 10^6 cells. This is even higher than the dose which is toxic for freshly isolated hepatocytes (500 nmol per 10^6 cells) or Ehrlich ascites tumour cells (1000 nmol per 10^6 cells). Clearly the cell number determines how much of the added HNE can be detoxified or scavenged (by cellular glutathione), whereas the absolute molar concentration of HNE is of minor importance. The most suitable parameter for an objective comparison of toxicity would be the protein content, i.e. nmol HNE per mg cell protein, but such numbers have not been reported in the literature. The importance of the appropriate reference system is clearly evident if, for instance, the HNE-metabolizing activity of two cell types is compared. For rat hepatocytes (1.78 mg protein per 10^6 cells) the consumption rate is 103 nmol per min per 10^6 cells. For MH_1C_1 cells, the consumption rate is only 40 nmol per min per 10^6 cells. The smaller consumption in MH_1C_1 is probably due to their much smaller cell size (0.25 mg protein per 10^6 cells). When rates were based on equal amounts of cellular protein, both cell types were nearly comparable with 73 and 55 nmol HNE min^{-1} (mg protein)$^{-1}$ (Ferro et al, 1988).

The mechanisms by which HNE and other 2-alkenals damage cells in a manner leading to cell death are not clear. High concentrations of hydroxyalkenals or, more precisely, a high ratio of HNE to cellular protein (about 300 nmol HNE (mg protein)$^{-1}$ or more), produce a multitude of effects such as depletion in glutathione, disturbance of calcium homeostasis, inhibition of respiration, inhibition of DNA, RNA and protein synthesis, initiation of lipid peroxidation, and inhibition or activation of some specific enzymes (see Table 4). The time sequence of these events has not been studied and it is therefore not possible to make a guess on the possible initial critical event, e.g. glutathione depletion, which could have triggered all subsequent disturbances.

Table 4. Effects of 4-hydroxyalkenals on enzymes.

Effects	Dose*	Reference
Inhibition		
S-Adenosylmethionine decarboxylase	100 μM (HPE)	Dianzani (1982)
Calcium sequestering activity of rat liver microsomes	40 μM (HNE)	Benedetti et al (1984)
Cathepsin B_1 (30% inhibition)	200 μM (HPE)	Dianzani (1979)
DNA polymerase α	370 μM (HNE)	Wawra et al (1986)
DNA polymerase β	290 μM (HNE)	Wawra et al (1986)
Galactosyltransferase	100 μM (HNE)	Marinari et al (1987)
Glucose-6-phosphatase	100 μM (HNE)	Poli et al (1982), Koster et al (1986)
Glyceraldehyde phosphate dehydrogenase	30 μM (HPE)	Schauenstein et al (1977)
Hexokinase	40 mM (HPE)	Schauenstein et al (1977)
Lactate dehydrogenase	40 mM (HPE)	Schauenstein et al (1977)
O^6-Methylguanine–DNA methyltransferase	100 μM (HNE)	Krokan et al (1985)
5′-Nucleotidase	3 mM (HNE)	Paradisi et al (1985)
Protein kinase C	90 mM (HNE)	Poli et al (1988)
RNA synthesis	320 μM (HPE)	Schauenstein et al (1977)
Sialyltransferase	50 μM (HNE)	Marinari et al (1987)
Stimulation		
Adenylate cyclase†	3 μM (HNE)	Paradisi et al (1985)
Phospholipase C‡	0.1 nM (HOE)	Rossi et al (1990)

HNE, 4-hydroxynonenal; HPE, 4-hydroxypentenal; HOE, 4-hydroxyoctenal.
* Approximate dose which gives half maximal effect.
† Stimulation is transient and followed by an inhibition.
‡ 4-Hydrohexenal and HNE stimulate at 1 nM and 1 μM respectively.

For mice or rats, 4-hydroxyalkenals were extremely toxic: for example, the lethal dose (LD_{50}) for HNE in mice, if given intraperitoneally, is 68 mg (kg body weight)$^{-1}$ (Schauenstein et al, 1977). 4-Hydroxyhexenal or HNE injected intravenously (13 to 18 mg kg^{-1}) as a phospholipid emulsion produces severe liver damage in rats very similar to that seen after poisoning with carbon tetrachloride or with senecionine alkaloids (Segall et al, 1985).

A connection has also been proposed (Turner et al, 1985) between the HNE content in oil samples collected during the outbreak (1981) of the toxic Spanish oil syndrome and the toxicity of these oil samples. In a bioassay screening programme, a correlation was found between the toxicity of the oil sample and the HNE content. In one study, not related to lipid peroxidation, HNE was found to be one of the most toxic principles in some red algae killing reef-dwelling fish (Paul and Fenical, 1980). For *Eupomacentrus leucostictus*, a dose of 8 μg of HNE per litre of water is lethal.

IV. EFFECTS ON GROWTH CAPABILITY OF MAMMALIAN CELLS

4-Hydroxyalkenals appear to possess the capacity to block cell proliferation under conditions where lethal effects as measured by trypan blue exclusion or LDH release are low or absent. The growth-inhibitory activity of HNE, other 4-hydroxyalkenals and 2-alkenals have been observed independently by a number of laboratories (Schauenstein et al, 1977; Schauenstein and Esterbauer, 1979; Khoschsorur et al, 1981; Schauenstein, 1982; Hauptlorenz et al, 1985; Krokan et al, 1985; Brambilla et al, 1986; Cajelli et al, 1987; Poot et al, 1987, 1988a,b; Kaneko et al, 1987, 1988). Typical response curves for HNE-mediated growth inhibition in six different cell lines are shown in Figure 2.

Although the methodologies which led to these dose responses are not fully comparable, it seems that the sensitivity towards HNE strongly depended on the cell type. For example, the HNE concentration required for a 50% reduction of colony-forming efficiency in bronchial fibroblasts was only about 2 μM (Krokan et al, 1985), whereas about 250 μM (estimated from Figure 2) was needed for 50% inhibition of growth of Ehrlich ascites tumour cells. The reasons for this different sensitivity are not clear, but may reside, among others, in the different capacity to metabolize and detoxify the aldehyde. Ehrlich ascites tumour cells (5×10^6 cells ml^{-1}) incubated in PBS with 160 μM HNE decreased the aldehyde concentration in the incubation medium by about 50% within 90 min (Schauenstein, 1982), whereas cultured human fibroblasts consumed within 24 h only about 60% of the initially added aldehydes (25 μM) (Kaneko et al, 1987). In addition, the cell density and the ratio of HNE to cellular protein are certainly important factors (see above).

The growth inhibition mediated by HNE and similar aldehydes seems to be a specific effect independent of the general toxicity of this class of compounds. This may be concluded from several experiments where both

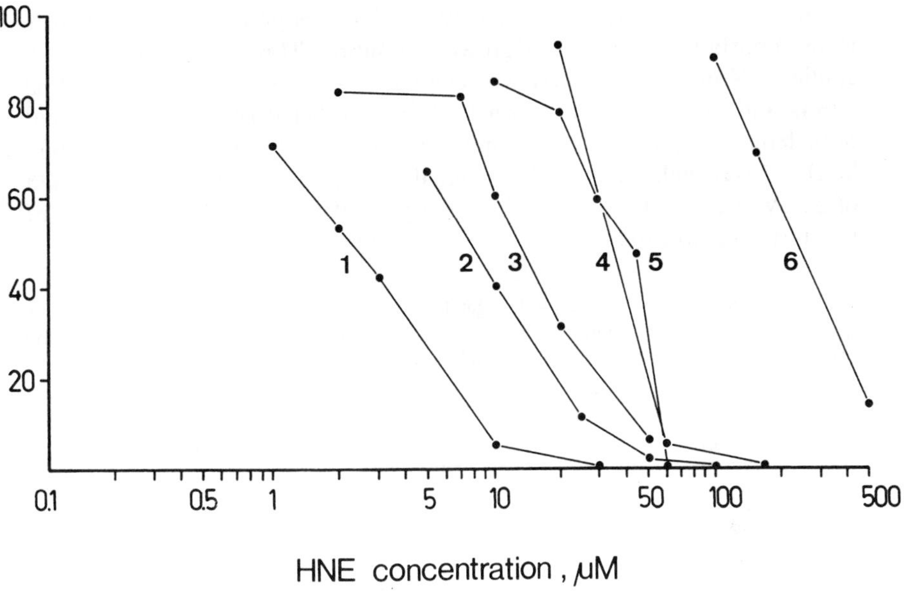

Figure 2. Effect of HNE on growth capability of mammalian cells. 1, Bronchial fibroblasts treated for 1 h in serum-free medium (Krokan et al, 1985); 2, human umbilical vein endothelial cells treated for 3 h in Earle's solution (Kaneko et al, 1988); 3, cultured Ehrlich ascites tumour cells, HNE added to the culture medium supplemented with 6% serum (Hauptlorenz et al, 1985); 4, Chinese hamster ovary cells treated for 1.5 h in serum-free medium (Brambilla et al, 1986); 5, V79 Chinese hamster lung cells treated for 1.5 h in serum-free medium (Cajelli et al, 1987); 6, peritoneal Ehrlich ascites tumour cells treated for 0.5 h in phosphate-buffered saline, then reimplanted (Khoschsorur et al, 1981; Schauenstein, 1982).

cell death and growth capability after aldehyde treatment were measured (Table 5). Chinese hamster ovary cells incubated for 90 min in serum-free medium with 60 μM HNE were fully viable (only 2% dead cells), but they did not proliferate and form colonies in culture (Brambilla et al, 1986). Similar results were obtained with homologous 4-hydroxyalkenals (4-hydroxypentenal, 4-hydroxyhexenal, 4-hydroxyoctenal, 4-hydroxyundecenal). The difference in concentration leading to lethal toxic effects and growth inhibition, however, was more pronounced for long-chain aldehydes. For instance, 4-hydroxyhexenal and 4-hydroxyundecenal had similar toxicity towards Chinese hamster ovary cells (ID_{50}, 100 and 120 μM); however, the concentrations reducing the plating efficiency to 50% differed widely being

80 and 5 μM respectively. This suggests that the lipophilicity of the aldehyde is an important parameter for growth inhibition. This is also supported by studies of Kaneko et al (1988), who compared for a series of 2-alkenals lethal effects and growth inhibition on human endothelial cells and found that both lethal effects and growth inhibition increased with increasing chain length. 2-Hexenal, 2-heptenal, 2-octenal, and 2-nonenal at a concentration of 25 μM decreased the growth capability to 106%, 90%, 33% and 10% of the untreated controls.

Table 5. HNE toxicity and growth inhibition of Chinese hamster ovary cells (CHO) (Brambilla et al, 1986), human umbilical vein endothelial cells (EC) (Kaneko et al, 1988) and Ehrlich ascites tumour cells (EATC) (Khoschsorur et al, 1981; Schauenstein, 1982).

	Incubation time (min)	HNE (μM)	Cell death[a] (%)	Cell growth[b] (%)
CHO	120	7	0	100
	120	20	1	93
	120	60	2	7
	120	170	96	0
	120	500	100	0
EC	180	5	5	65
	180	10	20	39
	180	25	45	11
	180	50	95	0.9
	180	100	99	0.2
EATC	30	160	2	90
	60	160	3	30
	120	160	10	15
	30	500	6	15
	60	500	15	0
	120	500	40	0

[a] Percentage trypan blue-positive cells at the end of incubation.
[b] CHO: plating efficiency at day 7. EC: cell number after two days of culturing in percentage of control. Ehrlich ascites tumour cells: HNE-treated cells injected intraperitoneally into mice, number of tumour cells determined at day 7–10 in percentage of control.

The concentration difference for HNE-mediated cell death and growth inhibition was somewhat less pronounced in endothelial cells compared to Chinese hamster ovary cells (Table 5). On the other hand, in the case of Ehrlich ascites tumour cells, HNE appears to block cell proliferation specifically, also at concentrations where virtually no acute lethal effects were produced. In this context it should be mentioned that 4-hydroxy-alkenals were extensively studied (review: Schauenstein et al, 1977) as carcinostatic agents, and some transplanted animal tumours could in fact be cured by local application of these aldehydes.

The chemical reactions and molecular mechanisms involved in HNE-mediated growth inhibition are largely unknown. Possible explanations which were offered are a specific inhibition of the DNA polymerase system (Schauenstein et al, 1977; Wawra et al, 1986) or disturbance of the cell cycle (Poot et al, 1988b). [^{3}H]-Thymidine incorporation into Ehrlich ascites tumour cells is strongly inhibited by 4-hydroxyalkenals at concentrations of 25–110 μM; for HNE the ID_{50} is in the range 25–50 μM (review: Schaur et al, 1991). Inhibition of thymidine incorporation is not an indirect effect caused by inhibition of thymidine transport across the cell membrane, but is in fact an inhibition of de novo synthesis of DNA (Hauptlorenz et al, 1985). A possible target for HNE is DNA polymerase α, which is the most active of the three eukaryotic DNA polymerases and possesses a functional SH-group, which could be blocked by HNE. This would be consistent with the sensitivity of DNA polymerase α (but not of polymerase β) to other SH-blocking aldehydes such as acrolein (Munsch et al, 1974), benzaldehyde and protocatechualdehyde (Suzuki et al, 1981) and to the general sulphydryl-reactive agent *N*-ethylmaleinimide (Krokan et al, 1979). As shown by Wawra et al (1986), isolated DNA polymerase α from neonatal rat liver is in fact inhibited by HNE, the concentration of the aldehyde required for 50% inhibition being 210 μM. This is about 5-fold the concentration (ID_{50} = 40 μM) required to obtain 50% inhibition of DNA synthesis in intact hepatocytes from adult or neonatal rat liver or in Yoshida hepatoma cells. Moreover, one must consider that in intact cells a concentration of 40 μM HNE is undoubtedly rapidly decreased to much lower levels by the action of detoxifying enzymes. This rather large difference in the effective concentrations required for the inhibition of DNA synthesis and for the inhibition of isolated DNA polymerase α makes it rather unlikely that DNA polymerase α is the primary target for HNE-dependent inhibition of proliferation.

It was also clearly shown (Poot et al, 1988a) that the inhibition of DNA synthesis seen in fibroblasts is not due to a depletion in cellular glutathione. Fibroblasts incubated with 20 μM HNE, which inhibited DNA synthesis to 70%, had no lowered glutathione content as compared to untreated controls, and with 40 μM HNE the glutathione content was two-fold higher than in controls. This is consistent with other observations (Poot et al, 1987, 1988b), according to which a single HNE pulse leads in fibroblasts only to a transient decrease of glutathione which then results in an overshoot of glutathione by stimulation of its de novo synthesis. Poot et al (1988b) have also studied the effect of HNE on the cell cycle of cultured fibroblasts (papillary dermal human diploid fibroblast-like cells, HDFL; amniotic fluid fibroblast-like cells, AFFC) and the results of these studies make it reasonable to assume that the reduction of DNA synthesis and inhibition of cell proliferation produced by HNE results at least in part from a severe disturbance of the cell cycle. The novel and elegant method used is bromodeoxyuridine–Hoechst flow cytometry, allowing the assay of the kinetics of cell proliferation by determining the duration of the transit from one cell cycle phase to another and the percentage of cells present in a distinct cell cycle phase, e.g. $G_0 + G_1$, S, G_2, G_1', G_2'. Quiescent fibroblasts were trypsinized and plated into culture flasks in MEM medium (10% fetal calf serum) supplemented with 2–20 μM HNE. These concentrations of HNE led to a dose-dependent arrest of cells in the G_2 phase of the first cell cycle and in the G_1 phase of the second cell cycle. A dose of 10 μM HNE increased after 70 h of culture the percentage of cells in G_2 from 5.1% (control) to 14.6%, and in G_1' from 22.5% to 32.1%. Other effects caused by this treatment with HNE were a marked delay in the onset of proliferation due to a retardation of the transit of the cells from G_0 to G_1, and a prolongation of the G_1 phase. In controls the duration of G_1 was 31.1 h, and in HNE (10 μM) treated cells 39.1 h. Moreover, HNE appears to provoke the exit of cells from G_1 and G_2 phase into the non-proliferating quiescent state G_0. The net result of the complex pattern of disturbed cell proliferation is of course a reduction of de novo synthesis of DNA and an inhibition of the rate of proliferation.

HNE produced more or less the same effects if added during quiescence, during early activation (i.e. transit from G_0 to G_1). This can be taken as evidence that the effects are not dependent on certain events or targets associated with the cell cycle. The effects of HNE on the kinetics of the cell cycle also strongly suggest that the HNE-dependent inhibition of proliferation is not due to an inhibition of DNA replication.

In view of the hypothesis (Burton et al, 1983; Burdon and Rice-Evans, 1989) that lipid peroxidation or products produced in this process (e.g. hydroxyalkenals) play a role in gross regulation of the cell cycle, it would be important to further investigate this problem with the aim of clarifying the underlying biochemical and molecular mechanisms. Some controversy exists about whether HNE-mediated inhibition of DNA synthesis and proliferation is reversible. Furthermore, it seems to be not clear whether the effects are produced by free HNE or by HNE bound to serum proteins. In the fibroblast studies by Poot et al (1988a,b) and also in other studies (Kaneko et al, 1987), HNE was added to the cell culture together with fetal calf serum or serum. Fetal calf serum contains SH proteins which rapidly bind HNE and we would assume that 99% or more of the added aldehyde was in fact bound within 1 h to the fetal calf serum proteins. Thus, it could be that the observed effects are in fact caused by HNE-modified proteins. Finally, with the finding that HNE can affect DNA (genotoxic effects) at very low concentrations and influence transcription of specific genes (Barrera et al, 1987; Cajone et al, 1989), the possibility should be considered that the HNE-mediated inhibition of proliferation results from an HNE-mediated modulation of the activity of genes involved in the regulation of the cell cycle. In this context it is also worth mentioning that rapidly dividing tumour cells do not undergo lipid peroxidation (review: Masotti et al, 1988) and it is intriguing to speculate that this has an important biochemical background. Ehrlich ascites tumour cells, which were forced by high concentrations of an iron–histidine complex to undergo lipid peroxidation and produce HNE intracellularly, were viable (no trypan blue cells), but completely lost their growth capability (Tillian et al, 1989).

V. GENOTOXIC EFFECTS

As described by Winter et al. (1986), 4-hydroxynonenal and 4-hydroxyhexenal react in vitro at pH 7.4 with deoxyguanosine to form cyclic adducts (Figure 3). The proposed reaction mechanism involves a Michael addition at C-3 of the aldehyde followed by cyclization to a mixture of two pairs of diastereoisomers. In the presence of peroxides a different reaction takes place and the product formed with HNE is $1,N^2$-ethenodeoxyguanosine (Sodum and Chung, 1988). The products were separated by HPLC and their structures were ascertained by mass spectroscopy and proton NMR. The yield of adduct formation was low despite long reaction times and very high concentrations of aldehyde (about 0.1 M) and deoxyguanosine. Such high aldehyde concentrations are orders of magnitude above the value expected to occur in vivo (Esterbauer and Zollner, 1989; Esterbauer et al, 1990a), and consequently the number of potentially hydroxyalkenal-modified guanosine residues in cellular DNA, if they existed at all, must be low. This is supported by an in vivo study (Grasse et al, 1985) in which rats received tritiated 4-hydroxyhexenal (3 mg, 2–9 μCi) injected directly into the hepatic portal vein. No significant binding of the aldehyde to DNA or RNA could be detected 2 or 16 h postinjection, whereas significant amounts of the aldehydes were bound to the liver protein (0.43 pmol (mg protein)$^{-1}$). Assuming an average molecular weight of 100 000 for the proteins, this figure means that about 1 out of 100 protein molecules possessed a covalently bound aldehyde. A rough calculation shows that the number of 4-hydroxyhexenal-modified DNA bases, if they occurred at all, must have been less than 1 in 50 000. In isolated DNA or RNA incubated with [^{14}C]4-hydroxypentenal (20–200 μM) no detectable amount of radioactivity was present

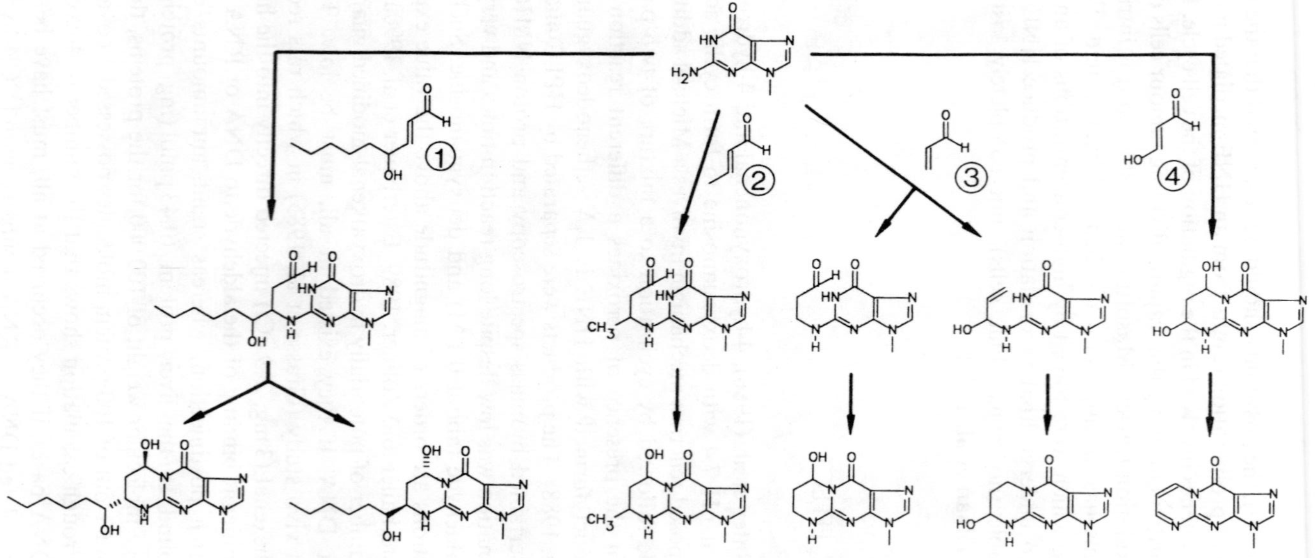

Figure 3. Reaction of guanosine with HNE (1), crotonaldehyde (2), acrolein (3) and malonaldehyde (4).

(Schauenstein and Esterbauer, 1979). The fact that so far a covalent binding of 4-hydroxyalkenals to DNA has been demonstrated neither in vitro nor in vivo should not be interpreted as evidence that binding does not occur in vivo, but rather as a methodological problem. The specific radioactivity of labelled 4-hydroxyalkenals is low, and therefore the sensitivity of the determination of low levels of DNA-bound aldehydes may be insufficient. In addition, it should be considered that such a binding would be at least in part reversible, and thus the aldehyde may have been lost during isolation of DNA and the subsequent work-up procedure. A similar problem with insufficient sensitivity was encountered in demonstrating the binding of MDA or β-substituted acroleins to DNA in vitro and in vivo (review: Basu et al, 1988).

Acrolein and crotonaldehyde, both aldehydes with the same functional group, —CH=H—CHO, as 4-hydroxyalkenals, alkylate deoxyguanosine in vitro in a similar way to 4-hydroxyalkenals, and give cyclic adducts between N^1 and N^2 of the guanosine base, i.e. tetrahydropyrimidopurinones (Chung et al, 1984) (Figure 3). Also, MDA, which exists in neutral aqueous solution mainly in the tautomeric form as β-hydroxyacrolein, reacts in a similar way with guanosine (Basu et al, 1988). In this case, the primary adduct is stabilized by elimination of water to a pyrimidopurinone (PyP) adduct. MDA in addition to the PyP adduct can give various other adducts with guanine, adenosine and cytidine bases, but the PyP adduct is so far the only one detected in MDA-treated DNA.

4-Hydroxyalkenals are mutagenic. This was first demonstrated by Marnett et al (1985) in a study where 28 carbonyl compounds (alkanals, 2-alkenals, dicarbonyls, 4-hydroxyalkenals) were screened for mutagenicity in *Salmonella typhimurium* TA 104. This strain is said to be more sensitive to carbonyl mutagenicity than any other characterized strain of *Salmonella*. The liquid preincubation procedure was used in this study (incubation of bacteria in phosphate-buffered aldehyde solution at 37°C for 20 min) followed by plating on minimal-glucose agar and scoring the revertants after 48 h. Clear dose–mutagenicity response curves were obtained for formaldehyde, acrolein, crotonaldehyde, methacrolein, 2-hexenal, 2-heptenal, 2,4-hexadienal, glutaraldehyde, glyoxal, methylglyoxal and ketoxal. The toxicity of the aldehydes strongly increased with increasing lipophilicity, i.e. increasing chain length; thus nonanal was about 300 times more toxic than acetaldehyde, and 2-nonenal was 400 times more toxic than acrolein. The high toxicity of these long-chain aldehydes may have precluded the detection of their mutagenicity. The toxicity was also a problem in the 4-hydroxyalkenal screening, the maximum nontoxic doses for 4-hydroxypentenal, 4-hydroxyoctenal and HNE being 4, 2.8 and 1.0 mM, and there being no significant increase of mutated bacteria at these doses. Addition of glutathione (20 mM) after the liquid preincubation period significantly decreased the toxicity of 4-hydroxyalkenals and other 2-alkenals, and under these conditions 4-hydroxypentenal gave a dose-dependent increase of mutations. The effect of glutathione also shows, according to Marnett et al (1985), that toxicity and mutagenicity are not related to each other, but are based on different chemical reactions. As mentioned previously, glutathione rapidly reacts, also non-enzymatically, with 4-hydroxyalkenals and the resulting conjugates are not toxic or are much less toxic. In addition, excess glutathione could have reactivated protein SH-groups blocked by the aldehyde, as reported for some enzymes (Schauenstein and Esterbauer, 1979).

The problem of detecting mutagenicity in the presence of high toxicity was described again by Cooper et al (1987), who tested crotonaldehyde, muconaldehyde and 4-hydroxynonenal in *Salmonella typhimurium* TA 100. From their data it appears that 4-hydroxynonenal is much more toxic in TA 100 than TA 104 strains. In TA 100 the maximal nontoxic dose was 0.05 mM, whereas in TA 104 a maximum nontoxic dose of 1 mM was reported (Marnett et al, 1985). At nontoxic doses of HNE no increase in mutation frequency was seen with TA 100. At concentrations which significantly decreased the survival count, significantly more apparent revertants were present on the plate (Table 6), but the authors concluded that these were not true mutants but rather pinpoint survivors. Critical comments on this conclusion were made by Zeiger (1988). It is interesting that the toxicity of 4-hydroxynonenal strongly depended on the type of agar plate used for plating; thus 0.1 mM or 0.15 mM HNE did not decrease survival counts in trypticase soy agar plates, but on minimal-agar plates the same dose reduced survival to less than 1% (Cooper et al, 1987). This indicates that the 4-hydroxynonenal toxicity is to some extent reversible, if the bacteria are transferred into a rich medium after aldehyde treatment. The nature of the protective substance is not known, but the effect is probably similar to the protective action seen with glutathione (Marnett et al, 1985).

The mutagenicity of HNE was also tested in V79 Chinese hamster ovary cells (Cajelli et al, 1987). At concentrations between 10 and 45 μM and with an exposure time of 90 min in serum-free medium, the aldehyde induced a dose-dependent increase of the 6-thioguanine-resistant (TG^r) cells. The spontaneous mutation frequency was 8.5–14.4 (mean 11.7) TG^r mutants per 10^6 surviving cells; at 45 μM HNE the mutation frequency ranged from 29.8 to 85.2 (mean 55.0) TG^r mutants per 10^6 survivors. Ethylmethanesulphonate used as a positive control gave at 2.5 mM on average 84 mutants per 10^6 survivors. According to Bradley et al (1981), a substance should be considered as mutagenic for V79 Chinese hamster ovary cells if the induced mutation frequency is at least three times higher than the spontaneous one. 45 μM HNE gave on average a 4.7-fold increase in mutation frequency.

Table 6. Cytotoxicity and mutagenicity of 4-hydroxyalkenals in *Salmonella typhimurium* TA 104 and TA 100.

Concentration (mM)	Survivors (%)	Revertants per plate
4-Hydroxypentenal,[a] TA 104		
1.0	100	37
4.0	100	340
10.0	100	775
4-Hydroxynonenal,[b] TA 100		
0.00	100	149
0.10	91	90
0.15	102	64
0.20	62	4432[c]
0.25	5	4644[c]
0.30	5	5976[c]

Table compiled from data reported by Marnett et al. (1985) and Cooper et al (1987).
[a] 20 min preincubation followed by a chase of 10 mM glutathione.
[b] 30 min preincubation, no glutathione chase, survivals tested in trypticase soy agar plates.
[c] The apparent revertants were assumed to be pinpoint survivors.

From the selection of 6-thioguanine resistance it was concluded (Cajelli et al, 1987) that HNE has induced a mutation at the locus of DNA coding for hypoxanthine–guanine phosphoribosyltransferase (HGPRT). The mutagenicity tests were limited also in this study by the cytotoxicity of the aldehydes at concentrations above 45 μM. Doses of 10, 20, 30 and 45 μM HNE reduced the relative survivals to 86%, 78%, 60% and 47%, but at 60 μM only 1% of the cells survived. This abrupt increase of toxicity suggests that the V79 Chinese hamster cells have, like other mammalian cells, the capacity to detoxify certain amounts of the aldehyde before toxicity results. The study with V79 cells also indicates that mammalian cells are probably better models for screening mutagenic and genotoxic effects of long-chain 4-hydroxyalkenals and other lipophilic aldehydes than the Ames test with *S. typhimurium* strains. In *S. typhimurium* TA 104, for example, the detection of mutagenicity of 4-hydroxynonenal was precluded by its high toxicity, and in the TA 100 strain possible mutagenicity was overlapped by cytotoxic effects. Similarly, in *S. typhimurium* assay amongst the alkanals, only formaldehyde was mutagenic, whereas all other long-chain alkanals tested were not mutagenic. In V79 cells, however, propanal, butanal and hexanal clearly increased the mutation frequency at concentrations of 10–30 mM (Brambilla et al, 1988).

The mechanism by which 4-hydroxyalkenals lead to mutation in V79 cells or in bacteria is not clear. In addition to a direct alkylation of guanosine bases in DNA, the aldehydes could also have acted indirectly by triggering other events which then caused DNA damage or by reducing the natural DNA repair capacity of the cells. It has been shown (Krokan et al, 1985) that 4-hydroxyalkenals (4-hydroxypentenal, 4-hydroxyoctenal, HNE) and other α,β-unsaturated aldehydes markedly inhibited in cultured human bronchial fibroblasts the important DNA repair enzyme O^6-methylguanine–DNA methyltransferase (O^6-MeGua–DNA methyltransferase). In intact cells treated with 100 μM HNE, the inhibition of this enzyme was 35%; in cell extracts the same dose led to an inhibition of 70%. O^6-MeGua–DNA methyltransferase is an enzyme with a functional SH-group in its active site, and it is reasonable to assume that the inhibition is produced by blockage of the SH-group. Another repair system, uracil–DNA glycosylase, was not affected by hydroxyalkenals at doses up to 100 μM. The doses of 4-hydroxyalkenals (100–300 μM) required to reduce O^6-MeGua–DNA methyltransferase activity were also highly toxic to the cells and decreased cell survival (colony-forming efficiency) below 1%. This decrease in colony-forming efficiency appears to be unrelated to the inhibition of the DNA repair system, since two other aldehydes (formaldehyde, acetaldehyde) also markedly decreased colony-forming efficiency, but had no effect on O^6-MeGua–DNA methyltransferase activity.

The effect of 4-hydroxyalkenals on the DNA repair mechanism appears to be complex and involves both an inhibition of specific repair enzymes and the induction of the gross DNA repair synthesis. The was first shown by Griffin and Segall (1986) in primary cultures of rat hepatocytes. In these cells, prolonged exposure (20 h) to 4-hydroxyalkenals (4-hydroxyhexenal, HNE) and 2-alkenals (hexenal, nonenal) elicited a dose-related increase of unscheduled DNA synthesis as measured autoradiographically by the number of nuclear grains resulting from [methyl-^{3}H]thymidine erroneously incorporated into the DNA by the repair mechanism. At doses of 480 and 600 nmol HNE per 10^6 cells (480–600 μM) the net grain count per nucleus was 13.8 and 15.2, whereas the control had negative nuclear grains (-2.88 ± 0.3 per nucleus); 95% of the cells were positive for unscheduled DNA synthesis. The authors have concluded that these results indicate that the alkenals had formed covalent bonds with hepatic DNA to a certain extent and that 4-hydroxyalkenals might have a carcinogenic potential. Propanal also showed the ability to induce unscheduled DNA synthesis in primary cultures of rat hepatocytes (Brambilla et al, 1988); with this aldehyde a clear increase of the net grain count per nucleus (11.3 ± 11.6) was, however, only obtained at the high concentration of 33 mM and with 20 h exposure time.

Numerous data show that 4-hydroxyalkenals and other 2-alkenals produce in mammalian cells a number of more or less unspecific genotoxic responses which might be described by the general term 'genotoxic effects'. Such genotoxic effects were assayed by measurement of sister chromatid exchange (SCE), chromosomal aberrations, DNA fragmentation and micronuclei. The capability of a series of 4-hydroxyalkenals (4-hydroxypentenal, 4-hydroxyhexenal, 4-hydroxyoctenal, HNE, 4-hydroxyundecenal) to induce DNA fragmentation and SCE in cultured Chinese hamster ovary cells was screened by Brambilla et al (1986). The alkaline elution assay or the alkaline denaturation followed by hydroxyapatite separation of single- and double-stranded DNA were used to detect DNA fragmentation, and the assays showed that the aldehydes at concentrations of 170 μM significantly increased the amount of DNA fragmentation. This concentration, however, was also highly cytotoxic towards the cells; for example, a 90-min exposure of Chinese hamster ovary cells to 170 μM HNE decreased cell viability measured by trypan blue exclusion to 4% and colony-forming efficiency to 0%. In the presence of such a high toxicity, the observed increase of DNA fragmentation could therefore simply be a result of cell death-associated activation of endonucleases (Williams et al, 1974). HNE also produced significantly ($p < 0.01$) increased DNA fragmentation at 60 μM, where cell viability was 98% and colony-forming efficiency 3%. No DNA fragmentation was seen at subtoxic doses of 20 μM HNE. These doses, however, caused statistically significant increases ($p < 0.01$) in the number of SCEs. At doses of 0, 5, 10 and 20 μM HNE, 11.8 ± 0.9, 11.8 ± 0.3, 13.3 ± 0.6 and 13.6 ± 0.8 SCEs per cell were present. With 20 μM HNE the cell viability was 99% and plating efficiency was 98%. The increase of SCE frequency was somewhat more pronounced with hydroxypentenal, hydroxyhexenal and hydroxyoctenal than with the other hydroxyalkenals tested. It is worth mentioning that acetaldehyde was also very effective in inducing SCE; doses of 30 μM and 100 μM gave on average 14.8 and 38.9 SCEs per cell, and the control value was 9.4 SCEs per cell (Brambilla et al, 1986).

This also confirms a series of previous studies (reviews: Brambilla et al, 1988, 1989) which showed that acetaldehyde increases the SCE frequency in bone marrow cells of mice, Chinese hamster lymphocytes, human lymphocytes and CHO cells.

Although the induction of DNA fragmentation in CHO cells by 4-hydroxyalkenals must be viewed with some caution because of the high cytotoxicity, the increased frequency of SCE at subtoxic concentrations shows that 4-hydroxyalkenals can produce genotoxic effects on mammalian cells. This was further confirmed in primary cultures of rat hepatocytes (Eckl and Esterbauer, 1989; Esterbauer et al, 1990b). Primary cultures of rat hepatocytes were treated in serum-free MEM medium for 3 h with 0 (control), 0.1, 1.0, 10 and 100 μM 4-hydroxynonenal or 2-nonenal or nonanal. After complete removal of the aldehydes by washing, the cells were incubated again for 48 h with fresh medium. The genotoxic and cytotoxic effects caused by the 3-h preincubation in the presence of the aldehyde was scored by the determination of percentages of mitotic cells, chromosomal aberrations, SCEs and micronuclei. A brief summary of the results obtained with 4-hydroxynonenal is given in Table 7. Included in this table are also the data reported by Griffin and Segall (1986) for unscheduled DNA synthesis and lactate dehydrogenase release. These two independent investigations clearly showed that doses of 100 μM HNE and above led to severe damage of the hepatocytes. Cultures treated for 3 h with 100 μM HNE or above showed immediately thereafter a large number of damaged cells (granular structure, pycnotic nuclei, loss of contact with the surface of the dishes, lactate dehydrogenase release); the cells did not recover, and after 25 h all the cells were dead. The strong increase of unscheduled DNA synthesis suggests that the cells tried to repair damaged DNA, yet without success with respect to their survival. In hepatocyte cultures treated with 10 μM HNE or less, no cytotoxic effects were observed. The mitotic index remained at the same level as in controls, which indicates that these concentrations do not affect cell proliferation. Moreover, no membrane damage occurred at 10 μM or below, as evidenced by the absence of lactate dehydrogenase release. Nevertheless, this subtoxic concentration of HNE clearly increased the number of micronuclei and the frequency of SCE (Esterbauer et al, 1990b). The parameter most sensitive to HNE treatment is SCE, which showed statistically significant increases (+39%) at concentrations as low as 0.1 μM. This dose is 100 to 1000 times lower than the genotoxic dose reported for other cell types, e.g. *Salmonella* and CHO cells. Moreover, this dose is two orders of magnitude lower than the cytotoxic dose. It therefore appears that primary cultures of rat hepatocytes are better targets for testing the genotoxic effects of 4-hydroxyalkenals and probably also of other aldehydes than other cell types. Rat hepatocytes have a high capacity to detoxify 4-hydroxynonenal and this may explain why the genotoxic effect reached a plateau at 0.1–1 μM. Higher aldehyde concentrations externally added are obviously rapidly reduced to lower levels by the combined action of the HNE-metabolizing systems of rat hepatocytes. On the other hand, at concentrations of 100 μM and above, the detoxification process is probably overwhelmed and HNE produces a number of lesions which finally lead to cell death. 2-Nonenal and nonenal were not cytotoxic at 100 μM, the maximum dose tested. At 100 μM 2-nonenal slightly increased the number of micronuclei, but chromosomal aberrations were not significantly altered. Nonanal had no detectable genotoxic effects. The effect of 2-nonenal and nonanal on SCE have not yet been investigated in this study. The basal level of endogenous 4-hydroxynonenal in the liver is in the genotoxic range of 0.1–1.0 μM (Table 1), and the concentration can significantly increase in

Table 7. Cytotoxic and genotoxic effects of HNE in primary cultures of rat hepatocytes. Table compiled from data reported by Esterbauer et al. (1990b) and Griffin and Segall (1986).

Concentration μM	Mitotic index[a]	LDH release[b]	Unscheduled DNA synthesis[c]	Micronuclei[d]	Chromosomal aberrations[e]	SCE[f]
0	1.94	100	−2.88	6.2	0.050	0.78
0.1	2.03	100	n.m	7.4	0.043	1.09*
1.0	1.75	100	n.m.	9.3*	0.18	1.05*
10.0	1.70	100	n.m.	9.8*	0.25	1.11*
60.0	n.m.	125*	−1.3			
100.0	toxic[g]	n.m.	n.m.			
120.0	n.m.	150*	0.25*			
240.0	n.m.	600*	4.69*			
600.0	n.m.	900*	15.2*			

[a] Percentage of mitotic cells.
[b] Lactate dehydrogenase release, percentage of control.
[c] Grains per nucleus.
[d] Number of cells with micronuclei per mitotic index 1.00.
[e] Aberrations per diploid cell (=42 chromosomes).
[f] Sister chromatid exchange per chromosome.
[g] All cells dead after 24 h.
* Significant difference to control with $P > 0.05$ or lower, control is 0 μM HNE.
n.m.; not measured.

conditions of oxidative stress such as vitamin E deficiency (reviews: Esterbauer and Zollner, 1989; Esterbauer et al, 1990a). Reactive aldehydes such as HNE or similar aldehydes produced by lipid peroxidation could therefore present a constant source of low-level DNA damage. However, so far no data are available which would suggest that such damage is persistent and leads to mutation.

ACKNOWLEDGEMENT

The authors' work has been supported by the Association for International Cancer Research (AICR) UK.

REFERENCES

Algar EM & Holmes RS (1989a) Kinetic properties of murine liver aldehyde dehydrogenases. *Prog. Clin. Biol. Res.* **82:** 94–103.

Algar EM & Holmes RS (1989b) Purification and properties of mouse stomach aldehyde dehydrogenases. Evidence for a role in the oxidation of peroxidic and aromatic aldehydes. *Biochim. Biophys. Acta* **995:** 168–173.

Alin P, Danielson UH & Mannervik B (1985) 4-Hydroxyalk-2-enals are substrates for glutathione transferase. *FEBS Lett.* **179:** 267–270.

Antonenkov VD, Pirozhkov SV & Panchenko LF (1987) On the role of microsomal aldehyde dehydrogenase in metabolism of aldehydic products of lipid peroxidation. *Biochim. Biophys. Acta* **224:** 357–360.

Barrera G, Martinotti S, Fazio V et al (1987) Inhibition of the expression of the c-myc oncogene in K562 cells. *Toxicol. Pathol.* **15:** 238–240.

Basu AK, O'Hara SM, Valladier P, Stone K, Mols O & Marnett LJ (1988) Identification of adducts formed by reaction of guanine nucleosides with malondialdehyde and structurally related aldehydes. *Chem. Res. Toxicol.* **1:** 53–59.

Benedetti A, Comporti M & Esterbauer H (1980) Identification of 4-hydroxynonenal as a cytotoxic product originating from the peroxidation of liver microsomal lipids. *Biochim. Biophys. Acta* **620:** 281–296.

Benedetti A, Fulceri R & Comporti M (1984) Inhibition of calcium sequestration activity of liver microsomes by 4-hydroxyalkenals originating from the peroxidation of liver microsomal lipids. *Biochim. Biophys. Acta* **793:** 489–493.

Bradley MO, Bhuyan B, Francis MC, Langenbach L, Petersen A & Huberman E (1981) Mutagenesis by chemical agents in V79 Chinese hamster cells: A review and analysis of the literature. *Mutat. Res.* **87:** 81–142.

Brambilla G, Sciaba L, Faggin P et al (1986). Cytotoxicity, DNA fragmentation and sister-chromatid exchange in Chinese hamster ovary cells exposed to the lipid peroxidation product 4-hydroxynonenal and homologous aldehydes. *Mutat. Res.* **171:** 169–176.

Brambilla G, Martelli A, Cajelli E, Canonero R & Marinari UM (1988) Lipid peroxidation products and carcinogenesis: Preliminary evidence of n-alkanal genotoxicity. In Nigam SK, McBrien DCH & Slater TF (eds), *Eicosanoids, Lipid Peroxidation and Cancer,* pp 243–251. Berlin, Heidelberg: Springer Verlag.

Brambilla G, Martelli A & Marinari UM (1989) Is lipid peroxidation associated with DNA damage? *Mutat. Res.* **214:** 123–127.

Burdon RH & Rice-Evans C (1989) Free radicals and the regulation of mammalian cell proliferation. *Free Radicals Res. Commun.* **6:** 345–358.

Burton GW, Cheesman KH, Ingold KU & Slater TF (1983) Lipid antioxidants and products of lipid peroxidation as potential tumor protective agents. *Biochem. Soc. Trans.* **11:** 261–262.

Cadenas E, Müller A, Brigelius R, Esterbauer H & Sies H (1983) Effects of 4-hydroxynonenal on isolated hepatocytes. *Biochem. J.* **214:** 479–487.

Cajelli E, Ferraris A & Brambilla G (1987) Mutagenicity of 4-hydroxynonenal in V79 Chinese hamster lung cells. *Mutat. Res.* **44:** 169–171.

Cajone F, Salina M & Benelli-Zazzera A (1989) 4-Hydroxynonenal induces a DNA-binding protein similar to the heat-shock factor. *Biochem. J.* **262:** 977–979.

Canuto RA, Muzio G, Biocca ME & Dianzani MU (1989) Oxidative metabolism of 4-hydroxy-2,3-nonenal during diethyl-nitrosamine-induced carcinogenesis in rat liver. *Cancer Lett.* **46:** 7–13.

Celotto C (1985) Microsomal and mitochondrial lipid peroxidation. PhD Thesis, University of Graz, Austria.

Cerutti PA (1985) Prooxidant states and tumor promotion. *Science* **227:** 375–381.

Chung FL, Young R & Hecht SS (1984) Formation of cyclic 1 N2-propanodeoxy-guanosine adducts in DNA upon reaction with acrolein or crotonaldehyde. *Cancer Res.* **44:** 990–995.

Comporti M (1985) Biology of disease. Lipid peroxidation and cellular damage in toxic liver injury. *Lab. Invest.* **53:** 599–623.

Comporti M (1989) Three models of free radical-induced cell injury. *Chem. Biol. Interact.* **72:** 1–56.

Cooper KO, Witz G & Witmer CM (1987) Mutagenicity and toxicity studies of several alpha, beta-unsaturated aldehydes in the *Salmonella typhimurium* mutagenicity assay. *Environ. Mutagen.* **9:** 289–295.

Curzio M (1988) Interaction between neutrophils and 4-hydroxyalkenals and consequences on neutrophil motility. *Free Radical Res. Commun.* **5:** 55–66.

Danielson UH, Esterbauer H & Mannervik B (1987) Structure–activity relationships of 4-hydroxyalkenals in the conjugation catalysed by mammalian glutathione transferase. *Biochem. J.* **247:** 707–713.

Dekant W, Lash LH & Anders MW (1988) Fate of glutathione conjugates and bioactivation of cysteine S-conjugates by cysteine conjugate β-lyase. In Sies H & Ketterer B (eds) *Glutathione Conjugation. Mechanisms and Biological Significance,* pp 415–450. London, New York: Academic Press.

Dianzani MU (1979) Biological activity of methylglyoxal and related aldehydes. *Ciba Found. Ser.* **67:** 245–270.

Dianzani MU (1982) Biochemical effects of saturated and unsaturated aldehydes. In McBrien DCH & Slater TF (eds) *Free Radicals, Lipid Peroxidation and Cancer,* pp 129–158. London: Academic Press.

Eckl P & Esterbauer H (1989) Genotoxic effects of 4-hydroxynonenals. *Adv. Biosci.* **76:** 141–157.

Esterbauer H (1982) Aldehydic products of lipid peroxidation. In McBrien DCH & Slater TF (eds) *Free Radicals, Lipid Peroxidation and Cancer*, pp 101–128. London: Academic Press.

Esterbauer H (1985) Lipid peroxidation products: Formation, chemical properties and biological activities. In Poli G, Cheeseman KH, Dianzani MU & Slater TF (eds) *Free Radicals in Liver Injury*, pp 29–47. Oxford, Washington DC: IRL Press.

Esterbauer H & Cheeseman KH (1990) Determination of aldehydic lipid peroxidation products: Malonaldehyde and 4-hydroxynonenal. *Methods Enzymol.* **186:** 407–421.

Esterbauer H & Zollner H (1989) Methods for determination of aldehydic lipid peroxidation products. *Free Radicals Biol. Med.* **7:** 197–203.

Esterbauer H, Zollner H & Scholz N (1975) Reaction of glutathione with conjugated carbonyls. *Z. Naturforsch.* **30:** 466–473.

Esterbauer H, Ertl A & Scholz N (1976) The reaction of cysteine with alpha,beta-unsaturated aldehydes. *Tetrahedron* **32:** 285–289.

Esterbauer H, Zollner H & Lang J (1985) Metabolism of the lipid peroxidation product 4-hydroxynonenal by isolated hepatocytes and by liver cytosolic fractions. *Biochem. J.* **228:** 363–373.

Esterbauer H, Jürgens G, Quehenberger O & Koller E (1987) Autoxidation of human low density lipoprotein: loss of polyunsaturated fatty acids and vitamin E and generation of aldehydes. *J. Lipid Res.* **28:** 495–509.

Esterbauer H, Zollner H & Schaur RJ (1988) Hydroxyalkenals: Cytotoxic products of lipid peroxidation. *ISI Atlas Sci.* **1:** 311–317.

Esterbauer H, Zollner H & Schaur RJ (1990a) Aldehydes formed by lipid peroxidation; Mechanisms of formation, occurrence and determination. In Vigo-Pelfrey C (ed.) *Membrane Lipid Oxidation*, Vol. I, pp 239–283. Boca Raton, Florida: CRC Press.

Esterbauer H, Eckl P & Ortner A (1990b) Possible mutagens derived from lipids and lipid precursors. *Mutat. Res.* **238:** 223–233.

Ferro M, Marinari UM, Poli G et al (1988) Metabolism of 4-hydroxynonenal by the rat hepatoma cell line MC1C1. *Cell Biochem. Funct.* **6:** 245–250.

Glasgow WC, Harris TM & Brash AR (1986) A short-chain aldehyde is a major lipoxygenase product in arachidonic acid-stimulated porcine leukocytes. *J. Biol. Chem.* **261:** 200–204.

Grasse LD, Lame MW & Segall HJ (1985) In vivo covalent binding of trans-4-hydroxy-2-hexenal to rat liver macromolecules. *Toxicol. Lett.* **29:** 43–49.

Griffin DS & Segall HJ (1986) Genotoxicity and cytotoxicity of selected pyrrolizidine alkaloids. A possible alkenal metabolite of the alkaloids and related alkenals. *Toxicol. Appl. Pharmacol.* **86:** 227–234.

Grosch W (1987) Reactions of hydroperoxides—products of low molecular weight. In Chan HWS (ed.) *Autoxidation of Unsaturated Lipids*, pp 95–139. London, New York: Academic Press.

Haenen GRMM, Tai Tin Tsoi JNL, Vermeulen NPE, Timmermann H & Bast A (1987) Hydroxy-2,3-trans-nonenal stimulates microsomal lipid peroxidation by reducing the glutathione-dependent protection. *Arch. Biochem. Biophys.* **259:** 449–456.

Hakki SFI & Nodes JT (1979) Metabolism of 4-hydroxy-2,3-pentene-1-al in sub-mitochondrial fractions of mouse liver. *Chem. Biol. Interact.* **25:** 363–368.

Hauptlorenz S, Esterbauer HW, Pümpel R, Schauenstein E & Puschendorf B (1985) Effects of the lipid peroxidation product 4-hydroxynonenal and related aldehydes on proliferation and viability of cultured Ehrlich ascites tumor cells. *Biochem. Pharmacol.* **34:** 3803–3809.

Hurst JS, Slater TF, Lang J, Jürgens G, Zollner H & Esterbauer H (1987) Effects of the lipid peroxidation product 4-hydroxynonenal on the aggregation of human platelets. *Chem. Biol. Interact.* **61:** 109–124.

Ishikawa T, Esterbauer H & Sies H (1986) Role of cardiac glutathione transferase and of the glutathione S-conjugate export system in biotransformation of 4-hydroxynonenal in the heart. *J. Biol. Chem.* **261:** 1576–1586.

Jensson H, Guthenberg C, Alin P & Mannervik B (1986) Rat glutathione transferase 8–8, an enzyme efficiently detoxifying 4-hydroxyalk-2-enals. *FEBS Lett.* **203:** 207–209.

Kagan VE (1988) *Lipid Peroxidation in Biomembranes*. Boca Raton, Florida: CRC Press.

Kaneko THS, Nakano SI & Matsuo M (1987) Lethal effects of a linoleic acid hydroperoxide and its autoxidation products, unsaturated aliphatic aldehydes, on human diploid fibroblasts. *Chem. Biol. Interact.* **63:** 127–137.

Kaneko T, Kaji K & Matsuo M (1988) Cytotoxicities of a linoleic acid hydroperoxide and its related aliphatic aldehydes toward cultured human umbilical vein endothelial cells. *Chem. Biol. Interact.* **67:** 295–304.

Kappus H (1985) Lipid peroxidation: Mechanisms, analysis, enzymology and biological relevance. In Sies H (ed.) *Oxidative Stress*, pp 273–310. London, New York: Academic Press.

Khoschsorur G, Schaur JR, Schauenstein E, Tillian HM & Reiter M (1981) Intracelluläre Wirkung von Hydroxyalkenalen auf tierischen Tumoren. *Z. Naturforsch.* **36:** 572–578.

Koster JF, Slee RG, Montfoort A, Lang J & Esterbauer H (1986) Comparison of the inactivation of microsomal glucose-6-phosphatase by in situ lipid peroxidation-derived 4-hydroxynonenal and exogenous 4-hydroxynonenal. *Free Radical Res. Commun.* **1:** 273–287.

Krokan H, Schaffer P & DePamphilis ML (1979) Involvement of eukaryotic deoxyribonucleic acid. *Biochemistry* **181:** 4431–4443.

Krokan H, Grafstrom RC, Sundqvist K, Esterbauer H & Harris CC (1985) Cytotoxicity, thiol depletion and inhibition of *O*-6-methylguanine-DNA methyltransferase by various aldehydes in cultured human bronchial fibroblasts. *Carcinogenesis* **150:** 369–378.

Lame MW & Segall HJ (1986) Metabolism of the pyrrolizidine alkaloid metabolite trans-4-hydroxy-2-hexenal by mouse liver aldehyde dehydrogenases. *Toxicol. Appl. Pharmacol.* **82:** 94–103.

Mannervik B & Danielson U (1988) Glutathione transferases—structure and catalytic activity. *CRC Crit. Rev.* **23:** 291–334.

Marinari UM, Pronzato MA, Cottalasso D et al (1987) Inhibition of liver Golgi glycosylation activities by carbonyl products of lipid peroxidation. *Free Radical Res. Commun.* **3:** 319–324.

Marnett LJ, Hurd HK, Hollstein MC, Levin DE, Esterbauer H & Ames BH (1985) Naturally occurring carbonyl compounds are mutagens in the Salmonella tester strain TA 104. *Mutat. Res.* **148:** 25–34.

Masotti L, Casali E & Galeotti T (1988) Lipid peroxidation in tumor cells. *Free Radicals Biol. Med.* **4:** 377–386.

Mitchel YD & Petersen DR (1987) The oxidation of alpha-beta unsaturated aldehydic products of lipid peroxidation by rat liver aldehyde dehydrogenases. *Toxicol. Appl. Pharmacol.* **87:** 403–410.

Mitchel YD & Petersen DR (1989) Oxidation of aldehydic products of lipid peroxidation by rat liver microsomal aldehyde dehydrogenase. *Arch. Biochem. Biophys.* **269:** 11–17.

Munsch N, de Recondon AM & Frayssinet C (1974) In vitro binding of 3H-acrolein to regenerating rat liver DNA polymerase. *Experientia* **30:** 1234–1236.

Norsten-Höög C (1989) Mechanism of ethanol toxicity studied by gas chromatography–mass spectrometry. Thesis, Karolinkska Institute, Stockholm.

Paradisi L, Panagini C, Parola M, Barrera G & Dianzani MU (1985) Effect of 4-hydroxynonenal on adenylate cyclase and 5′-nucleotidase activities in rat liver plasma membranes. *Chem. Biol. Interact.* **53:** 209–215.

Paul VJ & Fenical W (1980) Toxic acetylene-containing lipids from the red marine algae Liagora farinosa Lamouroux. *Tetrahedr. Lett.* **21** 327–330.

Poli G, Chiarpotto E, Biasi F, Pavia R, Albano E & Dianzani MU (1982) Enzymatic impairment induced by biological aldehydes. *Res. Commun. Chem. Pathol. Pharmacol.* **38:** 71–76.

Poli G, Dianzani MU, Cheeseman KH, Slater TF, Lang J & Esterbauer H (1985) Separation and characterization of the aldehydic products of lipid peroxidation stimulated by carbon tetrachloride or ADP-iron in isolated rat hepatocytes and rat liver microsomal suspensions. *Biochem. J.* **227:** 629–638.

Poli G, Albano E, Dianzani MU et al (1988) Carbon tetrachloride-induced inhibition of protein kinase-C in isolated rat hepatocytes. *Biochem. Biophys. Res. Commun.* **153:** 591–597.

Poot M, Verkerk A, Koster JF, Esterbauer H & Jongkind JF (1987) Influence of cumene hydroperoxide and 4-hydroxynonenal on the glutathione metabolism during in vitro ageing of human skin fibroblasts. *Eur. J. Biochem.* **162:** 287–291.

Poot M, Verkerk A, Koster JF, Esterbauer H & Jongkind JF (1988a) Reversible inhibition of DNA and protein synthesis by cumene hydroperoxide and 4-hydroxynonenal. *Mech. Ageing Dev.* **43:** 1–9.

Poot M, Esterbauer H, Rabinovitch PS & Hoehn H (1988b) Disturbance of cell proliferation by two model compounds of lipid peroxidation contradicts causative role in proliferative senescence. *J. Cell. Physiol.* **137:** 421–429.

Rao MS & Reddy JK (1987) Peroxisome proliferation and hepatocarcinogenesis. *Carcinogenesis* **8:** 631–636.

Rossi MA, Fidale F, Garramone A, Esterbauer H & Dianzani MV (1990) Effect of 4-hydroxyalkenals on hepatic phosphatidylinositol-4,5-bisphosphate-phospholipase C. *Biochem. Pharmacol.* **39:** 1715–1719.

Schauenstein E (1982) Effects of low concentrations of aldehydes on tumour cells and tumour growth. In McBrien DCH & Slater TF (eds) *Free Radicals, Lipid Peroxidation and Cancer*, pp 159–171. London, New York: Academic Press.

Schauenstein E & Esterbauer H (1979) Formation and properties of reactive aldehydes. *Ciba Found. Ser.* **67:** 225–244.

Schauenstein E, Esterbauer H & Zollner H (1977) *Aldehydes in Biological Systems: Their Natural Occurrence and Biological Activities.* London: Pion Ltd.

Schaur RJ, Zollner H & Esterbauer H (1991) Biological effects of aldehydes with particular attention to 4-hydroxynonenal and malonaldehyde. In Vigo-Pelfrey C (ed.) *Membrane Lipid Oxidation*, Vol. II, 141–163. Boca Raton: CRC Press.

Segall HJ, Wilson DW, Dallas JL & Haddon WF (1985) Trans-4-hydroxy-2-hexenal. A reactive metabolite from the macrocyclic pyrrolizidine alkaloid senecionine. *Science* **229:** 472–475.

Selley ML, Bartlett MR, McGuiness JA, Hapel AJ, Ardlie NG & Lacey MJ (1989) Determination of the lipid peroxidation product trans-4-hydroxy-2-nonenal in biological samples by high performance liquid chromatography and combined capillary gas chromatography–negative ion chemical ionization mass spectrometry. *J. Chromatogr.* **488:** 329–340.

Slater TF (1984) Free-radical mechanisms in tissue injury. *Biochem. J.* **222:** 1–15.

Sodum RS & Chung FL (1988) 1,N^2-ethenodeoxyguanosine as a potential marker for DNA adduct formation by trans-4-hydroxy-2-nonenal. *Cancer Res.* **48:** 320–323.

Stremler KE, Stafforini DM, Prescott SM, Zimmerman GA & McIntyre TM (1989) An oxidized derivative of phosphatidylcholine is a substrate for the platelet-activating factor acetylhydrolase from human plasma. *J. Biol. Chem.* **264:** 5331–5334.

Suzuki K, Miyaki M, Umeda M, Nishimura M & Ono T (1981) Differential inactivation of DNA polymerase alpha and beta by aldehyde compounds. *Biochem. Biophys. Res. Commun.* **100:** 1626–1633.

Tillian HM, Hammer A, Kink E, Schaur RJ & Schauenstein E (1989) Iron-induced lipid-peroxidation and inhibition of proliferation in Ehrlich ascites tumor cells. *J. Cancer Res. Clin. Oncol.* **115:** 79–83.

Tomita I, Yoshino K & Sano M (1987) Formation of aliphatic aldehydes in the plasma and liver of vitamin E deficient rats. In Hayaishi O & Mino M (eds) *Clinical and Nutritional Aspects of Vitamin E*, pp 277–280. Amsterdam: Elsevier Science Publishers.

Turner WE, Hill RH Jr, Hannon WH, Bernert JT Jr, Killbourne EM & Bayse DD (1985) Bioassay screening for toxicants in oil samples from the toxic-oil syndrome outbreak in Spain. *Arch. Environ. Contam. Toxicol.* **14:** 261–271.

Van Kuijk FJGM (1988) Detection of lipid peroxidation associated with retinal degeneration using gas chromatography mass spectroscopy. Thesis, University of Nijmegen, The Netherlands.

Wawra E, Zollner H, Schaur RJ et al (1986) The inhibitory effect of 4-hydroxy-nonenal on DNA-polymerases alpha and beta from rat liver and rapidly dividing Yoshida ascites hepatoma. *Cell. Biochem. Funct.* **4:** 31–36.

Williams JR, Little JB & Shipley WV (1974) Association of mammalian cell death with a specific endonucleolytic degradation of DNA. *Nature (Lond.)* **252:** 754–755.

Winter CK, Segall HJ & Haddon WF (1986) Formation of cyclic adducts of deoxyguanosine with the aldehydes trans-4-hydroxy-2-hexenal and trans-4-hydroxy-2-nonenal in vitro. *Cancer Res.* **46:** 5682–5686.

Winter CK, Segall HJ & Jones AD (1987) Distribution of trans-4-hydroxy-2-hexenal and tandem mass spectrometric detection of its urinary mercapturic acid in the rat. *Drug Metab. Dispos.* **15:** 608–612.

Witz G (1989) Biological interactions of alpha,beta-unsaturated aldehydes. *Free Radicals Biol. Med.* **7:** 333–349.

Yoshino K, Matsuura T, Sano M, Saito SI & Tomita I (1986) Fluorimetric liquid chromatographic determination of aliphatic aldehydes arising from lipid peroxides. *Chem. Pharm. Bull.* **34:** 1694–1700.

Zeiger E (1988) Comments on the mutagenicity of alpha,beta-unsaturated aldehydes in Salmonella and the adequacy of experimental data to demonstrate a hypothesis. *Environ. Mol. Mutagen.* **11:** 161–162.

14

Oxidation of Low-density Lipoprotein in vitro

JANUSZ M. GEBICKI,* GÜNTHER JÜRGENS and HERMANN ESTERBAUER
Institute of Biochemistry, University of Graz, Schubertstrasse 1, A-8010 Graz, Austria
* *On leave from the School of Biological Sciences, Macquarie University, Sydney, Australia*

I. Introduction 371
II. Methods to determine oxidation of LDL 372
III. Modification of LDL in the absence of cells 377
IV. Modification of LDL in the presence of cells 381
V. The role of endogenous antioxidants in retarding the oxidation of LDL . 390

I. INTRODUCTION

The application of ultracentrifugal separation to the isolation of blood lipoproteins (Gofman et al, 1950) quickly led to the discovery of their lability on storage (Ray et al, 1954). The components which are now identified as the low-density lipoprotein (LDL) proved to be particularly susceptible (Schuh et al, 1978). A detailed review of the state of knowledge on blood lipoproteins achieved by 1960 (Gurd, 1960) contained an account of some chemical changes detected in LDL during isolation and storage. They included bleaching of carotenoid pigments and spectral changes characteristic of conjugated dienes formed during peroxidation of unsaturated lipids.

A clue to the conditions promoting oxidative changes in LDL was provided quite early by the work of Ray et al (1954), who identified copper ions as a powerful catalyst of the process. Several reports of changes induced during oxidation of LDL then followed, providing, by 1980, a rough picture

Oxidative Stress: Oxidants and Antioxidants
ISBN 0-12-642762-3

of the chemical events which might follow careless handling or prolonged storage of blood lipoproteins. But it was only the last decade which saw a series of remarkable studies demonstrating that LDL oxidation is not only a storage nuisance but also an important in vivo phenomenon, with potential to explain the chain of events leading to the development of atherosclerotic occlusions. These studies, summarized in recent reviews (Steinberg et al, 1989; Esterbauer et al, 1990a), all led to a common focus: massive uptake of LDL by macrophages.

The route to this focus involved three major discoveries. First, Goldstein et al (1979) identified a macrophage 'scavenger' receptor which allows entry of acetylated LDL and converts the macrophages into foam cells. In the second, Steinberg's laboratory (Henriksen et al, 1981) demonstrated that LDL can be modified by incubation with endothelial cells to a form recognized by the macrophage scavenger receptor. Finally, Chisholm and his collaborators (Morel et al, 1982, 1983) reported that the cell-induced modification of LDL involves peroxidation of its constituent lipids. Much subsequent work confirmed and extended these observations which, taken together, provided the mechanism by which a physiologically plausible series of processes could lead to the build-up of a potentially fatal atherosclerotic plaque. In these processes, the key event triggering localized deposition of cholesterol esters, characteristic of the developing plaque, is the oxidation of LDL.

Studies of the conditions and agents capable of initiating oxidation of LDL can be grouped into two categories: those concerned with oxidation by cells, and the rest, where no cells were present. Both lines of work proceeded in parallel much of the time and both led to the general conclusion that cells and reagents potentiating the oxidation do so largely by accelerating the normal autoxidative changes in isolated LDL (Steinbrecher, 1988; Esterbauer et al, 1990a).

II. METHODS TO DETERMINE OXIDATION OF LDL

In the course of oxidation the chemical, physicochemical, functional and biological properties of LDL become progressively altered (review: Esterbauer et al, 1990a). The onset and progression of LDL oxidation in vitro can be followed by measuring the increase of TBARS, lipid hydroperoxides, conjugated dienes, aldehydes and fluorescent proteins or lipids (see Figure 1). Other possibilities include measurement of the disappearance of the endogenous antioxidant and polyunsaturated fatty acids, fragmentation of the apolipoprotein B to smaller peptides, and increase of the relative electrophoretic mobility of LDL. The biological assays used most frequently for the evaluation of the extent of oxidative modification of LDL are its uptake rate by cultured macrophages and its cytotoxicity towards cultured cells, e.g. fibroblasts. Finally, polyclonal and monoclonal antibodies against certain types of oxidatively modified LDL are available and can be used for immunocytochemical analysis. With this technique it was shown that MDA- and 4-hydroxynonenal-modified LDL occur in atherosclerotic arteries of the hyperlipidaemic Watanabe rabbit (Haberland et al, 1988; Palinski et al, 1989, 1990; Jürgens et al, 1990: Rosenfeld et al, 1990).

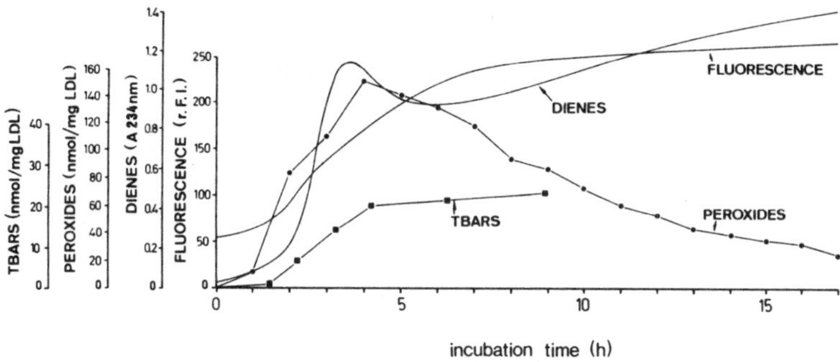

Figure 1. Time-course of Cu^{2+}-stimulated oxidation of LDL measured by the change of lipid hydroperoxides, dienes (234 nm absorption), 430 nm fluorescence (ex. 360 nm) and TBARS. LDL (0.25 mg ml^{-1}) was incubated in PBS + 1.66 μM Cu^{2+}. From Esterbauer et al (1990a), with permission.

The most commonly used assay in LDL oxidation studies, both in the presence and absence of cells, is the determination of thiobarbituric acid reactive substance (TBARS). The values found in oxidized LDL are in the range of about 1–15 nmol MDA equivalents per mg total LDL (review: Esterbauer et al, 1990a). The high variability of the TBARS values found by the different laboratories probably depends on methodological differences, but certainly also on different compositions of LDL used in the studies. It is known (Esterbauer et al, 1987, 1988, 1989a, 1990a,b) that in LDL the major sources of malonaldehyde (MDA) are arachidonic and docosahexaenoic acids. Since the concentration of these fatty acids in LDL can vary considerably from donor to donor, it follows also that significant variations in the TBARS values of oxidized LDL can be expected. Nevertheless, all reports agree that the TBARS values must reach a certain threshold level so that LDL is recognized and taken up by the scavenger receptor. This threshold level is about 2–4 nmol MDA per mg total LDL (Esterbauer et al, 1990a). In copper-stimulated oxidation of LDL, the formation of TBARS is preceded by a lag phase of about 40–140 min, depending on the LDL, during which the TBARS are low or undetectable. Thereafter, the TBARS value rapidly increases for about 1–2 h. At later stages the TBARS remain more or less constant or slowly increase (Esterbauer et al, 1989b). The kinetics of TBARS formation during cell-mediated LDL oxidation have so far not been studied thoroughly. In fact, most of the researchers only use a 24-h incubation period without intermediate checks of the stages of oxidation. From the few existent time-course studies (Fong et al, 1987; Morel et al, 1984; Heinecke et al, 1984; Hiramatsu et al, 1987) it appears that also in cell-mediated oxidation TBARS values do not continuously increase but approach a maximum.

Very useful parameters for monitoring the rate of LDL oxidation are the conjugated dienes (Esterbauer et al, 1989c). If the LDL lipids are oxidized, the polyunsaturated fatty acids (18:2, 20:4, 22:6) will be converted to fatty acid hydroperoxides with conjugated double bonds, showing a UV absorption maximum at 234 nm. Since oxidized LDL remains fully soluble in buffer, the increase of the 234-nm diene absorption can be measured directly in the LDL solution, i.e. without extraction of the LDL lipids. As with the TBARS time-course, the increase of the diene absorption is always preceded by a lag phase during which the 234-nm absorption remains constant or only slightly increases. Thereafter, the 234-nm absorption rapidly increases (propagation period) more or less in parallel with the TBARS values to a maximum value. In succession the 234-nm absorption decreases and then increases again (Figure 2). This agrees with the general mechanism of lipid peroxidation reactions according to which the conjugated lipid hydroperoxides are labile intermediates and decompose to a great variety of products, including aldehydes. Many of these products also have a strong UV absorption in the 210–240-nm region, which explains the second rise of the 234-nm absorption in the 'diene' time-profile.

The decomposition of the lipid hydroperoxides can clearly be seen when the time-course of LDL oxidation is followed by an assay measuring the lipid hydroperoxides in LDL (El-Saadani et al, 1989). There are three successive phases: a lag phase, a phase of propagation and a phase of decomposition. This time-profile also indicates that the chemical composition of LDL steadily changes during the process of oxidation (e.g. low to high to low peroxide value), and that an oxidized LDL with constant and definable properties does not in fact exist. The properties will always be different, depending on the time of oxidation.

Another useful method of following LDL oxidation is the measurement of the increase of the fluorescence at 430 nm with 360-nm excitation (Schuh et al, 1978; Jürgens et al, 1987; Steinbrecher, 1987). This fluorescence is almost entirely associated with apolipoprotein B, and various studies suggest that the formation of the fluorophor is due to binding of aldehydic lipid peroxidation products to free amino groups of the protein (review: Esterbauer et al, 1990a). All of the four methods (TBARS, diene, hydroperoxides, fluorescence) so far discussed give, in copper-stimulated LDL oxidation, similar time-profiles for the lag phase and the propagation phase; during the decomposition phase these methods, however, respond differently (Figure 1).

It has been shown that besides MDA many other aldehydes are formed during oxidation of LDL by decomposition of lipid hydroperoxides (Esterbauer et al, 1987, 1988, 1990a; Steinbrecher et al, 1989). Clearly identified were 4-hydroxynonenal, 4-hydroxyoctenal, 4-hydroxyhexenal, propanal, butanal, pentanal, hexanal, octanal and 2,4-heptadienal.

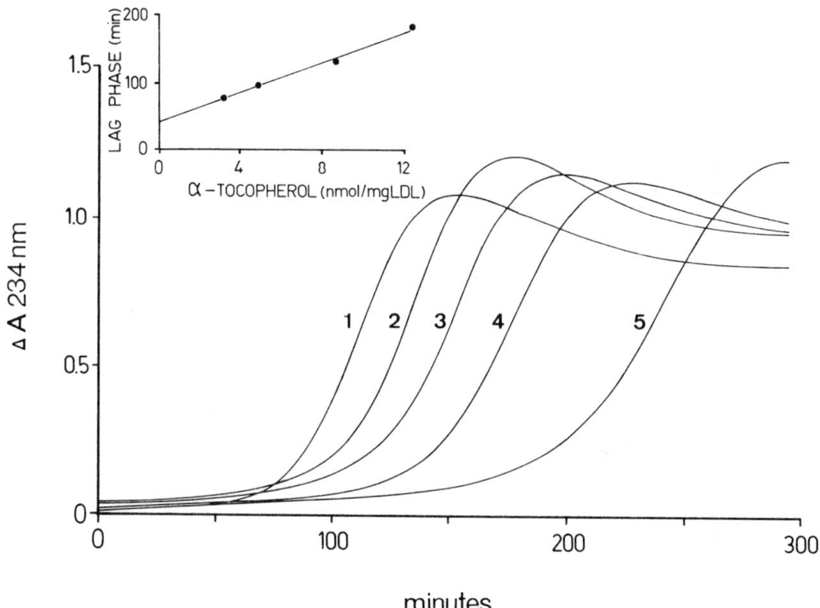

Figure 2. Determination of the oxidation resistance of LDL. The LDL samples (1 to 5) with increasing vitamin E content were prepared by supplementation of plasma with all-rac-α-tocopherol. To the LDL solution (0.25 mg per ml of PBS) was added 1.66 μM $CuSO_4$ and the oxidation was continuously followed by 234-nm diene absorption. The insert shows the dependence of the lag phase from the α-tocopherol content of the LDL. From Esterbauer et al (1990b), with permission.

The major aldehydes are always MDA (16–25 nmol (mg LDL)$^{-1}$), hexanal (10–12 nmol (mg LDL)$^{-1}$) and 4-hydroxynonenal (5–6.5 nmol (mg LDL)$^{-1}$). The contents of the other aldehydes are in the range 0.5–2.0 nmol (mg LDL)$^{-1}$. Methods are available to determine these aldehydes, either by HPLC or GC. The procedures, however, are rather time-consuming and are therefore only occasionally used in LDL oxidation studies. The time-courses of formation of the aldehydes also show lag phases and propagation phases, which correlate with TBARS and hydroperoxides. In later stages of oxidation the concentration can either decrease, remain constant or increase, depending on the type of aldehyde.

Measurements of the disappearance of the antioxidants during the lag phase, or the disappearance of the fatty acids, are important to achieve a better insight of the mechanism of LDL oxidation. However, for routine studies they have no advantage over the other much more simple methods. The same holds for degradation of apo B. The apolipoprotein B of native LDL has a molecular weight of about 500 000 and gives a single band in SDS electrophoresis. In oxidized LDL, this protein is degraded into smaller peptides to such an extent that SDS gel electrophoresis shows a smear with only a few distinct bands (Schuh et al, 1978; Fong et al, 1987). By densitometry, the time-dependence of the disappearance of the 500-kD band was quantified (unpublished). The disappearance temporarily correlated with the appearance of TBARS and conjugated dienes and lipid hydroperoxides, which suggests that the protein is fragmented by the attack of lipid alkoxy and lipid peroxy radicals.

LDL has a negatively charged surface and migrates to the anode in agarose gel electrophoresis under non-denaturing conditions. Oxidation renders LDL more negatively charged and accordingly its electrophoretic mobility increases. A measurable index for this is the relative electrophoretic mobility (REM), which is the ratio of the migration distance of oxidized to native LDL. Depending on the degree of oxidation, the REM increases from 1.00 (native LDL) to about 2.5–3.0 (Esterbauer et al, 1990a). The mechanism by which oxidation increases the negative charge of the surface is not yet clear. One possibility is that the binding of aldehydic products of lipid peroxidation to lysine residues (formation of Schiff bases) reduces the number of positive charges and consequently leads to an increase of net negative charge. Another possibility is that reactive oxygen species generated during oxidation of LDL convert histidine and proline residues to negatively charged aspartic or glutamic acid residues (Stadtman, 1989). A close linear relationship exists between the REM of oxidized LDL and its uptake rate by macrophages. Many researchers therefore use the REM as a preliminary index for predicting the uptake by the macrophage scavenger receptor. The receptor appears to be able to recognize a certain pattern of negative charges on the apo B. It should be noted, however, that a large number of negative charges (which can also be introduced by chemical modifications) by itself is not necessarily associated with increased uptake by the scavenger; rather, the density of charges at certain defined regions of the apo B seems to be an important determinant (review: Jürgens et al, 1987).

The most significant assay for characterizing the biological properties of oxidized LDL is probably its behaviour in the macrophage uptake experiments (review: Lang and Esterbauer, 1990). Mouse peritoneal macrophages, which are mainly used for such assays, express only a few receptors for LDL, and as a consequence the uptake of unmodified LDL is low in the range of about 1 μg LDL protein degraded per mg macrophage protein. In oxidized LDL this degradation is several-fold higher, in the range of about 10–15 μg protein. The experiments with macrophage uptake also revealed that the formation of the apo B epitopes, which bind to the scavenger receptor, occurs predominantly, if not exclusively, during the period of lipid hydroperoxides decomposing to secondary oxidation products (Bedwell and Jessup, 1987; Bedwell et al, 1989; Babiy et al, 1990). It follows that all earlier events occurring during oxidation of LDL (loss of antioxidants, increase of hydroperoxides) are, although critical requirements, not directly involved in the generation of the epitopes recognized by the macrophage scavenger receptor. LDL, being heavily oxidized by γ-irradiation, for example, was not a good substrate for the scavenger receptor, but when a previously γ-irradiated LDL was additionally treated with copper ions, which led to decomposition of the hydroperoxides, it was avidly taken up by macrophages.

Recently, it was reported (Parthasarathy et al, 1990) that native HDL inhibits the uptake of oxidatively modified LDL by macrophages. For example, endothelial cell-modified LDL with 56 nmol TBARS per mg LDL protein was degraded by macrophages with a rate of 14.5 μg protein per 5 h per mg cell protein. When LDL was oxidized by endothelial cells in the presence of 200 μg HDL ml^{-1}, somewhat higher TBARS were obtained (67 nmol mg^{-1}), but the rate of macrophage uptake was significantly decreased to 5.1 μg. In vivo, this could be of great significance, since it suggests that HDL, which is known to lower the risk of coronary heart diseases, diminishes the deleterious effects of oxidized LDL. This experiment also shows that oxidation indices (e.g. TBARS) may in some cases not be sufficient to predict the biological properties of oxidized LDL.

III. MODIFICATION OF LDL IN THE ABSENCE OF CELLS

Oxidation of LDL in the absence of cells or purposely added reagents is accompanied by changes characteristic of lipid peroxidation (Gurd, 1960). This can be a slow process, requiring days at low temperature (Lee, 1980), but it was recognized quite early by Ray et al (1954) that the presence of trace metal impurities is a determining factor in the rate of oxidation. This first systematic investigation of the effects of a wide range of metal ions on the ultracentrifugal homogeneity of human and rabbit LDL identified Cu^{2+} as the agent responsible for oxidative degradation. A number of agents, including EDTA, inhibited the oxidation, and this has led to its widespread use in the preparation of LDL. Cu^{2+} has also remained the main oxidant of LDL used in cell-free systems ever since.

While most of the early work on the oxidation of LDL concentrated on its avoidance, the actions of some deliberately added oxidizing agents were also studied. Nishida and Kummerow (1960) reported preferential denaturation of lipoproteins by methyl linoleate hydroperoxide, while another study (Nichols et al, 1961) showed extensive losses of polyunsaturated fatty acids in lipoproteins dialysed against low concentrations of Cu^{2+}. Similar losses were detected after exposure of human and chimpanzee LDL to H_2O_2, with cholesterol esters most affected and phospholipids least (Clark et al, 1969). The first direct observation of the effect of oxidation on the LDL protein was reported by Schuh et al (1978). Autoxidation of buffered LDL solutions was accompanied by extensive degradation of apoprotein B, which could be prevented by the absence of O_2 or by addition of EDTA, butylated hydroxytoluene (BHT) or propyl gallate. There was evidence of lipid hydroperoxides, and the authors suggested a causal relationship between lipid oxidation and apoprotein degradation. They speculated, farsightedly, that the protein damage could result from the action of free radicals generated during oxidation of the LDL lipids, and produced some evidence for the presence of low levels of oxidized lipoprotein in normal circulating blood.

A certain pattern can be detected in the many studies of oxidation of LDL in cell-free solutions which followed these early reports. LDL isolated by ultracentrifugation in the presence of EDTA was dialysed thoroughly, sometimes under an inert atmosphere, and then oxidized by incubation in PBS solution for several hours. Often micromolar concentrations of Cu^{2+} or, more rarely, Fe ions, were added to accelerate the process. Oxidation was monitored usually by assay of TBARS and by increase of electrophoretic mobility in an agarose gel. The former is a widely used semiquantitative indicator of lipid peroxidation, while the latter is characteristic of LDL oxidized or modified by treatment with breakdown products of lipid hydroperoxides (Jürgens et al, 1987). Additional indices of LDL oxidation were employed by different laboratories, but virtually all used TBARS and mobility measurements, which therefore form a useful index for comparison of results (Esterbauer et al, 1990a; Jürgens et al, 1987). The value of even simple TBARS assays was demonstrated by the Chisholm group (Morel et al, 1983), who showed for the first time that the known toxic action of LDL on cultured skin fibroblasts was due to the presence of oxidized lipids. Malonaldehyde, one of the products expected under the conditions employed, was not responsible for the toxicity.

Detailed studies of the chemical events which accompany spontaneous or Cu^{2+}-catalysed oxidation of LDL were carried out by Esterbauer et al (1988, 1989c, 1990a), showing that exposure of LDL to oxidizing conditions led to a loss of antioxidants and polyunsaturated fatty acids from the lipoprotein particle and to the appearance of lipid hydroperoxides and a range of aldehydes. Increase in electrophoretic mobility could be reproduced by incubation of native LDL with a range of aldehydes, 4-hydroxynonenal being the most effective (Jürgens et al, 1986, 1987; Esterbauer et al, 1987). Further studies showed that two Cu^{2+} per LDL particle were sufficient to induce maximum rates of oxidation. Formation of lipid hydroperoxide was preceded by a lag period; during this time the natural antioxidant disappeared. Addition of ascorbate prolonged the lag but did not change the order in which the antioxidants eventually disappeared (Esterbauer et al, 1989a,b).

The ability of other metal ions or free radical-generating systems to initiate these processes was also studied (Esterbauer, 1990a); in agreement with earlier work (Ray et al, 1954), Cu^{2+} was the only metal ion capable of accelerating oxidation of the LDL. Iron was tested extensively, because it is generally agreed that a transition metal catalyst is absolutely required in the initiation of LDL oxidation, and iron is the most likely adventitious contaminant of aqueous solutions (Halliwell and Gutteridge, 1989). However, iron proved ineffective, whether in the Fe^{2+} or Fe^{3+} form (Esterbauer et al, 1990a; Ray et al, 1954). This is not surprising, as there is no convincing evidence that either of these stable forms of iron is capable of initiating oxidation of polyunsaturated lipids directly. But then, neither is there any for Cu^{2+}, which nevertheless accelerates LDL oxidation. The only currently plausible explanation for the effects of Fe and Cu on LDL in buffered solutions depends on the assumption that even the most carefully isolated LDL contains traces of hydroperoxides which are decomposed by these metals (Chan, 1987; Halliwell and Gutteridge, 1989) to release lipid free radicals:

$$\text{LDL—OOH} + M^{(n+1)+} \longrightarrow \text{LDL—OO}^{\bullet} + M^{n+} + H^{+} \qquad (1)$$

In this reaction Cu^{2+} is reduced to Cu^{+} or Fe^{3+} to Fe^{2+}. The reduced form of the metal can then catalyse further hydroperoxide decomposition:

$$\text{LDL—OOH} + M^{n+} \longrightarrow \text{LDL—O}^{\bullet} + OH^{-} + M^{(n+1)+} \qquad (2)$$

The peroxyl (—$OO^{\bullet}$) or alkoxyl (—$O^{\bullet}$) lipid free radicals have the power to oxidize unsaturated lipids in the LDL particle:

$$\text{LDL—H} + \text{LDL—O}^{\bullet} \longrightarrow \text{LDL}^{\bullet} + \text{LDL—OH} \qquad (3)$$

$$\text{LDL—H} + \text{LDL—OO}^{\bullet} \longrightarrow \text{LDL}^{\bullet} + \text{LDL—OOH} \tag{4}$$

The symbol H indicates a labile H atom of the LDL lipids. With oxygen, the usual propagation reactions then take place:

$$\text{LDL}^{\bullet} + O_2 \longrightarrow \text{LDL—OO}^{\bullet} \tag{5}$$

followed by reaction 4 with the regenerated $LDL^{\bullet}$ free radical initiating further oxidation chains. There is experimental support for the occurrence of these reactions in the finding of lipid hydroperoxide and hydroxy derivatives in Cu^{2+}-oxidized LDL (Lenz et al, 1990).

Reaction 1 is much slower than reaction 2, so that the higher stable oxidation states of Cu or Fe are poor decomposers of lipid hydroperoxides (Chan, 1987; Halliwell and Gutteridge, 1989). However, other factors besides simple reaction rates are likely to be crucial in determining the ability of an ion to decompose hydroperoxides; they include concentration, presence of any chelators, lipophilicity and the nature of solvents (Pokorny, 1987). The stability of iron in solution is especially significant and causes a particular problem with Fe^{2+}. In the aqueous phosphate-buffered solutions commonly employed, the half-life of this species is less than 15 s and even in pure water it autoxidizes rapidly unless the pH is below about 3. It is likely that in studies of metal-assisted peroxidation of LDL (Esterbauer et al, 1990a; Ray et al, 1954) failure to detect oxidation even in solutions containing EDTA and H_2O_2 was caused by inadvertent oxidation of Fe^{2+} before it could generate sufficient oxidizing products (Halliwell and Gutteridge, 1989). In fact, LDL exposed to a slow infusion of acidified Fe^{2+} in the presence of H_2O_2 was recently successfully peroxidized in Esterbauer's laboratory (Gebicki and Esterbauer, unpublished).

The only other metal ion able to accelerate autoxidation of LDL is vanadyl (Dickson and Stern, 1990). At the relatively high concentration of 250 μM it caused a doubling of the spontaneous oxidation rate. Curiously, no lag period before the onset of peroxidation was reported, either in the presence or absence of the vanadyl ion.

The suggestion that free radicals may be responsible for the initiation of oxidation of LDL in the presence of cultured cells (Morel et al, 1982) has prompted much research, summarized in Section IV. However, a study carried out in the absence of cells has demonstrated which oxygen free radical could oxidize LDL and which could not. Using selected radicals generated by γ-radiation, Jessup and her coworkers (Bedwell et al, 1987, 1989) showed that the hydroxyl ($HO^{\bullet}$) and hydroperoxyl ($HO_2^{\bullet}$) free radicals were good oxidants of LDL, while superoxide ($O_2^{\bar{\bullet}}$) was quite poor, although it caused oxidation of the α-tocopherol held in the particle. An interesting finding was that the radical-oxidized LDL was not taken up by the mouse macrophage scavenger receptor. This suggests that either the presence of lipid hydroperoxides per se is insufficient to effect this uptake or that radiation-generated radicals prevented formation of the epitope characteristic of oxidized LDL and normally recognized by the scavenger receptor. Some unpublished experiments (Babiy and Gebicki) suggest that the former is the case. A study in which radiation was also used to produce free radicals (Babiy et al, 1990) led to experiments which confirmed the inability of mouse macrophages to recognize peroxidized LDL. However, when the oxidized LDL was incubated with Cu^{2+} in the absence of air, it was recognized by the scavenger macrophage receptor just like LDL oxidized by air incubation or by Cu^{2+}. This result suggests that decomposition of lipid hydroperoxides is a necessary prerequisite for this recognition.

Other agents able to oxidize LDL include oxygen- and sulphur-centred free radicals generated during metal-catalysed oxidation of thiols (Parthasarathy, 1987). Long-life free radicals are also likely to be involved in the modification of LDL by cigarette smoke extract to a form recognized by the macrophage scavenger receptor (Yokode et al, 1988). Not surprisingly, this subject is currently receiving much attention (Harats et al, 1989). Peroxyl free radicals, generated by the thermal decomposition of 2,2′-azobis(2-amidinopropane)dihydrochloride (AAPH), have been shown to be capable of initiating the oxidation of LDL (Sato et al, 1990). Finally, Sparrow et al (1988) have demonstrated for the first time that treatment of LDL with lipoxygenase in the presence of phospholipase A_2 leads to the same modification as exposure to Cu^{2+}, including uptake and degradation by macrophages.

IV. MODIFICATION OF LDL IN THE PRESENCE OF CELLS

The original seminal observation from Steinberg's laboratory (Henriksen et al, 1981) that exposure to cultured endothelial cells caused physical and biological alteration of LDL was followed by numerous studies on the nature of these alterations and of conditions which can bring them about. The changes in LDL properties observed originally were an increase in density and a ready uptake by macrophages via a mechanism not used for the native lipoprotein. The next significant finding was that incubation of LDL with endothelial or smooth muscle cells caused it to become toxic to proliferating fibroblasts and that the toxicity resulted from oxidation of the LDL (Morel et al, 1983).

Table I. Chronological list of selected studies related to oxidative modifications of LDL.

LDL modifier	Medium	Parameters measured	Effective inhibitor	Reference
Autoxidation + copper ions	PBS	Density Homogeneity	EDTA, ascorbate, glucosaminic acid, hesperidin, rutin, amino acids, BSA, glucoascorbate	Ray et al (1954)
Autoxidation	PBS	UV absorption		Gurd (1960)
Methyl linoleate hydroperoxide	PBS	Denaturation		Nishida and Kummerow (1960)
H_2O_2		Homogeneity		Clark et al (1969)
Cu^{2+}	Water	FA, Chol, aggregation		Nichols et al (1961)
Autoxidation	PBS	Apo B, Fl, TBARS	EDTA, N_2, BHT, propyl gallate	Schuh et al (1978)
Autoxidation	PBS	TBARS	EDTA, GSH	Lee (1980)
Rabbit aortic EC	F-10	Mph, density, REM		Henriksen et al (1981)
Oxy radicals		TBARS, CT	BHT, tryptophan	Morel et al (1982)
Oxy radicals	PBS	TBARS, CT	BHT, SOD	Evensen et al (1983)
Rabbit aortic EC, human umbilical EC, pig/guinea-pig SMC	F-10	Mph, density, REM		Henriksen et al (1983)
Autoxidation, MDA	PBS, M-199	TBARS, CT	N_2, EDTA, GSH	Hessler et al (1983)
Oxy radicals	PBS, M-199	TBARS, CT	SOD, Cat, dimethylfuran, mannitol	Morel et al (1983)
Oxy radicals from EC and SMC	M-199	TBARS, REM, CT, sterols	BHT, GSH, vitamin E	Morel et al (1984)
Human arterial SMC + Cu + Fe	DME	TBARS, REM, Mph	DF, BHT	Heinecke et al (1984)
Rabbit aortic EC, human umbilical EC $\pm$ Cu	F-10	TBARS, Mph, REM, density	EDTA, BHT, vitamin E, DME	Steinbrecher et al (1984)
Free radicals from monocytes or neutrophils	M-199	TBARS, CT, PC	EDTA, BHT, vitamin E, GSH	Cathcart et al (1985)

Phospholipase A_2	F-10, DME	Apo B, TBARS, Mph, PL	*p*-Bromophenacetyl bromide	Parthasarathy et al (1985)
O_2^- from human or monkey arterial SMC + Cu + Fe	F-10	TBARS, Mph, O_2^-	EDTA (+ Cu), DTPA, DF, BHT, BHA, SOD	Heinecke et al (1986)
Hydroxynonenal, MDA	tris-saline	REM		Jürgens et al (1986)
Reagent and rabbit EC H_2O_2 ± Cu	PBS, F-10	Apo B, TBARS, Mph, H_2O_2	SOD (Cu + H_2O_2)	Montgomery et al (1986)
Rabbit aortic EC ± Cu	F-10	TBARS, Mph, REM	Probucol	Parthasarathy et al (1986a)
Macrophage or Cu	F-10	TBARS, Mph, REM	BHT, MEM, acetyl LDL, EC modified LDL	Parthasarathy et al (1986b)
Free radicals or lipid peroxides	M-199	REM, Chol, PL, NH_2, lipid composition, TBARS, density, ketoprostaglandin F, lactate dehydrogenase	BHT, vitamin E, SOD (for Cu) serum, HDL, cell death	van Hinsbergh et al (1986)
HDL, from human EC or Cu				
$HO^\bullet$, $HO_2^\bullet$, Cu	PBS	Lipid peroxides, TBARS, REM, Mph		Bedwell and Jessup (1987)
Autoxidation	PBS	REM, FA, tocopherols, Fl, aldehydes		Esterbauer et al (1987)
O_2^- from monkey arterial SMC + cystine + Cu or Fe	MEM	TBARS, Apo B, Mph, REM, O_2^-	EDTA, DTPA, SOD	Heinecke et al (1987), Heinecke (1987)
O_2^- from activated human monocytes ± Fe	RPMI	TBARS, Mph, REM, O_2^-	EDTA, DF, BHT, SOD, Cat	Hiramatsu et al (1987)
Thiols + Cu or Fe	F-10, PBS	TBARS, Mph	EDTA, BHT	Parthasarathy (1987)
Lipid peroxide decomposition products	PBS	Mph, REM, Fl, amino acids		Steinbrecher (1987)

Table I. (continued)

LDL modifier	Medium	Parameters measured	Effective inhibitor	Reference
Rabbit aortic EC or Cu	F-10	TBARS, Mph, REM, Apo B, NH_2, Chol, TG, DL		Steinbrecher et al (1987)
Lipoxygenase + phospholipase A_2	Borate, buffered saline	TBARS, Mph, REM, Apo B, fibroblast recognition		Sparrow et al (1988)
O^-_2 from rabbit arterial EC and SMC, monkey arterial SMC, human skin fibroblasts	DME, F-10	TBARS, REM, Mph, H_2O_2, O^-_2, SOD, Cat, GSH peroxidase	EDTA, DF, DME, M-199, RPMI, phenol red, ascorbate, serum, SOD, low O_2	Steinbrecher (1988)
Cigarette smoke extract	PBS	TBARS, REM, Mph, Apo B	SOD	Yokode et al (1988)
$HO^\bullet$, $HO_2^\bullet$, O^-_2 + Cu	PBS	Apo B, Mph, REM, vitamin E, lipid peroxides		Bedwell et al (1989)
Activated human monocytes	RPMI	TBARS, CT, O^-_2	BHT, vitamin E, SOD	Cathcart et al (1989)
Cigarette smoking + Bovine aortic SMC	F-10	TBARS, Mph, Apo B		Harats et al (1989)
Copper ion	PBS	Apo B structure, F1, NH_2, heparin binding		McLean and Hagaman (1989)

Products from autoxidizing fatty acids or aldehydes	PBS	TBARS, REM, Chol, PL, Fl, FA, NH_2, antiserum binding		Steinbrecher et al (1989)
Cu or Fe	PBS	NMR spectra		Barenghi et al (1990)
V^{4+}	PBS	TBARS, Fl, Apo B	DTA, DF, ethanol, 2-propanol	Dickson and Stern (1990)
Cu	PBS	Mph, Fl, UV absorption	Probucol, 17β–oestradiol	Huber et al (1990)
Fe + macrophages or Cu or autoxidation	F-10, PBS	Mph, vitamin E, lipid peroxides	Tocopherol, flavonoid (morin)	Jessup et al (1990)
MDA, acetylation, methylation, rabbit aorta EC	F-10, PBS	TBARS, REM, Mph, Apo B, PC, UV spectra	HDL	Parthasarathy et al (1990)
Autoxidation	PBS	ESR	N_2	Krilov et al (1988)

Abbreviations: Apo B, apoprotein B degradation; BHA, butylated hydroxyanisole; BHT, butylated hydroxytoluene; BSA, bovine serum albumin; Cat, catalase; Chol, cholesterol; CT, cell toxicity; DF desferrioxamine; DME, Dulbecco's modified medium; DTPA, diethylene triamine pentaacetic acid; EC, endothelial cells; REM, relative electrophoretic mobility; FA, fatty acid; Fl, fluorescence; GSH, reduced glutathione; HDL, high-density lipoprotein; LDL, low-density lipoprotein; MDA, malonaldehyde; Mph, macrophage uptake or degradation; PBS, phosphate-buffered saline; PC, phosphatidylcholine; PL, phospholipid; RPMI, cell medium; SMC, smooth muscle cells; SOD, superoxide dismutase; TG, triglyceride.

The focusing of attention on the two phenomena, oxidation of LDL by some cells and its avid uptake by macrophages, led to intensive studies of both these phenomena. A selected chronological summary of studies of the modification of LDL in vitro is presented in Table 1. The summary is not intended to include every published result but rather to give an overview of the LDL-modifying agents, the parameters measured and effective inhibitors of the modification as they were first used. It is clear that the two separate discoveries of the biological significance of modified LDL (Henriksen et al, 1979, 1981; Morel et al, 1982) were quickly followed by work which unified the phenomena of modification and the toxicity of the altered LDL. This was achieved by focusing on the central significance of cell-produced free radicals (Morel et al, 1983, 1984). The mechanism of conversion of normal LDL by cells to a form recognized by the macrophage scavenger receptor has not yet been established. However, sufficient information exists for some processes to be strong candidates for inclusion in the chain of events leading to the formation of modified LDL.

It can be accepted that, put at the most basic level, the presence of the following components is necessary for a successful modification of LDL by cells: (1) viable cells; (2) oxygen; (3) transition metal ions; and (4) reducing agent or agents. There is good evidence for each of these requirements.

The need for the modifying cells to be viable while in contact with the LDL seems self-evident, but one study addressed this directly and found that only live endothelial cells were capable of modifying the LDL (van Hinsbergh et al, 1986). Similarly, since oxidation is virtually synonymous with modification of LDL, the need for oxygen can be taken for granted. There is in fact one report (Steinbrecher, 1988) showing that LDL incubated for 20 h with endothelial cells under 1% O_2 was not degraded by macrophages, even though it had a considerably elevated content of TBARS and increased electrophoretic mobility. When the O_2 level was raised to 2%, the rate of macrophage degradation was almost the same as for LDL incubated under normal air.

There is considerable evidence that transition metal ions are involved in cell-induced modifications of LDL. Many studies have shown inhibitory effects of EDTA (Steinbrecher et al, 1984; Cathcart et al, 1985; Heinecke, 1987; Heinecke et al, 1987; Hiramatsu et al, 1987; Steinbrecher, 1988), diethylene triamine pentaacetic acid (DTPA) (Heinecke et al, 1986, 1987; Steinbrecher, 1988) and desferrioxamine (DF) (Hiramatsu et al, 1987; Steinbrecher, 1988) on this process and the stimulating effect of Cu and Fe has also been consistently reported (Steinbrecher et al, 1984; Steinbrecher 1987; Heinecke et al, 1984, 1987; Heinecke 1987; Hiramatsu et al, 1987). In a study of successful modification of LDL in an RPMI medium devoid of metals, it was suggested that the monocytes themselves could be a source of ions (Cathcart et al, 1989).

The nature of the last component required in a cell system able to modify LDL, a reducing agent, is still uncertain. Initial studies (Cathcart et al, 1985; Chisholm et al, 1983), which used cytotoxicity of LDL exposed to monocytes and neutrophils as an index of modification, suggested that free radicals were the responsible agents. Both these lines of cells generate superoxide radicals and H_2O_2 which could oxidize the LDL directly or through the generation of more reactive species. Many subsequent studies provided support for this hypothesis. The strongest evidence for the involvement of $O_2^{\bar{\cdot}}$ came from observations of inhibition of the LDL-modifying action of a wide variety of cells by superoxide dismutase (SOD). Thus, Hiramatsu et al (1987) reported almost complete inhibition of LDL oxidation by monocytes when SOD was added at the beginning of the incubation. Later additions were less effective. Oxidation of LDL by monkey and human arterial smooth muscle cells in the presence of Fe or Cu was inhibited by SOD to about 10% of control values (Heinecke et al, 1986) and subsequently Heinecke et al (1987) and Heinecke (1987) confirmed that SOD was a very effective inhibitor of the conversion of LDL by the same line of cells. This report also showed that the conversion was uniquely dependent on L-cysteine.

An important contribution to these studies was the observation that oxidation of LDL by aortic endothelial cells, arterial smooth muscle cells, or skin fibroblasts, was only partly inhibitable by SOD (Steinbrecher, 1988). The author suggested that the total oxidation of LDL achieved was the sum of two components, one due to $O_2^{\bar{\cdot}}$ and the other independent of its presence. The effect of this second component varied with the conditions of incubation. This study also examined the relationship between the amount of $O_2^{\bar{\cdot}}$ generated by several lines of cells and the degree of LDL modification achieved. The results demonstrated that the greatest modification was produced by cells generating the highest amounts of $O_2^{\bar{\cdot}}$. The successful cells (rabbit aortic endothelial and smooth muscle cells, monkey aortic smooth muscle cells and human skin fibroblasts) contained less SOD and catalase than bovine aortic endothelial cells, which were unable to modify the LDL.

Further support for the role of $O_2^{\bar{\cdot}}$ in modification of LDL by cells came from studies which showed that monocytes or neutrophils activated with phorbol myristoyl acetate (Hiramatsu et al, 1987), opsonized zymosan (Cathcart et al, 1989) or lymphokine (Cathcart et al, 1985, 1986) were much more effective than the resting cells. Activation triggers an oxidative burst which releases large quantities of $O_2^{\bar{\cdot}}$ from these cells (see Chapter 15). The somewhat variable results obtained with cells phagocytosing latex particles were explained by variability in the degree of spontaneous activation occurring during the isolation of the monocytes (Cathcart et al, 1985).

The contrary view, that $O_2^{\bar{\cdot}}$ plays little, if any, role in cell-induced LDL oxidation, has also received experimental support. Some studies (van Hinsbergh et al, 1986; Montgomery et al, 1986) failed to affect the course of endothelial cell-induced oxidation of LDL, assayed by TBARS and electrophoretic mobility, by the addition of SOD. These results, and the variability in the quantitative effects of SOD observed in other studies, may perhaps be reconciled with evidence supporting a role for $O_2^{\cdot-}$ by the findings of Cathcart et al (1989), who investigated the time-dependence of the inhibition by SOD of modification of LDL by activated monocytes. The enzyme was most effective in preventing the modification when added to the cells together with the LDL and became gradually less effective when added at later times. The authors concluded that $O_2^{\cdot-}$ was essential for the initiation of the oxidation of the LDL but that further oxidation and conversion of the LDL to a cytotoxin required propagation not inhibitable by SOD, with further reactions depending on a variety of factors such as medium composition, and presence of transition metals and inhibitors (Steinbrecher, 1988). These factors are likely to vary in their effectiveness in different experiments and during the course of a typical 20–24 h incubation of LDL with cells.

The role of $O_2^{\cdot-}$ as an initiator of LDL oxidation in its own right is still unclear. The radical is unable to oxidize polyunsaturated lipids either in solution (Gebicki and Bielski, 1981) or in LDL particles (Bedwell et al, 1989). It could provide the necessary reducing power and H_2O_2 to drive the formation of hydroxyl radicals in a Fenton reaction (Halliwell and Gutteridge, 1989), but so far there is only negative evidence for these radicals in cell-induced modification of LDL. Catalase, mannitol and ethanol could not inhibit LDL oxidation (Morel et al, 1983; Heinecke et al, 1986; van Hinsbergh et al, 1986) and, while this evidence is insufficient to exclude the possible formation of $HO^{\cdot}$, it certainly lends it no support.

It appears more likely at this stage that in the initiation of oxidation $O_2^{\cdot-}$ acts as a reductant of any Cu^{2+} or Fe^{3+} present, converting them to a form able to decompose any lipid hydroperoxides present in the LDL. Sufficient hydroperoxides are likely to be present, especially in the later stages of the cell–LDL incubation, formed by the usual autocatalytic oxidation which occurs in the absence of cells. In fact, the commonly observed delay before any cell-induced modification may be due to the need for the LDL to become slightly peroxidized in cell-independent reactions, prior to the onset of more rapid oxidation requiring lipid hydroperoxides as initiating reactants. The reduced Cu or Fe formed by $O_2^{\cdot-}$ or other reducing agents released from the cells (Steinbrecher, 1988) can decompose the hydroperoxides to lipid free radicals, which then initiate oxidation of new lipid molecules. Under favourable conditions, several molecules of lipid may be oxidized for every initiation. Availability of reducing agents may be assisted by the cytotoxicity of oxidized LDL, which could lead to increases in cell permeability during the long contact period.

Based on these considerations, the mechanism of initiation of LDL oxidation by cells can be summarized as follows:
(a) formation of reduced transition metal ions

$$\text{viable cells} \longrightarrow \text{reducing agent} \xrightarrow{Fe^{3+},Cu^{2+}} Fe^{2+},Cu^{+}$$

(b) initiation of lipid peroxidation through pre-existing hydroperoxides in LDL (LDL—OOH)

$$\text{LDL—OOH} \xrightarrow{Fe^{2+},Cu^{+}} \text{LDL—O}^{\bullet} \xrightarrow{\text{LDL—H}} \text{LDL—OH} + \text{LDL}^{\bullet}$$

(c) propagating chain reaction and conversion of the PUFAs in the LDL lipids (LDL—H) to lipid hydroperoxides (LDL—OOH).

$$\text{LDL}^{\bullet} + O_2 \longrightarrow \text{LDL—OO}^{\bullet} \xrightarrow{\text{LDL—H}} \text{LDL}^{\bullet} + \text{LDL—OOH}$$

(with the $\text{LDL}^{\bullet}$ product returning to the start of the cycle)

Further LDL-modifying steps, which lead to recognition by the scavenger receptor and cytotoxicity, involve decomposition of the hydroperoxides and derivatization of the apoprotein B by some of the products (Esterbauer et al, 1990a).

Some alternative oxidizing pathways have been investigated, but not supported by experimental findings. Thus, carotene, bilirubin and diazabicyclooctane (Steinbrecher, 1988), effective singlet oxygen scavengers, failed to inhibit modification of LDL induced by endothelial cells. The involvement of cyclooxygenase of lipoxygenase in this process has been questioned, because aspirin, indomethacin and 5,8,11,14-eicosatetraenoic acid were without effect (van Hinsbergh et al, 1986). Other studies (Sparrow et al, 1988), however, showed that enzymatic modification of LDL by purified lipoxygenase plus phospholipase A_2 mimics cell-mediated oxidative modification. Additional studies by Steinberg's group appear now to support strongly an essential role of lipoxygenases in oxidative modification of LDL by cells (Parthasarathy, personal communication).

Whichever mechanism will ultimately prove to operate in the initiation of modification of LDL by cells, it is clear already that the final product has significant biological properties: acceptance by the macrophage scavenger receptor, cytotoxicity and chemotactic properties. It is widely believed that these properties can be elicited in vivo, and can lead to the onset and development of atherosclerosis in susceptible species (Steinberg et al, 1989).

V. THE ROLE OF ENDOGENOUS ANTIOXIDANTS IN RETARDING THE OXIDATION OF LDL

Early on it was observed that inclusion of high contents of vitamin E (100 nmol (mg LDL)$^{-1}$) into the medium of the culture prevented the oxidative modification of LDL by cells (Steinbrecher et al, 1984). The LDL from the vitamin E-supplemented cultures neither had increased TBARS values nor was it taken up by macrophages more rapidly than the LDL control. It was also shown that autoxidation of LDL in oxygen-saturated buffer (without addition of metal ions) only occurs when LDL is largely depleted of its endogenous vitamin E and carotenoids (Esterbauer et al, 1987). This finding led us to a systematic investigation into the importance of the endogenous antioxidants in preventing or retarding LDL oxidation in vitro (Esterbauer et al, 1990a,b, 1991). To test the oxidation resistance of LDL a standardized and highly reproducible procedure was developed; it allows the exact determination of the length of the lag period preceding the onset of rapid oxidation of LDL (Esterbauer et al, 1989c). This procedure is briefly as follows. To an EDTA-free solution of LDL in oxygen-saturated phosphate-buffered saline (PBS), pH 7.4 (0.25 mg total LDL (ml PBS)$^{-1}$), $CuCl_2$ is added as prooxidant to give a final concentration of 1.66 μM Cu^{2+}. The rate of LDL oxidation is then followed by measuring in a UV spectrometer the increase of the 234-nm diene absorption of the LDL solution. A selection of typical experiments is given in Figure 2. From the diene versus time profile the duration of the lag phase can easily be deduced. The LDL of a large number of healthy donors ($n = 73$) was investigated, the mean values for the lag phases being 74.6 ± 21.8 min; occasionally, much shorter and also much longer lag phases were found (range 44–150 min), and the first suspicion was that this might be due to variations in the antioxidant content of the LDLs. The major antioxidant contained in the LDL is α-tocopherol. Its average content is 2.6 nmol (mg total LDL)$^{-1}$, which corresponds to 6.5 molecules per LDL particle. The other substances with potential antioxidant activity identified in LDL are γ-tocopherol, β-carotene, lycopene, cryptoxanthine, zeaxanthine, lutein and phytofluene. Each of them is present in amounts of only about one-tenth that of α-tocoperol. When the loss of these antioxidants during an experiment with oxidation was followed, the same sequence of events was always found. The antioxidants disappearing first were α-tocopherol and γ-tocopherol. Thereafter, the carotenoids decreased to zero with cryptoxanthine as the first and β-carotene as the last one. Shortly after the time when LDL was depleted of vitamin E and carotenoids, the absorption of dienes rapidly began to rise, indicative of the onset of a propagating lipid peroxidation chain reaction. This sequence suggests that the lag phase and hence the oxidation resistance of LDL is mainly determined by its content of vitamin E and carotenoids. The resistance against oxidation (= lag phase) of an LDL sample then ought to correlate with the antioxidant content, or, in other words, the lag phase should be predictable from the antioxidant status. Much to our surprise no clearly significant correlation was found for the LDL from healthy donors in the age range 20–30 years. For example, the α-tocopherol content was correlated with the lag phase with $r^2 = 0.36$ ($n = 73$). This correlation was not much better if the other antioxidants were included in the statistical analysis. In a related study, Babiy et al (1990) found essentially no correlation between LDL–α-tocopherol and its oxidizability by γ-irradiation.

This weak statistical correlation does not, in our opinion, contradict the findings according to which the oxidation of LDL is always preceded by sequential loss of its antioxidants, but rather suggests that the oxidation resistance of a particular LDL depends on more than one (i.e. antioxidant content) variable. What this additional variable factor could be is presently unknown. Possible explanations could be pre-existing hydroperoxides in LDL, different PUFA content, additional unknown antioxidants, and different binding sites for the prooxidative Cu^{2+} ions.

The finding that vitamin E or carotenoids disappear before the onset of lipid peroxidation is, by itself, not proof that they act as antioxidants. Such kinetic behaviour could merely result from a lower resistance to oxidation of these compounds as compared to the PUFAs. To prove that α-tocopherol is a true antioxidant in LDL, various supplementation studies were performed by our group. The α-tocopherol content of LDL could be increased several-fold by preincubating plasma with vitamin E prior to the isolation of LDL (Esterbauer et al, 1991). The oxidation resistance of these samples of LDL linearly ($r^2 = 0.95$ or better) increased with the content of α-tocopherol. Interestingly, the plasma taken from different donors behaved differently with respect to the effectiveness of vitamin E in prolonging the lag phase. In addition to these in vitro experiments, we have also made an ex vivo study (Esterbauer et al, 1990b): volunteers (male, female; 20–30 years; clinically healthy) took RRR–α-tocopherol daily for three weeks (150, 225, 800 or 1200 IU; two persons per dose, four persons placebo). The α-tocopherol content of LDL from the volunteers and its oxidation resistance was determined two days prior to the supplementation, in five intervals during the supplementation, and one week afterwards. The major findings of this ex vivo study are: (a) The α-tocopherol of LDL and also its oxidation resistance significantly increase during the three weeks of supplementation. (b) The relative oxidation resistance (ROR), i.e. mean lag phase measured during the supplementation period; the lag phase prior to supplementation was 1.17, 1.56, 1.35, and 1.75 for the doses of 150, 225, 800, and 1200 IU vitamin E respectively. The ROR of the placebo group was 0.93. (c) The given effectiveness of vitamin E in increasing the oxidation resistance varied from person to person. For four persons the lag phases correlated with α-tocopherol highly significantly at $r^2 = 0.74–0.89$; with four other persons this correlation was less ($r^2 = 0.48–0.60$); and with the final four persons no correlation existed at all ($r^2 = 0.06–0.28$). (d) If all values for α-tocopherol and lag phase obtained in this study are treated statistically, the correlation coefficient of $r^2 = 0.51$ is obtained.

The data from the in vitro and ex vivo studies show that vitamin E is an important but not the only parameter determining the oxidation resistance of LDL. Additional studies (e.g. supplementation with β-carotene) should be made to elucidate the importance of the other antioxidants for the oxidation resistance of LDL.

ACKNOWLEDGEMENTS

The authors' work has been supported by the Association of International Cancer Research, and by the Austrian Science Foundation, project no. P6176B.

REFERENCES

Babiy A, Gebicki JM & Sullivan DR (1990) Vitamin E content and low density lipoprotein oxidizability induced by free radicals. *Atherosclerosis* **81:** 175–182.

Barenghi L, Bradamante S, Giudici GA & Vergani C (1990) NMR analysis of low-density lipoprotein oxidatively-modified in vitro. *Free Radical Res. Commun.* **8:** 175–183.

Bedwell S & Jessup W (1987) Effects of oxygen-centred free radicals on low density lipoprotein structure and metabolism. *Biochem. Soc. Trans.* **15:** 259–260.

Bedwell S, Dean RT & Jessup W (1989) The action of defined oxygen centred free radicals on human low-density lipoprotein. *Biochem. J.* **262:** 707–712.

Bermes EW & McDonald HJ (1972) LDL sensitivity to oxidation. *Ann. Clin. Lab. Sci.* **2:** 226–232.

Carew TE, Schwenke DC & Steinberg D (1987) Antiatherogenic effect of probucol unrelated to its hypocholesterolemic effect: Evidence that antioxidants in vivo can selectively inhibit low density lipoprotein degradation in macrophage-rich fatty streaks and slow the progression of atherosclerosis in the Watanabe heritable hyperlipidemic rabbit. *Proc. Natl. Acad. Sci. USA* **84:** 7725–7729.

Cathcart MK, Morel DW & Chisholm GM (1985) Monocytes and neutrophils oxidize low density lipoprotein making it cytotoxic. *J. Leukocyte Biol.* **38:** 341–350.

Cathcart MK, Chisholm GM, McNally AK & Morel DW (1986) Oxidation of low-density lipoprotein by activated human monocytes requires initiation by superoxide and propagation by other radicals. *J. Cell. Biol.* **103:** 505a.

Cathcart MK, McNally AK, Morel DW & Chisholm GM (1989) Superoxide anion participation in human monocyte-mediated oxidation of low-density lipoprotein and conversion of low density lipoprotein to a cytotoxin. *J. Immunol.* **142:** 1963–1969.

Chan HWS (1987). In Chan HWS (ed.) The mechanism of autoxidation. *Autoxidation of Unsaturated Lipids*, pp 1–16. London: Academic Press.

Chisholm GM, Hessler JR, Morel DW & Lewis LJ (1981) In vitro autoxidation renders low density lipoprotein toxic to cells in culture. *Arteriosclerosis* **1:** 359a.

Chisholm GM, Morel DW & Cathcart MK (1983) Macrophage and neutrophil oxidation of low density lipoprotein and LDL induced cell injury. *Arteriosclerosis* **3:** 472c.

Clark DA, Foulds EL & Wilson FH (1969) Effects of hydrogen peroxide on lipoprotein and associated lipids. *J. Lipid Res.* **4:** 1–8.

Dickson C & Stern A (1990) Tetravalent vanadium-mediated oxidation of low density lipoprotein. *Int. J. Biochem.* **22:** 501–506.

El-Saadani M, Esterbauer H, El-Sayed M, Goher M, Nassar AY & Jürgens G (1989) A spectrophotometric assay for lipid peroxides in serum lipoproteins using a commercially available reagent. *J. Lipid Res.* **30:** 627–630.

Esterbauer H, Jürgens G, Quehenberger O & Koller E (1987) Autoxidation of human low density lipoprotein: loss of polyunsaturated fatty acids and vitamin E and generation of aldehydes. *J. Lipid Res.* **28:** 495–509.

Esterbauer H, Quehenberger O & Jürgens G (1988) Oxidation of human low density lipoprotein with special attention to aldehydic lipid peroxidation products. In Rice-Evans C & Halliwell B (eds) *Free Radicals: Methodology and Concepts*, pp 243–268. London: The Richelieu Press.

Esterbauer H, Striegl G, Puhl H et al (1989a) The role of vitamin E and carotenoids in preventing oxidation of low density lipoproteins. *Ann. NY Acad. Sci.* **570:** 254–267.

Esterbauer H, Rotheneder M, Striegl G et al (1989b) Vitamin E and other lipophilic antioxidants protect LDL against oxidation. *Fat Sci. Technol.* **91:** 316–324.

Esterbauer H, Striegl G, Puhl H & Rotheneder M (1989c) Continuous monitoring of in vitro oxidation of human low density lipoprotein. *Free Radical Res. Commun.* **6:** 67–75.

Esterbauer H, Dieber-Rotheneder M, Waeg G, Striegl G & Jürgens G (1990a) Biochemical, structural and functional properties of oxidized low-density lipoprotein. *Chem. Res. Toxicol.* **3:** 77–92.

Esterbauer H, Dieber-Rotheneder M, Waeg G, Puhl H & Tatzber F (1990b) Endogenous antioxidants and lipoprotein oxidation. *Biochem. Soc. Trans.* **18:** 1059–1061.

Esterbauer H, Dieber-Rotheneder M, Striegl G & Waeg G (1991) Role of vitamin E in preventing the oxidation of low density lipoprotein. *Am. J. Clin. Nutr.* **53:** 3145–3215.

Evensen SA, Galdat KS & Nilsen E (1983) LDL-induced cytotoxicity and its inhibition by antioxidant treatment in cultured human endothelial cells and fibroblasts. *Arteriosclerosis* **49:** 23–30.

Fong LG, Partharasarathy S, Witztum JL & Steinberg D (1987) Nonenzymatic oxidative cleavage of peptide bonds in apoprotein B100. *J. Lipid Res.* **28:** 1466–1477.

Gebicki JM & Bielski BHJ (1981) Comparison of the capacities of the perhydroxyl and the superoxide radicals to initiate chain oxidation of linoleic acid. *J. Am. Chem. Soc.* **103:** 7020–7022.

Gofman JW, Lindgren F, Elliott H et al (1950) The role of lipids and lipoproteins in atherosclerosis. *Science* **111:** 166–171.

Goldstein JL, Ho YK, Basu SK & Brown MS (1979) Binding site of macrophages that mediates uptake and degradation of acetylated low density lipoprotein, producing massive cholesterol deposition. *Proc. Natl. Acad. Sci. USA* **76:** 333–337.

Gurd PRN (1960) Some naturally occurring lipoprotein systems. In Hanahan DJ (ed.) *Lipid Chemistry*, pp 260–325. New York: John Wiley & Sons.

Haberland ME, Fong D & Cheng L (1988) Malondialdehyde-altered protein occurs in atheroma of Watanabe heritable hyperlipidemic rabbits. *Science* **241:** 215–218.

Halliwell B & Gutteridge JMC (1989) *Free Radicals in Biology and Medicine*, 2nd ed, pp 18, 44, 210. Oxford: Clarendon Press.

Harats D, Ben-Naim M, Dabach Y, Hollander G, Stein O & Stein Y (1989) Cigarette smoking renders LDL susceptible to peroxidative modification and enhanced metabolism by macrophages. *Atherosclerosis* **79:** 245–252.

Heinecke JW (1987) Free radical modification of low-density lipoprotein: Mechanisms and biological consequences. *Free Radicals Biol. Med.* **3:** 65–73.

Heinecke JW, Rosen H & Chait A (1984) Iron and copper promote modification of low density lipoprotein by human arterial smooth muscle cells in culture. *J. Clin. Invest.* **4:** 1890–1894.

Heinecke JW, Baker L, Rosen H & Chait A (1986) Superoxide mediated modification of low density lipoprotein by arterial smooth muscle cells. *J. Clin. Invest.* **77:** 757–761.

Heinecke JW, Rosen H, Suzuki LA & Chait A (1987) The role of sulfur-containing amino acids in superoxide production and modification of low density lipoprotein by arterial smooth muscle cells. *J. Biol. Chem.* **62:** 10098–10103.

Henriksen T, Evenson SA & Carlander B (1979) Injury to human endothelial cells in culture induced by low density lipoproteins. *Scand. J. Clin. Lab. Invest.* **39:** 361–368.

Henriksen T, Mahoney EM & Steinberg D (1981) Enhanced macrophage degradation of low density lipoprotein previously incubated with cultured endothelial cells: Recognition by receptors for acetylated low density lipoproteins. *Proc. Natl. Acad. Sci. USA* **78:** 6499–6503.

Henriksen T, Mahoney EM & Steinberg D (1983) Enhanced macrophage degradation of biologically modified low density lipoprotein. *Arteriosclerosis* **3:** 149–159.

Hessler JR, Morel DW, Lewis LJ & Chisholm GM (1983) Lipoprotein oxidation and lipoprotein-induced cytotoxicity. *Arteriosclerosis* **3:** 215–222.

Hiramatsu K, Rosen H, Heinecke JW, Wolfbaur G & Chait A (1987) Superoxide initiates oxidation of low density lipoprotein by human monocytes. *Arteriosclerosis* **7:** 55–60.

Huber L, Scheffler E, Poll T, Ziegler R & Dresel HA (1990) 17 beta estradiol inhibits LDL oxidation and cholesteryl ester formation in cultured macrophages. *Free Radical Res. Commun.* **8:** 167–173.

Jessup W, Jürgens G, Lang J, Esterbauer H & Dean RT (1986) Interaction of 4-hydroxynonenal-modified low-density lipoproteins with the fibroblast apolipoprotein B/E receptor. *Biochem. J.* **234:** 245–248.

Jessup W, Rankin SM, De Whalley CV, Hoult JRS, Scott J & Leake DS (1990) Alpha-tocopherol consumption during low density-lipoprotein oxidation. *Biochem. J.* **265:** 399–405.

Jürgens G, Lang J & Esterbauer H (1986) Modification of human low density lipoprotein by the lipid peroxidation product 4 hydroxynonenal. *Biochim. Biophys. Acta* **875:** 103–114.

Jürgens G, Hoff HF, Chisholm GM & Esterbauer H (1987) Modification of human serum low density lipoprotein by oxidation-characterization and pathophysiological implications. *Chem. Phys. Lipids* **45:** 315–336.

Jürgens G, Ashy A & Esterbauer H (1990) Detection of new epitopes formed upon oxidation of low density lipoprotein, lipoprotein (a) and very low density lipoprotein. *Biochem. J.* **265:** 605–608.

Kita T, Nagano Y, Yokode M et al (1987) Probucol prevents the progression of atherosclerosis in Watanabe heritable hyperlipidemic rabbit, an animal model for familial hypercholesterolemia. *Proc. Natl. Acad. Sci. USA* **84:** 5928–5931.

Krilov D, Pifat G & Herak JN (1988) Electron spin resonance spin–trapping study of peroxidation of human low density lipoprotein. *Can. J. Chem.* **66:** 1957–1960.

Lang G & Esterbauer H (1990) In Vigo-Pelfrey C (ed.) *Membrane Lipid Oxidation.* Boca Raton, FL: CRC Press.

Lee DM (1980) Malondialdehyde formation in stored plasma. *Biochem. Biophys. Res. Commun.* **95:** 1633–1672.

Lee HS & Csallany AS (1987) Measurement of free and bound malondialdehyde in vitamin-E-deficient and vitamin-E-supplemented rat-liver tissues. *Lipids* **22:** 104–107.

Lenz L, Hughes H, Mitchell JL et al (1990) Lipid hydroperoxy and hydroxy derivatives in copper-catalysed oxidation of low density lipoprotein. *J. Lipid Res.* **31:** 1043–1050.

McLean LR & Hagaman KA (1989) Effect of probucol on the physical properties of low-density lipoproteins oxidized by copper. *Biochemistry* **28:** 321–327.

Montgomery RR, Nathan CF & Cohn ZA (1986) Effects of reagent and cell-generated hydrogen peroxide on the properties of low density lipoprotein. *Proc. Natl. Acad. Sci. USA* **83:** 6631–6635.

Morel DW, Hessler JR & Chisholm, GM (1982) Low density lipoprotein (LDL) cytotoxicity induced by lipid peroxidation. *J. Cell Biol.* **95:** 262a.

Morel DW, Hessler JR & Chisholm GM (1983) Low density lipoprotein toxicity induced by free radical peroxidation of lipids. *J. Lipid Res.* **24:** 1070–1076.

Morel DW, DiCorleto PE & Chisholm GM (1984) Endothelial and smooth muscle cells alter low density lipoprotein in vitro by free radical oxidation. *Arteriosclerosis* **4:** 357–364.

Nichols AV, Rehnborg CS & Lindgren FT (1961) Gas chromatographic analysis of fatty acids from dialyzed lipoproteins. *J. Lipid Res.* **2:** 203–207.

Nishida T & Kummerow FA (1960) Interaction of serum lipoproteins with the hydroperoxide of methyl linoleate. *J. Lipid Res.* **1:** 450–458.

Palinski W, Rosenfeld ME, Ylä-Herttuala S et al (1989) Low density lipoprotein undergoes oxidative modification in vivo. *Proc. Natl. Acad. Sci. USA* **86:** 1372–1376.

Palinski W, Ylä-Herttuala S, Rosenfeld ME et al (1990) Antisera and monoclonal antibodies specific for epitopes generated during oxidative modification of low density lipoprotein. *Arteriosclerosis* **10:** 325–335.

Parthasarathy S (1987) Oxidation of low-density lipoprotein by thiol compounds leads to its recognition by the acetyl LDL receptor. *Biochim. Biophys. Acta* **917:** 337–340.

Parthasarathy S, Steinbrecher UP, Barnett J, Witztum JL & Steinberg D (1985) Essential role of phospholipase A2 activity in endothelial cell-induced modification of low density lipoprotein. *Proc. Natl. Acad. Sci. USA* **82:** 3000–3004.

Pathasarathy S, Young SG, Witztum JL, Pittman RC & Steinberg D (1986a) Probucol inhibits oxidative modification of low density lipoprotein. *J. Clin. Invest.* **7:** 641–644.

Parthasarathy S, Printz DJ, Boyd D, Joy L & Steinberg D (1986b) Macrophage oxidation of low-density lipoprotein generates a modified form recognised by the scavenger receptor. *Arteriosclerosis* **6:** 505–510.

Parthasarathy S, Fong LG, Otero D & Steinberg D (1987) Recognition of solubilized apoproteins from delipidated, oxidized low density lipoprotein (LDL) by acetyl-LDL receptor. *Proc. Natl. Acad. Sci. USA* **84:** 537–540.

Parthasarathy S, Wieland E & Steinberg DA (1989) A role for endothelial cell lipoxygenase in the oxidative modification of low density lipoprotein. *Proc. Natl. Acad. Sci. USA* **86:** 1046–1050.

Parthasarathy S, Barnett J & Fong LG (1990) High-density lipoprotein inhibits the oxidative modification of low density lipoprotein *Biochim. Biophys. Acta* **1044:** 275–283.

Pokorny J (1987) Major factors affecting the autoxidation of lipids. In Chan HWS (ed.) *Autoxidation of Unsaturated Lipids*, pp 141–206. London: Academic Press.

Quehenberger O, Koller E, Jürgens G & Esterbauer H (1987) Investigation of lipid peroxidation in human low density lipoprotein. *Free Radical Res. Commun.* **3:** 233–242.

Ray RB, Davisson EO & Crespi HL (1954) Experiments on the degradation of lipoproteins from serum. *J. Phys. Chem.* **58:** 841–846.

Rosenfeld ME, Palinski W, Ylä-Herttuala S, Butler S & Witztum JL (1990) Distribution of oxidation specific lipid–protein adducts and apolipoprotein B in atherosclerotic lesions of varying severity from WHHL rabbits. *Arteriosclerosis* **10:** 336–349.

Sato K, Niki E & Shimasaki H (1990) Free radical-mediated chain oxidation of low density lipoprotein and its synergistic inhibition by vitamin E and C. *Arch. Biochem. Biophys.* **279:** 402–405.

Schuh J, Fairclough GF & Hashemeyer RH (1978) Oxygen-mediated heterogenicity of apo-low-density lipoprotein. *Proc. Natl. Acad. Sci. USA* **75:** 3173–3177.

Sialal I, Vega GL & Grundy SM (1990) Physiologic levels of ascorbate inhibit the oxidative modification of low density lipoprotein. *Atherosclerosis* **82:** 185–191.

Sparrow CP, Parthasarathy S & Steinberg D (1988) Enzymatic modification of low density lipoprotein by purified lipoxygenase plus phospholipase A2 mimics cell-mediated oxidative modification. *J. Lipid Res.* **29:** 745–753.

Stadtman ER (1989) Oxygen radical-mediated damage of enzymes: Biological implications. In Hayaishi O, Niki E, Kondo M & Yoshikawa T (eds) *Medical, Biochemical and Chemical Aspects of Free Radicals*, pp 11–19. Amsterdam: Elsevier Science Publishers BV.

Steinberg D, Parthasarathy S, Carew TE, Khoo JC & Witztum JL (1989) Beyond cholesterol: Modifications of low-density lipoprotein that increase its atherogenecity. *N. Engl. J. Med.* **320:** 915–924.

Steinbrecher UP (1987) Oxidation of human low density lipoprotein results in derivatisation of lysine residues of apo-protein B by lipid peroxide decomposition products. *J. Biol. Chem.* **262:** 3603–3608.

Steinbrecher UP (1988) Role of superoxide in endothelial-cell modification of low-density lipoproteins. *Biochim. Biophys. Acta* **959:** 20–30.

Steinbrecher UP, Parthasarathy S, Leake DS, Witztum JL & Steinberg D (1984). Modification of low density lipoprotein by endothelial cells involves lipid peroxidation and degradation of low density lipoprotein phospholipids. *Proc. Natl. Acad. Sci. USA* **81:** 3883–3887.

Steinbrecher UP, Witztum JL, Parthasarathy S & Steinberg D (1987) Decrease in reactive amino groups during oxidation or endothelial cell modification of LDL. *Arteriosclerosis* **7:** 135–143.

Steinbrecher UP, Lougheed M, Kwan WC & Dirks M (1989) Recognition of oxidized low density lipoprotein by scavenger receptor of macrophages results from derivatisation of apolipoprotein B by products of fatty acid peroxidation. *J. Biol. Chem.* **264:** 15216–15223.

van Hinsberg VW, Scheffer M, Havekes L & Kemper HJ (1986) Role of endothelial cells and their products in the modification of low density lipoproteins. *Biochim. Biophys. Acta* **878:** 49–64.

Yokode M, Kita T, Arai H, Kawai C, Narumiya S & Fujiwara M. (1988) Cholesteryl ester accumulation in macrophages incubated with low density lipoprotein pretreated with cigarette smoke extract. *Proc. Natl. Acad. Sci. USA* **85:** 2344–2348.

Zawadzki Z, Milne RW & Marcel YL (1989) An immunochemical marker of low density lipoprotein oxidation. *J. Lipid Res.* **30:** 885–891.

15

The Phagocytes and the Respiratory Burst

M. BAGGIOLINI and M. THELEN
Theodor-Kocher Institute, University of Bern, PO Box 99, CH-3000 Bern 9, Switzerland

I. The cells 399
II. The respiratory burst 406
III. Activation of the respiratory burst 407
IV. Biochemistry of the NADPH oxidase 411
V. Phagocyte priming 413

I. THE CELLS

The three main phagocytic cells of the animal organism, neutrophil leukocytes, eosinophil leukocytes and mononuclear phagocytes, have several properties in common. They are formed in the bone marrow, are equipped with granules containing a variety of storage enzymes, migrate from the blood into the tissues in response to chemoattractant stimuli, and are able to phagocytose. These properties are required for the fulfilment of an essential function of the phagocytes, the defence of the host organism against invading microbes and parasites. Microorganisms are usually phagocytosed and killed within a phagocytic vacuole, while larger parasites, like worms, are attacked by many phagocytes simultaneously which bind firmly to the surface of the target organism in the same way as to the surface of particles to be phagocytosed. Killing depends on the generation of superoxide and derivatives and on the release of microbicidal and digestive enzymes and of cytotoxic peptides.

Oxidative Stress: Oxidants and Antioxidants
ISBN 0-12-642762-3

A. Neutrophil leukocytes

The neutrophils constitute the largest population of circulating phagocytes and function as the first line of defence against bacteria. They form in the bone marrow by a complex differentiation process lasting about 14 days. They are then released into the blood where they circulate for a few hours before migrating into the tissues where they fulfil defensive or inflammatory functions. Their circulation time is drastically shortened in infection and inflammation, when the demand for functional neutrophils in the tissues is high.

1. *Maturation*

The most prominent event in neutrophil maturation is the formation of the storage granules which are the morphological hallmark of the mature cell. A first type of granule is formed in promyelocytes. These granules contain myeloperoxidase, and their formation can thus be studied with high accuracy by ultrastructural cytochemistry (Bainton et al, 1971). Myeloperoxidase staining is observed in the rough endoplasmic reticulum, where the enzyme is produced, in the Golgi apparatus, where it is packaged, and in the granules, which constitute its final storage site (Bainton et al, 1971). It is interesting to note that the entire endoplasmic reticulum, including the perinuclear cisterna, stains positive during the whole duration of the promyelocyte stage, indicating that myeloperoxidase is produced throughout the cell. This lack of compartmentation suggests that the other enzymes known to be localized in the azurophil granules are produced and packaged concomitantly.

On the basis of their ultrastructure and the intensity of the myeloperoxidase staining, two types of granules can be distinguished in promyelocytes. One type is characterized by a crystalloid inclusion and round or football-shaped cross-section. The other is somewhat larger, has a round or ovoid cross-section and a dense matrix often surrounded by a lighter halo, and lacks crystalloid inclusions (Brederoo et al, 1983; Scott and Horn, 1970b). The peroxidase staining is usually more intense in the crystalloid-bearing granules, suggesting that they contain more of this enzyme.

The second type of granule, the specific, is formed at the myelocyte stage (Brederoo et al, 1983). These granules are peroxidase-negative and, therefore, the only structures staining for peroxidase in myelocytes are the azurophil granules. Although their contents are different it is assumed that the mechanism of formation is similar for both types of granule. Ultrastructural studies show that newly formed granules bud from the inner cisternae of the Golgi apparatus as larger vesicles which are only partially filled with matrix material and are surrounded by an undulated membrane. As they leave the Golgi region, the forming granules acquire a more uniform content and a smoother surface (Bainton et al, 1971; Brederoo et al, 1983; Scott and Horn, 1970b).

When granule formation ceases, the rough endoplasmic reticulum is gradually eliminated and the Golgi apparatus decreases markedly in size. As a consequence, the cell loses the ability to form new granules. Despite the involution of the rough endoplasmic reticulum, mature neutrophils retain some protein synthesis capabilities and even show active production of certain export proteins (Granelli-Piperno et al, 1979). It is unclear whether such synthetic activities are relevant for the function of the mature cell or simply represent a vestigial process.

Before leaving the bone marrow, the neutrophil acquires the ability to respond to chemotactic stimuli and to phagocytose, and the capacity to turn on the respiratory burst. Little information is available on the processes of functional differentiation that precede the release of the cells into the circulation. In an interesting study, Lichtman and Weed (1972) showed that the mature neutrophil has higher deformability, spreading, adhesiveness, migratory capacity and phagocytic activity than its bone marrow precursors. It was also shown that the ability of the cells to activate the respiratory burst does not develop until the metamyelocyte stage (Zakhireh and Root, 1979).

2. *The granules*

Cytochemical staining for myeloperoxidase permits us to distinguish between azurophil granules, which are positive, and specific granules, which are negative. Both granules also differ in density and mass, and can be separated by centrifugation in order to study their biochemical composition. The resolution of azurophil or primary and specific or secondary granules was obtained for several species, and the most extensive biochemical studies were performed on preparations from rabbit (Baggiolini et al, 1969, 1970) and human neutrophils (Bretz and Baggiolini, 1974; Spitznagel et al, 1974) obtained by differential sedimentation and isopycnic equilibration.

The two types of granules differ almost completely in their composition, their only common constituent being lysozyme (Table 1). The azurophil granules are very rich in lytic enzymes, including the acid hydrolases that are characteristic for lysosomes, lysozyme, and three serine proteases with a neutral pH optimum. Myeloperoxidase is the only nonhydrolytic enzyme. In terms of actual protein elastase, cathepsin G, myeloperoxidase and lysozyme are the major constituents. The specific granules contain a collagenase, lysozyme, lactoferrin and vitamin B_{12}-binding proteins. While it is clear from their enzymatic equipment that the azurophil granules are involved in both microbial killing and digestion of biological material, it is more difficult to speculate about the possible function of the specific granules. It is hoped that the discovery of new components will help us to understand the functional role of these granules which constitute the major storage compartment of the neutrophil.

Table I. Contents of azurophil and specific granules of human neutrophils.

Class	Azurophil granules	Specific granules
Microbicidal enzymes	Myeloperoxidase	
	Lysozyme	Lysozyme
Neutral proteases	Elastase	
	Cathepsin G	
	Proteinase 3	Collagenase
Acid (lysosomal) hydrolases	N-Acetyl-β-glucosaminidase	
	β-Glucuronidase	
	β-Glycerophosphatase	
	α-Mannosidase	
	Cathepsin B	
	Cathepsin D	
Other		Lactoferrin
		Vitamin B_{12}-binding proteins

Subcellular fractionation studies and exocytosis experiments have demonstrated the existence of smaller storage organelles that are unrelated to the azurophil and specific granules. These structures contain some acid hydrolases, including a relatively large proportion of cathepsin B and D and of proteinase 3, which also occur in the azurophil granules, and gelatinase (Baggiolini et al, 1980; Dewald et al, 1982). Such organelles cannot be positively identified in the microscope and, therefore, it has not been possible to establish the mode and time of their formation. In a recent study, Hibbs and Bainton (1989) described the immunohistochemical colocalization of gelatinase and lactoferrin, suggesting that gelatinase is a component of the specific granules. These conclusions contrast with the results of our own studies which revealed a dissociation between gelatinase and specific granule components (Baggiolini et al, 1980; Dewald et al, 1982). Other subcellular compartments that are subject to exocytosis were shown to contain alkaline phosphatase (Borregaard et al, 1987) and, as recently shown, tetranectin (Borregaard et al, 1990). Small organelles, and in some instances the specific granules, are considered as the stores of receptors, e.g. CR1, CR3 and the fMet-Leu-Phe receptor (Fearon and Collins, 1983; Fletcher et al, 1982; O'Shea et al, 1985), and of adhesion proteins (Bainton et al, 1987; Singer et al, 1989), which are expressed on the phagocyte surface upon stimulation.

The membrane of the specific granules appears to function as an intracellular store of cytochrome b_{558}, the terminal element of the respiratory burst oxidase (Borregaard et al, 1983; Segal and Jones, 1979).

B. Eosinophil leukocytes

Like neutrophils, the eosinophil leukocytes mature in the bone marrow, where they acquire numerous storage granules. Their progenitor cell, however, is distinct from that of neutrophils and the granules are larger, are rich in proteins that stain very strongly with eosin and normally contain large crystalloids.

1. Maturation and the granules

In maturing human eosinophils the granules are formed by a mechanism similar to that described for human neutrophils (Breton-Gorius and Reyes, 1976; Scott and Horn, 1970a). Newly formed granules are mostly spherical and contain a pale, uniform matrix which subsequently undergoes condensation, leading to the appearance of the characteristic crystalloid structures. Using cytochemical techniques, Bainton and Farquhar (1970) have made a detailed study of the process of formation and packaging of eosinophil granule proteins in rat and rabbit bone marrow. Similar observations were made subsequently in human cells (Breton-Gorius and Reyes, 1976; Zucker Franklin, 1980). Two sets of granules arise in the maturing eosinophil, primary granules at the promyelocyte stage and secondary granules, later on, at the myelocyte stage (Scott and Horn, 1970a). The secondary granules are the characteristic crystalloid containing ones. They outnumber by far the primary granules, which are formed only in small numbers and can be recognized by their morphologically uniform content and more regular shape. It was recently reported that the Charcot–Leyden crystal protein is localized exclusively in a minor population of large, crystalloid-free granules which appear to correspond to the primary granules (Dvorak et al, 1988). This protein, which forms the Charcot–Leyden crystals found in eosinophilia, is a lysophospholipase that was originally believed to be a granule membrane component (Weller et al, 1982, 1984).

Like the azurophil granules of neutrophils, the main granule population of eosinophils contains a peroxidase (which differs from myeloperoxidase) (Archer and Broome, 1963) and a number of hydrolases, including acid phosphatase (Ghidoni and Goldberg, 1967), proteinases (Davis et al, 1984), phospholipase D (Kater et al, 1976) and arylsulphatase B. The latter enzyme was believed to have a role in the cleavage and inactivation of the 'slow-reacting substance of anaphylaxis', which was shown in the 1970s to consist of sulphidoleukotrienes (Weller and Austen, 1983; Weller et al, 1980). Eosinophil granules contain three characteristic cationic proteins: MBP, the major basic protein (Gleich et al, 1976), which constitutes the crystalloid, and two proteins with ribonuclease activity, the ECP (eosinophil cationic protein) (Olsson and Venge, 1974; Olsson et al, 1977) and the EDN (eosinophil-derived neurotoxin) (Durack et al, 1981) or EXP (eosinophil protein X) (Peterson and Venge, 1983), located in the granule matrix. All three cationic proteins have been implicated in the killing of parasites (e.g. *Schistosomula mansoni*) and in cell damage (Gleich and Adolphson, 1986). EDN or EXP are also considered to be the cause of paralysis, known as the Gordon phenomenon, when injected into the brain (Durack et al, 1981; Gleich and Adolphson, 1986). The toxicity of the cationic proteins does not correlate with their ribonuclease activity. Eosinophil granules appear to contain peroxisomal enzymes like catalase (Iozzo et al, 1982) and, as reported for rat eosinophils, enzymes involved in lipid β-oxidation (Yokota et al, 1983, 1984).

C. Monocytes and macrophages

1. *Maturation*

Monocytes originate from the same stem cells as neutrophils and undergo a similar process of maturation in the bone marrow, characterized by the formation of two distinct populations of granules. Granules containing peroxidase, arylsulphatase, and acid phosphatase are produced in promonocytes which are large, immature cells similar to promyelocytes (Bainton et al, 1971). At this stage, peroxidase and acid phosphatase can be detected by cytochemistry in the rough endoplasmic reticulum, the Golgi apparatus and the forming granules. More mature cells, the bone marrow monocytes, produce granules which are peroxidase-negative. These observations are essentially in keeping with the results of an earlier study on the maturation of rabbit and guinea-pig monocytes (Nichols et al, 1971). Like the neutrophils, monocytes thus appear to have two types of cytoplasmic storage granules, one containing acid hydrolases and peroxidase and the other lacking peroxidase. With respect to time of appearance and enzymatic content, the promonocyte granules correspond to the azurophil or primary granules of neutrophil leukocytes (Bainton and Farquhar, 1968; Bainton et al, 1971; Baggiolini et al, 1970; Bretz and Baggiolini, 1974). There is no definitive information about the enzymatic content of the second type of granules, which could not be separated from primary granules by subcellular fractionation. Since they are free of peroxidase, as shown by cytochemistry, it is tempting to assume, by analogy with neutrophil leukocytes, that they also lack lysosomal hydrolases.

Despite these similarities, monocytes differ from neutrophils in several respects. They are long-lived cells, a fact that is suggested by their ultrastructural appearance: in addition to the characteristic cytoplasmic granules, they have a well-developed rough endoplasmic reticulum and Golgi apparatus and numerous mitochondria. In contrast to neutrophils, monocytes retain the capability to form secondary granules following release from the bone marrow (Nichols and Bainton, 1973).

2. *From monocytes to macrophages*

Kinetic studies have established that the blood monocytes are the precursors of the macrophages (Nichols and Bainton, 1973; van Furth et al, 1975). Monocyte maturation with the production of peroxidase-positive and -negative granules, and the subsequent differentiation of the monocytes into macrophages, were studied by Bainton and Golde (1978) in cultures of human bone marrow. Differentiating cells acquire numerous mitochondria, enlarge their endoplasmic reticulum and Golgi apparatus, and develop numerous villosities of the plasma membrane. A decrease in the number of peroxidase-positive granules accompanies these changes. The granules are discharged into phagocytic vacuoles and are not replaced. As demonstrated by staining for acid phosphatase, these cells produce large amounts of lysosomal enzymes. Lead phosphate staining is found in the rough endoplasmic reticulum, in the Golgi apparatus, in numerous small vesicles, presumably corresponding to primary lysosomes, and in secondary lysosomes.

Macrophages are found as resident cells in many different tissues. They are particularly frequent in the liver (where they are called Kupffer cells), in the lung and in the haematopoietic organs. A precise appreciation of their wide distribution in the body was provided by studies in mice, using the monoclonal antibody F4/80 which detects a macrophage-specific surface protein (Gordon, 1985). These immunocytochemical studies demonstrate a very intimate contact between macrophages and tissue cells. The dendritic appearance of the macrophages and their frequent association with endothelial cells illustrate what histologists called the reticuloendothelial system. It is important to realize that tissue macrophages form highly ramified networks in the basal layers of epithelial barriers (Hume et al, 1984) and are thus placed in a critical location for early contact with foreign antigens and invading microorganisms. While presumably unable to efficiently kill microorganisms, the tissue macrophages are likely to function as antigen presenting cells as well as producers of cytokines that activate the defence cells in the blood (Hume et al, 1983b). In addition to protective actions against non-self, tissue macrophages appear to have trophic and homeostatic functions as suggested by the observations of Hume et al (1983a) in the developing mouse retina. Tissue macrophages normally derive from immigrating monocytes, although mitotic division of resident macrophages, in particular in the lung, has been shown to occur.

Tissue macrophages must be distinguished from the inflammatory macrophages like those recruited into the peritoneal cavity of mice with irritants or certain bacteria. These are younger cells with major secretory activities and the capacity to undergo a vigorous respiratory burst response, and are chiefly involved in antimicrobial defence, granuloma formation and chronic inflammation and tissue damage.

II. THE RESPIRATORY BURST

The sudden rise of oxygen consumption in phagocytosing neutrophils was first described in the 1930s and was thought to be involved in providing energy for the phagocytic process. Its role as a source of microbicidal oxidants was recognized much later, and its critical importance for antibacterial defence became evident with the discovery of chronic granulomatous disease, a condition associated with genetic defects of the respiratory burst (Babior, 1978, 1987). The reactive oxygen metabolites that kill bacteria also cause cell and tissue damage, and these unwanted consequences of phagocyte intervention have promoted efforts to study how the burst is initiated and regulated, with the hope of finding therapeutic means to influence its activity.

O_2^- is produced by the NADPH oxidoreductase, commonly called NADPH oxidase, a membrane-bound enzyme complex that catalyses the transfer of single electrons from NADPH in the cytosol to extracellular oxygen. The binding site for NADPH is exposed at the cytosolic face while the binding site for oxygen appears to be located within the outer lipid layer of the plasma membrane (Babior et al, 1981). The electron transfer is vectorial and O_2^- is thus released into the pericellular space or within phagocytic vacuoles. NADPH is mainly supplied by the hexose monophosphate shunt (Rossi, 1986). Its affinity for the oxidase is 10 to 50 times higher than that of NADH, which does not function as electron donor under physiological conditions (Rossi, 1986).

In resting cells the NADPH oxidase is inactive and disassembled into subunits that are located in different intracellular compartments. It is rapidly reconstituted and activated upon stimulation of the cells by chemotactic agonists or phagocytosis. Its activity can be assessed as the superoxide dismutase-sensitive reduction of cytochrome *c* which is commonly used to measure O_2^- (Markert et al, 1984; Thelen et al, 1988), while the dismutation product H_2O_2 is detected by the oxidation of chromogenic, fluorogenic or luminogenic H donors catalysed by added peroxidase (Hyslop and Sklar, 1984; Pick and Mizel, 1981; Ruch et al, 1983; Wymann et al, 1987a). Under rigorous conditions, luminol-dependent chemiluminescence is proportional to the rate of H_2O_2 formation and thus directly reflects the activity of the NADPH oxidase (Wymann et al, 1987a). Nonmitochondrial oxygen consumption is a reliable measure of the respiratory burst in neutrophils and eosinophils. In monocytes, however, inhibition of mitochondrial electron transport (by sodium azide) can affect the respiratory burst (Thelen et al, 1988), as shown by the sequence:

$$4NADPH + 4O_2 \longrightarrow 4H^+ + 4O_2^- \longrightarrow 2O_2 + 2H_2O_2 \longrightarrow 3O_2 + 2H_2O$$

O_2^- and H_2O_2 are subject to dismutation. When side reactions are excluded, for each mole of oxygen that is reduced one recovers 1 mole of O_2^- or 1/2 mole of H_2O_2. Since 3/4 of the oxygen is regenerated, only 1/4 of the actual consumption is detected. For further discussion of oxygen metabolites generated by neutrophils, see Hamers and Roos (1985).

III. ACTIVATION OF THE RESPIRATORY BURST

Despite impressive advances during the past ten years, the mechanism of assembly and activation of the NADPH oxidase is still only partially understood. Most studies have dealt with neutrophils stimulated with chemotactic agonists, phagocytosable particles and a number of agents that bypass receptors. We shall describe here signalling pathways and activation steps that have been characterized in these cells. The information thus far available suggests that similar mechanisms are operating in other phagocytes as well.

A. Agonists and receptors

Under physiological conditions the respiratory burst is induced by selective chemotactic agonists that bind to plasma membrane receptors. Several of these agonists have been characterized and shown to arise from different sources and under different pathophysiological situations. The complement fragment C5a is a very potent stimulus which is formed in plasma and exudates upon complement activation via the classical or the alternative pathway (Fernandez et al, 1978). *N*-Formyl methionyl peptides, e.g. fMet-Leu-Phe, are N-terminal cleavage products of bacterial proteins (Schiffmann et al, 1975) arising at sites of bacterial infection. Two bioactive lipids, platelet-activating factor (PAF) and leukotriene B_4 (LTB_4), are of special interest since they are formed by the phagocytes themselves and, therefore, are likely to act in auto- as well as paracrine fashion (Baggiolini et al, 1988; Jenkins et al, 1980; Snyderman and Pike, 1984). Most recently, a new class of neutrophil activating peptides was described, including NAP-1/IL-8 (Baggiolini et al, 1989) and MGSA (Moser et al, 1990), which arise in the tissues, and NAP-2, which derives from precursor proteins released from activated platelets (Walz et al, 1989).

Although structurally unrelated and acting via distinct receptors, these chemotactic agonists initiate common signal transduction events and elicit the respiratory burst by a similar mechanism. Nevertheles, the magnitude and duration of the response varies considerably, suggesting that different regulatory events may be involved. fMet-Leu-Phe and C5a elicit strong and relatively long-lasting burst responses (Wymann et al, 1987b) while LTB_4, PAF and NAP-1 are considerably less potent (Baggiolini et al, 1989; Omann et al, 1987; Wymann et al, 1987b). For a given agonist, the intensity of the response is related to the number of receptors occupied. With *N*-formyl peptides maximum rates of superoxide production are reached when approximately 10% of the receptors are occupied (Skalar et al, 1985). The response is initiated by the binding, and decays rapidly when the agonist is displaced by an antagonist or by binding with an antibody (Sklar et al, 1985). These observations indicate that the response depends on the persistence of agonist binding and suggests that the active form of the oxidase is labile (see below).

B. GTP-binding proteins

The involvement of a GTP-binding protein in the NADPH oxidase activation was originally suggested by the observation that preincubation of the neutrophils with *Bordetella pertussis* toxin prevents the respiratory burst response induced by receptor agonists (Ohta et al, 1985). The toxin catalyses the ADP-ribosylation of the α-subunit of GTP-binding proteins of the inhibitory (G_i) type in the α-β-γ conformation, and the resulting covalent modification is thought to prevent the interaction between receptor and G protein (Okajima et al, 1985). Further evidence for the involvement of a GTP-binding protein came from the observation that the respiratory burst is induced and/or enhanced by fluoride (Curnutte et al, 1979) and, in electropermeabilized cells, by the nonhydrolysable analogue of GTP, GTP-γ-S (Nasmith et al, 1989). Human neutrophils possess a GTP-binding protein with a distinct α-subunit, G_{i2}-α, which is linked to myristic acid at its N-terminus (Buss et al, 1987). It has been suggested that thanks to this lipophilic adduct the α-subunit remains bound to the plasma membrane (Jones et al, 1990; Mumby et al, 1990) following dissociation from the more hydrophobic β-γ moiety (Sternweis, 1986). On the other hand, however, Rudolph et al (1989) have proposed that the G_{i2}-α may actually detach from the plasma membrane during stimulation.

C. Phospholipase C and the formation of second messengers

Upon receptor–ligand interaction and coupling of the G-protein to the receptor a phosphatidylinositol-specific phospholipase C becomes activated (Smith et al, 1986). This enzyme cleaves phosphatidylinositol 4,5-bisphosphate, generating two second messengers, 1,4,5-inositol trisphosphate (IP_3) and diacylglycerol. IP_3 is released into the cytosol, binds to specific receptors on intracellular Ca^{2+} storage organelles (Volpe et al, 1988), and induces Ca^{2+} release and a rise in cytosolic free Ca^{2+} ($[Ca^{2+}]_i$) (Pozzan et al, 1983). The other second messenger, diacylglycerol, remains associated with the membrane and activates protein kinase C. Recently, stimulus-dependent activation of phospholipase D generating phosphatidic acid, which is then dephosphorylated, was described as an additional source of diacylglycerol (Billah et al, 1989).

D. The role of Ca^{2+}

Real-time recordings show that the stimulus-dependent rise in $[Ca^{2+}]_i$ is preceded by a short, distinct lag which increases with decreasing agonist concentration and which is thought to reflect the time required for the generation of threshold concentrations of IP_3 (von Tscharner et al, 1986). Receptor stimulation also leads to Ca^{2+} influx through the plasma membrane (Meldolesi and Pozzan, 1987), which contributes to the overall $[Ca^{2+}]_i$ change, but is not essential for the response since the burst can be induced in the absence of free extracellular Ca^{2+} (Pozzan et al, 1983). IP_3 does not appear to act on the plasma membrane, and it has been suggested that the $[Ca^{2+}]_i$ rise due to the release of storage pool Ca^{2+} could control the influx through Ca^{2+}-dependent cation channels that can be demonstrated by electrophysiological techniques (van Tscharner et al, 1986).

Neutrophils can be depleted of mobilizable intracellular Ca^{2+} through the combined use of a Ca^{2+} ionophore and intra- and extracellular chelators (Di Virgilio et al, 1984; Grzeskowiak et al, 1986). In such cells, no $[Ca^{2+}]_i$ rise is observed upon stimulation with agonists or ionophores, indicating that the mobilizable storage pool is indeed depleted (Dewald et al, 1988). In addition, phospholipase C activation is prevented, as indicated by the lack of formation of IP_3 and diacylglycerol (Grzeskowiak et al, 1986). Under these conditions receptor agonist stimulation does not lead to activation of protein kinase C and of the NADPH oxidase. A respiratory burst response is obtained, however, when protein kinase C is activated by phorbol esters or exogenous diacylglycerol (Dewald et al, 1988; Grzeskowiak et al, 1986). These observations indicate that the kinase is required for the induction of the respiratory burst. Ca^{2+} depletion prevents the burst, presumably because under these conditions phospholipase C remains inactive and no diacylglycerol is formed. In contrast to the formation of diacylglycerol, the transient rise in $[Ca^{2+}]_i$ does not appear to be essential since the respiratory burst can be induced with receptor agonists even in Ca^{2+}-depleted cells, provided that phorbol esters or similar agents are added to activate protein kinase C.

E. Protein kinase C

It has long been known that phorbol myristate acetate (PMA) elicits the respiratory burst in neutrophils (Repine et al, 1974), and it was shown more recently that the same response is induced by permeant diacylglycerols (Dewald et al, 1984). A role of protein kinase C and protein phosphorylations in the induction of the respiratory burst is also suggested by the effect of protein kinase inhibitors. Staurosporine and sphingosine bases, which powerfully inhibit protein kinase C and other kinases, prevent protein phosphorylations and the respiratory burst following stimulation of neutrophils with phorbol esters, diacylglycerols and receptor agonists (Dewald et al, 1989; Wilson et al, 1986).

High-resolution recordings of H_2O_2 production show that the onset of the respiratory burst is much more rapid, and the apparent rate of activation of the NADPH oxidase considerably higher, when the cells are stimulated with agonists rather than with phorbol esters or diacylglycerols (Wymann et al, 1987b). The response is relatively more rapid when PMA is combined with a Ca^{2+} ionophore, e.g. ionomycin, but is still markedly delayed as compared with that induced by receptor agonists, indicating that receptor-mediated activation cannot be mimicked by simply raising $[Ca^{2+}]_i$ and turning on the kinase.

F. Other signal transduction events

In human neutrophils stimulated with a chemotactic agonist, Ca^{2+} is mobilized after a distinct lag (von Tscharner et al, 1986). This lag is shorter than the onset time of the respiratory burst (Wymann et al, 1987b), which means that $[Ca^{2+}]_i$ rises before the NADPH oxidase is activated. When the cells are pretreated with PMA or another protein kinase C activator, however, the onset time of the burst is drastically shortened. When neutrophils are first activated with a low concentration of PMA, leading to low-rate production of H_2O_2, and are then stimulated with fMet-Leu-Phe or another agonist, a sudden rise in the rate of H_2O_2 production is observed, that ensues without a measurable lag and clearly precedes the rise in $[Ca^{2+}]_i$. Thus, under these conditions, the receptor-mediated respiratory burst response results from a Ca^{2+}-independent mechanism (Wymann et al, 1987b). Such a signal transduction event has also been evidenced in Ca^{2+}-depleted cells, which rapidly respond to agonists (despite their inability to raise $[Ca^{2+}]_i$), provided that protein kinase C is activated by pretreatment with PMA (Grzeskowiak et al, 1986). Additional evidence comes from the effects of 17-hydroxywortmannin, a fungal metabolite which was originally shown to inhibit the respiratory burst induced by chemotactic agonists or phagocytosis, but not that induced by PMA or diacylglycerols (Baggiolini et al, 1987; Dewald et al, 1988). When Ca^{2+}-depleted neutrophils are pretreated with threshold concentrations of PMA and then stimulated with fMet-Leu-Phe, the (Ca^{2+}-independent) respiratory burst response to the agonist is blocked by 17-hydroxywortmannin. The inhibitor, on the other hand, has no influence on Ca^{2+} mobilization (in normal cells) and on the function of G-proteins, phospholipase C and protein kinase C (Dewald et al, 1988).

Taken together, these observations indicate that two distinct signal transduction sequences must be operating for the activation of the respiratory burst by chemotactic agonists. One sequence is Ca^{2+}-dependent and leads to the activation of protein kinase C, while the other is Ca^{2+}-independent and does not involve phospholipase C or protein kinase C. The effects of Ca^{2+} depletion and 17-hydroxywortmannin indicate that both sequences must act in concert in order to induce the burst.

IV. BIOCHEMISTRY OF THE NADPH OXIDASE

A. Components and assembly

Neutrophils which were previously stimulated with PMA or phagocytosable material retain their respiratory burst activity following disruption. From the homogenate the NADPH oxidase can be partially purified and the activity remains associated with the particulate fraction (McPhail et al, 1976). NADPH oxidase was successfully reconstituted in vitro using cytosol, membranes, unsaturated fatty acids, magnesium and NADPH (Bromberg and Pick, 1984; Heyneman and Vercauteren, 1984; Curnutte, 1985; McPhail et al, 1985). Sodium dodecyl sulphate can replace arachidonic acid or *cis*-unsaturated fatty acids (Babior, 1988; Bromberg and Pick, 1985). Recently, remarkable progress has been made in the understanding of the molecular composition and activation of the respiratory burst enzyme. In resting neutrophils the NADPH oxidase is disassembled and its components are located in the plasma membrane, the cytoskeleton and the cytosol (Curnutte, 1985; McPhail et al, 1985; Babior et al, 1988). The membrane contains a 66-kDa flavoprotein as intermediate electron carrier (Doussiere and Vignais, 1985; Markert et al, 1985; Segal, 1989), and a *b*-type cytochrome, termed cytochrome b_{558}, because of its absorption maximum at 558 nm (Segal and Jones, 1978, 1979). The cytochrome is a heterodimer consisting of 92-kDa and 22-kDa subunits, which were recently cloned (Royer Pokora et al, 1986; Parkos et al, 1988). The midpoint potential of the haem, which is bound to the small subunit (Nugent et al, 1989), is unusually low ($E_o' = -245$ mV) (Cross et al, 1987) and apt to reduce molecular oxygen to superoxide ($E_o' = -160$ mV) (Prince and Gunson, 1987). A low molecular weight (22-kDa), *ras*-related GTP-binding protein, termed rap-1, is associated with thecytochrome (Quinn et al, 1989). A proton channel, possibly compensating for the vectorial electron transport through the oxidase (Henderson et al, 1987, 1988), and a 45-kDa flavoprotein, which is evidenced through the binding of the NADPH oxidase inhibitor diphenyl iodonium (Cross and Jones, 1986; Yea et al, 1990), are also membrane-bound. Four cytosolic components (commonly called cytosolic factors) with isoelectric points of 3.1, 6.1, 7 and approximately 10 (Curnutte et al, 1989) have been identified. The primary structures of two of them, p47 and p67, have been deduced from the corresponding cDNA, and both factors obtained by recombinant techniques could substitute for the cytosol in the cell-free reconstitution of NADPH oxidase activity (Lomax et al, 1989; Volpp et al, 1989; Leto et al, 1990). In vitro studies show the p47 is a substrate for purified protein kinase C (Kramer et al, 1988). Marked phosphorylation of p47 and its translocation to the plasma membrane together with p67 is observed in intact neutrophils following stimulation (Okamura et al, 1988; Heyworth et al, 1989; Clark et al, 1990). A 66-kDa NADPH-binding protein was also identified in the cytosol (Doussiere et al, 1986; Umei et al, 1987; Smith et al, 1989). On stimulation, this additional cytosolic factor could associate with the membrane and become part of the electron transport chain from NADPH to oxygen.

B. Chronic granulomatous disease

Chronic granulomatous disease (CGD) is a rare inherited disorder of the phagocytic cells (neutrophils, eosinophils, monocytes and macrophages) which affects single or multiple components of the NADPH oxidase (Curnutte and Babior, 1987). The defective phagocytes produce little or no superoxide and secondary reactive oxygen metabolites. As a consequence patients with CGD suffer from recurrent bacterial and fungal infections, often starting within the first year of life. X-linked and autosomal transmitted forms of the disease are known (Curnutte and Babior, 1987). In general, the cytochrome b_{558} is lacking in the X-linked form (Segal et al, 1983), but present and functional in the autosomal recessive form (Curnutte et al, 1988; Volpp et al, 1988). The recent discovery of exceptions to this rule has led to the definition of subclasses on the basis of cytochrome expression (Okamura et al, 1988). The X-linked, cytochrome b_{558}-negative form arises from a mutation of the gene coding for the larger subunit (β-chain) of the molecule (Royer Pokora et al, 1986). The mRNA of the smaller subunit (α-chain) is expressed in various tissues, but appears to be unstable in the absence of the β-chain (Parkos et al, 1988; Newburger et al, 1988). The rare X-linked, cytochrome b_{558}-positive CGD results from a Pro/His mutation of the β-subunit gene, yielding a protein with lower catalytic activity (Clark et al, 1989). The autosomal cytochrome b_{558}-positive form is due to defects of the cytosolic factors p47 and p67 (Volpp et al, 1988). Of 94 CGD patients studied, 88% lacked the 47-kDa phosphoprotein and 12% the 67-kDa phosphoprotein (Clark et al, 1989). An autosomal cytochrome b_{558}-negative form has also been described. It is caused by a mutation of the α-subunit gene (Weening et al, 1985). In most cases of X-linked disease, the level of membrane-associated flavoprotein is about half-normal (Segal, 1989). The total amount of the 66-kDa NADPH-binding flavoprotein, however, is unaffected, suggesting that other types of flavoproteins may be part of the active oxidase.

V. PHAGOCYTE PRIMING

The responsiveness of phagocytes to stimulation, and the consequent production of O_2^- and other oxygen-derived reactants, is markedly enhanced by different types of pretreatment or conditioning of the cells. This process is called priming, which means preparing for (enhanced) response. Phagocyte-priming agents include cytokines, bacterial endotoxins, bioactive lipids, protein kinase C activators and ionophores. Little is known about the mechanism of priming, but the variety of agents as well as the diversity of their effects are clear indications that enhanced responsiveness can be mediated by different intracellular processes. In fact, among the agents listed, one can distinguish those that induce a respiratory burst by themselves, e.g. PAF, phorbol esters, diacylglycerols and ionophores, from those that do not, like interferon-γ and colony-stimulating factors. Another difference lies in the time required for priming: the effect of PAF and protein kinase C activators on neutrophils and monocytes (Baggiolini et al, 1988; Dewald and Baggiolini, 1985; Dewald et al, 1984; Thelen et al, 1988) is virtually immediate, but it takes a few hours for the priming of neutrophils by granulocyte–macrophage colony-stimulating factor (Weisbart et al, 1985) and days for the priming of mononuclear phagocytes by interferon-γ (Garotta et al, 1986; Thelen et al, 1988).

The priming effects of phorbol esters, diacylglycerols and ionophores may represent, at least in part, a facilitation of signal transduction, presumably related to a rapid and persistent activation of protein kinase C, as suggested by the fact that both types of pretreatment shorten the onset time and prolong the duration of the respiratory burst response (Dewald et al, 1984; Wymann et al, 1987b). The mechanism of the effect of PAF and, to a minor extent, LTB_4 (Dewald and Baggiolini, 1985) remains unexplained. Both bioactive lipids induce by themselves a weak burst response with kinetic characteristics similar to these induced by C5a or fMet-Leu-Phe (Wymann et al, 1987b). It is unlikely that their priming effect is related to the generation of second messengers, like diacylglycerol or cytosolic free calcium, since the same is observed upon stimulation of the cells with chemotactic peptides, including NAP-1 and homologues, which are devoid of priming activity. Phagocyte priming by PAF and LTB_4 can be considered as a way to amplify host defence, since both bioactive lipids are produced by the phagocytes themselves and are likely to act in auto- or paracrine fashion to enhance the release of microbicidal products (Baggiolini et al, 1988). PAF may also be involved in the process of neutrophil priming by granulocyte–macrophage colony-stimulating factor, since pretreatment with this cytokine induces the rapid production of sufficient amounts of PAF to enhance the respiratory burst response of the cells to C5a or fMet-Leu-Phe (Wirthmueller et al, 1989).

The extensive literature describing effects of cytokines on neutrophils, including priming, has been reviewed recently (Steinbeck and Roth, 1989) and will be summarized here only briefly. Of the cytokines acting on myeloid precursor cells, interleukin-3 and granulocyte, and granulocyte–macrophage colony-stimulating factors, only the two latter ones act on mature neutrophils as well and prime them to give a strongly enhanced respiratory burst response to chemotactic peptides and phagocytosis. The colony-stimulating factors do not induce the burst directly or a rise in the levels of second messengers like calcium or diacylglycerol. Their priming effect is delayed, as already pointed out, and may involve protein synthesis. Pretreatment with the classical inflammatory cytokines, interleukin-1α and -1β, and tumour necrosis factor-α, also prime neutrophils to enhanced burst responses. Unlike interleukin-1, tumour necrosis factor-α directly induces the formation of O_2^- and H_2O_2. The priming effect of interferon-γ was studied extensively in macrophages. In these cells, the NADPH oxidase has about only 1/10th of the affinity for its substrate NADPH than in circulating monocytes. Treatment with interferon-γ restores high affinity and enhances burst responsiveness (Cassatella et al, 1985). Similar results are obtained with murine macrophages (Tsunawaki and Nathan, 1984). Interferon-γ also enhances the respiratory burst capacity of human monocytes—by a mechanism that does not involve changes in the K_m of the oxidase (Thelen et al, 1988)—and of neutrophils. In the latter instance, priming is dependent on protein synthesis (Berton et al, 1986; Cassatella et al, 1988).

Endotoxin (LPS) was probably the first priming agent to be studied. Effects were reported in neutrophils (see Guthrie et al (1984) for references), and newer information about the mechanism of priming in general came from studies on the LPS effect on macrophages (Aderem, 1988). LPS priming of macrophages correlates with the potentiation of protein kinase C-dependent responses, such as enhanced arachidonic acid liberation and metabolism, and the induction and myristoylation of a specific protein kinase C substrate (Aderem et al, 1988). A similar effect was recently observed in human neutrophils pretreated with tumour necrosis factor (Thelen et al, 1990).

Phagocyte priming is of practical importance in the context of cytokine therapy. It must be borne in mind that colony-stimulating factors that are administered to promote myelopoiesis affect the responsiveness of circulating granulocytes and can thus enhance antimicrobial defence as well as the likelihood of inflammatory side effects (Steinbeck and Roth, 1989). A similar situation may be encountered upon administration of interferon-γ. Priming of the phagocytes, by a mechanism that must still be clarified, could be the reason for the beneficial effect of this cytokine in CGD (Ezekowitz et al, 1988; Sechler et al, 1988).

ACKNOWLEDGEMENTS

This work was supported in part by the Swiss National Science Foundation, grant 31-25700-88. M.T. is a recipient of a fellowship (Th 399/1-1) from the

Deutsche Forschungsgemeinschaft. We thank Sabine Imer for editorial assistance.

REFERENCES

Aderem AA (1988) *J. Cell Sci.* **Suppl 9:** 151–167.
Aderem AA, Albert KA, Keum MM, Wang JKT, Greengard P & Cohn ZA (1988) *Nature (Lond.)* **332:** 362–364.
Archer RK & Broome J (1963) *Acta Haematol. (Basel)* **29:** 147–156.
Babior BM (1978) *N. Engl. J. Med.* **298:** 659–668.
Babior BM (1987) *Trends Biochem. Sci.* **12:** 241–243.
Babior BM (1988) *Hematol. Oncol. Clin. North. Am.* **2:** 201–212.
Babior BM, Curnutte JT & Okamura N (1988) *Blood* **72(5) (supplement (1):** 141a.
Babior GL, Rosin RE, McMurrich BJ, Peters WA & Babior BM (1981) *J. Clin. Invest.* **67:** 1724–1728.
Baggiolini M, Hirsch JG & de Duve C (1969) *J. Cell Biol.* **40:** 529–541.
Baggiolini M, Hirsch JG, & de Duve C (1970) *J. Cell Biol.* **45:** 586–597.
Baggiolini M, Schnyder J, Bretz U, Dewald B & Ruch W (1980) *Ciba Found Symp.* **75:** 105–121.
Baggiolini M, Dewald B, Schnyder J, Ruch W, Cooper PH & Payne TG (1987) *Exp. Cell Res.* **169:** 408–418.
Baggiolini M, Dewald B & Thelen M (1988) *Prog. Biochem. Pharmacol.* **22:** 90–105.
Baggiolini M, Walz A & Kunkel SL (1989) *J. Clin. Invest.* **84:** 1045–1049.
Bainton DF & Farquhar MG (1968) *J. Cell Biol.* **39:** 299–304.
Bainton DF & Farquhar MG (1970) *J. Cell Biol.* **45:** 54–73.
Bainton DF & Golde DW (1978) *J. Clin. Invest.* **78:** 1555–1569.
Bainton DF, Ullyot JL & Farquhar MG (1971) *J. Exp. Med.* **134:** 907–934.
Bainton DF, Miller, LJ, Kishimoto TK & Springer TA (1987) *J. Exp. Med.* **166:** 1641–1653.
Berton G, Zeni L, Cassatella MA & Rossi F (1986) *Biochem. Biophys. Res. Commun.* **138:** 1276–1282.
Billah MM, Eckel S, Mullmann TJ, Egan RW & Siegel MI (1989) *J. Biol. Chem.* **264:** 17069–17077.
Borregaard N, Heiple JM, Simons ER & Clark RA (1983) *J. Cell Biol.* **97:** 52–61.
Borregaard N, Miller LJ & Springer TA (1987) *Science* **237:** 1204–1206.
Borregaard N, Christensen L, Bjerrum OW, Birgens HS & Clemmensen I (1990) *J. Clin. Invest.* **85:** 408–416.
Brederoo P, van der Meulen J & Mommaas-Kienhuis AM (1983) *Cell Tissue Res.* **234:** 469–496.
Breton-Gorius J & Reyes F (1976) *Int. Rev. Cytol.* **46:** 251–321.
Bretz U & Baggiolini M (1974) *J. Cell Biol.* **63:** 251–269.
Bromberg Y & Pick E (1984) *Cell Immunol.* **88:** 213–221.
Bromberg Y & Pick E (1985) *J. Biol. Chem.* **260:** 13539–13545.
Buss JE, Mumby SM, Casey PJ, Gilman AG & Sefton BM (1987) *Proc. Natl. Acad. Sci. USA* **84:** 7493–7497.
Cassatella MA, Della Bianca V, Berton G & Rossi F (1985) *Biochem. Biophys. Res. Commun.* **132:** 908–914.

Cassatella MA, Cappelli R, Della Bianca V, Grzeskowiak M, Dusi S & Berton G (1988) *Immunology* **63:** 499–506.
Clark RA, Malech HL, Gallin JI et al (1989) *N. Engl. J. Med.* **312:** 647–652.
Clark RA, Volpp BD, Leidal KG & Nauseef WM (1990) *J. Clin. Invest.* **85:** 714–721.
Cross AR & Jones OT (1986) *Biochem. J.* **237:** 111–116.
Cross AR, Jones OT, Harper AM & Segal AW (1981) *Biochem. J.* **194:** 599–607.
Curnutte JT (1985) *J. Clin. Invest.* **75:** 1740–1743.
Curnutte JT & Babior BM (1987) *Adv. Hum. Genet.* **16:** 229–297.
Curnutte JT, Babior BM & Karnovsky ML (1979) *J. Clin. Invest.* **63:** 637–647.
Curnutte JT, Berkow RL, Roberts RL, Shurin SB & Scott PJ (1988) *J. Clin. Invest.* **81:** 606–610.
Curnutte JT, Scott PJ & Mayo LA (1989) *Proc. Natl. Acad. Sci. USA* **86:** 825–829.
Davis WB, Fells GA, Sun XH, Gadek JE, Venet A & Crystal RG (1984) *J. Clin. Invest.* **74:** 269–278.
Dewald B, Bretz U & Baggiolini M (1982) *J. Clin. Invest.* **70:** 518–525.
Dewald B & Baggiolini M (1985) *Biochem. Biophys. Res Commun.* **128:** 297–304.
Dewald B, Payne TG & Baggiolini M (1984) *Biochem. Biophys. Res. Commun.* **125:** 367–373.
Dewald B, Thelen M & Baggiolini M (1988) *J. Biol. Chem.* **263:** 16179–16184.
Dewald B, Thelen M, Wymann MP & Baggiolini M (1989) *Biochem. J.* **264:** 879–884.
Di Virgilio F, Lew DP & Pozzan T (1984) *Nature (Lond.)* **310:** 691–693.
Doussiere J & Vignais PV (1985) *Biochemistry* **24:** 7231–7239.
Doussiere J, Laporte F & Vignais PV (1986) *Biochem. Biophys. Res. Commun.* **139:** 85–93.
Durack DT, Ackerman SJ, Loegering DA & Gleich GJ (1981) *Proc. Natl. Acad. Sci. USA* **78:** 5165–5169.
Dvorak AM, Letourneau L, Login GR, Weller PF & Ackerman SJ (1988) *Blood* **72:** 150–158.
Ezekowitz RA, Dinauer MC, Jaffe HS, Orkin SH & Newburger PE (1988) *N. Engl. J. Med.* **319:** 146–151.
Fearon DT & Collins LA (1983) *J. Immunol.* **130:** 370–375.
Fernandez HN, Henson PM, Otani A & Hugli TE (1978) *J. Immunol.* **120:** 109–115.
Fletcher MP, Seligmann BE & Gallin JI (1982) *J. Immunol.* **128:** 941–948.
Garotta G, Talmadge KW, Pink JR, Dewald B & Baggiolini M (1986) *Biochem. Biophys. Res. Commun.* **140:** 948–954.
Ghidoni JJ & Goldberg AF (1967) *Am. J. Clin. Pathol.* **45:** 402–405.
Gleich GJ & Adolphson CR (1986) *Adv. Immunol.* **39:** 177–253.
Gleich GJ, Loegering DA, Mann KG & Maldonado JE (1976) *J. Clin. Invest.* **57:** 633–640.
Gordon S (1985) *J. Cell Sci.* **4:** 267–286.
Granelli-Piperno A, Vasalli J-D & Reich E (1979) *J. Exp. Med.* **149:** 284–289.
Grzeskowiak M, Della Bianca V, Cassatella MA & Rossi F (1986) *Biochem. Biophys. Res. Commun.* **135:** 785–794.
Guthrie LA, McPhail LC, Henson PM & Johnston RB Jr (1984) *J. Exp. Med.* **160:** 1656–1671.
Hamers MC & Roos D (1985) In Sies H (ed.) *Oxidative Stress*, pp 351–381. London: Academic Press.
Henderson LM, Chappell JB & Jones OT (1987) *Biochem. J.* **246:** 325–329.
Henderson LM, Chappell JB & Jones OT (1988) *Biochem. J.* **255:** 285–290.
Heyneman RA & Vercauteren RE (1984) *J. Leukocyte Biol.* **36:** 751–759.

Heyworth PG, Shrimpton CF & Segal AW (1989) *Biochem. J.* **260:** 243–248.
Hibbs MS & Bainton DF (1989) *J. Clin. Invest.* **84:** 1395–1402.
Hume DA, Perry VH & Gordon S (1983a) *J. Cell Biol.* **97:** 253–257.
Hume DA, Robinson AP, MacPherson GG & Gordon S (1983b) *J. Exp. Med.* **158:** 1522–1536.
Hume DA, Perry VH & Gordon S (1984) *Anat. Rec.* **210:** 503–572.
Hyslop PA & Sklar LA (1984) *Anal. Biochem.* **141:** 280–286.
Iozzo RV, MacDonald GH & Wight TN (1982) *J. Histochem. Cytochem.* **30:** 697–701.
Jenkins CS, Ali Briggs EF, Zonneveld GT, Sturk A & Clemetson KJ (1980) *Thromb. Haemostat.* **42:** 1490–1502.
Jones TLZ, Simonds WF, Merendino Jr JJ, Brann MR & Spiegel AM (1990) *Proc. Natl. Acad. Sci. USA* **87:** 568–572.
Kater LA, Goetzl EJ & Austen KF (1976) *J. Clin. Invest.* **57:** 1173–1180.
Kramer IM, Verhoeven AJ, van der Bend RL, Weening RS & Roos D (1988) *J. Biol. Chem.* **263:** 2352–2357.
Leto TL, Lomax KJ, Volpp BD et al (1990) *Science* **248:** 727–730.
Lichtman MA & Weed RI (1972) *Blood* **39:** 301–316.
Lomax KJ, Leto TL, Nunoi H, Gallin JI & Malech HL (1989) *Science* **245:** 409–412.
Markert M, Andrews PC & Babior BM (1984) *Methods Enzymol.* **105:** 358–365.
Markert M, Glass GA & Babior BM (1985) *Proc. Natl. Acad. Sci. USA* **82:** 3144–3148.
McPhail LC, DeChatelet LR & Shirley PS (1976) *J. Clin. Invest.* **58:** 774–780.
McPhail LC, Shirley PS, Clayton CC & Snyderman R (1985) *J. Clin. Invest.* **75:** 1735–1739.
Meldolesi J & Pozzan T (1987) *Exp. Cell Res.* **171:** 271–283.
Moser B, Clark-Lewis J, Zwahlen R & Baggiolini M (1990) *J. Exp. Med.* **171:** 1797–1802.
Mumby SM, Heukeroth RO, Gordon JI & Gilman AG (1990) *Proc. Natl. Acad. Sci. USA* **87:** 728–732.
Nasmith PE, Mills GB & Grinstein S (1989) *Biochem. J.* **257:** 893–897.
Newburger PE, Ezekowitz RA, Whitney C, Wright J & Orkin SH (1988) *Proc. Natl. Acad. Sci. USA* **85:** 5215–5219.
Nichols BA & Bainton DF (1973) *Lab. Invest.* **29:** 27–33.
Nichols BA, Bainton DF & Farquhar MG (1971) *J. Cell Biol.* **50:** 498–503.
Nugent JHA, Gratzer W & Segal AW (1989) *Biochem. J.* **264:** 921–924.
Ohta H, Okajima F & Ui M (1985) *J. Biol. Chem.* **260:** 15771–15780.
Okajima F, Katada T & Ui M (1985) *J. Biol. Chem.* **260:** 6761–6768.
Okamura N, Malawista SE, Roberts RL et al (1988) *Blood* **72:** 811–816.
Olsson I & Venge P (1974) *Blood* **44:** 235–246.
Olsson I, Venge P, Spitznagel JK & Lehrer RI (1977) *Lab. Invest.* **36:** 493–500.
Omann GM, Traynor AE, Harris AL & Sklar LA (1987) *J. Immunol.* **138:** 2626–2632.
O'Shea JJ, Brown EJ, Seligmann BE, Metcalf JA, Frank MM & Gallin JI (1985) *J. Immunol.* **134:** 2580–2587.
Parkos CA, Dinauer MC, Walker LE, Allen RA, Jesaitis AJ & Orkin SH (1988) *Proc. Natl. Acad. Sci. USA* **85:** 3319–3323.
Peterson CG & Venge P (1983) *Immunology* **50:** 19–26.
Pick E. & Mizel D (1981) *J. Immunol. Methods* **46:** 211–226.

Pozzan T, Lew DP, Wollheim CB & Tsien RY (1983) *Science* **211:** 1413–1415.
Prince RC & Gunson DE (1987) *Trends Biochem. Sci.* **12:** 86–87.
Quinn MT, Parkos CA, Walker L, Orkin SH, Dinauer MC & Jesaitis AJ (1989) *Nature (Lond.)* **342:** 198–200.
Repine JE, White JG, Clawson CC & Holmes BM (1974) *J. Lab. Clin. Med.* **83:** 911–920.
Rossi F (1986) *Biochim. Biophys. Acta* **853:** 65–89.
Royer Pokora B, Kunkel LM, Monaco AP et al (1986) *Nature (Lond.)* **322:** 32–38.
Ruch W, Cooper PH & Baggiolini M (1983) *J. Immunol. Methods* **63:** 347–357.
Rudolph U, Koesling D, Hinsch K-D et al (1989) *Mol. Cell. Endocrinol.* **63:** 143–153.
Schiffmann E, Corcoran BA & Wahl SM (1975) *Proc. Natl. Acad. Sci. USA* **72:** 1059–1062.
Scott RE & Horn RG (1970a) *J. Ultrastruct. Res.* **23:** 16–28.
Scott RE & Horn RG (1970b) *Lab. Invest.* **23:** 202–215.
Sechler JM, Malech HL, White CJ & Gallin JI (1988) *Proc. Natl. Acad. Sci. USA* **85:** 4874–4878.
Segal AW (1989) *J. Clin. Invest.* **83:** 1785–1793.
Segal AW & Jones OT (1978) *Nature (Lond.)* **276:** 515–517.
Segal AW & Jones OTG (1979) *Biochem. J.* **182:** 181–188.
Segal AW, Cross AR, Garcia RC et al (1983) *N. Engl. J. Med.* **308:** 245–251.
Singer II, Scott S, Kawka DW & Kazazis DM (1989) *J. Cell Biol.* **109:** 3169–3182.
Sklar LA, Hyslop PA, Oades ZG et al (1985) *J. Biol. Chem.* **260:** 11461–11467.
Smith CD, Cox CC & Snyderman R (1986) *Science* **232:** 97–100.
Smith RM, Curnutte JT & Babior BM (1989) *J. Biol. Chem.* **264:** 1958–1962.
Snyderman R & Pike MC (1984) *Annu. Rev. Immunol.* **2:** 257–281.
Spitznagel JK, Dallegri F, Leffell MS et al (1974) *Lab. Invest.* **30:** 774–785.
Steinbeck MJ & Roth JA (1989) *Rev. Infect. Dis.* **11:** 549–568.
Sternweis PC (1986) *J. Biol. Chem.* **261:** 631–637.
Thelen M, Wolf M & Baggiolini M (1988) *J. Clin. Invest.* **81:** 1889–1895.
Thelen M, Rosen A, Nairn AC and Aderem A (1990) *Proc. Natl. Acad. Sci. USA* **87:** 5603–5607.
Tsunawaki S & Nathan CF (1984) *J. Biol. Chem.* **259:** 4305–4312.
Umei T, Takeshige K & Minakami S (1987) *Biochem. J.* **243:** 467–472.
van Furth R, Langevoort HL & Schaberg A (1975) In van Furth R (ed.) *Mononuclear Phagocytes in Immunity, Infection, and Pathology*, pp 1–15. Oxford: Blackwell.
Volpe P, Krause KH, Hashimoto S et al (1988) *Proc. Natl. Acad. Sci. USA* **85:** 1091–1095.
Volpp BD, Nauseef WM & Clark RA (1988) *Science* **242:** 1295–1297.
Volpp BD, Nauseef WM, Donelson JE, Moser DR & Clark RA (1989) *Proc. Natl. Acad. Sci USA* **86:** 7195–7199.
von Tscharner V, Prodhom B, Baggiolini M & Reuter H (1986) *Nature (Lond.)* **324:** 369–372.
Walz A, Dewald B, von Tscharner V & Baggiolini M (1989) *J. Exp. Med.* **170:** 1745–1750.
Weening RS, Corbeel L, de Boer M, Lutter R, van Zwieten R & Roos D (1985) *J. Clin. Invest.* **75:** 915–920.
Weisbart RH, Golde DW, Clark SC, Wong GG & Gasson JC (1985) *Nature (Lond.)* **314:** 361–363.
Weller PF & Austen KF (1983) *J. Clin. Invest.* **71:** 114–123.

Weller PF, Wasserman SI & Austen KF (1980) In Mahmoud AA, Austen KF & Simon AS (eds) *The Eosinophil in Health and Disease*, pp 115–130. New York: Grune and Stratton.
Weller PF, Bach DS & Austen KF (1982) *J. Immunol.* **128:** 1346–1349.
Weller PF, Bach DS & Austen KF (1984) *J. Biol. Chem.* **259:** 100–105.
Wilson E, Olcott MC, Bell RM, Merrill AH Jr & Lambeth JD (1986) *J. Biol. Chem.* **261:** 12616–12623.
Wirthmueller U, De Weck AL & Dahinden CA (1989) *J. Immunol.* **142:** 3213–3218.
Wymann MP, von Tscharner V, Deranleau DA & Baggiolini M (1987a) *Anal. Biochem.* **165:** 371–378.
Wymann MP, von Tscharner V, Deranleau DA & Baggiolini M (1987b) *J. Biol. Chem.* **262:** 12048–12053.
Yea CM, Cross AR & Jones, OTG (1990) *Biochem. J.* **265:** 95–100.
Yokota S, Deimann W, Hashimoto T & Fahimi HD (1983) *Histochemistry* **78:** 425–433.
Yokota S, Tsuji H & Kato K (1984) *J. Histochem. Cytochem.* **32:** 267–274.
Zakhireh B & Root RK (1979) *Blood* **54:** 429–439.
Zucker Franklin D (1980) In Mahmoud AA, Austen KF & Simon AS (eds) *The Eosinophil in Health and Disease*, pp 43–59. New York: Grune and Stratton.

16

Oxidative Stress in Platelets

BERNHARD BRÜNE, FRANK VON APPEN and VOLKER ULLRICH
Fakultät für Biologie, Universität Konstanz, Postfach 5560, D-7750 Konstanz, Germany

I. Introduction 421
II. Biochemistry associated with platelet activation and inactivation 425
III. Oxidative stress conditions 429
IV. Conclusions 438

I. INTRODUCTION

An 'oxidative stress' situation has been defined (Sies, 1985) as an alteration of the steady-state concentrations of components of cellular redox systems in favour of the oxidized form. Of these, the levels of antioxidants like ascorbate and vitamin E and the ratios of glutathione to glutathione disulphide, NAD(P)H to $NAD(P)^+$ and protein thiol groups to protein disulphides proved significant for the cellular redox status. Oxidative stress often occurs as a consequence of toxic actions of various drugs and chemicals or their metabolites. Therefore, the preferred subject of study has so far mostly been the parenchymal liver cell, and modelling of oxidative stress conditions was achieved by application of hydroperoxides, GSH-depleting or oxidizing agents and by irradiation with UV light. Various parameters for the evaluation of oxidative stress have been established, i.e. increased intracellular calcium levels and increased energy consumption, lactate dehydrogenase (LDH) release from the cell, the formation of membrane protrusions (bleb formation) and numerous events of lipid peroxida-

Oxidative Stress: Oxidants and Antioxidants
ISBN 0-12-642762-3

tion. Here we would like to report on some aspects of oxidative stress in human platelets.

Upon activation, platelets enter into massive arachidonic acid metabolism, leading to liberation of prostaglandins; this serves to recruit and activate further platelets and thus enhances the response. The oxygenation of arachidonic acid in the cyclooxygenase and the 12-lipoxygenase pathways produces endo- and hydroperoxides which should establish a considerable oxidative challenge to the cells. Furthermore, in inflammatory conditions platelets may be subjected to external oxidative stress by exposure to hydrogen peroxide and superoxide radicals or arachidonic acid hydroperoxides released by granulocytes or macrophages. Thus, platelets are likely to face physiological oxidative stress conditions, particularly in situations which require modulation of platelet function, suggesting a regulatory function of oxidative signals. We shall, therefore, put special emphasis on the physiological effects of oxidative stress in platelets and would like to propose the underlying events as models for other cell activation processes.

A. Platelets

Unstimulated platelets circulate as biconvex discs about 2–3 μm in diameter in the bloodstream. They are formed by fragmentation of the cytoplasm of megakaryocytes in bone marrow. The count of these anucleated bodies in the peripheral blood is about 200 000–300 000 μl^{-1}, and the lifespan of human platelets is usually 7–10 days; they are removed by the reticuloendothelial system or by incorporation into haemostatic plugs. Their major function in vivo is to restrict the flow of blood from vascular lesions by adhering to discontinuities in blood vessels, protruding pseudopods, recruiting and adhering to other platelets, secreting vasoactive substances, providing a surface for the conversion of procoagulant to coagulant protein and contracting to form a compact clot in the vascular system.

These physiological responses are elicited by a chemically heterogeneous group of compounds including collagen, platelet-activating factor, vasopressin, adrenaline, ADP, serotonin, von Willebrand factor, thrombin, prostaglandin endoperoxides and thromboxane A_2. Platelets are activated by binding of these agonists to their receptors on the platelet surface, initiating an ordered sequence of various biochemcial changes which have been associated with the activation of phospholipases, activation of G-proteins, degradation of inositol phospholipids, protein phosphorylation, redistribution of calcium, arachidonic acid metabolism, actin–myosin interaction, sodium–proton exchange and receptor exposure for fibrinogen on the platelet surface.

Human platelets can be readily obtained free from other blood components by several centrifugational steps using anticoagulated blood samples. The physiological responses of platelets to stimuli can be studied using the photometric measurement of platelet aggregation (Born, 1962). This method is based on the observation that the optical density of a stirred platelet suspension falls as aggregates develop and returns towards normal as they disperse, providing a quantitative, kinetic measurement of platelet aggregation and disaggregation.

The possibility of studying physiological aggregation and secretion responses of human platelets under various experimental conditions and of correlating these variations with the underlying biochemical parameters provides a tool with which to investigate the in vivo situation. Therefore, human platelets represent a useful model for studying cell regulatory functions, especially under various oxidative stress conditions.

B. Cell morphology

The platelet plasma membrane is rich in glycoproteins and serves as a physical barrier between platelet cytoplasm and the external media. Several functions are common to the membranes of other cell types, e.g. active transport of ions and metabolites. In order to play the critical role in haemostasis, the platelet membrane surface has specialized receptors for a variety of platelet agonists and antagonists, as well as the transducing apparatus to transmit information. This allows platelets to respond to stimulating or antagonistic receptor interactions. In contrast to a resting platelet, the membrane of activated platelets becomes cohesive, so that platelets can form large aggregates.

Next to the glycocalyx membrane (Phillips and Agin, 1977a; O'Farrell, 1975) in the equatorial plane of a non-activated platelet one finds a bundle of microtubules forming a peripheral ring around the platelet. These peripheral microtubules are part of the cytoskeleton and help to maintain the platelet in its characteristic discoid shape (White, 1979). During the activation process, platelets are able to change their morphology dramatically (Allen et al, 1979). The actin–myosin interaction that occurs during activation is involved in the shape change reaction and the subsequent aggregation and secretion process (Gordon et al, 1977; Perry, 1979).

During platelet activation, mainly the contents of two different types of granular organelles are released. The contents of the various storage granules can be distinguished by density (Bentfeld and Bainton, 1975; Day et al, 1969; Siegel and Lüscher, 1967). The dense granules or dense bodies contain high concentrations of amines (serotonin and histamine), adenine nucleotides (ATP and ADP) and bivalent cations (Mg^{2+} and Ca^{2+}) (Da Prada and Picotti, 1979). Together with 5-hydroxytryptamine, approximately two-thirds of the amounts of nucleotides in platelets are stored within these granules. They are only exchanging very slowly with the cytoplasmic pool of ATP and therefore are not available as a source of energy. After secretion, these constituents are released and affect vascular tone and permeability as well as the development of a haemostatic plug.

The other main type of storage organelle is the α-granule (Holmsen and Weiss, 1979) so called because it was the first platelet organelle to be observed under the light microscope. These organelles largely contain proteins, some of which are specific to platelets, and others which can also be found in the plasma and other cell types. Proteins which appear to be specific to platelets are platelet factor 4, β-thromboglobulin, low-affinity platelet factor 4 and platelet-derived growth factor. Proteins which are not exclusively found in α-granules include fibrinogen, factor VIII-related antigen and fibrinonectin. Platelets also contain lysosomal enzyme storage organelles (Holmsen and Weiss, 1979; Li et al, 1973). Finally, the smallest and most numerous granules in platelets are the glycogen granules. This reflects the fact that most of the energy required is provided by glycolysis rather than by oxidative phosphorylation (Detwiler, 1972), which is consistent with the small number of mitochondria present.

Two other structural elements should be mentioned. The plasma membrane invaginations of the open canalicular system greatly increase the platelet surface and the contents of granules are released into this system. Closely associated with the canalicular system is the dense tubular system, analogous to the endoplasmic reticulum of other cells (White, 1972). This network is the main site of arachidonate metabolism and contains the inositol 1,4,5-trisphosphate-releasable Ca^{2+} pool.

C. Platelet functions

In order to initiate the haemostatic process, platelets have to adhere to damaged blood vessel walls through their affinity for subendothelial collagen, and to accumulate on the initial adherent layer, a process which is then called aggregation (Baumgartner, 1973; Baumgartner et al, 1976). Aggregation occurs simply as a special case of adhesion, when platelets stick to other platelets rather than to a different surface.

Platelet aggregation requires platelet–platelet collision and the presence of extracellular cofactors like calcium and fibrinogen to form intercellular bridges between the platelet surfaces (Philips and Agin, 1977b; Phillips et al, 1980; Mustard et al, 1978; Bennett and Vilaire, 1979). Platelet aggregation can be induced by a variety of soluble and particulate stimuli, each interacting through a specific receptor. Parallel to the aggregatory process, secretion of the different storage organelles takes place. The mobilization of calcium ions, fibrinogen and ADP from the intracellular stores is important to promote further aggregation. After contact activation, platelets release substances to activate more platelets by way of a positive feedback in order to develop a thrombus. Normally, fibrillar collagen is the main stimulus for platelet–vascular adhesion (Zucker and Borelli, 1962; Baumgartner, 1977). After activation of platelets they lose their discoid shape, rounding up and forming long pseudopodia. This is associated with a decrease of light transmission before a large increase of light transmission in parallel with the aggregatory response takes place. Pseudopodia formation is necessary to facilitate cell–cell contacts, which are essential for platelet aggregation. There is also a release of products of the arachidonate cascade, but these compounds are directly synthesized upon activation and are not stored in organelles.

II. BIOCHEMISTRY ASSOCIATED WITH PLATELET ACTIVATION AND INACTIVATION

The haemostatic function of platelets needs to be precisely controlled to prevent other haemorrhage or thrombosis. This is achieved by compounds which are able to promote or inhibit platelet aggregation originating from the vascular endothelium and from activated platelets themselves (Holmsen, 1977). They all trigger certain intracellular signal transduction pathways. For almost all combinations of activating and inactivating agents, pronounced synergism and antagonism have been observed (Kinlough-Rathbone et al, 1977; Grant and Scrutton, 1980; Huang and Detailer, 1981). The intracellular signal transduction cascades are briefly surveyed below.

A. The phosphatidylinositol cycle and calcium

Numerous platelet agonists have been reported to cause rises in cytoplasmic Ca^{2+} levels (Rink et al, 1982) and inositol phosphate generation (Rittenhouse, 1982; Lapetina et al, 1981a). Agonists rapidly initiate degradation of the inositol phospholipids in platelets by activation of phospholipase C (PLC) (Lapetina, 1982; Nishizuka, 1983). There are several indirect lines of evidence suggesting that PLC is coupled to the receptors by a G-protein, but the exact nature of this G-protein is still unknown (Cockcroft and Gomperts, 1985; Cockcroft, 1987).

The initial activation of PLC cleaves phosphatidylinositol 4,5-bisphosphate into inositol phosphates and diacylglycerol (Michell, 1975). The latter serves mainly to activate protein kinase C (Nishizuka, 1983), while the cleavage product inositol 1,4,5-trisphosphate (IP_3) is capable of mobilizing calcium from the intracellular stores (Berridge, 1987), which results in a rapid cytosolic calcium increase. For further information about regulation of platelet PLC, involvement of G-proteins and the inositol phosphates see Berridge and Irvine (1989), Shears (1989), Abdel-Latif (1986), Cockcroft and Stutchfield (1988), Fain et al (1988) and Majerus et al (1988).

Activation of protein kinase C is associated with the predominant phosphorylation of its preferred substrate, a 47-kDa cytosolic protein. Although this protein has been isolated, its function is still unknown (Nishizuka, 1984; Imaoka et al, 1983). As seen in many other cells, activation of protein kinase C is associated with cell activation, as well as with inhibitory functions (Nishizuka, 1988; Kikkawa et al, 1989). Besides stimulation of protein kinase C, the release of intracellular Ca^{2+} increased by the second messenger IP_3 results in further activation of processes which have been associated with platelet aggregation. The activation of the Ca^{2+} calmodulin-dependent myosin light-chain kinase with subsequent phosphorylation of myosin light chain (Adelstein, 1983) and stimulation of Ca^{2+} dependent phospholipases, like phospholipase A_2 which cleaves arachidonate from lipid pools (Bills et al, 1976; Bell et al, 1979; Billah and Lapetina, 1982), are important.

B. The arachidonic acid cascade

The various agonists of platelet aggregation have different capabilities in stimulating platelet responses (Charo et al, 1977). Activation by weak agonists like ADP and collagen depends on enhancement of the stimulus by platelet-derived mediators, i.e. eicosanoids produced in the arachidonate cascade. In unstimulated platelets virtually all the arachidonic acid is present in esterified form. A prerequisite, as well as being the rate-determining step, for the generation of arachidonic acid metabolites is the cleavage of arachidonic acid from its lipid source (Marcus et al, 1969; Neufeld and Majerus, 1983; Lands and Samuelsson, 1968).

Arachidonic acid is esterified to glycerol at the glycerol sn-2 position, while phosphatidylethanolamine and phosphatidylcholine contain about 70% of the total arachidonate present in platelets (Cohen and Derksen, 1967). Once platelets are activated by various agonists, the metabolic balance shifts to favour hydrolysis of arachidonic acid and formation of eicosanoids.

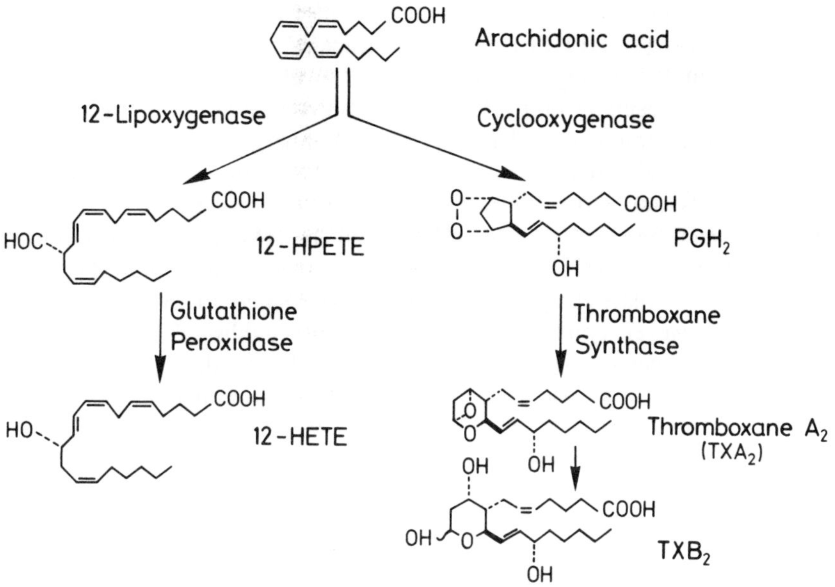

Figure 1. Oxidative metabolism of arachidonic acid in platelets.

Two major pathways exist in platelets for oxygenation of free arachidonic acid (Figure 1): the cyclooxygenase-mediated formation of prostaglandin endoperoxides PGG_2/PGH_2 with the subsequent formation of thromboxane A_2 (TXA_2) and the lipoxygenase-mediated formation of 12-hydroxyperoxy- and 12-hydroxy-5,8,10,14-eicosatetraenoic acid (Pace-Asciak and Asotra, 1989; Needleman et al, 1986; Nugteren, 1975). Cyclooxygenase and thromboxane synthase are located in the dense tubular system (Haurand and Ullrich, 1985), while the lipoxygenase is cytosolic and translocated to the membrane during activation (Baba et al, 1989). The prostaglandin endoperoxides are obligatory and short-lived intermediates for the generation of the stable prostaglandins PGE_2, $PDF_{2\alpha}$ and PGD_2 and for another class of eicosanoids, the thromboxanes. TXA_2 is formed by a P450 (haem-thiolate-containing) enzyme, the thromboxane synthase, and the biologically active compound TXA_2, with its short half-life of about 30 s, is converted to the chemically stable and biologically inactive hydrogen product TXB_2. TXA_2 is one of the most potent platelet agonists and normally functions in a self-amplification mechanism to promote the aggregatory response (Hamberg and Samuelsson, 1974; Hamberg et al, 1975; Svensson et al, 1976; Hamberg and Hamberg, 1980; Bryant et al, 1982; Lösche et al, 1984; MacIntyre et al, 1985; Friedman and Detwiler, 1975).

Platelet 12-lipoxygenase catalyses the formation of 12-hydroperoxy-eicosatetraenoic acid (12-HPETE) from free arachidonate. This hydroperoxide is tightly coupled through GSH peroxidase to glucose metabolism via the hexose monophosphate shunt (HMPS), leading to the reduction product 12-hydroxyeicosatetraenoic acid (12-HETE) (Hamberg and Hamberg, 1980). The HMPS is increased ten-fold above resting level in arachidonate-stimulated platelets, and a stoichiometric relationship of about two moles of 12-HETE formed per mole of CO_2 released by the HMPS has been reported (Bryant et al, 1982). This result implies that the consumption of GSH during early events of platelet aggregation is mainly related to the lipoxygenase pathway. However, conflicting results have been published concerning the share of lipoxygenase pathway and prostaglandin synthesis in GSH consumption and glucose oxidation. Thus, in another approach, about 70% of the flux through the HMPS has been attributed to the cyclooxygenase reaction (Lösche et al, 1984).

C. Cyclic nucleotides

Platelets are equipped to respond to hormones or agents that activate the membrane-bound adenylate cyclase or the cytosolic form of guanylate cyclase. The subsequent elevation of intracellular cyclic nucleotides (cAMP, cGMP) has in both cases been associated with inhibition of platelet activation (Friedmann and Detwiler, 1975; Schultz et al, 1977; MacIntyre et al, 1985; Böhme et al, 1977), although the mechanism of this inhibition is only poorly understood. The activation of HcAMP- or cGMP-dependent protein kinases induces the phosphorylation of specific platelet proteins. The most prominent of these are a 50-kDa protein and some low molecular weight proteins (24 kDa and/or 22 kDa) (Haslam et al, 1979; Takai et al, 1982) that have been related to inhibition of processes that decrease the Ca^{2+} concentration in the cytosol, thus inhibiting Ca^{2+}-dependent phosphorylation and Ca^{2+}-dependent activation mechanisms (Haslam et al, 1978; Fox et al, 1979; Takai et al, 1981; Yamanishi et al, 1983; Kawahara et al, 1984; Knight and Scrutton, 1984; Bushfield et al, 1985; Pannochia and Hardisty, 1985; Sage and Rink, 1985).

The stimulation of adenylate cyclase in intact cells appears to represent the mechanism by which various prostaglandins, such as PGI_2, PGE_1, PGD_2 and adenosine (A_2-receptors) raise the intracellular concentration of cAMP. Besides activation, the enzyme is also regulated by various inhibitory factors such as adrenaline (α_2-adrenoceptors), ADP, vasopressin, PAF and thrombin. Although an increase of cyclic AMP is associated with inhibition of platelet aggregation, inhibition of adenylate cyclase in intact platelets by aggregating agents (e.g. thrombin) does not seem to be a prerequisite for aggregation (Aktories and Jakobs, 1984).

For the past ten years the effect of cAMP on platelet activity has been explained by an enforced uptake of Ca^{2+} into the dense tubular system (Käser-Glanzmann et al, 1977), although this hypothesis cannot explain the inhibitory effect of cAMP in response to A23187-induced activation, where cAMP inhibits platelet activation without inhibiting cytosolic Ca^{2+} increase (Pannochia and Hardisty, 1985). These observations might suggest another effect of cAMP, distal to calcium haemostasis and protein phosphorylation. Several recent reports show an inhibitory effect of cAMP on the agonist-induced formation of inositol phosphates, 1,2-diacylglycerol and phosphatidic acid in parallel with an increased formation of phosphatidylinositol (Lapetina, 1986; Lapetina et al, 1981b; Billah and Lapetina, 1983).

The synthesis of cGMP from GTP is catalysed by the soluble form of guanylate cyclase, the predominant form of this enzyme in platelets. Activation is achieved by a whole variety of compounds, termed nitrovasodilators (nitroso compounds, nitrates and nitrate esters), which are able to release or generate nitric oxide (NO) in biological fluids, because NO was shown to activate soluble guanylate cyclase and to elevate tissue GMP levels (reviews: Waldman and Murad, 1987; Ignarro and Kadowitz, 1985; Walter, 1984; Shulz et al, 1989; Murad, 1988; Garbers, 1989). Another potent inhibitor of platelet aggregation, the endothelium-derived relaxing factor (EDRF), has been shown to be identical to NO and to cause activation of soluble guanylate cyclase and to accumulate intracellular cGMP (Chapter 20) (reviews: Furchgott and Vanhoutte, 1989; Moncada et al, 1988, 1989; Ignarro, 1989; Marletta, 1989; Collier and Vallance, 1989). The activation of soluble guanylate cyclase has also been associated with the inhibition of platelet activation. In analogy to cAMP this inhibitory activity is also associated with the 50-kDa, 24-kDa and/or 22-kDa protein phosphorylation by cGMP-dependent protein kinases, redistribution of cytosolic calcium, and the inhibition of the phosphatidylinositol response. Although activation of guanylate cyclase and inhibition of platelet aggregation by nitroso compounds are relatively well documented, there is little information about the physiological regulation of this enzyme and the biochemical events leading to platelet inactivation.

III. OXIDATIVE STRESS CONDITIONS

Considering a decreased ratio of GSH to GSSG, and of protein sulphydryls to protein disulphides, and/or decreased levels of antioxidants as an 'oxidative stress' situation for human platelets, we will discuss the underlying biochemical alterations, especially in relation to normal platelet responses. In order to investigate the mechanisms, various groups of compounds that elicit an oxidative challenge by different modes of action are at hand. Among them are model compounds previously used to study mechanisms of oxidative cell injury, like diamide or the redox cycler menadione, and also oxygen radicals and fatty acid hydroperoxides, which are physiologically present in the platelet environment and which are believed to play a crucial regulatory role during processes like inflammation or platelet aggregation itself.

A. Physiological oxidative stress in activated platelets

Unstimulated platelets maintain an intracellular concentration of 3–5 mM glutathione, with glutathione disulphide being about 5% of total glutathione. It has been shown that upon stimulation of platelet-rich plasma with arachidonate, collagen or thrombin, there is a transient decrease in the level of intracellular glutathione down to 20% (Hofmann et al, 1980; Thomas et al, 1986a). This oxidation of cellular glutathione is reversed to normal levels within a few minutes by a ten-fold increase in the net flux through the HMPS, which provides NADPH for re-reduction of glutathione disulphide via GSSG reductase (Bryant et al, 1982; Lösche et al, 1984a; Thomas et al, 1985, 1986a). This oxidative stress has been related to a stimulation of the arachidonic cascade, which leads to the formation of large amounts of eicosanoid hydroperoxides by both 12-lipoxygenase and cyclooxygenase pathways. The reduction rate of these eicosanoid metabolites initially exceeds the capacity of the HMPS to supply the amount of reducing equivalents which would be required to maintain the cellular redox balance unchanged.

Such a dramatic pertubation in the cellular redox status during normal cell function has so far not been seen with other cell systems. Therefore, the platelet might be an excellent system for study of suggested regulatory functions of oxidative changes. Several ways may exist by which the altered redox balance modifies cellular functions. Thus, a shift in the ratio of reduced to oxidized coenzymes could be the basis of allosteric regulations. An increased cellular level of GSSG may alter protein function by disulphide exchange reactions with critical protein sulphydryl groups. After exhaustion of the GSH-dependent reducing and radical-scavenging system, the remaining oxidants may act by themselves as effective regulatory messengers. The latter may apply for activation of thromboxane synthesis in stimulated platelets. Platelet cyclooxygenase was found to be fully active only at a certain level of hydroperoxides ('hydroperoxide tone'). In the time-course of platelet activation, thromboxane synthesis occurs simultaneously with the decrease in GSH concentration and ceases when the GSH concentration returns to normal levels (Thomas et al, 1986a). Impaired recovery of glutathione levels after stimulation results in prolonged thromboxane synthesis as observed in platelets of diabetic individuals (Thomas et al, 1986a). On the other hand, millimolar concentrations of extracellular glutathione and other thiols with moderate reducing capacity inhibit human platelet aggregation induced by ADP, collagen or arachidonic acid (Thomas et al, 1986b). These results suggest that GSH depletion by excessive production of hydroperoxides in the 12-lipoxygenase reaction is a prerequisite for activation of the cyclooxygenase during thromboxane synthesis. This is confirmed by the finding that thromboxane synthesis and aggregation is accelerated when a GSH-depleting agent is added to resting platelets simultaneously with an agonist (see Section III.B).

The short-term oxidative signal originating from the arachidonate cascade may also have some functions in feedback regulation of the early signal transduction events in stimulated platelets. PLC, diglyceride lipase, protein kinase C and IP_3-sensitive calcium channel are all reported to be inhibitable by oxidative modifications in platelets. For isolated protein kinase C from bovine brain and the sarcoplasmic calcium channel interesting results have been presented suggesting a biphasic regulation of these proteins (Abramson and Salama, 1989; Zaidi et al, 1989; Gopalakoishna and Anderson, 1989): under mild oxidative conditions both proteins are modified in such a way that they become active independent of their usual activators, i.e. calcium/phospholipids and IP_3, respectively. Prolonged exposure to the oxidant (Ca^{2+} channel) or the presence of the original activators (protein kinase C) then leads to inhibition. To what extent these findings are relevant for regulation of platelet function still needs to be elucidated.

B. Glutathione-oxidizing compounds

Some investigations used the thiol-oxidizing agent diamide (azo-diacarboxylic acid-bis-dimethylamide) to influence the aggregation behaviour, release and metabolism of arachidonic acid. A concentration of 0.5 mM diamide induces a reversible aggregation in platelet-rich plasma (Lösche et al, 1985). If diamide is applied simultaneously with activators like collagen, ADP or arachidonic acid, the onset of platelet aggregation is accelerated. Then, however, the aggregation becomes reversible and further activation of platelets by any agonist is completely inhibited. This inhibitory action is also seen at lower concentrations of diamide. After prolonged incubation with 0.1 mM diamide for up to 60 min the normal aggregatory response is restored, probably due to the regeneration of glutathione (Hofmann et al, 1983a; Lösche et al, 1984b; Spangenberg et al, 1984, 1987). It was found that diamide inhibits the formation of arachidonate metabolites in thrombin- or collagen-stimulated platelets. This is explained by an inhibitory effect on the release of arachidonate from membrane phospholipids, since the metabolism of exogenously added arachidonate is not diminished (Caruso et al, 1984; Lösche et al, 1984b; Hill et al, 1989a,b).

Exposure of thrombocytes to diamide also results in oxidation of sulphydryl groups present in the cytoskeleton and several other proteins. The distribution and level of organization of one of the major cytoskeletal proteins, actin, has been studied most extensively (Spangenberg et al, 1984, 1987; Hofmann et al, 1983b; Misselwitz et al, 1988). Diamide causes a redistribution of actin, reflected in an increase in cytoskeletal F-actin and a concomitant decrease in cytosolic actin. In cells where diamide treatment shows a total inhibition of the aggregatory response, the cytoskeletal-associated F-actin remains elevated. In contrast to untreated cells, where the network of filaments is subjacent to the cell membrane, the cytoskeleton in diamide-treated platelets shows electron-dense zones in the more central parts of the cytoplasm. A disturbance of cytoskeleton–membrane interaction by disulphide-linked polymer formation can be assumed, and this reorganization, associated with the inhibition of functional responses, stresses the dynamic nature of membrane–cytoskeletal interactions in shape change and aggregatory responses.

Concentrations of 0.5–5.0 mM diamide also cause a progressive decrease in polymerized tubulin in parallel to depleted acid-soluble and protein thiol groups (Steiner, 1985; Hofmann et al, 1983c). This is interesting because the collagen-induced aggregation especially shows a positive correlation with the pool of polymerized tubulin. It can be concluded from these results that the degree of polymerization of certain cytoskeletal platelet proteins, after the addition of diamide, parallels diamide-induced disaggregation and inhibition of aggregation.

The metabolism of the redox-active quinone, menadione (2-methyl-1,4-naphthoquinone) has also been associated with depletion of intracellular glutathione, and modification of protein thiols (Mirabelli et al, 1989). A dose-dependent increase of cytoskeletal-associated protein has been observed. These alterations were found to be associated with direct oxidative modifications of actin to form aggregates. At the same time platelets exhibited a sustained increase in cytosolic calcium after treatment with high concentrations of menadione. The increased levels of calcium may lead to the activation of a leupeptin-inhibitable protease which specifically degrades the actin-binding protein, which has the property of crosslinking actin filaments to form gels, and they also affect attachment of actin to the cell membrane. In all cases oxidative stress causes cytoskeletal alterations which affect cell structure by changing the normal cytoskeletal organization and by disturbing the normal cytoskeleton–plasma membrane interaction.

Depletion of intracellular glutathione by CDNB (1-chloro-2,4-dinitrobenzene) (Hill et al, 1989a,b), a substrate for GSH *S*-transferase, first used by Wahlländer and Sies (1979), shows that platelets respond normally at high agonist concentrations. These platelets are sensitized to oxidant agents such as diamide, which elicits a faster cytoskeletal protein oxidative polymerization. It seems likely that glutathione acts mainly as a reducing cofactor in human platelets to protect membrane and cytoskeletal protein thiol groups from oxidation, and these results also suggest that glutathione itself is not directly required for normal platelet aggregation or secretion.

C. Effects of hydroperoxides

The oxygenation of free arachidonate during the initial phase of platelet aggregation results in the formation of the prostaglandin endoperoxide PGG_2 and 12-HPETE. A glutathione-dependent peroxidase reduces the hydroperoxides efficiently to the corresponding hydroxy products PGH_2 and 12-HETE. In general, the peroxidase is quite efficient, and under normal, resting conditions the end-product of the lipoxygenase reaction is the alcohol (Brash, 1985). A dramatic decrease in the level of glutathione, seen after stimulation of platelets, is due to the consumption of GSH during reduction of hydroperoxides by GSH peroxidases.

However, under conditions of cell stimulation inflammatory reactions or glutathione depletion, so much of the hydroperoxy products (PGG_2/HPETE) are formed that increasing amounts escape reduction by GSH peroxidase. For example, activated platelets have been reported to release 12-HPETE and PGG_2/PGH_2 (Pace-Asciak, 1984, 1989). The latter can be utilized by the endothelium for the formation of prostacyclin (Marcus et al, 1980; Fitzgerald et al, 1984; Bordet and Lagarde, 1988). Cumene hydroperoxide, *t*-butylhydroperoxide or hydrogen peroxide influence the aggregation of human platelets in a way similar to diamide (Hofmann et al, 1983c). Addition of the compounds simultaneously with different platelet agonists to platelet-rich plasma accelerates the initial phase of aggregation, and results in a more sensitive response to the inducer; at later incubation times the aggregation becomes reversible, and this is followed by a switchover to disaggregation. A typical example for such a biphasic aggregatory response to hydrogen peroxide in combination with subthreshold concentrations of arachidonate is presented in Figure 2. Stimulation of platelets with 1.5 μM arachidonate induces complete and irreversible aggregation (Figure 2, trace

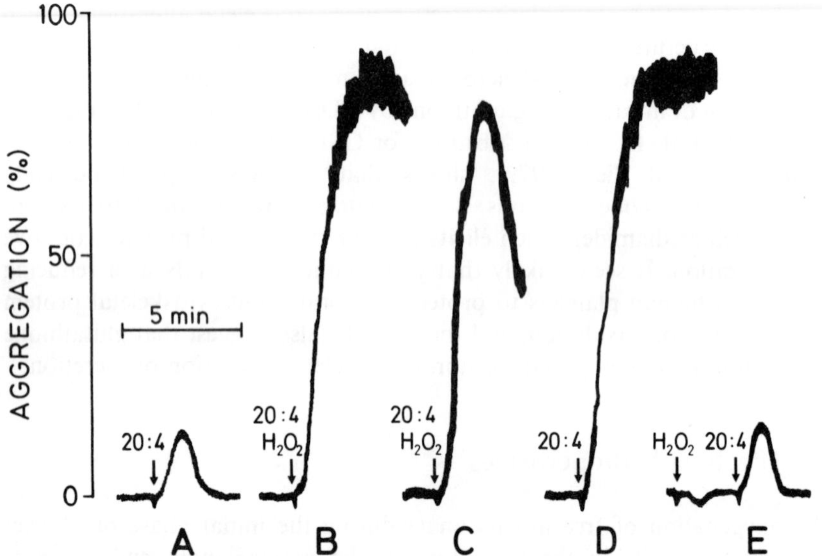

Figure 2. Effects of hydrogen peroxide on arachidonate induced platelet aggregation. Aliquots of arachidonate (20:4) and hydrogen peroxide were added to stirred suspensions of washed platelets at the times indicated by arrows. The aggregation curves represent typical light transmission–time traces monitored with an Elvi dual channel aggregometer. Final concentrations: A, 0.1 μM arachidonate; B, 0.1 μM arachidonate and 10 μM hydrogen peroxide simultaneously; C, 0.1 μM arachidonate and 100 μM hydrogen peroxide simultaneously; D, 1.5 μM arachidonate; E, 1.5 μM arachidonate after 2 min preincubation with 100 μM hydrogen peroxide. (Adapted from Hecker, G., Kupferschmidt, R. J. and Ullrich V., submitted).

D). Lower concentrations of arachidonate elicit only a weak reversible aggregatory response (trace A). Simultaneous addition of hydrogen peroxide enhances the aggregatory response (traces B and C). At higher concentrations of hydrogen peroxide, however, aggregation remains reversible (trace C). Preincubation with hydrogen peroxide for 2 min prevents arachidonate-induced aggregation (traces D and E). Several studies, including our own experiments, have shown that exogenous HPETEs are potent inhibitors of platelet aggregation. Preincubation with 12-HPETE (Figure 3) and 15-HPETE inhibited the arachidonic acid-induced aggregation of washed human platelets in a concentration-dependent fashion ($IC_{50} = 1$–$3\ \mu M$, for 12- and 15-HPETE) (Aharony et al, 1981, 1982; Vedelago and Mahadevappa, 1988). The activation process of other platelet inducers is inhibited as well, depending on the concentration of the antagonist as well as on the concentration of the agonist.

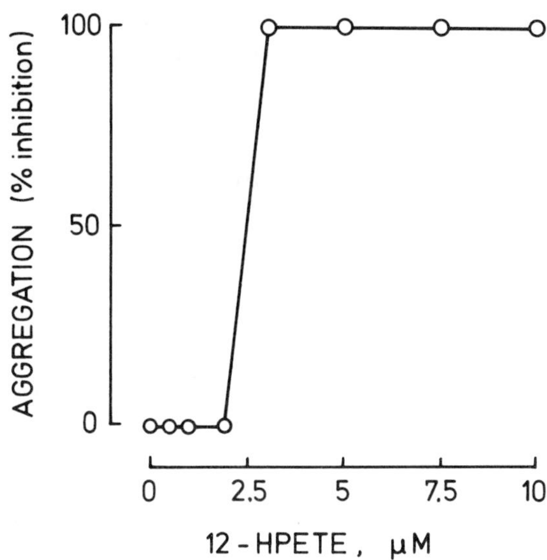

Figure 3. Inhibition of platelet aggregation by 12-HPETE. Washed human platelets (3–5 × 10^8 cells ml^{-1}) were incubated with 12-HPETE for 30 s before addition of the platelet agonist (4 μM arachidonate). After 60 s the aggregatory response was calculated and expressed as percentage inhibition caused by 12-HPETE in comparison to the maximal change in light transmission brought about by the agonist alone.

The exact inhibitory mechanism of the fatty acid hydroperoxides is still not fully elucidated. In this context inhibition of phospholipases or activation of soluble guanylate cyclase are being discussed (Vedelago and Mahadevappa, 1988; Hidaka and Asano, 1977). We would like to present some data in combination with already existing evidence to show that activation of platelet guanylate cyclase is the underlying mechanism for this inhibitory effect.

It has been known for a long time that fatty acids and unsaturated fatty acid hydroperoxides are able to stimulate purified or partially purified guanylate cyclase of different origin (Waldman and Murad, 1987; Glass et al, 1977; Gerzer et al, 1983, 1986; Thomas and Ramwell, 1986). The biochemical alterations caused by the lipoxygenase products eventually leading to the inhibition of aggregation are less well documented. Addition of 30 μM 12-HPETE or 15-HPETE to a suspension of washed platelets showed that after 30 s, 80 ± 6% of the HPETE had been reduced to the corresponding HETE; after 60 s, 97 ± 5% of the HPETE was metabolized to the reduced compound. Although this experiment demonstrates the huge capacity of platelets to reduce hydroperoxides, we were able to demonstrate that even lower concentrations of hydroxyperoxyeicosatetraenoic acids are able to stimulate guanylate cyclase in intact platelets. The results are shown in Table 1.

Table 1. Cyclic GMP formation in washed human platelets after the addition of different hydroxy(peroxy)eicosatetraenoic acids (10 μM). cGMP was determined by radioimmunoassay; samples were taken 2 min after addition of hydroperoxides (means ± SD, n = 3).

	cGMP (pmol 5 × 10^9 platelets)
Control, without addition	0.078 ± 0.006
12-HPETE	0.176 ± 0.012
15-HPETE	0.122 ± 0.002
PGG_2	0.091 ± 0.006
H_2O_2	0.160 ± 0.024
12-HETE	0.085 ± 0.007

Additional experiments revealed that only the HPETEs are able to activate guanylate cyclase, while the corresponding HETEs are inactive. 12-HPETE elevated intracellular cGMP levels 2–4-fold above controls, while 12-HETE showed no significant increase. Besides inhibition of aggregation and secretion, we also have evidence that HPETEs are able to inhibit the agonist-induced cytosolic calcium increase (Table 2) as well as the formation of phosphatidic acid. These observations are considered to be in close analogy to the alterations brought about by the classical activators of soluble guanylate cyclase in human platelets (Section II.C; Matsuoka et al, 1989; Mellion et al, 1981; Morgan and Newby, 1989; Gerzer et al, 1988).

It has previously been reported that processes involving oxidation–reduction and/or free radicals are considered important in the modulation of guanylate cyclase activity. Protein sulphydryl groups have also been implicated in influencing preparations of guanylate cyclase as well as in modifying responsiveness of the enzyme towards nitric oxide, nitroprusside and other compounds able to release nitric oxide (Waldman and Murad, 1987; Ignarro and Kadowitz, 1985; Schulz et al, 1989; Walter, 1984; Murad, 1988). Dithiothreitol prevents the spontaneous activation of crude guanylate cyclase preparations under aerobic conditions and also inhibits activation of spleen soluble guanylate cyclase by prostaglandin endoperoxides and fatty acid hydroperoxides. In contrast, diamide, glutathione disulphide and other thiol oxidants have been shown to inhibit guanylate cyclase activity and block enzyme activation by *N*-nitroso and other oxidative compounds in in vitro assays (Tsai et al, 1981).

Table 2. Inhibition of cytosolic calcium increase in aspirin-treated platelets by 12-hydroperoxyeicosatetraenoic acid (12-HPETE) after activation with 0.5 U ml^{-1} thrombin as an agonist in the presence of 1 mM extracellular calcium. Thrombin alone raised the intracellular calcium from a resting level of 90 nM to a stimulated level of 420 nM; set as 100% calcium increase.

12-HPETE (μM)	Cytosolic calcium increase (%)
0	100
5	85
10	70
15	60
20	50
30	41
40	36
50	35

Cytosolic calcium was measured using the Fura-2 method. 12-HPETE was preincubated for 2 min (typical experiment).

This information suggests that sulphydryl group modification in human platelets caused by hydroperoxy fatty acids might be responsible for the enhanced synthesis of GMP, whereas a direct reaction of nitric oxide with a sulphydryl group of soluble guanylate cyclase seems to be a less likely explanation for the activation process. The exact relationship between glutathione, mixed disulphide formation and sulphydryl group oxidation needs to be established using an intact cell system. If a situation which is defined as oxidative stress for platelets increases GMP, the balance between the activating and inhibiting signals must be known in order to influence platelet aggregation via the formation of this second messenger.

D. Effects of antioxidants

Naturally occurring antioxidants like vitamine E or C and selenium (Salonen, 1989; McCarty, 1986; Blackwell et al, 1985; McIntosh et al, 1987; Seeger et al, 1988; Toivanen, 1987; Mower and Steiner, 1982; Musca et al, 1982) as well as synthetic antioxidants like butylated hydroxytoluene (BHT) or nordihydroguaretic acid (NDGA) (Alexandre et al, 1986; Muranov et al, 1986; Takashi, 1985) have been used to influence platelet activation. All these studies report an inhibitory effect of antioxidants on platelet aggregation. Generally, the arachidonate-, thrombin- or collagen-induced aggregation was inhibited to a greater extent than the calcium ionophore A-23187-induced activation. In most cases, these observations were discussed in terms of inhibition of cyclooxygenase, leading to impaired synthesis of TXA_2. These results have also been connected to in vivo studies, performed with platelets isolated by gel filtration, which demonstrated that platelets from selenium-deficient rats aggregate to a significantly greater extent than platelets from selenium supplemented rats. Platelets from the deficient rats also produced more TXB_2 than platelets from control animals (Schoene et al, 1986).

Besides inhibition of aggregation and formation of TXB_2, antioxidants also prevent platelet membrane phosphatide breakdown, ATP secretion and the cytosolic calcium increase in aspirin-treated thrombocytes stimulated by thrombin. In addition, the same investigators found that the calcium release from the dense tubular system is affected and that the phorbol ester TPA-induced aggregation, which takes place without raising calcium at all, is also antioxidant-sensitive. Therefore, inhibition of calcium fluxes from the outside of the cell or calcium release from intracellular stores cannot alone account for the observed effects.

Antioxidants seem to desensitize platelets towards subsequent stimulation with different agonists although the final events which are responsible for the inhibitory effect are not known.

IV. CONCLUSIONS

In resting human platelets oxidative stress induces a biphasic response. An initial proaggregatory effect is later on reversed towards disaggregation and inhibition of further stimulation. Evidence for the initial stimulating action of oxidative stress is given by a low-amplitude aggregation in response to GSH-oxidizing agents and by an enhanced and accelerated agonist-induced aggregation in cases where oxidants and agonsits are added simultaneously. In fact, agonist-induced stimulation itself causes transient oxidative stress in platelets as evidenced by a reversible decrease in intracellular GSH content. This physiological oxidative stress is elicited by metabolites of the arachidonate cascade, i.e. the lipoxygenase product 12-HPETE and the endoperoxide products of the cyclooxygenase reaction. The physiological function of this unusual transient perturbation of the intracellular redox balance is still poorly understood, but some evidence exists for roles in the modulation of regulatory mechanisms during platelet activation. It is now widely accepted that the prostaglandin H synthase requires a certain 'hydroperoxide tone' for full activity. The reversible consumption of cellular GSH in the peroxidase-catalysed reduction of lipoxygenase products provides a means of maintaining transiently the required hydroperoxide tone during prostaglandin and thromboxane synthesis. On the other hand, several enzymes of the phosphoinositide cascade are reported to be sensitive to oxidants or to contain essential oxidant-sensitive thiol residues. The same applies for the arachidonate liberating enzymes diglyceride lipase and phospholipase A_2. Thus, there is room for an oxidative feedback regulation which terminates the receptor-triggered signal transduction pathway leading to release of arachidonate and Ca^{2+}.

However, to ensure normal platelet function the oxidative stress during platelet activation needs to be precisely controlled. Any superimposed exogenous oxidative stress eventually leads to inhibition of irreversible aggregation in vitro. Although a physiological relevance of this effect is not proven so far, it might be a model for inflammatory conditions in vivo, where platelets are exposed to external oxidative stress generated by granulocytes and macrophages. The oxidative inactivation of phospholipases and IP_3-sensitive calcium channels may be in part responsible for inhibition of receptor-triggered aggregation. We presented evidence that excessive fatty acid hydroperoxides activate soluble guanylate cyclase in human platelets. This enzyme is also directly activated by nitric oxide, which was found to be identical with another potent physiological inhibitor of platelet aggregation, the endothelium-derived relaxing factor (EDRF). An increased level of cGMP signals inhibition of platelet aggregation, possibly via stimulation of specific protein phosphorylation, leading to enhanced sequestration of cytosolic calcium and to inhibition of the phosphoinositide cascade. There is an analogous inhibition of platelet aggregation by an elevated level of intracellular cAMP. Thus, at least two different inhibitory intracellular signalling systems are employed in physiological regulation of platelet function: firstly, the receptor-triggered activation of adenylate cyclase upon stimulation with prostaglandins such as PGI_2 or PGE_1, and, secondly, the activation of guanylate cyclase achieved by direct interaction with EDRF or excessive fatty acid hydroperoxides.

In normal platelet activation the initial perturbation of the cellular redox balance is counterbalanced within a few minutes by an increased flux through the hexose monophosphate shunt with concomitant cessation of thromboxane synthesis. If, however, the restoration of the normal redox status is impaired by addition of GSH-oxidizing or -depleting agents, the aggregation becomes reversible and further stimulation is inhibited. The mechanisms by which this inhibitory action of prolonged oxidative stress is mediated have been investigated more thoroughly. Specific oxidation of protein sulphydryls has been demonstrated and disulphide crosslinking of cytoskeletal proteins was correlated with disaggregation and impaired membrane cytoskeleton interaction.

In summary, platelets can be considered as a suitable cell model to analyse the effects of 'physiological oxidative stress'. In fact, the term 'redox regulation' may be more appropriate to indicate the reversibility of those processes in contrast to the more toxic effects observed with redox cyclers and other initiators of lipid peroxidation.

REFERENCES

Abdel-Latif AA (1986) *Phamacol. Rev.* **38:** 227–272.

Abramson JJ & Salama G (1989) *J. Bioenerg. Biomembr.* **21:** 283–294.

Adelstein RS (1983) *J. Clin. Invest* **72:** 1863–1866.

Aharony D, Smith JB & Silver MJ (1981) *Thromb. Haemost.* **46:** 265 (abstract 0834).

Aharony D, Smith JB & Silver MJ (1982) *Biochim. Biophys. Acta* **718:** 193–200.

Aktories K & Jakobs KH (1984) *Eur. J. Biochem.* **145:** 333–338.

Alexandre A, Doni MG, Padoin E & Deana R (1986) *Biochem. Biophys. Res. Commun.* **139:** 509–514.

Allen RD, Zacharski LR, Widirstky ST, Rosenstein R, Zaitlin LM & Burgess DR (1979) *J. Cell Biol.* **83:** 126–142.

Baba A, Sakuma S, Okamoto H, Inoue T & Iwata H (1989) *J. Biol. Chem.* **264:** 15790–15795.

Baumgartner HR (1973) *Microvasc. Res.* **5:** 167–179.

Baumgartner HR (1977) *Thromb. Haemost.* **37:** 1.

Baumgartner HR, Muggli R, Tschopp TB & Turitto VT (1976) *Thromb. Haemost.* **35:** 124–138.

Bell RL, Kennerly DA, Stanford N & Majerus PW (1979) *Proc. Natl. Acad. Sci. USA* **76:** 3238–3241.

Bennett JS & Vilaire G (1979) *J. Clin. Invest.* **64:** 1393–1401.

Bentfeld ME & Binton DF (1975) *J. Clin. Invest.* **56:** 1635–1649.

Berridge MJ (1987) *Annu. Rev. Biochem.* **58:** 159–193.

Berridge MJ & Irvine RF (1989) *Nature (Lond.)* **341:** 197–205.

Billah MM & Lapetina EG (1982) *J. Biol. Chem.* **257:** 5196–5200.

Billah MM & Lapetina EG (1983) *Proc. Natl. Acad. Sci. USA* **80:** 965–968.

Bills TK, Smith JB & Silver MJ (1976) *Biochim. Biophys. Acta* **424:** 303–314.

Blackwell G-J, Radomski M & Moncada S (1985) *Thromb. Res.* **37:** 103–114.

Böhme E, Graf H, Hill HU & Argenow W (1977) *Naunyn-Schmiedeberg's Arch. Pharmacol.* **297:** R12.

Bordet JC & Lagarde M (1988) *Biochem. Pharmacol.* **37:** 3911–3914.

Born GVR (1962) *J. Physiol. (Lond.)* **162:** 67P–86P.

Bosia A, Spangenberg P, Ghigo D et al (1985) *Thromb. Res.* **37:** 423–434.

Brash AR (1985) *Circulation* **72:** 702–707.

Bryant RW, Simon TC & Bailey JM (1982) *J. Biol. Chem.* **257:** 14937–14943.

Bushfield M, McNicol A & MacIntyre DE (1985) *Biochem. J.* **232:** 267–271.

Caruso D, Galli G, Till U, Spangenberg P, Loesche W & Paoletti R (1984) *Thromb. Res.* **36:** 9–16.

Charo IF, Feinmann RD & Detwiler TC (1977) *J. Clin. Invest.* **60:** 866–873.

Cockcroft S (1987) *Trends Biochem. Sci.* **12:** 75–78.

Cockcroft S & Gomperts BD (1985) *Nature (Lond.)* **314:** 534–536.

Cockcroft S & Stutchfield, J (1988) *Phil. Trans R. Soc. Lond. (Biol.)* **320:** 247–265.
Cohen P & Derksen A (1967) *Br. J. Haematol* **17:** 359–371.
Collier J & Vallance P (1989) *TIPS* **10:** 427–431.
Da Prada M & Picotti AB (1979) *Br. J. Pharmacol.* **65:** 653–662.
Day H-J, Holmsen H & Hovig T (1969) *Scand. J. Haematol. Suppl* **7:** 3–35.
Detwiler TC (1972) *Biochim. Biophys. Acta* **256:** 163–179.
Fain JN, Wallace MA & Wojcikiewicz RJH (1988) *FASEB J.* **2:** 2569–2574.
Fitzgerald GA, Smith B, Pedersen AK & Brash AR (1984) *N. Engl. J. Med.* **310:** 1065–1068.
Fox JEB, Say AK & Haslam RJ (1979) *Biochem. J.* **184:** 651–661.
Friedmann F & Detwiler TC (1975) *Biochemistry* **14:** 1315–1320.
Furchgott RF & Vanhoutte PM (1989) *FASEB J.* **3:** 2007–2018.
Garbers DL (1989) *J. Biol. Chem.* **264:** 9103–9106.
Gerzer R, Hamet P, Ross AH, Lawson JA & Hardman JG (1983) *J. Pharmacol. Exp. Ther.* **226:** 180–186.
Gerzer R, Brash AR & Hardman JG (1986) *Biochim. Biophys. Acta* **886:** 383–389.
Gerzer R, Karrenbrock B, Siess W & Heim JH (1988) *Thromb. Res* **52:** 11–21.
Glass DB, Frey W, Carr DW & Goldberg ND (1977) *J. Biol. Chem.* **252:** 1279–1285.
Gopalakoishna R & Anderson WB (1989) *Proc. Natl. Acad. Sci. USA* **86:** 6758–6762.
Gordon DJ, Boyer JL & Korn ED (1977) *J. Biol. Chem* **252:** 8300–8309.
Grant JA & Scrutton MC (1980) *Br. J. Haematol.* **44:** 109–125.
Hamberg M & Hamberg G (1980) *Biochem. Biophys. Res Commun.* **95:** 1090–1097.
Hamberg M & Samuelsson B (1974) *Proc. Natl. Acad. Sci. USA* **71:** 3400–3404.
Hamberg M, Svensson J & Samuelsson B (1975) *Proc. Natl. Acad. Sci. USA* **72:** 2994–2998.
Haslam RJ, Davidson MML, Davis T, Lynham JA & McClenaghan MD (1978) *Adv. Cyclic Nucleotide Res.* **9:** 533–552.
Haslam RJ, Lynham JA & Fox JB (1979) *Biochem. J.* **178:** 397–406.
Haurand M & Ullrich V (1985) *J. Biol. Chem.* **260:** 15059–15067.
Hidaka H & Asano T (1977) *Proc. Natl. Acad. Sci USA* **74:** 3657–3661.
Hill TD, White JG & Rao GH (1989a) *Thromb. Res.* **53:** 457–465.
Hill TD, White JG & Rao GH (1989b) *Thromb. Res.* **53:** 447–455.
Hofmann B, Hofmann J & Till U (1983a) *Biomed. Biochim. Acta* **42:** K13–19.
Hofmann B, Danz R, Hofmann J et al (1983b) *Biomed. Biochim. Acta* **42:** 489–501.
Hofmann J, Lösche W, Till U et al (1980) *Artery* **8:** 431–436.
Hofmann J, Lösche W, Hofmann B et al *Biomed. Biochim. Acta* **42:** 479–487.
Holmsen H (1977) *Thromb. Haemost.* **38:** 1030–1041.
Holmsen H & Weiss HJ (1979) *Annu. Rev. Med.* **30:** 119–134.
Huang EM & Detwiler TC (1981) *Br. J. Haematol* **44:** 109–125.
Ignarro LJ (1989) *FASEB J.* **3:** 31–36.
Ignarro LJ & Kadowitz PJ (1985) *Annu. Rev. Pharmacol. Toxicol.* **25:** 171–191.
Imaoka T, Lynham JA & Haslam RJ (1983) *J. Biol. Chem.* **258:** 11404–11414.
Käser-Glanzmann R, Jakábová M, George JH & Lüscher EF (1977) *Biochim. Biophys. Acta* **466:** 429–440.
Kawahara Y, Yamanishi J & Fukuzaki H (1984) *Thromb. Res.* **33:** 203–209.
Kikkawa U, Kishimoto A & Nishizuka Y (1989) *Annu. Rev. Biochem.* **58:** 31–44.
Kinlough-Rathbone RL, Packham MA & Mustard JF (1977) *Thromb. Res.* **11:** 567–580.
Knight DE & Scrutton MC (1984) *Nature (Lond.)* **309:** 66–68.
Lands WEM & Samuelsson B (1968) *Biochim. Biophys. Acta* **164:** 426–429.

Lapetina EG (1982) *Trends Pharmacol. Sci.* **3:** 115–118.
Lapetina EG (1986) *FEBS Lett.* **195:** 111–114.
Lapetina EG, Billah MM & Cuatrecasas P (1981a) *J. Biol. Chem.* **256:** 5037–5040.
Lapetina EG, Billah MM & Cuatrecasas P (1981b) *Nature (Lond.)* **292:** 367–369.
Li CY, Lam KW & Yam LT (1973) *J. Histochem. Cytochem.* **21:** 1–12.
Lösche W, Pescarmona GP, Hofmann J et al (1984a) *Biomed. Biochim Acta* **43:** 1325–1328.
Lösche W, Michel E, Thielmann K & Till U (1984b) *Folia Haematol. (Leipz.)* **111:** 769–773.
Lösche W, Bosia A, Caruso D et al (1984) *Biomed. Biochim Acta* **43:** S362–S365.
Lösche W, Burgess-Wilson N, Michel E, Hoptinstall S & Till U (1985) *Thromb. Res.* **40:** 869–874.
MacIntyre DE, Bushfield M & Shaw AM (1985) *FEBS Lett.* **188:** 383–388.
Majerus PW, Connolly TM, Bansal VS, Inhorn RC, Ross TS & Lips DL (1988) *J. Biol. Chem.* **263:** 3051–3054.
Marcus AJ, Ullman HL & Safier LB (1969) *J. Lipid Res.* **10:** 108.
Marcus A, Weksler B, Jaffe FA & Broekman MJ (1980) *J. Clin. Invest.* **66:** 979–986.
Marletta MA (1989) *TIBS* **14:** 488–492.
Matsuoka J, Nakahata N & Nakanishi H (1989) *Biochem. Pharmacol.* **38:** 1841–1847.
McCarty MF (1986) *Med. Hypotheses* **19:** 345–357.
McIntosh GH, Bulman FH, Looker JW, Russell GR & James M (1987) *J. Nutr. Sci. Vitaminol. (Tokyo)* **33:** 299–312.
Mellion BT, Ignarro LJ, Ohlstein EH, Pontecorve EG, Hyman AL & Kadowitz PJ (1981) *Blood* **57:** 946–951.
Michell RH (1975) *Biochim. Biophys. Acta* **415:** 81–147.
Mirabelli F, Salis A, Vairetti M, Bellomo G, Thor H & Orrenius S (1989) *Arch. Biochem. Biophys.* **270:** 478–488.
Misselwitz F, Domogatsky SP, Repin VS, Spangenberg P & Till U (1988) *Thromb. Res.* **50:** 627–636.
Moncada S, Palmer RMJ & Higgs EA (1988) *Hypertension* **12:** 365–372.
Moncada S, Palmer RMJ & Higgs EA (1989) *Biochem. Pharmacol.* **38:** 1709–1715.
Morgan RO & Newby AC (1989) *Biochem. J.* **258:** 447–454.
Mower R & Steiner M (1982) *Prostaglandins* **24:** 137–148.
Murad F (1988) *Biochem. Soc. Trans.* **16:** 490–492.
Muranov KO, Gashev SB, Smirnov LD, Shvedova AA, Ritor VB & Kagan VE (1986) *Byll. EKSP Biol. Med.* **101:** 337–339.
Musca A, Cordova C, Violi F, Perrone A, Alessandi C & Salvadori F (1982) *Clin. Ther.* **102:** 273–276.
Mustard JF, Packham MA, Kinlough-Rathbone RL, Perry DW & Regoeczi E (1978) *Blood* **52:** 453–466.
Needleman P, Turk J, Jakschik BA, Morrison AR & Lefkowith JB (1986) *Annu. Rev. Biochem.* **55:** 69–102.
Neufeld EJ & Majerus PW (1983) *J. Biol. Chem.* **258:** 2461–2467.
Nishizuka Y (1983) *Philos. Trans. R. Soc. Lond. [Biol.]* **302:** 101–112.
Nishizuka Y (1984) *Nature (Lond.)* **308:** 693–698.
Nishizuka Y (1988) *Nature (Lond.)* **334:** 661–665.
Nugteren DH (1975) *Biochim. Biophys. Acta* **380:** 299–307.
O'Farrell PH (1975) *J. Biol. Chem.* **250:** 4007–4021.
Pace-Asciak CR (1984) *J. Biol. Chem.* **259:** 8332–8337.

Pace-Asciak CR (1989) *Biochim. Biophys. Acta* **793:** 485–488.
Pace-Asciak CR & Asotra S (1989) *Free Radicals Biol. Med.* **7:** 409–433.
Pannochia A & Hardisty RM (1985) *Biochem. Biophys. Res. Commun.* **127:** 339–345.
Perry SV (1979) *Biochem. Soc. Trans* **7:** 593.
Phillips DR & Agin PP (1977a) *Biochem. Biophys. Res. Commun.* **75:** 940–947.
Phillips DR & Agin PP (1977b) *J. Clin. Invest.* **60:** 535–545.
Phillips DR, Jennings LK & Prasana MR (1980) *J. Biol. Chem.* **255:** 11629–11632.
Rink T, Smith SW & Tsien RY (1982) *FEBS Lett.* **148:** 21–26.
Rittenhouse SE (1982) *Cell Calcium* **3:** 311–322.
Sage SO & Rink TJ (1985) *FEBS Lett.* **188:** 135–140.
Salonen JT (1989) *Ann. Med.* **21:** 59–62.
Schoene NW, Morris VC & Levander OA (1986) *Nutr. Res.* **6:** 75–83.
Schultz KD, Schultz K & Schultz G (1977) *Nature (Lond.)* **265:** 750–751.
Schulz S, Chinkers M & Garbers DL (1989) *FASEB J.* **3:** 2026–2035.
Seeger W, Moser U & Roka L (1988) *Naunyn-Schmiedeberg's Arch. Pharmacol.* **338:** 74–81.
Shears SB (1989) *Biochem. J.* **260:** 313–324.
Siegel A & Lüscher EF (1967) *Nature (Lond.)* **215:** 745–747.
Sies H (ed.) (1985) *Oxidative Stress,* pp 1–8. London: Academic Press.
Spangenberg P, Heller R, Bosia A, Arese P & Till U (1984) *Thromb. Res.* **36:** 609–618.
Spangenberg P, Till U, Gschmeissner S & Crawford N (1987) *Br. J. Haematol.* **67:** 443–450.
Steiner M (1985) *Thromb. Haemost.* **53:** 176–179.
Svensson J, Hamberg M & Samuelsson B (1976) *Acta Physiol. Scand.* **98:** 285–294.
Takai Y, Kaibuchi K, Matsubara T & Nishizuka Y (1981) *Biochem. Biophys. Res. Commun.* **101:** 61–67.
Takai Y, Kaibuchi K, Sano K & Nishizuka Y (1982) *J. Biochem.* **91:** 403–406.
Takashi O (1985) *Food Chem. Toxicol.* **23:** 937–940.
Thomas G & Ramwell P (1986) *Biochem. Biophys. Res. Commun.* **139:** 102–108.
Thomas G, Skrinska V, Lucas FV & Schumacher OP (1985) *Diabetes* **34:** 951–954.
Thomas G, Lucas FV, Schumacher OP & Skrinska V (1986a) *Prostaglandins Leukotrienes Med.* **22,** 117–128.
Thomas G, Skrinska VA & Lucas FV (1986b) *Thromb. Res.* **44:** 859–866.
Toivanen, J.L. (1987) *Prostaglandins Leukotrienes Med.* **26:** 265–280.
Tsai SC, Adamik R, Manganiello VC & Vaughan M (1981) *Biochem. Biophys. Res. Commun.* **100:** 637–643.
Vedelago HR & Mahadevappa VG (1988) *Biochem. Biophys. Res. Commun.* **150:** 177–184.
Wahlländer A & Sies H (1979) *Eur. J. Biochem.* **96:** 441–446.
Waldman S.A. & Murad F (1987) *Pharmacol. Rev.* **39:** 163–196.
Walter U (1984) *Adv. Cyclic Nucleotide Protein Phosphorylation Res.* **17:** 249–258.
White JG (1972) *Am. J. Pathol.* **66:** 295–306.
White JG (1979) *Am. J. Clin. Pathol.* **71:** 363–378.
Yamanishi J, Kawahara Y & Fukuzuki H (1983) *Thromb. Res.* **32:** 183–188.
Zaidi NF, Lagenaur CF, Abramson JJ, Pessah I & Salama G (1989) *J. Biol. Chem.* **264:** 21725–21736.
Zucker MB & Borelli J (1962) *Proc. Soc. Exp. Biol. Med.* **109:** 779–787.

17

Vasodilation and Oxygen Radical Scavenging by Nitric Oxide/EDRF and Organic Nitrovasodilators

EIKE NOACK and MICHAEL MURPHY*

*Institut für Pharmakologie and *Institut für Physiologische Chemie I, Heinrich-Heine-Universität Düsseldorf, Moorenstrasse 5, D-4000 Düsseldorf 1, Germany*

I. Introduction 446
II. From EDRF to nitric oxide (NO)—some historical remarks 448
III. Techniques for quantification of NO 451
IV. Biochemistry of NO synthesis 454
V. Biochemical properties of NO and EDRF 460
VI. NO depression or imbalance in cardiovascular diseases 469
VII. Drug-dependent vasorelaxation—donors of NO 472
VIII. Future developments 476
IX. Conclusions 478

ABBREVIATIONS

ACh	acetylcholine
DTT	dithiothreitol
EDRF	endothelium-derived relaxing factor
FMLP	*N*-formyl-L-methionyl-L-leucyl-L-phenylalanine
INF-γ	interferon-γ
L-NIO	*N*-imminoethyl-L-ornithine

Oxidative Stress: Oxidants and Antioxidants
ISBN 0-12-642762-3

L-NMMA (guanidino)-*N*-monomethyl-L-arginine
L-NNA (guanidino)-*N*-nitro-L-argine
LPS bacterial lipopolysaccharide
SOD superoxide dismutase
TNF tumour necrosis factor

I. INTRODUCTION

An endothelium-derived relaxing factor (EDRF) plays a vital role in the physiological regulation of vasomotor tone, platelet aggregation (Furchgott and Zawadzki, 1980; Azuma et al, 1986; Furlong et al, 1987; Pohl and Busse, 1990), cell communiction (Moncada et al 1989), neutrophil activity (Schmidt et al, 1989), and perhaps ischaemia and other pathological conditions. The definitive identification of nitric oxide (NO) as an EDRF (distinct from the prostaglandins) is based on the following combined pieces of evidence: (1) NO is produced by the cells which produce EDRF; (2) stimulants which trigger NO production parallel those which induce EDRF release; (3) NO and EDRF show identical chemical properties (with the exceptions discussed below); (4) NO and EDRF trigger identical responses in target cells via indistinguishable mechanisms; and (5) the quantity of NO produced accounts for the effects of EDRF (Ignarro et al, 1987b; Palmer et al, 1987, 1988a,b; Kelm et al, 1988; Moncada et al, 1988; Buga et al, 1989; Chen et al, 1989).

The most important action of NO in many target tissues is the stimulation of soluble guanylate cyclase. This raises the cyclic GMP (cGMP) level, activates cGMP-dependent enzymes and eventually leads to the altered properties of target cells. The stimulation of guanylate cyclase is the basis for the vasodilatory effect of NO in vascular smooth muscle, the antiaggregatory and antiadhesive effects in platelets, and the antiadhesive effect in polymorphonuclear leukocytes. The relaxant effect of NO on vascular smooth muscle is analogous to that of the nitrovasodilators, which liberate NO and activate the soluble cytosolic guanylate cyclase (Schröder et al, 1985; Noack et al, 1986; Schröder and Noack, 1987; Feelisch and Noack, 1987a). Elevated cGMP may also account for an increased permeability of kidney epithelial cells and the effects of NO on neurotransmission in the brain (Garthwaite et al 1988), and at the inhibitory non-adrenergic, noncholinergic nerves of the gastrointestinal tract (Boeckxstaens et al, 1990; Bult et al, 1990).

NO can have effects besides the activation of guanylate cyclase by binding to most iron-containing enzymes, including those of the respiratory chain, as evidenced by electron paramagnetic resonance (Collier and Vallance, 1989). Inhibition of enzymes (aconitase, NADH–ubiquinone oxidoreductase and succinate–ubiquinone oxidoreductase) might account for the contribution of NO to the antimicrobial and antitumour actions of macrophages (Hibbs et al, 1988; Marletta et al, 1988; Tayeh and Marletta, 1989), although other modes of actions are possible (Iyengar et al, 1987). It is not yet clear whether an effect on cGMP levels or a more direct inhibition of DNA synthesis (perhaps via ribonucleotide reductase inhibition or ADP-ribosyltransferase activation (Brüne and Lapetina, 1989)) is responsible for the antiproliferative activity of NO in mesangial, adenocarcinoma and smooth muscle cells (Garg and Hassid, 1989a,b).

NO is a colourless gas under normal conditions, with a low water solubility.

NO is paramagnetic and decomposes quickly in oxygenated solutions, with a half-life of only a few seconds (Moncada et al, 1988). The loss is thought to be due to an exothermic reaction with oxygen, forming nitrogen dioxide (NO_2) (reaction 1), a brown gas, and also a free radical, which disproportionates to nitrite (NO_2^-) and nitrate (NO_3^-) in neutral aqueous solution (reaction 2):

$$2NO + O_2 \longrightarrow 2NO_2 \quad (1)$$

$$2NO_2 + H_2O \longrightarrow NO_2^- + NO_3^- + 2H^+ \quad (2)$$

NO also has a high affinity for numerous metal complexes, and the physiological actions of NO are blocked by certain redox compounds (Moncada et al, 1986). Superoxide dismutase (SOD), however, enhances the relaxant effects of EDRF/NO, supporting the view of Gryglewski et al (1986) that NO is readily inactivated by superoxide (O_2^-) via an intermediate formation of peroxynitrite (reaction 3) and subsequent conversion to nitrate (reaction 4) (Blough and Zafiriou, 1985).

$$NO + O_2^- \longrightarrow ONOO^- \quad (3)$$

$$ONOO^- \longrightarrow NO_3^- \quad (4)$$

It has been suggested that generation of oxygen radicals at the time of reoxygenation may be responsible for reperfusion injury in the heart. NO may exhibit protective activity under these conditions due to its ability to scavenge O_2^-. In the same way, drugs which release NO may provide a beneficial scavenger function during ischaemia if given prophylactically. However, the peroxonitrate intermediate formed by reaction between O_2^-

and NO may yield hydroxyl radical upon breakdown to NO_2 (reaction 5) (Beckman et al, 1990). If this occurs in vivo as well, the potential toxicity of superoxide may not be mitigated at all by its reaction with NO.

$$ONOO^- + H^+ \longrightarrow NO_2 + OH \quad (5)$$

On the other hand, some cellular and subcellular damage during ischaemia may be related to the relative deficiency of local NO generation or availability. Under in vitro conditions the efficacy of NO is enhanced when the NO-deactivating effect of superoxide is blocked (Feelisch et al, 1989). Additional pathophysiological alterations may also result in a lower efficacy of NO. Thus the observation that the dilatatory response of atherosclerotic vessels to NO is impaired may at least partly be due to the direct interaction of oxidized low-density lipoprotein with guanylate cyclase, as was recently demonstrated in a simpler in vitro system by Schmidt et al (1990).

II. FROM EDRF TO NITRIC OXIDE (NO)—SOME HISTORICAL REMARKS

Endothelial cells cover the interior of the vessel wall like a tapestry. They provide a mechanical barrier throughout the circulatory system, effectively separating the soluble and cellular components of blood from the subendothelial cell layers and the vascular smooth muscle. Until 1980 it was generally believed that the mechanical protective function of the endothelium was its only physiological task besides its poorly understood roles in metabolism, coagulation, and substrate transport. Then Furchgott and Zawadzki (1980) published their seminal experimental findings on isolated rabbit aorta, which immediately intensified scientific work worldwide. Like many important findings that revolutionized a scientific field, the experiments were at the same time simple and ingenious. The authors isolated rabbit aortas and other arteries and removed the endothelial cells from some by rubbing the intimal surface with filter paper or by exposure to collagenase. After precontracting the isolated aortic strips with $10^{-7.7}$ M of (−)-noradrenaline, they added increasing concentrations of acetylcholine (ACh) or other cholinergic substances, which were well-known vasodilators. In the presence of an intact endothelium, the strips responded by relaxing. However, after removal of the endothelial cells, neither ACh nor the other agonists induced relaxation (Figure 1) and in most arteries a paradoxical contraction was seen in response to higher levels of ACh. It was therefore concluded that ACh triggers the release of a factor from the endothelial cells which relaxes blood vessels. This publication is one of the most often cited papers in this research field.

The chemical nature of this extremely labile but powerful vasoregulatory mediator was unknown, and it was merely termed the endothelium-derived relaxing factor. A number of local regulators, neurohumoral substances, hormones and autocoids (e.g. adenosine diphosphate, bradykinin, serotonin, substance P and thrombin) induce arterio- and venodilation by stimulating the release of EDRF. Increases in shear stress may also cause higher liberation rates of EDRF. Thus, for the first time it was recognized that the endothelial cell layer is a metabolically active system, synthesizing and releasing EDRF and other vasoactive substances from their luminal and abluminal surfaces, thereby influencing the vascular tone of the underlying vascular smooth muscle and other cell functions like thrombocyte aggregation (Furchgott, 1984).

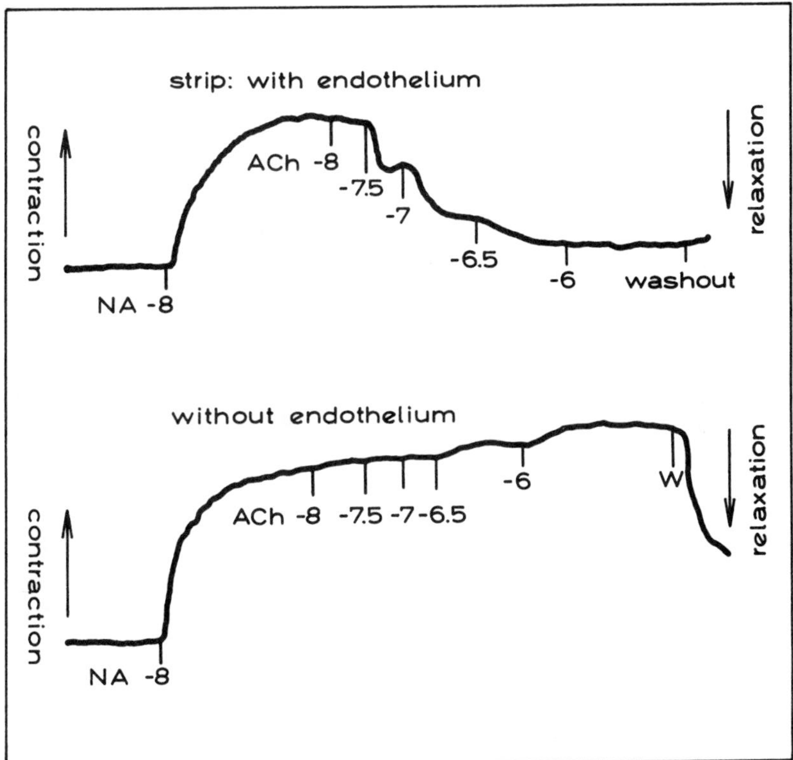

Figure 1. Loss of relaxing response of a rabbit aorta preparation to rising concentrations of acetylcholine after rubbing off the intimal surface. The vessel was precontracted with 10 pM noradrenaline (NA). Reprinted with permission from *Nature* Vol. 288, pp. 373–376. Copyright (c) 1990 Macmillan Magazines Ltd.

Despite substantial efforts, the chemical identity of EDRF remained elusive for several years. Several studies had demonstrated that EDRF is a labile substance with a half-life of only a few seconds and is readily inactivated by haemoglobin and O_2^-. This led Furchgott to propose in 1986 that EDRF may simply be NO. This hypothesis had the distinct advantage of explaining the similarity of EDRF action on vascular smooth muscle with that of the nitrovasodilators. Nevertheless, Vanhoutte (1987) still raised the question of whether there is more than one EDRF, suggesting that products of arachidonic acid (via the P450 cytochrome pathway) or ammonia may be likely candidates.

Progress came when Palmer et al (1987) were successful in chemically identifying EDRF as NO, using a chemiluminescence method. Shortly after this, our laboratory reported a further analytical method which allowed the quantitative and continuous measurement of small amounts of NO released from cultured endothelial cells (Kelm et al, 1988), from isolated perfused hearts (Kelm and Schrader, 1988) or from NO-carrying prodrugs (Feelisch and Noack, 1987a,b) using difference spectrophotometry which we had previously developed in the analysis of the bioactivation pathways of nitrovasodilators such as glyceryl trinitrate (Noack et al, 1986). Substantial evidence has accumulated that, at least in the coronary system and the large arteries, EDRF is indistinguishable from NO. This was a most exciting phase in both vascular physiology and cardiovascular pharmacology, bringing these two disciplines more closely together than ever before.

On a subcellular level, the action of NO in target cells is often mediated by activation of the soluble guanylate cyclase. NO activates guanylate cyclase by interacting with the ferrohaem centre of the enzyme (Ignarro et al, 1986a). Long before the discovery of the direct effect of NO, intracellular elevation of cGMP was a recognized effect of various nitrovasodilators. As early as 1973 cGMP was defined as a negative feedback signal which could counter a vasoconstriction of smooth muscle. Soluble guanylate cyclase was isolated from various tissues and is widely used in a bioassay for NO-generating substances and NO itself. The synthesis of NO may also induce a rise in cGMP within the generator cell itself, perhaps providing a negative feedback to turn off further NO release. The mechanism by which the rise in cGMP mediates further effects remains to be established. In smooth muscle it includes a decrease of cellular free calcium concentration after protein phosphorylation. The reuptake of calcium into intracellular storage sites such as the endoplasmic reticulum may be intensified (Karczewski et al, 1990).

Regarding the biochemical mechanisms of NO synthesis, it was found that the amino acid L-arginine is a precursor of NO synthesis (Palmer et al, 1988b; Schmidt et al, 1988a,b; Moncada et al, 1989; Vallance et al, 1989).

The NO-forming enzyme, termed nitric oxide synthetase, is NADPH-dependent and requires a divalent cation. It is not restricted to endothelial cells but may also occur in a variety of other tissues (Palacios et al, 1989). Synthesis of NO is blocked stereospecifically by some arginine analogues such as *N*-monomethyl-L-arginine (L-NMMA). These antagonists are now suitable tools for studying the biological function of NO in vivo.

III. TECHNIQUES FOR QUANTIFICATION OF NO

NO can be quantitatively determined by means of spectroscopy, chemiluminescence and various bioassay techniques. The assays may measure NO directly, the breakdown products of NO such as NO_2^- and NO_3^-, or the NO recovered from a reduction of NO_2^-. The reliability of the assays depends on their ability to measure all (or a constant fraction) of the NO originally produced, and to avoid measuring compounds not derived from the original NO. The determinations may be inaccurate, e.g. in the presence of oxygen radicals if they accelerate the breakdown of NO, or alter the ratio of NO_2^- to NO_3^- in favour of the latter. There is incomplete information as to what extent metals, thiols or other radicals interfere in the measurements.

A. Chemiluminescence with ozone

Palmer et al (1987) used chemiluminescence in combination with classical bioassays in their original identification of EDRF as NO. The technique is based on the homogeneous gas phase reaction of NO with ozone to yield electronically excited NO_2^* (reaction 6). This unstable intermediate emits light upon its return to ground state (reaction 7) which can be quantitated by a photodetector.

$$NO + O_3 \longrightarrow NO_2^* + O_2 \tag{6}$$

$$NO_2^* \longrightarrow NO_2 + h\nu \tag{7}$$

Channelling the incubation solution to a reflux vessel containing a boiling acetic acid–iodide mixture causes any NO_2^- to be converted to NO, but has no effect on NO_3^- or the pre-existent NO. The NO is then drawn out of the solution as a vapour and directed through the reaction/detection chamber to react with ozone. NO production in the incubation may often be underestimated, since NO converted to NO_3^- will never be measured. We adapted this protocol so that the incubation chamber is permanently blown through with argon in order to rapidly degas the NO from the solution (Feelisch and Noack, 1987b).

B. Spectrophotometry with oxyhaemoglobin

Small amounts of NO can be quantitatively and continuously determined using rapid NO-induced conversion of oxyhaemoglobin (HbO_2) to methaemoglobin (MetHb) and NO_3^- (reaction 8) (Doyle and Hoekstra, 1981; Doyle et al, 1983; Haussmann and Werringloer, 1987; Feelisch and Noack, 1987a). A continuous recording of the increasing MetHb concentration reflects the rate of NO formation.

$$HbO_2 + NO \longrightarrow MetHb + NO_3^- \quad (8)$$

The difference spectrum of MetHb versus HbO_2 is recorded by a double-beam spectrophotometer. HbO_2 is prepared by standard methods (Cassoly, 1981), and used at 4–10 μM in any ordinary buffer at neutral pH. The difference spectrum possesses an absorption maximum in the Soret region at 401 nm with an isosbestic point at 411 nm. The method is highly sensitive, and specific for NO. There is no interference with dissolved oxygen or inorganic nitro compounds like NO_2^- or NO_3^-. Selectivity and accuracy have been verified by the chemiluminescence technique (Feelisch and Noack, 1987b). The rate constant for reaction 8 is near $80 \times 10^6 \, M^{-1} s^{-1}$ at 37°C (Doyle, personal communication), so that the reaction with NO will be virtually complete within 100 ms in the presence of micromolar HbO_2.

One type of interference can occur when significant levels of H_2O_2 are generated in the assay mixture (Prasad et al, 1989), for example when measuring NO production from macrophages, from cells with impaired H_2O_2-metabolizing capabilities, or from sydnonimines that generate O_2^-. H_2O_2 quickly oxidizes MetHb to a compound with a lower and shifted absorption in the Soret region. The addition of catalase can prevent this but to be fully effective it must be added at levels nearly equal to the expected levels of MetHb. Unfortunately, the strong absorption of catalase in the Soret region also shifts after it reacts with H_2O_2, making the 401–411-nm measurement less reliable. This complication is avoided by measuring the change in the α- and β-bands of HbO_2 absorption (e.g. 578 nm maximum and 592 nm isosbestic point), where catalase absorption is small and little changed by its reaction with H_2O_2. The main drawback here is a 4–5-fold loss in sensitivity at these wavelengths.

C. Diazotization of sulphanilic acid

NO and NO_2^- accumulation in the culture medium of endothelial cells can be estimated spectrophotometrically after diazotization of sulphanilic acid and coupling with *N*-(1-naphthyl)-ethylene diamine (Ignarro et al, 1987a). This method is 100-fold more sensitive to NO than to NO_2^- and does not respond to NO_3^-. An aliquot of the medium (100–300 μl) is added to 25 μl 6.5 M HCl and 25 μl 37.5 mM sulphanilic acid. Ten minutes later 25 μl 12.5 mM *N*-(1-naphthyl)-ethylenediamine is added. After another 30 min, the absorbance is measured at 540 nm.

D. Application of ^{15}N-containing substances

NO generation can be measured from compounds which contain the nitrogen isotope ^{15}N, and the resulting ^{15}NO, $^{15}NO_2^-$ or $^{15}NO_3^-$ can be determined after conversion to nitrobenzene according to the method described by Iyengar et al (1987) using a GC–MS system.

E. Indirect measurement by bioassay or guanylate cyclase assay

NO may also be assessed by determining its vasorelaxant activity or its stimulation of guanylate cyclase. These assays are sensitive and have inherent physiological relevance, but suffer from a potential for artefactual effects of unrelated substances. In all cases the quantification is indirect via a calibration of the system to standard solutions of NO or NO-releasing agents (e.g. SIN-1 or glycerol trinitrate).

1. *Vasorelaxant activity*

There are several variations of the same basic technique in the literature (Lückhoff et al, 1987; Mülsch et al, 1987; Busse et al, 1988; Kelm et al, 1988). Arterial segments are collected and cut into rings, and the endothelial cell layer is removed by gently skimming with a razor blade or pipe cleaner. The rings are then mounted between two hooks, so as to be superfused with a test solution. Alternatively, the segments are cannulated from both sides after ligation of the side branches, stretched into their mean in situ length and mounted in an organ bath containing oxygenated Tyrode or Krebs-Ringer solution. All solutions are supplemented with 10 μM indomethacin in order to avoid disturbances by the prostaglandin system. The segments are precontracted by noradrenaline and allowed to equilibrate for about 60 min.

The outflow tubing from the source of NO (e.g. a column of cultured endothelial cells on beads or an arterial segment with an intact endothelial layer) then superfuses the ring or is directed to the inflow cannula of the detector segment. A short transit time is essential, since the half-life of NO is only a few seconds under these conditions. The tension of rings is monitored by a force transducer, or the outer diameter of segments is monitored by means of a photoelectric device.

2. *Guanylate cyclase stimulation*

A relaxant agent (e.g. ACh or bradykinin) is first added to precontracted arterial or venous rings in a bath chamber in order to assess the degree of relaxation (Ignarro et al, 1986b). After this relaxant response, the ring is rinsed, allowed to equilibrate for half an hour and then added to a suspension of purified, soluble guanylate cyclase (Gerzer et al, 1981) containing the relaxant agent. This is further incubated at 37°C for 10 min and the formation of cGMP is determined by the formation of [^{32}P]cGMP from [^{32}P]GTP.

Endothelium-derived NO is also released to the luminal side of vessels and is then able to increase cGMP in platelets which are passing by in close proximity. This increase normally inhibits both the aggregation and adhesion of platelets (Azuma et al, 1986; Furlong et al, 1987). The recent data of Pohl and Busse (1989) indicate that basally released EDRF/NO is able to increase cGMP in platelets even during a single passage through the coronary bed of isolated, saline-perfused rabbit hearts.

IV. BIOCHEMISTRY OF NO SYNTHESIS

The enzymatic activities which catalyse the production of NO/EDRF from arginine can be divided into two distinct subtypes best characterized in endothelial and macrophage cells. No single protein responsible for this conversion has been purified to homogeneity from any tissue, and the enzymatic activity is usually referred to simply as the NO synthetase enzyme. Characterization of the enzyme has been usually carried out in whole cells or lysates, but in some cases at least a partial purification has also been possible.

Evidence that NO is derived from arginine comes from [^{15}N-]guanidino-labelled arginine, which releases ^{15}N-labelled NO (Palmer et al, 1988a,b; Marletta et al, 1988; Hibbs et al, 1988; Schmidt et al, 1988b). Furthermore, evidence that the ordinary metabolites of arginine, such as urea, are not the

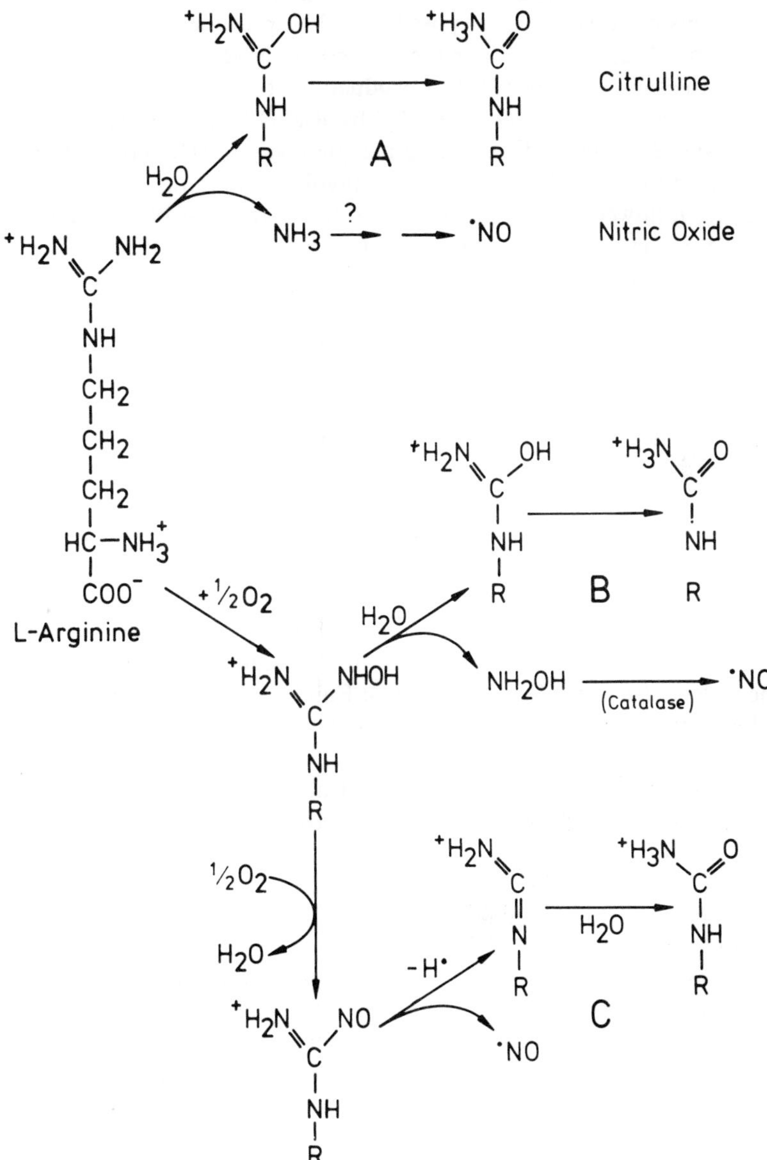

Figure 2. Biochemical pathways for NO production in vivo.

eventual substrate comes from [^{14}C]guanidino-labelled arginine, which is converted to ^{14}C-labelled citrulline (Palmer and Moncada, 1989; Hibbs et al, 1988). This conversion is not possible via the previously known enzymatic pathways. Three pathways for the conversion have been proposed (Figure 2), but the lack of expected NO production from the proposed ammonia intermediate of one (pathway A) (Hibbs et al, 1987) or from the proposed hydroxylamine intermediate of another (pathway B) (DeMaster et al, 1989) make them unlikely. A third proposal, involving an initial hydroxylation of arginine, followed by a release of NO and formation of a diimide intermediate to citrulline (pathway C) (Marletta et al, 1988; Tayeh and Marletta, 1989), remains a viable, albeit unsubstantiated, possibility.

A. Endothelial enzyme

NO synthetase has been best characterized in endothelial cells. This enzymatic activity depends on Ca^{2+} and NADPH and uses arginine as a substrate (Figure 3). The dependence on calcium is critical, and the stimulation of EDRF/NO production by this enzyme in vivo is consistently linked to agents which raise intracellular calcium (Furchgott, 1984). In vitro studies have indicated that the enzyme's K_m for Ca^{2+} is in the physiological range (Mayer et al, 1989; Carter et al, 1990). For endothelial cells, stimulants include ACh, bradykinin, ATP, melittin, substance P, A23187 and many other agents (reviews: Furchgott, 1984; Furchgott and Vanhoutte, 1989). It is therefore likely that most conditions which lead to an increase in cell calcium (e.g. limited cell injury, ischaemia or inactivation of calcium pumps) would also lead to a production of NO. Calcium inflow may be involved in the mechanism by which flow-induced shear stress stimulates NO production by endothelial cells (Cooke et al, 1990; Kelm et al, 1990). Lipopolysaccharides enhance the production of NO by endothelial cells (Salvemini et al, 1990), although this stimulant is usually associated with macrophage stimulation (see below).

A number of analogues of arginine inhibit the basal and stimulated synthesis of NO by the endothelial enzyme, including L-NMMA ($ED_{50} \approx 4\ \mu M$) (Palmer and Moncada, 1989; Palmer et al, 1988a,b; Rees et al, 1989a), *N*-nitro-L-arginine (L-NNA), L-argininosuccinate and L-canavanine (Schmidt et al, 1988b; Gold et al, 1989). The inhibitors are stereospecific and have different potencies but are usually reversible by excess (100 μM to 1 mM) arginine (Rees et al, 1989a; Gold et al, 1989; Palmer et al, 1988a,b). No inhibitors affected the responsiveness of target cells to other sources of NO, such as sodium nitroprusside or glycerol trinitrate (Rees et al, 1989b). This implies a simple competitive inhibition at the active site of the NO-synthesizing enzyme, and subsequent investigations of the inhibition characteristics in vitro have thus far yielded the same conclusion. An exception may be L-argininosuccinate, which reportedly caused an irreversible inhibition in bovine pulmonary aortic rings (Gold et al, 1989). However, numerous small variations between species in the sensitivity of tissues to the inhibitors and reversibility by arginine still may be due to cell-specific differences in the turnover of the compound via diffusion, transport or metabolism. For example, L-NMMA is more rapidly degraded than the *N*-nitro analogue. Typical literature values for the average concentration of arginine in human (55 μM), rat (100 μM) and mouse (84 μM) plasma indicate that these pools may influence results of whole animal studies unless a sufficient excess of inhibitor is used.

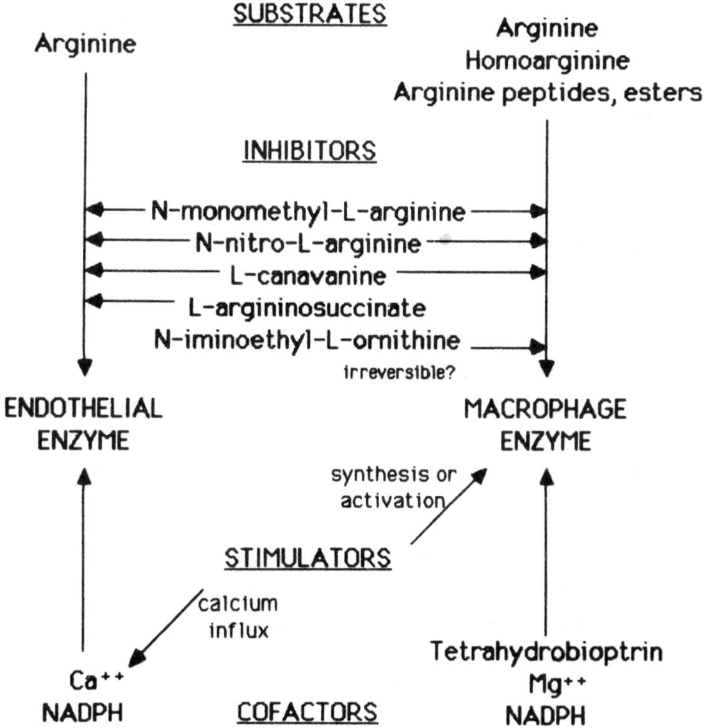

Figure 3. Comparison of characteristics of endothelial and macrophage NO synthetase enzymes.

Arginine is the sole compound proven to be a substrate for the endothelial enzyme (Rees et al, 1989a; Palmer and Moncada, 1989; Palmer et al, 1988b).

Some investigators had suggested that basic polyamino acids (such as poly-L-arginine or poly-L-lysine) served as partial substrates for the enzyme, since the addition of these, but not equal amounts of L-arginine, induced the release of NO from endothelial cells (Ignarro et al, 1989). These results have not been reproduced in lysates or homogenates, and thus direct evidence is lacking that polyamino acids are substrates for the enzyme. The high potency of poly-L-lysine, the reversibility by polyanions, the appearance of the phenomenon only in intact cells, and the slow desensitization of the cells, all support an alternative view that the basic polyamino acids have a nonspecific membrane effect, probably leading to a transient increase in intracellular calcium (Thomas et al, 1989). The stimulation of EDRF formation by phospholipase A_2 or lysophospholipids probably occurs via a similar membrane effect, although these treatments have been prematurely interpreted as giving rise to a non-NO type of EDRF (Saito et al, 1988).

B. Macrophage enzyme

The second type of enzyme converting arginine to NO is best characterized in macrophage cells; its properties differ considerably from those of the enzyme from endothelial cells. For example, macrophages apparently generate NO at levels high enough to be cytotoxic (Hibbs et al, 1988). The macrophage enzyme is inducible, does not depend on Ca^{2+} for its activity, utilizes NADPH and arginine (ED_{50} 150 μM) as its natural substrates, and has tetrahydrobiopterin and possibly Mg^{2+} as cofactors (Tayeh and Marletta, 1989) (Figure 3). FAD may also be a cofactor (Stuehr et al, 1990). As with the endothelial enzyme, ^{15}N and ^{14}C tracer studies have demonstrated conclusively that NO is derived from the guanidino nitrogen of arginine, and that citrulline is the other direct product of the reaction (Marletta et al, 1988; Iyengar et al, 1987; Hibbs et al, 1988).

The inducibility and Ca^{2+}-independence are the critical features of this enzyme. The production of NO by macrophages can be considered as a part of the antimicrobial arsenal of the macrophage, with its activity being induced by the same types of stimulants as induce the NADPH-oxidase responsible for O_2^- production (Ding et al, 1988; Iyengar et al, 1987). Most experiments so far employ a combination of interferon-γ (IFN-γ) plus either bacterial lipopolysaccharide (LPS) or tumour necrosis factor (TNF) to induce the NO synthetase activity. The induction requires about 12–24 hours for full effect (Drapier et al, 1988; Ding et al, 1988; Marletta et al, 1988; Hibbs et al, 1987; Iyengar et al, 1987). INF-γ alone, but not TNFα or β, could induce some NO synthetase activity, but the combinations were more potent. Interleukins 1β, 2, 3, or 4 or NF-α or -β did not induce NO synthesis (they also failed to induce O_2^- synthesis) and could not substitute for TNF (Drapier et al, 1988; Ding et al, 1988). Muramyl dipeptide could substitute for TNF, possibly by stimulating the endogenous synthesis of TNF (Drapier et al, 1988). Compared to other combinations of stimulants, INF-γ plus LPS may lead to more NO production and less O_2^- production (Ding et al, 1988). The significant hypotension usually seen following infusion of TNF is prevented by L-NMMA, suggesting that stimulation of NO synthesis is a major action of TNF at some sites in vivo (Kilbourn et al, 1990).

In contrast to the endothelial cell enzyme (Rees et al, 1989a) the macrophage enzyme has a range of substrates which can substitute for arginine, such as L-homoarginine, L-arginine methyl ester, L-arginamide and the L-arginyl-L-aspartate dipeptide (Iyengar et al, 1987). The latter three can be converted to arginine by macrophages, so it is still unclear whether they are true substrates. The production of L-monocitrulline from L-homoarginine shows that the latter is a true substrate for the enzyme (Hibbs et al, 1987). Canavanine, (guanidino)-*N*-nitro-L-arginine (L-NNA), *N*-nitro-L-arginine methyl ester and L-NMMA (Marletta et al, 1988; McCall et al, 1990) are inhibitors of the macrophage enzyme. Interestingly, L-NMMA is reported to slightly decrease the production of O_2^- from macrophages, suggesting that some regulatory connections between the NO and O_2^- pathways is possible. *N*-iminoethyl-L-ornithine (L-NIO) appears to cause an irreversible inhibition of the macrophage enzyme (McCall et al, 1990), but it is unclear whether this involves a suicide inactivation. Since the compound does not irreversibly inactivate the endothelial enzyme, it may become an effective tool for the differentiation between the two major types.

C. Other cell types

The macrophage type of the NO synthetase has been found in bone marrow (Kamoun et al, 1988), adenocarcinoma cells (Amber et al, 1988), and neutrophils (Schmidt et al, 1989). Adenocarcinoma cells were similar to macrophages, being stimulated by INF-γ plus TNF, but the further addition of LPS led to even greater NO synthetase activity (Amber et al, 1988). The NO-synthesizing activity in bone marrow cells shared the Ca^{2+} independence of the macrophage enzyme, but was reported to have the unusual property of being stimulated specifically by a source of O_2^- radicals (Kamoun et al, 1988). The enzyme was induced in the bone marrow cells by the lyophilized, dialysed supernatant of a mixed leukocyte culture after a 1–2-day lag period, perhaps due to the action of cytokines released by the leukocytes. The NO-synthesizing activity of neutrophils was stimulated by either *N*-formyl-L-methionyl-L-leucyl-L-phenylalanine (FMLP), platelet-activating factor or leukotriene B_4. The synthesis was inhibitable by L-canavanine, but, in contrast to the long induction required by macrophages, the neutrophils showed quick induction (Schmidt et al, 1989).

Brain homogenates also contain a NO-synthesizing activity, the nature of which is still slightly unclear. In one report, it was like the endothelial enzyme in showing a Ca^{2+} dependence, but, surprisingly, L-homoarginine, L-arginyl-L-aspartate (Knowles et al, 1989) and L-arginine methyl ester (Knowles et al, 1989; Mayer et al, 1990) could replace arginine as substrates. Bredt and Snyder (1989a) suggested it was Ca^{2+} calmodulin-dependent. *N*-methyl-D-asparate, a neurotransmitter thought to trigger Ca^{2+} inflow, stimulated EDRF/NO synthesis in both neuroblastoma cells and rat cerebellum (Garthwaite et al, 1988; Bredt and Snyder, 1989b).

V. BIOCHEMICAL PROPERTIES OF NO AND EDRF

The similar chemical properties of NO and EDRF constitute a critical part of the evidence that they are in fact the same compound. The properties of each have been studied in detail separately, and in many cases in parallel. The bulk of information concerns the compounds which deactivate NO/EDRF, plus the consistent observation that SOD enhances their effects. However, there is some evidence that their properties are not always identical. The following section discusses this information.

A. Deactivation and inhibition

The ability of NO/EDRF to act as an inter- and intracellular signal depends on its having only a limited period of action. Unlike other physiological transmitters, there is no proposal that a specific deactivating enzyme or factor even exists for NO/EDRF. This is not surprising, since NO reacts readily with oxygen, HbO_2, and perhaps many normal constituents of cells. These reactions are favoured by the fact that NO is a free radical. The limited half-life of NO is an advantage in terms of its roles as a local vasodilator and in the respiratory burst, but at the same time the variety of possible reactions makes it difficult to identify precisely the factors most important to its half-life in vivo. They also complicate the identification of mechanisms by which compounds may deactivate NO.

1. *Oxyhaemoglobin*

Oxyhaemoglobin (HbO_2) is the most commonly used compound thought to inactivate NO directly. The salient reaction (reaction 8) involves a direct attack by the NO on the bound oxygen, and parallels the same reaction with oxymyoglobin. At a typical level of 10 μM in the media or reaction solutions, the disappearance of NO would proceed at a pseudo-first order rate of 850 s^{-1} (half-life of 0.8 ms). This is three to four orders of magnitude faster than the loss of NO in ordinary oxygenated buffer (reactions 1 and 2) and easily accounts for the effectiveness of HbO_2 in inhibiting the action of NO.

Methaemoglobin (MetHb) and metmyoglobin do not undergo reactions with NO, but do show a weak reversible binding. It may be possible for MetHb to alter the effects of NO in perfusion systems via binding and facilitating dispersion, but there have been consistent reports of inhibition in static systems (Ignarro et al, 1980; Ignarro and Gruetter, 1980; Gruetter et al, 1981). The mechanism was unclear, but contamination by HbO_2 was a possible problem. The inability of pure MetHb to affect NO is usually reported by others.

Deoxyhaemoglobin binds NO in a practically irreversible manner, at a rate as rapid as that of the reaction between NO and HbO_2. In atmospheric or oxygen-saturated solutions, though, virtually no deoxyhaemoglobin exists, and so the effect of haemoglobin under these conditions cannot be attributed to direct binding, although this was occasionally inferred due to poor techniques (e.g. Ignarro et al, 1987a,b). Moreover, the NO–haemoglobin complex is unstable in oxygenated solutions. It converts to MetHb in a reaction that may involve oxygen, and which appears faster than the dissociation of the complex (Chiodi and Mohler, 1985). The use of deoxyhaemoglobin to bind NO has thus far been successfully employed only in the study of the NO release from anaerobic bacteria (Goretski and Hollocher, 1988).

A novel observation in the literature suggests that HbO_2 and MetHb have direct effects on vascular smooth muscle used as the target tissue in investigations of NO/EDRF (Bény et al, 1989). Whether this effect can be shown to extend to other tissues and other conditions is not yet known, but the possibility should not be discounted that large organic molecules can have indirect effects.

2. *Methylene blue*

Methylene blue is not usually considered to deactivate NO directly, but rather to act as a slowly reversible inhibitor of the ability of NO to stimulate guanylate cyclase (Gruetter et al, 1981). Its mechanism of action is not clearly known. Considerable evidence indicates that the guanylate cyclase enzyme contains a haem group, and that the enzyme is activated when NO is bound to the haem (Craven and DeRubertis, 1976; Ignarro et al, 1982; Wood and Ignarro, 1987). Apparently the haem must be in the reduced (ferrous) form to bind NO, and only the (ferrous) haem–NO form activates the enzyme. Neither the ferric nor the ferrous form of haem without NO activates the enzyme, but, interestingly, protoporphyrin IX (haem lacking iron) and perhaps perferryl haem (analogous to the compound I of catalase) activate the enzyme in the absence of NO.

Methylene blue can be both an oxidant and a reductant of haem groups. Its ability (after cell-mediated reduction) to reduce the ferric form of haemoglobin to the ferrous form is well known, and is even used therapeutically in the treatment of idiopathic methaemoglobinaemia (Metz et al, 1976). It is difficult to view a reduction of guanylate cyclase by methylene blue as a potential mechanism of action (Martin et al, 1985), since this would theoretically yield the active (ferro)enzyme, but the oxidation of guanylate cyclase is a potential mechanism. It has been suggested that methylene blue may interact directly with EDRF/NO (Martin et al, 1985). This would not be surprising since it is a powerful redox compound and NO is a free radical, but this would not explain the slow reversibility of inhibition following washout of methylene blue. To further complicate matters, Palmer and Moncada (1989) reported that methylene blue inhibited the formation of citrulline (i.e. actually inhibited the synthesis of NO) in endothelial cell homogenates. The mechanism of methylene blue and the role of the redox state of guanylate cyclase in enzyme activity both need further clarification.

3. *Compounds claimed to release superoxide*

NO reacts with O_2^- to yield nitrate, and the apparent similar sensitivity of NO and EDRF to O_2^- is important support for the proposal that they are identical (Hutchinson et al, 1987; Ignarro et al, 1987b; Palmer et al, 1987). An unusual assortment of agents which deactivate NO/EDRF is claimed to act by generating O_2^- (Gryglewski et al, 1986; Moncada et al, 1986), but all the agents have other effects and properties which may also contribute to this deactivation. It is worth noting that virtually all effective inhibitors of EDRF are antioxidants and free radical quenchers (Griffith et al, 1984). It should also be remembered that H_2O_2 (always produced when O_2^- is present) directly stimulates NO production by endothelium (Rubanyi and Vanhoutte, 1986b).

The reaction rate between NO and O_2^- at 37°C in neutral buffer is around $56 \times 10^6\ M^{-1}\ s^{-1}$ (Saran et al, 1990). In order for O_2^- to bring the half-life of NO to around 7 s (i.e. equalling the spontaneous loss of NO/EDRF in oxygenated buffer, pseudo-first order $0.1\ s^{-1}$; Griffith et al (1984), Kelm et al (1988) and Moncada et al (1988)), it must be at a concentration of 1.8 nM.

This is at least 100-fold higher than normal cytosolic O_2^- concentrations (Tyler, 1975). Without rigorous removal of trace metals, such low concentrations of O_2^- show a pseudo-first order disappearance rate (Fielden et al, 1974), which in Krebs-Henseleit buffer is about $1.0\,s^{-1}$ (M.E. Murphy, C. Firnhaber and H. Sies, unpublished). Therefore, O_2^- would need to be generated at $1.8\,nM\,s^{-1}$ in buffer, or $>10^4$-fold faster inside cells (which, depending on the tissue, contain about $10\,\mu M$ endogenous SOD ($K_{SOD} + O_2^- = 3 \times 10^9\,M^{-1}\,s^{-1}$ (Fielden et al, 1974)), to maintain a steady-state O_2^- concentration capable of affecting the half-life of NO. For comparison, $2 \times 10^5\,ml^{-1}$ FMLP-activated neutrophils generate O_2^- at about $2\,nM\,s^{-1}$ (McCall et al, 1989; Schmidt et al, 1989). It is important to realize, however, that normal tissues are not likely to reach this level of O_2^- production or concentration (Freeman and Crapo, 1982).

Ferrous iron

Ferrous iron added to buffer or media deactivates NO and EDRF. It is known that Fe(II) can react in oxygenated solutions to generate O_2^- and ferric iron (Fe(III)) (reaction 9). Additionally, the H_2O_2 arising from the subsequent reactions of O_2^- can also oxidize the Fe(II), yielding Fe(III), hydroxide and hydroxyl radical. The rate of O_2^- generation from the amounts of Fe(II) used to deactivate NO/EDRF (e.g. $300\,\mu M$ $Fe(II)SO_4$) may be sufficient to explain the deactivation. However, ferrous iron is also capable of a rapid and strong binding of NO in buffers that contain anions, to form tetrahedral complexes such as NO_2FeCl_2 and $NO_2Fe(PO_4)_2$ (McDonald et al, 1965; Beattie, 1967; Stolze and Nohl, 1989). There is no evidence concerning the importance of this mechanism of deactivation by Fe(II).

$$Fe(II) + O_2 \rightleftharpoons Fe(III) + O_2^- \tag{9}$$

The potential binding of NO to ferrous iron may have further implications. Since the reaction of oxygen with Fe(II) is reversible, the formation of O_2^- by other mechanisms may lead to the reduction of trace Fe(III) to Fe(II) in solution. The ferrous iron might then bind to and thus deactivate NO. This deactivation would depend on O_2^-, and be inhibitable by SOD, but would not be due to the direct reaction between O_2^- and NO. The importance of this mechanism is likewise unknown, but would theoretically depend on the ratio between the concentrations of NO and Fe(III).

Pyrogallol

Pyrogallol (1,2,3-trihydroxybenzol) undergoes a spontaneous autoxidation in oxygenated solutions to yield a variety of coloured compounds. O_2^- is not only generated by the reaction, but also catalyses the reaction by reacting with pyrogallol (or perhaps its radical intermediates) to initiate more O_2^- production (Marklund and Marklund, 1974). Figure 4 indicates a possible series of reactions, but the exact chemistry is not clearly defined. Nevertheless, this self-catalytic O_2^- generation forms the basis for the use of pyrogallol in SOD assays. Pyrogallol is structurally related to hydroquinone, also purported to deactivate NO via O_2^- production (Hutchinson et al, 1987), and is also related to gossypol, whole mechanism is still in question (Förstermann et al, 1986; Bhardwaj and Moore, 1988).

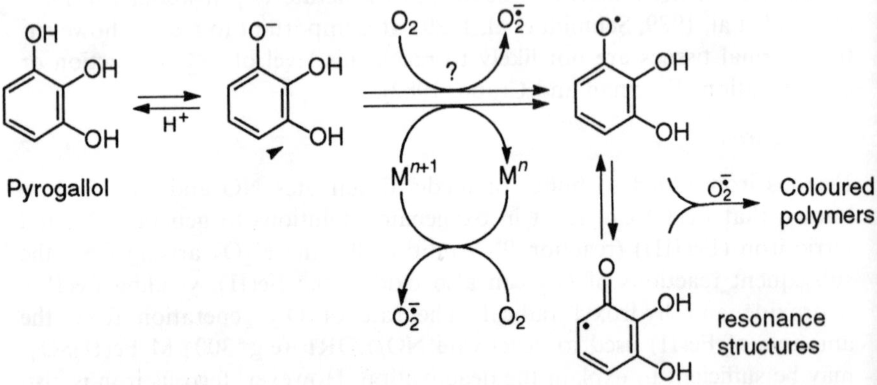

OH
OH
OH
Pyrogallol
H^+
O^-
OH
OH
O_2
$O_2^{\bar{}}$
?
M^{n+1}
M^n
$O_2^{\bar{}}$
O_2
$O^\bullet$
OH
OH
O
OH
OH
$O_2^{\bar{}}$
Coloured polymers
resonance structures

Figure 4. Hypothetical reaction scheme for pyrogallol autoxidation in oxygenated solutions.

Pyrogallol added to buffer or media deactivates NO/EDRF, and this is attributed to the O_2^- generated by pyrogallol. However, the possibility of a direct reaction of NO with pyrogallol, or with a radical intermediate of its autoxidation, has not been studied. Pyrogallol, like hydroquinone, is actually an excellent free radical scavenger, and is even the recommended antioxidant for some lipid extraction procedures. SOD not only reduces the steady-state concentration of O_2^- in the presence of pyrogallol, but also indirectly decreases the rate of generation of the other radical intermediates. Thus, any reaction between NO and such intermediates would likewise be inhibited by SOD.

Dithiothreitol

Dithiothreitol (DTT) is a small dithiol compound designed and employed as a potent reducing agent that can be tolerated by cells (Cleland, 1964). Its reducing potential is responsible for its ability to autoxidize and generate O_2^- in oxygenated solutions. The ability of DTT to deactivate NO/EDRF in buffer and media is attributed to this O_2^- generation, but amazingly the reference (Misra, 1974) often cited to support this theory specifically demonstrates that O_2^- production does not occur at pH 7.4 and is catalysed by trace metals (Figure 5). This would generate the reduced state of the metal, such as Fe(II), whose chemistry is outlined above. The binding of NO by complexes of Fe(II) and thiols is also possible (McDonald et al, 1965). Furthermore, the oxidation of a thiol would generate at least a transient thiol radical intermediate. The role of trace metals and/or radical thiol intermediates in the deactivation of NO/EDRF by DTT has not been investigated.

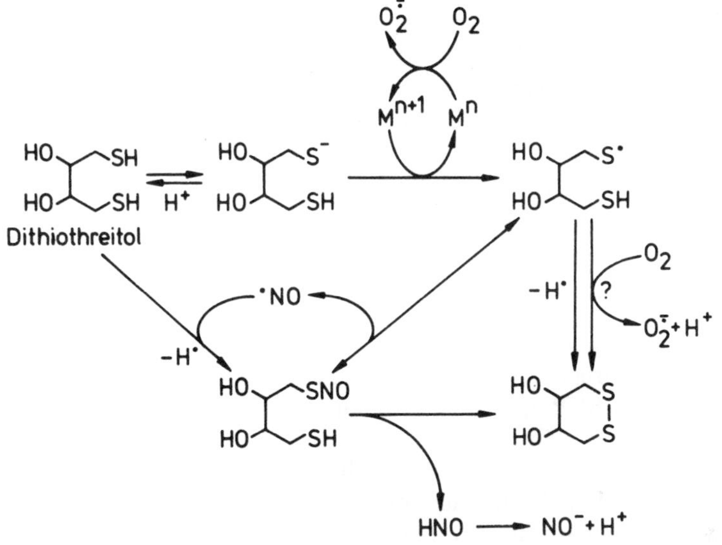

Figure 5. Theoretical pathway leading to generation of radical thiol intermediates and O_2^- anions from dithiothreitol which may deactivate NO/EDRF.

Nitrosothiols are products of a direct reaction between thiols and NO and are fairly stable in oxygenated solutions (Ignarro and Gruetter, 1980; Ignarro et al, 1981; Ignarro, 1990). The half-life of *S*-nitrosocysteine at 37°C in air is over one hour (Ignarro et al, 1981). The half-life of nitrosodithiothreitol may be shorter than this (Ignarro and Gruetter, 1980), but if the direct reaction between DTT and NO proceeds readily enough, this transient binding could account for the deactivating effects of DTT. This would be most effective in perfused systems, but a slow reversibility of NO binding could still facilitate dispersion of NO in closed systems. The ability of cysteine to cause the release of NO from guanylate triphosphate and subsequently stimulate guanylate cyclase shows lower than expected activation of guanylate cyclase at high cysteine levels, perhaps also reflecting the NO-binding effects of excess thiol.

The release of NO from thiols may involve disulphide formation (Hart, 1985), perhaps releasing hyponitrous acid (HNO) which deprotonates to nitrosyl anion (NO^-) at neutral pH (Park, 1988) (Figure 5). Neither the formation of NO^- via thiols nor its ability to activate guanylate cyclase has been investigated, but is is worth noting that thiols increase the NO_2^-/NO_3^- ratio arising from NO (Feelisch et al, 1989; Stuehr et al, 1990), and that NO_2^- (but not NO_3^-) is an expected product of NO^- (Seddon et al, 1973; Saran et al, 1990). The disulphide formation and NO release is slower for nitrosoglutathione ($3 \times 10^{-4}\ M^{-1}\ s^{-1}$) than for the cysteine or protein adducts ($0.1\ M^{-1}\ s^{-1}$) (Park, 1988), and it is therefore interesting that agents which specifically deplete glutathione enhance vasodilation (Needleman et al, 1973). A need for a second thiol to allow NO release from nitrosothiols may explain why nitrosothiols are ineffective in activating guanylate cyclase pretreated with ethacrynic acid or cystine (Ignarro et al, 1980), but a complete inactivation of guanylate cyclase by these two strong thiol oxidants cannot be excluded (Craven and DeRubertis, 1978; Needleman et al, 1973). It may be worth noting that DTT is a common (or even necessary) component of in vitro guanylate cyclase assays (Ignarro et al, 1981).

DTT slightly decreases the response of guanylate cyclase to pure NO (Ignarro and Gruetter, 1980), but increases the stimulation due to other agents, such as SNP and other nitrosothiols (Ignarro and Gruetter, 1980; Ignarro et al, 1980; Craven and DeRubertis, 1978). DTT and other thiols can also reverse the oxidation of thiol residues of guanylate cyclase, restoring the enzyme to full activity (Brandwein et al, 1981). Furthermore, DTT and other thiols directly stimulate guanylate cyclase, especially under low oxygen conditions (Niroomand et al, 1989). DTT also has direct effects on the responsiveness of target tissues to NO that may be independent of any direct interaction with NO (Needleman and Johnson, 1973; Axelsson et al, 1982), and may protect or maintain NO synthetase in a high-activity form (Stuehr et al, 1990). The role of these changes in the effects of DTT have not usually been controlled for, and leaves doubt about the mechanism by which DTT inhibits the action of NO in most studies.

B. Enhancement by superoxide dismutase

Superoxide dismutase (SOD) is the only agent consistently reported to stabilize NO/EDRF. The effectiveness of SOD has been demonstrated in perfused and closed systems involving a variety of tissues and cells. Interestingly, in only a few cases has the production of O_2^- been simultaneously estimated (e.g. McCall et al, 1989). The rate of O_2^- generation, and estimates of its concentration, are necessary to distinguish whether a direct reaction between NO and O_2^- is causing the NO deactivation, whether other reactions are involved. For example, the half-life of NO perfused through a heart was about 120 ms (i.e. about 50-fold faster than in the buffer) (Kelm and Schrader, 1990). The major end-product of the NO was NO_2^-, which argues against O_2^- driving the breakdown, since NO_3^- would be expected. A similar finding led Furchgott and Jothianandan (1990) to conclude that reactions 3 and 4 (Section I) are wrong. Nevertheless, SOD protected the NO at least to some extent in both studies. Ignarro (1990) has also noted these discrepancies. The protection of NO/EDRF by SOD also seems too good in other cases, since it reportedly worked even in the absence of cells (Warner et al, 1989). The conclusion of the investigators was that Krebs-Henseleit buffer must actually contain a high steady-state level of O_2^-, but the improbability of this demands a further investigation of the effects of SOD.

C. Purported differences between NO and EDRF

The recognition that NO and EDRF are identical is not entirely universal, with two basic challenges still standing. The first involves the differences between the chemical properties of NO and EDRF, and the second involves identification of compounds (nitrosothiols) with properties similar to those of NO and which have potent EDRF activity. The response in support of NO has been extensive, but not yet irrefutable.

Some authors report that EDRF differs from NO in that it can be bound to anion exchange columns, survives storage at −70°C and is stabilized in acid solutions (Murray et al, 1986). NO shares none of these properties. Furthermore, there are observations that a subset of target tissues respond to NO while being unaffected by the EDRF released from cultured bovine endothelial cells (Gräser et al, 1990; Shikano and Berkowitz, 1987; Shikano et al, 1988). It would be shortsighted to regard these observations as artefacts. More probably, they reflect the uncharacterized interference by other endothelial-derived relaxing and/or contracting factors, or involve some bound or reversibly altered form of NO, such as nitrosothiols or NO^-. It is worth noting that the odd properties of EDRF only appear under specific experimental conditions.

Nitrosothiols have been well studied and in some ways represent a legitimate alternative possibility as EDRF. They can be formed by reactions of many types of thiols with NO or with other nitrogen compounds (Ignarro, 1990). The nitrosothiols can also activate guanylate cyclase

(Ignarro and Gruetter, 1980), but this property may just be due to the NO that they spontaneously release (Noack and Feelisch, 1991; Section V.C). Also, critical evidence is lacking that nitrosothiols are produced by the EDRF-generating cells, although it is worth noting that supposedly 'direct' measurements of NO are often made following treatments with strong reducing conditions, in which nitrosothiols would give an identical release of NO (Myers et al, 1990). We are currently investigating whether endogenous thiols are required for NO synthesis by cultured porcine endothelial cells. Our data so far suggest that the depletion of glutathione and/or protein thiols has no direct effect on NO production (Murphy and Sies, unpublished results).

D. Interference by other factors

In response to the same agents that stimulate NO generation, endothelial cells can affect the relaxation of smooth muscle by other direct and indirect mechanisms (review: Furchgott and Vanhoutte, 1989). This concern applies as well to other pairs of NO-generating and NO-responsive cells. In addition, the NO-generating cells are often responsive to their own NO, with NO-dependent mechanisms controlling the synthesis of release of the other factors (Yoon and Deykin, 1990; Boulanger and Lüscher, 1990). These interactions complicate the interpretation of studies involving NO synthesis and action, but are not insurmountable problems.

Endothelial cells produce contractions in isolated arteries and veins by releasing endothelium-derived contracting factors (De Mey and Vanhoutte, 1983; Furchgott, 1988; Furchgott and Vanhoutte, 1989). One of these may be endothelin, a vasoconstricting 21-amino acid peptide synthesized by aortic endothelial cells and released simultaneously with NO. Superoxide may be another contracting factor (Vanhoutte and Katusic, 1988). The endothelial cells of veins also produce a vasoconstrictor which may be a cyclooxygenase product, while the cells from arteries produce a vasoconstrictor that may be thromboxane A_2. In addition, NO may inhibit thromboxane A_2 synthesis (Darius et al, 1984). Taken all together, these changes may explain the endothelium-dependent constriction of smooth muscle when NO is deactivated by HbO_2 (Martin et al, 1985), or when its synthesis is inhibited with arginine analogues.

Prostacyclins and other products of the lipoxygenase and cyclooxygenase pathways are also released by stimulated endothelium. Although some constrict veins, others dilate arteries, as well as having antiaggregatory and antiadhesive effects on platelets. These mediators were well studied before the recognition of NO as an EDRF, and to avoid their effects, the lipoxygenase inhibitor indomethacin (10 μM) is used in nearly all studies in which target tissue response to NO/EDRF is measured. Finally, endothelial cells produce a diffusible factor termed the endothelium-derived hyperpolarizing factor (EDHF) with properties distinct from those of NO. As its name implies, the target smooth muscle cells experience a hyperpolarization of their membranes and relax. The release of a non-NO EDHF may be stimulated by bradykinin, substance P and thimerosal (Bény, 1990), consistent with our findings that thimerosal does not induce NO synthesis (Murphy and Sies, unpublished results). However, conflicting evidence that NO can have direct hyperpolarizing effect also exists (Tare et al, 1990).

VI. NO DEPRESSION OR IMBALANCE IN CARDIOVASCULAR DISEASES

Experimental work suggests that NO released by the endothelium is one of the most important factors controlling the vascular tone of resistance vessels and large arteries such as the coronary arteries. When the endothelium is damaged by atherosclerotic, toxic or mechanical injury of the vessel wall, there is no longer a vasodilator response. The absence or damage of endothelium can even lead to vasoconstriction, due to a local deficiency of NO and/or the production of naturally occurring vasoconstricting substances such as endothelin. The dysfunction of EDRF production by endothelial cells, or at least a deterioration of local tonus homeostasis, appears to be an early sign of vascular disease such as atherosclerosis and hypertension and may favour cerebral or coronary vasospasm (Rubanyi and Vanhoutte, 1986b).

Evidence for a fundamental role of NO in blood flow regulation and blood pressure came from experiments with endothelium-containing vascular rings of the aorta and other arteries. If such rings are brought to contraction with a vasoconstrictor, addition of a selective inhibitor of endothelial NO synthesis or a scavenger of NO rapidly produces an additional augmentation of contraction by blocking the action of basally released NO. Later, specific inhibitors of the enzymatic formation of NO in endothelium (Section IV.A) allowed an analysis of the physiological role of endothelium-derived NO in vivo, and this has indicated that it plays a significant role in regulating the tone of resistance vessels and other vessels. Rees et al (1989b) intravenously administered the antagonist L-NMMA in the anaesthetized rabbit and observed a stereospecific, dose-dependent (3–100 mg kg^{-1}) elevation of systemic blood pressure. The hypertension was long-lasting (15–90 min), and as expected an excess of arginine rapidly reversed the effect. L-NMMA also partially inhibited the hypotensive action of ACh, supporting the assumption that ACh exerts its in vivo haemodynamic activity by triggering NO release.

Physicochemical factors may stimulate NO release (Furchgott and Vanhoutte, 1989), including pulsatile characteristics of blood flow (Pohl et al, 1986) and a reduced oxygen content of the blood (Ku, 1982; Mügge et al, 1989). These act in concert with circulating biogenic amines and vasoactive compounds which bind to specific receptors at the endothelial surface and trigger the release of NO. In addition, NO contributes to the control of renin release, with changes in hydromechanical forces influencing the release of NO from the endothelium of the renal vascular system (Rubanyi et al, 1986). Thus, these endothelial cells could act as sensors which respond to changes in physical factors such as an increase in shear stress, inducing proportional changes in renin release (Vidal et al, 1988).

A. Coronary artery disease

Coronary artery disease is characterized by an imbalance between oxygen supply and demand in the working myocardium due to the diminished blood flow resulting from a mechanical hindrance. The most important processes which narrow the vessel lumen are plaque fissuring, coronary vasospasm and platelet thrombus formation. There is experimental evidence that normal endothelium function is a prerequisite for preventing such vascular complications (review: Lüscher, 1989). It has a protective function against increased platelet adhesion and aggregation as well as against vascular hyperreactivity. Endothelial-derived NO may explain these effects, as well as regulate the normal coronary vascular tone (Amezcua et al, 1989). The evidence of a direct correlation between endothelium dysfunction and coronary artery occlusion justifies therapeutic efforts for a substitution of a local deficiency of NO by administration of drugs like glycerol trinitrate or isosorbide dinitrate. Nitrovasodilator compounds also increase the content of cGMP in isolated platelets, probably by the same mechanism (Men'shikov et al, 1987), although NO may normally act synergistically with endothelium-derived prostacyclins to inhibit platelet aggregation (Moncada et al, 1987).

In addition, endothelium-dependent relaxations are often decreased in atherosclerotic arteries in response to ACh and other vasoactive substances, even though endothelium-independent vasodilators remain fully active (Freiman et al, 1986; Bossaller et al, 1987). Elevated vascular tone resulting from impaired relaxation (a dynamic coronary stenosis) may become crucial if an additional local injury or alteration occurs. Damaged endothelium regenerates quickly but afterwards may show paradoxical reactions to vasorelaxants, resulting in vasoconstriction. Thus, infusion of ACh into the left anterior descending coronary artery causes contractions in patients with coronary heart disease (Ludmer et al, 1986). Epidemiological evidence indicates that hypertension favours atherosclerotic alterations of the coronary arteries. One reason may be that acute hypertension selectively and profoundly attenuates endothelium-dependent responses to serotonin, as shown in the canine coronary circulation by Lamping and Dole (1987).

B. Hypertension

A decrease in endothelium-dependent relaxation has been demonstrated in vessels from hypertensive animals (Lüscher et al, 1988) and in a variety of experimental models of hypertension (Winquist et al, 1984). Combined with the evidence above that NO is a major component of the normal control of blood pressure, it is tempting to assume that disturbances in the biochemical sequences of NO synthesis and activities at the vascular site may be involved in certain forms of hypertension or may be one of the fundamental factors in the process of hypertension development. Some forms of hypertension may also be due to a relative deficit of NO as a consequence of increased liberation of constricting factors, such as endothelin, by the endothelial cells. Therapeutic progress in the future management of essential hypertension may thus include manipulation of vascular NO formation.

C. Ischaemia

An impairment of regional NO release may be a significant pathophysiological factor in myocardial ischaemia. For example, strong hypoxia impairs NO-related vascular relaxation. The presence of oxygen is essential for the synthesis of NO (Section IV.A) and long-lasting hypoxia may even result in local NO deficiency in coronary vessels after acute myocardial ischaemia (Ku, 1982; Lefer et al, 1990). Rubanyi and Vanhoutte (1985) demonstrated an endothelial-dependent vasoconstriction under comparable conditions in the coronary arteries of canines, which may be due to increased endothelin release. There is also indirect evidence that oxygen-derived free radicals may aggravate ischaemia, since agents which are thought to scavenge or inhibit the production of oxygen radicals often reduce reperfusion-induced ventricular fibrillation and arrhythmias (Hearse and Tosaki, 1987). However, whether O_2^- scavenging by NO (or a lack of normal scavenging) might be a factor in ischaemia or reperfusion would depend on the local concentrations of half-lives of both NO and O_2^-.

VII. DRUG-DEPENDENT VASORELAXATION—DONORS OF NO

A. Implications for therapy

As outlined above, NO was only recently found to be the EDRF synthesized and released from various tissues. Of its different physiological activities, the vasorelaxant effect is clinically the most important, not only by allowing new insights into the pathogenesis of cardiovascular diseases but also by providing a fascinating basis for the expansion and development of new rational therapies in the treatment of cardiovascular diseases. Although nitrovasodilator drugs such as nitroglycerol, sodium nitroprusside or isosorbide dinitrate have been widely used for decades, it was realized only recently that all these compounds exert their pharmacodynamic action by an identical final step of bioactivation, i.e. the generation of NO (Feelisch and Noack, 1987a,b; Noack et al, 1986; Schröder and Noack 1987). This confirmed the early thesis of Katsuki et al (1977), who favoured NO as the terminal activator of guanylate cyclase. Thus nitrovasodilators are prodrugs preserving NO in their molecular structure, carrying it safely to the vascular site, and exhibiting their biological activity after NO liberation. However, the pathways of bioactivation differ between the nitrovasodilators.

B. Chemistry

Nitrovasodilator compounds differ considerably chemically. Figure 6 presents the structures of some important compounds.

The substances most often used therapeutically are the nitric acid esters of organic alcohols such as nitroglycerol (glyceryl trinitrate, GTN) and derivatives of isosorbide. Esters of nitrous acid, such as isopentylnitrite, also have pharmacological activity, and were already clinically used in the last century. Of the sydnonimine derivatives, only molsidomine is therapeutically used. The furoxans are a new group of vasodilators which were further developed from sydnonimines. Currently, *S*-nitrosothiols and NO itself are merely of scientific interest since their half-life in vivo is too short to be of practical use. Last but not least, sodium nitroprusside, an inorganic compound, spontaneously releases NO.

C. Bioactivation pathways

The pharmacodynamic activity of organic nitrates primarily depends on the physicochemical properties of these drugs. There is a close correlation between the lipophilicity and the coronary vasodilatory action in the isolated, pressure-constant perfused Langendorff heart preparation (Noack, 1984). This is due to the requirement for organic nitrates to penetrate the plasmalemma of the vascular smooth muscle cells before bioconversion can take place. Stereochemical aspects also play an additional, modulatory role in the case of isosorbide derivatives such as ISDN, where the *exo* conformation of the nitrate ester group shows higher pharmacological activity than the *endo* position.

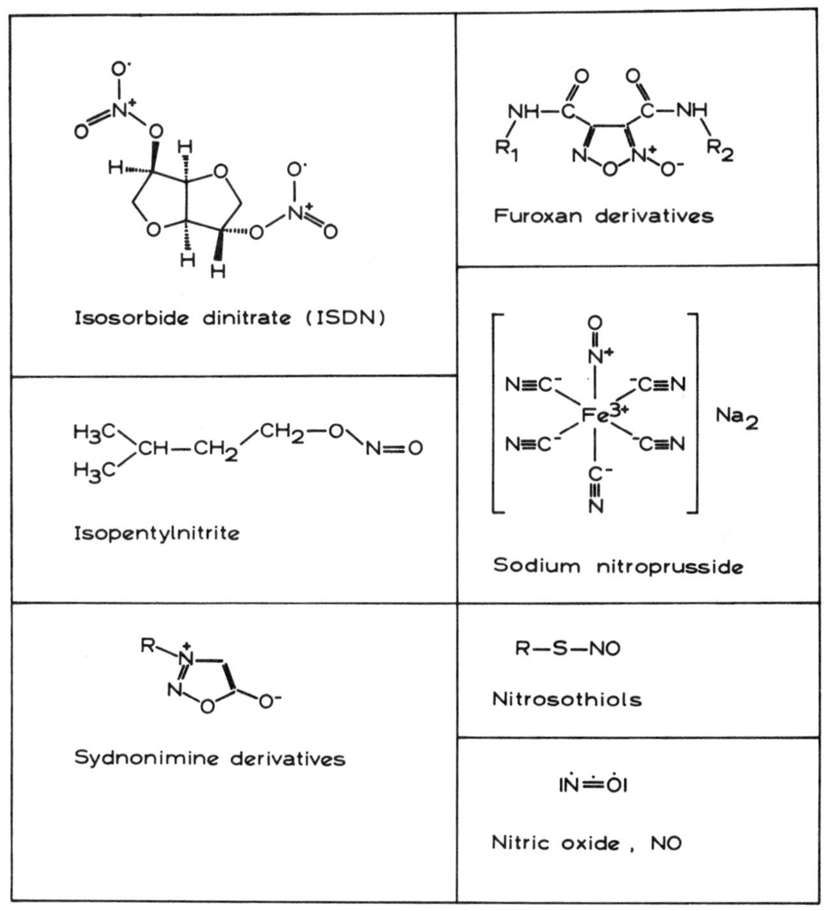

Figure 6. Representative nitrovasodilators.

Two types of bioactivation pathway, one enzymatic and one non-enzymatic, lead to the generation of NO from organic nitrates. They may compete with one another in vivo, depending on the local enzymatic presence and conditions. The enzymatic pathway involves glutathione *S*-transferase, which catalyses the degradation of organic nitrates to inorganic nitrite, requiring the oxidation of two moles of GSH. In 1883 Hay proposed that the nitrite anion formed upon enzymatic degradation might be the ultimate active metabolite. By simultaneous estimation of NO and NO_2^- formation, we excluded the possibility that NO_2^- participates in the pharmacological activity of organic nitrates (Feelisch and Noack, 1987a). This concept was also contradicted by the finding that NO_2^- only showed a vasodilatory effect at considerably higher doses than the organic nitrates. Finally, the basal NO_2^- concentration within the smooth muscle is already in the micromolar range. Thus, the intracellular activity of the glutathione *S*-transferase may be regarded pharmacologically as disadvantageous, destroying part of the drug. There may be other enzymatic degradation pathways like a cytochrome P450-catalysed reductive denitration in the presence of NADPH (Servent et al, 1989) or a microsomal monooxygenase system which produces minor amounts of NO.

The pathway through which organic nitrates liberate NO and activate guanylate cyclase requires the presence of cysteine, a thiol-containing amino acid, in broken cell preparations and under in vitro conditions. However, incubation with other thiols resulted in decomposition of organic nitrates without the formation of NO (Feelisch and Noack, 1987c). Therefore, the thiol-catalysed degradation of organic nitrates does not necessarily give rise to guanylate cyclase activation.

On the other hand, the cysteine-mediated in vitro reaction may reflect processes at the active centre of the guanylate cyclase itself, where cysteine molecules are known to be in close proximity. It is not known whether these cysteinyl residues closely associated with the active site of guanylate cyclase may also catalyse the generation of NO.

However, we propose that non-enzymatic bioactivation is the essential mechanism for the pharmacological efficacy of organic nitrates in vivo, especially in the case of high tissue concentrations.

The cysteine-induced reaction is characterized by a simultaneous formation of NO_2^- and NO at a constant ratio of 14:1, regardless of the structure of the organic nitrate. Therefore, a common thiol-specific intermediate, possibly a thionitrate ($RSNO_2$), may result from an intermolecular *trans* esterification of the nitrate residue to the respective thiol, which thereafter degrades with a fixed ratio of NO_2^- and NO after intramolecular rearrangement. The resulting rate of NO formation directly correlates with the extent of guanylate cyclase activation (Noack et al, 1986; Feelisch and Noack, 1987a). These in vitro findings were recently confirmed in strips of isolated rabbit aorta and bovine pulmonary artery. Kawamoto et al (1990) examined glyceryl trinitrate biotransformation and cGMP elevation in vascular smooth muscle before the onset of GTN-induced relaxation and found results consistent with the hypothesis that the magnitude of GTN biotransformation is the important determinant of subsequent tissue relaxation. The rate of degradation of GTN and of release of NO directly correlated with the extent of guanylate cyclase activation and cGMP formation (Schröder et al, 1985).

Isopentylnitrite (amyl nitrite) and other organic nitrites spontaneously release NO in aqueous solution at a slow rate. If cysteine or any other thiol-containing compound is present, the thiolate ion rapidly reacts to form the corresponding *S*-nitrosothiol (Patel and Williams, 1990), from which NO is ultimately liberated at a rate dependent on the chemical stability of the nitrosothiol.

Sodium nitroprusside contains NO bound in a complex form. It may spontaneously decompose in aqueous solutions and release NO directly, but the process is probably more complex. For example, there is no linear correlation between nitroprusside concentration and NO formation, with some evidence also suggesting a direct interaction between nitroprusside and guanylate cyclase as a mechanism of activation. Furthermore, it is suggested that a one-electron reduction of nitroprusside drives the release of NO. Ascorbate or thiols may reduce the nitroprusside (Ignarro et al, 1980; Wood and Ignarro, 1987), but O_2^- also acts as a reductant and accelerates the apparent rate of NO release (Misra, 1984). It is interesting to speculate that this may make nitroprusside-like vasodilators even more potent at tissue sites (e.g. reperfused or atherosclerotic vessels) where the O_2^- concentration may be elevated. However, nitroprusside is reported to be a specific inhibitor of SOD (Misra, 1984). It is worth noting, though, that it showed an identical K_i for several unrelated forms of SOD, thus suggesting a mechanism other than simple inhibition.

In contrast to the organic nitrates, sydnonimines degrade in the absence of thiol-containing compounds. The bioactivation pathway of these pharmacologically active substances (e.g. SIN-1, the metabolite of molsidomine) was recently clarified in detail (Bohn and Schönafinger, 1989; Feelisch et al, 1989; Noack and Feelisch, 1989). NO is released from the open-ring A forms of sydnonimines, which form in a base-catalysed reaction. Since this reaction is fairly slow at neutral pH, the overall NO release is nonlinear with time but concentration-dependent over a wide range. All sydnonimines also generate NO_2^- and NO_3^- at an approximately equimolar ratio. Molecular oxygen is a prerequisite for the final step of NO release from the open-ring A form, acting as a one-electron oxidant (being converted to O_2^-), while the radical cation A form of the sydnonimine then releases NO. Metals and metal complexes which can undergo one-electron oxidations can substitute for oxygen, in some cases yielding O_2^- in later reactions with O_2. Superoxide from either source may catalyse further steps of sydnonimine decomposition. The reaction between NO and O_2^- may falsify the determination of NO formation from sydnonimines.

Furoxans are chemically related to sydnonimines but liberate only NO upon addition of thiols. Unlike the case with the organic nitrates, all thiols drive this degradation (Feelisch and Noack, 1989). Furoxans rapidly form *S*-nitrosothiols, the concentration of which parallels the rate of NO generation.

S-Nitrosothiols were suggested to be direct activators of the guanylate cyclase (Ignarro et al, 1979). We synthesized a series of nitrosothiols and correlated the individual NO liberation rates with the extent of enzyme activation (Noack and Feelisch, 1991); NO liberation and biological activity were directly dependent on the chemical instability of the compounds. This indicated that NO mediates the reaction at the enzyme site and not the intact nitrosothiols.

D. Nitrate tolerance

The fact that nitrovasodilators possess different bioactivation pathways for the generation of NO provides the key for a better understanding of the clinical phenomenon of tolerance. This occurs almost exclusively with organic nitrates because these cannot set free NO directly like the sydnonimines or organic nitrites. The tendency for a nitrate to induce tolerance correlates with the number of nitrate groups in the molecule, lipophilicity and degradation rate (Noack, 1990). Since bioactivation may specifically depend on the presence of cysteine for the intermediary formation of thionitrates, a shortage of this essential cofactor may be crucial for nitrate-tolerance development. Blood vessels that relax in response to even low concentrations of organic nitrates (e.g. veins) may have a greater metabolic capacity for degradation (enzymatic as well as non-enzymatic). Whether this is due to a larger cysteine-regenerating capacity has to be confirmed by additional investigations.

VIII. FUTURE DEVELOPMENTS

A. Oxygen radical scavenging by donors of NO?

The beneficial effects of nitrovasodilators in the therapy of angina have most often been ascribed to haemodynamic changes such as a decrease of preload, bringing indirect relief to the working heart by covering its oxygen demand. The mechanism of action of the various nitrovasodilators would logically be to release NO in vessels, perhaps compensating for a local deficiency of NO (e.g. as a consequence of atherosclerotic alterations). However, protecting ischaemic myocardial tissue from irreversible damage as a consequence of oxygen free radicals may also be important, especially during long-term prophylactic treatment. Thus, in ischaemic heart disease, nitrovasodilators such as nitroglycerol may be advantageous because they release NO as scavengers of O_2^-, and may thus represent a type of oxygen radical-scavenging therapy analogous to that proposed for SOD and catalase (Greenwald, 1990). So far, clinical studies have been unable to confirm this pharmacotherapeutic aspect due to technical limitations. McCall et al (1989) suggested that NO may significantly reduce the levels of O_2^-, so that neutrophils may leave the vascular system without damaging the endothelium, at least at lower degrees of stimulation. But, as mentioned earlier, whether the concentration of either NO or O_2^- is significantly affected by a reaction with the other cannot be decided without a better knowledge of the actual steady-state concentrations and rates of alternative reactions of each.

B. ACE inhibitors

Some in vitro experiments suggest that compounds with a sulphydryl moiety, such as the angiotensin-converting enzyme (ACE) inhibitors captopril and zofenopril, may be cardioprotective by interacting with oxygen free radicals (Mak et al, 1989; Przyklenik and Kloner, 1987; Westlin and Mullane, 1988). It remains uncertain whether these findings can be transferred to in vivo conditions, since rather high in vitro concentrations of captopril had to be used (IC_{50} values of 8–10 μM versus therapeutic levels near 1 μM). Preincubation of cultured endothelial cells (10^6 ml^{-1}) with SH-containing ACE inhibitors (captopril, epi-captopril, and free-SH form of zofenofril) before free radical addition resulted in an equipotent inhibition of lipid peroxidation ($IC_{50} = \sim 25$ μM; Mak et al (1989)). Furthermore, both captopril and its isomer SQ 14,534, which is 100-fold less potent as an ACE inhibitor but equipotent in scavenging superoxide anions, equally improved postischaemic contractile dearrangements when administered at reperfusion (5 mg kg^{-1}).

C. Sydnonimines

As mentioned, during the generation of NO from the sydnonimines such as molsidomine (the prodrug of SIN-1), oxygen is converted to superoxide radicals while the A form of the sydnonimines releases NO. The O_2^- may reduce their pharmacological efficacy by reacting with NO, and would certainly nullify any potential therapeutic benefit of these vasodilators as O_2^- scavengers. Since the relationship between superoxide and NO generation may depend on chemical structure, the sydnonimines that appear to yield a higher ratio of NO to O_2^- may find wider application in the future.

D. New nitrates

New nitrovasodilators will be introduced into therapy which are free of nitrate tolerance development, which may involve the exhaustion of the cellular cysteine pool (Noack, 1990). Tolerance may even cause an undesirable positive feedback on the need for NO, since basal EDRF release from endothelial cells is also lowered by cysteine depletion (Minor et al, 1989). This clinical goal may be achieved by creating compounds which contain not only the structural prerequisites for NO liberation (e.g. nitrates and nitrites) but also sulphydryl groups for intramolecular *trans* esterification reactions and intermediary formation of thionitrites. NO could thus be generated independently of the intracellular status of thiol compounds such as cysteine and glutathione (Noack and Feelisch, 1991). The direct application of new nitrosothiols, more stable than those available now, would also avoid the problem of tolerance. Recently a captopril derivative, *S*-nitrosocaptopril, has been suggested to possess such properties (Cooke et al, 1989).

IX. CONCLUSIONS

In summary, both endogenously related NO and the commonly used nitrovasodilators (prodrugs of NO) have potent effects on blood vessels and other tissues, but in addition may have oxygen radical-scavenging capabilities. Thus, NO may potentially play a role in the biological as well as pharmacological fight against the adverse reactions of oxygen-derived free radicals, which may contribute to the pathophysiology of human diseases. NO, therefore, is a very simple nitrogen derivative which nature has utilized to modulate the function of vascular smooth muscle, but which is also likely to influence numerous other biochemical hazards and activities.

ACKNOWLEDGEMENT

This work was supported in parts by the Deutsche Forschungsgemeinschaft (SFB 242, Coronary Heart Disease, Teilprojekt C 3 Noack, Düsseldorf) and the Ernst-Jung-Stiftung für Medizin, Hamburg (Murphy).

REFERENCES

Amber IJ, Hibbs JB Jr, Taintor RR & Vavrin Z (1988) Cytokines induce an L-arginine-dependent effector sytem in nonmacrophage cells. *J. Leukocyte. Biol.* **44:** 58–65.

Amezcua JL, Palmer RMJ, de Souza BM & Moncada S (1989) Nitric oxide synthesized from L-arginine regulates vascular tone in the coronary circulation of the rabbit. *Br. J. Pharmacol.* **97:** 1119–1124.

Axelsson KL, Andersson RGG & Wikberg JES (1982) Vascular smooth muscle relaxation by nitro compounds: reduced relaxation and cGMP elevation in tolerant vessels and reversal of tolerance by dithiothreitol. *Acta Pharmacol. Toxicol.* **50:** 350–357.

Azuma H, Ishikawa M & Sekizaki S (1986) Endothelium-dependent inhibition of platelet aggregation. *Br. J. Pharmacol.* **88:** 411–415.

Beattie IR (1967) Nitric oxide. In *Supplement to Mellor's Comprehensive Treatise on Inorganic and Theoretical Chemistry*, Vol. VIII, Suppl. II, Section XXV, pp. 217–240. London: Longmans, Green and Co. Ltd.

Beckman JS, Beckman TW, Chen J, Marshall PA & Freeman BA (1990) Apparent hydroxyl radical production by peroxynitrite: Implications for endothelial injury from nitric oxide and superoxide. *Proc. Natl. Acad. Sci. USA* **87:** 1620–1624.

Bény J-L. (1990) The endothelium dependent relaxation linked to smooth muscle hyperpolarization is independent from the EDRF(NO) relaxation. *Arch. Int. Pharmacodyn. Ther.* **305:** 231.

Bény J-L, Brunet PC & Van der Bent V (1989) Hemoglobin causes both endothelium-dependent and endothelium-independent contraction of the pig coronary arteries, independently of an inhibition of EDRF effects. *Experientia* **45:** 132–134.

Bhardwaj R & Moore PK (1988) Endothelium-derived relaxing factor and the effects of acetylcholine and histamine on resistance blood vessels. *Br. J. Pharmacol.* **95:** 835–843.

Blough NV & Zafiriou OC (1985) Reaction of superoxide with nitric oxide to form peroxonitrite in alkaline aqueous solution. *Inorg. Chem.* **24:** 3502–3504.

Boeckxstaens GE, Bult H, Pelckmans PA, Jordaens FH, Herman AG & Van Maercke YM (1990) Nitric oxide release in response to stimulation of non-adrenergic non-cholinergic nerves. *Arch. Int. Pharmacodyn. Ther.* **305:** 232.

Bohn H & Schönafinger K (1989) Oxygen and oxidation promote the release of nitric oxide from sydnonimines. *J. Cardiovasc. Pharmacol.* **14 (supplement 11):** S6–S12.

Bossaller C, Habib GV, Yamamoto H, Williams C, Wells S & Henry PD (1987) Impaired muscarinic endothelium-dependent relaxation and cyclic guanosine 5′-monophosphate formation in atherosclerotic human coronary artery and rabbit aorta. *J. Clin. Invest.* **79:** 170, 174.

Boulanger C & Lüscher TF (1990) Release of endothelin from the porcine aorta. *J. Clin. Invest.* **85:** 587–590.

Brandwein HJ, Lewicki JA & Murad F (1981) Reversible inactivation of guanylate cyclase by mixed disulfide formation. *J. Biol. Chem.* **256:** 2958–2962.

Bredt DS & Snyder SH (1989a) Isolation of nitric oxide synthetase, a calmodulin-requiring enzyme. *Proc. Natl. Acad. Sci. USA* **87:** 682–685.

Bredt DS & Snyder SH (1989b) Nitric oxide mediates gluatamate-linked enhancement of cGMP levels in the cerebellum. *Proc. Natl. Acad. Sci. USA* **86:** 9030–9033.

Brüne B & Lapetina EG (1989) Activation of a cytosolic ADP-ribosyltransferase by nitric oxide-generating agents. *J. Biol. Chem.* **264:** 8455–8458.

Buga GM, Gold ME, Wood KS, Chaudhuri G & Ignarro LJ (1989) Endothelium-derived nitric oxide relaxes nonvascular smooth muscle. *Eur. J. Pharmacol.* **161:** 61–72.

Bult H, Boeckxstaens GE, Pelckmanns PA, Jordaens FH, Van Maercke YM & Herman AG (1990) Nitric oxide as an inhibitory non-adrenergic non-cholinergic neurotransmitter. *Nature (Lond.)* **345:** 346–347.

Busse R, Lückhoff A, Mülsch A & Bassenge E (1988) Modulation of EDRF release assessed by guanylate cyclase assay. In Bevan JA, Majewski H, Maxwell RA & Story DF (eds) *Vascular Neuroeffector Mechanisms*, pp 111–118. Oxford, Washington DC: IRL Press.

Carter TD, Bjaaland T & Pearson JD (1990) Endothelial cell (EC) calcium homeostasis and regulation of EDRF synthesis. *Arch. Int. Pharmacodyn. Ther.* **305:** 236.

Cassoly R (1981) Preparation of hemoglobin hybrids carrying different ligands: valency hybrids and related compounds. *Methods Enzymol.* **76:** 106–113.

Chen W-Z, Palmer RMJ & Moncada S (1989) Release of nitric oxide from rabbit aorta. *J. Vasc. Med. Biol.* **1:** 2–6.

Chiodi H & Mohler JG (1985) Effects of exposure of blood hemoglobin to nitric oxide. *Environ. Res.* **37:** 355–363.

Cleland WW (1964) Dithiothreitol, a new protective reagent for SH groups. *Biochemistry* **3:** 480–482.

Collier J & Vallance P (1989) Second messenger role for NO widens to nervous and immune systems. *Trends Pharmacol. Sci.* **10:** 428–431.

Cooke JP, Andon N, DelDuca D, Dzau V & Loscalzo J (1989) S-nitrosocaptopril: a novel vasodilator resistant to nitrate tolerance. *Circulation* **80 (supplement II-559):** 2226.

Cooke JP, Rossitch E Jr, Andon N, Loscalzo J & Dzau VJ (1990) Flow activates a specific endothelial potassium channel to release an endogenous nitrovasodilator. *Arch. Int. Pharmacodyn. Ther.* **305:** 238.

Craven PA & DeRubertis FR (1976) Chromatographic separation of cyclic guanosine 3′,5′-monophosphate from guanylate cyclase. *Anal. Biochem.* **72:** 455–459.

Craven PA & DeRubertis FR (1978) Restoration of the responsiveness of purified guanylate cyclase to nitrosoguanidine, nitric oxide, and related activators by heme and hemeproteins. *J. Biol. Chem.* **253:** 8433–8443.

Darius H, Ahland B, Rücker W, Klaus W, Peskar BA & Schrör K (1984) The effects of molsidomine and its metabolite SIN-1 on coronary vessel tone, platelet aggregration, and eicosanoid formation in vitro—inhibition of 12-HPETE biosynthesis. *J. Cardiovasc. Pharmacol.* **6:** 115–121.

DeMaster EG, Raij L, Archer SL & Weir EK (1989) Hydroxylamine is a vasorelaxant and a possible intermediate in the oxidative conversion of L-arginine to nitric oxide. *Biochem. Biophys. Res. Commun.* **163:** 527–533.

De Mey JG & Vanhoutte PM (1983) Anoxia and endothelium-dependent reactivity of the canine femoral artery. *J. Physiol. (Lond.)* **335:** 65–74.

Ding AH, Nathan CF & Stuehr DJ (1988) Release of reactive nitrogen intermediates and reactive oxygen intermediates from mouse peritoneal macrophages. *J. Immunol.* **141:** 2407–2412.

Doyle MP & Hoekstra JW (1981) Oxidation of nitrogen oxides by bound dioxygen in hemoproteins. *J. Inorg. Biochem.* **14:** 351–358.

Doyle MP, Pickering RA & Cook BR (1983) Oxidation of oxymyoglobin by nitric oxide through dissociation from cobalt nitrosyls. *J. Inorg. Biochem.* **19:** 329–338.

Drapier J-C, Wietzerbin J & Hibbs JB Jr (1988) Interferon-γ and tumor necrosis factor induce the L-arginine-dependent cytotoxic effector mechanism in murine macrophages. *Eur. J. Immunol.* **18:** 1587–1592.

Engler R (1987) Consequences of activation and adenosine-mediated inhibition of granulocytes during myocardial ischemia. *Fed. Proc.* **46:** 2407.

Feelisch M & Noack E (1987a) Correlation between nitric oxide formation during degeneration of organic nitrates and activation of guanylate cyclase. *Eur. J. Pharmacol.* **139:** 19–30.

Feelisch M & Noack E (1987b) Nitric oxide (NO) formation from nitrovasodilators occurs independently of hemoglobin or non-heme iron. *Eur. J. Pharmacol.* **142:** 465–469.

Feelisch M & Noack E (1987c) Molecular prerequisites of thiol-containing compounds like cysteine for their stimulatory action on the degradation of organic nitrates and activation of soluble guanylate cyclase. *Naunyn-Schmiedeberg's Arch. Pharmacol.* **335 (supplement):** 46.

Feelisch M & Noack E (1989) Thiol-induced generation of nitric oxide (NO) accounts for the vasodilatory action of furoxans. *Naunyn-Schmiedeberg's Arch. Pharmacol.* **339 (supplement):** R67.

Feelisch M, Ostrowski J & Noack E (1989) On the mechanism of NO release from sydnonimines. *J. Cardiovasc. Pharmacol.* **14 (supplement 11):** S13–S22.

Fielden EM, Roberts PB, Bray RC et al (1974) The mechanism of action of superoxide dismutase from pulse radiolysis and electron paramagnetic resonance. *Biochem. J.* **139:** 49–60.

Förstermann U, Goppelt-Strübe M, Frölich JC & Busse R (1986) Inhibitors of acyl-coenzyme A:lysolecithin acytltransferase activate the production of endothelium-derived vascular relaxing factor. *J. Pharmacol. Exp. Ther.* **238:** 352–359.

Freeman BA & Crapo JD (1982) Biology of disease. Free radicals and tissue injury. *Lab. Invest.* **47:** 412–426.

Freiman PC, Mitchell GG, Heistad DD, Armstrong ML & Harrison DG (1986) Atherosclerosis impairs endothelium-dependent vascular relaxation to acetylcholine and thrombin in primates. *Circ. Res.* **58:** 783–789.

Furchgott RF (1984) The role of endothelium in the response of vascular smooth muscle to drugs. *Annu. Rev. Pharmacol. Toxicol.* **24:** 175–197.

Furchgott RF (1988) Studies on relaxation of rabbit aorta by sodium nitrite: Basis for the proposal that the acid-activatable component of the inhibitory factor from retractor penis is inorganic nitrite and the endothelium-derived relaxing factor is nitric oxide. In Vanhoutte PM (ed.) *Mechanisms of Vasodilation*, pp 401–414. New York: Raven Press.

Furchgott RF & Jothianandan D (1990) Oxidation of nitric oxide (NO) to nitrite (NO_2^-) by oxygen (O_2) and by superoxide anion (O_2^-): relevance to identification of EDRF. *Arch. Int. Pharmacodyn. Ther.* **305:** 247.

Furchgott RF & Vanhoutte PM (1989) Endothelium-derived relaxing and contracting factors. *FASEB J.* **3:** 2007–2018.

Furchgott RF & Zawadzki JV (1980) The obligatory role of endothelial cells in the relaxation of arterial smooth muscle by acetylcholine. *Nature (Lond.)* **288:** 373–376.

Furlong B, Henderson AH, Lewis MJ & Smith JA (1987) Endothelium-derived relaxing factor inhibits in vitro platelet aggregation. *Br. J. Pharmacol.* **90:** 687–692.

Garg UC & Hassid A (1989a) Nitric oxide-generating vasodilators and 8-bromo-cyclic guanosine monophosphate inhibit mitogenesis and proliferation of cultured rat vascular smooth muscle cells. *J. Clin. Invest.* **83:** 1774–1777.

Garg UC & Hassid A (1989b) Inhibition of rat mesangial cell mitogenesis by nitric oxide-generating vasodilators. *Am. J. Physiol.* **257:** F60–F66.

Garthwaite J, Charles SL & Chess-Williams R (1988) Endothelium-derived relaxing factor release on activation of NMDA receptors suggests role as intercellular messenger in the brain. *Nature (Lond.)* **336:** 385–388.

Gerzer R, Hofmann F & Schulz G (1981) Purification of a soluble, sodium-nitroprusside-stimulated guanylate cyclase from bovine lung. *Eur. J. Biochem.* **116:** 479.

Gold ME, Wood KS, Buga GM, Byrns RE & Ignarro LJ (1989) L-arginine causes whereas L-argininosuccinic acid inhibits endothelium-dependent vascular smooth muscle relaxation. *Biochem. Biophys. Res. Commun.* **161:** 536–543.

Goretski J & Hollocher TC (1988) Trapping of nitric oxide produced during denitrification by extracellular hemoglobin. *J. Biol. Chem.* **263:** 2316–2323.

Gräser T, Vedernikow YP & Vanin AF (1990) Endothelium-derived relaxing factor may not be identical to nitric oxide. *Arch. Int. Pharmacodyn. Ther.* **305:** 250.

Greenwald RA (1990) Superoxide dismutase and catalase as therapeutic agents for human diseases. *Free Radicals Biol. Med.* **8:** 201–209.

Griffith TM, Edwards DH, Lewis MJ, Newby AC & Henderson AH (1984) The nature of endothelium-derived vascular relaxant factor. *Nature (Lond.)* **308:** 645–647.

Gruetter CA, Gruetter DY, Lyon JE, Kadowitz PJ & Ignarro LJ (1981) Relationship between cyclic guanosine 3′:5′-monophosphate formation and relaxation of coronary arterial smooth muscle by glyceryl trinitrate, nitroprusside, nitrite and nitric oxide: effects of methylene blue and methemoglobin. *J. Pharmacol. Exp. Ther.* **219:** 181–186.

Gryglewski RJ, Palmer RMJ & Moncada S (1986) Superoxide is involved in the breakdown of endothelium-derived vascular relaxing factor. *Nature (Lond.)* **320:** 454–456.

Hart TW (1985) Some observations concerning the S-nitroso and S-phenylsulphonyl derivatives of L-cysteine and glutathione. *Tetrahedron Lett.* **26:** 2013–2016.

Haussmann HJ & Werringloer J (1987) Mechanism and control of the denitrosation of N-nitrosodimethylamine. In Bartsch H, O'Neill K & Schulte-Herman R (eds) *Relevance of N-nitroso Compounds to Human Cancer: Exposure and Mechanisms*, Vol 84, pp 109–112. Lyon: IARC Sci. Publ.

Hay M (1883) The chemical nature and physiological action of nitroglycerin. *Practitioner* **30:** 422–433.

Hearse DJ & Tosaki A (1987) Free radicals and reperfusion-induced arrhythmias: protection by spin trap agent PBN in the rat heart. *Circ. Res.* **60:** 375.

Hibbs JB Jr, Taintor RR & Vavrin Z (1987) Macrophage cytotoxicity: Role for L-argninine deiminase and imino nitrogen oxidation to nitrite. *Science* **235:** 473–476.

Hibbs JB Jr, Taintor RR, Vavrin Z & Rachlin EM (1988) Nitric oxide: a cytotoxic activated macrophage effector molecule. *Biochem. Biophys. Res. Commun.* **157:** 87–94.

Hutchinson PJA, Palmer RMJ & Moncada S (1987) Comparative pharmacology of EDRF and nitric oxide on vascular strips. *Eur. J. Pharmacol.* **141:** 445–451.

Ignarro LJ (1990) Biosynthesis and metabolism of endothelium-derived nitric oxide. *Annu. Rev. Pharmacol. Toxicol.* **30:** 535–560.

Ignarro LJ & Gruetter CA (1980) Requirement of thiols for activation of coronary arterial guanylate cyclase by glyceryl trinitrate and sodium nitrite. *Biochim. Biophys. Acta* **631:** 221–231.

Ignarro LJ, Edwards JC, Gruetter DY, Barry BK & Gruetter CA (1979) Possible involvement of S-nitrosothiols in the activation of guanylate cyclase by nitroso compounds. *FEBS Lett.* **110:** 275–278.

Ignarro LJ, Barry BK, Gruetter DY et al (1980) Guanylate cyclase activation by nitroprusside and nitrosoguanidine is related to formation of S-nitrosothiol intermediates. *Biochem. Biophys. Res. Commun.* **94:** 93–100.

Ignarro LJ, Lippton H, Edwards JC et al (1981) Mechanism of vascular smooth muscle relaxation by organic nitrates, nitrites, nitroprusside and nitric oxide: evidence for the involvement of S-nitrosothiols as active intermediates. *J. Pharmacol. Exp. Ther.* **218:** 739–749.

Ignarro LJ, Degnan J, Baricos WH, Kadowitz PJ & Wolin MS (1982) Activation of purified guanylate cyclase by nitric oxide requires heme. Comparison of heme-deficient, heme-reconstituted and heme-containing forms of soluble enzyme from bovine lung. *Biochim. Biophys. Acta* **718:** 49–59.

Ignarro LJ, Adams JB, Horwitz PM & Wood KS (1986a) Activation of soluble guanylate cyclase by NO-hemoproteins involves NO-heme exchange. Comparison of heme-containing and heme-deficient enzyme forms. *J. Biol. Chem.* **261:** 4997–5002.

Ignarro LJ, Harbison RG, Wood KS & Kadowitz PJ (1986b) Activation of purified soluble guanylate cyclase by endothelium-derived relaxing factor from intrapulmonary artery and vein: stimulation by acetylcholine, bradykinin and arachidonic acid. *J. Pharmacol. Exp. Ther.* **237:** 893–900.

Ignarro LJ, Buga GM, Wood KS, Byrns RR & Chaudhuri G (1987a) Endothelium-derived relaxing factor produced and released from artery and veins is nitric oxide. *Proc. Natl. Acad. Sci. USA* **84:** 9265.

Ignarro LJ, Byrns RE, Buga GM & Wood KS (1987b) Endothelium-derived relaxing factor from pulmonary artery and vein possesses pharmacological and chemical properties identical to those of nitric oxide radical. *Circ. Res.* **61:** 866–879.

Ignarro LJ, Byrns RE & Wood KS (1988) Biochemical and pharmacological properties of endothelium-derived relaxing factor and its similarity to nitric oxide radical. In Vanhoutte PM (ed.) *Mechanisms of Vasodilation*, pp 427–435. New York: Raven Press.

Ignarro LJ, Gold ME, Buga GM et al (1989) Basic polyamino acids rich in arginine, lysine, or ornithine cause both enhancement of and refractoriness to formation of endothelium-derived nitric oxide in pulmonary artery and vein. *Circ. Res.* **64:** 315–329.

Iyengar RD, Stuehr DJ & Marletta MA (1987) Macrophage synthesis of nitrite, nitrate, and N-nitrosamines: precursors and role of the respiratory burst. *Proc. Natl. Acad. Sci. USA* **84:** 6369.

Kamoun PP, Schneider E & Dy M (1988) Superoxide-induced deimination of arginine in hematopoietic cells. *FEBS Lett.* **226:** 285–286.

Karczewski P, Kelm M, Hartmann M & Schrader J (1990) Bedeutung von Phospholamban bei der Endothel-vermittelten Relaxation der Rattenaorta. *Z. Kardiol.* **79 (supplement 1):** 212.

Katsuki S, Arnold W, Mittal C & Murad F (1977) Stimulation of guanylate cyclase by sodium nitroprusside, nitroglycerin and nitric oxide in various tissue preparations and comparison to the effects of sodium azide and hydroxylamine. *J. Cycl. Nucleotide Res.* **3:** 23–35.

Kawamoto JH, McLaughlin BE, Brien JF, Marks GS & Nakatsu K (1990) Biotransformation of glyceryl trinitrate and elevation of cyclic GMP precede glyceryl trinitrate-induced vasodilation. *J. Cardiovasc. Pharmacol.* **15:** 714–719.

Kelm M & Schrader J (1988) Nitric oxide release from the isolated guinea pig heart. *Eur. J. Pharmacol.* **155:** 317–321.

Kelm M & Schrader J (1990) Control of coronary vascular tone by nitric oxide. *Circ. Res.* **66:** 1561–1575.

Kelm M, Feelisch M, Spahr R, Piper H-M, Noack F & Schrader J (1988) Quantitative and kinetic characterization of nitric oxide and EDRF released from cultured endothelial cells. *Biochem. Biophys. Res. Commun.* **154:** 236–244.

Kelm M, Feelisch M, Deußen A, Schrader J & Strauer BE (1990) Flow- and pressure-dependent release of endothelium-derived nitric oxide (NO). *Arch. Int. Pharmacodyn. Ther.* **305:** 257.

Kilbourn R, Gross SS, Griffith OW, Levi R & Lodato RF (1990) Vascular shock induced by tumor necrosis factor is mediated by overproduction of nitric oxide. *Arch. Int. Pharmacodyn. Ther.* **305:** 258.

Knowles RG, Palacios M, Palmer RM & Moncada S (1989) Formation of nitric oxide from L-arginine in the central nervous system: A transduction mechanism for stimulation of the soluble guanylate cyclase. *Proc. Natl. Acad. Sci. USA* **86:** 5159–5162.

Ku DD (1982) Coronary vascular reactivity after acute myocardial ischemia. *Science* **218:** 576–578.

Lamping KG & Dole WP (1987) Acute hypertension selectively potentiates constrictor responses of large coronary arteries to serotonin by alternating endothelial function in vivo. *Circ. Res.* **61:** 904–913.

Lefer AM, Tsao PS, Lefer DJ & Johnson III G (1990) Time course and mechanism of inhibition of coronary EDRF release during myocardial ischemia and reperfusion. *Arch. Int. Pharmacodyn. Ther.* **305:** 262.

Lückhoff A, Busse R, Winter I & Bassenge E (1987) Characterization of vascular relaxant factor released from cultured endothelial cells. *Hypertension* **9:** 295–303.

Ludmer PL, Selwyn AP, Shook TL et al (1986) Paradoxical vasoconstriction induced by acetylcholine in atherosclerotic coronary arteries. *N. Engl. J. Med.* **315:** 1046–1051.

Lüscher TF (1989) Endothelium-derived relaxing and contracting factors: potential role in coronary artery disease. *Eur. Heart. J.* **10:** 847–857.

Lüscher TF, Diederick D, Weber E, Vanhoutte PM & Bühler FR (1988) Endothelium-dependent responses in carotid and renal arteries of normotensive and hypertensive rats. *Hypertension* **11:** 573–578.

Mak IT, Dickens BF & Weglicki WB (1989) Hydroxyl radical scavenging and attenuation of free radical injury in endothelial cells by SH-containing ACE inhibitors. *Circulation* **80 (supplement II-242):** 966.

Marklund S & Marklund G (1974) Involvement of the superoxide anion radical in the autoxidation of pyrogallol and a convenient assay for superoxide dismutase. *Eur. J. Biochem.* **47:** 469–474.

Marletta MA, Yoon PS, Iyengar R, Leaf CD & Wishnok JS (1988) Macrophage oxidation of L-arginine to nitrite and nitrate: nitric oxide is an intermediate. *Biochemistry* **27:** 8706–8711.

Martin W, Villani GM, Jothianandan D & Furchgott RF (1985) Selective blockade of endothelium-dependent and glyceryl trinitrate-induced relaxation of hemoglobin and by methylene blue in the rabbit aorta. *J. Pharmacol. Exp. Ther.* **232:** 708–716.

Mayer B, Schmidt K, Humbert P & Böhme E (1989) Biosynthesis of endothelium-derived relaxing factor: a cytosolic enzyme in porcine aortic endothelial cells Ca^{2+}-dependently converts L-arginine into an activator of soluble guanylyl cyclase. *Biochem. Biophys. Res. Commun.* **164:** 678–685.

Mayer B, Evers B, Wilke P & Böhme E (1990) Partial purification and characterization of an EDRF-forming enzyme from bovine brain. *Arch. Int. Pharmacodyn. Ther.* **305:** 265.

McCall TB, Boughton-Smith NK, Palmer RMJ, Whittle BJR & Moncada S (1989) Synthesis of nitric oxide from L-arginine by neutrophils. Release and interaction with superoxide anion. *Biochem. J.* **261:** 293–296.

McCall TB, Feelisch M, Palmer RMJ & Moncada S (1990) Characterization of the L-arginine: nitric oxide pathway in neutrophils and the macrophage cell line J774. *Arch. Int. Pharmacodyn. Ther.* **305:** 265.

McCord JM (1985) Oxygen-derived free radicals in postischemic tissue injury. *N. Engl. J. Med.* **312:** 159.

McCord JM (1987) Oxygen-derived radicals: a link between reperfusion injury and inflammation. *Fed. Proc.* **46:** 2402.

McDonald CC, Phillips WD & Mower HF (1965) An electron spin resonance study of some complexes of iron, nitric oxide, and anionic ligands. *J. Am. Chem. Soc.* **87:** 3319–3326.

Men'shikov MYu, Baldenkov GN, Negresku EV, Mazaev AV & Tkachuk VA (1987) Calcium-blocking action of nitrocompounds on human platelets: correlation with the change in the content of cyclic GMP. *Biochemistry (USSR)* **52:** 371–377.

Metz EN, Balcerzak SP & Sagone AL Jr (1976) Mechanisms of methylene blue stimulation of the hexose monophosphate shunt in erythrocytes. *J. Clin. Invest.* **58:** 797–802.

Minor RL, Meyers PR & Bates JN (1989) Basal EDRF release is reduced by depletion of cysteine from endothelial cells. *Circulation* **80 (supplement II-281):** 1121.

Misra HP (1974) Generation of superoxide free radical during the autoxidation of thiols. *J. Biol. Chem.* **249:** 2151–2155.

Misra HP (1984) Inhibition of superoxide dismuatse by nitroprusside and electron spin resonance observations on the formation of a superoxide-mediated nitroprusside nitroxyl free radical. *J. Biol. Chem.* **259:** 12678–12684.

Moncada S, Palmer RMJ & Gryglewski RJ (1986) Mechanism of action of some inhibitors of endothelium-derived relaxing factor. *Proc. Natl. Acad. Sci. USA* **83:** 9164–9168.

Moncada S, Palmer RMJ & Higgs EA (1987) Prostacyclin and endothelium-derived relaxing factor: biological interactions and significance. In Verstraete J, Linen HR & Arnout J (eds) *Thrombosis and Haemostasis*, pp 597–618. Leuven: International Society on Thrombosis and Haemostasis and Leuven University Press.

Moncada S, Radomski MW & Palmer RMJ (1988) Endothelium-derived relaxing factor. Identification as nitric oxide and role in the control of vascular tone and platelet function. *Biochem. Pharmacol.* **37:** 2495–2501.

Moncada S, Palmer RMJ & Higgs EA (1989) Biosynthesis of nitric oxide from L-arginine. A pathway for the regulation of cell function and communication. *Biochim. Pharmacol.* **38:** 1709–1715.

Mügge A, Förstermann U & Lichtlen PR (1989) Endothelial functions in cardiovascular diseases. *Z. Kardiol.* **78:** 147–160.

Mülsch A, Böhme E & Busse R (1987) Stimulation of soluble guanylate cyclase by endothelium-derived relaxing factor from cultured endothelial cells. *Eur. J. Pharmacol.* **135:** 247–250.

Murray JJ, Fridovich I, Makhoul RG & Hagen P-O (1986) Stabilization and partial characterization of endothelium-derived relaxing factor from cultured bovine aortic endothelial cells. *Biochem. Biophys. Res. Commun.* **141:** 689–696.

Myers PR, Minor RL Jr, Guerra R Jr, Bates JN & Harrison DG (1990) Vasorelaxant properties of the endothelium-derived relaxing factor more closely resemble S-nitrosocysteine than nitric oxide. *Nature (Lond.)* **345:** 161–163.

Needleman P & Johnson EM Jr (1973) Mechanism of tolerance development to organic nitrates. *J. Pharmacol. Exp. Ther.* **184:** 709–715.

Needleman P, Jakschik B & Johnson EM, Jr (1973) Sulfhydryl requirement for relaxation of vascular smooth muscle. *J. Pharmacol. Exp. Ther.* **187:** 324–331.

Niroomand F, Rössle R, Mülsch A & Böhme E (1989) Under anaerobic conditions soluble guanylate cyclase is specifically stimulated by glutathione. *Biochem. Biophys. Res. Commun.* **161:** 75–80.

Noack E (1984) Investigation on structure–activity relationship in organic nitrates. *Methods Find. Exp. Clin. Pharmacol.* **6:** 583–586.

Noack E (1990) Mechanisms of nitrate tolerance—influence of the metabolic activation pathway. *Z. Kardiol.* **79 (supplement 3):** 51–55.

Noack E & Feelisch M (1989) Molecular aspects underlying the vasodilator action of molsidomine. *J. Cardiovasc. Pharmacol.* **14 (supplement 11):** S1–S5.

Noack E & Feelisch M (1991) Molecular mechanisms of nitrovasodilator bioactivation. *Basic Res. Cardiol.* (in press).

Noack E, Schröder H & Feelisch M (1986) Continuous determination of nitric oxide formation during non-enzymatic degradation of organic nitrates and its correlation to guanylate cyclase activation. *Naunyn-Schmiedeberg's Arch. Pharmacol.* **332 (supplement):** 125.

Palacios M, Knowles RG, Palmer RMJ & Moncada S (1989) Nitric oxide from L-arginine stimulates the soluble guanylate cyclase in adrenal glands. *Biochem. Biophys. Res. Commun.* **165:** 802–809.

Palmer RMJ & Moncada S (1989) A novel citrulline-forming enzyme implicated in the formation of nitric oxide by vascular endothelial cells. *Biochem. Biophys. Res. Commun.* **158:** 348–352.

Palmer RMJ, Ferrige AG & Moncada S (1987) Nitric oxide release accounts for the biological activity of endothelium-derived relaxing factor. *Nature (Lond.)* **327:** 524–526.

Palmer RMJ, Ashton DS & Moncada S (1988a) Vascular endothelial cells synthesize nitric oxide from L-arginine. *Nature (Lond.)* **333:** 664–666.

Palmer RMJ, Rees DD, Ashton DS & Moncada S (1988b) L-arginine is the physiological precursor for the formation of nitric oxide in endothelium-dependent relaxation. *Biochem. Biophys. Res. Commun.* **153:** 1251–1256.

Park JW (1988) Reaction of S-nitrosoglutathione with sulfhydryl groups in protein. *Biochem. Biophys. Res. Commun.* **152:** 916–920.

Patel HMS & Williams DLH (1990) Nitrosation by alkyl nitrites. Part 6. Thiolate Nitrosation. *J. Chem. Soc. Perkin Trans* **2:** 37–42.

Peach MJ, Singer HA, Izzo NJ Jr & Loeb AL (1987) Role of calcium in endothelium-dependent relaxation of arterial smooth muscle. *Am. J. Cardiol.* **59:** 35A–43A.

Pohl U & Busse R (1989) EDRF increases cyclic GMP in platelets during passage through the coronary vascular bed. *Circ. Res.* **65:** 1798–1803.

Pohl U & Busse R (1990) Endothelium-dependent modulation of vascular tone and platelet function. *Eur. Heart J.* **11 (supplement B):** 35–42.

Pohl U, Busse R, Kuon E & Bassenge E (1986) Pulsatile perfusion stimulates the release of endothelial autacoids. *J. Appl. Cardiol.* **1:** 215–235.

Przyklenik K & Kloner RA (1987) Acute effects of hydralazine and enalapril on contractile function of postischemic stunned myocardium. *Am. J. Cardiol.* **60:** 934–936.

Rees DD, Palmer RMJ, Hodson HF & Moncada S (1989a) A specific inhibitor of nitric oxide formation from L-arginine attenuates endothelium-depedent relaxation. *Br. J. Pharmacol.* **96:** 418–424.

Rees DD, Palmer RMJ & Moncada S (1989b) Role of endothelium-derived nitric oxide in the regulation of blood pressure. *Proc. Natl. Acad. Sci. USA* **86:** 3375–3378.

Rubanyi GM & Vanhoutte PM (1985) Hypoxia releases a vasoconstrictor substance from the canine vascular endothelium. *J. Physiol.* **364:** 45–56.

Rubanyi GM & Vanhoutte PM (1986a) Superoxide anions and hyperoxia inactivate endothelium-derived relaxing factor(s). *Am. J. Physiol.* **250:** H822–H827.

Rubanyi GM & Vanhoutte PM (1986b) Oxygen-derived free radicals, endothelium, and responsiveness of vascular smooth muscle. *Am. J. Physiol.* **250:** H815–H821.

Rubanyi GM, Romero JC & Vanhoutte PM (1986) Flow induced release of endothelium-derived relaxing factor. *Am. J. Physiol.* **250:** H1145.

Saito T, Wolf A, Menon NK, Saeed M & Bing RJ (1988) Lysolecithins as endothelium-dependent vascular smooth muscle relaxants that differ from endothelium-derived relaxing factor (nitric oxide). *Proc. Natl. Acad. Sci. USA* **85:** 8246–8250.

Salvemini D, Korbutt R, Änggard E & Vane J (1990) Immediate release of a nitric oxide-like factor from bovine aortic endothelial cells by *Escherichia coli* lipopolysaccharide. *Proc. Natl. Acad. Sci. USA* **87:** 2593–2597.

Saran M, Michel C & Bors W (1990) Reaction of NO with O_2^-. Implications for the action of endothelial derived relaxation factor (EDRF). *Free Radical Res. Commun.* (in press).

Schmidt HHHW, Klein MM, Niroomand F & Böhme E (1988a) Is arginine a physiological precursor of endothelium-derived nitric oxide? *Eur. J. Pharmacol.* **148:** 293–295.

Schmidt HHHW, Nau H, Wittfoht W et al (1988b) Arginine is a physiological precursor of endothelium-derived nitric oxide. *Eur. J. Pharmacol.* **154:** 213–216.

Schmidt HHHW, Seifert R & Böhme E (1989) Formation and release of nitric oxide from human neutrophils and HL-60 cells induced by a chemotactic peptide, platelet activating factor and leukotriene B4. *FEBS Lett.* **244:** 357–360.

Schmidt K, Graier WF, Kostner GM & Kukovetz WR (1990) Activation of soluble guanylate cyclase by EDRF (NO) is inhibited by oxidized low density lipoprotein (LDL-ox). *Arch. Int. Pharmacodyn.* **305:** 257 (abstract 106).

Schröder H & Noack E (1987) Structure–activity relationship of organic nitrates for activation of guanylate cyclase. *Arch. Int. Pharmacodyn.* **290:** 235–246.

Schröder H, Noack E & Müller R (1985) Evidence for a correlation between nitric oxide formation by cleavage of organic nitrates and activation of guanylate cyclase. *J. Mol. Cell. Cardiol.* **17:** 931–934.

Seddon WA, Fletcher JW & Sopchyshyn FC (1973) Pulse radiolysis of nitric oxide in aqueous solution. *Can. J. Chem.* **51:** 1123–1130.

Servent D, Delaforge M, Ducrocq C, Mansuy D & Lenfant M (1989) Nitric oxide formation during microsomal hepatic denitration of glyceryl trinitrate: involvement of cytochrome P-450. *Biochem. Biophys. Res. Commun.* **163:** 1210–1216.

Shikano K & Berkowitz BA (1987) Endothelium-derived relaxing factor is a selective relaxant of vascular smooth muscle. *J. Pharmacol. Exp. Ther.* **243:** 55–60.

Shikano K, Long CJ, Ohlstein EH & Berkowitz BA (1988) Comparative pharmacology of endothelium-derived relaxing factor and nitric oxide. *J. Pharmacol. Exp. Ther.* **247:** 873–881.

Stewart DJ, Pohl U & Bassenge E (1988) Free radicals inhibit endothelium-dependent dilation in the coronary resistance bed. *Am. J. Physiol.* **255:** H765–769.

Stolze K & Nohl H (1989) Detection of free radicals as intermediates in the methemoglobin formation from oxyhemoglobin induced by hydroxylamine. *Biochem. Pharmacol.* **38:** 3055–3059.

Stuehr DJ, Kwon NS & Nathan CF (1990) FAD and GSH participate in macrophage synthesis of nitric oxide. *Biochem. Biophys. Res. Commun.* **168:** 558–565.

Tare M, Parkington HG, Coleman HA, Neild TO & Dusting GJ (1990) Nitric oxide hyperpolarizes arterial smooth muscle. *Arch. Int. Pharmacodyn. Ther.* **305:** 285.

Tayeh MA & Marletta MA (1989) Macrophage oxidation of L-arginine to nitric oxide, nitrite, and nitrate. *J. Biol. Chem.* **264:** 19654–19658.

Thomas G, Hecker M & Ramwell PW (1989) Vascular activity of polycations and basic amino acids: L-arginine does not specifically elicit endothelium-dependent relaxation. *Biochem. Biophys. Res. Commun.* **158:** 177–180.

Tyler DD (1975) Polarographic assay and intracellular distribution of superoxide dismutase in rat liver. *Biochem. J.* **147:** 493–504.

Vallance P, Collier J & Moncada S (1989) Nitric oxide synthesised from L-arginine mediates endothelium dependent dilatation in human veins in vivo. *Cardiovasc. Res.* **23:** 1053–1057.

Vanhoutte PM (1987) The end of the quest? *Nature (Lond.)* **327:** 459–460.

Vanhoutte PM & Katusic ZS (1988) Endothelium-derived contracting factor: endothelium and/or superoxide anion? *Trends Pharmacol. Sci.* **9:** 229–230.

Vidal MJ, Romero JC & Vanhoutte PM (1988) Endothelium-derived relaxing factor inhibits renin release. *Eur. J. Pharmacol.* **149:** 401–402.

Warner TD, de Nucci G & Vane JR (1989) Comparison of the survival of endothelium-derived relaxing factor and nitric oxide within the isolated perfused mesenteric arterial bed of the rat. *Br. J. Pharmacol.* **97:** 777–782.

Westlin W & Mullane K (1988) Does captopril attenuate reperfusion-induced myocardial dysfunction by scavenging free radicals? *Circulation* **77 (supplement 1):** I-30–I-39.

Winquist RJ, Bunting PB, Baskin EP & Wallace AA (1984) Decreased endothelium-dependent relaxation in New Zealand genetic hypertensive rats. *J. Hypertension* **2:** 541–545.

Wood KS & Ignarro LJ (1987) Hepatic cyclic GMP formation is regulated by similar factors that modulate activation of purified hepatic soluble guanylate cyclase. *J. Biol. Chem.* **262:** 5020–5027.

Yoon PS & Deykin D (1990) Nitric oxide abolishes NADPH oxidase-catalyzed superoxide production in human polymorphonuclear leukocytes. *Arch. Int. Pharmacodyn. Ther.* **305:** 296.

Westb[illegible] ... mitochondrial dysfunction by an endogenous free radical? [illegible] (Suppl [illegible]) [illegible]

Winquist RJ, Smith [illegible], Baskin [illegible] & Wallace [illegible] (198[illegible]) [illegible] dependent relaxation [illegible] hypertension [illegible] 537–545.

Wood KS & Ignarro LJ (1987) Hepatic cyclic GMP formation is regulated by similar factors that modulate activation of purified hepatic soluble guanylate cyclase. J. Biol. Chem. 262: 5020–5027.

[illegible] & Dennis [illegible] (1990) Nitric oxide inhibits NADPH oxidase-catalyzed superoxide production in human polymorphonuclear leukocytes. [illegible] 305: 25[illegible]

TOWARDS CLINICAL MEDICINE

18

Ischaemia–Reperfusion

BASSAM OMAR,* JOE McCORD* and JAMES DOWNEY†
**Webb-Waring Lung Institute, University of Colorado, Denver, Colorado 80262, USA*
†Department of Physiology, The University of South Alabama, Mobile, Alabama 36688, USA

I. Introduction 493
II. General aspects of ischaemia–reperfusion 494
III. What are the sources of free radicals in the reperfused organ? 496
IV. Ischaemia–reperfusion injury in the heart 503
V. Ischaemia–reperfusion injury in other organs 515
VI. Concluding remarks 519

I. INTRODUCTION

Ischaemia is a source of tissue necrosis in a wide variety of pathological conditions, and there are similarities in the ischaemic syndromes seen in many organs, including heart, intestine, brain, kidney, and lung. Permanent deprivation of blood flow is lethal to any tissue and the prudent therapy for ischaemia unquestionably is reperfusion. While reperfusion is necessary to reverse the progression towards ischaemic death, reperfusion is also thought to be accompanied by its own component of injury. Although we are far from understanding the exact intracellular targets of injury during ischaemia and reperfusion, evidence has accumulated implicating cytotoxic oxygen-derived free radicals as mediators of at least part of this injury, based on the ability of free radical scavengers to diminish injury in animal models of transient ischaemia. More recently free radicals have been directly demonstrated in reperfused hearts using electron spin resonance (ESR) spec-

Oxidative Stress: Oxidants and Antioxidants
ISBN 0-12-642762-3

troscopy and spin trapping agents. In studying the role of antioxidant therapy in ischaemia and reperfusion, quantifying injury has, unfortunately, been a major obstacle in many of the animal models, and conclusions regarding effectiveness of the various antioxidants have varied widely. While interest in ischaemic injury to the heart has been especially intense, the many conflicting reports in the literature continue to confuse the issue. This chapter will attempt to summarize the current state of our knowledge of ischaemia–reperfusion injury and the role that free radicals play in it. We will attempt to identify those concepts that have been firmly established and those that remain speculation, and, most importantly, try to identify why so many of the studies have been contradictory.

II. GENERAL ASPECTS OF ISCHAEMIA–REPERFUSION

A. Pathogenesis of ischaemic injury

The aetiology of ischaemia is diverse in clinical medicine. It can result from atherosclerosis, thromboembolism, or external pressure on vessels (resulting from tumours), or can be iatrogenic, as can occur during surgery, when the blood flow to an organ must be interrupted as in organ transplantation. Ischaemia, from whatever source, is accompanied by a lack of both oxygen and substrate, and therefore a lack of aerobic energy production. Anaerobic ATP generation is not sufficient to keep pace with the metabolic requirements and, thus, ATP content of the tissue falls rapidly. The lack of ATP initiates a cascade of damaging effects, including inability to maintain membrane ionic balance. There is an influx of sodium into the cells (Renlund et al, 1982) which, upon reperfusion, could exchange with calcium (Bersohn et al, 1982), leading to the activation of injurious enzymes and disruption of the membrane, causing further abnormalities in ionic homeostasis. When cytosolic calcium rises, mitochondria actively concentrate calcium within their matrix, which leads to swelling and dysfunction (Vlessis and Mela-Riker, 1989; Stone et al, 1989). Moreover, the ischaemic tissue becomes prone to attack by inflammatory cells upon reperfusion, and this can further exacerbate the ischaemic damage. In many organs, such as the kidney, intestine and lung, there is also widespread damage to the endothelium. This may lead to fluid imbalances and directly affect capillary function. Endothelial injury may also affect vessel patency such that reperfusion flow is poor, thus prolonging ischaemia. In these organs the capillary endothelium is a primary site of ischaemic damage. Even brief periods of ischaemia cause large increases in permeability to macromolecules, hindering the function of these organs. The final magnitude of injury is a function of all of these processes, and no one event can be singly equated with cell death.

The status of intracellular enzymes during ischaemia is crucial. First, a potentially detrimental event is the rupture of lysosomes (Dickens et al, 1988; Kalra et al, 1989) and the release of proteases, which can lead to the degradation of cytosolic proteins. A second injurious event is the activation of various calcium-dependent proteases and phospholipases as a result of calcium influx (Katz et al, 1985; Kako, 1986; Kishimoto et al, 1981). The activation of phospholipases has direct damaging effects on membranes and results in the release of free fatty acids and lysophospholipids. These are noxious by themselves, and can initiate arachidonic acid metabolism, with the resulting formation of various products and accompanying generation of oxy radicals (Torok et al, 1984; Chan and Fishman, 1984; Yagi, 1987). The activation of proteases, on the other hand, can degrade vital enzymes and the cytoskeleton of the cells and may induce proteolytic changes which in themselves will increase free radical production on reperfusion (Roy and McCord, 1983).

If ischaemia is not reversed by reperfusion in a timely manner, the tissue will be irreversibly injured by the ischaemia alone. Although some free radicals may be generated early in ischaemia and may contribute to cell death, they are not thought to be the major toxic event during total ischaemia, because molecular oxygen is unavailable at that time. Rather, intracellular ATP seems to correlate best with whether the injury is reversible or irreversible (Jennings et al, 1978).

B. Reperfusion a double-edged sword?

Reperfusion is essential for the salvage of ischaemic tissue. By current definitions, reperfusion will salvage cells that were reversibly injured by the ischaemia and not those that were irreversibly injured. The question arises as to whether any reversibly injured cells are further injured by events associated with reperfusion itself. It is this population of cells that is the subject of debate as to whether it is really killed by reperfusion, whether it represents an appreciable enough fraction of the total cell population to warrant intervention, and as to how such injury might be eliminated. Generally, three types of reperfusion inury have been postulated (Fantone, 1990). First, reperfusion injury may arise from inside the parenchymal cell, caused by either free radical production or calcium entry. In the heart muscle, cell death during reperfusion is known to involve a rapid influx of calcium into the cells (Bourdillon and Poole-Wilson, 1981, 1982; Shen and Jennings, 1972) but the mechanism for this influx is not known. While little of this calcium enters through the slow calcium channels, much of it has been ascribed to gross membrane defects (Poole-Wilson et al, 1984; Buja et al, 1988). Calcium influx on reperfusion may also result from a rapid exchange with sodium which had entered during the ischaemic phase through the sodium–calcium exchanger on the cell membrane (Bersohn et al, 1982; Grinwald, 1982; Renlund et al, 1982). Second, reperfusion injury may be caused by endothelial cell swelling, interstitial oedema or neutrophil plugging, any of which could limit reflow of blood to the ischaemic area and thereby protract the ischaemia. The consequences of this 'no-reflow' phenomenon have been investigated in great detail (Harris, 1975; Humphrey et al, 1980; Kloner et al, 1974; Powell et al, 1976; Tranum-Jensen et al, 1981). Third, reperfusion injury may be caused by the infiltration of activated neutrophils which initiate an inflammatory response and further tissue destruction. All these sources of injury could conceivably involve oxygen-derived free radicals.

III. WHAT ARE THE SOURCES OF FREE RADICALS IN THE REPERFUSED ORGAN?

Although spin trap studies have now established that free radicals are produced in reperfused tissue, their sources are still a matter of speculation. Similarly, it has not been firmly established which of the reduced oxygen intermediates (if any) is responsible for the injury seen with ischaemia–reperfusion. In the following section we will review sources of oxy radical production.

A. Xanthine oxidase

Xanthine oxidase has been proposed to be an important source of oxygen-derived free radicals in reperfused tissue (Chambers et al, 1985; Granger et al, 1981; McCord et al, 1985; Ytrehus et al, 1986; Hearse et al, 1986). Xanthine oxidase exists in vivo primarily as a dehydrogenase form, which uses NAD^+ as an electron acceptor during the oxidation of xanthine and hypoxanthine (Battelli et al, 1972). Studies in the intestine indicate that ischaemia may promote the conversion of xanthine dehydrogenase to xanthine oxidase such that after only a few minutes of ischaemia virtually all of the enzyme is in the oxidase form (Parks and Granger, 1986). The elevated cytosolic calcium with ischaemic cells has been proposed to activate a calcium-dependent protease that catalyses this conversion (McCord,

1985a). Some conversion results from sulphydryl oxidation as well and is reversible by strong reducing agents (Engerson et al, 1987). Ischaemia causes the degradation of ATP into hypoxanthine and xanthine, which form the substrates for xanthine oxidase. The missing substrate required for the reaction, molecular oxygen, is introduced upon reperfusion. Therefore, in the first few minutes of reperfusion xanthine oxidase will produce measurable amounts of uric acid and superoxide anion and H_2O_2 as byproducts (Roy and McCord, 1983; McCord, 1985b; Huizer et al, 1989). The importance of conversion of xanthine dehydrogenase to oxidase has been less clear in the heart. We found that only 10% of the enzyme was in the oxidase form in non-ischaemic dog heart while 32% was in the oxidase form after 30 min of ischaemia (Chambers et al, 1985). A similar conversion was seen in ischaemic rat heart with 25% conversion after 5 min of ischaemia and no further increase thereafter (Downey et al, 1987). When Kehrer et al (1987) examined conversion in the rat heart they found that about 32% of the enzyme was in the oxidase form in fresh hearts and no further conversion could be found after 60 min of hypoxia and reperfusion. Moreover, although Bindoli et al (1988) reported a limited increase (9%) in the reversible xanthine oxidase following ischaemia and reperfusion in the rat heart, these authors suggested that the 20% xanthine oxidase present in fresh rat hearts could contribute to the background production of free radicals. Indeed, recent studies by Engerson et al (1987) indicate that the conversion to the oxidase form is probably too slow to be important. Nevertheless, the 8–32% of oxidase that is clearly present in *healthy* dog or rat heart could become a significant source of free radicals when fuelled by the hypoxanthine resulting from breakdown of purine nucleotides in ischaemic tissue, *even if no additional conversion occurred.* If the percentage of oxidase rises in injured tissues as a result of sulphydryl oxidation or proteolysis, then the insult would become two- to ten-fold greater.

Much of the evidence supporting the involvement of xanthine oxidase in myocardial reperfusion injury has been obtained indirectly from the protection afforded by allopurinol to the reperfused heart. While some have found allopurinol to be protective in the isolated ischaemic or hypoxic rat heart model (Grisham et al, 1986b; Brown et al, 1988), this has not been a universal finding (Kehrer et al, 1987).

The cardiac xanthine oxidase data are further complicated by important species differences. Both the rat and the dog contain at least several orders of magnitude higher amounts of xanthine oxidase in their hearts than rabbit, pig or man (Grum et al, 1986; Parks and Granger, 1986; Eddy et al, 1987). We and others have used the rabbit heart as a model of the xanthine oxidase-deficient human heart. Enzyme-inhibiting doses of allopurinol have not been found to be protective in rabbit heart (McCord et al 1989; Downey et al,

1986). A very high dose of 1 mM in the perfusate was reported to protect the rabbit heart (Myers et al, 1985a), but that was more likely by direct radical scavenging (Moorhouse et al, 1987) than enzyme inhibition. On the basis of the rabbit experiments it would seem that xanthine oxidase inhibitors would not be protective to human hearts. Although human heart has undetectable amounts of xanthine oxidase (or dehydrogenase) by biochemical assay (Grum et al, 1989; Eddy et al, 1987), fluorescent antibodies to xanthine oxidase do bind to coronary endothelium from humans (Bruder et al, 1983), suggesting that some enzyme is present. How much enzyme is actually present or whether it is enough to contribute to injury is currently unknown. Human intestine and kidney have considerably more xanthine oxidase than human heart (Parks and Granger, 1986), and in those organs xanthine oxidase inhibitors may still have therapeutic possibilities.

B. Arachidonate pathways

Investigations into the metabolism of arachidonic acid by cyclooxygenase and by lipoxygenase have revealed that intermediate peroxy compounds and hydroxyl radicals are produced in these reactions (Samuelsson, 1983; Kuehl et al, 1979; Mullane et al, 1987; Kontos, 1987; Kukreja et al, 1986). Experimental evidence also indicates that a potent oxidant, possibly the hydroxyl radical, results from the peroxidative conversion of endoperoxide PGG_2 to PGH_2. This oxidant regulates prostaglandin and thromboxane production by multiple mechanisms, which include deactivation of cyclooxygenase, peroxidase, and prostacyclin synthetase (Penfield and Dale, 1985; Kontos, 1987; Kukreja et al, 1986; Oyanagui, 1976). Free radical scavengers have been shown to modulate the conversion of prostaglandin endoperoxides to more stable, classical prostaglandins, such as PGE_2, PGF_2, PGI_2 (prostacyclin), and thromboxanes. Arachidonic acid can also be metabolized by enzymatic oxygenation via a specific lipoxygenase. This pathway may involve either a one- or two-step dioxygenation of substrates. Monohydroperoxy and dihydroperoxy fatty acids are primary products of these reactions (Samuelsson, 1983; Borgeat et al, 1983). The potential role that the products of hydroperoxy fatty acids play in the mediation of tissue injury involves at least two mechanisms. First, like other peroxides, both monohydroperoxy and dihydroperoxy fatty acids can serve as substrates for the production of hydroxyl or hydroperoxyl radicals via metal-catalysed reactions. These compounds can then interact with other target molecules and alter the biochemical or functional properties of these molecules. Second, metabolism of hydroperoxy compounds produces a variety of biologically active molecules that can modulate both immune and inflammatory-cell functions (Gardner, 1989; Piretti and Pagliuca, 1989). This results in chemotaxis and granulocyte which can worsen the damage initiated by the arachidonate pathway metabolites.

Agents that block arachidonate metabolism, such as ibuprofen (Jugdutt, 1985), nafazatrom (Simpson et al, 1987), and BW755C (a dual inhibitor of both cyclooxygenase and lipoxygenase enzymes) (Klein et al, 1988) have been reported to protect the myocardium against ischaemia–reperfusion injury. These agents, however, also inhibit leukocyte activation (Mullane et al, 1987; Flynn et al, 1984), and their protective effect, therefore, may have been secondary to the restriction of neutrophil infiltration at reperfusion. Other investigators, however, find that nonsteroidal antiinflammatory drugs aggravate acute myocardial ischaemia (Berti et al, 1988). Moreover, steroid therapy (which indirectly inhibits both cyclooxygenase and lipoxygenase pathways) has been shown to cause delayed healing and ventricular aneurysm following acute myocardial infarction (Bulkley and Roberts, 1974). Arachidonate metabolism, however, seems to play a more definitive role in brain ischaemia, where a beneficial effect of steroidal and nonsteroidal antiinflammatory therapy has been documented (Hall and Travis, 1988; Kontos et al, 1985).

C. Catecholamines

The autoxidation of catecholamines has been proposed to cause myocardial damage during ischaemia–reperfusion (Singal et al, 1982, 1983). This proposal has been supported by indirect evidence whereby high, non-physiological levels of catecholamines in the blood have been found to cause ultrastructural and functional damage to the heart (Wheatley et al, 1985; Ferrans et al, 1970; Opie et al, 1979). Matsuki et al (1989), however, found no protective effort of chronic chemical sympathectomy with 6-hydroxydopamine on myocardial infarct size in rabbits. Jewett et al (1989) recently reviewed studies on the role of catecholamine autoxidation in ischaemia–reperfusion and suggested that this process is extremely low or non-existent at physiological pH and therefore unlikely to be a primary source of oxygen radicals in pathological states. They also suggested that the oxidation of catecholamines observed in biological systems is likely to be due to catalysed oxidations involving superoxide anion generated by substrate-dependent processes, enzyme-catalysed peroxidative processes, and possibly trace metal catalysis.

D. Mitochondria

The production of partially reduced oxygen species by the mitochondrion has been known for nearly two decades (Boveris and Chance, 1973; Loschen et al, 1973; Dionisi et al, 1975; Boveris and Cadenas, 1975; Boveris et al, 1976; Turrens and Boveris, 1980). The mechanism of this free radical production, however, has only recently been examined. The rate of superoxide production by the mitochondrion increases when the concentration of oxygen is increased or the respiratory chain becomes largely reduced (Turrens et al, 1982b). The first process, for example, takes place in the lungs of animals exposed to high oxygen concentrations (Crapo et al, 1983; Turrens et al, 1982a). The second process occurs when coupled mitochondria consume oxygen in state 4 (that is, in the absence of ADP and in the presence of substrate and oxygen), when the electron flow is limited by lack of substrate. Mitochondria produce superoxide anions at two sites in the electron transport chain. The first site is the ubiquinone to cytochrome c_1 step, which passes through the intermediate ubisemiquinone (Cadenas et al, 1977; Boveris et al, 1976; Turrens et al, 1985). Ubisemiquinone is capable of reducing oxygen to superoxide, which dismutates spontaneously to form hydrogen peroxide. The second site of superoxide anion formation is the NADH dehydrogenase (Turrens and Boveris, 1980).

Several studies have demonstrated that mitochondria become uncoupled upon exposure of heart tissue to ischaemia (Jennings and Ganote, 1976; Watanabe et al, 1985; Piper, 1989). It has also been observed that state 3 respiration in mitochondria isolated from the ischaemic rat and rabbit heart is much lower compared with those isolated from control organs (Turrens and McCord, 1989, 1990). In addition, ischaemic tissue contains low levels of ADP, depending on the length of ischaemia (Nishida et al, 1987), which would keep the mitochondria in state 4, thus generating H_2O_2 at maximal rates. Several reports indicate that the inhibition of the mitochondrial respiratory chain, may be caused by the uptake of calcium (Parr et al, 1975; Saris and Akerman, 1980; Vlessis and Mela-Riker, 1989). Mitochondria have been shown to accumulate calcium electrogenically, using the H^+ gradient for this purpose instead of producing ATP, even in the presence of ADP. It is possible that calcium overload of mitochondria may activate phospholipases or proteases, either of which may cause uncoupling and inhibition of respiration.

E. Neutrophils

Activated nicotinamide adenine dinucleotide (NADPH) oxidase is an important source of free radicals in the cell membranes of phagocytes. When polymorphonuclear leukocytes and macrophages are stimulated by particulate or soluble inflammatory mediators, they experience a respiratory burst that is characterized by increased oxygen consumption and glucose metabolism secondary to activation of the hexose monophosphate shunt (Babior, 1978; Klebanoff, 1980). In stimulated neutrophils, after initiation of the respiratory burst, more than 90% of the consumed oxygen can be accounted for by the generation of superoxide. Most H_2O_2 released during the stimulation of phagocytic cells appears to be directly derived from the dismutation of $O_2^{\bar{\cdot}}$. Production of $O_2^{\bar{\cdot}}$ and H_2O_2 by neutrophils is enhanced after the cells adhere to surfaces or after they are primed with a chemically different stimulus (Dahinden et al, 1983; Fletcher et al, 1982). The oxygen concentration at sites of inflammation may also be a rate-limiting factor in the generation of $O_2^{\bar{\cdot}}$ by phagocytes (Edwards et al, 1984).

Hydroxyl radicals are also formed when stimulated phagocytes produce $O_2^{\bar{\cdot}}$ and H_2O_2. Electron spin trapping studies have shown that as neutrophils, monocytes, or macrophages are stimulated, they generate significant quantities of hydroxyl radicals, presumably through a Fenton-like reaction (McCord and Day, 1978). Hydroxyl radicals can also be produced in vitro from H_2O_2 in the presence of lactoferrin, an iron-binding protein found in specific granules of neutrophils (Ambruso and Johnston, 1981). Since myeloperoxidase, an enzyme found in lysosomal granules, is capable of catalysing the formation of hypochlorous acid (HOCl), singlet oxygen ($^1O_2$), and long-acting chloramine compounds, it appears that compounds within these granules could enhance the production of specific free radicals in tissues and may do so either independently or in conjunction with the production of superoxide anions and hydrogen peroxide.

Hernandez et al (1987) have demonstrated an important role for neutrophils in ischaemia–reperfusion microvascular intestinal injury. Using either antiserum to neutrophils or a monoclonal antibody to the neutrophil's CD 11/18 complex to prevent neutrophil adhesion, these investigators showed a near-complete attenuation of microvascular permeability caused by ischaemia–reperfusion of the small intestine. The fact that oxygen radical scavengers and allopurinol also provided a comparable extent of protection to that of neutrophil depletion led these investigators to suggest that xanthine oxidase-derived reactive oxygen metabolites play an important role in eliciting ischaemia–reperfusion-induced neutrophil infiltration.

The involvement of neutrophils in myocardial reperfusion injury has been investigated as well. Although neutrophils are always associated with infarcted myocardium, a cause-and-effect relationship has been difficult to prove. Simpson et al (1988a,b) attempted to block neutrophil-mediated damage using the anti-Mo-1-monoclonal antibody, a reagent which prevents neutrophils from interacting with vascular and myocardial cells during the phase of reperfusion (Simon et al, 1986). Administration of anti-Mo-1 antibody (Simpson et al, 1988a) or neutrophil depletion (Simpson et al,

1988b) resulted in a significant decrease of infarct size in the canine heart subjected to 90 min of ischaemia and reperfusion for 6 h. A disturbing feature of the Simpson et al (1988a) study, however, was that a significant limitation of infarct size could not be shown against a third group receiving a control IgG_1 antibody. A protective effect of anti-Mo-1 antibody against myocardial stunning, however, could not be documented (Schott et al, 1989).

F. Myoglobin and iron-mediated radical generation

Several lines of evidence indicate that haemoproteins promote lipid hydroperoxide-dependent lipid peroxidation in vitro (Tappel, 1953; Kaschnitz and Hatefi, 1975). This could be very important for the heart, due to its high content of myoglobin. Grisham (1985) showed that myoglobin, in the presence of hypoxanthine and xanthine oxidase, catalyses the peroxidation of arachidonic acid. Oxy (ferrous) myoglobin was found to be the most effective catalyst for arachidonic acid peroxidation when compared to metmyoglobin, haemoglobin or ADP-iron chelates. Galaris et al (1989) suggested that myoglobin and H_2O_2 promote peroxidation of unsaturated fatty acids and, thus, may cause damage to cellular constituents. However, these authors showed that lipid peroxidation is inhibited in the presence of ascorbate. They therefore suggested that the redox cycling of myoglobin may act as an important electron 'sink' and defence mechanism against peroxides during oxidative challenge to muscle. Hochstein et al (1989) have demonstrated that ascorbate (100 μM) in the perfusion medium prevents the leak of LDH from isolated rat hearts after 15 min of global ischaemia.

The Haber–Weiss reaction (whereby the hydroxyl radical is produced from the reaction of $O_2^{\cdot-}$ and H_2O_2) does not occur to any significant extent in the absence of iron or other redox-active transition metals (McCord and Day, 1978; Halliwell, 1978). The hydroxyl radical ($^{\cdot}OH$) is assumed to represent the ultimate oxidant involved in hydrogen atom abstraction from polyunstaturated fatty acids. However, the involvement of $^{\cdot}OH$ as an initiator of lipid peroxidation has recently been questioned, due to the inability of $^{\cdot}OH$ traps to inhibit lipid peroxidation (Minotti and Aust, 1987; Thomas et al, 1985; Bucher et al, 1983). Alternatively, Samokyszyn and Aust (1990) proposed the involvement of iron-oxo intermediates as direct initiators of lipid peroxidation. Minotti and Aust (1990) demonstrated that the rate of lipid peroxidation was absolutely dependent on relative Fe(III)/Fe(II) ratios, with maximum rates occurring at a ratio of 1:1.

Several studies have demonstrated a protective effect of iron chelators on ischaemic tissues. Myers et al (1985a) showed that deferoxamine reduced the creatine kinase release from isolated rabbit hearts. Several other studies demonstrated a protective effect with iron chelators in many organs, such as the skin (Angel et al, 1986), kidneys (Fuller et al, 1987), liver (Parnham et al, 1985) and brain (Ikeda et al, 1989). Not all studies, however, were able to demonstrate protection with deferoxamine (Maxwell et al, 1989).

IV. ISCHAEMIA–REPERFUSION INJURY IN THE HEART

Much of our knowledge about ischaemia–reperfusion has been obtained in the heart model. In the ischaemic heart, reperfusion early after the onset of myocardial ischaemia promotes survival of tissue which would otherwise have died. Such an intervention has the effect of preserving pump function. Several strategies have been devised for effecting early reperfusion of the ischaemic heart, including emergency angioplasty and the thrombolytic agents. Clinical trials have confirmed that both mortality and myocardial pump function are improved by early reperfusion (ISAM Study Group, 1986; ISIS-2 Collaborative Group, 1988; TIMI Study Group, 1985) and few doubt that benefit was derived from infarct size reduction (Campbell et al, 1988; Darius et al, 1986), although it is still not possible to directly measure infarct size in man.

Are some potentially viable cells killed by reperfusion per se or does reperfusion simply cause cells previously killed by ischaemia to suddenly undergo abrupt morphological changes, possibly through an osmotic effect (Jennings and Reimer, 1983)? This question has been difficult to answer, since the only known way to test whether ischaemic tissue is dead or not is to see if it recovers on reperfusion, a test which in itself would invoke reperfusion injury should it exist. Reoxygenation is one component of reperfusion which has been suggested to cause reperfusion injury. The most compelling evidence in favour of reoxygenation injury comes from the work of Hearse et al (1973, 1975), who demonstrated that the reintroduction of oxygen to a hypoxic heart was accompanied by an abrupt disruption of the tissue as assessed by both release of cytosolic enzymes and ultrastructure. Among the several possible explanations which they considered was the production of oxygen-derived free radicals (Hearse and Humphrey, 1975). Several investigators subsequently found that a variety of free radical scavengers, including SOD, could reduce reoxygenation and reperfusion injury, which argued in favour of a free radical mechanism (Guarnieri et al, 1978; Shlafer et al, 1982; Woodward and Zakaria, 1985; Gardner et al, 1983; Gauduel and Duvelleroy, 1984). Three reduced intermediates of oxygen have been implicated in this injury, the superoxide radical, hydrogen peroxide, and the hydroxyl radical. All three are cytotoxic and, although hydrogen peroxide is not a radical species, it is the primary precursor for the hydroxyl radical. It is believed that hypoxic conditions during ischaemia set the stage for oxy radical production when oxygen is reintroduced during reperfusion. Indeed, reperfusion of the heart after a suitable period of ischaemia will be accompanied by a sudden disruption of the tissue (Fukuhara, 1985; Hearse et al, 1975), similar to that seen in the hypoxia/reoxygenation model. Although reperfusion involves more than just reoxygenation, reoxygenation injury could still be a significant part of reperfusion injury. Recent studies with spin trap probes confirm that a variety of free radicals are generated in the reperfused heart (Blasig et al, 1986; Arroyo et al, 1987; Garlick et al, 1987; Bolli et al, 1988; Weglicki et al, 1988). Moreover, if hearts are perfused with SOD, all trapped radical signals are attenuated, suggesting that all are derived secondarily from superoxide (Zweier et al, 1987, 1989).

A. Ischaemia–reperfusion causes specific lesions in the heart

In the heart, three levels of ischaemic injury are currently recognized and all three have been proposed to involve, at least in part, a free radical mechanism. The first detectable level of injury is the generation of reperfusion arrhythmias. Reperfusion after an ischaemic period of only several minutes can result in ventricular tachycardia or fibrillation (Bernier et al, 1986). Second, increasing the length of the ischaemic period to between 5 and 15 min will result in a prolonged deficit in contractility following reperfusion. This state has been called the 'stunned' myocardium. In the stunned myocardium all of the ischaemic myocytes are still viable and the heart will completely recover its function, although such recovery may require several days (Bolli, 1988; Farber et al, 1988; Przyklenk and Kloner, 1986). Finally, when the ischaemic period is extended to 20 min or longer, some of the heart cells will be irreversibly injured and become necrotic (Reimer et al, 1977). The greater the depth and duration of the ischaemic insult, the more widespread the cell death. There seems to be little question that free radicals are contributory to both reperfusion arrhythmias and to stunning. As will be discussed below, it is their role in necrosis that has yet to be established.

Hearse and coworkers have demonstrated the involvement of free radicals in the genesis of reperfusion arrhythmias in isolated and in vivo rat hearts. They demonstrated oxy radical generation with the use of spin traps (Hearse and Tosaki, 1987a,b), and by ameliorating arrhythmias using a variety of agents such as mannitol (Bernier and Hearse, 1988), SOD (Riva et al, 1987) and catalase (Bernier et al, 1989). Other investigators have also shown a protective effect against reperfusion arrhythmias with radical scavengers (Woodward and Zakaria, 1985; Nejima et al, 1989; Yamakawa et al, 1989; Watanabe et al, 1989).

Much of the evidence for the involvement of free radicals in myocardial stunning has been provided by Bolli et al (1988, 1989), who have presented direct evidence, with the use of spin traps, that oxy radicals contribute to postischaemic myocardial dysfunction in dogs. They further demonstrated the attenuation of stunning with the administration of a variety of radical scavengers, including *N*-2-mercaptopropionyl glycine (MPG) (Myers et al, 1986), dimethylthiourea (DMTU) (Bolli et al, 1987), and SOD plus catalase (Myers et al, 1985b). Their work points to xanthine oxidase as a significant source of oxygen radicals implicated in stunning (Charlat et al, 1987), while others suggest that leukocytes may be an important cause of myocardial stunning (Westlin and Mullane, 1989; Engler and Covell, 1987). Other investigators have also documented a protective effect of radical scavengers on the stunned myocardium (Gross et al, 1986; Farber et al, 1988).

B. The controversy surrounding antioxidant therapy and tissue necrosis

It has been clearly established that oxygen-derived free radicals can kill heart cells when administered exogenously (Basu and Karmazyn, 1987; Blaustein et al, 1986). Similarly, it has been established that reperfused hearts do produce free radicals. What remains to be shown is whether enough free radicals are produced upon reperfusion to contribute to necrosis. Virtually all of the evidence supporting the theory that oxygen radicals promote cell death in reperfused myocardium has been obtained indirectly from experiments in which oxidants were seen to limit infarct size in the reperfused heart. A variety of antioxidants have been tested against all three levels of myocardial ischaemia–reperfusion injury. Anti-free radical therapies include: the enzymatic scavengers such as SOD and catalase, xanthine oxidase inhibitors, organic scavengers such as DMSO (Ganote et al, 1982) and MPG (Fuchs et al, 1985), and iron chelators (Menasche et al, 1987). Unfortunately, the reports have been very mixed regarding all of these agents and that is why the free radical hypothesis of necrosis remains controversial. Only when it has been proven that an anti-free radical intervention can unambiguously limit necrosis will the free radical hypothesis gain full acceptance. In the following, we summarize the available data for two of the antioxidant approaches and attempt to determine if current evidence supports either as being capable of limiting infarction. The bulk of the available data concerning antioxidant therapy and infarct size limitation is confined to either the xanthine oxidase inhibitors or the enzymatic scavengers, SOD and catalase. For this reason we will focus on these two interventions.

C. The xanthine oxidase inhibitors

We have already dissussed xanthine oxidase as a potential source of oxygen-derived free radicals during ischaemia–reperfusion. Allopurinol, a competitive inhibitor of xanthine oxidase, has been shown to be cardioprotective in some species against both reperfusion arrhythmias and stunning. Hearse and coworkers (Manning et al, 1984; Hearse et al, 1986) have domonstrated that allopurinol can reduce reperfusion-induced arrhythmias in both isolated and in vivo rat hearts. Bolli and coworkers (Charlat et al, 1987) have established a protective effect for allopurinol in the in vivo canine heart against stunning as well. The effect of allopurinol on canine infarct size, however, has varied (Reimer and Jennings, 1985; Werns et al, 1986). The causes of this discrepancy are not clear, but one point is that the pretreatment period, which has been longer in the positive studies, appears to be important for allopurinol to be converted into the noncompetitive inhibitor, oxypurinol (Werns, 1990). Indeed, when both allopurinol and oxypurinol were administered acutely with no pretreatment, only oxypurinol was protective (Matsuki et al, 1990). Post-treatment also seems to be important. Puett et al (1987) failed to demonstrate limitation of infarct size with oxypurinol when the inhibitor was given as a single bolus at reperfusion (half-life of about 3 h in dogs). In an almost identical protocol, infarct size was limited by oxypurinol when its plasma levels were maintained all through the reperfusion period (Matsuki et al, 1990), suggesting that xanthine oxidase may generate free radicals long into the reperfusion period. In all of the positive studies with allopurinol cited above, infarct size was determined by staining with triphenyltetrazolium chloride (TTC) within 24 h of reperfusion. As will be seen in the next section, SOD seems to delay but not prevent necrosis as assessed by this method. Thus, while failure of the negative studies with xanthine oxidase inhibitors might be ascribed to an improper administration schedule, it may also be true that these agents are simply incapable of causing a sustained limitation of infarct size. That can only be tested in models where the reperfusion time is extended.

Although allopurinol is thought to be protective primarily via xanthine oxidase inhibition, several studies suggest different potential mechanisms for its action. Godin and Bhimji (1987) found pretreatment with a relatively high dose of allopurinol ($75\,mg\,kg^{-1}$) protected the xanthine oxidase deficient rabbit hearts against ischaemia–reperfusion injury. These investigators suggested that allopurinol pretreatment may protect by enhancing the antioxidant capacity of myocardial tissues. Myers et al (1985a) also demonstrated a protective effect of a high allopurinol dose (1 mM) on the rabbit heart against hypoxia/reoxygenation, thus raising further questions regarding the mechanism of action of allopurinol. Moorhouse et al (1987) found allopurinol and oxypurinol to be scavengers of the hydroxyl radical, with rate constants comparable to those seen for any organic molecule from which a hydrogen atom may be abstracted (e.g. glucose). They proposed on the basis of this observation that the protective effect of these compounds against reperfusion injury may be due, in part, to radical scavenging rather than xanthine oxidase inhibition. Furthermore, Werns et al (1989) reported that amflutizole, shown to inhibit cardiac xanthine oxidase, did not attenuate myocardial stunning in the xanthine oxidase-containing dog heart, which suggested that the mechanism of protection by allopurinol against myocardial stunning is not via xanthine oxidase inhibition.

D. Enzymatic scavengers: SOD and catalase

SOD and catalase have been employed extensively in several models, either singly or in combination. Woodward and Zakaria (1985) documented a protective effect of SOD on reperfusion arrhythmias in the rat; other reports supported their findings (Riva et al, 1987). Several investigators (Gross et al, 1986; Buchwald et al, 1989; Myers et al, 1985a) demonstrated a protective effect of SOD and catalase on the functional recovery of the stunned canine myocardium. There is consensus among investigators about the protective effect of SOD and catalase on both arrhythmias and stunning. Unfortunately, this is not the case with infarct size limitation.

Table 1 is adapted from a recent review by Engler and Gilpin (1989) and summarizes 14 infarct size trials using SOD with and without catalase. Jolly et al (1984) found smaller infarct size in dogs that received SOD and catalase at the time of reperfusion. Chambers et al (1985) later reported a positive effect in open chest, nephrectomized dogs receiving SOD only. Werns et al (1985) and Ambrosio et al (1986) confirmed that SOD alone can reduce infarct size. Finally, SOD plus catalase was reported to limit infarct size in a rabbit (Downey et al, 1987) and a porcine model (Naslund et al, 1986) of ischaemia–reperfusion. Except for the study by Werns et al (1988), infarct size was evaluated by TTC 24 h or less following reperfusion with no collateral flow measurement being done in the dog studies. Unfortunately it takes at least 24 h, preferably several days, of coronary artery reperfusion before degenerative changes indicative of infarction become apparent at the light-microscopic level. In the TTC staining method, dehydrogenase

TABLE 1. Summary of the findings of different SOD trials in the literature using the infarct size model.

Study	Catalase present	Species	Col fl as covariate	Ischaemic duration	Reperfusion duration	Detection method
Positive studies						
Ambrosio et al (1986)		Dog	No	90	48 h	Gross
Werns et al (1988)		Dog	Yes	90	6, 24 h	TTC
Naslund et al (1986)		Pig	N/A	60	5 h	TTC
Chambers et al (1985)	*	Dog	No	60	4 h	TTC
Jolly et al (1984)	*	Dog	No	90	20 h	TTC
Werns et al (1985)		Dog	No	90	6 h	TTC
Downey et al (1987)	*	Rabbit	N/A	45	4 h	TTC
Negative studies						
Uraizee et al (1987)		Dog	Yes	40	96 h	Hist
Gallagher et al (1986)	*	Dog	No	180	24 h	TTC
Nejima et al (1989)		Dog	Yes	90	7 d	Hist
Patel et al (1988)		Dog	Yes	120	4, 48 h	TTC
Shirato et al (1988)		Rabbit	N/A	45	3, 24, 72 h	TTC
Miura et al (1988)	*	Rabbit	N/A	45	72 h	Hist
Richard et al (1988)	*	Dog	Yes	90	48 h	Hist

Adapted from Engler and Gilpin, *Circulation* **79:** 1137–1142, 1989.
Abbreviations: Col fl, collateral flow; Hist, histology; TTC, tetrazolium staining; N/A, not applicable to this species; * indicates that catalase was present with SOD. Ischaemic times are in minutes. Reperfusion times are h for hours and d for days.

enzymes and NADH in surviving cells react with tetrazolium salts to produce a highly coloured and insoluble formazan pigment (Klein et al, 1981; Nachlas and Schnitka, 1963). Because reperfusion quickly washes these components out of dead cells, it has been found in untreated hearts that TTC staining just several hours after reperfusion yields essentially the same infarct size as that seen with histology several days after reperfusion (Horneffer et al, 1987).

The lower half of Table 1 shows negative studies with these agents. Uraizee et al (1987) found no protective effect with SOD when the coronary artery was reperfused for four days and infarcts were sized by histology and analysed according to their collateral flow. The same investigators (Richard et al, 1988) repeated the study using a 90-min, instead of 40-min, occlusion period with a combination of SOD and catalase, and again no protection was seen. Nejima et al (1989) also failed to demonstrate protection with SOD alone, using histology as the endpoint and including collateral flow in their analysis.

Some studies using TTC as the endpoint failed to show protection. Patel et al (1988) used SOD alone at two doses, 5 and 15 mg kg^{-1}, and examined infarct size following 6 and 48 h of reperfusion using TTC. They included collateral flow in the analysis and again no effect on infarct size was seen. Gallagher et al (1986) have also reported a negative study using TTC, although collateral flow was not measured.

One possible explanation for the differences in findings among the studies is that the earlier studies did not take an important determinant of infarct size, collateral flow, into account (Reimer et al, 1977; Miura et al, 1987). Figure 1 shows that the wide variation in collateral flow from dog to dog, ranging from less than 5% to as high as 80% of the preocclusion value, makes it difficult to achieve a normal distribution with a small number of animals per group. Therefore, it is likely that some of the early studies were compromised by unaccounted differences in collateral flow between groups. The importance of measuring collateral flow is illustrated in the study by Ambrosio et al (1986), who found significantly smaller infarcts in the SOD-treated group when infarct size as a fraction of the risk zone was compared. Nevertheless, a plot of percentage infarction against collateral flow shows that the SOD-treated animals tended to have higher collateral flows. Their interpretation of this plot was that only the drug-treated animals with low collateral flow benefited from the SOD. When we performed a one-way analysis of variance, with collateral flow as covariate, on their published data, no significance was found. It is possible that there was no real drug effect in the Ambrosio et al study and that only chance variations in collateral flow between the groups led them to conclude that infarct size had been reduced by SOD. The problem is compounded by the small group size of only eight. Thus, four studies from Table 1 (Ambrosio et al, 1986; Chambers et al, 1985; Jolly et al, 1984; Werns et al, 1985), should be cautiously interpreted due to the lack of collateral flow analysis.

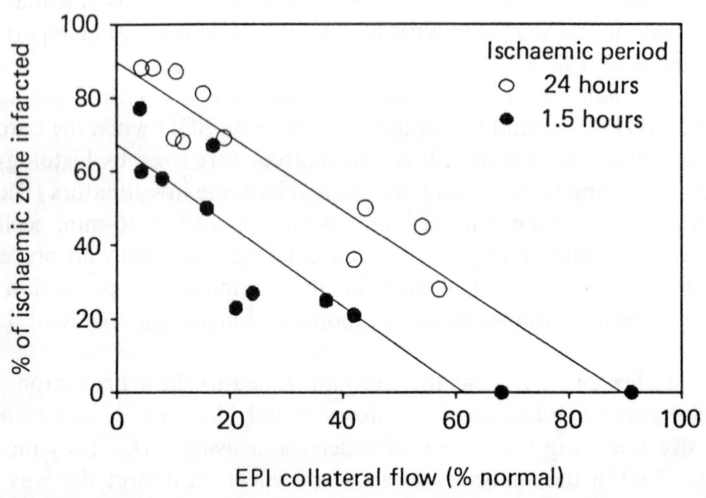

Figure 1. The extent of infarction of the ischaemic zone is plotted against collateral flow for both permanently occluded (open circles) and reperfused (solid circles) dog hearts. Note that infarct size is inversely related to collateral flow at the subepicardium (EPI) and that early reperfusion shifts that relationship down in a parallel fashion. A cardioprotective drug should also cause a downward shift of the relationship. The wide range of collateral flows seen in dogs causes significant variability in infarct size (Downey, 1990).

E. SOD may only delay necrosis

The collateral flow analysis argument, however, can hardly be applied to rabbit (Downey et al, 1987) or pig (Naslund et al, 1986) trials since neither of these species has significant coronary collateral circulation (Del Maestro et al, 1981; Maxwell et al, 1987). Furthermore, the dog study of Werns et al (1988) does use the collateral flow analysis. A second possible source of the discrepancy could be the method used to estimate the infarct size. All of the positive studies involved TTC staining 24 h or less after reperfusion. The most logical explanations are that protection afforded by SOD is simply not sustained, or that a second, slowly progressing, non-radical-mediated component of injury is revealed, once the radical-mediated component has been suppressed. Shirato et al (1988) recently examined SOD alone in a rabbit model with three different reperfusion times. After 3 and 24 h of reperfusion, TTC indicated very small infarcts, but at 72 h no differences were found by TTC. One disturbing feature of this work was the direct comparison of infarcts, by histology, with those by TTC in the 24-h reperfusion group. Figure 2 shows that TTC consistently underestimated infarct size in the SOD group but not the control group. SOD treatment seems to retard the rate of loss of dehydrogenase enzyme and cofactor from necrotic tissue, perhaps by preserving capillary integrity (Granger et al, 1986). Based on the above studies we must conclude that there is little evidence that SOD, with or without catalase, can actually limit infarct size. Furthermore, none of the antioxidants to date has been shown to limit infarct size in a model where the heart was analysed 48 h or longer after reperfusion.

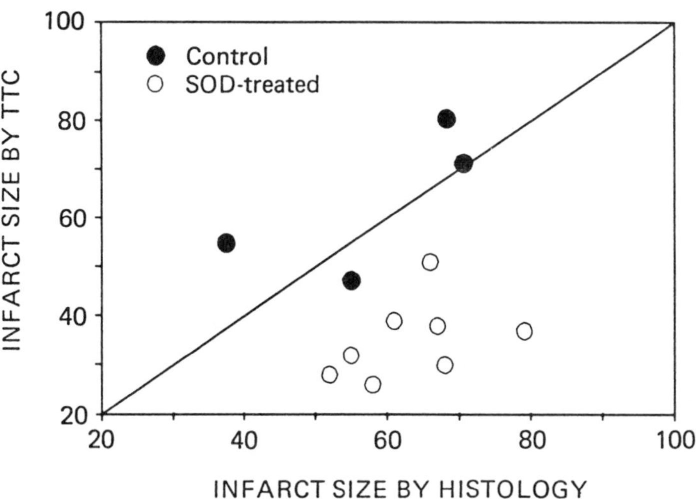

Figure 2. Ischaemic–reperfused rabbit hearts were analysed by both histology and TTC. The infarct size is expressed as a percentage of the area at risk. Similar values were obtained by both methods in untreated animals (closed circles). In SOD-treated animals (open circles), however, TTC consistently underestimated the infarct size measured by histology. It is possible that SOD may cause dead tissue to retain the ability to react with TTC (Shirato et al, 1989).

F. SOD loses its effect at high doses

There are several possible explanations for why SOD failed to limit infarct size in the animal models. For example, SOD may not have access to critical compartments in the heart, or superoxide may simply not contribute to necrosis. Earlier in this chapter we pointed out that the schedule for administration of a xanthine oxidase inhibitor may be crucial to the outcome of the study. A similar stipulation may exist for SOD. Several studies showed that the amount of protection SOD confers to the myocardium is dose-dependent. Woodward and Zakaria (1985) showed that SOD at concentrations of 5, 10 and 20 U ml^{-1} decreased the incidence of ventricular fibrillation in isolated rat hearts by 38%, 64% and 76% respectively. Aoki et al (1988) showed that SOD at 5 and 10 mg kg^{-1} reduced the creatine kinase loss from in situ rat hearts subjected to left anterior descending artery ligation and reperfusion by 68% and 88% respectively. Burton (1985) found that 10 and 20 μg ml^{-1} of SOD enhanced the recovery of developed tension of isolated perfused interventricular septal preparations from the rabbit by 39% and 69% respectively. All these limited dose–response studies happened to fall on the ascending limb of the dose–response relationship. Only Bernier et al (1989) showed that protection against reperfusion arrhythmias in the rat is lost at higher doses of SOD.

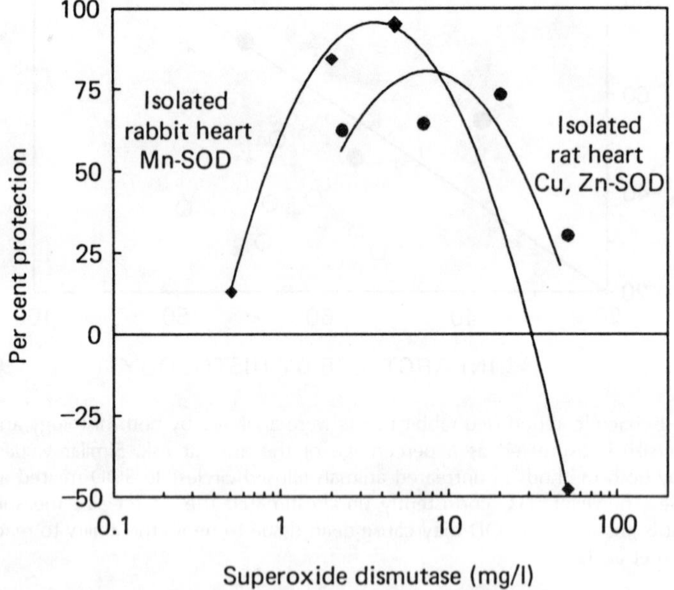

Figure 3. Dose–response curves for the effect of SOD on the isolated heart in terms of percentage protection relative to controls for each model, versus the logarithm of SOD concentration. Note the bell-shaped characteristics of both curves despite the differences in animal species (rat versus rabbit), endpoint (enzyme release versus developed tension), model (hypoxia versus ischaemia) and SOD species used (Cu/Zn-SOD versus Mn-SOD) (Omar and McCord, 1990b; Omar et al, 1990a).

We have carried out dose–response studies using both rabbit and rat hearts. Using isolated buffer-perfused rabbit hearts (Omar and McCord, 1990) subjected to 1 h of global ischaemia and 1 h of reperfusion, lower Mn-SOD doses (0.5, 2 and 5 mg l^{-1}) caused significantly higher recovery of developed tension while a higher (50 mg l^{-1}) Mn-SOD dose resulted in a lower recovery of developed tension compared to controls. Using isolated rat hearts (Omar et al, 1990) subjected to 40 min of hypoxia and 10 min of reoxygenation, lower Cu/Zn-SOD doses (2.3, 7 and 20 mg l^{-1}) resulted in a lower creatine kinase release, while a higher (50 mg l^{-1}) Cu/Zn-SOD dose resulted in enzyme release that was not significantly different from controls. Both of these dose–response curves are replotted in Figure 3 to demonstrate their bell-shaped characteristics. We examined the effect of a very high dose of SOD on infarct size in the in situ rabbit heart (Omar et al, 1990). Cu/Zn-SOD, administered as a single bolus of 50 mg kg^{-1} 10 min prior to reperfusion, resulted in an infarct size significantly *larger* than that seen in the control rabbits.

G. Pretreatment may also be important

We examined the effect of preischaemic equilibration of the isolated rabbit heart with human recombinant SODs on its tolerance to ischaemia–reperfusion (Omar and McCord 1991). After either 15 or 50 min of equilibration the hearts were subjected to 1 h of ischaemia followed by reperfusion. Fluid weeping from the ventricular surface was collected and assumed to represent cardiac lymph. Only 15 min of preischaemic equilibration with the positively charged Mn-SOD was sufficient to load the lymph and protect function. This period, however, was insufficient for either the negatively charged Cu/Zn-SOD or the large molecular size and negatively charged PEG-SOD to equilibrate with the lymph. This effect was reflected by the lack of protection given by both of these forms of SOD. However, when the equilibration period was raised to 50 min, a sufficient amount of time for the Cu/Zn-SOD to equilibrate fully with the interstitium, Cu/Zn-SOD did show good protection. Polyethylene glycol conjugated SOD (PEG-SOD) never equilibrated with the lymph and never protected. Figure 4 illustrates the strong correlation between SOD activity in the lymph and recovery of developed tension for the various SOD treatments and preischaemic equilibration times.

In reviewing the SOD trials listed in Table 1, none of the technically acceptable studies (those in which the reperfusion time was 48 h or longer and collateral flow was accounted for) met the dose and schedule requirements dictated by our isolated heart studies. It is interesting to note that the only trials in which pretreatment and dose even approached adequacy were the early positive TTC studies from Lucchesi's laboratory (Werns et al, 1985, 1988; Jolly et al, 1984).

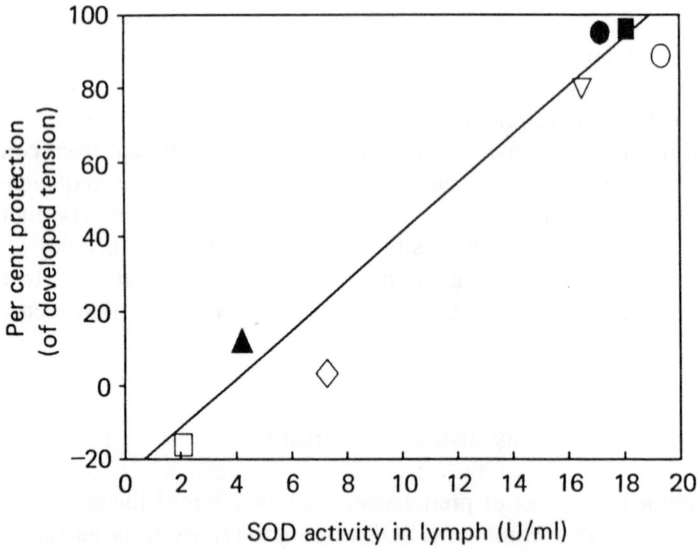

Figure 4. Percentage protection by SOD versus SOD activity in cardiac lymph. SOD activity in the lymph correlates well with the protection it affords to the myocardium (R = 0.97). Various SODs and equilibration times (as described in the text) are designated by the following symbols: □, PEG at 15 min; ▲, Cu/Zn at 15 min; ◇, PEG at 50 min; ▽, Cu/Zn at 50 min; ●, Mn at 15 min; ■, Mn at 50 min; ○, sheep Cu/Zn at 15 min (Omar and McCord, 1990a).

H. Post-treatment may be a requirement

In our in vitro studies SOD was present in the perfusate throughout. Therefore, we do not know when the drug can be withdrawn. Recent studies with PEG-SOD by Lucchesi's group (Tamura et al, 1988; Chi et al, 1989) suggest that prolonged post-treatment is indeed a requirement. PEG-SOD has a plasma half-life in excess of a day. However, we (Ooiwa et al, 1991) saw no limitation of infarct size in our rabbit model with PEG-SOD given at 1000 U kg^{-1} (the same dose as that of Tamura et al) when histology was used to size the infarcts. Tanaka et al (1989) also failed to find protection with PEG-SOD in dog hearts using histology. PEG-SOD probably fails to reduce infarct size because of its inability to enter the interstitial space. Even though PEG-SOD may not salvage myocardium, post-treatment may still be a requirement for a positive outcome, especially if neutrophils are the source of injurious free radicals. *None of the studies in Table 1 maintained any significant post-treatment.*

V. ISCHAEMIA–REPERFUSION INJURY IN OTHER ORGANS

So far we have discussed in detail ischaemia–reperfusion in the heart. Several studies indicate that other organs undergo a similar phenomenon when they are exposed to a transient period of ischaemia followed by reperfusion. Ischaemia–reperfusion in the gastrointestinal tract, lung, kidneys and central nervous system will be discussed next.

A. Gastrointestinal tract

Reperfusion of the ischaemic intestine results in accelerated tissue damage that adds considerably to the already existing ischaemic insult (Haglund and Lundgren, 1978). Granger et al (1980) have shown that 1 h of local arterial hypotension (30 mmHg arterial pressure), followed by normotensive reperfusion, causes a significant increase in intestinal capillary permeability. In another study (Granger et al, 1981), neither antihistamine, indomethacin nor methylprednisolone pretreatment was able to attenuate the increase in capillary permeability following 1 h of partial ischaemia followed by reperfusion in the cat intestine. However, SOD, and to a lesser extent catalase (Parks and Granger, 1984), significantly diminished the permeability increase. Further experiments showed that the infusion of xanthine and xanthine oxidase (a superoxide-generating system) caused an increase in capillary permeability in normal intestine comparable to that seen following ischaemia (Grogaard et al, 1982) which was abolished by SOD pretreatment. Younes et al (1987) reported that glutathione is consumed, while glutathione disulphide is formed, in the intestinal mucosa following ischaemia–reperfusion. The same investigators also observed a significant increase in lipid peroxidation products following reperfusion of the ischaemic intestine, which was attenuated by SOD.

One source of reactive oxygen species in the reperfused intestine has been suggested to be xanthine oxidase. It has been shown that allopurinol and oxypurinol are almost as effective as SOD and catalase in ameliorating the increase in vascular permeability (Granger et al, 1981) and morphological changes (Morris et al, 1987) seen with intestinal ischaemia–reperfusion.

Furthermore, a low-molybdenum, high-tungsten diet, which leads to production of inactive demolybdo-xanthine oxidase, protects against the increased postischaemic intestinal microvascular permeability (Parks et al, 1986). Although the conversion of xanthine dehydrogenase to xanthine oxidase in the ischaemic intestine has been originally reported to occur within 1 min (Roy and McCord, 1983), more recent reports suggest a much slower conversion rate (Parks et al, 1988). Conversion, however, might not be necessary, since the fresh intestine is already rich in xanthine oxidase (Parks and Granger, 1986).

Another potential source of oxygen radicals in the ischaemic intestine is the NADPH oxidase of neutrophils. Grisham et al (1986a) reported an increase in neutrophil infiltration of the ischaemic intestine upon reperfusion which was attenuated by SOD and allopurinol. Otamiri (1989) has also shown that either SOD plus allopurinol or hydrocortisone pretreatment results in the attenuation of the postischaemic increase of mucosal permeability, malondialdehyde content and myeloperoxidase activity (a neutrophil granulocyte marker) in the mucosa.

Reperfusion of the ischaemic liver results in the release of enzymatic indicators of hepatocellular injury which is attenuated by SOD and catalase administration (Adkison et al, 1986; Atalla et al, 1985). Treatment with the antioxidants α-tocopherol (Marubayashi et al, 1986) or coenzyme Q_{10} (Marubayashi et al, 1984) increases the survival and accelerates ATP synthesis after reperfusion of the ischaemic liver. Moreover, administration of glutathione (GSH) prior to hepatic ischaemia accelerates the recovery of ATP synthesis and decreases lipid peroxide formation upon reperfusion (Marubayashi et al, 1984). Allopurinol has been found to be as effective as oxy radical scavengers in ameliorating the reperfusion injury of the ischaemic liver (Adkison et al, 1986; Atalla et al, 1985), suggesting xanthine oxidase as an important source of these radicals. The role of neutrophils in hepatic ischaemia–reperfusion injury is not yet well defined.

B. Lungs

Several lines of evidence indicate that reperfusion of the ischaemic lung results in some free radical generation. Koyama et al (1987) demonstrated that reperfusion of an isolated perfused dog lung lobe subjected to 6 h of ischaemia resulted in progressive injury, as assessed by increase in lung weight. This injury was markedly attenuated by SOD, indicating a free radical mechanism. Other models of hyperoxic lung injury have also been studied and shown to involve the production of reactive oxygen species (Repine et al, 1987; Jamieson et al, 1986).

Several sources of oxy radical production in the lung have been implicated. One important source seems to be xanthine oxidase. The presence of xanthine oxidase activity has been demonstrated in the lung of several animal species, including the dog and the rat (Parks and Granger, 1986). Cheronis et al (1987) have shown that the addition of tungsten to the diet of rats for three weeks prevented lung injury caused by hyperoxia. The slow conversion of xanthine dehydrogenase to xanthine oxidase in the lungs (Engerson et al, 1987) raised some questions about the amount of damage contributed by this source of free radicals. However, such conversion might not be necessarily required, as discussed above, since the initial xanthine oxidase content itself is high and may contribute to oxidative stress. Another source of free radicals is the mitochondrion. Lung mitochondria have been shown to increase their hydrogen peroxide production dramatically as the oxygen tension rises (Turrens et al, 1982b). Lung mitochondria, however, generate less hydrogen peroxide than liver mitochondria under the same conditions (Boveris and Chance, 1973), which may be an adaptive response to the high oxygen tensions normally present in the lung.

A third potential source of free radicals in the lung is the neutrophil. A number of reports support the involvement of neutrophils in the pathogenesis of the hyperoxia-induced acute oedematous lung injury (Fox et al, 1981a,b; Tate and Repine, 1983). Neutrophils have been shown to accumulate along damaged endothelial cells in lungs from rats exposed to hyperoxia (Fox et al, 1981a; Barry and Crapo, 1985). Furthermore, neutrophils were shown to make potent toxins, damage cultured lung endothelial cells and cause acute oedematous injury when added to the perfusate of isolated lungs (Shasby et al, 1982). The mechanism of neutrophil-mediated lung injury has been suggested to involve a synergistic action of both lysosomal proteases and oxygen-derived free radicals (Ward et al, 1986; Repine et al, 1987). Some oxygen derivatives such as hydrogen peroxide have been shown to potentiate the effects of proteases, resulting in a modification of protein substrates such that they become much more susceptible to proteolysis (Fligiel et al, 1984).

C. Kidneys

Indirect evidence for oxygen radical-mediated damage in ischaemic–reperfused kidneys comes from the protective effects of the xanthine oxidase inhibitor allopurinol in dogs (Chatterjee and Berne, 1976), rabbits (Hansson et al, 1982), and rats (Paller et al, 1984; Bayati et al, 1985). Linas et al (1987) have demonstrated that DMTU and GSH are consumed in the ischaemic–reperfused kidney, which indirectly implicates cytotoxic oxygen species.

Green et al (1986) have demonstrated a protective effect with deferoxamine against lipid peroxidation on rabbit kidneys subjected to warm ischaemia for 60 or 120 min followed by reperfusion for 60 min or 24 h. Protective effects of SOD against ischaemia–reperfusion-induced kidney damage have been demonstrated in rabbits (Hansson et al, 1983), rats (Paller et al, 1984; Schneider et al, 1987), and dogs (Ouriel et al, 1985). SOD treatment improved renal function and circulation, decreased tubular injury, and prevented lipid peroxidation in cortical mitochondria (Paller et al, 1984). SOD also suppressed the increase in urinary protein excretion associated with immunological nephritis in rats (Adachi et al, 1986). Information on the involvement of neutrophils in ischaemic renal injury is inconclusive. Hellberg et al (1988) reported that neutrophil depletion with antineutrophil serum (ANS) caused a modest increase in the immediate reperfusion glomerular filtration rate in the rat. Klausner et al (1989) reported that ANS-treated rats developed less azotaemia 24 h after ischaemia than did non-neutropaenic controls. In contrast, Paller (1989) and Thornton et al (1989) failed to show any protection from either ANS or Mo-1 antibody against ischaemia–reperfusion in rabbit and rat kidneys. The protection against oxygen-induced tissue damage is of relevance to the kidneys especially during transplantation, when ischaemia is followed by reperfusion (Parks et al, 1983). The improvement of renal function and survival rate after kidney transplantation in dogs receiving allopurinol (Toledo-Pereyra et al, 1976; Owens et al, 1974) is likely to be due to inhibition of superoxide formation, although other mechanisms, as discussed above, cannot be ruled out.

D. Central nervous system

Brain ischaemia in experimental models is associated with distinct free radical pathology which affects the predominant membrane lipids in the ischaemic tissues. Kirsch et al (1987) recently demonstrated an ESR signal using the spin adduct phenyl-*t*-butylnitrone (PBN) in rat brains exposed to global ischaemia followed by reperfusion. Snelling et al (1987) and Davis et al (1987) showed significant improvement in postischaemic cerebral blood flow and somatosensory evoked potential recovery upon SOD administration in models of reversible global and focal ischaemia in the cat. Hall and coworkers (Hall and Braughler, 1986; Hall and Wolf, 1986) found that pretreatment with high oral doses of vitamin E and selenium prevented any post-traumatic decrease in white matter perfusion in the cat. The same investigators (Hall and Braughler, 1986) also found that ascorbic acid, administered as a single bolus intravenously, significantly retards post-traumatic spinal cord hypoperfusion. Other investigators have reported a protective role for vitamin E in cerebral ischaemia models in the rat (Yamamoto et al, 1983), dog (Fujimoto et al, 1984) and cat (Hall, 1987). Non-glucocorticoid 21-aminosteroids have been demonstrated to inhibit lipid peroxidation in the CNS (Braughler et al, 1987). One such agent, U74006F, has been reported to improve postischaemic neuronal survival in gerbils (Berry et al, 1987) and reduce infarction and improve glucose utilization in cat brain following cerebral ischaemia (Silvia et al, 1987; Hall and Yonkers, 1988). The biochemical, physiological and pharmacological evidence for the involvement of oxygen radicals and lipid peroxidation in CNS trauma and stroke has been recently reviewed (Braughler and Hall, 1989; Hall and Braughler, 1989). In the feline regional cerebral ischaemia model, wherein one middle cerebral artery is occluded, there is a period of approximately two hours during which the situation is reversible; reperfusion before two hours in this model results in no discernible infarction and no clinical deficits (Flamm et al, 1980). Three hours of occlusion, however, will result in progressive decline in blood flow (90–95%) which appears to coincide with irreversibility and amplification of morphological damage (Ransohoff et al, 1980). The mechanism behind this hypoperfusion and its consequent damage is unknown, but there is some evidence that it may be free radical mediated. It has been postulated (Demopoulos et al, 1980) that lipid peroxides produced from lipid radical reactions selectively inhibit the endothelial synthesis of PGI_2, which counteracts the proaggregating properties of thromboxane A_2 in platelets. This will result in platelet aggregation and numerous platelet-induced microocclusions which exacerbate hypoperfusion.

VI. CONCLUDING REMARKS

A great amount of evidence has implicated oxidative stress in ischaemic–reperfusion injury. It has been rather difficult, however, to reach a consensus regarding the extent of this oxidative stress and the relevance of antioxidant therapy. The failure of antioxidants to provide consistent protection in animal models in vivo, which more closely mimic clinical disease states, has led to scepticism regarding the importance of reperfusion-induced oxidative stress. The most prominent example of this controversy is the reduction of myocardial infarct size by SOD. We have pointed out that the inadequacy of the traditional methods of quantifying infarct size, especially regarding the use of TTC staining and in the absence of collateral flow measurement, has plagued this science for a number of years. Moreover, the dose and administration schedule of SOD are important variables that have been overlooked by many investigators. Consideration of these factors sheds some light on the controversy, and explains at least some of the negative reports. Although the work in vitro has established a theoretical framework for the free radical hypothesis, in the final analysis this question will only be answered by measuring the protection derived in a whole animal by antioxidant interventions.

REFERENCES

Adachi T, Fukuta M, Ito Y, Hirano K, Sugiura M & Sugiura K (1986) *Biochem. Pharmacol.* **35:** 341–345.

Adkison D, Hollwarth ME, Benoit JN, Parks DA, McCord JM & Granger DN (1986) *Acta Physiol. Scand.* **126 (supplement 548):** 101–107.

Ambrosio G, Becker LC, Hutchins GM, Weisman HF & Weisfeldt ML (1986) *Circulation* **74:** 1424–1433.

Ambruso DR & Johnston RB Jr (1981) *J. Clin. Invest.* **67:** 352–360.

Angel MF, Narayanan K, Swartz WM et al (1986) *Br. J. Plast. Surg.* **39:** 469–472.

Aoki N, Bitterman H, Brezinski ME & Lefer AM (1988) *Br. J. Pharmacol.* **95:** 735–740.

Arroyo CM, Kramer JH, Dickens BF & Weglicki WB (1987) *FEBS Lett.* **221:** 101–104.

Atalla SL, Toledo-Pereyra LH, Mackenzie GH & Cederna JP (1985) *Transplantation* **40:** 584–590.

Babior BM (1978) *N. Engl. J. Med.* **298:** 659–668.

Barry BW & Crapo JD (1985) *Am. Rev. Respir. Dis.* **132:** 548–555.

Basu DK & Karmazyn M (1987) *J. Pharmacol. Exp. Ther.* **242:** 673–685.

Battelli MG, DellaCorte E & Stirpe F (1972) *Biochemistry* **126:** 747–749.

Bayati A, Frodin L, Kallskog O, Hellberg O & Wolgast M (1985) *Acta Physiol. Scand.* **124**: 383.

Bernier M & Hearse DJ (1988) *Am. J. Physiol.* **254:** H862–H870.

Bernier M, Hearse DJ & Manning AS (1986) *Circ. Res.* **58:** 331–340.

Bernier M, Manning AS & Hearse DJ (1989) *Am. J. Physiol.* **256:** H1344–H1352.

Berry KP, Braughler JM & Hall ED (1987) *Neuroscience* **13:** 1494 (abstract).

Bersohn MM, Philipson KD & Fukushima JY (1982) *Am. J. Physiol.* **242:** C288–295.

Berti F, Rossoni G, Magni F et al (1988) *J. Cardiovasc. Pharmacol.* **12:** 438–444.

Bindoli A, Cavallini L, Rigobello MP, Coassin M & Di Lisa F (1988) *Free Radicals Biol. Med.* **4:** 163–167.

Blasig IE, Ebert B & Lowe H (1986) *Studia Biophys.* **116:** 35–42.

Blaustein AS, Schine L, Brooks WW, Franburg BL & Bing OH (1986) *Am. J. Physiol.* **250:** H595–H599.

Bolli R (1988) *J. Am. Coll. Cardiol.* **12:** 239–249.

Bolli R, Zhu WX, Hartley CJ et al (1987) *Circulation* **76:** 458–468.

Bolli R, Patel BS, Jeroudi MO, Lai EK & McCay PB (1988) *J. Clin. Invest.* **82:** 476–485.

Bolli R, Jeroudi MO, Patel BS et al (1989) *Proc. Natl. Acad. Sci. USA* **86:** 4695–4699.
Borgeat P, Frateau de Laclos B & MacLouf J (1983) *Biochem. Pharmacol.* **32:** 381–387.
Bourdillon PDV & Poole-Wilson PA (1981) *Cardiovasc. Res.* **15:** 121–130.
Bourdillon PDV & Poole-Wilson PA (1982) *Circ. Res.* **50:** 360–368.
Boveris A & Cadenas E (1975) *FEBS Lett.* **54:** 311–314.
Boveris A & Chance B (1973) *Biochem. J.* **134:** 707–716.
Boveris A, Cadenas E & Stoppani AOM (1976) *Biochem. J.* **156:** 435–444.
Braughler JM & Hall ED (1989) *Free Radicals Biol. Med.* **6:** 289–301.
Braughler JM, Pregenzer JF, Chase RL, Duncan LA, Jacobsen EJ & McCall JM (1987) *J. Biol. Chem.* **262:** 10438–10440.
Brown JM, Terada LS, Grosso MA et al (1988) *Mol. Cell. Biochem.* **84:** 173–175.
Bruder G, Heid HW, Jarasch ED & Mather IH (1983) *Differentiation* **23:** 218–225.
Bucher JR, Tien M & Aust SD (1983) *Biochem. Biophys. Res. Commun.* **111:** 777.
Buchwald A, Klein HH, Lindert S et al (1989) *J. Cardiovasc. Pharmacol.* **13:** 258–264.
Buja LM, Hagler HK & Willerson JT (1988) *Cell Calcium* **9:** 205–217.
Bulkley BH & Roberts WC (1974) *Am. J. Med.* **56:** 244–250.
Burton KP (1985) *Am. J. Physiol.* **248:** H637–H643.
Cadenas E, Boveris A, Ragan CI & Stopanni AOI (1977) *Arch. Biol. Biophys.* **180:** 248–257.
Campbell CA, Alker KJ & Kloner RA (1988) *J. Am. Coll. Cardiol.* **11:** 54A.
Chambers DE, Parks DA, Patterson G et al (1985) *J. Mol. Cell Cardiol.* **17:** 145–152.
Chan PH & Fishman RA (1984) *Fed. Proc.* **43:** 210–213.
Charlat ML, O'Neill PG, Egan JM et al (1987) *Am. J. Physiol.* **252:** H566–H577.
Chatterjee SN & Berne TV (1976) *Am. J. Surg.* **131:** 658–659.
Cheronis JC, Rodell TC & Repine JE (1987) *Clin. Res.* **35:** 170A.
Chi L, Tamura Y, Hoff PT et al (1989) *Circ. Res.* **64:** 665–675.
Crapo JD, Freeman BA, Barry BE, Turrens JF & Young SL (1983) *Physiologist* **26:** 170–176.
Dahinden CA, Fehr J & Hugli T (1983) *J. Clin. Invest.* **72:** 113–121.
Darius H, Yanagisawa A, Brezinski ME & Lefer AM (1986) *Fed. Proc.* **45:** 808 (abstract).
Davis RJ, Bulkley GB & Traystman RJ (1987) *Fed. Proc.* **46:** 799.
Del Maestro RF, Björk J & Arfors KE (1981) *Microvasc. Res.* **22:** 255–270.
Demopoulos HB, Flamm ES, Pietronigro DD & Seligman ML (1980) *Acta Physiol. Scand* **492 (supplement):** 91–119.
Dickens BF, Mak IT & Weglicki WB (1988) *Mol. Cell. Biochem.* **82:** 119–123.
Dionisi D, Galeotti T, Terrehove T & Azzi A (1975) *Biochim. Biophys. Acta* **403:** 292–300.
Downey JM (1990) *Annu. Rev. Physiol.* **52:** 487–504.
Downey J, Chambers D, Miura T, Yellon D & Jones D (1986) *Circulation* **74 (supplement II):** 372.
Downey JM, Miura T, Eddy LJ et al (1987) *J. Mol. Cell. Cardiol.* **19:** 1053–1060.
Eddy LJ, Stewart JR, Jones HP, Engerson TD, McCord JM & Downey JM (1987) *Am. J. Physiol.* **253:** H709–H711.
Edwards SW, Hallett MB & Campbell AK (1984) *Biochem.* **217:** 851–854.
Engerson TD, McKelvey TG, Rhyne DB, Boggio EB, Snyder SJ & Jones HP (1987) *J. Clin. Invest.* **79:** 1564–1570.
Engler R & Covell JW (1987) *Circ. Res.* **61:** 20–28.

Engler R & Gilpin E (1989) *Circulation* **79:** 1137–1142.

Fantone JC (1990) In Schaik TV (ed.) *Clinical Ischemic Syndromes: Mechanisms and Consequences of Tissue Injury*, pp 137–145. St Louis, MO: C.V. Mosby Company.

Farber NE, Vercellotti GM, Jacob HS, Pieper GM & Gross GJ (1988) *Circ. Res.* **63:** 351–360.

Ferrans VJ, Hibbs RGT, Weily HS, Weilbaecher DG, Walsh JJ & Burch GE (1970) *J. Mol. Cell. Cardiol.* **1:** 11–22.

Flamm ES, Seligman ML & Demopoulos HB (1980) In Cotrell JE & Turndorf H (eds) *Anesthesia and Neurosurgery*, p. 248. St Louis, MO: C.V. Mosby Company.

Fletcher MP, Seligmann BE & Gallin JI (1982) *J. Immunol.* **128:** 941–948.

Fligiel SEG, Lee EC, McCoy JP, Johnson KJ & Varani J (1984) *Am. J. Pathol.* **115:** 418–425.

Flynn PJ, Becker WK, Vercellotti GM et al (1984) *Inflammation* **8:** 33–44.

Fox RB, Hoidal JR, Brown DM & Repine JE (1981a) *Am. Rev. Respir. Dis.* **123:** 521–523.

Fox RB, Shasby DM, Harada RN & Repine JE (1981b) *Chest* **80 (supplement):** 3S–4S.

Fuchs J, Mainka L & Zimmer G (1985) *Arzneimittelforsch.* **35:** 1394–1402.

Fujimoto S, Mizoi K, Yoshimoto T & Suzuki J (1984) *Surg. Neurol.* **22:** 449–454.

Fukuhara T (1985) *Jpn. Circ. J.* **49:** 432–445.

Fuller BJ, Lunec J, Healing G, Simpkin S & Green CJ (1987) *Transplantation* **43:** 604–606.

Galaris D, Cadenas E & Hochstein P (1989a) *Arch. Biochem. Biophys.* **273:** 497–504.

Galaris D, Eddy L, Arduini A, Caderas E & Hochstein P (1989b) Mechanisms of reoxygenation injury in myocardial infarction: implications of a myoglobin redox cycle. *Biochem. Biophys. Res. Commun.* **160:** 1162–1168.

Gallagher KP, Buda AJ, Pace D, Gerren RA & Shlafer M (1986) *Circulation* **73:** 1065–1076.

Ganote CE, Simms M & Safavi S (1982) *Am. J. Pathol.* **109:** 270–276.

Gardner HW (1989) *Free Radicals Biol. Med.* **7:** 65–86.

Gardner TJ, Stewart JR, Casale AS, Downey JM & Chambers DE (1983) *Surgery* **94:** 423–427.

Garlick PB, Davies MJ, Hearse DJ & Slater TF (1987) *Circ. Res.* **61:** 757–760.

Gauduel Y & Duvelleroy MA (1984) *J. Mol. Cell. Cardiol.* **16:** 459–470.

Godin DV & Bhimji S (1987) *Biochem. Pharmacol.* **36:** 2101–2107.

Granger DN, Sennett M, McElearney P & Taylor AE (1980) *Gastroenterology* **79:** 474–480.

Granger DN, Rutili G & McCord JM (1981) *Gastroenterology* **81:** 22–29.

Granger DN, Hollwarth ME & Parks DA (1986) *Acta Physiol. Scand.* **126 (supplement 548):** 47–63.

Green CJ, Healing G, Simkin S, Lunec J & Fuller BJ (1986) *Comp. Biochem. Physiol.* **85B:** 113–117.

Grinwald PM (1982) *J. Mol. Cell. Cardiol.* **14:** 359–365.

Grisham MB (1985) *J. Free Radicals Biol. Med.* **1:** 227–232.

Grisham MB, Hernandez LA & Granger DN (1986a) *Am. J. Physiol.* **251 (GILP 14):** G567–G574.

Grisham MB, Russell WJ, Roy RS & McCord JM (1986b) In Rotilio G (ed.) *Superoxide and Superoxide Dismutase in Chemistry, Biology and Medicine*, pp 571–575. Amsterdam: Elsevier.

Grogaard B, Parks DA, Granger DN, McCord JM & Forsberg JO (1982) *Am. J. Physiol.* **242:** G448–G454.

Gross GJ, Farber NE, Hardman HF & Warltier DC (1986) *Am. J. Physiol.* **250:** H372–H377.
Grum CM, Ragsdale RA, Ketai LH & Shlafer M (1986) *Biochem. Biophys. Res. Commun.* **141:** 1104–1108.
Grum CM, Gallagher KP, Kirsh MM & Shlafer M (1989) *J. Mol. Cell. Cardiol.* **21:** 263–267.
Guarnieri C, Ferrari R, Visioli O, Caldarera CM & Nayler WG (1978) *J. Mol. Cell. Cardiol.* **10:** 893–906.
Haglund U & Lundgren O (1978) *Fed. Proc.* **37:** 2729–2733.
Hall ED (1987) In Tsuchiya M, Asano M, Mishima Y & Oda M (eds) *Microcirculation: An update*, pp 553–556. Amsterdam: Excerpta Medica.
Hall ED & Braughler JM (1986) *CNS Trauma* **3:** 281–293.
Hall ED & Braughler JM (1989) *Free Radicals Biol. Med.* **6:** 303–313.
Hall ED & Travis MA (1988) *Brain Res.* **451:** 350–352.
Hall ED & Wolf DL (1986) *J. Neurosurg.* **64:** 951–961.
Hall ED & Yonkers PA (1988) *Stroke* **19:** 340–344.
Halliwell B (1978) *FEBS Lett.* **92:** 321–326.
Hansson R, Gustafsson B, Jonsson O et al (1982) *Transplant. Proc.* **14(1):** 51–58.
Hansson R, Jonsson O, Lundstam S, Pettersson S, Scherst NT & Waldenström J (1983) *Clin. Sci.* **65:** 605–610.
Harris P (1975) *Eur. J. Cardiol.* **3:** 157–163.
Hearse DJ & Humphrey SM (1975) *J. Mol.Cell. Cardiol.* **7:** 463–482.
Hearse DJ & Tosaki A (1987a) *Circ. Res.* **60:** 375–383.
Hearse DJ & Tosaki A (1987b) *J. Cardiovasc. Pharmacol.* **9:** 641–650.
Hearse DJ, Humphrey SM & Chain EB (1973) *J. Mol. Cell. Cardiol.* **5:** 395–407.
Hearse DJ, Humphrey SM, Nayler WG, Slade A & Border D (1975) *J. Mol. Cell. Cardiol.* **7:** 315–324.
Hearse DJ, Manning AS, Downey JM & Yellon DM (1986) *Acta Physiol. Scand.* **126 (supplement 548):** 65–78.
Hellberg POA, Källskog Ö, Wolgast M & Öjteg G (1988) *Acta Physiol. Scand.* **134:** 313–315.
Hernandez LA, Grisham MB, Twohig B, Arfors KE, Harlan JM & Granger DN (1987) *Am. J. Physiol.* **253(HCP 22):** H699–H703.
Horneffer PJ, Healy B, Gott VL & Gardner TJ (1987) *Circulation* **76:** v39–v42.
Huizer T, De Jong JW, Nelson JA et al (1989) *J. Mol. Cell. Cardiol.* **21:** 691–695.
Humphrey SM, Gavin JB & Herdson PB (1980) *J. Mol. Cell. Cardiol.* **12:** 1397–1406.
Ikeda Y, Ikeda K & Long DM (1989) *J. Neurosurg.* **71:** 233–238.
ISAM Study Group (1986) *N. Engl. J. Med.* **314:** 1465.
ISIS-2 Collaborative Group (1988) *Lancet* **2:** 349.
Jamieson D, Chance B, Cadenas E & Boveris A (1986) *Annu. Rev. Physiol.* **48:** 703–719.
Jennings RB & Ganote CE (1976) *Circ. Res.* **38 (supplement 1):** 1.
Jennings RB & Reimer KA (1983) *Circulation* **68 (supplement):** 125–136.
Jennings RB, Hawkins HK, Lowe JE, Hill MC, Klottman S & Reimer KA (1978) *Am. J. Pathol.* **92:** 187–214.
Jewett SL, Eddy LJ & Hochstein P (1989) *Free Radicals Biol. Med.* **6:** 185–188.
Jolly SR, Kane WJ, Bailie MB, Abrams GD & Lucchesi BR (1984) *Circ. Res.* **54:** 277–285.
Jugdutt BI (1985) *Circulation* **72(4):** 907–914.
Kako KJ (1986) *Can. J. Cardiol.* **2:** 184–194.
Kalra J, Chaudhary AK & Prasad K (1989) *J. Mol. Cell. Cardiol.* **21:** 1125–1136.

Kaschnitz RM & Hatefi Y (1975) *Arch. Biochem. Biophys.* **171:** 292–304.
Katz AM, Freston JW, Messineo FC & Herbette LG (1985) *J. Mol. Cell. Cardiol.* **17 (supplement 2):** 11–20.
Kehrer JP, Piper HM & Sies H (1987) *Free Radicals Res. Commun.* **3:** 69–78.
Kirsch JR, Phelan AM, Lange DG & Traystman RJ (1987) *Pediatr. Res.* **21:** 202A.
Kishimoto A, Kajikawa N, Tabuchi H, Shiota M & Nishizuka Y. (1981) *J. Biochem.* **90:** 889–892.
Klausner JM, Paterson IS, Goldman G et al (1989) *Am. J. Physiol.* **256:** F794–F802.
Klebanoff SJ (1980) *Ann. Intern. Med.* **93:** 480–489.
Klein HH, Puschmann S, Schaper J & Schaper W (1981) *Virchows Arch. (A)* **393:** 287–297.
Klein HH, Pich S, Bohle RM et al (1988) *J. Cardiovasc. Pharmacol.* **12:** 338–344.
Kloner RA, Ganote CE & Jennings RB (1974) *J. Clin. Invest.* **54:** 1496–1508.
Kontos HA (1987) *Am. Rev. Respir. Dis.* **136:** 474–477.
Kontos HA, Wei EP, Ellis EF et al (1985) *Circ. Res.* **57:** 142–151.
Koyama I, Toung TJK, Rogers MC, Gurtner GH & Traystman RJ (1987) *J. Appl. Physiol.* **63(1):** 111–115.
Kuehl FA, Humes J & Torchiana ML (1979) *Adv. Inflamm. Res.* **1:** 419–430.
Kukreja RC, Kontos HA, Hess ML & Ellis EF (1986) *Circ. Res.* **59:** 612–619.
Linas SL, Shanley PF, White CW, Parker NP & Repine JE (1987) *Am. J. Physiol.* **253:** F692–F701.
Loschen G, Azzi A & Flohe L (1973) *FEBS Lett.* **33:** 84.
Manning AS, Coltart DJ & Hearse DJ (1984) *Circ. Res.* **55:** 545–548.
Marubayashi S, Dohi K, Yamada K & Kawasaki T (1984) *Biochim. Biophys. Acta* **797:** 1–9.
Marubayashi S, Dohi K, Ochi K & Kawasaki T (1986) *Surgery* **99:** 184–192.
Matsuki T, Cohen MV, Holt G, Ayling J, Hearse DJ & Downey JM (1989) *Am. J. Physiol.* **256:** H1322–H1327.
Matsuki T, Shirato C, Cohen MV & Downey JM (1990) *Can. J. Cardiol.* **6:** 123–129.
Maxwell MP, Hearse DJ & Yellon DM (1987) *Cardiovasc. Res.* **21:** 737–746.
Maxwell MP, Hearse DJ & Yellon DM (1989) *J. Cardiovasc. Pharmacol.* **13:** 608–615.
McCord JM (1985a) *N. Engl. J. Med.* **312:** 159–163.
McCord JM (1985b) In Oberley L (ed.) *Superoxide Dismutase*, Vol. III, *Pathological States*, pp 143–150. Boca Raton: CRC Press.
McCord JM & Day ED Jr (1978) *FEBS Lett.* **86:** 139–142.
McCord JM, Roy RS & Schaffer SW (1985) *Adv. Myocardiol.* **5:** 183–189.
McCord JM, Omar BA & Russell WJ (1989) In Hayaishi E, Kondo M & Yoshikawa T (eds) *Medical, Biochemical, and Chemical Aspects of Free Radicals*, pp 1113–1118. Amsterdam: Elsevier Science Publishers.
Menasche P, Grousset C, Gauduel Y, Mouas C & Piwnica A (1987) *Circulation* **76:** V180–V185.
Minotti G & Aust SD (1987) *Arch. Biochem. Biophys.* **253:** 257.
Minotti G & Aust SD (1990) *J. Biol. Chem.* **262:** 1098.
Miura T, Yellon DM, Hearse DJ & Downey JM (1987) *J. Am. Coll. Cardiol.* **9:** 647–654.
Miura T, Downey JM, Hotta D & Iimura O (1988) Effects of superoxide dismutase plus catalase on myocardial infarct size in rabbits. *Can. J. Cardiol.* **4:** 407–411.
Moorhouse PC, Grootveld M, Halliwell B, Quinlan JG & Gutteridge JMC (1987) *FEBS Lett.* **213:** 23–28.

Morris JB, Haglund U & Bulkley GB (1987) *Gastroenterology* **187:** 9–17.
Mullane KM, Salmon JA & Kraemer R (1987) *Fed. Proc.* **46:** 2422–2433.
Myers CL, Weiss SJ, Kirsh MM & Shlafer M (1985a) *J. Mol. Cell. Cardiol.* **17:** 675–684.
Myers ML, Bolli R, Lekich RF, Hartley CJ & Roberts R (1985b) *Circulation* **72:** 915–921.
Myers ML, Bolli R, Lekich RF, Hartley CJ & Roberts R (1986) *J. Am. Coll. Cardiol.* **8:** 1161–1168.
Nachlas M & Schnitka T (1963) *Am. J. Pathol.* **42:** 379–406.
Naslund U, Haggmark S, Johansson G, Marklund SL, Reiz S & Oberg A (1986) *J. Mol. Cell. Cardiol.* **18:** 1077–1084.
Nejima J, Knight DR, Fallon JT et al (1989) *Circulation* **79:** 143–153.
Nishida T, Shibata H, Koseki M et al (1987) *Biochim. Biophys. Acta* **890:** 82–88.
Omar BA & McCord JM (1990) *Free Radicals Biol. Med.* **9:** 473–478.
Omar BA & McCord JM (1991) *J. Mol. Cell. Cardiol.* (in press).
Omar BA, Gad NM, Jordan MC et al (1990) *Free Radicals Biol. Med.* **9:** 465–471.
Ooiwa H, Stanley A, Felaneous-Bylund AC, Wilborn W & Downey JM (1991) *J. Mol. Cell. Cardiol.* (in press).
Opie LH, Thandroyen FT, Muller C & Bricknell OL (1979) *J. Mol. Cell. Cardiol.* **11:** 1073–1094.
Otamiri T (1989) *Surgery* **105:** 593–597.
Ouriel K, Smedira NG & Ricotta JJ (1985) *J. Vasc. Surg.* **2:** 49–53.
Owens ML, Lazarus HM, Wolcott MW, Maxwell JG & Taylor JB (1974) *Transplantation* **17:** 424–427.
Oyanagui Y (1976) *Biochem. Pharmacol.* **25:** 1465–1472.
Paller MS (1989) *J. Lab. Clin. Med.* **113:** 379–386.
Paller MS, Hoidal JR & Ferris TF (1984) *J. Clin. Invest.* **74:** 1156–1164.
Parks DA & Granger DN (1984) *Gastroenterology* **862:** 1207 (abstract).
Parks DA & Granger DN (1986) *Acta Physiol. Scand.* **126 (supplement 548):** 87–99.
Parks DA, Bulkley GB & Granger DN (1983) *Surgery* **94(3):** 428–431.
Parks DA, Henson JL & Granger DN (1986) *Physiologist* **29:** 101.
Parks DA, Williams TK & Beckman JS (1988) *Am. J. Physiol.* **254:** G768–G774.
Parnham MJ, Leyck S, Dereu N, Winkelmann J & Graf E (1985) *Adv. Inflamm. Res.* **10:** 397–398.
Parr DR, Wimhurst JM & Harris EF (1975) *Cardiovasc. Res.* **9:** 366–372.
Patel BS, Jeroudi MO, O'Neill PG, Roberts R & Bolli R (1988) *Circulation* **78:** II-373 (abstract).
Penfield A & Dale MM (1985) *FEBS Lett* **181:** 335–338.
Piper HM (1989) *Klin. Wochenschr.* **67:** 465–476.
Piretti MV & Pagliuca G (1989) *Free Radicals Biol. Med.* **7:** 219–221.
Poole-Wilson PA, Harding DP, Bourdillon PDV & Tones MA (1984) *J. Mol. Cell. Cardiol.* **16:** 175–187.
Powell WJ, Dibona DR, Flores J & Leaf A (1976) *Circulation* **54:** 603–615.
Przyklenk K & Kloner RA (1986) *Circ. Res.* **58:** 148–156.
Puett DW, Forman MB, Cates CU et al (1987) *Circulation* **76:** 678–686.
Ransohoff J, Flamm ES & Demopoulos HB (1980) In Cotrell JE & Turndorf H (eds) *Anesthesia and Neurosurgery*, p. 361. St Louis, MO: C.V. Mosby Company.
Reimer KA & Jennings RB (1985) *Circulation* **71:** 1069–1075.
Reimer KA, Lowe JE, Rassmussen MM & Jennings RB (1977) *Circulation* **56:** 786–794.

Renlund DG, Gerstenblith G, Lakatta EG, Jacobusd WE, Kallman CH & Weisfeldt ML (1982) *Circulation* **66II:** 158 (abstract).
Repine JE, Cheronis JC, Rodell TC, Linas SL & Patt A (1987) *Am. Rev. Respir. Dis.* **136:** 483–485.
Richard VJ, Murry CE, Jennings RB & Reimer KA (1988) *Circulation* **78:** 473–480.
Riva E, Manning AS & Hearse DJ (1987) *Cardiovasc. Drugs Ther.* **1:** 133–139.
Roy RS & McCord JM (1983) In Greenwald R & Cohen G (eds) *Oxy Radicals and Their Scavenger Systems:* Vol II. *Cellular and Molecular Aspects*, pp 145–153. New York: Elsevier Science.
Samokyszyn VM & Aust SD (1990) In Schaik TV (ed.) *Clinical Ischemic Syndromes: Mechanisms and Consequences of Tissue Injury*, pp 187–201. St Louis, MO: C.V. Mosby Company.
Samuelsson B (1983) *Science* **220:** 568–575.
Saris NE & Akerman EO (1980) In Sanadi DR (ed.) *Current Topics in Bioenergetics*, Vol. 10, pp 103–179. New York: Academic Press.
Schneider J, Friderichs, E & Giertz H (1987) *Free Radicals Biol. Med.* **3:** 21–26.
Schott RJ, Nao BS, McClanahan TB et al (1989) *Circ. Res.* **65:** 1112–1124.
Shasby DM, Van Benthuysen KM, Tate RM et al (1982) *Am. Rev. Respir. Dis* **125:** 443–447.
Shen AC & Jennings RB (1972) *Am. J. Pathol.* **67:** 441–452.
Shirato C, Miura T & Downey JM (1988) *FASEB J.* **2:** A918 (abstract).
Shirato C, Miura T, Ooiwa H, Toyofuku T, Wilborn WH & Downey JM (1989) *J. Mol. Cell. Cardiol.* **21:** 1187–1193.
Shlafer M, Kane PF & Kirsh MM (1982) *J. Thorac. Cardiovasc. Surg.* **83:** 830–839.
Silvia RC, Piercy MF, Hoffmann WE, Chase RL, Braughler JM & Tang AH (1987) *Neuroscience* **13:** 1495 (abstract).
Simon RH, DeHart PD & Todd RF (1986). *J. Clin. Invest.* **78:** 1375.
Simpson PJ, Mickelson JK & Lucchesi BR (1987) *Fed. Proc.* **46:** 2413–2421.
Simpson PJ, Todd RF, Fantone JC, Mickelson JK, Griffin JD & Lucchesi BR (1988a) *J. Clin. Invest.* **81:** 624–629.
Simpson PJ, Fantone JC, Mickelson JK, Gallagher KP & Lucchesi BR (1988b) *Circ. Res.* **63:** 1070–1079.
Singal PK, Kapur N, Dhillon KS, Beamish RE & Dhalla NS (1982) *Can. J. Physiol. Pharmacol.* **60:** 1390–1397.
Singal PK, Beamish RE & Dhalla NS (1983) *Adv. Exp. Med. Biol.* **161:** 391–401.
Snelling LK, Ackerman AD, Dean JM, North MC & Traystman RJ (1987) *Anesthesiology* **67:** A153.
Stone D, Darley-Usmar V, Smith DR & O'Leary V (1989) *J. Mol. Cell. Cardiol.* **21:** 963–973.
Tamura Y, Chi L, Driscoll EM Jr et al (1988) *Circ. Res.* **63:** 944–959.
Tanaka M, FitzHarris GP, Stoler RC, Jennings RB & Reimer KA (1989) *Circulation* **80: (supplement II)**: II-296 (abstract).
Tappel AL (1953) *Arch. Biochem. Biophys.* **44:** 378–395.
Tate RM & Repine JE (1983) *Am. Rev. Respir. Dis.* **128:** 552–559.
Thomas CE, Morehouse LA & Aust SD (1985) *J. Biol. Chem.* **260:** 3275–3280.
Thornton MA, Winn R, Alpers CE & Zager RA (1989) *Am. J. Pathol.* **135:** 509–515.
TIMI Study Group (1985) *N. Engl. J. Med.* **312:** 932.
Toledo-Pereyra LH, Simmons RL & Najarian JS (1976) *Ann. Surg.* **189:** 780–782.
Torok B, Roth E, Tigyi A, Zsoldos T, Matkovics B & Szabo L (1984) *Acta Chir. Hung.* **25:** 185–192.

Tranum-Jensen J, Janse MJ, Fiolet JWT, Krieger WJG, D'Alnoncourt CN & Durrer D (1981) *Circ. Res.* **49:** 364–381.
Turrens JF & Boveris A (1980) *Biochem. J.* **191:** 421–427.
Turrens JF & McCord JM (1991) In Crastes de Paulet A, Douste Blazy L & Paoletti R (eds) *Free Radicals, Lipoproteins and Membrane Lipids*, pp 65–71. New York: Plenum Publishing.
Turrens JF & McCord JM (1990) In Schaik TV (ed.) *Clinical Ischemic Syndromes: Mechanisms and Consequences of Tissue Injury*, pp 203–212. St Louis, MO: C.V. Mosby Company.
Turrens JF, Freeman BA, Levitt JG & Crapo JD (1982a) *Arch. Biochem. Biophys.* **217:** 401–410.
Turrens JF, Freeman BA & Crapo JD (1982b) *Arch. Biochem. Biophys.* **217:** 411–419.
Turrens JF, Alexandre A & Lehninger AL (1985) *Arch. Biochem. Biophys.* **237:** 408–414.
Uraizee A, Reimer KA, Murry CE & Jennings RB (1987) *Circulation* **75:** 1237–1248.
Vlessis AA & Mela-Riker L (1989) *Am. J. Physiol.* **256:** C1196–C1206.
Ward PA, Johnson KJ & Till GO (1986) *Acta Physiol. Scand.* **126 (supplement 548):** 79–85.
Watanabe F, Hashimoto T & Tagawa K (1985) *J. Biochem.* **97:** 1229–1234.
Watanabe N, Inoue M & Morino Y (1989) *Biochem. Pharmacol.* **38:** 3477–3483.
Weglicki WB, Arroyo CM, Kramer JH et al (1988) In Cerutti PA, Fridovich I & McCord JM (eds) *Oxy-Radicals in Molecular Biology and Pathology*, pp 357–364. New York: Alan R. Liss, Inc.
Werns SW (1990) In Schaik TV (ed.) *Clinical Ischemic Syndromes: Mechanisms and Consequences of Tissue Injury*, pp 213–226. St Louis, MO: C.V. Mosby Company.
Werns SW, Shea MJ, Driscoll EM et al (1985) *Circ. Res.* **56:** 895–898.
Werns SW, Shea MJ, Mitsos SE et al (1986) *Circulation* **73:** 518–524.
Werns SW, Simpson PJ, Mickelson JK, Shea MJ, Pitt B & Lucchesi BR (1988) *J. Cardiovasc. Pharmacol.* **11:** 36–44.
Werns S, Ventura A, Li GC & Lucchesi BR (1989) *Circulation* **80(4):** II-295 (abstract).
Westlin W & Mullane KM (1989) *Circulation* **80:** 1828–1836.
Wheatley AM, Thandroyen FT & Opie LH (1985) *J. Mol. Cell. Cardiol.* **17:** 349–359.
Woodward B & Zakaria MN (1985) *J. Mol. Cell. Cardiol.* **17:** 485–493.
Yagi K (1987) In Kim CH, Tedeschi H, Diwan JJ & Salerno JC (eds) *Advances in Membrane Biochemistry and Bioenergetics*, pp 553–560. New York: Plenum Publishing Corp.
Yamakawa T, Kadowaki Y, Garcia-Alves M, Yokoyama M, Iwashita Y & Nishi K (1989) *J. Mol. Cell. Cardiol.* **21:** 441–452.
Yamamoto M, Shima T, Uozumi T, Sogabe T, Yamada K & Kawasaki T (1983) *Stroke* **14:** 977–982.
Younes M, Mohr A, Schoenberg MH & Schildberg FW (1987) *Res. Exp. Med.* **187:** 9–17.
Ytrehus K, Myklebust R & Mjos OD (1986) *Cardiovasc. Res.* **20:** 597–603.
Zweier JL, Rayburn BK, Flaherty JT & Weisfeldt ML (1987) *J. Clin. Invest.* **80(6):** 1728–1734.
Zweier JL, Kuppusamy P, Williams R et al (1989) *J. Biol. Chem.* **264:** 18890–18895.

19

The Lens and Oxidative Stress

ABRAHAM SPECTOR
Biochemistry and Molecular Biology Laboratory, Department of Ophthalmology, College of Physicians and Surgeons of Columbia University, New York, New York, USA

I. Introduction 529
II. Oxidative stress upon the lens 534
III. H_2O_2 535
IV. Protection against oxidative stress 539
V. Hypothesis: oxidative stress causes cataract 544
VI. Protection from oxidative stress 551
VII. Concluding remarks 554

I. INTRODUCTION

A. Morphology and cell biology

The lens of the eye has the apparently simple function of focusing light entering the eye on the retina. It performs this task by changing the curvature of its surface and by having a relatively uniform refractive index which is definitively different from the surrounding environment. While the function of the lens is simple, the requirements for carrying out its task are complex. It must be transparent, contain material which will contribute to its unusual refractive index and be pliable. Furthermore, it should be functional for the life of the individual. It should be noted, however, that presbyopia, which is usually attributable to a hardening of the tissue, is a ubiquitous complaint of the middle-aged.

Oxidative Stress: Oxidants and Antioxidants
ISBN 0-12-642762-3

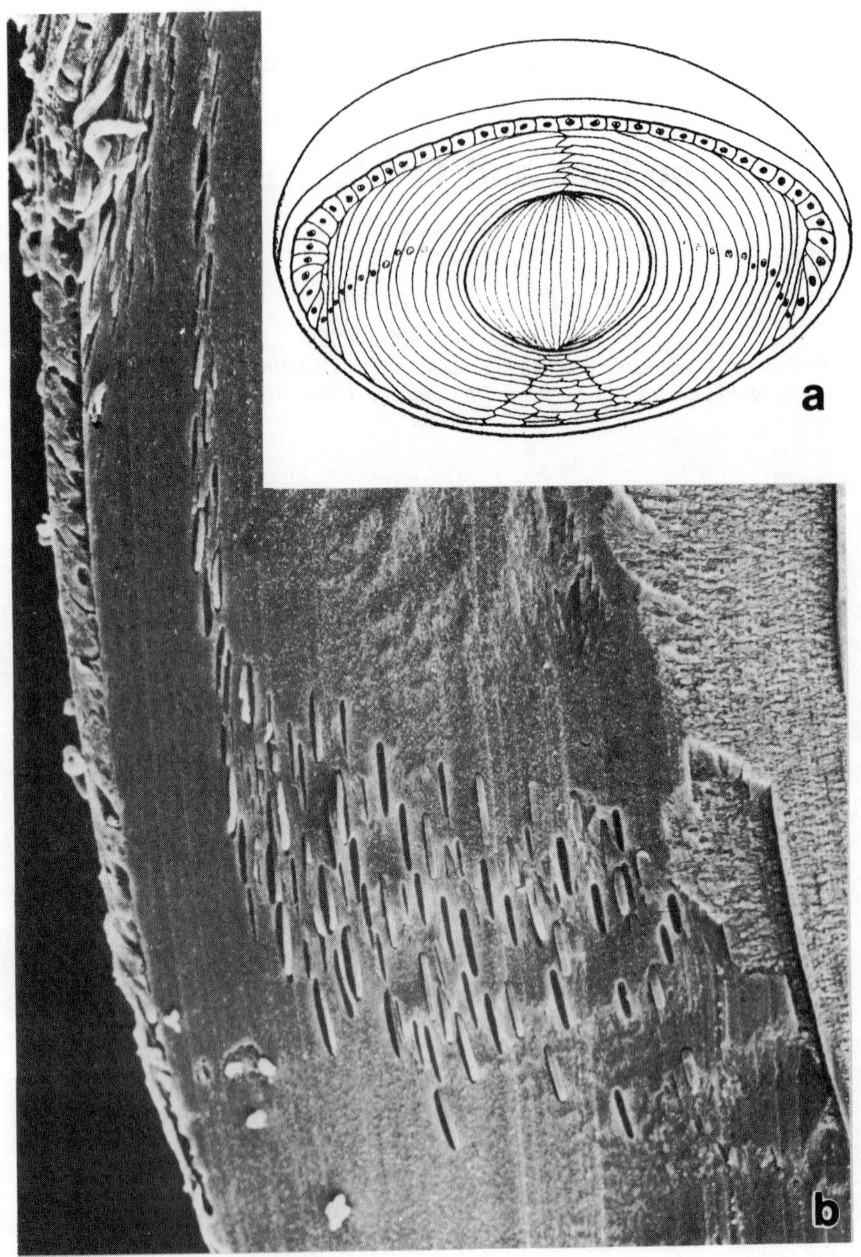
a
b

Nature has resolved most of the requirements mentioned above by designing a unique system (Harding and Crabbe, 1984). The lens is encapsulated by a thin, transparent porous extracellular basal lamina containing elastin, collagen and mucopolysaccharides. Immediately beneath the capsule on the anterior side is a single layer of epithelial cells which can be divided into three groups: a nondividing cell population which is present at the anterior pole, dividing cells which surround the anterior pole, and, finally, the equatorial zone in which cells have begun to differentiate terminally, changing in shape from typical cuboidal epithelial cells to fibre cells (see Figures 1 and 2). In this differentiation process, there is a dramatic elongation of the cells, with processes developing towards both the anterior and posterior poles. Mature human lens fibres may be 7–10 mm in length but only 2 μm in thickness and 10–12 μm in width. With the initiation of the terminal differentiation process, there is a marked increase in protein synthesis. The synthesis is primarily directed to the production of the crystallins, the so-called structural lens proteins which fill the fibres, and the lens fibre membrane proteins, especially MIP26. With cell maturation, there is a gradual loss of all intracellular structures, including the cell nucleus, mitochondria and microsomes. Thus, mature fibres have lost much of their metabolic capacity; they cannot synthesize new protein or generate ATP. It is estimated that human lens fibres which are approximately two layers in from the capsule at the anterior pole have already lost their nuclei and much of their metabolic function. At the equatorial pole, there may be as many as 200 layers before the nuclei are completely lost (Kuszak and Rae, 1982).

The fibres are in close proximity, being held together by ball and socket and ridge and valley structures. Thus, there is little extracellular space (see Figure 2). Since the lens fibre membrane contains a high concentration of communicating junctions, described by some investigators as connections, there is extensive communication between fibres.

There are a number of significant consequences that result from this organization. No cells are lost from the tissue. The lens continues to lay down new fibre cells all through life, with the older cells being displaced

---

Figure 1. (**a**) Schematic representation of a section through the polar axis of a newborn human crystallin lens. The lens is essentially a ball of crescent-like cells arranged end-to-end, covered on its anterior half by the lens epithelium. The lens epithelium is the basal layer of this stratified epithelial organ (schematic modified from Kuszak (1990)). Because of the inverted developmental scheme of the lens, as cells are formed they are laid onto the existing cell mass and every cell formed is retained throughout life. (**b**) SEM micrograph of the bow region of the lens showing the progressive elimination of fibre cell nuclei as a function of terminal differentiation. (Micrograph courtesy of J.R. Kuszak.)

toward the interior of the tissue. The process slows down dramatically after the first few years of life, so that in the mature individual it may take decades for a cell to move into the differentiating zone. The oldest fibres of the organ, representing the fetal lens, are present in the centre of the structure, and the youngest fibres are at the periphery. Since most lens fibres have lost their ability to synthesize new protein, there is a concomitant aging of cell fibres and the macromolecules they contain. The lens contains no vasculature and must depend on the thin band of active peripheral tissue to pump in the required nutrients and to pump out waste products. The overall maintenance of the homeostasis of this system is dependent on the epithelial cell layer and the zone of newly formed fibres still capable of performing some metabolic functions. The transparency of the tissue is enhanced by the small number of cell layers containing intracellular bodies, the high concentration of protein (35–40% on a wet weight basis in the human) and the uniformity of the refractive index. The stability of the tissue is protected by its isolation, being surrounded by the aqueous humor on its anterior side and the vitreous body on the posterior side.

For most of the lens fibres, damage is irreversible. It does not appear to be possible to repair damage to a macromolecular constituent throughout much of the tissue. Since the transparency of a fibre cell is dependent on maintaining the native structure of the protein, it must be concluded that transparency is heavily dependent on the outer zone of actively metabolizing tissue which maintains the environment and detoxifies noxious components.

The epithelial cell layer is capable of synthesizing new protein, a function which is dependent on the integrity and fidelity of the DNA present in the cell. The lens has a limited number of viable cells (estimated to be somewhat greater than 350 000 for the human lens) (J. R. Kuszak, personal communication). Irreversible damage to the DNA can lead to a deficient cell. Each epithelial cell can be considered to be maintaining the homeostasis of the section of fibres immediately beneath it. Thus, mutation to one of the cell's vital housekeeping genes or a gene which contributes to the constant environment of the fibre such as a Na/K-ATPase gene, could lead to opacification of the fibre region subtending the cell and the death of the cell.

Figure 2. (**a**) SEM photomicrographic montage showing surface morphology of the low cuboidal lens epithelium overlying the lens fibre cell mass. Occasional breaks in the lateral membrane of these cells afford a view of their nuclei (arrows). (**b**) SEM photomicrographic montage of surface morphology of the germ cells of the lens epithelium (tall columnar cells) and the onset of fibre cell elongation (tall columnar cells rotating about their axis and then elongating bidirectionally). (Micrographs courtesy of J.R. Kuszak.)

B. Aging

With aging, a number of changes occur in the lens. The tissue yellows, suggesting the accumulation of pigment (Said and Weale, 1959; Spector et al, 1975). There is a shift in the size of the crystallins to higher molecular weight, which contributes to some loss of transparency (Spector, 1985; Sigelman et al, 1974) (it has been shown that this modification is primarily noncovalent in the aging but normal lens) (Spector, 1984), non-enzymatic racemization of protein aspartate occurs in a time-dependent manner (Masters et al, 1977, 1978; Garner and Spector, 1978), a low level of glycation is observed (Spector, 1985), and deamidation occurs (Harding and Crabbe, 1984). There is exposure of buried protein thiol groups and possibly other reactive groups (Garner and Spector, 1980a,b) and, although not well documented, some decrease in the activity of certain enzymes has been reported.

These observations suggest that, with aging, the system changes. The proteins are modified, causing changes in conformation which expose reactive groups. These conformational changes are probably due to the slow non-enzymatic post-translational modifications observed in the system, some of which are noted above. It can be concluded that the older tissue, as a result of these changes, is more susceptible to oxidative insult than its younger counterpart. What then is the oxidative stress which is imposed on this vulnerable system?

II. OXIDATIVE STRESS UPON THE LENS

A. Oxygen tension

The lens is exposed to a lower oxygen tension than most tissues as a result of its nonvascularity. The major source of oxygen on the anterior surface of the lens is due to the oxygen which passes through the cornea and aqueous humor. This process results in a gradient of oxygen tension from 160 mm at the corneal surface to approximately 10 mm at the surface of the lens. A similar tension has been reported on the posterior surface as a result of oxygen diffusion from the highly vascularized retina and other tissues bordering the vitreous body. This low concentration of oxygen is sufficient to support the oxygen-dependent metabolism observed in the lens.

B. Oxygen reduction

Oxygen is a poor oxidant since it contains two electrons with unpaired spins.

However, under certain conditions it can become a powerful oxidant in one of two ways. Oxygen can absorb energy and go to an excited singlet state with paired electrons or it can be involved in single-electron reactions in which spin restrictions do not apply.

Oxygen can be reduced in four one-electron reactions to H_2O. In this process, three oxidants are generated, superoxide, hydrogen peroxide and hydroxyl radical. While the superoxide and hydroxyl radicals are much more potent oxidants than H_2O_2, they are highly unstable and quickly react in the near vicinity of the place where they are generated. In contrast, H_2O_2 is relatively stable and can react at sites far removed from its initiation site. Because of its nonionic structure, it is not hindered by membrane barriers.

III. H_2O_2

A. Concentration and source

The aqueous humor fills the anterior chamber between the cornea and the anterior surface of the lens. It supplies nutrients to the lens and receives waste products eliminated from the tissue. Examination of the aqueous humor and lenses from primates and humans indicates significant levels of H_2O_2 (Pirie, 1965; Matsuda et al, 1981; Bhuyan and Bhuyan, 1977; Spector and Garner, 1981; Giblin et al, 1984). In primates and humans, values averaging from 24 to 34 μM were observed. In most experiments, the H_2O_2 was determined by the 2,6-dichlorophenolindophenol procedure of Mapson (1945). While these levels are remarkably high, they have been confirmed by an electrochemical technique (Huang and Hu, 1990) and a colorimetric method based on the oxidation of ferrous ion (Hildebrandt et al, 1978; A. Spector, unpublished). In some cases, it was shown that addition of catalase eliminated the H_2O_2. While the source of the H_2O_2 is not certain, it is probable that the high concentration of ascorbate, approximately 1 mM, found in the aqueous humor and the lens contributes significantly to the generation of the normal H_2O_2 levels.

1. Ascorbate

It has been proposed that ascorbic acid reacts non-enzymatically with O_2 to give rise to H_2O_2 and dehydroascorbate (Pirie, 1965). Giblin et al. (1984) have shown that rat and frog aqueous humor have concentrations of ascorbate approximately ten-fold less than the most characteristic mammalian levels and concomitantly 5–10-fold less H_2O_2 than is found in mammals. Furthermore, elevating ascorbate two-fold by intraperitoneal injection in rabbits led to an approximately two-fold increase in H_2O_2. Lowering the ascorbate level in guinea-pigs with an ascorbate-deficient diet led to a 40% drop in both ascorbate and H_2O_2 concentrations.

However, because of the presence of light energy in the aqueous humor and the lens, absorbing chromophores such as riboflavin (Kinsey and Frohman, 1951), tryptophan degradation products (Zigman, 1985) and trace levels of metal ions such as copper, other reactions leading to H_2O_2 and radical formation may also occur. The remarkable levels of H_2O_2 found in the aqueous humor and lenses of some cataract patients (in some cases greater than 100 μM) (Spector and Garner, 1981) cannot be explained simply by elevated ascorbate concentrations. Even ascorbate levels twice the physiological range gave H_2O_2 levels 2–5-fold less than observed in some of the patients with cataract.

2. *Photochemical reactions*

The cornea absorbs light of wavelengths below 295 nm (Spikes and Livingston, 1969). Light above this wavelength passes through the cornea to the lens. Such light energy could be absorbed by chromophores such as β-carbolines (Dillon et al, 1976) and 3-hydroxykynurenine and its glucosides produced by the degradation of tryptophan (van Heyningen, 1973) or riboflavin (RF), and generate superoxide (Jernigan et al, 1981).

$$\mathrm{RF} + \mathrm{O_2} \xrightarrow{h\nu} \text{photooxidized RF} + \mathrm{O_2^{\bullet -}}$$

However, in the presence of substantial levels of reducing compounds such as ascorbate, a somewhat different set of reactions may occur. The excited riboflavin may acquire an electron from ascorbate (AH^-).

$$\mathrm{RF} \xrightarrow{h\nu} \mathrm{RF^*}$$

$$\mathrm{RF^*} + \mathrm{AH^-} \longrightarrow \mathrm{RF^\bullet} + \mathrm{A^{\bullet -}} + \mathrm{H^*}$$

* = excited state

The excited hydrogen atom may react with O_2 to form a hydroperoxyl radical which rapidly dissociates to produce superoxide (Schulte-Frohlinde et al, 1981). The riboflavin radical may also react with oxygen to produce superoxide and ground state riboflavin.

$$\mathrm{RF^\bullet} + \mathrm{O_2} \longrightarrow \mathrm{RF} + \mathrm{O_2^{\bullet -}}$$

If this process occurs in the lens, the superoxide could be detoxified by superoxide dismutase (SOD), producing H_2O_2. In the aqueous humor where SOD is probably not present, a more likely reaction is for the superoxide to react with ascorbate to again give H_2O_2 as well as the ascorbate semiquinone (Cabbelli and Bielski, 1983). The semiquinone can then react to form the more stable ascorbate, either by disproportionation or by reaction with H_2O_2, yielding the more potent hydroxyl radical.

$$A^{\bullet -} + H_2O_2 \longrightarrow A + {}^{\bullet}OH + OH^-$$

The overall result of these reactions involving riboflavin and ascorbate is the consumption of O_2, production of H_2O_2, and formation of the ascorbate semiquinone and perhaps other radicals such as $^{\bullet}OH$. The presence of low concentrations of metal ion accelerates the consumption of oxygen by ascorbate, increasing the formation of H_2O_2.

B. Oxidant stability

It should be noted that in generating an oxidative stress upon a system, the time necessary for the oxidizing species to reach its target in relationship to the lifetime of the species must be considered. For potent oxidants such as hydroxyl radical and superoxide, the target must be in the immediate vicinity of the oxidant. This conclusion is supported by experiments reported by Zigler et al. (1985, 1989). Damage to lenses in organ culture caused by H_2O_2 could be eliminated by addition of Fe^{2+}-EDTA to the culture medium. Under these conditions, the H_2O_2 was converted to $^{\cdot}OH$ in a Fenton reaction (Fenton, 1894).

$$H_2O_2 + Fe^{2+} \longrightarrow OH^- + {}^{\bullet}OH + Fe^{3+}$$

The short-lived radical could not get to a target site before decaying to a nonreactive species. In contrast, H_2O_2 is relatively stable, and could move into the cell and easily react directly or generate other oxidants close to the target sites.

C. Source of more reactive oxidants

There are two sets of data which suggest that even when H_2O_2 gets into the lens or is generated in the lens, it may not be a damaging agent but rather is converted to a more toxic oxidant. It has recently been shown that H_2O_2

causes significant damage to lens epithelial cell DNA when epithelial cells are incubated in media containing low levels of H_2O_2 (Spector et al, 1989; Kleiman et al, 1990). When *o*-phenanthroline (de Mello et al, 1985), a powerful iron chelator, was added to the incubation medium, the H_2O_2 effect as measured by single-strand breaks was eliminated. The results suggested a Fenton reaction as described previously. If this were the case, then $^{\cdot}OH$ quenchers should markedly decrease the H_2O_2 effect. This was found to be the case with potassium iodide and dimethylsulphoxide, preventing most of the damage.

Furthermore, if an inhibitor of SOD, diethyldithiocarbamate (DDC) (Heikkila et al, 1976), is added to the lens epithelial cells in the presence of H_2O_2, the extent of DNA damage is substantially increased (Kleiman et al, 1990). It can also be shown that DDC in the presence of H_2O_2 plus a hydroxyl radical quencher KI, has no effect, indicating no direct superoxide involvement in producing detectable DNA damage under these conditions. These observations suggest that the superoxide may be contributing to cycling the limited concentration of Fe^{3+} back to the Fe^{2+} state.

$$O_2^{\cdot -} + Fe^{3+} \longrightarrow Fe^{2+} + O_2$$

The metal ion then again participates in a Fenton reaction, the overall result being

$$O_2^{\cdot -} + H_2O_2 \longrightarrow O_2 + OH^- + {}^{\cdot}OH$$

This reaction is known as the Haber–Weiss reaction.

Of course, in the lens, which contains high levels of reductants such as ascorbate and glutathione, the recycling of the metal ion can also be carried out with either of these reductants.

$$2Fe^{3+} + 4\ \text{ascorbate} + O_2 \longrightarrow 4\ \text{dehydroascorbate} + 2Fe^{2+} + H_2O_2 + 2H^+$$

$$2Fe^{3+} + 4GSH + O_2 \longrightarrow 2GSSG + 2Fe^{2+} + H_2O_2 + 2H^+$$

Thus, in this case, the H_2O_2 necessary for $^{\cdot}OH$ formation is generated at the metal ion site.

It is apparent from the above discussion that, in spite of the low O_2 tension in the vicinity of the lens, the tissue is under considerable oxidative stress. By generation of oxidants in other parts of the eye, by photochemical reactions, by metal-catalysed reactions, and possibly by intrinsic metabolism, the lens is constantly exposed to oxidative challenge. Furthermore, it is also apparent that, with aging, the constituents of the tissue are increasingly susceptible to oxidation.

IV. PROTECTION AGAINST OXIDATIVE STRESS

A. Glutathione

1. *Concentration and localization*

How does the tissue protect itself from oxidative stress? One of the major constituents of lens epithelial cells is glutathione. (A comprehensive review of glutathione in the lens and other ocular tissues was recently published (Rathbun, 1989).) It is present at high concentrations in the outer region of the organ and then gradually decreases towards the centre of the tissue (Reddy and Giblin, 1984; Harding and Crabbe, 1984; Pau et al, 1990). Reddy and Giblin (1984) have reported values of 64 μmol g^{-1} in the epithelium of rabbit lenses, approximately five-fold greater than the outer fibre layers. In the bovine lens, values ranging from 3 to 14 μmol g^{-1} have been reported (Harding and Crabbe, 1984; Kuck et al, 1982). The glutathione levels in human lenses have been reported to range from approximately 2 to 5 μmol g^{-1} (Reddy and Han, 1976; Dickinson et al, 1968; Kinoshita and Merola, 1973; Anderson et al, 1979). However, in a recent paper, Pau et al (1990) reported remarkably high total glutathione values of approximately 28 μmol g^{-1} in the cortex plus epithelium of three fresh human lenses from people of approximately 80 years of age. It appears that the glutathione levels decrease with time, being one half the fresh value in lenses 20 hours postmortem. It is probable that the variation in reported values is partially due to the decrease in glutathione levels observed with aging. Harding (1970) has reported, in the very young human lens, glutathione values of approximately 4 μmol g^{-1} which gradually decrease to values less than 2 μmol g^{-1} in normal lenses taken from individuals in their sixties. The concentration of glutathione also drops sharply with the development of cataract (Harding and Crabbe, 1984), particularly in subcapsular cataracts (Pau et al, 1990).

The lens also contains glutathione analogues in which the thiol group or the CH_2SH group of the cysteinyl residue are replaced by a methyl group (ophthalmic acid and norophthalmic acid, respectively) (Waley and van Heyningen, 1962). Ophthalmic acid is present at concentrations of approximately 10% of glutathione, and norophthalmic acid at approximately 1% (Waley, 1969). The physiological function of these compounds is not known, although ophthalmic acid has been shown to inhibit insulin degradation (Offord et al, 1979).

2. *Redox state*

Glutathione is normally found predominantly in the reduced state (Akerboom and Sies, 1981) and, in spite of its high concentration in the lens, this is also true in this tissue (Kinoshita and Masurat, 1957). Lens epithelial cells have a high capacity for maintaining glutathione in the reduced state (Spector et al, 1985, 1987). When the glutathione level in lens epithelial cells was doubled as a result of incubation with glutathione ethyl ester, the cells easily maintained the total glutathione in the reduced state (Spector et al, 1987). Indeed, even after oxidative stress generated by direct exposure to H_2O_2 (see Table 1), within 4 min no glutathione disulphide could be detected even though 86% was in the disulphide form 0.5 min after the H_2O_2 insult (Spector et al, 1985).

Table 1. GSH recovery from H_2O_2 stress in lens epithelial cells.

Post-H_2O_2 (min)	GSH (nmol per 10^5 cells)	GS^- (nmol per 10^5 cells)
0.5	0.5 ± 0.1	3.0 ± 0.2
2.0	1.5 ± 0.2	1.1 ± 0.2
4.0	2.9 ± 0.3	Trace
6.0	3.0 ± 0.1	Trace
Control	3.7 ± 0.2	0

From Spector et al. (1987).

The remarkable ability of the cell to maintain or return glutathione to the reduced state is linked to the active hexose monophosphate shunt activity (Kinoshita, 1955), stimulated supposedly as a result of the drop in NADPH (Giblin et al, 1981) caused by its utilization to reduce glutathione disulphide. Thus, glucose is the ultimate source of reducing potential, generating NADPH through the oxidation of glucose 6-phosphate via glucose-6-phosphate dehydrogenase, 6-phosphoglucono-δ-lactone lactonase and 6-phosphogluconate dehydrogenase. The glutathione disulphide GSSG is returned to the reduced state via glutathione reductase utilizing NADPH.

3. *Glutathione as antioxidant*

Glutathione exerts a powerful antioxidative effect by being involved in a number of reactions.

1. It acts as a cofactor of GSH peroxidase to detoxify H_2O_2.

$$H_2O_2 + 2GSH \xrightarrow[\text{peroxidase}]{\text{GSH}} GSSG + 2H_2O$$

2. It can detoxify free radicals.

$$GSH + R^{\bullet} \longrightarrow GS^{\bullet} + RH$$
$$GS^{\bullet} + GS^{\bullet} \longrightarrow GSSG$$

3. It can reduce protein disulphides.

$$GSH + PS\text{—}SP \longrightarrow PS\text{—}SG + PSH$$
$$PS\text{—}SG + GSH \longrightarrow PSH + GSSG$$

$$2GSH + PSSP \longrightarrow 2PSH + GSSG$$

4. The high concentration of GSH present in the lens makes it probable that if thiol oxidation is induced by a particular oxidizing agent, it is more likely to target GSH rather than the relatively low level of protein thiol.
5. Finally, there is one other interesting protective function. In the development of sugar-induced cataract, glycation occurs (Stevens et al, 1978). Sugar aldehydes react with protein amino groups to form Schiff bases which may later be stabilized by Amadori reactions. It has been shown that GSH at concentrations found in the lens will inhibit the glycation reaction (Huby and Harding, 1988). It has been proposed that glycation causes an unfolding of the protein which leads to thiol oxidation and protein–protein disulphide formation. Thus, GSH may be acting in this system on at least three different levels, preventing glycation and thiol oxidation and reducing protein disulphides. It would appear that as long as the GSH concentration in the lens remains high, glycation is not an important event. Thus, the loss of GSH contributes to conditions suitable for significant glycation and subsequent protein thiol oxidation.

4. Mixed disulphides

Under conditions where significant levels of GSSG are present, formation of mixed protein–S–SG might be expected (for a discussion, see Brigelius (1985)). Indeed, such mixed disulphides have been detected in cataractous lenses (Srivastava and Beutler, 1973; Truscott and Augusteyn, 1977; Anderson and Spector, 1978), and it has been proposed that the formation of the mixed disulphide is due to an exchange reaction with GSSG (Reddy and Han, 1976; Spector et al, 1986; Lou et al, 1986). It is of interest that Lou et al. (1986) found that, in normal lens, mixed cysteine–protein disulphides were more prevalent than GSH–protein disulphides by a factor of ten, but with development of cataract, the GSH–protein mixed disulphides increased dramatically, with little change in the abundance of the cysteine–protein mixed disulphide.

One of the characteristics of cataract development is the early precipitous drop in GSH concentration, the cause of which is unknown (Reddy, 1990). Glutathione is synthesized and catabolized in the lens (Kinsey and Merriam, 1950; Cliffe and Waley, 1958; Rathbun, 1980). Thus, drops in synthesis, breakdown, utilization and/or diffusion from the lens must all be considered. In the normal lens, diffusion from the tissue is not significant and synthesis and catabolism occur at comparable rates (Reddy, 1990).

B. Ascorbate as an antioxidant

It is generally assumed that the function of ascorbate in the aqueous humor and the lens is to act as an antioxidant and protect the lens. Yet the previous discussion suggests that, in the presence of riboflavin and light energy, deleterious components may be generated. Indeed, Wolff et al (1987) demonstrated that ascorbate in the presence of riboflavin was phototoxic to lens epithelial cells. This result differs somewhat from those of Varma et al (1979) and Jernigan et al (1981), where riboflavin toxicity was partially eliminated by ascorbate. However, Wolff et al (1987) were working with lens cell cultures and Varma et al (1979) and Jernigan et al (1981) with organ culture, and their conditions differed. In the latter study, it was clear that damage to lens transport systems did not occur in the absence of oxygen and that it is probably H_2O_2 that is causing the damage.

The early literature suggests that ascorbate may prevent cataract, since a drop in ascorbate levels was associated with the development of lens opacity (Bietti, 1935; Bellows, 1936; Bouton, 1939). Recently, work from Trevithick's group has suggested that supplementing the diet with vitamins C and E decreases the risk of cataract by 50% (Robertson et al, 1989). Furthermore, Frei et al (1989) have shown that, in plasma, ascorbate is effective in protecting lipids against peroxidation but protein thiols are not protected and perhaps other groups sensitive to oxidation but not measured may also be vulnerable under their conditions (see Chapter 9). Thus, based upon present information, the role of ascorbate as an effective antioxidant for the lens remains unclear.

C. Vitamin E as an antioxidant

Vitamin E has been suggested to act as an antioxidant defence for the lens.

Varma et al (1982) showed that in vitro lens lipid oxidation resulting from photochemical insult was decreased by vitamin E. Bhuyan et al (1982) reported that vitamin E prevented the increase in H_2O_2 in the aqueous humor which resulted from ingestion of 3-aminotriazole. There is also some evidence from a small retrospective epidemiological study that vitamin E may be beneficial in contributing to the prevention of cataract (Robertson et al, 1989). Some reports suggest that cataract development associated with diabetes (Ross et al, 1982; Trevithick et al, 1989) or other genetic factors (Varma et al, 1984) may be retarded by elevated vitamin E intake.

Unfortuntely, it is difficult to demonstrate definitively a rigorous relationship between increased vitamin E levels and a decrease in cataract development. However, the cited articles, while preliminary and controversial, suggest that vitamin E therapy for the prevention of cataract deserves further study.

D. Enzyme defences

Glutathione peroxidase and catalase are the two enzymes which metabolize H_2O_2 in the lens. GSH peroxidase was first shown to be present in the lens by Pirie (1965) and was confirmed by Holmberg (1968) and Bergad et al (1982). Catalase is present in the lens in very low levels and appears to be concentrated in the epithelial layer (Bhuyan and Bhuyan, 1978). However, it would seem that these enzymes are present at sufficient concentrations to handle the H_2O_2 levels which, in the normal human lens, are reported to be approximately 35 µM (Spector and Garner, 1981). It is generally believed that catalase is relatively inactive in metabolizing H_2O_2 at low H_2O_2 concentrations. However, it should be noted that Flohé et al (1972) demonstrated that glutathione, peroxidase and catalase metabolize H_2O_2 at similar rates. Thus, the enzyme present at the highest concentration in a given cellular region would be the major contributor to H_2O_2 detoxification. GSH peroxidase, in contrast to catalase, is also capable of reducing hydroperoxides such as are found as a result of lipid oxidation (Srivastava, 1976). However, it should be noted that the enzyme has an apparent K_m for H_2O_2 of approximately 45 µM, while the affinity for hydroperoxides is about tenfold lower (Bergad et al, 1982). GSH peroxidase activity decreases in the lens with age and is not present at high concentration (Rathbun et al, 1986; Rathbun and Bovis, 1986).

When rats were fed a long-term selenium-deficient diet, lens GSH peroxidase activity was decreased by 85%. Cataract development was not observed under these conditions (Lawrence et al, 1974). However, second-generation selenium-deficient rats do exhibit cataract (Sprinker et al, 1971). There are conflicting reports as to whether a lens selenium deficiency is related to human cataract (Swanson and Truesdale, 1971; Lakoma and Ekland, 1979).

Superoxide dismutase (SOD) has been shown to be present in the lens at low concentration (Bhuyan and Bhuyan, 1978; Augusteyn, 1981), with the highest activity in the epithelial region. As with other enzymes, the activity has been reported to decrease markedly in cataract.

Thioredoxin/thioredoxin reductase is another enzyme system that may contribute to the lens' defence against oxidative stress. Operating in the micromolar range and utilizing the reducing power of NADPH, this system effectively reduces some protein disulphides and quenches free radicals (Laurent et al, 1964; Moore et al, 1964; Holmgren, 1968, 1985; Spector et al, 1988). The effectiveness of thioredoxin (Tx) in reducing protein disulphides results from its containing two thiols, one of which has an unusual pK which increases its ability to interact with disulphides (Kallis and Holmgren, 1980).

$$\mathrm{Tx(SH)_2 + Protein(S{-}S) \longrightarrow Tx(S{-}S) + Protein(SH)_2}$$

Thioredoxin reductase (Tx red), a flavoprotein containing flavin adenine dinucleotide, then catalyses the reduction of Tx, utilizing NADPH as a reductant.

$$\mathrm{NADPH + H^+ + Tx(S{-}S) \xrightarrow{Tx\ red} NADP^+ + Tx(SH)_2}$$

The actual biological function of this system is not clear, since it is involved in a number of unrelated reactions such as the reduction of ribonucleotide diphosphates to the deoxyribonucleotide diphosphates, the regulation of fructose-1,6-diphosphatase in the photosynthetic process, the regulation of insulin activity, and possibly the reversal of certain oxidative damage to protein. Furthermore, the Tx system is believed to be required to maintain methionine sulphoxide reductase activity. This enzyme reduces peptide methionine sulphoxide to peptide methionine. Methionine sulphoxide reductase is present in the lens and acts to maintain the peptide methionine groups when the system is under oxidative stress (Spector et al, 1982).

While Tx has not been found in the lens, Tx red activity has been observed. It has also been shown that *E. coli* Tx (mol wt ≈ 12 000) gets into lens epithelial cells and contributes to the ability of such cells to withstand H_2O_2 insult (Spector et al, 1988).

V. HYPOTHESIS: OXIDATIVE STRESS CAUSES CATARACT

It is apparent from the above discussion that the lens, in spite of its array of defences against oxidative stress, is susceptible to oxidative damage. The continuous presence of oxidative stress, the presence of a limited cell population which continues to differentiate throughout life, the absence of viable repair systems and defences against oxidative stress throughout much of the tissue, the age-dependent changes in activity of protective enzymes and increased vulnerability of cell constituents all contribute to an increased probability of oxidative damage. It is from such considerations that the hypothesis was developed that oxidative stress initiates or is an early event in the development of cataract. We now examine the evidence which indicates that oxidative insult to lens constituents has occurred.

A. Thiol oxidation

1. Disulphide formation

Oxidation of lens protein thiols accompanies the formation of human cataract (Dische and Zil, 1951; Harding, 1972). Some of these disulphides are intermolecular and stabilize large protein aggregates (Spector and Roy, 1978) that contribute to the loss of transparency. Such disulphide bonds are present to an insignificant extent in the normal lens. While less than 10% of the protein thiol is present as disulphides in the normal lens, more than 70% is found in such linkage in human cataract. The disulphide-linked aggregates involve both cytosol and membrane components (Figure 3). The data suggest that in cataract involving the inner or nuclear region of the lens, as much as 50% of the cytosol protein is linked to the membrane by reducible bonds, probably disulphides (Spector, 1984; Spector and Garner, 1980; Garner and Spector, 1980b). Upon reduction, the cytosolic protein was released from the membrane fraction. Further work has shown that the major cytosol constituent involved in the disulphide linkage is γ-crystallin, a lens protein which is present at higher concentration in the inner region of the lens than in the cortex and which is relatively rich in thiol groups. The above analysis of cytosolic protein–membrane protein linkage was performed upon the water-insoluble fraction, the fraction that would normally contain membrane components. But if with the development of cataract the membrane were to break down, then it might be expected that the high molecular weight disulphide-linked aggregates associated with membrane fragments would be found separated from the intact insoluble membrane fraction. Utilizing antibodies to intrinsic membrane components and to cytosol protein, it was shown that such entities do exist in the cataractous condition. This is confirmed by observing a loss of lipid from the membrane fraction but not from the lens. Thus, in certain types of cataract, there appears to be a general breakdown of membrane structure (Figure 3). It has

A. Nuclear Cataract

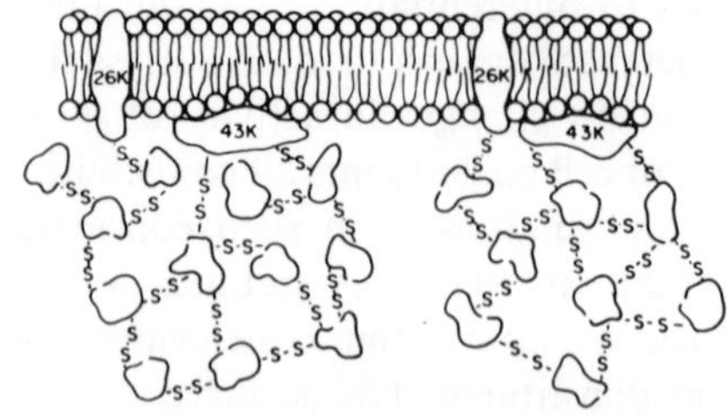

B. Cortical Cataract

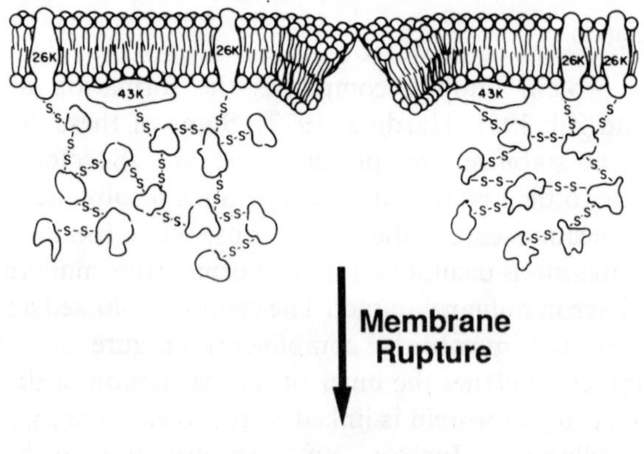

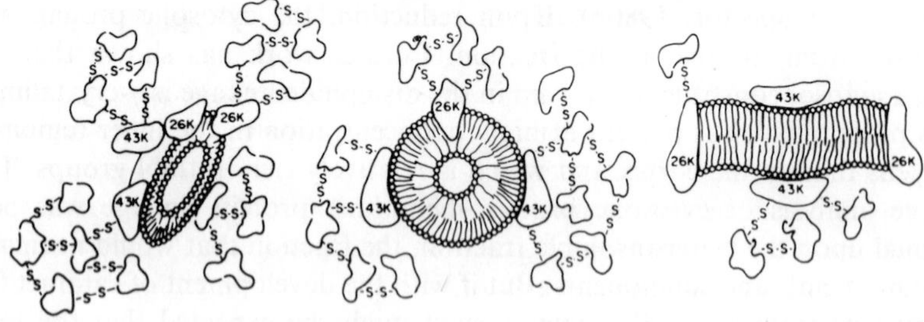

Figure 3. Schematic representation of lens membrane–cytosol protein aggregates. (**A**) Depiction of aggregates in the nuclear (inner) region of the lens. Intrinsic and extrinsic membrane protein is disulphide-linked to cytosol protein units, which are in turn disulphide-linked to each other. Such giant aggregates scatter light and contribute to the loss of transparency. (**B**) Aggregates in the outer cortical region of the lens. While nuclear fibre cell membranes appear to be rigid and do not break with aggregate formation, in the cortical region the formation of the aggregates causes membrane rupture and the appearance of membrane fragments linked to cytosol protein as well as the nuclear region type of aggregates.

been found that this breakdown is not observed in cataracts involving the inner region of the tissue. In such cataracts, the high molecular weight protein aggregates remain disulphide-linked to the membrane. Only in cataract of the outer region do the large protein aggregates linked to the membrane appear to break off from the membrane structure and localize in the cytoplasm. This difference in behaviour may be related to changes in the structure of the lens fibre membrane. In both types of aggregate formation, the oxidation appears to begin at the membrane, since all aggregates contain membrane components.

In most experimentally induced cataracts, there is a loss of both total and reduced glutathione (Rathbun, 1989); this is also observed in human cataract (Pau et al, 1990; Harding, 1970; Mach, 1966). The loss in glutathione is probably caused by a number of factors, including a decrease in glutathione synthesis, formation of mixed disulphides and increased linkage of glutathione to protein due to elevation of glutathione disulphide concentration.

2. *Exposure of thiol groups*

Examining the soluble protein in normal human lenses as a function of age, it was found that there is a change in the conformation of the protein with aging, the thiol groups of the protein from young individuals being essentially buried (Garner and Spector, 1980a). This was determined by isolating the protein under N_2 in the presence of iodoacetamide. Under these conditions, exposed thiols would react with the thiol reagent, forming carboxyamidomethylcysteine. Buried thiols were then exposed by utilizing denaturing agents and modified with methyl iodide, forming methylcysteine. With this approach, it is possible to differentiate between buried and exposed thiols. When a similar analysis was carried out with a group of normal lenses in the 60–65-year-old range, more than 50% of the thiols were exposed. However, no oxidation could be detected in the thiol groups except in the intrinsic membrane fraction. A similar study of protein methionine again indicated no oxidation, except for the membrane fraction and high molecular weight fraction, where up to approximately 25% of the methionine appeared oxidized. In young lenses 6–12 years of age, no methionine oxidation was observed.

In contrast to these observations, in human cataract almost all the thiols are exposed, and 75–100% of the thiols are oxidized. More than 50% of the methionines are also oxidized. It is interesting to note that membrane oxidation appears to be greater than in other parts of the fibre cell. The pattern that emerges from this study is that the gradual change in protein conformation makes the system more vulnerable to oxidative attack. What initiates the conformational changes is not apparent. However, it is probable that racemization, deamidation and glycation contribute to the process.

B. Na/K-ATPase and Ca^{2+}-ATPase

The observation that oxidation is occurring at the cell membrane has led to the examination of pump systems, particularly Na/K-ATPase. It is well known that an early change in cataract development is an increase in cellular sodium concentration (Mercantonio et al, 1980; Marlat et al, 1981; Maraini and Pasino, 1983). With increases in human lens opacification, there is a concomitant decrease in Na/K-ATPase measured by steady-state hydrolysis of saturating levels of Mg^{2+}-ATP (Kobayashi et al, 1983). A similar relationship with K^+ influx measured by $^{86}Rb^+$ uptake has been reported by Maraini and Pasino (1983). That ATPase is susceptible to oxidative insult was shown by exposing lenses in organ culture to oxidizing conditions. Utilizing either *t*-butylhydroperoxide or hydrogen peroxide, K^+ influx can be markedly inhibited (Giblin et al, 1982; Garner et al, 1983). However, under these conditions the steady state hydrolysis of Mg^{2+}-ATP is normal (Garner et al, 1983). Thus, when subjected to oxidative stress, the Na/K-ATPase appears to be oxidized, causing uncoupling of ion translocation and energy utilization. The wasteful utilization of ATP might be expected to decrease lens ATP levels; indeed, a small but significant decrease of about 15% was observed which could not be explained by a suppression in glucose metabolism (Garner et al, 1983). Examination of the impact of H_2O_2 on Na/K-ATPase kinetics indicated that H_2O_2 altered the interaction of the pump with ATP from negative to positive kinetic cooperativity (Garner et al, 1984). When a small group of normal and cataractous human lenses was subjected to kinetic analysis of Na/K-ATPase, measuring the steady-state hydrolysis of Mg^{2+}-ATP, it was found that in the cataract population a significant fraction of the epithelial cells from these lenses had either no Na/K-ATPase activity or demonstrated positive cooperativity similar to that found with the H_2O_2-treated preparations (Garner and Spector, 1986). All of the apparently normal lenses were active and did not have positive cooperativity kinetics.

Ion derangement is not limited to Na^+ and K^+. There have been a number of reports delineating changes in Ca^{2+} concentrations related to cataract formation (Hightower and Reddy, 1982; Duncan and Bushell, 1975). Recently, it has been reported that Ca^{2+}-ATPase is highly sensitive to oxidation by low levels of H_2O_2 (Borchman et al, 1989). Thus, it is apparent that not only membrane structural elements but also pump systems are susceptible to oxidative stress.

C. DNA and H_2O_2 insult

Damage to DNA should also be considered. But the data linking specific DNA mutations to cataract formation are not as well defined as is damage to lens protein. There are a large number of genetically linked diseases, such as Down's syndrome, galactosaemia, Marfan syndrome, and Cockayne syndrome, in which congenital cataract is observed (Bardelli et al, 1989). In some cases these diseases involve mutations which cause metabolic aberrations leading to cataract, such as galactosaemia, while in other cases the specific DNA lesion is not known. In at least one situation, Cockayne syndrome, it is suspected that repair of DNA damage has been affected. Fibroblasts isolated from the skin of Cockayne syndrome patients have been shown to be more sensitive to UV than normal cells (Schmickel et al, 1977; Friedberg, 1985) and there is a general sensitivity of the Cockayne syndrome patient to sunlight. The UV exposure is believed to cause DNA damage which is either not repaired or repaired incorrectly.

It may be argued that loss of lens transparency is not the result of somatic mutation but of some other effect of the traumatizing agent. However, it can be demonstrated that lens epithelial cell DNA can be damaged by a number of conditions, including UV radiation, H_2O_2 and ionizing radiation, and it has been concluded that such damage can lead to cataract (Cogan et al, 1952; von Sallmann, 1950, 1951; Worgul and Rothstein, 1975; Worgul, 1989; Kleiman et al, 1990). In certain cases, it has been suggested that the development of opacification can be related to the specific epithelial cells that have mutated (Worgul and Rothstein, 1975; Worgul, 1989). Probably the most carefully examined factor causing lens DNA damage is H_2O_2 (Kleiman et al, 1990; Spector et al, 1989). At concentrations of H_2O_2 only slightly above normal, 50 μM, significant single-strand breaks have been observed. With bovine lens epithelial cell cultures, essentially no double-strand breaks, pyrimidine dimers or thymine glycol have been observed, even at concentrations of 200 μM. The H_2O_2 damage has been shown to be directly caused by H_2O_2 interaction with metal ion to generate hydroxyl radical. Repair of the single-strand breaks is rapid, with almost complete recovery being observed in 30 min. When similar experiments were conducted with UVB radiation (the major UV region causing DNA damage which reaches the lens), the results were more complex. The primary damage to DNA was found to be the formation of pyrimidine dimers but DNA–DNA linkage and thymine glycol formation were observed. Furthermore, the repair process required hours rather than minutes. It is strange that the lens, which is subjected to UVB radiation throughout life, should appear to have a less efficient system for repairing UVB damage than for H_2O_2 repair (Spector et al, 1990).

Thus, the oxidation of thiols, leading to large aggregates capable of scattering light and contributing to cell breakdown, the massive oxidation of methionine residues, the loss of vital pump function, probably as a result of oxidation, the observed loss of GSH and enzyme activity and appearance of GSSG and mixed disulphides, and the observed sensitivity of epithelial cell DNA to oxidation, all support the contention that oxidative stress may cause cataract.

D. Compounds producing oxidative stress cause cataract

While a number of compounds causing oxidative stress have been investigated, only two which may have physiological relevance and are related to cataract are considered in this discussion.

1. H_2O_2

H_2O_2 levels have been determined in the aqueous fluid and lenses of a small number of patients with maturity-onset cataract. Elevated H_2O_2 levels ranging from 45 μM to 663 μM were found in the aqueous fluid of 7 out of 17 patients. In the cataractous lenses, elevated H_2O_2 levels ranging from 60 μM to 153 μM were found in 10 out of 30 patients (Spector and Garner, 1981). High H_2O_2 levels in cataract patients have recently been observed in two other laboratories (Morris and Varma, 1990; Huang and Hu, 1990).

Elevated H_2O_2 levels have been shown to cause cataract in organ culture (Garner et al, 1982; Giblin et al, 1987) and to generally cause a pattern of oxidative damage similar to that found in the cataractous lens (Zigler et al, 1989). Furthermore, in studies with lens epithelial cell cultures, H_2O_2 stress produces changes in metabolic parameters which are similar to those seen in cataract (Spector et al, 1988; Anderson and Spector, 1978). Such findings implicate H_2O_2 as one of the compounds probably producing oxidative stress to the lenses and capable of causing cataract in vivo.

2. Hyperbaric O_2

It is apparent that exposure of the lens to a high concentration of O_2 or to hyperbaric oxygen will cause cataract. In patients subjected to hyperbaric oxygen, opacification of the inner region of the lens has been reported (Cohen, 1986). In one report, 14 out of 15 patients developed nuclear opacities (Palmquist et al, 1984). Experiments with mice have shown that multiple exposure to 100% O_2 will cause nuclear opacity (Schocket et al, 1972), and exposure of rabbit lenses in organ culture to hyperbaric oxygen can cause some loss of nuclear transparency and a drop in GSH in the nuclear region (Giblin et al, 1988). It was found that the drop in GSH did not occur in the epithelium. It is well known that the level of glutathione may be as much as ten times higher in the epithelium than in the nuclear region, and yet no increase in GSSG was observed in the epithelium under most hyperbaric conditions. In in vitro experiments, it was shown that concentrations of GSH comparable to those found in the lens generated H_2O_2 at concentrations of 10–20 μM in the presence of hyperbaric O_2. That this H_2O_2 production is metal-dependent was demonstrated by repression but not elimination of the H_2O_2 formation by the addition of 2 μM EDTA. Thus, it is conceivable that some aspects of O_2-induced cataract may be a result of H_2O_2 formation.

VI. PROTECTION FROM OXIDATIVE STRESS

A. Elevation of GSH

There may be a number of ways to protect the lens from oxidative insult, such as increasing the concentration of GSH in the system or increasing the activity of the enzymes which metabolize components which may cause oxidation. On the basis of this premise, lens epithelial cells in culture had their GSH levels increased by as much as 87% (Spector et al, 1987) by utilizing GSH ethyl ester, which effectively moves through membrane barriers, in contrast to free GSH (Puri and Meister, 1983). This approach appeared reasonable, since lowering GSH levels or preventing the reduction of GSSG by inhibiting GSSG reductase made the system more susceptible to oxidative stress (Reddy and Giblin, 1984; Giblin and McCready, 1983). When the GSH was elevated in epithelial cell preparations, the system was able to maintain the GSH in the reduced state. Three parameters were utilized to characterize the cells' resistance to oxidative stress: ATP levels, indicative of overall metabolism; glyceraldehyde-3-phosphate dehydrogenase activity, reflecting a cytosol enzyme sensitive to oxidative stress; and Na/K-ATPase activity, an indicator of membrane viability. Increased resistance to H_2O_2 insult was observed with only the cytosol-localized glyceraldehyde-3-phosphate dehydrogenase. In a typical experiment, a loss of 85% in enzymatic activity would be expected after H_2O_2 insult (0.3 mM H_2O_2 for 3 min); with elevated GSH in the cell, only a 25% inhibition was observed. However, the elevated GSH concentration significantly slowed the recovery from oxidative stress based on ATP levels and Na/K-ATPase activity. It would appear that increasing the GSH concentration seriously alters the redox equilibrium of the cell and seriously compromises certain aspects of the metabolic process.

B. GSH peroxidase mimics

The question remains of how to effectively protect the cell from oxidative stress. As mentioned previously, of the major oxidants found in biological systems, few are stable. Most of the common potent reactive oxidants, such as superoxide and the hydroxyl radical, have short half-lives, and thus react at the source of their generation. In many cases, they are formed as a result of reaction with H_2O_2 as, for example, the formation of hydroxyl radical in a Fenton-type reaction. Thus, if H_2O_2 can be maintained at normal low levels, oxidative stress would be significantly diminished.

There are certain requirements for maintaining low H_2O_2 concentrations in a biological system. The removal of the hydroperoxide must be rapid, for otherwise it would move to sites where it could be damaging. The protective agent must not alter significantly the oxidation reduction setpoint of the cell for, as shown elsewhere with GSH, increasing the reducing potential is deleterious. The agent should be present at low concentration to avoid toxic side reactions. The agent should act catalytically so that small quantities will be able to detoxify relatively large quantities of hydroperoxide without being consumed. Ideally, the compound would use the metabolic apparatus of the cell to maintain a steady state. Theoretically, elevating the concentration of an existing enzyme, such as GSH peroxidase, might meet the above requirements. However, the localization of the enzyme to certain parts of the cell may leave other parts of the cell exposed to H_2O_2 attack. An agent that is present throughout the cell would be ideal. Thus, attempts have been made to synthesize pseudoenzymes or enzyme mimics which can perform in a similar manner to GSH peroxidase and move easily through the cell. Such efforts were stimulated by reports that ebselen, a selenocompound, has GSH peroxidase activity and other antioxidative effects (Müller et al, 1984, 1985; Wendel et al, 1984; Parnham and Kindt, 1984; Hurst et al, 1989). Ebselen also appears to inhibit cyclooxygenase activity and is presently being considered as an antiinflammatory compound. The studies also show that it utilizes GSH.

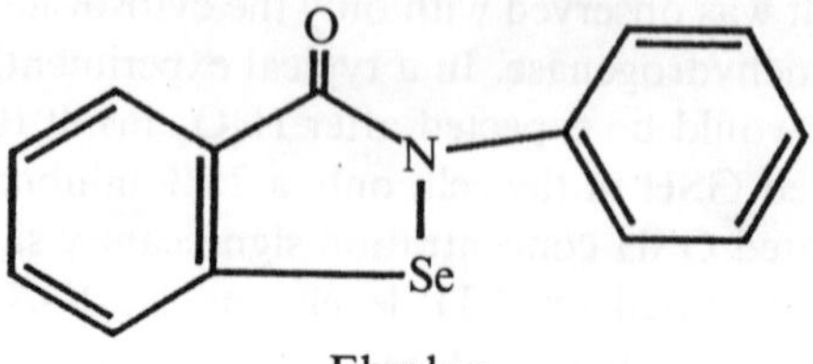

O
N
Se

Ebselen

GSH peroxidase is a selenoprotein whose X-ray crystal structure has been determined with 0.2-nm resolution (Epp et al, 1983). Thus, the general structure of the active centre of the enzyme is known. From such analyses, as well as biochemical information, a reaction mechanism for the enzyme has been proposed. The reduced enzyme is believed to have the selenium in the selenol form SeH. The reduced selenol is readily oxidized by H_2O_2 to selenenic acid (SeOH).

$$E\text{—}SeH + H_2O_2 \longrightarrow E\text{—}SeOH + H_2O$$

The selenenic acid is then reduced by GSH.

$$E\text{—}SeOH + GSH \longrightarrow E\text{—}SeSG + H_2O$$
$$E\text{—}SeSG + GSH \longrightarrow E\text{—}SeH + GSSG$$

Utilizing the proposed mechanism, the information available on the structure of the enzyme and recent observations on selenium chemistry (Reich and Jasperse, 1987), two compounds have been synthesized which have high GSH peroxidase activity in the 1–2 μM range (Wilson et al, 1989).

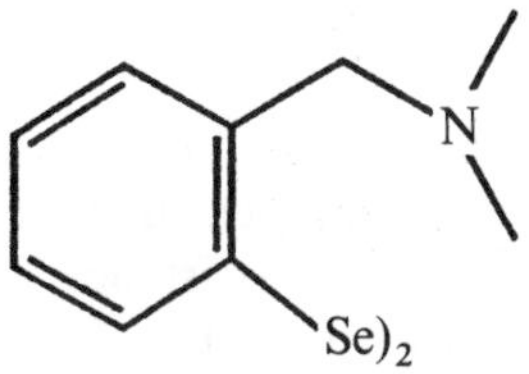

N
Se)$_2$

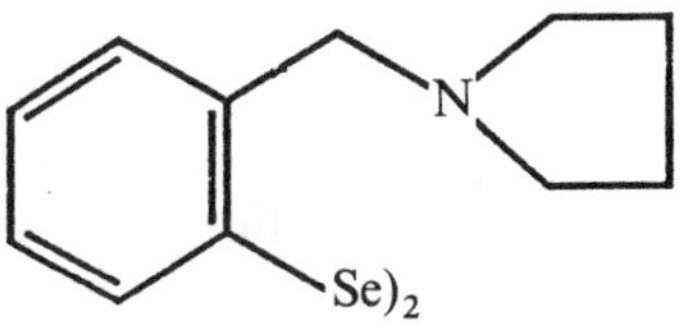

N
Se)$_2$

By coupling the GSH peroxidase reaction with the reduction of GSSG by GSSG reductase and NADPH, it could be shown that the compounds utilized GSH. The GSH peroxidase mimics rapidly metabolized H_2O_2 without loss of catalytic activity. Unpublished results from this laboratory suggest that these compounds rapidly pass into cell cultures and increase the GSH peroxidase activity of the cell. Experiments have now been initiated to determine the effectiveness of these compounds in protecting the lens from oxidative stress. Preliminary results indicate that lens epithelial cell cultures are protected from H_2O_2 stress by these GSH peroxidase mimics.

Comparisons between ebselen and the compounds mentioned above indicate that the new selenium compounds are approximately 10-fold more active as GSH peroxidase mimics than ebselen. The new compounds were designed to increase the acidity of the selenium. 7-Nitroebselen has comparable activity to the compounds discussed above (Parnham et al, 1989). The introduction of the nitro electron-withdrawing group also increases the selenium acidity. The nitro group also enhances the activity of the compounds reported by Wilson et al (1989). Diphenyl diselenides have activities comparable to that of ebselen (Wilson et al, 1989).

VII. CONCLUDING REMARKS

It is apparent that the lens is subject to oxidative stress and becomes less capable of handling such stress with aging. There is also a large body of work supporting the concept that oxidation is a major factor in the development of cataract in the aged human population. However, it cannot be stated with certainty that oxidation is a cause of maturity-onset cataract. Only by diminishing oxidative stress and observing a concomitant decrease in cataract development can it be definitively demonstrated that cataract and oxidation are linked. Perhaps the emerging GSH peroxidase mimics and other compounds now being developed will allow us to adequately test the hypothesis.

ACKNOWLEDGEMENTS

The work from the author's laboratory was supported by the National Institutes of Health, Alcon Laboratories and Research to Prevent Blindness. The author is grateful for their support. The editorial assistance and manuscript preparation by Elaine Bluberg is gratefully acknowledged.

REFERENCES

Akerboom TPM & Sies H (1981) *Methods Enzymol.* **77:** 373–382.
Anderson EI & Spector A (1978) *Exp. Eye Res.* **26:** 407–417.
Anderson EI., Wright DD & Spector A (1979) *Exp. Eye Res.* **29:** 233–243.
Augusteyn RC (1981) In Duncan G (ed.) *Mechanisms of Cataract Formation in the Human Lens*, pp 72–111. New York: Academic Press.
Bardelli AM, Lasorella G & Vanni M (1989) *Ophthalmic Ped. Genet.* **10:** 293–298.
Bellows J (1936) *Arch. Ophthalmol.* **15:** 78–83.
Bergad PL, Rathbun UB & Linder W (1982) *Exp. Eye Res.* **34:** 131–144.
Bhuyan KC & Bhuyan DK (1977) *Biochim. Biophys. Acta* **497:** 641–651.
Bhuyan KC & Bhuyan DK (1978) *Biochim. Biophys. Acta* **542:** 28–38.
Bhuyan KC, Bhuyan DK & Podos SM (1982) *Ann. NY Acad. Sci.* **393:** 169–171.
Bietti G (1935) *Bull. Oculist* **14:** 3–33.
Borchman D, Paterson CA & Delamere NA (1989) *Invest. Ophthalmol Vis. Sci.* **30:** 1633–1637.

Bouton SM (1939) *Arch. Intern. Med.* **63:** 930–945.
Brigelius R (1985) In Sies H (ed.) *Oxidative Stress*, pp 243–272. London: Academic Press.
Cabelli DE & Bielski BHJ (1983) *J. Phys. Chem.* **87:** 1809–1812.
Cliffe FE & Waley SG (1958) *Biochem. J.* **69:** 649–655.
Cogan DG, Donaldson DD & Reese AB (1952) *Arch. Ophthalmol.* **47:** 55–70.
Cohen GH (1986) *Postgrad. Med.* **79:** 89.
de Mello Filho AC & Meneghini R (1985) *Biochim. Biophys. Acta* **847:** 82–89.
Dickinson JC, Durham DG & Hamilton PB (1968) *Invest. Ophthalmol.* **7:** 551–563.
Dillon J, Spector A & Nakanishi K (1976) *Nature (Lond.)* **254:** 422–423.
Dische Z & Zil H (1951) *Am. J. Ophthalmol.* **34:** 104–113.
Duncan G & Bushell AR (1975) *Exp. Eye Res.* **20:** 223–230.
Epp O, Ladenstein R & Wendel A (1983) *Eur. J. Biochem.* **133:** 51–69.
Fenton HJM (1894) *J. Chem. Soc.* **65:** 899–910.
Flohé L, Loschen G, Günzler WA & Eichele E (1972) *Hoppe-Seyler's Z. Physiol. Chem.* **353:** 987–999.
Frei B, England L & Ames BN (1989) *Proc. Natl. Acad. Sci. USA* **86:** 6377–6381.
Friedberg EC (1985) *DNA Repair*, pp 536–539. New York: W.H. Freeman Co.
Garner WH & Spector A (1978) *Proc. Natl. Acad. Sci. USA* **75:** 3618–3620.
Garner MH & Spector A (1980a) *Proc. Natl. Acad. Sci. USA* **77:** 1274–1277.
Garner MH & Spector A (1980b) *Exp. Eye Res.* **31:** 361–369.
Garner WH, Garner MH & Spector A (1983). *Proc. Natl. Acad. Sci. USA* **80:** 2044–2048.
Garner MH & Spector A (1986) *Exp. Eye Res.* **42:** 339–348.
Garner MH, Garner WH & Spector A (1982) *Invest. Ophthalmol. Vis. Sci.* **22:** 34.
Garner MH, Garner WH & Spector A (1984) *J. Biol. Chem.* **259:** 7712–7718.
Giblin FJ & McCready JP (1983) *Invest. Ophthalmol. Vis. Sci.* **24:** 113–118.
Giblin FJ, Neis DE & Reddy VN (1981) *Exp. Eye Res.* **33:** 289–298.
Giblin FJ, McCready JB & Reddy VN (1982) *Invest. Ophthalmol. Vis. Sci.* **22:** 330–335.
Giblin FJ, McCready JP, Kodama T & Reddy VN (1984) *Exp. Eye Res.* **38:** 87–93.
Giblin FJ, McCready JP, Schrimscher L & Reddy VN (1987) *Exp. Eye Res.* **45:** 77–91.
Giblin FJ, Schrimscher L, Chakrapani B & Reddy VN (1988) *Ivest. Ophthalmol. Vis. Sci.* **29:** 1312–1319.
Harding JJ (1970) *Biochem. J.* **117:** 957–960.
Harding JJ (1972) *Exp. Eye Res.* **13:** 33–40.
Harding JJ & Crabbe JC (1984) In Davison H (ed.), *The Eye*, pp 207–492. New York: Academic Press.
Heikkila RE, Cabbat FS & Cohen G (1976) *J. Biol. Chem.* **251:** 2182–2185.
Hightower KR & Reddy VN (1982) *Exp. Eye Res.* **34:** 413–421.
Hildebrandt AG, Roots I, Tjoe M & Heinemeyer G (1978) *Methods Enzymol.* **LII:** 342–350.
Holmberg N (1968) *Exp. Eye Res.* **7:** 570–580.
Holmgren A (1968) *Eur. J. Biochem.* **6:** 475–484.
Holmgren A (1985) *Annu. Rev. Biochem.* **54:** 237–271.
Huang Q-L & Hu T-S (1990) *Invest. Ophthalmol. Vis. Sci.* **31:** 350.
Huby R & Harding JJ (1988) *Exp. Eye Res.* **47:** 53–59.
Hurst JS, Patterson CA, Bhattacherjee P & Pierce WM (1989) *Biochem. Pharmacol.* **38:** 3357–3363.

Jernigan HJJ, Fukui HN, Goosey JD & Kinoshita JH (1981) *Exp. Eye Res.* **32:** 461–466.
Kallis GB & Holmgren A (1980) *J. Biol. Chem.* **255:** 10261–10265.
Kinoshita JH (1955) *Arch. Ophthalmol* **54:** 360–368.
Kinoshita HJ & Masurat T (1957) *Arch. Ophthalmol.* **57:** 266–274.
Kinoshita JH & Merola LO (1973) In *The Human Lens in Relation to Cataract Ciba Symposium Foundation*, pp 173–184. North Holland: Elsevier Excerpta Medica.
Kinsey VE & Frohman CE (1951) *Arch. Ophthalmol.* **46:** 536–541.
Kinsey VE & Merriam FC (1950) *Arch. Ophthalmol.* **44:** 370–380.
Kleiman NJ, Wang R-R & Spector A (1990) *Mutat Res.* **240:** 35–45.
Kobayashi S, Roy D & Spector A (1983) *Curr. Eye Res.* **2:** 327–334.
Kuck JFR, Yu N-T & Askren CC (1982) *Exp. Eye Res.* **34:** 23–37.
Kuszak JR (1990) In Tasman W & Jaegar E (eds) *Duane's Clinical Ophthalmology*, Vol. 1, pp 1–9, Chapter 71 A. Philadelphia: J. P. Lippincott.
Kuszak JR & Rae JL (1982) *Exp. Eye. Res.* **35:** 499–519.
Lakoma E-L & Ekland P (1979) *Nuc. Act. Tech. Life Sci.* **1978:** 333–344.
Laurent TC, Morre EC & Reichard P. (1964) *J. Biol. Chem.* **239:** 3436–3444.
Lawrence RA, Sunde R, Schwartz G & Hoekstra W (1974) *Exp. Eye Res.* **18:** 563–569.
Lou MF, McKellar R & Chyan O (1986) *Exp. Eye Res.* **42:** 607–616.
Mach H (1966) *Klin. Mbl. Augénheilk* **149:** 897–904.
Mapson LW (1945) *Biochem. J.* **39:** 228–236.
Maraini G & Pasino M (1983) *Exp. Eye Res.* **33:** 543–550.
Marlat P, Bracchi PG & Maraini G (1981) *Ophthalmol Res.* **13:** 293–301.
Masters PM, Bada JL & Zigler Jr JS (1977) *Nature (Lond.)* **268:** 71–73.
Masters PM, Bada JL & Zigler Jr JS (1978) *Proc. Natl. Acad. Sci. USA* **75:** 1204–1208.
Matsuda H, Giblin FJ & Reddy VN (1981) *Exp. Eye Res.* **33:** 253–265.
Mercantonio SM, Duncan G, Davies PN & Bushell AR (1980) *Exp. Eye Res.* **31:** 227–237.
Moore EC, Reichard P & Thelander L (1964) *J. Biol. Chem.* **239:** 3445–3452.
Morris SM & Varma SD (1990) *Invest. Ophthalmol. Vis. Sci.* **31:** 349.
Müller A, Cadenas E, Graf P & Sies H (1984) *Biochem. Pharmacol.* **33:** 3225–3229.
Müller A, Gabriel H & Sies H (1985) *Biochem. Pharmacol.* **34:** 1185–1189.
Offord RE, Phillippe J, Davis JG, Halloon PA & Berger M (1979) *Biochem. J.* **182:** 249–251.
Palmquist BM, Philipson B & Barr PO (1984) *Br. J. Ophthalmol.* **68:** 113.
Parnham MJ & Kindt P (1984) *Biochem. Pharmacol.* **33:** 3247–3250.
Parnham MJ, Biedermann CH, Bittner N, Dereu N, Leyck S & Wetzig H (1989) *Agents Actions* **27:** 306–308.
Pau H, Graf P & Sies H (1990) *Exp. Eye Res.* **50:** 17–20.
Pirie A (1965) *Biochem. J.* **96:** 244–253.
Puri RN & Meister A (1983) *Proc. Natl. Acad. Sci. USA* **80:** 5258–5260.
Rathbun WB (1980) In Srivastava SK (ed.) *Red Blood Cell and Lens Metabolism*, pp 169–173. New York: Elsevier, North Holland.
Rathbun WB (1989) In Dolphin D, Poulson R & Avramovic O (eds) *Coenzymes and Cofactors*. Vol. III. *Glutathione*, pp 467–509. New York: John Wiley & Sons, Inc.
Rathbun WB & Bovis MG (1986) *Curr. Eye Res.* **5:** 381–385.
Rathbun WB, Bovis MG & Holleschau AM (1986) *Curr. Eye Res.* **5:** 195–199.
Reddy VN (1990) *Exp. Eye Res.* **50:** 771–778.

Reddy VN & Giblin F (1984) In Nugent J & Whelan J (eds) *Human Cataract Formation Ciba Foundation Symposium*, pp 65–83. London: Pitman.
Reddy VN & Han RF (1976) *Doc. Ophthalmol. Proc. Ser.* **8:** 153–160.
Reich NJ & Jasperse CP (1987) *J. Am. Chem. Soc.* **109:** 5549–5551.
Robertson JM, Donner AP & Trevithick JR (1989) *Ann. NY Acad. Sci.* **570:** 372–382.
Ross WM, Creighton MO, Stewart-DeHaan PJ, Sanwal M, Hirst M & Trevithick JR (1982) *Can. J. Ophthalmol.* **17:** 61–66.
Said F & Weale RA (1959) *Gerontology* **3:** 213–231.
Schmickel RD, Chu EHY, Trosko JE & Chang C-C (1977) *Pediatrics* **60:** 135–139.
Shocket SS, Esterson J, Bradford B, Michaelis M & Richards RD (1972) *Israel J. Med. Sci.* **8:** 1596.
Schulte-Frohlinde D, Anker R & Bothe E (1981) In Rodgers MAJ & Powers EL (eds) *Oxygen and Oxy-radicals in Chemistry and Biology*, pp 61–67. New York: Academic Press.
Sigelman J, Trokel SL & Spector A (1974) *Arch. Ophthalmol.* **92:** 437–442.
Spector A (1984) *Invest. Ophthalmol. Vis. Sci.* **25**: 130–146.
Spector (1985) In Maisel H (ed.) *The Ocular Lens*, pp 405–438. New York: Marcel Dekker, Inc.
Spector A & Garner MH (1980) In Srivastava SK (ed.) *The Red Blood Cell and Lens Metabolism*, pp 233–236. New York: Elsevier, North Holland.
Spector A & Garner WH (1981) *Exp. Eye Res.* **33:** 673–681.
Spector A & Roy D (1978) *Proc. Natl. Acad. Sci. USA* **75:** 3244–3248.
Spector A, Roy D & Stauffer J (1975) *Exp. Eye Res.* **21:** 24–29.
Spector A, Scotto R, Weissbach H & Brot N (1982) *Biochem. Biophys. Res. Commun.* **108**: 429–434.
Spector A, Huang R-RC & Wang G-M (1985) *Curr. Eye Res.* **4:** 1289–1295.
Spector A, Wang G-M & Huang R-RC (1986) *Curr. Eye Res.* **5:** 47–51.
Spector A, Huang R-RC, Wang G-M, Schmidt C, Yan G-Z & Chifflet S (1987) *Exp. Eye Res.* **45:** 453–465.
Spector A, Yan G-Z, Huang R-RC, McDermott MJ, Gascoyne PRC & Pigiet V (1988) *J. Biol. Chem.* **263:** 4984–4990.
Spector A, Kleiman NJ, Huang R-RC & Wang RR (1989) *Exp. Eye Res.* **49:** 685–698.
Spector A, Kleiman N & Wang R-R (1990) *Invest. Ophthalmol. Vis. Sci.* **31:** 436.
Spikes JD & Livingston R (1969) *Adv. Radiat. Biol.* **3:** 29–121.
Sprinker L, Harr J, Newberne P, Whanger P & Weswig P (1971) *Nutr. Rev. Int.* **4:** 335–340.
Srivastava SK (1976) *Exp. Eye Res.* **22:** 577–585.
Srivastava SK & Beutler E (1973) *Exp. Eye Res.* **17:** 33–42.
Stevens VJ, Rouzer CA, Monnier VM & Cerami A (1978) *Proc. Natl. Acad. Sci. USA* **75:** 2918–2922.
Swanson A & Truesdale A (1971) *Biochem. Biophys. Res. Commun.* **45:** 1488–1496.
Trevithick JR, Linklater HA, Mitton KP, Dzialoszynski T & Sanford SE (1989) *Ann. NY Acad. Sci.* **570:** 358–371.
Truscott RJW & Augusteyn RC (1977) *Exp. Eye Res.* **25:** 139–148.
van Heyningen R (1973) *Exp. Eye Res.* **17:** 137–147.
Varma SD, Kumar S & Richards RD (1979) *Proc. Natl. Acad. Sci. USA* **76:** 3504–3506.
Varma SD, Beachy NA & Richards RD (1982) *Photochem. Photobiol.* **36:** 623–626.

Varma SD, Chand D, Sharma YR, Kuck JF & Richards RD (1984) *Curr. Eye Res.* **3:** 35–57.
von Sallmann L (1950) *Trans. Am. Ophthalmol. Soc.* **48:** 228–242.
von Sallmann L (1951) *Arch. Ophthalmol.* **45:** 149–164.
Waley SG (1969) In Davson H (ed.) *The Eye*, pp 299–379. New York: Academic Press.
Waley SG & van Heyningen R (1962) *Biochem. J.* **83:** 274–283.
Wendel A, Fausel M, Safayhi H, Tiegs G & Otter R (1984) *Biochem. Pharmacol.* **33:** 3241–3245.
Wilson SR, Zucker PA, Huang R-RC & Spector A (1989) *J. Am. Chem. Soc.* **111:** 5936–5939.
Wolff SP, Wang G-M & Spector A (1987) *Exp. Eye Res.* **45:** 777–789.
Worgul BV (1989) *Lens Eye Toxicity Res.* **6:** 559–571.
Worgul BV & Rothstein H (1975) *Ophthalmic Res.* **7:** 21–32.
Zigler JS, Jernigan HMJ, Garland D & Reddy VN (1985) *Arch. Biochem. Biophys.* **241:** 163–172.
Zigler JS, Huang Q-L & Du X-Y (1989) *Free Radicals Biol. Med.* **7:** 499–505.
Zigman S (1985) In Maisel H (ed.) *The Ocular Lens*, pp 301–347. New York: Marcel Dekker, Inc.

20

Photooxidative Stress in the Skin

JÜRGEN FUCHS* and LESTER PACKER†

* *Zentrum der Dermatologie und Venerologie, Klinikum der Johann Wolfgang, Goethe Universität, 6000 Frankfurt/M 70, Germany.*

† *Department of Molecular and Cell Biology, University of California, Berkeley, and Membrane Bioenergetics Group, Lawrence Berkeley Laboratory, Berkeley, CA 94720, USA*

I. Introduction 559
II. Photosensitization 561
III. Photooxidative skin damage 567
IV. Skin diseases 570
V. Antioxidants 574
VI. Conclusions 578

I. INTRODUCTION

Skin is a biological interface with the environment and functions as the first line of defence against noxious external stimuli such as ultraviolet and visible irradiation, prooxidant chemicals, infection, and ionizing radiation. Under some conditions, e.g. inflammation or ischaemia, endogenous sources of free radicals can also become significant. Reactive oxygen species (ROS) may play an important role in skin aging, tumour promotion, cutaneous autoimmune disease, and phototoxicity/photosensitivity. It has been reported that there is no direct evidence relating free radicals or ROS with acute or chronic ultraviolet effects in skin (Epstein, 1977). Recently, however, it was pointed out that considerable circumstantial evidence suggests that ROS are responsible for some of the deleterious effects of ultraviolet light upon skin

Oxidative Stress: Oxidants and Antioxidants
ISBN 0-12-642762-3

(Black, 1987a). ROS can be formed by UVC (200–280 nm) through water photolysis, and by UVB (280–320 nm) and UVA (320–400 nm) through photodynamic action and photodissociation of several types of molecules, e.g. hydrogen peroxide. Evidence for the existence of ROS and other free radicals in skin in vitro and in vivo is difficult to obtain because of the high reactivity of these species, their low steady-state concentrations and the heterogeneity of the organ. The epidermis is composed of different cell layers (stratum corneum, s. granulosum, s. spinosum, s. basale), each having distinct metabolic activities. In the epidermis the keratinocyte is the dominating cell species (90%); melanocytes and Langerhans cells comprise about 10% of the total cell population. In the dermis noncellular elements like collagen and elastin fibres predominate over cells, such as fibroblasts (major fraction) and mast cells (minor fraction). Epidermal cells, dermal cells and cell infiltrates (leukocytes) can be sources or targets of biological oxidants in skin during normal or altered metabolic activity.

Direct evidence for free radical formation in skin has been obtained by low-temperature electron paramagnetic resonance (EPR) spectroscopy (Norins, 1962; Pathak and Stratton, 1968). Irradiation of human skin with wavelengths greater than 320 nm enhances the endogenous paramagnetic melanin signal in pigmented skin samples. Non-pigmented human skin samples give no endogenous melanin free radical signal, and irradiation with similar wavelengths does not produce any detectable free radicals. Ultraviolet radiation with wavelengths shorter than 320 nm enhances the melanin signal in pigmented skin and also produces other radical species in pigmented and in non-pigmented skin (Pathak and Stratton, 1968). EPR spectroscopy at ambient temperature has revealed that UVA irradiation of mouse skin results in the formation of the endogenous ascorbyl radical (Buettner et al, 1987). Epidermal application of free radical-generating compounds such as 4-hydroxyanisole (Riley, 1970) and anthralin (Schroot and Brown, 1986; Fuchs and Packer, 1989) results in generation of persistent organic radicals in skin. Indirect experimental evidence for the formation of ROS and free radicals in epidermal and dermal cells as well as in intact skin has been obtained by detection of free radical reaction products, e.g. lipid peroxidation products (thiobarbituric acid reactive substances, TBARS) (Meffert and Reich, 1969; Niwa et al, 1987), by spin-trapping glutathione thiyl radicals in keratinocytes with 5,5-dimethyl-1-pyrroline-*N*-oxide (DMPO) (Schreiber et al, 1989), and alkyl and alkoxyl radicals in keratinocytes by DMPO (Taffe et al, 1987). Unidentified thiyl radicals arising after photosensitization in intact skin were also spin-trapped by DMPO (Li, personal communication).

II. PHOTOSENSITIZATION

Photosensitization is a cutaneous reaction to electromagnetic irradiation. The reaction is distinguished from normal photodermatitis by an amplification of the physiological response due to endogenous or exogenous chemical substances. Drug-induced phototoxicity must be distinguished from drug-induced photoallergy. Epstein was a pioneering photobiologist who first differentiated between phototoxic and photoallergic skin reactions in humans (Epstein, 1939). The reaction pathways of photosensitizers usually proceed via the triplet state and can be characterized as type I and type II reactions. In the type I reaction, the activated photosensitizer directly reacts with a substrate molecule via electron or hydrogen atom transfer and gives rise to free radical formation. In the type II reaction (photodynamic reaction), the electronically excited photosensitizer reacts with oxygen, thereby forming ROS as reactive intermediates. Most photosensitized reactions are oxygen-dependent, and may cause ROS-mediated molecular damage at various cellular sites (microsomes, mitochondria, cytoskeleton, lysosomes, plasma membrane) (Miyachi et al, 1986; Torinuki et al, 1980).

A. Endogenous photosensitizers

Photodermatoses can be classified into genetic, metabolic, idiopathic, chemical, degenerative and neoplastic disorders. Endogenous photosensitizers may play a role in all of these entities and comprise porphyrins, melanins and photosensitizing molecules present in natural substances (Ames, 1983; Straight & Spikes, 1985). In addition, endogenous metabolites like flavins, NADH, NADPH, and nucleosides can sensitize the transmission of photon energy from solar radiation to oxygen, resulting in formation of superoxide anion radical (Cunningham et al, 1985; Ballou et al, 1969; Massey et al, 1969). Tryptophan and *N*-formyl kynurenine (McCormick et al, 1976) are endogenous sources of hydrogen peroxide upon irradiation with near-ultraviolet light.

1. Melanins

Melanins are redox polymers containing high concentrations of *o*-quinones (oxidizing groups) and *o*-hydroquinones (reducing groups). According to the presumed chemical precursor, melanins can be classified into eumelanins (dopa/tyrosine) phaeomelanins (dopa/cysteine) and allomelanins (catechol or other polyhydroxy aromatics). A free radical property of melanin was reported in 1954 (Commoner et al, 1954), and confirmed by others (Blois et al, 1964; Mason et al, 1960). Melanins can undergo reversible redox processes and are able to scavenge persistent radicals such as diphenylpicrylhydrazyl and nitroxides (Sarna et al, 1985a), as well as highly reactive oxygen species (Sarna et al, 1986). Although the main function of melanin is considered to be protection against ultraviolet irradiation, some results show that melanins may also act as photosensitizers. ROS are formed during irradiation of melanogenic precursors as well as during irradiation of melanins (Tomita et al, 1984; Joshi et al, 1987). Irradiation of melanogenic precursors may induce radical photoreactions in vivo that play an important role in acute solar responses such as erythema, as well as in chronic sequellae such as premature aging and carcinogenesis (Chedekel and Zeise, 1988).

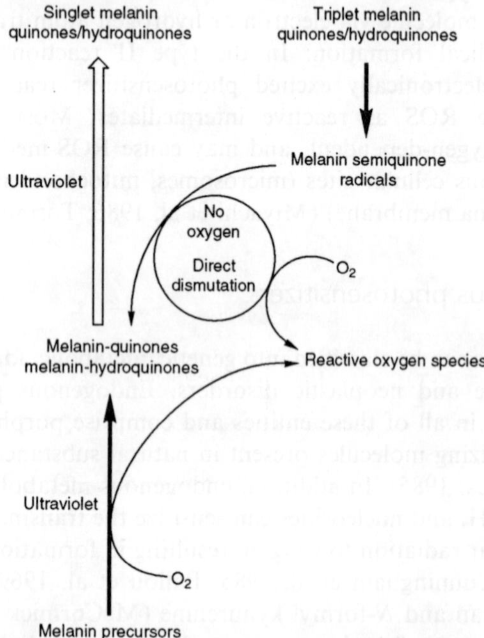

Figure 1. Melanin photochemistry.

Melanin may act as a photoprotective agent at long ultraviolet and visible wavelengths, while it acts as a photosensitizer in the short ultraviolet range. The chromophore responsible for photoinduced oxygen consumption in melanins is not identical to the major melanin chromophore responsible for absorption of visible light. Quantum yields for oxygen consumption in eumelanin are low with visible light but increase sharply with light of shorter wavelengths (Sarna and Sealy, 1984a,b). Rates of both free radical production and oxygen consumption in melanins can be greatly enhanced in the presence of other photosensitizers. Whereas oxygen consumption and free radical production induced by direct irradiation appear to be closely related for eumelanins and phaeomelanins, sensitized reactions reveal a greater reactivity of phaeomelanins. This might be important in view of the putative link between phaeomelanin and the skin of red-haired individuals and the incidence of skin cancer in these persons (Sarna et al, 1985b). The ultraviolet carcinogenesis hypothesis of cutaneous melanoma has some inconsistencies, and it was suggested that an unknown xenobiotic is involved in the induction of melanoma (Rampen and Fleuren, 1987). This unknown xenobiotic could work as a photosensitizer interacting with phaeomelanin. An extremely high content of phaeomelanin is reported in dysplastic melanocytic nevi (a melanoma precursor), and it was postulated that ultraviolet irradiation, through the generation of ultraviolet photoproducts upon exposure of phaeomelanin, results in the transformation of dysplastic melanocytes (Yamada et al, 1989).

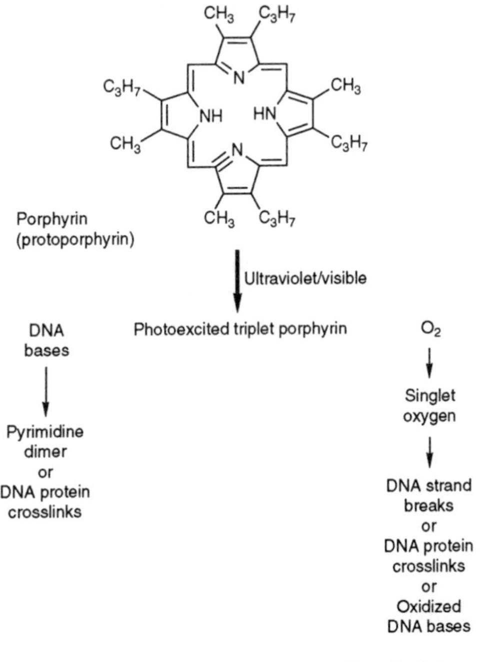

Figure 2. Porphyrin photochemistry.

2. Porphyrins

Porphyrins are important endogenous photosensitizers in photodermatoses such as porphyria cutanea tarda, porphyria variegata and erythropoietic protoporphyria. The mechanisms of cutaneous photosensitization by porphyrins may be mediated in type I and type II photoreactions. Skin damage in mice induced by porphyrin photosensitization occurs under aerobic but not under anoxic conditions (Gomer and Razum, 1984), indicating involvement of ROS in the mechanism of phototoxicity. Porphyrin-induced skin photosensitization reactions involve the production of singlet oxygen but not superoxide anion radical (Pathak and Carraro, 1987). β-Carotene prevented the lethal photosensitization induced in mice by haematoporphyrin (Mathews, 1964). The biological significance of oxidative injury as one of the key pathophysiological mechanisms in porphyrin photosensitivity is underlined by findings that the antioxidant β-carotene shows photoprotective effects in patients (Mathews-Roth et al, 1970, 1972, 1974; Mathews-Roth, 1987). Oral administration of β-carotene significantly increased the minimal erythema dose of solar radiation in humans (Mathews-Roth et al, 1972), and topical application of β-carotene significantly reduces UVB-induced lipid peroxidation in hairless mice (Pugliese and Lampley, 1985). Vitamin E prevents ultraviolet-induced in vitro lipid peroxidation of red blood cells from patients with erythropoietic protoporphyria (Goldstein and Harber, 1972).

3. *Flavins*

As mentioned above, photosensitization of flavins involves formation of ROS. Human bone marrow progenitor cells can be inactivated in vitro by photosensitizing the cells with riboflavin (Petkau et al, 1987), and flavins have been identified as endogenous photosensitizers in human diploid lung fibroblasts (Pereira et al, 1976). Mitochondrial energy metabolism (Aggarwal et al, 1978) and microsomal drug metabolism are inhibited by flavin-mediated photosensitization (Augusto and Packer, 1981). Riboflavin is present free and as a conjugate in skin. Although involvement of flavins in any known photodermatoses is not yet established, it is suggested that flavins are potential chromophores for evoking ultraviolet radiation effects in human skin (Joshi, personal communication).

B. Exogenous photosensitizers

A variety of drugs may cause photosensitivity reactions in skin. Different mechanisms are involved in the pathogenesis of photosensitivity reactions. Phototoxic reactions occur most frequently and are non-immunological. Photoallergic reactions involve photochemical modification of a haptene molecule to a reactive species that readily reacts with a macromolecule to yield a complete antigen. Several mechanisms are involved in photosensitivity due to drugs. They include formation of free radicals, activation of

ROS, photoactivation resulting in a toxic non-radical product and photoaddition of the drug to biological target molecules (Epstein and Wintroub, 1985; Epstein, 1977). A free radical mechanism is probably involved in the photosensitivity of various xenobiotics (Kochevar, 1987).

Table 1. Involvement of free radicals/reactive oxygen species in drug photosensitivity.

Acridine dyes	Martin and Logsdon (1987)
Benoxaprofene	Kochevar (1987)
Chlorpromazine	Chignellet et al (1985), Motten et al (1985)
Piroxicam	Western et al (1987)
Sulfonamides	Kochevar (1987)
Tartrazine	Merville et al (1984)
Tar products	Joshi and Pathak (1984), Kochevar et al (1982)
Tetrachlorosalicylanilides	Jenkins et al (1964)
Tetracyclines	Li et al (1987), Glette and Sandberg (1986), Shea et al (1986)
Thiazides	Hayakawa et al (1987), Matsuo et al (1986)

C. Photochemotherapy

In photochemotherapy, ultraviolet or visible irradiation is used in combination with a drug, administered either topically or systemically. Epidermodermal lymphocytes and granulocytes are considered to be target cells in photo- and photochemotherapy of various dermatological disorders. Furthermore, induction of normal protective mechanisms, alteration of metabolites and killing of microorganisms are possible mechanisms in therapeutic photodermatology (Parrish, 1981).

1. *8-Methoxypsoralen*

Psoralens are widely used in dermatological photochemotherapy of cutaneous disorders like T-cell lymphomas, mycosis fungoides, lichen ruber, psoriasis, vitiligo, urticaria pigmentosa, pityriasis lichenoides and polymorphous light eruption. The major photobiological effects of psoralens were thought to be mediated through UVA-induced stable covalent photoaddition of psoralens to chromosomal DNA (reaction type I) (Song and Tapley, 1979). Although in biological systems photosensitization with psoralen was oxygen-independent (Oginsky et al, 1959; Musajo and Rodighiero, 1972), it was reported that free radical formation may be involved in psoralen photochemistry (Pathak et al, 1961). It is now believed that psoralens also undergo type II reactions in vivo, in which the photoexcited psoralen reacts with oxygen and forms ROS, e.g. singlet oxygen and superoxide anion (Joshi and Pathak, 1984). The cell-killing effect of UVA-irradiated 8-methoxypsoralen is 2.5-fold increased in the presence of deuterated water, indicating involvement of singlet oxygen (Meffert et al, 1982). Skin photosensitization by psoralen ultraviolet A (PUVA) may be mediated by generation of singlet oxygen, superoxide anion radical and hydroxyl radicals (Pathak and Carraro, 1987), and may lead to lipid peroxidation. Indeed, PUVA therapy of guinea-pigs increases the formation of lipid peroxidation products in the skin and in serum; the changes are reversible after about four weeks (Akimov et al, 1984).

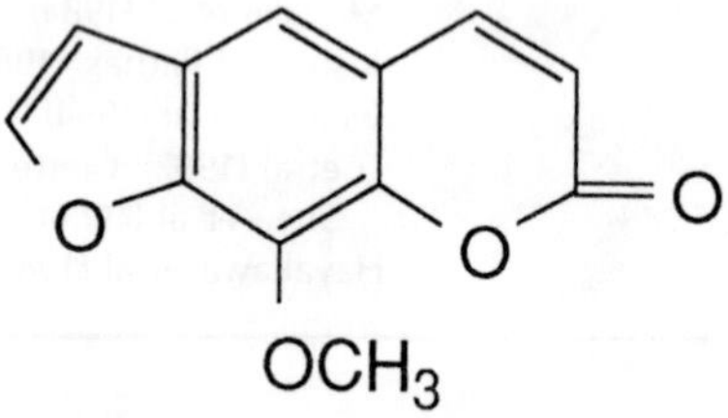

O O O OCH_3

Figure 3. 8-Methoxypsoralene.

2. *Haematoporphyrin*

Haematoporphyrin derivatives are used in the treatment of malignant cutaneous tumours (Dougherty, 1981). This new technique is based upon the tumour-localizing ability of haematoporphyrins and their photodynamic action. Exposure to activating visible light results in a relatively specific photosensitization of malignant tissues. Patients with psoriasis have also been successfully treated by topical application of haematoporphyrin derivatives and visible/UVA irradiation (McCullough et al, 1983; Meffert et al, 1989). Recently, successful therapy of psoriasis, unresponsive to other forms of therapy, was reported by systemic administration of tin-protoporphyrin and UVA irradiation (Emtestam et al, 1989). Light activation of haematoporphyrins results in generation of singlet oxygen (Keene et al, 1986; Blum and Fridovich, 1985), the apparent cytotoxic agent causing injury to tumour cells (Dougherty, 1986). Another reactive species involved in haematoporphyrin photochemotherapy could be thiyl radicals. Thiyl radicals were detected during light activation of haematoporphyrin derivatives, and singlet oxygen and hydroxyl radicals were indicated to be the initiating species (Buettner, 1984).

III. PHOTOOXIDATIVE SKIN DAMAGE

The reaction pattern and the time-course of acute ultraviolet photodamage, e.g. skin inflammation are wavelength-dependent. This may be due to differences in ultraviolet penetration into skin layers (UVA > UVB > UVC) and specific reaction cascades initiated by a specific wavelength band. Depending on the radiation dose, UVA and UVB can reach the dermis and the dermal vasculature. The inflammatory response to UVA, UVB and UVC irradiation may occur by separate molecular pathways differing with respect to inflammatory mediators and oxygen dependence. Prostaglandins are the major inflammatory mediators of UVB erythema but play a minor role in UVC and in UVA erythema. Major skin cell types, such as keratinocytes, Langerhans cells and fibroblasts, are damaged by UVB light exposure, and reactive oxidants generated by ultraviolet radiation are believed to participate in keratinocyte (Miyachi et al, 1983; Danno and Horio, 1987) and Langerhans cell photoinjury (Horio and Okamoto, 1987).

A. Lipid peroxidation

Irradition of skin with UVA and UVB can induce the formation of lipid peroxidation products. UVA light is capable of peroxidizing human skin surface lipids (Nazarro-Porro et al, 1986), and UVA irradiation of guinea-pigs significantly enhances lipid peroxidation in the skin (Akimov, 1987). Irradiation of extracted human skin surface lipids with sunlight or UVB causes a strong decline in desaturation of fatty acids (Horacek and Cernikova-Brunn, 1961). Exposure of human skin to an artificial ultraviolet light source (mercury lamp) results in a significant increase in lipid peroxidation products (Meffert and Reich, 1969). Products of lipid peroxidation (TBARS) are elevated in chronically sun-exposed human skin (Niwa et al, 1987).

1. *Time-course of ultraviolet lipid peroxidation in skin*

Analysis of time-course of lipid peroxidation in skin after UVB irradiation reveals that irradiation of hairless mice results in a significant elevation of lipid peroxidation products (TBARS) 3 h after irradiation. Immediately after ultraviolet exposure, accumulation of lipid peroxidation products in the epidermis is normal (Ogura et al, 1987); similar findings using UVB light were obtained by other authors (Maisuradze et al, 1987; Fuchs et al, 1989b). In skin of hairless mice, TBARS are detected in the epidermis after irradiation with UVB with a maximum at 9 h after irradiation, returning to preirradiation values after about 20 h. There is a nonlinear accumulation of TBARS with radiation dose, the maximal response occurring at 200 mJ cm^{-2}; at higher energy doses TBARS content decreases (Pugliese and Lampley, 1985).

2. Lipid peroxidation and ultraviolet-induced skin inflammation

The cytotoxic and proinflammatory potency of lipid peroxidation products are suggested to play an important role in UVB-induced skin inflammation (Meffert and Lohrisch, 1971; Ohsawa et al, 1984). Lipid peroxidation products (Ohsawa et al, 1984; Tanaka, 1979; Ogura, 1981) exert acute inflammatory effects in mammalian skin. In mice, subcutaneous administration of an aqueous extract of ultraviolet-irradiated linoleic acid induces necrosis of the epidermis and dermis followed by inflammatory responses and disruption of collagen and elastic fibres (Waravdekar et al, 1965). Hydroperoxides are important regulators of cyclooxygenase and lipoxygenase activity. Modulation of the lipid peroxide baseline levels will influence prostaglandin and leukotriene metabolism (Lands, 1985). Increased steady-state concentrations of lipid hydroperoxides in skin may stimulate cyclooxygenase activity and result in formation of proinflammatory prostaglandins and leukotrienes. Prostaglandins of the E series appear to be important mediators of UVB-induced skin inflammation in the early phase (<24 h) (Eaglstein et al, 1979). It is not known, however, whether ultraviolet light-induced lipid peroxidation is the cause of skin inflammation or a consequence of it. Careful kinetic studies are required to solve this problem.

3. Lipid peroxidation and photocarcinogenesis

It was suggested that lipid peroxidation may play a role in at least a part of the photocarcinogenic response in hairless mice. The level of dietary lipid intake and the degree of lipid saturation have pronounced effects on photoinduced skin cancer, with increasing levels of unsaturated fat intake enhancing cancer expression. The level of intake of these lipids is also manifested in the level of epidermal lipid peroxidation (Black et al, 1985).

B. Nucleic acid damage

DNA damage in skin has been extensively studied in vitro and in vivo using ultraviolet irradiation as the damaging agent (see Chapter 2). Far (200–290 nm = UVC) and near (290–400 nm = UVB + UVA) ultraviolet irradiation can cause DNA damage in skin. At 260 nm the principal photoproducts in most types of DNA are cyclobutyl pyrimidine dimers. Photodimerization of pyrimidine residues in DNA has been causally related to photocarcinogenesis (Jung, 1978; Doniger et al, 1981; Sutherland et al, 1980). At longer wavelengths it is likely that thymine dimers may not be the most biologically important lesions. In addition to pyrimidine dimers, pyrimidine–cytosine adducts are formed by UVB irradiation (280–320 nm), and non-dimer adducts (DNA–protein crosslinks) by UVA (320–400 nm), respectively.

1. *UVA damage*

UVA radiation can be lethal and mutagenic in mammalian cells (Wells and Han, 1984; Tyrell et al, 1984). UVA-induced single-strand breaks in fibroblasts are inhibited by catalase but cell lethality is not decreased by the antioxidant enzyme (Roza et al, 1985). UVA-induced photoproducts are cyclobutyl pyrimidine dimers, single-strand breaks and DNA–protein crosslinks. Cyclobutyl pyrimidine dimers are significantly induced in human skin by UVA irradiation at an erythemogenic dose (Freeman et al, 1987).

2. *UVB damage*

Cyclobutyl pyrimidine dimers are detected in the DNA of human skin following in vivo irradiation with suberythemal doses of UVB light (Sutherland et al, 1980). Dimer formation increases with increasing UVB dose and correlates with the erythemal response. Increased dimer formation is found in individuals with high UVB sensitivity (Freeman et al, 1986). The dimer photoproducts in human skin produced by simulated solar irradiation (Bruze et al, 1989) or UVB (Reusch et al, 1988) are rapidly removed, with approximately 30% remaining after 4 h, and 10% after 24 h, indicating efficient secondary antioxidant nucleic acid repair in human skin.

3. *Oxidative DNA damage*

Although the DNA molecule (maximal absorbance at 260 nm) has been identified as the chromophore for mutagenic and lethal effects of far-ultraviolet irradiation, it absorbs only 0.1% at 320 nm of that which is absorbed at 260 nm. Therefore, it was suggested that while DNA is not the critical chromophore for near-ultraviolet irradiation, it is the main target of it (Eisenstark, 1987). Several studies have shown involvement of ROS in DNA damage (Nishida et al, 1981; Breimer and Lindahl, 1985; Sestili et al, 1986). DNA transformation and mutagenesis occur more readily in the presence of oxygen (Cabrera-Juarez, 1981; Nishida et al, 1981), indicating involvement of ROS. Free radical DNA damage products are thymine glycol, thymidine glycol, 5-hydroxymethyluracil and 8-hydroxyguanine. These products can be detected in human urine and have the potential for screening of oxidative damage in vivo. Guanine bases can be inactivated by ultraviolet in a photosensitized reaction involving singlet oxygen, forming guanidine and trioxoimidazole.

It was suggested that an endogenous triplet sensitizer may mediate cyclobutyl pyrimidine dimer formation in the near-ultraviolet (Peak et al, 1984). However, it is not known whether dimer formation in skin results from a sensitized reaction or is a direct near-ultraviolet effect. Near-ultraviolet also has some mutagenic and lethal effects, even under anaerobic conditions (Peak et al, 1983). This shows that near-ultraviolet may also have a direct action on DNA.

IV. SKIN DISEASES

Ultraviolet and visible light can cause quantitatively and qualitatively abnormal reactions in susceptible patients. Skin diseases with abnormal reactions to light are categorized into disorders primarily caused by ultraviolet light (photodermatoses) and skin diseases precipitated and/or exacerbated by light. Photodermatoses can be classified into idiopathic diseases (unknown aetiology), metabolic and genetic diseases, and neoplastic and degenerative diseases. Skin diseases which are precipitated or exacerbated by ultraviolet light are very heterogeneous, and individual susceptibility shows large variations. Involvement of reactive oxidants in photooxidative stress is reported in a group of diseases with increased cellular sensitivity of DNA to ultraviolet light, in patients with lupus erythematosus and in photocarcinogenesis. The involvement of reactive oxidants in skin disease with abnormal reactions to light is speculative in all cases. Indirect evidence mainly stems from in vitro demonstrations which show that reactive oxidants and free radical species are involved in important mechanisms relevant to the specific disease. In several cases reaction products of reactive oxidants are increased in diseased skin, and skin antioxidant activity is found to be impaired. Finally, clinical symptoms can be ameliorated by antioxidant treatment of patients.

A. Diseases with increased cellular susceptibility

A group of heritable diseases can be linked with increased cellular sensitivity of DNA to several physicochemical agents. The clinical picture is distinct for each condition, but the diseases often overlap with respect to some of the following clinical abnormalities: high rate of malignancy, premature aging, skin photosensitivity, hyperpigmentation, teleangiectasia, atrophy and neurological dysfunctions. The underlying mechanism of increased cellular sensitivity to ultraviolet irradiation, γ-irradiation and increased rates of spontaneous chromosome breaks is not known in most cases, and participation of ROS is suspected in some cases. The clinically heterogeneous diseases comprise xeroderma pigmentosum. Blooms syndrome, Cockayne's syndrome, Fanconi's anaemia, ataxia teleangiectasia, dyskeratosis congenita and others. Although it is not known how all of the clinical features of these diseases are explained by increased cellular sensitivity of DNA, it is evident that this dysfunction is of importance for the pathophysiology of the specific syndrome.

1. *Xeroderma pigmentosum*

Xeroderma pigmentosum is an autosomal recessive disease, characterized by increased sun sensitivity, freckles and squamous cell carcinoma in sun-exposed areas. There is cellular hypersensitivity to ultraviolet irradiation and to certain chemicals (4-nitroquinoline-1-oxide, 8-methoxypsoralen + UVA, nitrogen mustard) in association with abnormal DNA repair. Genetic heterogeneity has been demonstrated among xeroderma pigmentosum patients and seven complementation groups of excision repair-defective cells have now been identified. Cultured fibroblasts of xeroderma pigmentosum patients have a markedly low catalase (Vuillaume et al, 1986) and superoxide dismutase (SOD) activity (Nishigori et al, 1989). The decrease in catalase and SOD activity in xeroderma pigmentosum cells could be an additive effect for inducing the carcinogenic process, particularly for xeroderma pigmentosum patients with normal excision repair (Vuillaume et al, 1986).

2. *Bloom's syndrome*

Bloom's syndrome is an autosomal recessive disease characterized clinically by growth retardation, immunodeficiency, solar sensitivity, teleangiectatic lesions in sun-exposed skin areas and proneness to the development of cancer. Cells from these patients are not unusually sensitive to ultraviolet. Clastogenic activity was detected in a concentrated ultrafiltrate of culture media from fibroblasts of patients with Bloom's syndrome (Emerit and Cerutti, 1981). Fibroblasts from patients with Bloom's syndrome have significantly lowered catalase activity (Vuillaume et al, 1986). The working hypothesis was proposed that Bloom's syndrome cells may be deficient in the detoxification of ROS and it is these species that lead to the formation of clastogenic components. Another hypothesis suggests that cells from Bloom's syndrome patients suffer from an elevated endogenous generation of ROS (Poote, personal communication).

B. Lupus erythematosus

Lupus erythematosus is an autoimmune disease of unknown origin, characterized by major alterations of both humoral and cellular immunity. The disease can involve multiple internal organ systems (systemic lupus erythematosus, SLE) or the skin only (chronic discoid lupus erythematosus, CDLE). There are several studies indicating involvement of reactive oxidants in the pathogenesis of lupus erythematosus. Photosensitivity is a prominent feature in patients with lupus erythematosus. UVB light may also be effective in causing photosensitivity in patients. Tocopherol provides significant systemic photoprotection in patients with cutaneous lupus erythematosus (Jordan and Wulf, 1950). The lymphocytes of SLE patients are sensitive to near-ultraviolet light, and this photosensitivity seems to be related to the presence of a chromosome-damaging agent (clastogenic factor) in these cells.

1. Clastogenic factors in lupus erythematosus

SLE patients have increased numbers of chromosome breaks (Tuschl et al, 1984), which are mediated by a serum factor. Lymphocytes of healthy subjects exposed to the factor also become sensitive to light of the same wavelengths (Emerit and Michelson, 1981). Generation of ROS in the medium of lymphocyte cultures results in production of chromosome breakage factors (clastogenic material) which induce chromosome breaks and sister chromatid exchanges (Emerit, 1982). Clastogenic material may be composed of lipid peroxidation products, such as TBARS, hydroxyalkenals, and conjugated dienes (Emerit et al, 1985; Khan and Emerit, 1985). Chromosome breakage in lymphocytes is inhibited by exogenous SOD in vitro and is also diminished by injection of SOD in lupus erythematosus patients (Michelson, 1986) and New Zealand Black (NZB) mice (Emerit et al, 1980), an animal model for systemic lupus erythematosus. The presence of clastogenic products in plasma is not specific for lupus erythematosus. Clastogenic products were first described in plasma from patients irradiated with ionizing radiation and are also found in patients with spontaneous chromosomal instability, as well as chronic inflammatory and autoimmune diseases such as progressive systemic sclerosis, dermatomyositis, periarteriitis nodosa, rheumatoid arthritis and Crohn's disease. The fact that patients with progressive systemic sclerosis, Crohn's disease and rheumatoid arthritis have no significantly increased photosensitivity, argues against a common mechanism of clastogenic factors-mediated photosensitivity.

C. Photocarcinogenesis

Although the involvement of sunlight in the induction of squamous cell carcinoma and basalioma is established, its participating role in the pathogenesis of melanoma is controversial, with the exception of lentigo maligna melanoma. The more common types of melanoma, superficial spreading melanoma and nodular melanoma, typically not located in sun-exposed skin areas, are not associated with solar skin damage around the tumour and do not occur more frequently among outdoor workers. However, melanoma incidence among white-skinned populations increases as their residence approaches the equator, and melanoma is very rare in dark-skinned populations.

The most effective wavelengths for ultraviolet carcinogenesis are in the UVB region. UVC is probably much less carcinogenic than UVB in vivo (Lill, 1983; De Gruijl et al, 1983), due to the strong absorption of UVC in the superficial skin layers. Skin tumours can be induced in hairless mice following multiple small exposures to UVB as well as by exposure to a large single fluence ($0.36\,J\,cm^{-2}$) of UVB radiation (Mathews-Roth, 1983). In hairless mice multiple small doses of UVB radiation are more effective in producing skin cancer than larger, less frequent ultraviolet doses. The rate of skin cancer induction in hairless mice by UVB light is enhanced by subsequent UVA irradiation (Staberg et al, 1983b). However, irradiation with multiple doses of combined UVB and UVA light (total fluence UVB $20\,J\,cm^{-2}$, total fluence UVA $60\,J\,cm^{-2}$) induces a slightly lower tumour incidence as compared to UVB alone (Staberg et al, 1983a). In another study (Willis et al, 1981) simultaneous UVB/UVA of hairless mice resulted in a significant increase in cancer induction.

These discrepant results may be due to time–dose relationships. Recently, it was reported that irradiation of skin of hairless mice with multiple doses of UVA irradiation from which the UVB light had been rigorously excluded did induce tumours (Strickland, 1986). However, the total UVA fluence applied in this study was very large ($2.5\,J\,cm^{-2}$ single exposure, $1170\,J\,cm^{-2}$ total exposure). Exposure of hairless mice to multiple small doses of the deeper-penetrating UVA irradiation with a total dose of $71\,J\,cm^{-2}$ (Staberg et al, 1983a,b) or a total dose of $40.5\,J\,cm^{-2}$ (Bech-Thomsen et al, 1988) respectively, did not result in an increased skin tumour incidence. UVA-induced skin tumours require orders of magnitude greater doses than tumour induction by UVB light.

Large fluences of UVA light are emitted by artificial suntanning devices. Since tanning requires much higher doses of UVA than UVB, the carcinogenic risks of tanning by UVA and UVB are reported to be of the same order of magnitude (Engel et al, 1988). Exposure of human skin to UVA light was considered to be completely harmless (Raab, 1979). Now it is evident that exposure to high-intensity UVA light sources or to solar UVA for prolonged time periods may significantly contribute to skin carcinogenesis. However, treatment of hairless mice with 42 J cm^{-2} UVA light before exposure to broad-band ultraviolet irradiation significantly delays skin tumour development (Bech-Thomsen et al, 1988). Pretreatment with UVA did not alter epidermal thickness or melanin content. UVA, in addition to its carcinogenic properties (photoaugmentation of UVB-induced skin tumours), may also act as an antipromoter in skin.

1. *Antioxidants in photocarcinogenesis*

Dietary antioxidants inhibit lipid peroxidation and photocarcinogenesis. It was suggested that the carcinogenic process in skin may involve free radical-mediated lipid peroxidation reactions (Black, 1987b). Antioxidants may play a role in both inhibition of the primary events involved in ultraviolet light-mediated tumour induction and subsequent development of precancerous lesions into tumours (Black and Chan, 1975). Cholesterol-α-oxide, a known carcinogen, is formed in skin of hairless mice upon ultraviolet irradiation and might be involved in the aetiology of ultraviolet-induced carcinogenesis (Black and Douglas, 1973). Systemic application of a mixture of α-tocopherol, glutathione and butylated hydroxytoluene (BHT) inhibits the photochemical conversion of skin cholesterol-α-oxide in hairless mice (Lo and Black, 1973). This suggests a possible prophylactic effect of systemic antioxidants on the formation of this carcinogen and the subsequent pathological conditions that may result from its formation.

V. ANTIOXIDANTS

A. Ultraviolet radiation effects on skin antioxidants

Low molecular weight lipophilic and hydrophilic antioxidants can be consumed during reactions with ROS if they are not regenerated (see Chapter 9), and antioxidant enzymes can be inactivated during oxidative exposure.

Catalase is photoinactivated by near-ultraviolet and visible light (Cheng and Packer, 1979; Freierabend and Engel, 1987; Mitchell and Anderson, 1965), SOD is readily inactivated by hydrogen peroxide (Bray and Cockle, 1974), and glutathione peroxidase is inhibited by superoxide anion radicals (Blum and Grossweiner, 1985). Very little is known about the acute and chronic responses of enzymatic and non-enzymatic antioxidants in skin after ultraviolet challenge. However, some studies indicate that after ultraviolet exposure the enzymatic and non-enzymatic antioxidant potential of skin decreases (Connor and Wheeler, 1987; Fuchs et al, 1989b).

1. *Enzymatic skin antioxidants*

Skin SOD activity is initially decreased after acute oxidant injury. A single exposure of mouse skin to UVB irradiation results in a significant decrease of SOD activity 24–48 h after exposure, returning to the normal level 72 h after irradiation. Three-hour postirradiation SOD activity is unaltered in skin (Miyachi et al, 1987). In guinea-pig skin no change in SOD activity was observed immediately after UVB exposure (Ogura et al, 1987). Immediately after irradiation of mouse skin with a single exposure of ultraviolet irradiation (>280 nm), a small but significant inhibition of glutathione reductase and catalase activity was observed; glutathione peroxidase and SOD were not affected (Fuchs et al, 1989b). Cu/Zn-SOD and Mn-SOD activities in sun-exposed and non-exposed skin of healthy individuals are not significantly different (Sugiura et al, 1985).

2. *Non-enzymatic skin antioxidants*

Tocopherol and ubiquinone are increased in chronically sun-exposed human skin (De Simone et al, 1987), but are decreased immediately after UVB irradiation in mouse skin (Fuchs et al, 1989b). In ultraviolet-irradiated skin of hairless mice the hydrophilic antioxidant glutathione decreases and concomitantly glutathione disulphide increases. A slight depletion of the total glutathione content was observed; the skin concentration of total ascorbic acid did not change. The ratio of dehydroascorbate to ascorbate was, however, not analysed in this study (Fuchs et al, 1989). Treatment of hairless mice with 8-methoxypsoralen and skin exposure to 5 J cm^{-2} UVA results in a depletion of cutaneous glutathione with a maximum decrease observed after 24–28 h (Wheeler et al, 1986). Irradiation of skin of hairless mice with 90 mJ cm^{-2} UVB results in a decrease in epidermal glutathione to 60% of its control level within 10 min and an increase in glutathione disulphide. This returns to control levels within 30 min (Connor and Wheeler, 1987).

3. One-electron reduction potential at the epidermal surface

UVB irradiation of skin of hairless mice inhibits the free radical scavenging activity at the epidermal surface, as analysed by scavenging of persistent piperidinoxyl-type nitroxide radical (Fuchs et al, 1989). In guinea-pig and human skin UVA and UVB inhibit nitroxide radical-scavenging activity at the epidermis. Ultraviolet-induced deactivation of thioredoxin reductase has been suggested as the cause of this inhibition (Schallreuter and Wood, 1989). Our results, however, show that thioredoxin reductase is not deactivated by UVA and UVB light and argue against thioredoxin reductase being the only mechanism in epidermis involved in reduction of nitroxides (Fuchs, 1988; Fuchs et al, 1989a). Photooxidation of epidermal ascorbate by ultraviolet irradiation (Buettner et al, 1987) could significantly contribute to loss of nitroxide radical-scavenging activity at the epidermal surface.

B. Photoprotection by antioxidants

Orally administered and topically applied antioxidants provide protection from ultraviolet irradiation-induced erythema and cytotoxicity. An important role in the development of UVB erythema is suggested to be due to photochemical reactions after irradiation, most probably the development of peroxidation processes (Potapenko et al, 1980, 1984). Agents capable of enhancing natural defence mechanisms, e.g. melanization and antiradical activity, are suggested to hold the greatest promise for future developments on systemic photoprotective agents (Black 1987b,c).

1. Tocopherol

Tocopherol, when applied on skin before and within 2 min of irradiation, greatly inhibits the erythemal response to UV light (280–365 nm). The antierythema efficiency of tocopherol applied 2 min after irradiation was not less than that observed with application before irradiation. This means that the antioxidant modulates the dark stages of erythema. The antioxidant treatment of the skin 5 h after irradiation or later has almost no influence on the erythema response (Potapenko et al, 1980). In contrast to inhibitors of cyclooxygenase such as indomethacin (Snyder, 1975; Kaidbey and Kurban, 1976), the local application of antioxidants such as tocopherol results in significant restriction of sunburn cell formation caused by UVB in human skin, when administered before irradiation (Msika et al, personal communication). In vitro, α-tocopherol inhibits ultraviolet light-induced peroxidation of squalene, one of the major skin surface lipid components (Ohsawa et al, 1984).

The photoprotective potency of the natural RRR-tocopherols in UVB-induced oedema of the hairless mouse (topical application of tocopherol prior to irradiation) is of the same order as their reactivity to peroxyl radicals: α-tocopherol (100%) > γ-tocopherol (72%) > β-tocopherol (42%) > δ-tocopherol (40%) (Potakar et al, personal communication). Topical application of β-carotene and tocopherol results in a significant decrease of TBARS after UVB exposure of hairless mice (Pugliese and Lampley, 1985). α-Tocopherol protects rabbit skin against PUVA erythema at a medium ultraviolet fluence of 3 MED (MED = minimal erythemal dose) only if applied before irradiation. If applied shortly after irradiation, its inhibitory effect is greatly diminished. α-Tocopherol application prior to PUVA treatment with a large ultraviolet fluence of 10 MED has virtually no inhibitory action (Potapenko et al, 1980). α-Tocopherol has a weak ultraviolet filter effect; however, it inhibits UVB-induced erythema irrespective of whether it is applied before or immediately after UVB irradiation. At high tocopherol concentrations on the skin surface ($>5 \times 10^{-9}$ mol cm^{-2}) the protective action is decreased, which is suggested to be due to prooxidant action at higher concentrations (Potapenko et al, 1984). However, a nonspecific, local irritating effect at high concentrations is an alternative explanation of this phenomenon.

2. *Carotenoids*

In animal models β-carotene and canthaxanthin have protected against ultraviolet-induced skin cancer as well as some chemically induced tumours. β-Carotene was successfully used in the treatment of patients with erythropoietic protoporphyria (Mathews-Roth et al, 1970; Baart De La Faille et al, 1972), porphyria cutanea tarda (Ippen, 1972), actinic reticuloid and solar urticaria (Kobza et al, 1973). Oral administration of β-carotene significantly increased the minimal erythema dose of solar radiation in humans (Mathews-Roth et al, 1972), and topical application of β-carotene significantly diminishes UVB-induced lipid peroxidation in hairless mice (Pugliese and Lampley, 1985). A photoprotective mechanism of β-carotene in skin involving scavenging of singlet oxygen was suggested. Although clinical effectiveness of β-carotene in the treatment of erythropoietic protoporphyria is suggested by several other case reports, controlled trials have not been performed (Pollitt, 1975).

3. *Dihydrolipoate*

The antioxidant dihydrolipoate (Scholich et al, 1989) provides local photoprotection in human skin when applied before or immediately after UVB irradiation; however, no photoprotection is observed when the drug is applied 4 h after irradiation (Fuchs and Milbradt, 1989).

4. *Superoxide dismutase*

Pretreatment of the epidermal surface of guinea-pigs before exposure to UVB irradiation with liposomal SOD prevents increases in skin lipid peroxidation and inhibits skin inflammation (Ogura et al, 1987). Exogenous SOD inhibits UVB phototoxic effects in cultured fibroblasts. Since intracellular penetration of non-liposomal SOD is slow, it is suggested that membrane-bound SOD inhibits phototoxic processes at the fibroblast plasma membranes (Emerit et al. 1981).

5. *Butylated-hydroxytoluene*

Oral administration of BHT to hairless mice resulted in a decreased light transmission of the stratum corneum of treated animals. Photoprotection is suggested to be related to changes in chemical properties of the stratum corneum rather than changes in physical parameters (Koone and Black, 1986). Structure–activity evaluations of chemically related phenols reveal that the antioxidant properties of BHT are not significant determinants of photoprotective activity (Black and Tigges, 1986). A large component of the photoprotective effect of BHT is thought to be alterations of the physicochemical properties of the stratum corneum, reducing the levels of ultraviolet irradiation reaching epidermal target sites (Black et al, 1984).

VI. CONCLUSIONS

There is little direct evidence that free radicals and reactive oxygen species are primarily involved in the pathophysiology of cutaneous photooxidative stress. However, most of the experimental data are circumstantial or indirect and based upon the effects which appear to be abolished by free radical scavengers and on the detection of free radical reaction products such as spin-trapped adducts, lipid peroxidation products and oxidative damage to DNA. This discrepancy is explained by the extreme difficulty in directly detecting low steady-state concentrations of highly reactive free radicals in biological material. In the future, improved analytical procedures may help to narrow the gap between direct and indirect experimental evidence.

Proctor and Reynolds (1984) suggested that a 'free radical disease' is associated with abnormal flux of free radicals or intermediates and that specific radical species or their characteristic reaction products must be demonstrated at the site of a lesion. In vitro demonstrations should show that free radical species are involved in important mechanisms relevant to the specific disease. Similar clinical symptoms should be produced by otherwise dissimilar aetiological agents which produce free radical species in common or which inhibit or deplete components of the natural antioxidant defence system. Furthermore, the pathogenesis of the disease should be influenced by administration of antioxidants or free radical quenchers.

In our analysis no photodermatological disorder qualifies, according to the criteria of Proctor and Reynolds (1984), as a 'free radical disease'. A variety of open clinical studies and case reports do, however, support the hypothesis that oxidative stress is a major significant determinant in the pathogenesis of certain photodermatoses and skin diseases triggered by solar irradiation. In contrast to distinct clinical skin diseases caused, initiated or exacerbated by ultraviolet irradiation, the role of ROS and other free radicals in the pathophysiological scenario is better characterized in photosensitivity reactions, photochemotherapy, and photocarcinogenesis.

REFERENCES

Aggarwal BB, Quintanilha AT, Cammack R & Packer L (1978) *Biochim. Biophys. Acta* **502:** 367–382.

Akimov VT (1987) *Vestn. Dermatol. Venerol.* **12:** 7–10.

Akimov VT, Lashmanova AP & Marzeeva GI (1984) *Vestn. Dermatol. Venerol.* **8:** 22–25.

Ames BN (1983) *Science* **221:** 1256–1264.

Augusto O & Packer L (1981) *Photochem. Photobiol.* **33:** 765–767.

Baart De La Faille H, Suurmond D, Went LN, Stevenick J & Van Schothorst AA (1972) *Dermatologica* **145:** 389–394.

Ballou D, Palmer G & Massey V (1969) *Biochem. Biophys. Res. Commun.* **36:** 898–904.

Bech-Thomsen N, Wulf H, Poulsen T & Lundgreen K (1988) *Arch. Dermatol.* **124:** 1215–1218.

Black HS (1987a) *Photochem. Photobiol.* **46:** 213–221.

Black HS (1987b) *Photodermatology* **4:** 187–195.

Black HS (1987c) *Fed. Proc.* **46:** 1901–1905.

Black HS & Chan JT (1975) *J. Invest. Dermatol.* **65:** 412–414.

Black HS & Douglas DR (1973) *Cancer Res.* **33:** 2094–2095.

Black HS & Tigges J (1986) *Photochem. Photobiol.* **43:** 403–408.

Black HS, Lenger W, Gerguis J & McCann V (1984) *Photochem. Photobiol.* **40:** 69–75.

Black HS, Lenger WA, Gerguis J & Thornby JI (1985) *Cancer Res.* **45:** 6254–6259.

Blois MS, Zahlan AB & Maling JE (1964) *Biophys. J.* **4:** 471–490.

Blum A & Grossweiner LI (1985) *Photochem. Photobiol.* **41:** 27–32.

Blum J & Fridovich I (1985) *Arch. Biochem. Biophys.* **240:** 500–508.
Bray RC & Cockle SA (1974) *Biochem. J.* **139:** 43–48.
Breimer LH & Lindahl T (1985) *Mutat. Res.* **150:** 85–89.
Bruze M, Emmett EA, Creasey J & Strickland PT (1989) *J. Invest. Dermatol.* **93:** 341–344.
Buettner GR (1984) *FEBS Lett.* **177:** 295–299.
Buettner GR, Motten AG, Hall RD & Chignell CF (1987) *Photochem. Photobiol.* **46:** 161–164.
Cabrera-Juarez E (1981) *Mutat. Res.* **83:** 301–306.
Chedekel MR & Zeise L (1988) *Lipids* **23:** 587–591.
Cheng LY & Packer L (1979) *FEBS Lett.* **97:** 124–128.
Chignell CF, Motten AG & Buettner GR (1985) *Environ. Health Perspect.* **64:** 103–110.
Commoner B, Townsend J & Pake GW (1954) *Nature (Lond.)* **174:** 689–691.
Connor MJ & Wheeler LA (1987) *Photochem. Photobiol.* **47:** 239–245.
Cunningham ML, Krinsky NI, Giovanazzi SM & Peak MJ (1985) *J. Free Radicals Biol. Med.* **1:** 381–385.
Danno K & Horio T (1987) *Photochem. Photobiol.* **43:** 683–690.
De Gruijl FR, Van der Meer JB & Van der Leun JC (1983) *Photochem. Photobiol.* **37:** 53–62.
De Simone C, Rusciani L, Venier A et al (1987) *J. Invest. Dermatol.* **89:** 317.
Doniger J, Jacobson ED, Krell K & DiPaolo JA (1981) *Proc. Natl. Acad. Sci. USA* **78:** 2378–2382.
Dougherty TJ (1981) *J. Invest. Dermatol.* **77:** 122–124.
Dougherty TJ (1986) *Semin. Surg. Oncol.* **2:** 24–37.
Eaglstein WH, Sakai M & Mizuno N (1979) *J. Invest. Dermatol.* **72:** 59–63.
Eisenstark A (1987) *Environ. Mol. Mutagen.* **10:** 317–337.
Emerit I (1982) *Prog. Mutat. Res.* **4:** 61–72.
Emerit I & Cerutti P (1981) *Proc. Natl. Acad. Sci. USA* **78:** 1866–1872.
Emerit I & Michelson AM (1980) *Acta Physiol. Scand.* **492 (supplement):** 59–65.
Emerit I & Michelson AM (1981) *Proc. Natl. Acad. Sci. USA* **78:** 2537–2540.
Emerit I, Feingold J, Levy A, Martin E & Housset E (1980) *J. Natl. Cancer Inst.* **64:** 513–516.
Emerit I, Khan SH & Cerutti PA (1985) *J. Free Radicals Biol. Med.* **1:** 51–57.
Emerit I, Michelson AM, Martin E & Emerit J (1981) *Dermatologica* **163:** 295–299.
Emtestam L, Berglund L, Angelin L, Drummond GS & Kappas A (1989) *Lancet* June 3. 1231–1233.
Engel A, Johnson ML & Haynes SG (1988) *Arch. Dermatol.* **124:** 72–79.
Epstein JH (1977) In Pryor WA (ed.) *Free Radicals in Biology*, pp 219–249. New York: Academic Press.
Epstein JH & Wintroub BU (1985) *Drugs* **30:** 42–57.
Epstein S (1939) *J. Invest. Dermatol.* **2:** 43–51.
Freeman SE, Gange RW, Matzinger EA and Sutherland BM (1986) *J. Invest. Dermatol.* **86:** 34–36.
Freeman SE, Gange RW, Sutherland JC, Matzinger EA and Sutherland BM (1987) *J. Invest. Dermatol.* **88:** 430–433.
Freierabend J and Engel S (1987) *Arch. Biochem. Biophys.* **251:** 567–576.
Fuchs J (1988) *J. Invest. Dermatol.* **91:** 92–93.
Fuchs J & Milbradt R (1989) Thioctsäure. In Borbe HO & Ulrich H (eds) *Neue biochemische, pharmakologische und klinische Erkenntnisse zur Thioctsäure*, pp 299–303. Frankfurt: pmi Verlag.

Fuchs J & Packer L (1989) *J. Invest. Dermatol.* **92:** 677–682.
Fuchs J, Mehlhorn RJ and Packer L (1989a) *J. Invest. Dermatol.* **93:** 633–640.
Fuchs J, Huflejt M, Rothfuss L, Carcamero G and Packer L (1989b) *J. Invest. Dermatol.* **93:** 769–773.
Glette J & Sandberg S (1986) *Biochem. Pharmacol.* **35:** 2883–2885.
Goldstein BD & Harber LC (1972) *J. Clin. Invest.* **51:** 892–902.
Gomer CJ & Razum NJ (1984) *Photochem. Photobiol.* **40:** 435–439.
Hayakawa K, Matsuo I, Fujita H & Ohkido M (1987) In Hayaishi O, Imamura S & Miyachi Y (eds) *The Biological Role of Reactive Oxygen Species in Skin*, pp 101–106. New York: Elsevier.
Horacek J & Cernikova-Brünn M (1961) *Arch. Klin. Exp. Dermatol.* **213:** 124–129.
Horio T & Okamoto H (1987) *J. Invest. Dermatol.* **88:** 699–702.
Ippen H (1972) *Hautarzt* **23:** 47–50.
Jenkins FP, Welti D & Baines D (1964) *Nature (Lond.)* **201:** 827–828.
Jordan P & Wulf K (1950) *Hautarzt* **1:** 233–234.
Joshi PC & Pathak MA (1984) *J. Invest. Dermatol.* **82:** 67–73.
Joshi PC, Carraro C & Pathak MA (1987) *Biochem. Biophys. Res. Commun.* **142:** 265–274.
Jung EG (1978) *Bull. Cancer (Paris)* **65:** 315–322.
Kaidbey KH & Kurban AK (1976) *J. Invest. Dermatol.* **66:** 153–156.
Keene JP, Kessel D, Land EJ, Redmond RW & Truscott TG (1986) *Photochem. Photobiol.* **43:** 117–120.
Khan SH & Emerit I (1985) *J. Free Radicals Biol. Med.* **1:** 443–449.
Kobza A, Ramsay CA & Magnus IA (1973) *Br. J. Dermatol.* **88:** 157–161.
Kochevar IE (1987) *Photochem. Photobiol.* **46:** 891–895.
Kochevar IE, Armstrong RB, Einbinder J, Walther RR & Harber LC (1982) *Photochem. Photobiol.* **36:** 65–69.
Koone MD & Black HS (1986) *J. Invest. Dermatol.* **87:** 343–347.
Lands WEM (1985) *J. Free Radicals Biol. Med.* **1:** 97–101.
Li ASW, Chignell CF & Hall RD (1987) *Photochem. Photobiol.* **46:** 379–382.
Lill PH (1983) *J. Invest. Dermatol.* **81:** 342–346.
Lo WB & Black HS (1973) *Nature (Lond.)* **246:** 489–491.
Maisuradze VN, Platanov AG, Gudz TI, Goncharenko EN & Kudriashov IUB (1987) *Biol. Nauki.* **5:** 31–35.
Martin JP & Logsdon N (1987) *Photochem. Photobiol.* **46:** 45–53.
Mason HS, Ingram DJE & Allen B (1960) *Arch. Biochem. Biophys.* **86:** 225–230.
Massey V, Strickland S, Mayhew SG et al (1969) *Biochem. Biophys. Res. Commun.* **36:** 891–897.
Mathews MM (1964) *Nature (Lond.)* **203:** 1092.
Mathews-Roth M (1983) *Photochem. Photobiol.* **37:** 509–511.
Mathews-Roth MM (1987) *Fed. Proc.* **46:** 1890–1893.
Mathews-Roth MM, Pathak MA, Fitzpatrick TB, Harber LC & Kass EH (1970) *N. Engl. J. Med.* **282:** 1231–1234.
Mathews-Roth MM, Pathak MA, Parrish J et al (1972) *J. Invest. Dermatol.* **59:** 349–353.
Mathews-Roth MM, Pathak A, Fitzpatrick TB, Harber CC & Kass EH (1974) *JAMA* **228:** 1004–1008.
Matsuo I, Fujita H, Hayakawa K & Ohkido M (1986) *J. Invest. Dermatol.* **87:** 637–641.
McCormick JP, Fisher JR, Pachlatko JP & Eisenstark A (1976) *Science* **191:** 468–469.

McCullough J, Weinstein G & Eaglstein WH (1983) *J. Invest. Dermatol.* **81:** 928–933.
Meffert H & Lohrisch I (1971) *Dermatol. Monatsschr.* **157:** 793–801.
Meffert H & Reich P (1969) *Dermatol. Monatsschr.* **155:** 948–954.
Meffert H, Böhm F, Röder B & Sönnichsen N (1982) *Dermatol. Monatsschr.* **168:** 387–393.
Meffert H, Pres H, Diezel W & Sönnichsen N (1989) *Dermatol. Monatsschr.* **175:** 28–34.
Merville MP, Decuyper J, Lopez M, Piette J & Van de Vorst A (1984) *Photochem. Photobiol.* **40:** 221–226.
Michelson AM (1986) In Johnson JE, Walford R, Harman D & Miquel J (eds) *Free Radicals, Aging and Degenerative Diseases*, pp 263–291. New York: Alan R. Liss Inc.
Mitchell RL & Anderson IC (1965) *Science* **150:** 74.
Miyachi Y, Horio T & Imamura S (1983) *Clin. Exp. Dermatol.* **8:** 305–310.
Miyachi Y, Imamura S, Niwa Y, Tokura Y & Takigawa M (1986) *J. Invest. Dermatol.* **86:** 26–28.
Miyachi Y, Imamura S & Niwa Y (1987) *J. Invest. Dermatol.* **89:** 111–112.
Motten AG, Buettner GR & Chignell CF (1985) *Photochem. Photobiol.* **42:** 9–15.
Musajo L & Rodighiero G (1972) In Giese A (ed.) *Photophysiology*, Vol. 7, pp 115–147. London: Academic Press.
Nazzaro-Porro M, Picardo M, Finotti E & Passi S (1986) *J. Invest. Dermatol.* **89:** 320.
Nishida KI, Takagi M & Yano K (1981) *J. Gen. Microbiol.* **27:** 447–485.
Nishigori C, Miyachi Y, Imamura S & Takebe H (1989) *J. Invest. Dermatol.* **92:** 491.
Niwa Y, Kanoh T, Sakane T, Soh H, Kawai S & Miyachi Y (1987) *J. Clin. Biochem. Nutr.* **2:** 245–251.
Norins AL (1962) *J. Invest. Dermatol.* **39:** 445–448.
Oginsky EL, Green GS, Griffith DG & Fowlks WL (1959) *J. Bacteriol.* **78:** 821–833.
Ogura R (1981) *Kurume Med. J.* **45:** 279–301.
Ogura R, Sugiyama M, Sakanashi T & Hidaka T (1987) In Hayaishi O, Imamura S & Miyachi Y (eds) *The Biological Role of Reactive Oxygen Species in Skin*, pp 55–63. New York: Elsevier.
Ohsawa K, Watanabe T, Matsukawa R, Yoshimura Y & Imaeda K (1984) *J. Toxicol. Sci.* **9:** 151–159.
Parrish JA (1981) *J. Invest. Dermatol.* **77:** 167–171.
Pathak MA & Carraro C (1987) In Hayaishi O, Imamura S & Miyachi Y (eds) *The Biological Role of Reactive Oxygen Species in Skin*, pp 75–94. New York: Elsevier.
Pathak MA & Stratton K (1968) *Arch. Biochem. Biophys.* **123:** 468–476.
Pathak MA, Allen B, Ingram DIE & Fellman JH (1961) *Biochim. Biophys. Acta.* **54:** 506–515.
Pathak MA, Kramer DM & Fitzpatrick TB (1974) In Fitzpatrick TB (ed.) *Sunlight and Man*, pp 335–368. Tokyo: University of Tokyo Press.
Peak JG, Peak MJ & Tuveson RW (1983) *Photochem. Photobiol.* **38:** 541–543.
Peak MJ, Peak JG, Moehring MP & Webb RB (1984) *Photochem. Photobiol.* **40:** 613–620.
Pereira OM, Smith JR & Packer L (1976) *Photochem. Photobiol.* **24:** 237–242.
Petkau A, Chelack WS, Palamar E & Gerrard J (1987) *Res. Commun. Chem. Pathol. Pharmacol.* **57:** 107–116.
Pollitt N (1975) *Br. J. Dermatol.* **93:** 721–724.

Potapenko AY, Abijev GA & Pliquett F (1980) *Bull. Exp. Biol. Med.* **89:** 611–615.
Potapenko AY, Abijev GA, Pitsov MY et al (1984) *Arch. Dermatol. Res.* **276:** 12–16.
Proctor PH & Reynolds ES (1984) *Physiol. Chem. Phys. Med. NMR* **16:** 175–195.
Pugliese PT & Lampley CB (1985) *J. Appl. Cosmetol.* **3:** 129–138.
Raab W (1979) *Z. Hautkr.* **55:** 497–513.
Rampen FHJ & Fleuren E (1987) *Med. Hypotheses* **22:** 341–346.
Reusch MK, Meager K, Leadon SA & Hanawalt PC (1988) *J. Invest. Dermatol.* **91:** 349–352.
Riley PA (1970) *J. Pathol.* **101:** 163–169.
Roza LGP, Van der Schans P & Lohman PH (1985) *Mutat. Res.* **146:** 89–98.
Sarna T & Sealy RC (1984a) *Photochem. Photobiol.* **39:** 69–74.
Sarna T & Sealy RC (1984b) *Arch. Biochem. Biophys.* **232:** 574–578.
Sarna T, Korytowski W & Sealy RC (1985a) *Arch. Biochem. Biophys.* **239:** 226–233.
Sarna T, Menon IA & Sealy RC (1985b) *Photochem. Photobiol.* **42**: 529–532.
Sarna T, Pilas B, Land EJ & Truscott TGI (1986) *Biochim. Biophys. Acta* **883:** 162–167.
Schallreuter KU & Wood JM (1989) *Free Radicals Biol. Med.* **6:** 519–532.
Schreiber J, Foureman GL, Hughes MF, Mason RP & Eling TE (1989) *J. Biol. Chem.* **264:** 7936–7943.
Scholich H, Murphy ME & Sies H (1989) *Biochem. Biophys. Acta.* **1101:** 256–261.
Schroot B & Brown C (1986) *Arzneimittelforsch./Drug Res.* **36:** 1253–1255.
Sestili P, Piedimonte G, Flaminio C & Cantoni O (1986) *Biochem. Int.* **12:** 493–501.
Shea CR, Wimberly J & Hasan T (1986) *J. Invest. Dermatol.* **87:** 338–342.
Snyder DS (1975) *J. Invest. Dermatol.* **64:** 322–325.
Song PA & Tapley JJ Jr (1979) *Photochem. Photobiol.* **29:** 1177–1197.
Staberg B, Wulf HC, Klemp P, Poulsen T & Brodthagen H (1983a) *J. Invest. Dermatol.* **81:** 517–519.
Staberg B, Wulf HC, Poulsen T, Klemp P & Brodthagen H (1983b) *Arch. Dermatol.* **119:** 641–643.
Straight RC & Spikes JD (1985) In Frimer AA (ed.) *Singlet Oxygen*, Vol. 4, pp 91–143. Boca Raton: CRC Press.
Strickland PT (1986) *J. Invest. Dermatol.* **87:** 272–275.
Sugiura K, Ueda H, Hirano K & Adachi T (1985) *Jpn. J. Dermatol.* **95:** 1541–1545.
Sutherland BM, Harber LC & Kochevar IE (1980) *Cancer Res.* **40:** 3181–3185.
Taffe BG, Takahashi N, Kensler TW & Mason RP (1987) *J. Biol. Chem.* **262:** 12143–12149.
Tanaka T (1979) *Vitamins (Kyoto)* **53:** 577–586.
Tomita Y, Hariu A, Kato C & Seiji M (1984) *J. Invest. Dermatol.* **82:** 573–576.
Torinuki W, Miura T & Seiji M (1980) *Br. J. Dermatol.* **102:** 17–27.
Tuschl H, Kovac R, Wolf A & Smolen JS (1984) *Mutat. Res.* **128:** 167–172.
Tyrrell RM, Werfelli P & Moraes EC (1984) *Photochem. Photobiol.* **39:** 183–189.
Vuillaume M, Calvayrac R, Best-Belpomme M et al (1986) *Cancer Res.* **46:** 538–544.
Waravdekar VS, Saslaw LD, Jones WA & Kuhns JG (1965) *Arch. Pathol.* **80:** 91–95.
Wells RL & Han A (1984) *Mutat. Res.* **129:** 251–258.
Western A, Van Camp JR, Bengasson R, Land EJ & Kochevar IE (1987) *Photochem. Photobiol.* **46:** 469–475.
Wheeler LA, Aswad A, Connor MJ & Lowe M (1986) *J. Invest. Dermatol.* **87:** 658–662.
Willis I, Menter JM & Whyte J (1991) *J. Invest. Dermatol.* **76:** 404–448.
Yamada K, Salopek T, Jimbow K & Ito S (1989) *J. Invest. Dermatol.* **92:** 544.

21

Interactions between Reactive Oxygen and Mediators of Sepsis and Shock

ALBRECHT WENDEL, MARCUS NIEHÖRSTER and GISA TIEGS
Faculty of Biology, University of Konstanz, Postfach 5560, D-7750 Konstanz, Germany

I. Bacterial endotoxins and their pathophysiological roles 585
II. Eicosanoids 587
III. Reactive oxygen species as initiators of endotoxin-induced cytokine release 589
IV. Conclusions 591

Abbreviations

LPS	Lipopolysaccharide = endotoxin
LTD_4	Leukotriene D_4
TNF	Tumour necrosis factor
α_1PI	α_1 Proteinase inhibitor
SOD	Superoxide dismutase

I. BACTERIAL ENDOTOXINS AND THEIR PATHOPHYSIOLOGICAL ROLES

Gram-negative bacteria contain certain cell wall components, the endotoxins, which share a common structural principle: they consist of a hydrophilic heteropolysaccharide component, covalently linked to a hydrophobic

Oxidative Stress: Oxidants and Antioxidants
ISBN 0-12-642762-3

lipid portion termed lipid A (Rietschel et al, 1984). Whenever these lipopolysaccharides are present in the circulation of mammals, characteristic symptoms arise such as fever, bidirectional and reversible changes in leukocyte counts, alterations in blood pressure, or ultimately shock and death. The existing large variations in the outer carbohydrate moieties of the different bacterial endotoxins would require a vast variety of different antibodies in order to counteract endotoxin-induced pathological reactions by means of passive immunization. Even though all endotoxins share a common lipid A core, this part of the molecule is masked in a way that does not permit immunological neutralization of the endotoxic potential. Consequently, alternative approaches to the therapy of endotoxic reactions require a refined mechanistic knowledge of the pathogenic events triggered by lipopolysaccharides (Rietschel and Brade, 1987). The chemical identities of the major mediators as well as their biokinetics are now needed for developing strategies against septic shock.

A most common outcome of Gram-negative bacterial infection is septic lung injury which may eventually lead to adult respiratory distress syndrome (ARDS). Since the lung contains a large pool of endotoxin-responsive leukocytes, it is not surprising that this organ is primarily exposed to the mediators activated by endotoxin. In spite of the clinical experience that the lung is the major target of endotoxic complications, it should be noted that the complex of events involves all cells, tissues and organs, including their humoral products (Newman, 1985). In other words, advanced stages of shock affect all major organs of the body, leading to the clinical situation of multiorgan failure.

Among the different leukocyte populations, be they resident or circulating cells, a wide variety of different types of mediators become activated and are released. These mediators include histamine, products of the arachidonic acid cascade, activation of the clotting and fibrinolytic systems, plasma complement activation, the formation of reactive oxygen species, the release of proteases from cells, and the release of cytokines. It becomes increasingly clear that the release of cytokines such as tumour necrosis factor (TNF) and members of the interleukin family represents the decisive point of no return in the pathogenic sequence of endotoxaemic organ injury (Beutler and Cerami, 1988). It is the purpose of this chapter to show some relationships between selected parts of this complex self-activation system and the role of reactive oxygen in its function as a permissive modulator. It is emphasized that these functions of reactive oxygen are different from the direct chemical effects, i.e. acutely toxic effects of activated oxygen on tissue, e.g. lipid peroxidation-mediated cytotoxicity.

II. EICOSANOIDS

Endotoxic complications can be ameliorated by preventive administration of glucocorticoids (Thomas and Smith, 1954). In recent years, interest originating from these early observations focused on individual metabolites of arachidonic acid, especially on cyclooxygenase and lipoxygenase products.

A. Cyclooxygenase products

The contribution of prostaglandin-like material to endotoxin shock has been demonstrated in many animal species such as dog, cattle, guinea-pig and rabbit (Flynn, 1985). Initial studies with animal bolus endotoxin models provided evidence that the prostaglandins E and F are increased in the plasma of these animals after endotoxin administration (Anderson et al, 1975). Subsequently, similarly designed studies strongly supported the hypothesis that intravenously administered endotoxin is a potent stimulus also for other prostanoid products such as prostacyclin and thromboxane, triggered by at least two different mechanisms. Prostacyclin (PGI_2) is probably the predominant prostanoid released in response to endotoxin exposure (Bult and Herman, 1983). The endotoxin-induced release of PGI_2 seems to depend on complement activation, since in rabbits inactivation of complement with cobra venom factor suppressed the PGI_2 increase (Rampart et al, 1981). Functionally, this PGI_2 response may be interpreted as a vasodilative counterregulation by the endothelium against the endotoxin-stimulated, leukotriene-mediated vasoconstriction (see below). Endotoxin-dependent release of E- and F-type prostaglandins has been consistently found in vivo and in vitro (Anderson et al, 1975; Kurland and Bockman, 1978). On the other hand, administration of PGE_1 or $PGF_{2\alpha}$ in vivo antagonized the cardiovascular effects in canine endotoxic shock and improved survival (Raflo et al, 1973). This apparent protection may be attributed to cAMP-enhancing effects of the prostaglandin.

Whether endotoxin-stimulated thromboxane A_2 release from platelets under in vivo conditions is an effect directly mediated by the lipopolysaccharide (LPS) cannot be decided at present. Kupffer cells treated with LPS release large amounts of thromboxane A_2 (Bowers et al, 1985). Furthermore, evidence is available that the release of the cytokine TNFα from endotoxin-stimulated Kupffer cells is suppressed by PGE_2 (Karck et al, 1988) while the synthesis and secretion of interleukin 1 is not affected (Scales et al, 1989).

B. Lipoxygenase products

The leukotrienes exert a variety of important biological actions in several cell types and contribute to the major effects of ischaemia and shock (Lefer, 1986). The main characteristics of the peptidoleukotrienes are vaso- and bronchoconstriction, while the dihydroxylated product LTB_4, i.e. a nonpeptide leukotriene, is a potent chemoattractant and chemotactic agent and thus recruits leukocytes into inflamed areas. Moreover, LTB_4 contributes to the adherence of mobile cells to endothelial membranes of the vasculature and thus eventually leads to obstruction of blood flow in these areas. Peptidoleukotrienes stimulate smooth muscle and are potent vasoconstrictors in nanomolar concentrations. The vasoconstrictive effects of peptidoleukotrienes are only exceeded by those of thromboxane. An early connection between shock and increased lipoxygenase pathway activity was established by Ogletree et al (1982), who observed an increased release of 5-hydroxyeicosatetraenoic acid during endotoxaemia in sheep, i.e. animals which are extremely sensitive to endotoxin. Two lines of independent experimental approach provided the link between this finding and a direct effect of leukotrienes in endotoxic situations. First, it was shown that endotoxin-induced lethality in mice can be prevented by leukotriene antagonists (Hagmann and Keppler, 1982). Second, it was demonstrated that mouse peritoneal macrophages incubated in the presence of endotoxin released LTC_4 in a dose-dependent manner (Lüderitz et al, 1983). With the improvement of the analytical procedure for detection of leukotrienes it became possible to demonstrate the production and excretion of peptidoleukotrienes into the bile in endotoxic shock in rats (Hagmann et al, 1985). The pharmacological inhibition profiles obtained with different agents interfering with eicosanoid metabolism suggested that LTD_4 might contribute the major activity as a shock mediator. By direct intravenous administration of LTD_4 to galactosamine-sensitized mice, an identical time-course, morphology and pathophysiology compared to endotoxin administration was obtained in a mouse liver injury model (Tiegs and Wendel, 1988).

The ability of leukotrienes to mimic effects of endotoxin indicates that these eicosanoids propagate the endotoxin stimulus by leading to the release of cytokines. Indeed, it was shown in vitro that leukotrienes augment interleukin 1 production by monocytes (Rola-Pleszczynski and Lemaire, 1985). These findings were extended by showing that leukotrienes also activate macrophages to release TNF (Gagnon et al, 1989). In vivo, in the galactosamine-sensitized mouse, lipoxygenase inhibitors protected against endotoxin-mediated organ injury, but not against TNFα-induced lesions (Tiegs et al, 1989). Concordantly, it was demonstrated in a similar animal model that mice protected by lipoxygenase inhibitors were unable to release TNF into the circulation (Schade et al, 1989).

III. REACTIVE OXYGEN SPECIES AS INITIATORS OF ENDOTOXIN-INDUCED CYTOKINE RELEASE

If phagocytic cells are directly exposed to endotoxin, they are not stimulated to release superoxide. However, when these cells were primed by endotoxin before exposure to established leukocyte stimuli such as FMLP or zymosan, a massive respiratory burst was evoked (Guthrie et al, 1984). In vivo, endotoxin-induced superoxide production requires the presence and activation of complement factors. The exposure of animals to endotoxin results in the production of C5a, a direct potent stimulus of superoxide release (Snyderman et al, 1969). An alternative mechanism for superoxide generation in vivo was considered to depend on LPS-induced LTD_4 production. Peptidoleukotrienes are highly vasoconstrictive compounds. They act on smooth muscle cells of vessel walls and may thereby cause a transient ischaemia followed by reflow during which superoxide production occurs (Lefer, 1986). It therefore seemed reasonable to suggest that reactive oxygen intermediates may also promote liver injury by an endogenously caused transient ischaemia (Arthur, 1988). Indeed, it has been shown that suppressors of an ischaemia/reflow syndrome, i.e. SOD or catalase or allopurinol, protected against endotoxin-induced liver injury in galactosamine-sensitized mice (Wendel et al, 1987).

Under physiological conditions, two molecules of superoxide dismutate to H_2O_2 which can then be enzymatically utilized by myeloperoxidase to yield highly reactive hypochlorous acid. Activated neutrophils release high amounts of myeloperoxidase into the extracellular fluid where the substrate H_2O_2 is present. During inflammation, both hypochlorous acid and superoxide kill bacteria and may destroy intact tissue. Besides these unspecific cytotoxic actions of reactive oxygen, superoxide induces a variety of specific interactions with systems regulating the inflammatory response. In addition to the long-known direct leukotactic effect of superoxide, it has been recently recognized that extracellular superoxide leads to a rapid and irreversible inactivation of the production of the endothelium-derived relaxing factor, i.e. NO (Gryglewski et al, 1986). In other words, this event triggered by superoxide may eventually lead to a restriction of the local blood flow by preventing the counterregulation via dilation of the vessels. By enhancing the peroxide tone in leukocytes, in turn the leukotriene production of leukocytes is further enhanced, which again promotes vasoconstriction induced by peptidoleukotrienes or leukotaxis induced by LTB_4 (Lands et al, 1971; Egan et al, 1983).

Not only the class of lipid autacoid activity is subject to modulation by reactive oxygen; evidence is now accumulating that activated oxygen regulates the release of cytokines. It was shown that hydrogen peroxide or sodium periodate increased the secretion of TNF from mouse peritoneal macrophages exposed to LPS (Chaudhri and Clark, 1989). The processing and release of TNF includes a step which is catalysed by a serine protease: inhibition of proteolytic activity by a low molecular weight serine protease inhibitor prevented selectively the release of TNF from stimulated human mononuclear cells while it had no effect on the release on interleukin 1 (Scuderi, 1989). Hence, TNF secretion depends on the activity of a protease of the elastase type. It is interesting that during sepsis and inflammation neutrophils release large amounts of elastase into the circulation (Jochum and Fritz, 1989). Moreover, the enzyme is released together with its physiological inhibitor, the α_1 protease inhibitor (α_1PI). This antiprotease activity can be inactivated by oxidation of a single methionine residue in the active centre of the molecule when exposed to superoxide or hypochlorous acid (Johnson and Travis, 1979). Thus, leukocytic elastase becomes catalytically active during inflammation. These findings link the oxidative burst with protease-dependent TNFα release. In line with these observations, it has been shown that a transmembrane 26-kDa TNFα precursor is formed (Kriegler et al, 1988) which may be proteolytically cleaved to the soluble bioactive 17-kDa TNFα. Indeed, the N-terminus of this soluble form of TNFα contains amino acid residues typical for cleavage products of elastase (valine in human TNFα, leucine in mouse TNFα).

Recently, evidence was obtained that these in vitro findings may also apply to the in vivo situation. In the animal model of endotoxin-induced hepatitis in galactosamine-sensitized mice (Wendel and Tiegs, 1986), pretreatment with α_1PI prevented TNFα release into the circulation and therefore protected against hepatic injury (Table 1). Accordingly, α_1PI did not protect when galactosamine-sensitized animals were challenged with TNFα instead of endotoxin. Table 1 shows also that pharmacological interventions directed against the formation or presence of reactive oxygen prevented TNFα release into the circulation and protected against hepatitis. In contrast, SOD or allopurinol did not protect against TNFα-induced liver injury. These findings indicate that the formation of reactive oxygen is required for the release or secretion of TNFα.

A hypothesis to explain the current state of interrelationships between mediators of shock and sepsis includes the following steps. In monocytes or macrophages LPS induces the synthesis of membrane-bound 26-kDa TNFα (Figure 1: square + triangle symbol). The bioactive 17-kDa form of TNFα (square symbol) is released by a proteolytically active, elastase-like serine protease. Active protease is provided by oxidative inactivation of α_1PI. Agents counteracting reactive oxygen species may interrupt the sequestration of bioactive TNFα by indirectly preventing the activation of the proteolytic activity required for the processing of a TNFα precursor (Figure 1).

Table 1. Effect of antiprotease or compounds interfering with reactive oxygen on liver injury and TNF-secretion in galactosamine-sensitized mice.

Pretreatment, 1 hour prior to LPS	Challenge	Hepatitis	Systemic TNF
Saline, iv	LPS	yes	+
42 mg/kg α_1PI, iv	LPS	no	−
42 mg/kg α_1PI, iv	TNF	yes	
10^6 U/kg SOD, iv	LPS	no	−
10^6 U/kg SOD iv	TNF	yes	
100 mg/kg Allopurinol, ip	LPS	no	−
100 mg/kg Allopurinol, ip	TNF	yes	

From Tiegs and Wendel (1988) and Niehörster et al (1990).

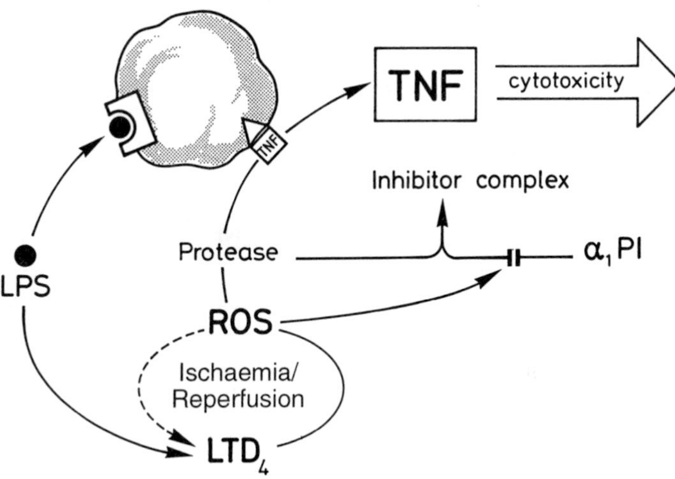

Figure 1. Possible interrelationships between the oxidative inactivation of α_1 proteinase inhibitor and the secretion of the cytokine TNFα.

IV. CONCLUSIONS

The role of reactive oxygen species as immediate and direct cytotoxic agents in acute intoxication or oxidative metabolism under overload conditions has been the focus of previous research interests. It now appears that, in a much lower concentration range, forms of extracellular reactive oxygen, i.e.

superoxide, H_2O_2 and hypohalous acids, have humoral functions modulating endogenous host defence systems (Figure 2).

This chapter compiles some of the known examples of oxidative activation of lipid mediators such as leukotrienes which, secondarily as potent vasoconstrictors, may initiate another postischaemic reactive oxygen burst. One of the most fascinating and new aspects of the proinflammatory role of steady low-level oxidative stress concerns the delicate balance of extracellular protease/antiprotease activity which may govern the ultimate step in cytokine release by processing bioactive TNFα. This piece of evidence might represent the missing link between the radical-orientated and the immunologically determined view of the complex pathophysiology in shock and inflammation (see Chapter 22).

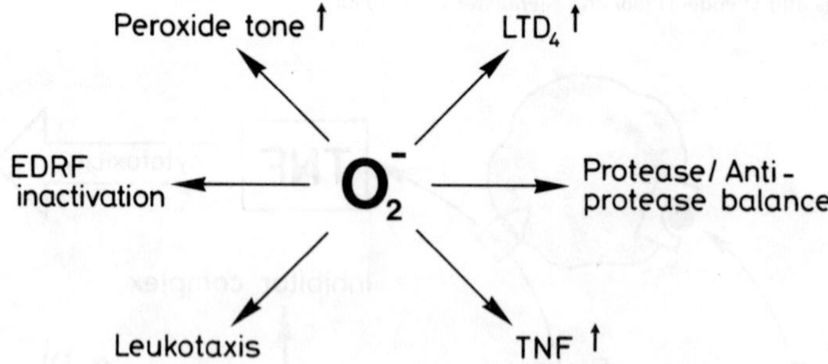

Figure 2. Biological activities of superoxide in inflammation and shock.

REFERENCES

Anderson FL, Jubiz LS, Tsagaris TJ & Kuidah H. (1975) *Am. J. Physiol.* **228:** 410–414.

Arthur MJP (1988) *J. Hepatology* **6:** 125–131.

Beutler B & Cerami A (1988) *Ann. Rev. Biochem.* **57:** 505–518.

Bowers GJ, MacVittie TJ, Hirsch EF, Conklin JC, Nelson RD, Roethel RJ & Fink MP (1985) *J. Surgical Res.* **38:** 501–508.

Bult H & Herman AG (1983) *Biochem. Pharmacol.* **32:** 2523–2528.

Chaudhri G & Clark JA (1989) *J. Immunol.* **143:** 1290–1294.

Egan RW, Tischler AN, Baptista EM, Ham EA, Soderman DD & Gale PH (1983). In Samuelsson B, Paoletti R and Ramwell P (eds) *Advances in Prostaglandin, Thromboxane, and Leukotriene Research*, Vol. II, pp 151–157. New York: Raven Press.

Flynn JT (1985) In Hinshaw LB (ed.) *Handbook of Endotoxin, Vol. 2: Pathophysiology of Endotoxin*, pp 237–285. Amsterdam: Elsevier Science Publishers.

Gagnon L, Filion LG, Dubois C & Rola-Pleszczynski M (1989) *Agents and Actions* **26:** 141–147.
Gryglewski RJ, Palmer RMJ & Moncada S (1986) *Nature* **320:** 454–456.
Guthrie LA, McPhail LC, Henson PM & Johnston RB Jr (1984) *J. Exp. Med.* **160:** 1656–1671.
Hagmann W & Keppler D (1982) *Naturwissenschaften* **69:** 594–595.
Hagmann W, Denzlinger C and Keppler D (1985) *FEBS Lett.* **180:** 309–313.
Jochum M & Fritz H (1989) In Faist E, Ninnemann JL and Green DR (eds) *Immune Consequences of Trauma, Shock and Sepsis*, pp 165–172. Berlin: Springer-Verlag.
Johnson D & Travis J (1979) *J. Biol. Chem.* **254:** 4022–4026.
Karck U, Peters T & Decker K (1988) *J. Hepatol.* **7:** 352–361.
Kriegler M, Perez C, DeFay K, Albert I & Lu SD (1988) *Cell* **53:** 45–53.
Kurland JI & Bockman R (1978) *J. Exp. Med.* **147:** 952–957.
Lands W, Lee R & Smith W (1971) *Ann. N.Y. Acad. Sci.* **180:** 107–122.
Lefer A (1986) *Biochem. Pharmac.* **35(2):** 123–127.
Lüderitz T, Rietschel ET & Schade UF (1983) *Immunobiology* **165:** 312–313.
Newman JH (1985) *Clinics in Chest Medicine* **6:** 371–384.
Niehörster M, Tiegs G, Schade UF & Wendel A (1990) *Biochem. Pharmacol.* **40:** 1601–1603.
Ogletree ML, Oates JA, Brigham KL & Hubbard WC (1982) *Prostaglandins* **4:** 459–468.
Raflo GT, Wangensteen SL, Glenn TM & Lefer AM (1973) *Eur. J. Pharmacol.* **24:** 86–95.
Rampart M, Bult H & Herman AG (1981) *Arch. Int. Pharmacodyn. Ther.* **249:** 328–329.
Rietschel ET, Zähringer U, Wollenweber H-W, Miragliotta G, Musehold J, Lüderitz T & Schade UF (1984) *Am. J. Emergency Med.* **2:** 60–69.
Rietschel ET & Brade H (1987) *Infection* **15:** 133–141.
Rola-Pleszczynski M & Lemaire I (1985) *J. Immunol.* **135:** 3958–3961.
Scales WE, Chensue SW, Otterness I & Kunkel SL (1989) *J. Leukocyte Biology* **16:** 116–131.
Schade UF, Ernst M, Reinke M & Wolter DT (1989) *Biochem. Biophys. Res. Commun.* **159:** 748–754.
Scuderi P (1989) *J. Immunol.* **143:** 168–173.
Snyderman R, Shin JK, Gewurz H & Mergenhagen SE (1969) *J. Immunol.* **103:** 413–422.
Thomas L & Smith R (1954) *Proc. Soc. Exp. Biol. Med.* **86:** 801–815.
Tiegs G, Wolter M & Wendel A (1989) *Biochem. Pharmacol.* **38(4):** 627–631.
Tiegs G & Wendel A (1988) *Biochem. Pharmacol.* **37:** 2569–2573.
Wendel A, Tiegs G & Werner C (1987) *Biochem. Pharmacol.* **36:** 2637–2639.
Wendel A & Tiegs G (1986) *Biochem. Pharmacol.* **35:** 2115–2118.

22

Involvement of Oxygen Radicals in Shock and Organ Failure

HEINZ REDL,* HARALD GASSER,* SETH HALLSTRÖM,* EVA PAUL,* SOHEYL BAHRAMI,* GÜNTHER SCHLAG* and ROGER SPRAGG†

**Ludwig Boltzmann Institute for Experimental and Clinical Traumatology, Donaueschingenstrasse 13, A-1200 Vienna, Austria*

†Department of Medicine, Pulmonary and Critical Care Division, University of California, Medical Center, San Diego, Ca, USA

I. Onset of shock and progression of organ failure 596
II. Involvement of oxygen radicals in shock 597
III. Sources of oxygen radicals in shock 600
IV. Possible targets and effects of radical action 605
V. Involvement of different organ systems 607
VI. Experiences with therapeutic radical scavenging 608
VII. Conclusion 610

The syndrome of multiorgan failure (MOF) occurs in critically ill patients and is the result of tissue trauma, ischaemia, hypoxia and sepsis. Despite intensive research efforts this syndrome is still associated with a high mortality rate. A mechanism responsible for post-traumatic organ failure is the formation of oxygen radicals and subsequent chain reactions leading to lipid peroxidation. This chapter focuses on the interaction between oxygen radicals produced during shock conditions and antioxidant defence systems with mention of other related situations such as ischaemia–reperfusion or inflammatory events. Many of the proposed interrelationships have only been measured indirectly or described in vitro and are applied hypothetically to the in vivo setting.

Oxidative Stress: Oxidants and Antioxidants
ISBN 0-12-642762-3

I. ONSET OF SHOCK AND PROGRESSION OF ORGAN FAILURE

Post-traumatic organ failure is related to the effects of a multitude of mediators released via activation cascade systems involving both humoral and blood cell components. Essentially, the activation sequence consists of two phases: the primary events are immediately post-trauma, and the secondary events start between three and seven days after trauma, and in many cases are the consequences of sepsis.

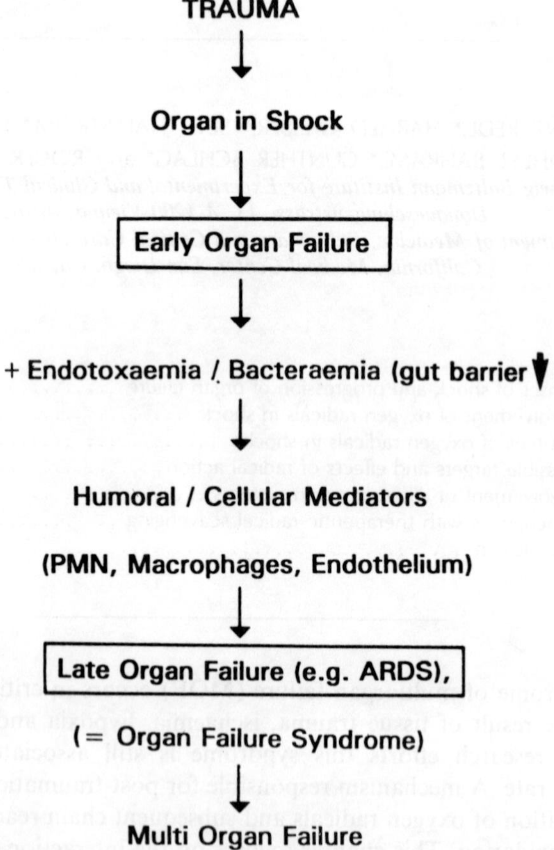

Figure 1. Reactions leading to post-traumatic cell damage, organ failure and multiorgan failure (MOF) syndrome.

Polytrauma is commonly associated with hypovolaemic-traumatic shock causing hypoperfusion of vital organs. Reperfusion of these organs occurs following resuscitation. Hypoperfusion–reperfusion events together with phagocyte activation caused by products of humoral cascade systems (e.g. complement) represent the non-bacterial inflammatory events. The result may be early organ failure, which together with the sepsis syndrome may progress to multiorgan system failure (Figure 1).

II. INVOLVEMENT OF OXYGEN RADICALS IN SHOCK

A. Measurement of reactive oxygen products

In the intensive care area, measurements in patients by Baldwin et al (1986) showed significantly elevated hydrogen peroxide levels in the breath condensate of patients with acute lung injury, consistent with neutrophil activation. These results were confirmed by Sznajder et al (1989), studying brain injury and sepsis.

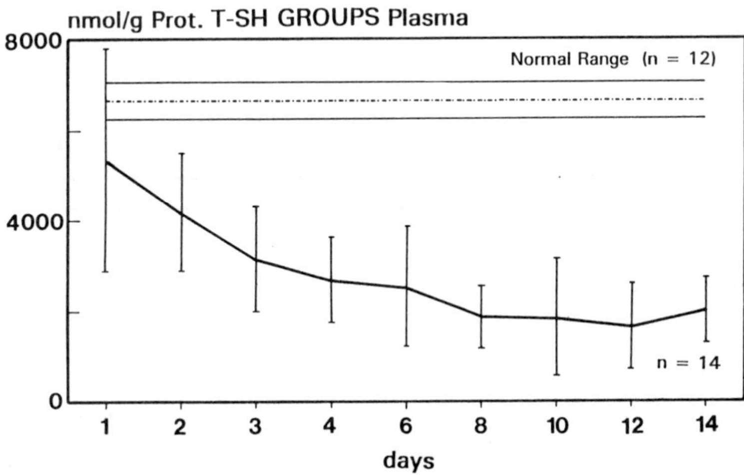

Figure 2. The decrease of DTNB reactive total sulphydryl groups in plasma from patient is shown for 14 consecutive days after burn injury. For comparison the range of plasma values from healthy volunteers is indicated in the graph (dotted line represents the mean value, $n = 12$). $\overline{x} \pm$ SD.

B. Analysis of antioxidants

Changes of antioxidant levels can be used as an index of oxidant production. Takeda et al (1984), studying critical-care surgical patients, found plasma tocopherol levels of 5–6 μg ml^{-1}, comparable to those seen in congenital vitamin E deficiency patients. In addition, levels of plasma thiobarbituric acid reactive substances (TBARS) were consistent with ongoing lipid peroxidation. Similar findings have been reported in patients with acute lung injury (Richard et al, 1990). Dithionitro benzoic acid (DTNB)-reactive sulphydryl groups were diminished in severe burns (Figure 2). Decreased plasma ceruloplasmin and increased TBARS were found in patients developing organ failure (Dauberschmidt et al, 1989). In the primate model of haemorrhagic-traumatic shock, tocopherol levels fell significantly during the shock period before retransfusion of shed blood and remained below baseline until the end of the study (Figure 3). There was a loss of plasma SH-groups, and plasma levels returned to baseline after the reinfusion procedure. In tissue samples from patients in cardiogenic shock, a decreased tissue content of total glutathione and an increase in glutathione disulphide were found (Corbucci et al, 1985).

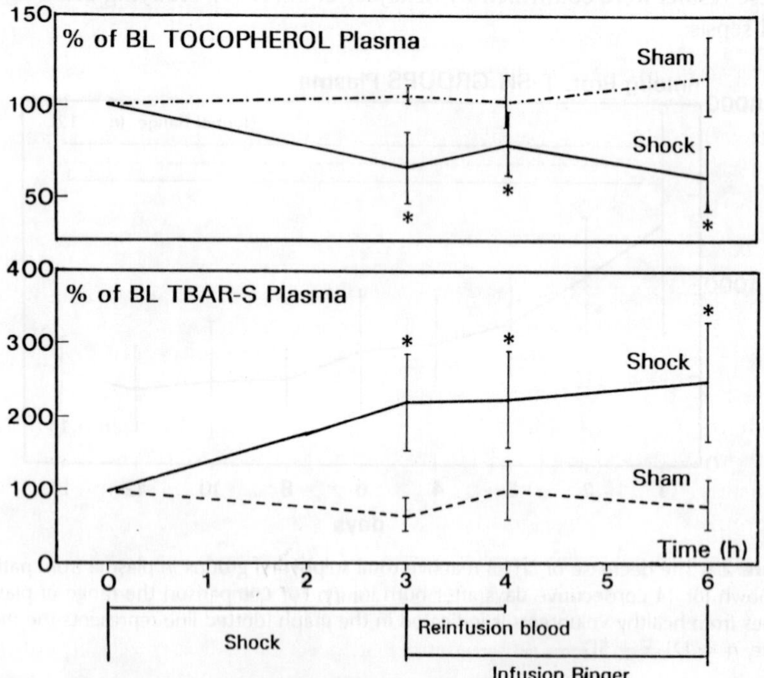

Figure 3. α-Tocopherol (top) and thiobarbituric acid reactive substance (TBARS) (bottom) plasma levels during haemorrhagic-traumatic shock and after retransfusion in baboons. Changes in concentration are given in percentage calculated from baseline values ± SD. Sham, ---- ($n = 3$); shock, —— ($n = 5$); 100% corresponds to 23.3 ± 9.3 μg (g plasma protein)$^{-1}$ and 1.32 ± 0.33 nmol (g protein)$^{-1}$, respectively.
*, significantly different from baseline. $p < 0.05$.

C. Lipid peroxidation products

The polyunsaturated ω-6 (i.e. linoleic acid, arachidonic acid, γ-linolenic acid) and ω-3 (i.e. α-linolenic acid, decosahexaenoic acid) acids contained in plasma and biomembranes are sensitive targets for oxygen radicals in vivo. They are oxidatively degraded by lipid peroxidation (Kappus, 1985).

1. *Trauma*

Shock-related studies of lipid peroxidation were performed during extracorporeal circulation (Westaby et al, 1986), in tissue of deceased polytrauma patients (Nerlich et al, 1986), or in polytrauma experiments (Figure 3). Reactive hydroxyalkenals such as HNE (4-hydroxynon-2-enal) are products of lipid peroxidation (Esterbauer et al, 1984) and increase in response to oxidative stress, for instance polytrauma (Lieners et al, 1989). We found a 1.5–4.5-fold increase of HNE in dog but not baboon plasma after 3 h of polytrauma and hypovolaemia (baseline values: range 0.6–2.2 nmol HNE per g plasma protein). The high reactivity of HNE can be traced by the formation of fluorescent products (Esterbauer et al, 1986), which were detected in plasma in the hypovolaemic-traumatic shock model (Lieners et al, 1989).

2. *Burns*

In rat burn shock Till et al (1983) studied the sequence of appearance of lipid peroxidation products and found that conjugated dienes appeared in burned skin and in the lung as well as in plasma. Increased plasma conjugated dienes were also found in postburn sheep (Demling et al, 1989), with the burn wound as the major source (Jin et al, 1986). Similarly, increases of TBARS were found by Nishigaki et al (1980) in rat burn shock. In addition to the measured peroxidation products, the use of scavengers such as SOD or allopurinol also suggested oxygen radical production (Saez et al, 1984).

3. *Sepsis and endotoxaemia*

Experimentally, organ failure can be induced by endotoxin (LPS) administration (i.v., i.p.). After endotoxin administration to rats TBARS increase transiently in plasma (Redl et al, 1990). In studies of mice receiving LPS, Peavy and Fairchild (1986) found TBARS and exhaled ethane. Increased conjugated dienes were found by Demling et al (1986) in ovine endotoxaemia in arterial but not venous plasma. Elevated TBARS were detected in ovine lung tissue after high but not low LPS doses (Wong et al, 1984). In intensive care patients, hydroperoxides were found to be increased in adult respiratory distress syndrome (ARDS) (Frei et al, 1988).

D. Scavenging of radicals

To test the hypothesis that oxygen radicals are involved in the pathogenesis of shock/trauma-related organ failure, antioxidants (Till et al, 1985) or spin traps (Novelli et al, 1986) have been used investigationally and therapeutically.

E. Similarities between shock and exogenously induced radical damage

A typical shock-induced insult of the lung is the increase in permeability of the alveolar–capillary membrane with subsequent oedema formation. The isolated lung system has been used to investigate H_2O_2-induced oedema formation (Redl et al, 1986). There was a rapid and profound fall in tissue ATP levels before the onset of lung oedema. This depletion might be due to arrest of ATP production as is also seen in isolated endothelial cells after peroxide exposure (Hyslop et al, 1988) with resultant disturbance of membrane-associated cation pumps. This H_2O_2-induced lung injury was accompanied by increased levels of TBARS in lung tissue. Using activated neutrophils as the radical source in combination with antioxidant or anti-protease regimens, Shasby et al (1983) demonstrated the potential for oxygen products from polymorphonuclear leukocytes (PMN) to cause lung oedema in the isolated organ. In vivo neutrophil activation is easily achieved with complement fragments. Specific complement activation was used to study neutrophil-dependent organ damage in vivo. Ward et al (1985) used systemic cobra venom factor to activate complement and found elevated conjugated dienes, hydroperoxides and fluorochromic substances but not elevated TBARS.

III. SOURCES OF OXYGEN RADICALS IN SHOCK

A. Xanthine oxidase

According to McCord (1985), haemorrhagic shock may be viewed as 'whole body ischaemia'. The consequence of acute ischaemia is inhibition of cellular

ATP synthesis by substrate (glucose) and oxygen deprivation. There is an increase in the levels of the ATP degradation products adenosine, inosine, and hypoxanthine (Saugstad and Ostrem, 1977; Redl et al, 1988). An involvement of oxygen radicals was first demonstrated by Granger et al (1981) in studies of small intestine ischaemia. Oxidant-injured tissue may leak sufficient xanthine oxidase (XO) into the systemic circulation to mediate injury to the microvasculature of unaffected tissue (Yokoyama and Parks, 1988). XO may be detectable within the bloodstream of patients with pathological conditions such as ARDS (Grum et al, 1987), extremity ischaemia (tourniquet)–reperfusion (Friedl et al, 1990) or ischaemia–reperfusion of the liver (Yokoyama et al, 1990). Furthermore, XO in plasma had increased catalytic activity after thermal injury, probably due to the enhancement of XO activity by histamine (Friedl et al, 1989). On the other hand, an increased substrate level (hypoxanthine) was found during hypovolaemic-traumatic shock (Redl et al, 1988).

XO action may also activate inflammatory cells. The proposed mechanism is that XO-derived oxidants produced at the time of reperfusion promote the formation of plasma compounds that attract and activate PMN (Petrone et al, 1980). However, these effects might not be due to chemotactic substances formed in plasma but to direct effects upon the endothelium, which becomes more adhesive. This effect was blocked by SOD or antiadherence molecule antibody (IB4), leading to the conclusion that superoxide mediates reperfusion-induced leukocyte–endothelial interactions (Suzuki et al, 1989), which was also true in intestinal ischaemia–reperfusion. Marzi et al (1990) used intravital microscopy during haemorrhagic shock to demonstrate increased PMN adherence to endothelium in the liver, which was attenuated by human recombinant SOD (hrSOD) administration upon retransfusion.

One probable explanation is suggested by the observation of Lewis et al (1988) that H_2O_2 induces platelet-activation factor (PAF) production by the endothelium. PAF remains bound and then causes leukocyte adherence via the PAF receptor on leukocytes. A similar mechanism has also been described for thrombin-stimulated endothelial cells, but not for cytokines (Zimmerman et al, 1990). Conversely, accumulated PMN aggregates produce circulatory interruption in the capillaries, which leads again to ischaemia. Capillary plugging by granulocytes apparently accounts for the no-reflow phenomenon (Schmid-Schönbein, 1987), probably together with endothelial swelling.

B. Phagocyte production

A role for neutrophils in shock-related organ damage is largely based on the following observations:

1. Neutrophils accumulate in tissues during shock and retransfusion, giving leukostasis of organs (Redl et al, 1984).
2. Animals (rats) with a large number of nitroblue tetrazolium (NBT)-positive (O_2^--producing) neutrophils did not survive haemorrhage, while those with low numbers did (Barroso-Aranda and Schmid-Schönbein, 1989).
3. Neutropenia offers protection in shock models.
4. Monoclonal antibodies against the leukocyte adhesion complex CD11/CD18 attenuate organ damage.

In response to activation by phagocytosis or soluble inflammatory mediators, neutrophils (macrophages) undergo a respiratory burst (Chapter 15; Baggiolini and Wymann, 1990). The amount of O_2^- and H_2O_2 produced is species- and stimulus-specific (Fantone and Ward, 1983). Macrophages release O_2^- and H_2O_2 similarly to PMN upon stimulation with IgA immune complexes in vitro (Warren et al, 1987) as well as ex vivo after IgA complex stimulation in vivo (Johnson et al, 1986). Further reactive metabolites, e.g $^{\bullet}OH$, are formed by the iron-catalysed Fenton mechanism. With the parallel release of oxygen radicals and myeloperoxidase from PMN, HOCl is formed, especially by PMN, where 25–40% of H_2O_2 is halogenated (Weiss and Slivka, 1982), while macrophages contain little or no myeloperoxidase.

The toxic effects of HOCl include oxidation of plasma membrane sulphydryls and oxidation of methionine and tryptophan residues, followed by ATP loss and cell lysis (Schraufstätter et al, 1990). Secondary products of lipid peroxidation, such as hydroxynonenal (HNE) formed by activated neutrophils (see Chapter 13), are thought to be relevant in leukocyte-dependent inflammatory reactions. HNE may produce positive feedback loops by exerting a chemotactic activity towards neutrophils (Curzio, 1988).

An increased adherence of PMN to endothelium is presumed to constitute the prerequisite for the local effectiveness of tissue-damaging mediators of PMN (proteinases, toxic radicals). Leukocyte stimulation increases adhesiveness to the endothelial surface via Fc and PAF receptors and by adhesion molecules (Wallis et al, 1986), e.g. the CD18 complex (Gamble et al, 1985).

1. *PMN stimuli during shock*

The most prominent PMN stimuli are anaphylatoxins (C3a, C5a), released after trauma due to complement activation (Heideman and Hugli, 1984). Ischaemic tissue causes massive complement activation, resulting in elevated levels of the anaphylatoxins C3a and C5a (Bengtson et al, 1987). Beside other stimuli, LTB_4 enhanced production of H_2O_2 by monocytes and O_2^- by PMN (Perlman et al, 1989). Plasma kallikrein itself was also thought to attract and activate PMN (Wachtfogel et al, 1983), but Zimmerli et al (1989) showed that kallikrein functions to prime PMN for superoxide production. Aside from cell-activating products of the humoral cascade systems, endotoxins and cytokines have direct PMN-activating properties.

The question is whether phagocytes are activated during shock and trauma. Evidence comes from the in vitro (ex vivo) activation capacity (chemiluminescence) of blood cells. The drawback is that only circulating cells are quantified. Nevertheless, enhanced whole blood chemiluminescence (CL) consistent with PMN activation has been demonstrated in samples from postoperative and septic patients (Inthorn et al, 1987) by whole blood CL (Redl et al, 1983). Enhanced CL was found when patients became septic (Inthorn et al, 1987), during febrile periods (Pauksens et al, 1989) and in PMN from ARDS patients (Zimmerman et al, 1983; Schmeling et al, 1988). For trauma patients, blood PMN CL was higher than in controls throughout the post-traumatic course, while PMN recovered in the lung lavage fluid had diminished activation capacity (Regel et al, 1987a). In contrast, alveolar macrophages were normal immediately post-trauma but demonstrated a continuous increase in CL during the following days (Dwenger et al, 1990). There was also increased CL during ovine endotoxaemia (Traber et al, 1987), not found in another sheep study (Regel et al, 1987b), where only PMN adherence function was increased. Whole blood photometric (Bellavite et al, 1983) and fluorometric (Lieners et al, 1989) measurements during early organ failure revealed significantly increased oxygen radical production during zymosan-induced rat peritonitis (van Bebber et al, 1989), but not during hypovolaemic-traumatic shock in dogs (Lieners et al, 1989).

Compared with ex vivo measurements, analysis of plasma markers of PMN activation such as elastase permits detection of products released by circulating cells. There is a positive correlation between trauma severity and elastase plasma levels in patients (Dittmer et al, 1986; Nuytinck et al, 1986). In our baboon hypovolaemic-traumatic shock model, there is a significant increase in plasma elastase levels both at the end of the shock period and after retransfusion (Redl et al, 1989). There is also a significant correlation between sepsis and PMN-elastase plasma levels (Duswald et al, 1985) and the severity of multiorgan failure (Nuytinck et al, 1986; Pacher et al, 1989), indicating a relationship between the activation of granulocytes and development of organ failure (Nuytinck et al, 1986; Lang et al, 1989). Thus, identification of circulating elastase does not necessarily indicate elastase-induced organ failure, but does suggest the involvement of PMN. Involvement of PMN in shock injury mechanisms was also shown by PMN depletion. Experimental studies with pharmacologically induced leukopenia showed that shock-related organ damage, e.g. in the lung, can be prevented (Schraufstätter et al, 1984). Granulocytes are also of decisive importance in the well-known 'no-reflow syndrome', as granulocyte depletion was found to prevent 'no reflow' and mortality in haemorrhagic shock (Barroso-Aranda et al, 1988) as well as haemorrhagic shock-induced injury of the gastric mucosa (Smith et al, 1987).

The role of PMN in late organ failure is substantiated by experiments with granulocyte depletion, in which protection from TNF-mediated lung injury is demonstrable (Stephens et al, 1988). However, neutropenia did not preclude the development of ARDS in selected patients (Maunder et al, 1986).

Adherent PMN release substantially more O_2^- than PMN in suspension (Dahinden et al, 1983). Antibodies with antiadherence properties (60.3, TS1/18, IB4) against the common CD-18 β-chain of these complexes possess a tissue-protective potential. As with the ischaemia–reperfusion situation, a protective effect of blocking PMN adherence was found in rabbits during haemorrhagic shock (Hernandez et al, 1987; Vedder et al, 1988). Vedder et al (1989) then used the same rabbit model of haemorrhagic shock in rabbits to determine whether neutrophils are important during reinfusion. At the start of resuscitation the monoclonal antibody 60.3, directed to the human neutrophil adherence molecule CD18, was infused to block neutrophil adherence. This led to significantly increased survival and attenuation of several organ injuries but did not diminish pulmonary injury and leukostasis. Further evidence for the central role of PMN adherence in microvascular damage leading to plasma leakage comes from studies in which anti-CD-18 MoAb 60.3 was protective against local endotoxin injection (Arfors et al, 1987), LPS-induced leukostasis (Cooper et al, 1989) or PMN/monocyte-related gut injury (Hernandez et al, 1987). In addition, IB4 was protective against lung reperfusion injury (Horgan et al, 1989) and phorbol myristic acid-induced lung oedema (Kaslovsky et al, 1991).

C. Other radical sources

Cyclooxygenase generates O_2^- (Kukreja et al, 1986). Ischaemia-induced disturbances of mitochondrial electron transport might also contribute to shock-induced O_2^- generation, e.g. in mitochondria ('mitochondrial overspill') (Boveris, 1977).

Oxygen therapy is a source of oxygen radical production. High inspired oxygen concentrations, as encountered by many critically ill patients cause pulmonary oedema and pulmonary fibrosis. Jacobson et al (1990) have shown that tocopherol deficiency predisposes animals to O_2 toxicity. Synergistic interaction between hyperoxia and PMN products might be critical for the development of acute lung injury (Krieger et al, 1985).

IV. POSSIBLE TARGETS AND EFFECTS OF RADICAL ACTION

A. Direct

The endothelium appears to be an early target of oxidant injury as demonstrated using activated adherent neutrophils (Sacks et al, 1978) or acellular systems that produce oxygen radicals. In particular, hydrogen peroxide (Martin, 1984) or peroxide-derived products induce cell lysis in vitro (Weiss et al, 1981) and increase permeability (Shasby et al, 1983; Malik, 1986) in the isolated lung (Redl et al, 1986). This increase in permeability was independent of cyclooxygenase products but was attenuated by lipoxygenase inhibitors (Burghuber et al, 1985).

Oxygen radical cytotoxicity occurring in vivo may be limited by the modulating action of endothelial cells (Fehr et al, 1985), by the protective role of glutathione in the endothelial cell, by intracellular antioxidative enzymes (Hoover et al, 1987) or by the scavenging action of erythrocytes (Forslid et al, 1989). Changes in microfilament organization following H_2O_2 injury (Hinshaw et al, 1988) are very similar to ischaemic alterations. There is a drop of ATP secondary to H_2O_2-mediated inhibition of ATP synthesis (Schraufstätter et al, 1986; Hinshaw et al, 1986) prior to cell lysis, which is in contrast to the injury caused by HOCl, where loss of ATP was clearly followed by cell death (Schraufstätter et al, 1990). Also, the DNA strand breaks seen after exposure of cells to H_2O_2 (Schraufstätter et al, 1988) were absent after exposure to HOCl (Schraufstätter et al, 1990). In the isolated perfused rabbit lung a rapid and profound fall in cellular ATP levels and energy charge following exposure to H_2O_2 was found that occurred before the onset of lung oedema; in addition, increased levels of TBARS were produced (Redl et al, 1986).

Increased free cytosolic Ca^{2+} was found after exposure of rat tubular epithelium to a sublethal exogenous oxidative stress. The increase preceded morphological changes, depletion of sulphydryl groups or TBARS formation (Swann et al, 1990). Increased cytosolic Ca^{2+} concentration might cause a positive feedback loop in the formation of lipid peroxides. Cotterill et al (1989) showed that a Ca^{2+} channel-blocking agent (verapamil) decreased the extent of lipid peroxidation in ischaemic kidneys, while added Ca^{2+} increased the oxidative damage. In hypovolaemic shock, Ca^{2+} channel blockers have been found protective with regard to cardiac function (Hess et al, 1983), and in septic shock nifedipine decreased intestinal mucosal injury (Bosson et al, 1988). Decreased Ca^{2+} activation of proteases, including conversion of xanthine dehydrogenase to XO, may be responsible for this protection. Verapamil also attenuates the lung vascular response to

LPS in sheep, probably through inhibition of PMN superoxide production (Parker et al, 1988).

Endothelial cells constantly produce small amounts of O_2^- (Ratych et al, 1987), perhaps as a physiological mechanism to control the action of endothelial-derived relaxation factor (EDRF) (NO, which is inactivated by O_2^-) (see Chapter 17). However, substantially more O_2^- may be produced by endothelial cells upon reperfusion. Cell injury is known to cause release of iron (Halliwell and Gutteridge, 1986). Thus, the conversion of O_2^- and H_2O_2 to the aggressive $^{\bullet}OH$ is likely to occur. Hydroxyl radical is a very important oxidant involved in neutrophil-mediated cell injury (Ward et al, 1983).

An essential role of elastase in endothelial cell damage has recently been demonstrated with combined endotoxin (LPS) + chemotactic factor stimulation of PMN (Smedley et al, 1986). Radical and elastase seem to be synergistic in some PMN-mediated injuries (Varani et al, 1989). In support of the hypothesis that not only the endothelial cell but also the underlying basement membrane is a target in oxidant-induced organ failure, Peters et al (1986) found that cellular fibronectin with the extra type III (ED1) domain was released from isolated rabbit oxidant-injured lungs in parallel with oedema formation. In vivo elevated levels of fibronectin bearing the ED1 domain were found intravascularly and in the lung lavage (Peters et al, 1988) with inflammatory injury (H_2O_2 or activated leukocytes) and in patients with trauma and sepsis.

B. Indirect

Although proteinase inhibitors are present in plasma, proteolytic damage may occur in the microenvironment in the contact area between PMN and endothelial cells. In that environment, α_1PI (Weiss and Regiani, 1984) may be inactivated by oxygen radicals. Oxidant inactivation of proteinase inhibitors was investigated in isolated rabbit lungs, in which stimulated neutrophils induce permeability oedema. Oedema formation was blocked by antioxidants or by addition of the oxidant-resistant antielastase (cathepsin G) inhibitor eglin (Neuhof, 1990). In human lungs free elastase activity together with oxidized and thus inactivated α_1PI was detected in lavage fluid (Cochrane et al, 1983).

Oxygen radicals are also implicated in enhanced generation of tumour necrosis factor (TNF), one of the central mediators in sepsis (see Chapter 21). Increased release of TNF from LPS-stimulated monocytes/macrophages under oxidant stress was inhibitable by radical scavengers and by iron chelators (Chaudhri and Clark, 1989). Similarly, interleukin 1-like activity was found after oxidative stress (Kasama et al, 1989).

V. INVOLVEMENT OF DIFFERENT ORGAN SYSTEMS

A. Lung

Endothelial swelling and PMN accumulation in human lungs occurs within 1 h after the onset of shock (Schlag and Regele, 1972) and was experimentally confirmed (Redl et al, 1984).

Following haemorrhage, during the 'low-flow syndrome' some areas of the lung might also become partially ischaemic. In ischaemia–reperfusion lung injury, Gurtner et al (1988) suggested that oxygen radicals are derived from cytochrome P450 during reperfusion. During shock (low-flow state) we noted a markedly decreased perfusion of various areas of the lung. However, compared with ischaemic periods (at least 12 h) that preceded an allopurinol-attenuable lung permeability increase, shock-related 'ischaemic' periods might not have been sufficient for this mechanism (Horgan et al, 1989). Thus, for shock, leukocyte accumulation is likely to be the more important mechanism. The high permeability rate of lung injury secondary to pancreatitis also appears to be due to neutrophil-dependent oxygen radicals (Guice et al, 1988).

Lung injury secondary to skin burns in rats was prevented by depletion of complement or neutrophils, by antioxidant enzymes, by iron chelators and by scavengers of hydroxyl radical (Till et al, 1983, 1985). These observations suggest that $^{\bullet}OH$ radicals derived from complement-stimulated neutrophils are the injurious agent.

Studies of bronchoalveolar lavage fluid recovered from patients who have experienced polytrauma reveal evidence of loss of surface activity of lung surfactant (Pison et al, 1990). While inhibitors may contribute to the loss of surface activity (Spragg et al, 1988), our recent observations (Gilliard et al, 1990) suggest that lipid peroxidation may also occur and be accompanied by loss of surface activity, which has been observed in some in vitro studies (Seeger et al, 1985).

B. Muscle

Haemorrhage is associated with an approximately 80% decrease of muscle pO_2 (Kessler et al, 1976). During the 'low-flow syndrome' (before reinfusion) no endothelial cell damage was documented; only after retransfusion was there massive endothelial cell oedema, but no plugging with leukocytes as seen in the lung or liver (Schlag and Redl, 1985).

C. Gut

Endotoxaemia occurs after trauma as well as after major surgery in the critically ill patient and is frequently without a septic focus. According to Fine et al (1972), the intestine is the source of endotoxin, probably because of shock-induced bacterial translocation (Baker et al, 1988).

One of the reasons for bacterial translocation is the breakdown of the intestinal mucosa barrier by hypoxia (Younes et al, 1984) by reperfusion injury, and by oxygen radicals released from granulocytes (Grisham et al, 1986). Endotoxin perpetuates this process by a positive feedback loop (Deitch et al, 1987). The relative contribution of ischaemia and reperfusion to mucosal injury was reviewed recently by Kvietys and Granger (1989).

In haemorrhagic–reinfusion shock experiments in rats, allopurinol as well as tungsten diet decreased bacterial translocation, indicating participation of the XO system (Deitch et al, 1988). In the baboon model of hypovolaemic-traumatic shock this bacterial translocation (Schlag et al, 1990) was evident during both hypovolaemia and reperfusion and was not inhibitable by SOD.

Haemorrhagic shock in baboons also resulted in gastric mucosal lesions, which were evident during ischaemia (where not prevented by allopurinol or SOD) and which were further aggravated during retransfusion (antioxidant-attenuable) (Von Ritter et al, 1988). Oxygen radicals contribute to the formation of gastric mucosal lesions in septic cats if animals become hypotensive. However, neither SOD nor catalase affected intestinal damage, even when reduced intestinal blood flow was found (Arvidsson et al, 1985).

VI. EXPERIENCES WITH THERAPEUTIC RADICAL SCAVENGING

A. Enzymes

As a therapeutic agent in haemorrhagic (traumatic) shock, human recombinant SOD (hrSOD) was found to diminish leukocyte adherence in the liver (Marzi et al, 1990). In a blinded placebo-controlled study we used hrSOD in our hypovolaemic-traumatic shock model in baboons. hrSOD was continuously infused for 5 h, starting 30 min after trauma induction at either 6, 30 or 120 mg kg^{-1}. We found no attenuation in plasma tocopherol and sulphydryl decrease as well as no attenuation of the increase of TBARS and conjugated dienes. With this SOD regimen we could not block organ damage, including intestinal injury.

Allopurinol combined with SOD, as well as SOD–albumin conjugates, also failed to prevent haemorrhagic shock-induced severe gastric mucosal lesions in the baboon, although minor to medium degree of injury was decreased, and tissue glutathione levels were higher (von Ritter et al, 1988) with SOD. SOD-Ficoll did not improve survival in haemorrhagic shock when infused during reinfusion in the rat (Lee et al, 1987).

In endotoxaemic or septic (like) organ failure, there are widely varying reports on the therapeutic efficacy of oxygen radical scavengers. There was no effect of SOD on respiratory failure in canine, pig, rabbit or rat models (Olson et al, 1987; Novotny et al, 1988; Broner et al, 1989; McKechnie et al, 1986), while further worsening occurred in sheep (Traber et al, 1985). On the other hand, positive effects were seen by Yoshikawa et al (1983), Kunimoto et al (1987) and Schneider (1990) with application of hrSOD ($28\,mg\,kg^{-1}\,h^{-1}$) in rat LPS-induced pulmonary oedema and in mice given LPS–galactosamine (Wendel et al, 1987). In contrast to SOD, catalase has been found to be protective for sheep (Seekamp et al, 1988), although a protective effect of (Ficoll)-catalase in goats could not be demonstrated (Maunder et al, 1988).

In a sepsis-like situation caused by zymosan-induced peritonitis, we found that lipid peroxidation was attenuated by application of enzyme conjugates (Van Bebber et al, 1991); however, the mortality with SOD/catalase was actually increased.

B. Iron chelators

Dogs treated with deferoxamine up to 4 h after haemorrhagic-traumatic shock induction had significantly increased survival rates (72 h) as compared to controls (Sanan et al, 1989). In a porcine haemorrhagic shock model, survival was significantly improved (Jacobs et al, 1990). Brackett et al (1990) found no effect in LPS or sepsis models of deferoxamine covalently bound to dextran or hydroxyl starch (Hallaway et al, 1989).

Non-corticoid 21-aminosteroids (e.g. U74006F), which have both chelating and radical-scavenging effects, appear to be especially beneficial in the treatment of head trauma (Hall et al, 1988a). In haemorrhagic shock, the decline in arterial pressure upon reinfusion was prevented by this drug (Hall et al, 1988b), with no beneficial effects (mortality) being found in rat endotoxin shock (Redl et al, 1990).

C. Natural antioxidants (non-enzymatic)

Tocopherol is a protective agent for PMN-mediated endothelial damage and also a potent modulator of endothelial function (i.e. PGI_2 production) (Boogaerts et al, 1984) and PMN adherence (Lafuze et al, 1984). There is a possibility of therapeutic usefulness in hypovolaemic–traumatic-shock and in endotoxaemia/sepsis (Kunimoto et al, 1987; McKechnie et al, 1986), although Broner et al (1989) has found no effect. High enteral doses of tocopherol are reported to prevent lung failure in rabbits with pulmonary microembolization, and in an uncontrolled study of ventilated patients at high risk a striking improvement in survival was found (Wolf and Lasch, 1984). Little is known of the therapeutic effects of ascorbic acid or glutathione, e.g. McKechnie et al (1986) and Kunimoto et al (1987).

D. Synthetic antioxidants and XOD blockers

From studies in the intestine it is clear that allopurinol and the long-lived metabolite oxypurinol have protective effects in ischaemia–reperfusion situations. Treatment of haemorrhagic shock with allopurinol dates back to Crowell's treatment of dogs in 1975 (Crowell and Nelson, 1975). Recently, Cederna et al (1990) reported protective effects of allopurinol during 1 h haemorrhagic shock in rats. Allopurinol combined with catalase was reported to have beneficial effects in dogs undergoing limited haemorrhagic shock (Bond et al, 1988).

Spin-trapping agents have been used in shock research to scavenge radicals for therapeutic purposes. Phenyl butyl nitrone was successfully used in Noble-Collip drum-shocked rats (McKechnie et al, 1986) to achieve long-term survival.

N-Acetylcysteine (NAC) is undergoing clinical trials to evaluate its usefulness in the therapy of intensive care patients (Goris, personal communication). NAC given to endotoxic sheep diminished the increase in lung permeability (Brigham, 1986). This result could be due either to decreased injury of lung endothelium by oxygen radicals or to attenuated eicosanoid release (Bernard et al, 1984). In porcine septic shock the effect of NAC on lung permeability was absent (Groeneveld et al, 1990). A selenoorganic substance, ebselen, has been tested in rat galactosamine-LPS hepatitis and had protective effects (Konz et al, 1987).

VII. CONCLUSION

Since shock-related organ failure evolves from a variety of starting points—ischaemia, reperfusion, non-bacterial or bacterial inflammation—several mechanisms are involved. In addition to the effects of xanthineoxidase after ischaemia—reperfusion, toxic oxygen species from phagocytes that accumulate in both intra- and extravascular tissue space are of central importance. A critical event is the contact (adhesion) of leukocytes to endothelial cells, which consequently are the significant targets for leukocyte products.

Damage of membranes by lipid peroxidation by exposure to mediators such as PAF and by proteolytic events leads to increased permeability, tissue oedema and organ dysfunction. Thus antioxidants, protease inhibitors and other agents that control phagocyte function are likely to contribute to the protection of the permeability barrier.

ACKNOWLEDGEMENTS

We are indebted to M. Grossauer for preparing the manuscript and Prof. H. Sies for valuable discussions. Some of the cited studies were performed with the support of Lorenz Böhler Fonds.

REFERENCES

Arfors KE, Lundberg C, Lindbom L, Lundberg L, Beatty PG & Harlan JM (1987) *Blood* **69:** 338–340.
Arvidsson S, Fält K, Marklund S & Haglund U (1985) *Circ. Shock* **16:** 383–393.
Baggiolini M & Wymann MP (1990) *TIBS* **15:** 69–72.
Baker JW, Deitch EA, Berg RD & Specian RD (1988) *J. Trauma* **28:** 896–906.
Baldwin SR, Simon RH, Grum CM, Ketai LH, Boxer LA & Devall LH (1986) *Lancet* **I:** 11–13.
Barroso-Aranda J & Schmid-Schönbein GW (1989) *Am. J. Physiol.* **257:** H846–H852.
Barroso-Aranda J, Schmid-Schönbein GW, Zweifach BW & Engler RL (1988) *Circ. Res.* **63:** 437–448.
Bellavite P, Dri P, Della Biance V & Serra MC (1983) *Eur. J. Clin. Invest.* **13:** 363–368.
Bengston A, Lannsjo W & Heideman M (1987) *Br. J. Anaesth.* **59:** 1093–1097.
Bernard GR, Lucht WD, Niedermeyer ME, Snapper JR, Ogletree ML & Brigham KL (1984) *J. Clin. Invest.* **73:** 1772–1784.
Bond RF, Haines GA & Johnson G III (1988) *Circ. Shock* **25:** 139–151.
Boogearts MA, van de Broeck J, Deckmyn H, Roelant C, Vermylen J & Verwilghen RL (1984) *Thromb. Haemost.* **51:** 89–92.
Bosson S, Fält K & Haglund U (1988) *Res. Exp. Med.* **188:** 357–365.
Boveris A (1977) *Adv. Exp. Med. Biol.* **78:** 61–82.
Brackett DJ, Lerner MR, Archer LT & Wilson MF (1990) *Circ. Shock* **31:** 35.
Brigham KL (1986) *Chest* **89:** 859–863.
Broner CW, Shenep JL, Stidham GL et al (1989) *Circ. Shock* **29:** 77–92.
Burghuber OC, Strife RJ, Zirrolli J et al (1985) *Am. Rev. Respir. Dis.* **131:** 778–785.
Cederna J, Bandlien K, Toledo-Pereyra C, Bergren C, MacKenzie G & Guttierez-Vega R (1990) *Transplant. Proc.* **22:** 444–445.
Chaudhri G & Clark IA (1989) *J. Immunol.* **143:** 1290–1294.
Cochrane CG, Spragg R & Revak SD (1983) *J. Clin. Invest.* **71:** 754–761.

Cooper JA, Neumann PH, Wright SD & Malik AB (1989) *Am. Rev. Respir. Dis.* **139:** A301.
Corbucci GG, Gasparetto A, Candiani A et al, (1985) *Circ. Shock* **15:** 15–26.
Cotterill LA, Gower JD, Fuller BJ & Green CJ (1989) *Transplantation* **48:** 745–751.
Crowell JW & Nelson KM Jr (1975) *Circ. Shock* **2:** 21–28.
Curzio M (1988) *Free Radical Res. Commun.* **5:** 55–66.
Dahinden C, Galanos C & Fehr J (1983) *J. Immunol.* **130:** 857–862.
Dauberschmidt R, Mrochen R, Griess B et al (1989) *Prog. Clin. Biol. Res.* **308:** 331–338.
Deitch EA, Berg R & Specian R (1987) *Arch. Surg.* **122:** 185–190.
Deitch EA, Bridges W, Baker J et al (1988) *Surgery* **104:** 191–198.
Demling RH, Lalonde C, Jin LJ, Ryan P & Fox R (1986) *J. Appl. Physiol.* **60:** 2094–2100.
Demling RH, Lalonde C, Fogt F, Zhu D & Liu Y (1989) *Crit. Care Med.* **17:** 1025–1030.
Dittmer H, Jochum M & Fritz H (1986) *Unfallchirurg* **89:** 160–169.
Duswald KH, Jochum M, Schramm W & Fritz H (1985) *Surgery* **98:** 892–899.
Dwenger A, Beychok C, Schweitzer G & Pape HC (1990) *Fresenius J. Anal. Chem.* **337:** 86–87.
Esterbauer H, Lang J, Zadravec S & Slater TF (1984) *Methods Enzymol.* **105:** 319–328.
Esterbauer H, Koller E, Slee RG & Koster JF (1986) *Biochem. J.* **239:** 405–409.
Fantone JC & Ward PA (1983) *Am. J. Pathol.* **107:** 397–418.
Fehr J, Moser R, Leppert D & Groscurth P (1985) *J. Clin. Invest.* **76:** 535–542.
Fine J, Caridis DT, Cuevas P, Ishiyama M & Reinhold R (1972) In Forscher BK, Lillehei RC & Stubbs S (eds) *Shock in Low- and High-Flow States*, pp 7–11. Amsterdam: Excerpta Medica.
Forslid J, Bjrksten B, Hagersten K & Hed J (1989) *Inflammation* **13:** 543–551.
Frei B, Stocker R & Ames BN (1988) *Proc. Natl. Acad. Sci. USA* **85:** 9748–9752.
Friedl HP, Till GO, Trentz O & Ward PA (1989) *Am. J. Pathol.* **135:** 203–217.
Friedl HP, Smith DJ, Till GO, Thomson PD, Louis DS & Ward PA (1990) *Am. J. Pathol.* **136:** 491–495.
Gamble JR, Harlan JM, Klebanoff SJ & Vadas MA (1985) *Proc. Natl. Acad. Sci. USA* **82:** 8667–8671.
Gilliard N, Heldt G, Merritt TA, Pappert D & Spragg RG (1990) *Am. Rev. Respir. Dis.* **141:** A635.
Granger DN, Rutili G & McCord JM (1981) *Gastroenterology* **81:** 22–29.
Grisham MB, Hernandez LA & Granger DN (1986) *Am. J. Physiol.* **251:** G567–G574.
Groeneveld ABJ, den Hollander W, Straub J, Nauta JJP & Thijs LG (1990) *Circ. Shock* **30:** 185–205.
Grum CM, Ragsdale RA, Ketai LH & Simon RH (1987) *J. Crit. Care* **2:** 22–26.
Guice KS, Oldham KT, Johnson KJ, Kunkel RG, Morganroth ML & Ward PA (1988) *Ann. Surg.* **208:** 71–77.
Gurtner GH, Traystman RJ & Toung TJK (1988) *J. Appl. Physiol.* **64:** 1757–1761.
Hall ED, Pazara KE & Braughler JM (1988a) *Stroke* **19:** 997–1002.
Hall ED, Yonkers PA & McCall JM (1988b) *Eur. J. Pharmacol.* **147:** 299–303.
Hallaway PE, Eaton JW, Panter SS & Hedlund BE (1989) *Proc. Natl. Acad. Sci USA* **86:** 10108–10112.
Halliwell B & Gutteridge JMC (1986) *Arch. Biochem. Biophys.* **246:** 501–514.

Heideman M & Hugli TE (1984) *J. Trauma* **24:** 1038–1043.
Hernandez LA, Grisham MB, Twohig B, Arfors KE, Harlan JM & Granger N (1987) *Am. J. Physiol.* **253:** H699–H703.
Hess ML, Warner MF, Smith JM, Manson NH & Greenfield LJ (1983) *Circ. Shock* **10:** 119–130.
Hinshaw DB, Sklar LA, Bohk B et al (1986) *Am. J. Pathol.* **123:** 454–464.
Hinshaw DB, Armstrong BC, Burger JM, Beals TF & Hyslop PA (1988) *Am. J. Pathol.* **132:** 479–488.
Hoover RL, Robinson JM & Karnovsky MJ (1987) *Am. J. Pathol.* **126:** 258–268.
Horgan MJ, Everitt J & Malik AB (1987) *Am. Rev. Respir. Dis.* **135:** A260.
Horgan MJ, Lum H & Malik AB (1989) *Am. Rev. Respir Dis.* **140:** 1421–1428.
Hyslop PA, Hinshaw DB, Halsey WA et al (1988) *J. Biol. Chem.* **263:** 1665–1675.
Inthorn D, Szczeponik T, Mühlbayer D, Jochum M & Redl H (1987) *Prog. Clin. Biol. Res.* **236B:** 51–58.
Jacobs D. Julsrud J, Hedlund B, Hallaway P & Bubrick M (1990) *Circ. Shock* **31:** 37.
Jacobson JM, Michael JR, Harfi MH Jr & Gurtner GH (1990) *J. Appl. Physiol.* **68:** 1252–1259.
Jin LJ, LaLonde C & Demling RH (1986) *J. Appl. Physiol.* **61:** 103–112.
Johnson KJ, Ward PA, Kunkel RG & Wilson BS (1986) *Lab. Invest.* **54:** 499–506.
Kappus H (1985) In Sies H (ed.) *Oxidative Stress*, pp 273–310. London: Academic Press.
Kasama T, Kobayashi K, Fukushima T et al (1989) *Clin. Immunol. Immunopathol.* **53:** 439–448.
Kaslovsky RA, Lum H, Gilboa N, McCandless BK, Wright SD & Malik AB (1991) *Circ. Res.* (in press).
Kessler M, Höper J & Krumme BA (1976) *Anesthesiology* **45:** 184–197.
Konz KH, Tiegs G & Wendel A (1987) *Prog. Clin. Biol. Res.* **236A:** 281–288.
Krieger BP, Loomis WH, Czer GT & Spragg RG (1985) *J. Appl. Physiol.* **58:** 1326–1330.
Kukreja RC, Kontos HA, Hess ML & Ellis EF (1986) *Circ. Res.* **59:** 612–619.
Kunimoto F, Morita T, Ogawa R & Fujita T (1987) *Circ. Shock* **21:** 15–22.
Kvietys PR & Granger DN (1989) In Marston A, Bulkley GB, Fiddian-Green RC & Haglund UH (eds) *Splanchnic Ischemia and Multiple Organ Failure*, pp 127–134. London: Edward Arnold.
Lafuze JE, Weisman SJ, Apert LA & Baehner RL (1984) *Pediatr. Res.* **18:** 536–540.
Lang H, Jochum M, Fritz H & Redl H (1989) *Prog. Clin. Biol. Res.* **308:** 701–706.
Lee ES, Greenburg AG, Muffuid P, Melcher ED & Velkey TS (1987) *J. Surg. Res.* **42:** 1–6.
Lewis MS, Whatley RE, Cain P, McIntyre TM, Prescott SM & Zimmerman GA (1988) *J. Clin. Invest.* **82:** 2045–2055.
Lieners C, Redl H, Molnar H, Fürst W, Hallström S & Schlag G (1989) *Prog. Clin. Biol. Res.* **308:** 345–350.
Malik AB (1986) In Novelli GP & Ursini F (eds) *Oxygen Free Radicals in Shock. Int. Workshop, Florenz 1985*, pp 83–86. Basel: S. Karger.
Martin WJ (1984) *Am. Rev. Respir. Dis.* **130:** 209–213.
Marzi I, Rehkopf A, Hower R, Bühren V & Trentz O (1990) *Circ. Shock* **31:** 38.
Maunder RJ, Hackman RC, Riff E, Albert RK & Springmeyer SC (1986) *Am. Rev. Respir. Dis.* **133:** 313–316.
Maunder RJ, Winn RK, Gleisner JM, Hildebrandt J & Harland JM (1988) *J. Appl. Physiol.* **64:** 697–704.

McCord JM (1985) *N. Engl. J. Med.* **312:** 159–163.
McKechnie K, Furman BL & Parratt JR (1986) *Circ. Shock* **19:** 429–439.
Nerlich ML, Seidel J, Regel G, Nerlich AG & Sturm JA (1986) *Langenbecks Arch. Chir.* **supplement**: 217–222.
Neuhof H (1990) In Schlag G & Redl H (eds) *Shock, Sepsis and Organ Failure, First Wiggers Bernard Conference*, pp 404–420. Heidelberg: Springer.
Nishigaki J, Hagihara M, Hiramatsu M, Izawa Y & Yagi K (1980) *Biochem. Med.* **24:** 185–189.
Novelli GP, Angiolini P, Consales G, Lippi R & Tani R (1986) In Novelli GP & Ursini F (eds) *Oxygen Free Radicals in Shock, International Workshop, Florenz 1985*, pp 119–124. Basel: S. Karger.
Novotny MJ, Laughlin MH & Adams HR (1988) *Am. J. Physiol.* **254:** H954–H962.
Nuytinck JKS, Goris RJA, Redl H, Schlag G & VanMunster PJJ (1986) *Arch. Surg.* **121:** 886–890.
Olson NC, Grizzle MK & Anderson DL (1987) *J. Appl. Physiol.* **63:** 1526–1532.
Pacher R, Redl H, Frass M, Petzl DH, Schuster E & Woloszczuk W (1989) *Crit. Care Med.* **17:** 221–226.
Parker RE, Hardin JR & Brigham KL (1988) *J. Appl. Physiol.* **65:** 2138–2143.
Pauksens K, Sjlin J & Venge P (1989) *Scand. J. Infect. Dis.* **21:** 277–284.
Peavy DL & Fairchild EJ II (1986) *Infec. Immunity* **52:** 613–616.
Perlman MB, Johnson A, Jubiz W & Malik AB (1989) *Circ. Res.* **64**: 62–73.
Peters JH, Ginsberg MH, Bohl BP, Sklar LA & Cochrane CG (1986) *J. Clin. Invest.* **78:** 1596–1603.
Peters JH, Ginsberg MH, Case CM & Cochrane CG (1988) *Am. Respir. Dis.* **138:** 167–174.
Petrone WF, English DK, Wong K & McCord JM (1980) *Proc. Natl. Acad. Sci. USA* **77:** 1159–1163.
Pison U, Obertacke U, Brand M et al (1990) *J. Trauma* **30:** 19–26.
Ratych RE, Chuknyiska RS & Bulkley GB (1987) *Surgery* **102:** 122–131.
Redl H, Lamche H & Schlag G (1983) *Klin. Wochenschr.* **61:** 163–164.
Redl H, Schlag G & Hammerschmidt DE (1984) *Acta Chir. Scand.* **150:** 113–117.
Redl H, Schlag G, Schiesser A, Bahrami S, Junger W & Spragg RG (1986) In Novelli, GP & Ursini F (eds) *Oxygen Free Radicals in Shock. International Workshop, Florenz 1985*, pp 180–184. Basel: S. Karger.
Redl H, Hallström S, Lieners C, Fürst W & Schlag G (1988) In Hörl WH & Heidland A (eds) *Proteases II—Potential Role in Health and Disease*, pp 449–455. New York: Plenum.
Redl H, Schlag G, Paul E & Davies J (1989) *Circ. Shock* **27:** 308.
Redl H, Lieners C, Bahrami S, Schlag G, van Bebber IPT & Goris RJA (1990) In Emeritt I, Packer L & Auclair C (eds) *Antioxidants in Therapy and Preventive Medicine*, Vol. 264, pp 17–27. New York: Plenum.
Regel G, Dwenger A, Seidel J, Nerlich ML, Sturm JA & Tscherne H (1987a) *Unfallchirurg* **90:** 99–106.
Regel G, Nerlich ML, Dwenger A, Seidl J, Schmidt C & Sturm JA (1987b) *J. Surg. Res.* **42:** 74–84.
Richard C, Lemonnier F, Thibault M, Couturier M & Auzepy P (1990) *Crit. Care Med.* **18:** 4–9.
Sacks T, Moldow CF, Craddock PR, Bowers TK & Jacob HS (1978) *J. Clin. Invest.* **61:** 1161–1167.
Saez JC, Ward PH, Günther B & Vivaldi E (1984) *Circ. Shock* **12:** 229–239.

Sanan S, Sharma G, Malhotra R, Sanan DP, Jain P & Vadhera P (1989) *Free Radical Res. Commun.* **6:** 29–38.
Saugstad OD & Ostrem T (1977) *Eur. Surg. Res.* **9:** 48–56.
Schlag G & Redl H (1985) *Crit. Care Med.* **13:** 1045–1049.
Schlag G & Regele H (1972) *Med. Welt.* **23:** 1755–1758.
Schlag G, Redl H, Hallström S, Radmore K & Davies K (1990) *Circ. Shock* **31:** 11–12.
Schmeling DJ, Drongoswki RA & Coran AG (1988) *Circ. Shock.* **24:** 293.
Schmid-Schönbein GW (1987) *Fed. Proc.* **46:** 2397–2401.
Schneider J, Friderichs E, Heintze K & Flohe L (1990) *Circ. Shock* **30:** 97–106.
Schraufstätter IU, Revak SD & Cochrane CG (1984) *ZJ. Clin. Invest.* **73:** 1175–1184.
Schraufstätter IU, Hinshaw DB, Hyslop PA, Spragg RG & Cochrane CG (1986) *J. Clin. Invest.* **77:** 1312–1320.
Schraufstätter IU, Hyslop PA, Jackson JH & Cochrane CG (1988) *J. Clin. Invest.* **82:** 1040–1050.
Schraufstätter IU, Browne K, Harris A et al (1990) *J. Clin. Invest.* **85:** 554–562.
Seeger W, Lepper H, Wolf HRD & Neuhof H (1985) *Biochim. Biophys Acta* **835:** 58–67.
Seekamp A, Lalonde C, Zhu D & Demling R (1988) *J. Appl. Physiol.* **65:** 1210–1216.
Shasby DM, Shasby SS & Peach MJ (1983) *Am. Rev. Respir. Dis.* **127:** 72–76.
Smedly LA, Tonnesen MG, Sandhaus RA et al (1986) *J. Clin. Invest.* **77:** 1233–1243.
Smith SM, Holm-Rutili L & Perry MA (1987) *Gastroenterology* **93:** 466–471.
Spragg RG, Richman P, Gilliard N & Merritt TA (1988) In Lachmann B (ed.) *Surfactant Replacement Therapy in Neonatal and Adult Respiratory Distress Syndrome*, pp 203–211. Berlin: Springer.
Stephens KE, Ishizaka A, Wu Z, Larrick JW & Raffin TA (1988) *Am. Rev. Respir. Dis.* **138:** 1300–1307.
Suzuki M, Inauen W, Kvietys PR et al (1989) *Am. J. Physiol.* **257:** H1740–H1745.
Swann JD, Jones TW, Maki A et al (1990) *Circ. Shock* **31:** 41.
Sznajder JI, Fraiman A, Hall JB et al (1989) *Chest* **96:** 606–612.
Takeda K, Shimada Y, Amano M, Sakai T, Okada T & Yoshiya I (1984) *Crit. Care Med.* **12:** 957–959.
Till GO, Beauchamp C, Menapace D et al (1983) *J. Trauma* **23:** 269–277.
Till GO, Hatherill JR & Ward PA (1985) *Circ. Shock* **16:** 65–66.
Traber DL, Adams T, Sziebvert L, Stein M & Traber L (1985) *J. Appl. Physiol.* **58:** 1005–1009.
Traber DL, Schlag G, Redl H, Strohmaier W & Traber LD (1987) *Circ. Shock* **22:** 185–193.
Van Bebber IPT, Boekholz WKF, Goris RJA et al (1989) *J. Surg. Res.* **47:** 471–475.
Van Bebber IPT, Lieners CFJ, Joldewijn EL, Redl H & Goris RJA (1991) *J. Surg. Res.* (in press).
Varani J. Ginsburg I, Schuger L et al (1989) *J. Pathol.* **135:** 435–438.
Vedder NB, Winn RK, Rice CL, Chi EY, Arfors KE & Harlan JM (1988) *J. Clin. Invest.* **81:** 939–944.
Vedder NB, Fouty BW, Winn RK, Harlan JM & Rice CL (1989) *Surgery* **106:** 509–516.
Von Ritter C, Hinder RA, Oosthuizen MMJ, Svensson LG, Hunter SJS & Lambrecht H (1988) *Digest. Dis. Sci.* **33:** 857–864.
Wachtfogel YT, Kucich U, James HJ et al (1983) *J. Clin. Invest.* **72:** 1672–1677.
Wallis WJ, Hickstein DD, Schwartz BR et al (1986) *Blood* **67:** 1007–1013.

Ward PA, Till GO, Kunkel R & Beauchamp C, (1983) *J. Clin. Invest.* **72:** 789–801.
Ward PA, Till GO, Hatherill JR, Annesley TM & Kunkel R (1985) *J. Clin. Invest.* **76:** 517–527.
Warren JS, Kunkel SL, Johnson KJ & Ward PA (1987) *Am. J. Pathol.* **129:** 578–588.
Weiss SJ & Regiani S (1984) *J. Clin. Invest.* **73:** 1297–1303.
Weiss SJ & Slivka A (1982) *J. Clin. Invest.* **69:** 255–262.
Weiss SJ, Young J, LoBuglio AF, Slivka A & Nimeh NF (1981) *J. Clin. Invest.* **68:** 714–721.
Wendel A, Tiegs G & Werner C (1987) *Biochem. Pharmacol.* **36:** 2637–2639.
Westaby S, Fleming J & Royston D (1986) *Trans. Am. Soc. Artif. Intern. Organs* **31:** 604–609.
Wolf HRD & Lasch HG (1984) *Intensivmed. Prax.* **21:** 149–153.
Wong C, Flynn J & Demling RH (1984) *Arch. Surg.* **119:** 77–82.
Yokoyama Y & Parks DA (1988) *Gastroenterology* **94:** 607.
Yokoyama Y, Beckman JS, Beckman TK et al (1990) *Am. J. Physiol.* **258:** G564–G570.
Yoshikawa T, Murakami M, Yoshida N, Seto O & Kondo M (1983) *Thromb. Haemost.* **50:** 869–872.
Younes M, Schoenberg MH, Jung H, Fredholm BB, Haglund U & Schildberg W (1984) *Res. Exp. Med.* **184:** 259–264.
Zimmerli W, Huber I, Bouma BN & Lämmle B (1989) *Thromb. Haemost.* **62:** 1121–1125.
Zimmerman GA, Renzetti AD & Hill HR (1983) *Am. Rev. Respir. Dis.* **127:** 290–300.
Zimmerman GA, McIntyre TM, Mehra M & Prescott SM (1990) *J. Cell Biol.* **110:** 529–549.

Abbreviations

ABP	aminobiphenyl
ACE	angiotensin-converting enzyme
ACh	acetylcholine
AET	2-aminoethylisothiouronium bromide-HBr
AFB_1	aflatoxin B_1
AFBO	aflatoxin 8,9-oxide
AFFC	amniotic fluid fibroblast-like cells
α_1PI	α_1 protease inhibitor
ANS	antineutrophil serum
AP	apurinic/apyrimidic
ApppGpp	adenosine 5′,5″-triphospho-guanosine-3″-diphosphate
ARDS	adult respiratory distress syndrome
Asc	ascorbate
AT	ataxia telangiectasia
BHA	2(3)-*tert*-butyl-4-hydroxy-anisole
BHA	butylated hydroxyanisole
BHT	butylated hydroxytoluene
BHTOOH	2,6-di-*tert*-butyl-4-hydro-peroxyl-4-methyl-2,5-cyclo-hexadienone
BLM	bleomycin
bp	base pair
BP	benzo(a)pyrene
BPDE	benzo(a)pyrene-7,8-diol-9,10-oxide
BzPO	benzoyl peroxide
CDLE	chronic discoid lupus erythematosus
CDNB	1-chloro-2,4-dinitrobenzene
CGD	chronic granulomatous disease
Chl	chlorophyll
CHO	Chinese hamster ovary
CL	chemiluminescence
CPP	cyclopento(c,d)pyrene
DDC	diethyldithiocarbamate
DES	diethylstilbestrol
DF	desferrioxamine
DMBA	7,12-dimethylbenz(a)-anthracene
DMPO	5,5-dimethyl-1-pyrroline-*N*-oxide
DMSO	dimethylsulphoxide
DMTU	dimethylthiourea
dRpase	deoxyribophosphodiester-ase
DTPA	diethylene triamine penta-acetic acid
DTT	dithiothreitol
EC	endothelial cell
ECP	eosinophil cationic protein
EC-SOD	extracellular superoxide dismutase
EDHF	endothelium-derived hyper-polarizing factor
EDN	eosinophil-derived neuro-toxin
EDRF	endothelium-derived relaxing factor
EET	epoxyeicosatrienoic acids
EPR	electron paramagnetic resonance
ESR	electron spin resonance
EXP	eosinophil protein X

2-FAA 2-fluorenylacetamide
fd_{ox} oxidized ferredoxin
fd_{red} reduced ferredoxin
FMLP *N*-formyl-L-methionyl-L-leucyl-L-phenylalanine
FMO flavin-containing monooxygenases
GAPDH glyceraldehyde-3-phosphate dehydrogenase
G6PDH glucose-6-phosphate dehydrogenase
$GS^{\cdot}$ thiyl radical
GSH glutathione
GSH-Px glutathione peroxidase
GSSG glutathione disulphide
GST glutathione transferase
GTN glyceryl trinitrate
HbO_2 oxyhaemoglobin
HDFL human diploid fibroblast-like cells
HETE 5-hydroxyeicosatetraenoic acid
HGPRT hypoxanthine–guanine phosphoribosyltransferase
HMPS hexose monophosphate shunt
HNE 4-hydroxynon-2-enal
HP-I hydroperoxidase I
HP-II hydroperoxidase II
HPE 4-hydroxypentenal
HPETE (*S*)-5-hydroperoxy-6-*trans*-8,11,14-*cis*-eicosatetraenoic acid
hrSOD human recombinant superoxide dismutase
IFN interferon
IP_3 inositol 1,4,5-trisphosphate
IPG isopropylideneguanosine
LDL low-density lipoprotein
L-NIO *N*-iminoethyl-L-ornithine
L-NMMA *N*-monomethyl-L-arginine
L-NNA (guanidino)-*N*-nitro-L-arginine
LPS lipopolysaccharide
LT leukotriene
LTB_4 leukotriene B_4
MAB *N*-methyl-4-aminobenzene
MDA malondialdehyde
MBP major basic protein
MED minimal erythemal dose
MetHb methaemoglobin
MFO mixed function oxygenase
MNNG *N*-methyl-*N*-nitrosoguanidine
MPG *N*-2-mercaptopropionyl glycine
MPO myeloperoxidase
1-NA 1-naphthylamine
NAC *N*-acetylcysteine
N-AcO-2-FAA *N*-acetoxy-*N*-(2-fluorenyl)-acetamide
NBT nitroblue tetrazolium
NCS neocarzinostatin
NDGA nordihydroguaiaretic acid
N-OH-2-FAA *N*-hydroxy-*N*-(2-fluorenyl)-acetamide
4-NQO 4-nitroquinoline-*N*-oxide
8-OH-dG 8-hydroxydeoxyguanosine
8-OH-G 8-hydroxyguanosine
8-OH-Gua 8-hydroxyguanine
O^6-Me-dG O^6-methyldeoxyguanosine
O^4-Me-dT O^4-methyldeoxythymidine
8-oxo-dG 8-oxo-7,8-dihydrodeoxyguanosine
PA primary acceptor
PAF platelet-activating factor
PAH polycyclic aromatic hydrocarbons
PAN peroxyacetyl nitrate
PBN phenyl-*t*-butylnitrone
PBS phosphate-buffered saline
PG prostaglandin
3′-PGA 3′-phosphoglycolaldehyde
PGI_2 prostacyclin
PHGSH-Px phospholipid hydroperoxide glutathione peroxidase
PHS prostaglandin H synthase
PIP peroxidation-inhibiting protein
PKC protein kinase C
PLA_2 phospholipase A_2
PLC phospholipase C
PMA phorbol myristate acetate
PMN polymorphonuclear leukocytes
ppGpp 5′-diphosphate-3′-diphosphate
ppm parts per million
PUFA polyunsaturated fatty acids
QR quinone reductase

REM	relative electrophoretic mobility
RF	riboflavin
ROR	relative oxidation resistance
ROS	reactive oxygen species
SCE	sister chromatid exchange
SLE	systemic lupus erythematosus
SOD	superoxide dismutase
STZ	streptozotocin
TBARS	thiobarbituric acid reactive substance
TBHQ	*tert*-butylhydroquinone
TBQ	*tert*-butylquinone
TCDD	2,3,7,8-tetrachlorodibenzo-*p*-dioxin
TG	5,6-dihydroxy-5,6-dihydrothymine
TNF	tumour necrosis factor
TPA	12-*O*-tetradecanoylphorbol-13-acetate
TTC	triphenyltetrazolium chloride
Tx	thioredoxin
TxA_2	thromboxane A_2
Tx red	thioredoxin reductase
UV	ultraviolet
vitE	vitamin E
VOC	volatile organic compounds
XO	xanthine oxidase

Index

Note: inorganic radicals and oxidants are indexed under their formula

N-Acetoxy aminobiphenyl, glutathione conjugation and, 172
N-Acetyl benzoquinone imine, glutathione conjugation and, 172, 173
2–Acetylaminofluorene, phenolic antioxidants enhancing tumourigenesis by, 263
Acetylcholine as a vasodilator, 448–9
  coronary heart disease and, 470, 471
N-Acetylcysteine in shock therapy, 610
Acid hydrolases in neutrophils, 402
Acid smog, 14
Acrolein, genotoxicity, 357
Actin, platelet, glutathione-oxidizing compounds and, 432, 432–3
Actinic reticuloid cell line, UV effects, 68–9
Acylation of lysophospholipids, re-, 326
Adducts, *see* Crosslinking
Adenine, 8–hydroxyguanine and, base pairing between, 109
Adenocarcinoma cells, NO synthesis in, 459
Adenosine 5′,5″-triphosphoguanosine-3″-diphosphate accumulation, UVA and, 73, 74
Adenylate cyclase, platelet, 428–9
Adriamycin, toxicity/carcinogenicity, 296–8, 299
Adult respiratory distress syndrome, endotoxin and, 586
Aflatoxin B_1, glutathione conjugation and, 178–80
Aging, lens changes with, 534, 547
Agonists, chemotactic, respiratory burst and, 407–8
AhpC, 161, 162, 164, 166
AhpF, 161, 162, 166
Air pollution, oxygen radicals and, 3–55, *see also* Atmosphere
Albumin
  antioxidant activity, 216, 218–19, 233
  bilirubin bound to, antioxidant activity, 221–2, 227, 228
Aldehyde(s) from lipid hydroperoxides, 337–69, 375–6
Aldehyde dehydrogenases, 4-hydroxyalkenals and, 340–2
Alkaloids, plant, 4-hydroxyalkenal production from, 340
2–Alkenals, 338, *see also specific alkenals*
  growth inhibition by, 350, 352
  metabolism, 342
  toxicity, 348–9
    geno-, 360–1
Alkenes, peroxyacetyl nitrate reacting with, 22
Alkoxyl radicals, 188
  instability, 306
  from lipid oxidation, 379
    aldehydes formed by β-cleavage reactions of, 337–69
  toxicity/detoxification, 188
Alkyl hydroperoxide reductase, 63, 73, 161–2
Alkylation
  by 2–alkenals, 342
  glutathione and, 85–6

Alkylperoxyl radicals, production/toxicity/detoxification, 188
Allergy, photo-, drug-induced, 561
Allopurinol
  ischaemia–reperfusion injury and effects of, 506–7, 515, 516
  shock and organ failure and effects of, 608, 610
α-granules of platelets, 424
Amines
  antioxidative mechanisms in plants involving, 33
  aromatic, oxidation to free radicals, 301–2
  peroxyacetyl nitrate reacting with, 22
Aminobiphenyl (ABP), glutathione conjugation and, 172
Aminochrome [2,3-dihydroindole-5,6-quinone], quinone reductase and, 201
21-Aminosteroids
  cerebral ischaemia and effects of, 519
  shock and organ failure and effects of, 609
Aminothiols, oxidation, 92, 93
Ammonia, plant damage involving, 28–9
AMP, cyclic, platelets and, 428–9, 439
AmpR gene, 163
Amyl nitrite (isopentylnitrite), 472, 473, 475
Anaphylatoxins, trauma-related release, 602
Angina, nitrovasodilator therapy, 476–7
Angiotensin-converting enzyme inhibitors, therapeutic uses, 477
Animals, pollutants affecting, 35–44
Anthracyclines, oxidation to free radicals, 296–300
Anthrones, free radical metabolites of, 305
Anthropogenic emissions, 6–9, 45
Anticarcinogen, *see* Carcinogen
Antiinflammatory effects of phenolic antioxidants, 248–51
Antioxidants
  carcinogenic, suspected, 259–62, 264, 574–8
  8-hydroxydeoxyguanosine formation and its inhibition in the detection of natural, 105
  in ischaemia–reperfusion injury, protection/therapy with, 505–6, 519
  LDL oxidation and, 376, 390–2
  in lens, 540–1, 542
  phenolic, see Phenolic antioxidants
  photochemical processes/light radiation and, 574–8
  in plasma, 213–43
    interaction between, 231–3
    site-specific protection by, 233
  platelets and effects of, 437–8
  prooxidant activity, *see* Prooxidant activity
  in serum, 230
  in shock and trauma-related organ failure, 597–8
  synthetic, 610
  vitamins, *see* Ascorbic acid, Vitamin E
Antioxidative systems
  in animals (in general), 69–70, 72
    diesel soot particles and, 40
  in microbes, 62–3, 69
  in plants, 27, 32–4
  UVA and, 62–3
α-Antiprote(in)ase, *see* Protease inhibitor
Aortic endothelial cells, LDL oxidative modifications involving, 382, 383, 384, 385
AP sites, *see* Apurinic/apyrimidinic sites
Apn1 yeast endonuclease, 139–40, 142–3
Apolipoprotein B, LDL oxidation measured by fluorescence associated with, 374–5, 376
Apurinic/apyrimidinic (AP) sites
  endonuclease/lyase, 128, 129, 131, 137–9, 139–41, 141–2, 157–9
  oxidized, 123, 124
  repair at, 111, 128–30, 137–9, 139–41, 141–2
Arachidonate halohydrin glutathione conjugates, 186
Arachidonic acid metabolism, 173, 182–5, 190, 426–8, 587–8
  cutaneous, UV effects, 568
  endotoxic shock and the products of, 587–8
  enzymes of, 173, 181–5, *see also specific enzymes*

glutathione conjugation and, 173, 181–5, 190
NO and the products of, 468–9
particulate components affecting, 44–5
phenolic antioxidants affecting, 251
platelet, 422, 426–8, 430–1, 432, 433–6, 437–8
reperfused organs and free radical generation and the, 498–9
Arginine
analogues, NO studies with, 451, 456–7, 459
NO synthesis from, 450–1, 454–6
Aromatic amines, oxidation to free radicals, 301–2
Aromatic hydrocarbons, 295–6
carcinogenicity and toxicity, 181, 197–8, 291–2, 295–6
glutathione conjugation and, 181
nitropolycyclic, 181
oxidation to free radicals, 295–6
polycyclic, 205, 291–2, 295–6
quinone metabolites and, 197–8, 296
quinone reductase induced by, 205
Aromatic isothiocyanates, anticarcinogenic effects, 206
Aromatic nitro compounds, *see* Nitroaromatic compounds
Arrhythmias, reperfusion, 504–5, 506, 507
Asbestos fibres, 41–2, 103
Ascites tumour cells, Ehrlich, 4-hydroxyalkenals in, effects, 346–8, 350, 351, 352, 353, 355
Ascorbic acid, 535–6, 542
as an antioxidant, 220–1, 225, 226, 227, 228, 230, 231–3, 235, 236, 542
erythrocytic, 235
in lenses, 535–6, 536, 537
leukocytic, 236
plasma, 220–1, 225, 226, 227, 228, 230, 231–3, 235
as a prooxidant, 224–5, 535–6, 536, 537
tocopherols and, interactions, 231–3
Astaxanthin as an antioxidant, 229
Ataxia telangiectasia, 70
Atherosclerosis, phenolic antioxidants in the prevention of, 252
Atmosphere, 4–17, *see also* Air
chemistry/reactions, 9–17
emissions into, *see* Emissions
ATP
in lens, 548
in shock and organ failure, 605
ATPases, lens, 548
Autoxidation, LDL, 380, 382, 383, 385
Autoxidized saturated fatty acids, 8-hydroxydeoxyguanosine formation in DNA by, 104
Azaserine, phenolic antioxidants affecting carcinogenicity of, 254
Azo-diacarboxylic acid-bis-dimethylamide, *see* Diamide
Azo dyes, 7,12–dimethylbenz(a)anthracene-induced toxicity and the protective effects of, 205–6
Azoreductase, quinone reductase as a, 202–3
Azurophil granules in neutrophils, 401–2

Bacteria
antioxidant enzymes in, 62–3, 71
antioxidant molecules in, 69
DNA damage in, 60
repair of, 62, 67–8, 109–11, 112, 127–30, 130–9, 142–4, 144–5, 157–9, 360
endotoxin, *see* Endotoxin
gene expression in, induction, 72–4
hydrogen peroxide in, *see* H_2O_2
4-hydroxyalkenals as mutagens in, 357–8, 359
killing/inactivation, 63–5, 73, 156–7
by hydrogen peroxide, 156–7
by neutrophil enzymes, 402
by NO, 458
respiratory burst and, 406
regulator genes, 163
transforming DNA, 60–1
translocation, in gut, 608
UVA effects on, 60, 60–1, 63–5, 71–2, 72–4
Bases (of nucleic acid)
radical-damaged, 120–2
repair, 127–35
sites free of, *see* Apurinic/apyrimidinic sites
Bay region diol epoxides, glutathione conjugation and, 178

Beck–Halliwell–Asada cycle, 27, 32
Benign lesions, progression to malignancy, 279–80
Benzanthracene, glutathione conjugation and, 172, 173, 178
Benzene, 293–5
 enzymatic conversion, 281
 free radicals from the oxidation of, 293–5
 superoxide dismutase and, 281, 294
 toxicity, 290–1, 293–5
  quinone reductase and, 200
Benzidine, 301, 302
 glutathione conjugation and, 180–1
 oxidation, 301
Benzidine diimine, 301
 glutathione conjugation and, 173, 180
Benzo(a)pyrene (BP)
 carcinogenicity/mutagenicity of (and its derivatives), 283, 286–7, 291–2, 295–6
  glutathione conjugation and, 172, 173, 178
  phenolic antioxidants affecting, 254, 258
Benzo(a)pyrene-7,8–diol-9,10-epoxide, toxicity/carcinogenicity, 296
Benzo(a)pyrene-7,8–diol-9,10-oxide (BPDE), glutathione conjugation and, 172, 173, 178
Benzo(a)pyrenediones, quinone metabolites and, 197–8
Benzoyl peroxide, free radical formation from, 307
β-cell damage by streptozotocin, 300–1
 β-cleavage reactions of alkoxy radicals arising from hydroperoxides of PUFAs, aldehydes formed by, 337–69
β-elimination reaction, 128, 129
β-lyase, 131
Bilirubin
 antioxidant activity, 221–2, 227, 228, 229, 230, 233
 conjugated, 222
Biomolecules, air pollutants reacting with, 17–29, *see also* Macromolecules
Bleomycin (BLM), 281–2
 DNA changes induced by, 123, 124, 281–2
  repair, 138–9, 145
 lung toxicity, 282
Blood
 cells in, antioxidant defences provided by, 234–7
 flow and pressure regulation, NO and, 469–70
 plasma, endogenous antioxidant defences, 213–43
 serum, endogenous antioxidant defences, 230
 whole, chemiluminescence, with post-trauma neutrophil activation, 603
Bloom's syndrome, 571–2
Bone marrow
 eosinophil maturation, 403–4
 macrophage maturation from monocytes in, 404
 neutrophil maturation in, 401
 NO synthesis in cells of, 459
 xenobiotic activation in, 290–1
Brain
 ischaemia, 518–19
  arachidonate metabolism and, 499
 NO synthesis in, 460
Burns, 599
 lipid peroxidation and, 599
 lung injury secondary to, 607
Buthionine-S,R-sulphoximine, 69
Butylated hydroquinone (2(3)-*tert*-butylhydroquinone; TBHQ), adverse effects, 261, 264
Butylated hydroxyanisole (2(3)-*tert*-butyl-4-hydroxyanisole; BHA), 245–67
 adverse actions, 259–66 *passim*, 286
 butylated hydroxytoluene and, synergism between, 266
 protective effects of, 246–59 *passim*
  menadione-induced singlet oxygen production and, 199
Butylated hydroxytoluene (BHT), 246–67 *passim*
 adverse actions, 259–66 *passim*, 307–10
 butylated hydroxyanisole and, synergism between, 266

free radical metabolites of, 307–10
photoprotective effects, 578
protective effects of, 246–59 *passim*
Butylated quinone (2(3)-*tert*-butylquinone; TBQ), adverse effects, 261, 264
t-Butylhydroperoxide, platelet aggregation and effects of, 433

Calcium
NO synthetase and, 456, 458
phenolic antioxidants affecting events regulated by, 248–50
platelet, 425–6, 436, 437, 438
reperfusion injury and, 495–6
respiratory burst in phagocytes and, 409–11
in shock and organ failure, 605
Calcium ATPases in lens, 548
Canalicular system platelets, 424
Cancer, *see sites and specific types of tumours* and Carcinogen; Chemo-therapeutic agents; Cytotoxic agents; Tumours
Canthaxanthin as an antioxidant, 229
Carbohydrates, *see also* Sugar
heated, DNA damage induced by, 100, 104
in plants, pollutants and the transport of, 32
Carbon
compounds containing, *see specific formula*
in plants, pollutants and the partitioning of, 32
Carbonyl reductase, quinone reduction by, 205
Carcinogen(s), 277–319, 574–5, *see also* Tumours
asbestos as a, 103
cigarette smoke as a, 38, 102
diesel soot particles as, 40
free radicals in the activation of, 277–317
lipid peroxidation in formation of, 258, 284
oxidation, 292–304
phenolic antioxidants as, 259–62, 264
protection against
novel substances conferring, identification, 206–7
phenolic antioxidants conferring, 253–9
2,3,7,8-tetrachlorodibenzo-*p*-dioxin (TCDD) as, 44
Carcinogenesis, 574–5
antioxidants implicated in, 259–62, 264, 574
DNA changes/damage etc. in the process of, 99, 102, 103–8, 112
8–hydroxydeoxyguanosine and, 102–108, 112
multi-stage nature of, 278–80
photochemical processes/light radiation in, 66, 563, 568, 574–5
skin, 563, 573–4
lipid peroxidation and, 568
Carcinomas, forestomach, phenolic antioxidants inducing, 259, 260
Cardiovascular disease, *see also* Heart
NO and, 447, 469–71, 476–7, 477–8
oxygen radicals and, 447, 471, 477
therapy, 476–8
β-Carotene
antioxidant role, 223–4, 229, 564, 577
in skin, 564, 577
Carotenoids
antioxidant role of, 223–4, 229, 390–1, 564, 577
LDL oxidation and the, 390–1
in skin, 564, 577
UVA damage prevented by, 64, 69
Catalase
ischaemia–reperfusion injury and, 507–10, 515, 516
in lenses, 543
in mammalian cells, UVA and, 72
in microbes
hydrogen peroxide and, 160
UVA and, 62–3, 71, 73
plasma (human), 215–16
in shock and organ failure, 609
UV-inactivated, 575
xeroderma pigmentosum and, 571
Catalysis, particle, 24–5
Cataracts, 544–51
antioxidants in the prevention of, 542–3

Cataracts (*cont.*)
  cortical, 546
  DNA damage and, 549
  glutathione and, 541, 542
  nuclear, 545, 546
  oxidative stress causing, 544–51
  sugar-induced, 541
  UV and, 58
Catechol, benzene-derived, 294–5
Catecholamines, ischaemia–reperfusion injury and, 499
Cationic proteins in eosinophils, 404
CD18 (neutrophil adherence molecule), haemorrhagic shock and, 604
Cell(s), *see specific (types of)* cells and Bacteria; Mammalian cells
Cell cycle, lipid peroxidation and the, 354
Central nervous system ischaemia, 499, 518–19
Cerebral ischaemia, *see* Brain
Ceruloplasmin, antioxidant activity, 216, 217
Charcot–Leyden crystals, 403–4
Chemiluminescence
  ozone, NO assay via, 451–2
  whole blood, with post-trauma neutrophil activation, 603
Chemoprotected state, quinone reductase as a marker of, 205–7
Chemotactic agonists, respiratory burst and, 407–8
Chemotherapeutic agents
  phase II enzymes and, 204
  photo-, 565–6
  toxicity/carcinogenicity, 285, 296–9
Chinese hamster ovary cells, 4-hydroxyalkenal effects, 347, 351–2, 358–9, 361
1–Chloro-2,4-dinitrobenzene
  as a glutathione transferase substrate, 176–7, 177, 433
  platelet glutathione depleted by, 433
Chlorophyll, bleaching by lipid peroxidation, 28
Chloroplasts, reactions in, 25–9
Chromatid exchange, sister, 4-hydroxyalkenal genotoxicity and, 361–4
Chromosomal aberrations, 4-hydroxyalkenal genotoxicity and, 361–4
Cigarette smoke
  damage caused by, 36–9, 102, 107
  LDL oxidative modifications caused by, 381, 384
Ciprofibrate, 8-hydroxydeoxyguanosine in DNA and effects of, 106–7
Clastogenic factors in lupus erythematosus, 572–3
Clofibrate, aldehyde dehydrogenases and, 341
CO, atmospheric oxidation, 11
Coenzyme Q_{10}, reduced (ubiquinol-10), antioxidant activity, 224, 229
Collateral flow, ischaemia–reperfusion injury and, 508, 509, 510
Colony-stimulating factors, phagocyte priming by, 414–15, 415
Complement, trauma-related release, 602
Conjugated dienes in monitoring of LDL oxidation rate, 374
Conjugation, *see also* Cross-linking
  bilirubin, 222
  glutathione, *see* Glutathione
Contraction, smooth muscle, phenolic antioxidants inhibiting, 252
Copper, oxidation induced by, 229, 378–80, 382
Copper–zinc superoxide dismutase
  in ischaemia–reperfusion injury, 512, 513
  in mammalian cells, 75
  in plants, 27, 33
Coronary artery disease, NO and, 470–1
Cortical cataracts, 546
Coumarins, anticarcinogenic effects, 206
Criegee zwitterions, 20
Crosslinking/adducts, *see also* Conjugation
  DNA–DNA
  interstrand, 126
  intrastrand, 125–6
  radical-induced, 125–6
    UV-induced, 569
  DNA-polycyclic aromatic hydrocarbon, 292
  DNA-protein
    radiation, wavelengths inducing, 60, 569

radical-induced, 126–7
Crotonaldehyde, genotoxicity/mutagenicity, 357, 358
Cruciferae, anticarcinogens in, 207
Cryptoxanthine, LDL oxidation and the antioxidant role of, 390
Crystallins
age-related changes, 534
disulphide formation, 545
UV irradiation-related changes, 58
Cumene hydroperoxide in platelets, effects, 433
Cutaneous tissue, *see* Skin
Cyclic nucleotides, *see* Nucleotides *and specific nucleotides*
8,5′-Cyclo-2′-deoxyguanosine formation, 125–6
Cyclohexa-2,5–dienone products, 262
Cyclooxygenase
cutaneous, UV effects, 568
glutathione conjugation and the, 173, 182–5
LDL oxidation and, 389
NO and the products of, 468–9
platelet, 427, 430, 431
reperfused organs and free radical generation and, 498, 499
shock and, 604
endotoxic, 587
Cyclopento(c,d)pyrene, activation, 286
CysB gene, 163
Cysteamine, UV radiation and, 70
Cysteine
diesel soot particles and, interactions, 40
NO release requiring, 465–6, 474
seleno-, peroxidases containing, 327, 328–9
Cytochrome b_{558} as an NADPH oxidase component, 412, 413
Cytochrome P450-dependent monooxygenases, *see* Oxygenases
Cytokines
endotoxic shock and, 458, 459, 586, 589–91, 606
NO and, 458–9
phagocyte priming by, 414–15, 415
reactive oxygen species and, 589–91
Cytoplasmic enzymes, xenobiotic/carcinogen activation involving, 283
Cytoskeleton, platelet, glutathione-oxidizing compounds and the, 432
Cytosolic protein–membrane protein linkage in lens, 545–7
Cytotoxic agents
phase II enzymes and the design of, 204
toxicity/carcinogenicity, 285, 296–9
Cytotoxicity
of 4-hydroxyalkenals, 345–50
of oxygen radicals, 605

Daunomycin, toxicity/carcinogenicity, 296–8, 299
Death/inactivation
of bacteria, *see* Bacteria
of mammalian cells, 65, 352
Defence mechanisms (cellular), *see also specific (types of) defence mechanisms*
in bacteria, 66–9, 71
to hydrogen peroxide, 157–9
in lung, damage to, 37
UVA and, 66–72
Deferoxamine (desferrioxamine)
ischaemia–reperfusion injury and effects of, 502–3
shock and organ failure and effects of, 609
Dehydroascorbic acid
erythrocytic, 235
leukocytic, 236
Dense granules/bodies of platelets, 423–4
Deoxyguanosine
4-hydroxyalkenal reaction with, 355
8-hydroxydeoxyguanosine formation from, 101
Deoxyhaemoglobin, NO and, 460
2-Deoxypentos-4-ulose formation, 123
2-Deoxyribonolactone formation, 123
Deoxyribophosphodiesterase, *E. coli*, DNA repair by, 139
Deoxyribose damage, radical-induced, 123–5
repair, 134–43
Dermatology, *see* Skin
Dermis, 560
Detoxification
alkoxyl/alkylperoxyl radicals, 188
hydroquinones, 202, 290

Detoxification (*cont.*)
  4-hydroxyalkenals, 188, 343
  in lens, 540, 541
    of free radicals, 541
    of hydrogen peroxide, 540
  in plants, 27, 32
Diacylglycerols, phagocyte priming by, 414
Diamide (azo-diacarboxylic acid-bis-dimethylamide), glutathione oxidation by, 87
  in platelets, 431–2
Diazotization, sulphanilic acid, NO assay via, 453
N,N-Dibutylnitrosamine, phenolic antioxidants enhancing tumourigenesis by, 263
Dicoumarol, quinone toxicity and quinone reductase and effects of, 199, 200, 203, 205
Dienes, conjugated, in monitoring of LDL oxidation rate, 374
Diesel soot particles, damage caused by, 39–41
3′-Diesterases, 139, 142
Diethyldithiocarbamate, lens epithelial cells and effects of, 538
Diethylnitrosamine (and its derivatives), phenolic antioxidants affecting carcinogenicity of, 255
Diethylstilboestrol
  quinone metabolite of, formation of/damage caused by, 299–300
  redox cycling by, 300
    quinone reductase-inhibited, 199–200
  toxicity/carcinogenicity, 299–300
Differentiating tissues, membrane destruction in, 331
Dihydrodiols, chemical-initiated formation, 283
Dihydrolipoate, photoprotection by, 577–8
7,12-Dihydroxymethylbenzanthracene, glutathione conjugation and, 178
1,4-Dihydroxynonene metabolism in hepatocytes, 343, 344
7,12-Dimethylbenz(a)anthracene
  azo dyes in protection from toxic effects of, 205–6
  glutathione conjugation and, 178
  phenolic antioxidants affecting carcinogenicity of (and its derivatives), 255
1,2-Dimethylhydrazine, phenolic antioxidants enhancing tumourigenesis by, 263
Dinitrobenzoic acid in shock/trauma-related organ failure, 598
Diol epoxides, Bay region, glutathione conjugation and, 178
Dioxins, polychlorinated
  carcinogenicity, 44
  sources, 9
Disulphides, protein, in lens, 541–2, 545–7
  formation, 541–2, 545–7
  reduction, 541
Disulphiram, reduction and clinical use, 93–4
Diterpenes as anticarcinogens, 206
2,6-Di-*t*-butyl-4-hydroperoxyl-4-methyl-2,5-cyclohexadienone formation, 308–10
Dithiocarbamate disulphide, reduction and clinical use, 93–4
1,2-Dithiol-3-thiones as anticarcinogens, 206
Dithiothreitol, superoxide release by, 464–6
Divalent metals, exonuclease III requirement for, 143
DNA, *see also* Genotoxicity
  crosslinking, *see* Crosslinking
  by cytotoxic agents, 299
  damage, 99–113 *passim*, 119–54, 357–64, 568–70
    by aromatic amines, 301
    by benzene oxidation products, 295
    consequences, 120–7
    by diethylstilboestrol-quinone, 299
    by 4-hydroxyalkenals, 357–64
    in lens cells, 533, 538, 549
    in mammalian cells, 60, 99–113 *passim*, 357–64
    in microbes, 60, 62
    by nitro compounds, 302
    by peroxide free radical metabolites, 307

by phenolic antioxidants, 264
in plants, 34
by polycyclic aromatic hydrocarbon oxidation products, 296
repair, *see below*
in respiratory systems of animals, 38
UVA-related, 59–60, 62, 569
UVB-related, 569
8-hydroxydeoxyadenosine in, 105
8-hydroxydeoxyguanosine in, *see* 8-Hydroxydeoxyguanosine
8-hydroxyguanine in, 107, 122, 134–5
peroxidation, 189
recombination, *see* Recombination
repair (of damage), 66–9, 109–11, 112–13, 127–54, 189
issues in, 145–8
in microbes, 62, 67–8, 109–11, 112, 127–40, 142–5, 157–9, 360
replication/synthesis
4-hydroxyalkenal effects, 353–4
misreading during, 108–9, 112
transcription, hydrogen peroxide stress and, 166–7
transforming (bacterial), UVA-related inactivation, 60–1
DNA glycosylases, 127–32
DNA methyltransferase, 4-hydroxyalkenal effect on, 360
DNA polymerases
4-hydroxyalkenal effects, 353
repair involving, 144–5
Drosophila apurinic/apyrimidinic (AP) endonuclease, 140–1
Drugs
glutathione oxidation and the role of, 86–7, 94, 95
hydrogen peroxide generation and, 94
P450-dependent monooxygenases and, 94, 95
photosensitivity induced by, 561, 564
vasodilatory, NO-donating, 472–6
Dysrhythmias, reperfusion, 504–5, 506, 507

Ebselen, 552–3, 610
EDRF, *see* endothelium-derived relaxing factor
Ehrlich ascites tumour cells, 4-hydroxyalkenals in, 346–8, 350–3, 355
Eicosanoids, *see* Arachidonic acid metabolism *and specific eicosanoids*
Elastase
endotoxin-induced cytokine release and, 590
in shock and organ failure, 590, 603, 606
Electron transport chain, *see* Respiratory chain
Electrophiles
4-hydroxyalkenals as, 342
quinones as, 197
xenobiotic, glutathione conjugation and, 178
Electrophoretic mobility, relative, in measurements of LDL oxidation, 376
Emissions, atmospheric, 4–9, 45
anthropogenic, 6–9, 45
natural, 4–6
Endonuclease
bacterial, 109–11, 133–4, 137–9, 139–41
DNA repair involving, 109–11, 113, 133–4, 137–9, 139–41
drosophila, 140–1
human cell, 111, 113, 141–2
8-hydroxydeoxyguanosine, 109–11, 112
8-hydroxyguanine, 111, 113, 134–5, 149
yeast, 139–40
Endonuclease III
bacterial, 111, 128–31
DNA repair involving, 128–31, 137–9, 139–40
Endonuclease IV
DNA repair involving, 137–9, 139, 142–3
yeast Apn1 endonuclease and, homology, 140
Endonuclease V, DNA repair involving, 133–4
Endothelial cells
LDL oxidative modifications involving, 372, 381, 382–5 *passim*
NO synthetase, 456–8

Endothelial cells (*cont.*)
  NO/EDRF-interfering factors in, 468–9
  shock-related organ failure and, 605, 606
  umbilical vein, 4-hydroxyalkenal effects, 347, 351, 352
Endothelium
  cardiovascular disease and the role of, 469–71
  shock-related organ failure and, 602, 605
Endothelium-derived hyperpolarizing factor, 469
Endothelium-derived relaxing factor, 445–89
  biochemical properties, 460–9
  deactivation and inhibition, 460–6
  historical background, 448–51
  interference by other factors with, 468–9
  NO and
    differences, 467–8
    relationships/identity, 429, 446, 448–51, 460–9
  platelet aggregation and, 429, 439
Endotoxin (lipopolysaccharide; LPS), 585–93
  cytokines and, 458, 459, 586, 589–91, 606
  lipid peroxidation and, 599–600
  NO and, 458, 459
  organ failure caused by, radical scavenging therapy, 609
  pathophysiological role, 585–6
  phagocyte priming by, 415
Energy balance upset by redox-cycling agents, 148
Enzyme(s), *see also specific enzymes*
  antioxidant
    in mammalian cells, 72
    in microbes, 62–3, 70–2
    in plasma, 214–16
    in shock and organ failure, 608–9
    UV and, 62–3, 70–2, 73, 571, 575
  carcinogen and other xenobiotic-activating, 280–4, 287–92 *passim*
  DNA repair, 109–11, 112–13, 127–45, 157–9, 360
  eosinophil, 403–4
  4-hydroxyalkenal effects on, 349, 360
  inactivation, UVA, 62–3, 571
  in ischaemia, 495
  lens
    defensive role, 543–4
    mimics of, 552–4
  membrane-associated, *see* Membranes
  monocyte/macrophage, 404, 405
  neutrophil, 400, 401–2
  phase II, preneoplastic nodules and tumours and, 204, 206
  skin diseases and, 571
Eosinophil(s), 403–4
  granules, 403–4
  maturation, 403–4
  other phagocytic cells and, properties in common, 399
  respiratory burst in, 407
Eosinophil-derived neurotoxin, 404
Eosinophil protein X, 404
Epidermis, 560
  one-electron reduction potential at surface of, 576
Epithelial lens cells, 533
  DNA damage, 533, 538, 549
  hydrogen peroxide stress in, 540
Epithelium, renal tubular, oxidative damage and the, 605
Epoxidation, benzo(a)pyrene-7,8-diol, 286
Epoxides, Bay region diol, glutathione conjugation and, 178
2-Epoxybenzoquinone, quinone reductase and the reduction of, 201
Epoxyeicosatrienoic acid glutathione conjugates, 186
Erythema, UV-related, 65–6, 576
Erythrocytes, *see* Red blood cells; Reticulocytes
Erythropoietic protoporphyria, 564, 577
Escherichia coli
  antioxidant molecules in, 69
  DNA damage in, 60
    repair of, 62, 67–8, 109–12, 128–30, 131–9, 142–3, 143–4
  gene expression in, induction, 73
  hydrogen peroxide in, 73, 136, 155–69

UVA effects in, 60, 62, 67–8, 69, 71
1,N^2,-Ethenodeoxyguanosine production, 355
Exonuclease III, DNA repair involving, 67, 71, 135–7, 139, 142–3, 157–9
Eye, *see* Lens

F-Actin, platelet, glutathione-oxidizing compounds and, 432
FAD-containing flavoprotein, quinone reduction by a, 195
FAPy glycosylase, 131–2, 149, *see also* Formamidopyrimidine
Fatty acids
autoxidized saturated, 8-hydroxy-deoxyguanosine formation in DNA by, 104
damaged, elimination, 325–6
hydroperoxy
phospholipase A_2 specificity for, 325
platelet aggregation and, 435
reperfused organs and free radical generation and, 498–9
LDL oxidation and the disappearance of, 376
peroxidation, *see* Peroxidation
polyunsaturated, *see* Polyunsaturated fatty acids
in thylakoid membranes, degradation, 26
α,β-unsaturated ketone-containing, glutathione conjugates derived from, 186–7
Fenton reaction, 59, 76, 101
Ferredoxin, reduced, oxygen reduction involving, 26
Ferritin, antioxidants and, 217, 218
Ferrous and ferric iron, *see* Iron
Fibroblasts
antioxidant molecules in, 69
DNA repair in, 68
heat treatment effects, 75
hydrogen peroxide effects, 75
4-hydroxyalkenal effects, 345–6, 347, 350, 351, 353
UVA effects, 65, 68
Fibronectin in shock and organ failure, 606
Flavin(s), photosensitization, 564
Flavin-containing monooxygenases, multisubstrate, properties, 89–91, 92, 93
Flavonoid antioxidants, 246
adverse effects, 261, 263–4
antiinflammatory action, 248
platelet aggregation inhibited by, 251
Flavoprotein(s)
carcinogen/other xenobiotics activated by, 280–1
in lens, 544
as an NADPH oxidase component, 412
quinone reduction by, 195, 197
Fluorescence at 430nm with 360nm excitation, LDL oxidation measured by increases in, 374–5
Food antioxidants
anticarcinogenic properties, 253
carcinogenic properties, 260, 262
Forestomach cancer, phenolic antioxidants inducing, 259, 264
Formaldehyde
glutathione reacting with, 94–5
photolysis, 10
Formamidopyrimidine (FAPy) in DNA, radical-induced formation, 120, *see also* FAPy
Fpg gene, 131, 149
Free radical(s), *see also specific (types of) radicals*
carcinogen activation and the role of, 277–317
carcinogen oxidation to, 292–305
endotoxin-induced cytokine release initiated by, 589–90
ischaemia–reperfusion and, 447, 471, 496–503, 503–4, 505, 516, 517, 517–18
LDL oxidation and, 379–81, 382, 383
in lens, 536–7, 538
phenolic antioxidants removing metabolites of, 258
scavenging, *see* Scavenging
in shock and organ failure, 589–90, 597–606, 608–9
involvement, 589–90, 597–600
sources, 600–4
targets and effects of, 605–6
in skin, 559–60, 562–3, 578–9
Free radical disease, 578–9

Furoxan derivatives, 472, 473, 476

G-protein, platelet, 425–6
Gallic acid esters
  adverse effects, 261
  protective effects of, 247–59 *passim*
γ-radiation
  DNA crosslinking induced by, 125–6
  8-hydroxydeoxyguanosine formation in DNA induced by, 103
  8-hydroxyguanine formation in DNA induced by, 122
Gastrointestinal tract
  ischaemia–reperfusion injury and the, 501, 515–16
  shock-related injury to, 605–6
Gene expression/activity
  4-hydroxynonenal-mediated modulation of, 355
  UVA-induced, 72–8
Genotoxicity
  of 4-hydroxyalkenals, 355–64
  of phenolic antioxidants, 261
Germany (West), anthropogenic emissions of main pollutants in, 7
Glucose, antioxidant activity, 222
Glutamate toxicity, 203
γ-Glutamyl cysteine synthetase, 69
Glutathione, 85–97, 171–94, 431–3, 539–42, *see also* Thiyl radicals
  analogues of, in lens, 539
  conjugation, 86, 94–5, 171–94, 541–2
    with 4-hydroxyalkenals, 340–1, 342, 344
  erythrocytic, 234–5
  eukaryotic, 69, 85–97
  in lenses, 539–42
    antioxidant role, 540–1
    concentration and localization, 539
    elevation, 551–2
    loss, 541, 547
    redox state, 539–40
  oxidation, 85–97, 431–3
    mechanisms, 87–94, 431–3
  platelet, 428, 430–1, 431–3
  prokaryotic, 69
Glutathione peroxidase
  extracellular human, 214–15
  lens, 540, 543–4
    mimics of, 552–4
  membranous, 326, 328–9
  phospholipase plus, antiperoxidative action exerted by, 329
  phospholipid hydroperoxide glutathione peroxidase and, relationship between, 328–9
  platelet, 428
  selenium dependence, 327, 328–9, 553
Glutathione reductase
  bacterial, hydrogen peroxide and, 160–1
  human/mammalian, phenolic antioxidant anticarcinogenicity and, 253
Glutathione transferases, 171–94
  human, 189–90
  4-hydroxynonenal-metabolizing, 342–3
  isoenzymes, 174–81
  nitrovasodilators and, 474
  platelet, 433
Glycation, cataract development and, 541
Glycerol, radiation effects prevented by, 63
Glyceryl trinitrate (GTN), 472, 475
Glycocalyx membrane of platelets, 423
Glycolysis, red cell, 235
Glycosylases, DNA, 127–32
GMP, cyclic
  NO and, 429, 439, 446, 450
  platelets and, 428, 429, 436, 439
GorA gene, 160–1, 164
G-protein, platelet, 425–6
Gram-negative bacterial endotoxin, *see* Endotoxin
Granules (storage)
  eosinophils, 403–4
  monocyte/macrophage, 404, 405
  neutrophil, 400–3
  platelet, 423–4
Granulomatous disease, chronic, NADPH oxidase in, 413
Growth inhibition by 4-hydroxyalkenals in mammalian cells, 350–5
GTP-binding proteins, respiratory burst and, 408–9
(Guanidino)-*N*-monomethyl-L-arginine (L-NMMA), NO synthesis inhibited by, 456, 457, 459

(Guanidino)-N-nitro-L-arginine (L-NNA), NO synthesis inhibited by, 456
Guanine modification, 99–116
Guanosine, aldehydes reacting with, 357, *see also* GMP; GTP
Guanosine 5′-diphosphate-3′-diphosphate synthesis, UVA and, 73, 74
Guanylate cyclase
  NO and, 429, 439, 446, 450, 454, 462, 466, 474
  platelet, 429, 435–6, 439
Gut, *see* Gastrointestinal tract

Haem, pools, 76, 78
Haem oxygenase, UVA and, 70, 76, 78
Haematoporphyrins, 566
Haemoglobin oxidation, 218, *see also* Methaemoglobin; Oxyhaemoglobin
Haemopexin, antioxidant activity, 216, 218
Haemoproteins, ischaemia–reperfusion injury and, 502–3
Haemorrhagic shock, 608, 609, 610
  oxygen radicals and, 604, 608, 609
Halohydrin glutathione conjugates, arachidonate, 186
Haptoglobin, antioxidant activity, 216, 218
Heart, *see also* Myocardium
  arrhythmias, 504–5, 506, 507
  ischaemia–reperfusion injury, 503–15, 519
    neutrophils and, 501–2
    NO and, 447, 471
  ischaemic and ischaemia–reperfusion injury, free radicals and, 447, 471, 500
Heat shock responses, stress responses and, 75
HeLa cells
  apurinic/apyrimidinic (AP) endonuclease, 141–2
  8-hydroxydeoxyguanosine in DNA of, 105
Hepatic tissue, *see* Liver
Hepatitis, endotoxin-induced, 590
Hepatocellular injury, ischaemia–reperfusion-related, 516
Hepatocytes, 4-hydroxyalkenals in
  effects, 346, 347, 361–2
  metabolism, 343–4
Hepatoma cells
  anticarcinogen tests in, 206–7
  4-hydroxyalkenals in, effects, 346, 347, 353
Hepoxilins, glutathione conjugation and, 187
12-HETE, platelets and, 428, 433–6
Heterogeneous atmosphere chemistry, 17
Hexose monophosphate shunt
  lens, 540
  platelet, 428
  red cell, 235–6
HNO_3 (nitric acid), atmospheric generation, 13
$HO_2^{\bullet}$ (hydroperoxyl) radical
  atmospheric reactions involving, 10, 13
  hydroperoxide oxidative decomposition giving rise to, 321
H_2O_2 (hydrogen peroxide), 155–69, 535–8, 549–50, 550
  atmospheric reactions involving, 13, 14
  in bacteria, 72, 73, 136, 155–69
    adaptive processes, 72, 155–69
    inactivation by, 73
    lesions cause by, 136
  biomolecules reacting with, 22
  in chloroplasts, detoxification, 27
  diesel soot particles and, in animals, interactions, 40
  dismutation, 407
  DNA damage induced by, 136, 137
  drug-dependent generation, 94
  iron and, 58–9
  ischaemia–reperfusion injury and, 501
  in lenses, 535–8, 549–50, 550
    concentration and source, 535–7
    conversion to more damaging oxidant, 537–8
    detoxification, 540, 543
    impact of/injury caused by, 540, 548, 549–50, 550
    stability, 537

H_2O_2 (hydrogen peroxide) (*cont.*)
  N-Methyl-*N*′-nitro-*N*-nitrosoguanidine and, phenolic antioxidants scavenging radical formed from, 258
  NO assays interfered with by, 452
  in platelets, effects, 433–4
  respiratory burst and the production of, 407, 410
  in respiratory system, effects of NO_2 in combination with, 38
  shock-related organ failure and, 602
12-HPETE, platelets and, 428, 433–6
$HSO_3^{\bullet}$, biomolecules reacting with, 18
HSO_3^- (sulphite)
  animal injury caused by, 40
  biomolecules reacting with, 18, 27
  diesel soot particles and, 40
  plant injury caused by, 27–8
H_2SO_4 (sulphuric acid), production, 12
Hydrogen-containing radicals/oxidants/inorganic compounds, *see specific formula*
Hydrolases, acid, in neutrophils, 402
Hydroperoxidase(s), xenobiotic/carcinogen activation involving, 282
Hydroperoxidase-I (HP-I), 71, 160
Hydroperoxidase-II (HP-II), 71
Hydroperoxides (organic), 319–36, 433–7
  aldehydes formed by β-cleavage reactions of alkoxyl radicals arising from, 337–69
  free radical metabolites of, 305–7
  LDL oxidation through pre-existing, 389
  in leaves of isoprene-emitting plants after ozone exposure, 16–17
  lipid
    decomposition, 374
    ischaemia–reperfusion injury and, 502
    PUFAs in LDL lipids converted to, 389
  in membranes, 319–36
    decomposition, 321
    elimination (by hydrolysis or reduction), 323, 324–9, 327
    formation, 320–4
    physiological effects, 329–32
  platelet, 430–1, 433–7
    effects, 433–7
  as tumour promoters, 305–7
Hydroperoxy fatty acids, *see* Fatty acids
12-Hydroperoxy-eicosatetraenoic acid (12-HPETE), platelets and, 428, 433–6
Hydroperoxyl radical, *see* $HO_2^{\bullet}$
Hydroquinones, 202, 290–1, 294
  detoxification, 202, 290
  oxidation, 202, 203
  quinone reduction to, 199
  toxicity, 290
Hydroxamic acids and their metabolites, 303
4-Hydroxyalkenals, 188, 337–69
  detoxification, 188, 343
  growth inhibition by, 350–5
  metabolism, 188, 340–5, 375
  toxicity, 188
    cyto-, 345–50
    geno-, 355–64
8-Hydroxydeoxyadenosine (in DNA), 105
8-Hydroxydeoxyguanosine (in DNA), 99–116
  in chloroplasts, 34
  formation, 34, 100–8, 111–12, 120
    discovery, 100–1
    in vitro, 101–5
    in vivo, 105–7
  in mammals, 99–105
  misreading of, during replication, 108–9, 112
  repair of, 109–11, 149
8-Hydroxydeoxyguanosine endonuclease, 109–11, 112
8-Hydroxydeoxyguanosine triphosphate, incorporation into RNA, 108
12-Hydroxyeicosatetraenoic acid (12-HETE), platelets and, 428, 433–6
8-Hydroxyguanine
  adenine and, base pairing between, 109
  in DNA, 107, 122, 134–5
  in RNA, 107
8-Hydroxyguanine (endo)nuclease, 111, 113, 134–5, 149
2-Hydroxyhydroquinone, quinone reductase and the production of, 201

Hydroxyl radical, *see* ˙OH
5-Hydroxymethyluracil formation in DNA, 122
4-Hydroxynonenal (4-hydroxynon-2-enal; HNE), 188, 338–64
  detoxification, 188
  growth inhibition by, 350–5
  metabolism, 188, 340–5
  physiological levels, 339
  shock and organ failure and, 602
  toxicity, 188
    cyto-, 345–50, 359
    geno-, 355–64
4-Hydroxynonenoic acid metabolism in hepatocytes, 343, 344
4-Hydroxyoctenal, effects on enzymes, 349
4-Hydroxypentenal
  cytotoxic effects, 359
  enzymic effects, 349
  mutagenic effects, 359
15-Hydroxyprostaglandin dehydrogenase, NAD^+-dependent, cigarette smoke inactivating, 39
2-Hydroxysemiquinone, quinone reductase and the production of, 201
Hyperbaric oxygen, cataracts caused by, 550–1
Hyperoxia-induced acute oedematous lung injury, 517
Hyperplasia, forestomach, phenolic antioxidants inducing, 259, 260, 264
Hypertension, NO and, 471
Hypochlorous acid in shock and organ failure, 589, 602, 605
Hypoperfusion, polytrauma and, 596–7
Hypovolaemic-traumatic shock, 596–7, 608, 609
Hypoxanthine glycosylase, 132

IlvY gene, 163
N-Iminoethyl-L-ornithine (L-NIO), NO synthesis inhibited by, 459
Infarct, myocardial, size, 507, 508, 509, 510, 511, 514, 519
Infection, damaged defence mechanisms in lung cells, 37, *see also* Sepsis
Inflammation
  phenolic antioxidants and their effects on, 248–51
  skin, UV-induced, 65–6, 568
Inflammatory cells, 406
  xanthine oxidase and, in shock, 601
  xenobiotic activation by, 287–92
1,4,5-Inositol triphosphate
  platelet, 425, 439
  respiratory burst in phagocytes and, 409
Insulin-producing β-cells, streptozotocin-related damage, 300–1
Interferon-γ
  NO and, 458, 459
  phagocyte priming by, 415
Interleukin-1, phagocyte priming by, 415
Intestine
  ischaemia–reperfusion injury and the, 501, 515–16
  shock-related injury to, 606
Ion pumps in lens, 548
Ionizing radiation, DNA damage/changes induced by, 103, 122, 125–6
Ionophores, phagocyte priming by, 414
Iron
  enzymes containing, NO and, 446–7
  ferrous, superoxide release by, 463
  hydrogen peroxide and, 58–9
  ischaemia–reperfusion injury and, 502–3
  lens cells and, 538
  proteins binding, antioxidant activity of, 216–18
  toxicity, 283
    defence against, 76
Iron chelators
  ischaemia–reperfusion injury and effects of, 502–3
  shock and organ failure and effects of, 609
Iron–oxo complexes, hydroperoxide generation from membrane phospholipids involving, 321–2
Irradiation, *see specific forms of radiation*
Ischaemia, 493–527
  free radicals and, 447, 471, 496–503
  general aspects of, 494–6

Ischaemia (*cont.*)
  injury caused by, 503–19
    aetiology/pathogenesis, 494–5
  NO and, 447, 471
  reversal by reperfusion, 493, 495
Ischaemia–reperfusion, 493–527
  damage caused by, 502–19
Islet cell damage by streptozotocin, 300–1
Isopentylnitrite, 472, 473, 475
Isosorbide derivatives, 472, 473
Isothiocyanates, aromatic, anticarcinogenic effects, 206

KatE gene, 71
KatF gene, 71
KatG gene, 62–3, 71, 161–6
Ketones, α,β-unsaturated
  fatty acids containing, glutathione conjugates derived from, 186–7
  prostaglandins containing, glutathione conjugates of, 186–7
Kidney
  8-hydroxydeoxyguanosine in DNA of, 106
  ischaemia–reperfusion injury, 517–18
  shock and injury to, 605
Killing, *see* Death

Lactoferrin, antioxidant activity, 216, 217, 218
Lamellar arrangement of membrane phospholipids, lipid peroxidation and the, 322
Leaf surface, pollutants and the, 16
Lens
  opacities, *see* Cataracts
  oxidative stress, 529–58
    protection from/against, 539–44, 551–4
Leukocytes, 585, 589–91, *see also specific types*
  antioxidant defences, 236–7
  common properties of different types, 399
  DNA-polycyclic aromatic hydrocarbon adducts in, 292
  endotoxin-related release of mediators from, reactive oxygen species and, 589–91
Leukotriene(s)
  cutaneous, UV effects, 568
  glutathione conjugation and, 181–2, 190
  respiratory burst and, 408, 414
Leukotriene-like compounds, glutathione conjugation and, 186–7
Leukotrienes, endotoxins and, 588
Light-driven reactions, *see* Photochemical reactions; Ultraviolet
Liliaceae, anticarcinogens in, 207
Lipid(s)
  in animal membranes, UVA-related damage, 61–2
  antioxidant activity, 226, 227
  peroxidation, *see* Peroxidation
Lipid-soluble antioxidants, 222–4, 227–30
  water-soluble and, interaction between, 231–3
Lipopolysaccharide, *see* Endotoxin
Lipoproteins, high-density, oxidized LDL and, 377
Lipoproteins, low-density, 371–97
  lipid peroxidation of, 389
    phenolic antioxidants preventing, 252
  lipid-soluble antioxidants in, 229
  oxidation, 229, 371–97
    antioxidants retarding, 390–2
    methods of determination, 372–7
    modifications (of LDL) as a result of, 377–89
Lipoxygenases
  cutaneous, UV effects, 568
  endotoxic shock and the products of, 588
  glutathione conjugation and, 181–2
  LDL oxidative modifications involving, 381, 384, 389
  membrane, peroxidation involving, 330–1
  NO and the products of, 468–9
  platelet, 427, 428, 430, 431
  reperfused organs and free radical generation and, 498, 499
Liver, *see also entries under* Hepat-
  endotoxic injury, α-protease inhibitor protecting against, 590
  glutathione transferase, 177, 188

4-hydroxyalkenals in, metabolism, 340, 341–2, 343–4
8-hydroxydeoxyguanosine in DNA of, 106, 106–7
ischaemia–reperfusion injury, 516
parenchymal cells, oxidative stress studies in, 421
phenolic antioxidants with adverse effects in, 261
preneoplastic nodules, quinone reductase activity in, 204
London smog, 14
Los Angeles smog, 14
Low flow syndrome, 607
Lung
bleomycin toxicity, 282
cells, defence mechanisms, damage to, 37
diesel soot particles in, carcinogenicity, 40
endotoxic injury, 586
fibrosis, asbestos causing, 42
hyperoxia-induced acute oedematous injury, 517
ischaemia–reperfusion injury, 516–17
ozone in, toxicity/carcinogenicity, 35–6, 284
phenolic antioxidants damaging, 266
shock and injury to, 586, 600, 607
Lupus erythematosus, 572–3
Lutein, antioxidant role of, 229
LDL oxidation and the, 390
Lycopene, antioxidant role of, 229
LDL oxidation and the, 390
Lymphocytes in systemic lupus erythematosus, 572
Lymphokines, *see* Cytokines
Lysophospholipids, reacylation, 326
Lysosomal enzyme in ischaemia, 495
Lysosomal hydrolases in neutrophils, 402
LysR gene, 163

Macromolecules, hydrogen peroxide-related damage to, 156, *see also* Biomolecules
Macrophages, *see also* Monocytes
asbestos fibres and, cooperative effects between, 43
inflammatory, 406
LDL uptake by, oxidatively modified, 377
maturation, 405–6
NO synthetase, 457, 458–9
role/function, 405–6
scavenger receptors, 372, 381
tissue, 405–6
xenobiotic activation by, 287–92 *passim*
Magnesium, exonuclease III requirement for, 143
Magnesium (Mg^{2+})-ATP in the lens, 548
Malignant tumours, progression from benign to, 279–80, *see also sites and types of tumours* and Tumours
Malondialdehyde
genotoxicity, 357
in LDL oxidation, formation, 373–5
Mammalian cells
antioxidant enzymes in, 72
antioxidant molecules in, 69–70
death/inactivation, 65, 352
DNA damage in, 60
DNA repair in, 68, 111, 113, 141–2
gene expression, induction, 74–8
genotoxicity, 355–7, 358–9, 360–4
4-hydroxyalkenals in
growth inhibition by, 350–5
toxicity, 345–50
8-hydroxydeoxyguanosine in DNA of, 105–7
LDL oxidative modifications involving, 381–9
radiation-sensitive, 570–2
UVA effects, 60, 65, 68, 69–70, 72, 74–8
Manganese-superoxide dismutase, 75
in ischaemia–reperfusion injury, 512, 523
Melanins, photosensitization and, 561–3
Melanoma, 563, 573
Membranes (animal), 319–36
destruction, peroxidative, 330–1
enzymes associated with, 323, 324–9
xenobiotic/carcinogen activation involving, 281–2
glutathione transferase bound to, 177
in lens, cytosolic protein linkage with proteins in, 545–7
platelet, 423–4

Membranes (animal) (*cont.*)
repair processes with oxidative damage, 325–7
signal transduction across, *see* Signal transduction
UVA-related damage, 60–1
Membranes (bacterial), hydrogen peroxide-related damage to, defence against, 162–3
Membranes (plant)
disruption by lipid peroxidation, 25
reactive oxygen species, generation/damage caused by/detoxification, 26–7
Menadione (2-methyl-1,4-naphthoquinone)-induced reactions, 200
in platelets, 432
quinone reductase effects on, 199
Mercaptoimidazoles, oxidation, 92
Mercapturic acid, 4-hydroxynonenal-derived, 344–5
Metabolism, active, in bacteria, hydrogen peroxide-related killing requiring, 157
Metal(s), *see also specific metals*
divalent, exonuclease III requirement for, 143
LDL oxidation involving, 378–80, 386, 389
NO affinity for complexes containing, 447
phenolic antioxidants binding, 246
proteins binding, antioxidant activity of, 216–18, 230
Metallic compounds in cigarette smoke, 38
Metalloproteins, DNA repair enzymes as, 143
Methaemoglobin
formation, 218
NO and, 460
Methane, atmospheric oxidation, 11–12
Methionine in lens proteins, oxidation, 547
8-Methoxypsoralen, 565–6
Methyl radical ($CH_3^{\bullet}$), organic hydroperoxides and the generation of, 306
N-Methyl-4-aminobenzene, glutathione conjugation and, 180
Methylation reactions in carcinogenesis, 306–7
7-Methylbenzanthracene, glutathione conjugation and, 178
3-Methylcholanthrene, glutathione conjugation and, 178
Methylene blue, NO and, 461–2
O^6-Methylguanine-DNA methyltransferase, 4-hydroxyalkenal effect on, 360
2-Methyl-1,4-naphthoquinone, *see* Menadione
N-Methyl-*N'*-nitro-*N*-nitrosoguanidine and hydrogen peroxide, phenolic antioxidants scavenging radical formed from, 258
N-Methyl-*N*-nitrosourea, phenolic antioxidants enhancing tumourigenesis by, 263
MetR gene, 163
Micellar arrangement of membrane phospholipids, lipid peroxidation and the, 322
Micronuclei 4-hydroxyalkenal genotoxicity and, 361–4
Microorganisms, *see also specific (types of) microorganisms*
antioxidant enzymes in, 62–3, 71
antioxidant molecules in, 69
DNA damage in, 60, 62
repair of, 62, 67–8, 109–11, 112, 127–40, 142–5, 157–9, 360
enzymes killing, 402
gene expression, induction, 72–4
hydrogen peroxide in, 72, 73, 136, 155–69
inactivation, 63–5
NO and the killing of, 458
respiratory burst and the killing of, 406
UVA effects on, 60, 62, 63–5, 69, 71–2, 72–4
Microsomes
aldehyde dehydrogenases in, 341–2
glutathione transferase in, 177, 188
hydrogen peroxide production in, 266
4-hydroxyalkenal metabolism, 341–2
phenolic antioxidants in, 264, 266
Mitochondria
free radicals and ischaemia–reperfusion injury and, 499–500, 517

liver, 8-hydroxydeoxyguanosine in DNA of, 107
lung, 517
Mitomycin C, toxicity/carcinogenicity, 285, 298, 298–9
Monocytes, 404–5
maturation to macrophages, 405–6
respiratory burst in, 407, 415
xenobiotic activation by, 287–92 *passim*
Monooxygenases, *see* Oxygenases
Monoterpenes, ozone reaction with, 16
Muconaldehyde, mutagenicity, 358
Muscle
contraction, phenolic antioxidants inhibiting, 252
shock and injury to, 605
smooth, *see* Smooth muscle
Mutagen(s)
diesel soot particles as, 39–40
4-hydroxyalkenals as, 357
8-hydroxydeoxyguanosine formation in DNA induced by, 104
Mutagenesis
bacterial defence against, 145–8, 162–3
DNA repair and, 145–8
8-hydroxydeoxyguanosine in DNA and, 104
in lens cells, 533, 549
phenolic antioxidants in, 253
inductive effects, 261
protective effects, 253
MutM gene, 149
Myelocyte granules, 400
Myeloperoxidase (MPO), 400
xenobiotic activation involving, 287–92 *passim*
Myocardium
infarct size, 507, 508, 509, 510, 511, 514, 519
ischaemia–reperfusion injury in, 497, 504–15 *passim*, 519
free radical sources in, 499
superoxide dismutase effects on, 512
stunned, 504–5, 507
Myoglobin, ischaemia–reperfusion injury and, 502–3

NADH-cytochrome b_5 reductase, carcinogens/other xenobiotics activated by, 280
NADPH, peroxyacetyl nitrate reacting with, 22
NADPH-cytochrome P450 (cytochrome *c*) reductase, carcinogens/other xenobiotics activated by, 280–1
NAD(P)H-dependent aldehyde dehydrogenases, 4-hydroxyalkenals and, 340–2
NADPH oxidase, 406–7, 411–13
in chronic granulomatous disease, 413
components and assembly, 411–12
ischaemic intestine and, 516
respiratory burst and, 406–7, 411–13
NAD(P)H:quinone reductase, *see* Quinone reductase
Naphthoquinones, 198
redox cycling, 24, 25, 41
Naphthol, toxicity, 198
Necrosis, tissue
antioxidants and, 505–6, 510–11
superoxide dismutase and, 510–11
Neocarzinostatin (NCS), DNA changes induced by, 123, 124
Nervous system, central, ischaemia, 499, 518–19
Neurotoxin, eosinophil-derived, 404
Neutrophil(s) (polymorphonuclear), 400–3, 500–2
asbestos and, 42
diesel soot particles and, 41
granules, 400–3
ischaemia–reperfusion injury and, 500–2, 517, 518
kidney, 518
lung, 517
maturation, 400–1
monocytes and, differences, 405
NO synthesis in, 459–60
other phagocytic cells and, properties in common, 399
plasma exposed to oxidants from activated, 230, 231
respiratory burst in, 406, 407–11, 414–15, 500–1
phenolic antioxidants and the, 250–1
in shock/trauma-related organ failure, 598, 601–4
xenobiotic activation by, 287–92 *passim*

Neutrophil-activating peptides, respiratory burst and, 408
Nfo gene, 138
NH_3, plant damage involving, 28–9
Nitrates, organic, 450, 472–6
  glutathione oxidation by, 87
  new, 478
  therapeutic uses, 476–7, 477–8
  tolerance to, 476, 478
Nitric oxide, *see* NO
Nitroaromatic compounds, 301–2
  oxidation to free radicals, 301–2
  quinone reductase as substrates for, 203
  redox cycling, 24, 25, 41
7-Nitroebselen, 553
Nitrogen, radicals/oxidants/inorganic compounds containing, *see specific formula*
Nitrogen-15, NO assay via substances containing, 453
Nitropolycyclic aromatic hydrocarbons, glutathione conjugation and, 181
2-Nitropropane, 8-hydroxydeoxyguanosine in DNA/RNA and effects of, 107
4-Nitroquinoline 1-oxide, toxicity, 303, 304
Nitroreductase activity of quinone reductase, 203
Nitrosothiols, NO/EDRF and, 465–6, 467–8, 473, 476, 478
Nitrovasodilators, 450, 472–6
  therapeutic uses, 476–7, 477–8
  tolerance to, 476, 478
NO (nitric oxide), 445–89
  atmospheric, 19
  biochemical properties, 460–9
  deactivation and inhibition, 460–6
  donors, as vasodilators, 472–6
  endothelium-derived relaxing factor and, *see* Endothelium-derived relaxing factor
  general properties, 446–7
  historical background, 448–51
  interference by other factors with, 468–9
  platelet aggregation and, 429, 436–7, 439
  quantification, 451–4
  synthesis, 454–60
$NO^•$ radical, atmospheric reactions involving, 12
NO synthetase, 454–60
  endothelial, 456–8
  macrophage, 457, 458–9
NO_2
  biomolecules reacting with, 19
  in cigarette smoke, damage caused by, 36–8
  photolysis, 9
$NO_3^•$ radical, atmospheric reactions involving, 10
NO_x (nitrogen oxides)
  anthropogenic emissions, 8
  biomolecules reacting with, 19
  at epidermal surface, reduction, 576
  natural emissions, 6
NodD gene, 163
Nodules, preneoplastic, quinone reductase activity in, 204
Non-steroid antiinflammatory drugs, myocardial ischaemia–reperfusion injury and effects of, 499
Nordihydroguaiaretic acid (NDGA), phospholipase A_2 inhibition by, 324, 325
Nth gene, 130, 131
Nuclear cataracts, 545, 546
Nucleases, DNA repair by, 67, 71, 109–11, 112–13, 132–43, *see also* Endonuclease; Exonuclease
Nucleic acids, *see* Bases; DNA; Nucleoside; Nucleotide; RNA
Nucleoside modification, generation, 99–116
Nucleotide(s), cyclic, *see also specific nucleotides*
  NO and, 429, 439, 446, 450
  platelets and, 428–9, 436, 439
Nucleotide 5′-aldehyde formation, 123
Nuv gene, 73–4

$^1O_2$ (singlet oxygen)
  in mammalian cells
    inactivation by, 65
    quinones and the formation of, 197, 199
  in microbes
    generation, 64

protection from, 64
in plants, 25
UVA and, 59, 64
$O_2^{-\bullet}$ (superoxide), 462–6, *see also* Superoxide dismutase
compounds claimed to release, 462–3
dismutation, 407, 501
endotoxin-induced cytokine release initiated by, 589
ferritin and the generation of, 217
hydroquinone oxidation involving, 202, 203
ischaemia–reperfusion injury and, 500
LDL oxidative modifications involving, 383, 384, 387–8
in lens, 536–7, 538
in microbes, 74
mitochondrial, 500
neutrophil, 501
phagocytic, 407, 501
plant damage involving, 25
prevention, 33–4
quinones and the generation of, 197, 202
respiratory burst and the production of, 406–7, 501
in respiratory systems of animals
diesel soot particles and, 41
formation, 38, 41
scavenging/depletion of, 447, 476–7
by NO, 447, 462–6, 476–7, 478
by NO donors, 476–7, 478
by phenolic antioxidants, 250
by superoxide dismutase, 202
shock and organ failure and, 589, 602, 604
sulphite reaction with, 28
UVA and the intercellular generation of, 59
O_3 (ozone), 20–1
animal injury involving, 35–6
atmospheric reactions involving, 9, 11, 13, 15
biomolecules reacting with, 20–1
carcinogenesis and, 284
chemiluminescence, NO assay via, 451–2
formation in atmosphere, 16
photolysis, 9, 15
plant injury involving, 28–9
in smog, 15, 16
Oestrogen, synthetic, *see* Diethylstilboestrol
$^{\bullet}OH$ (hydroxyl) radical
atmospheric reactions involving, 10, 11, 11–13
8-hydroxydeoxyguanosine formation involving, 105
ischaemia–reperfusion injury and, 501, 502
lens, generation, 537
NO yielding, 448
oxygen species with properties similar to, in plants, 26
quinones and the formation of, 197
in smog, generation, 15
One-electron reduction potential at epidermal surface, 576
Opacities, lens, *see* Cataracts
Ophthalmic acid, 539
Organ(s), 8-hydroxydeoxyguanosine in DNA of various, 105
Organ failure, 595–616
multiple, 595
post-traumatic, 595–616
progression, 596–7
Organic compounds
in atmosphere, chemical lifespan, 10–11
volatile, *see* Volatile organic compounds
Organic free radicals, chemical activation enhanced by, 286
Organic hydroperoxides and peroxides, *see* Hydroperoxides; Peroxides
Organic nitrates, *see* Nitrates
Ornithine decarboxylase induction
antioxidants inducing, 260
by tumour promoters, antioxidants suppressing, 258
Oxidases, *see* Oxygenases
Oxidative stress, definition, 421 (*see also*, Introductory Remarks, (xv–xxii)
6-Oxy-benzo(a)pyrene radical, formation and damage caused, 295
Oxygen, *see also* Hyperoxia
LDL oxidative modifications and the role of, 386
in lens
reduction, 534–5

Oxygen, in lens (*cont.*)
tension, 534
in respiratory bursts, consumption, 406–7
therapy
hyperbaric, cataracts caused by, 550–1
injury caused by, in critically ill patients, 604
Oxygen-containing oxidants/inorganic compounds, *see specific formula*
Oxygen radicals, *see also specific formulae of radicals and* Free radicals
air pollution and, 3–55
DNA damaged by, repair, 119–54
endotoxin-induced cytokine release initiated by, 589–90
8-hydroxydeoxyguanosine formation in DNA by, 99–116
ischaemia/ischaemia–reperfusion injury and, 447, 471, 500, 501, 503–4, 505, 516, 517, 517–18
LDL oxidation and, 379–81, 383
in lens, production, 536–7, 538
scavenging/depletion of, 447, 476–7, 608–9
by NO, 447, 462–6, 476–7, 478
by NO donors, 476–7, 478
by phenolic antioxidants, 250
by superoxide dismutase, 202
in shock and organ failure, 589–90, 597–610
involvement, 589–90, 597–600
sources, 600–4
targets and effects of, 605–6
Oxygen species, reactive
endotoxin-related release of leukocyte mediators and, 586, 589–91
glutathione transferase and the products of, 171–94
measurement, 597
in plants
damage caused by, 26
detoxification, 27
generation, 26
shock involving, 589–90, 597–610
in skin, 559–60
UV and the generation of, 58–66, 559–60
Oxygenases
carcinogen/other xenobiotics activated by, 280–1
cytochrome P450-dependent mixed function, 171
carcinogen/other xenobiotics activated by, 280–1
drugs and, 94, 95
flavin-containing, multisubstrate, properties, 89–91, 92, 93
Oxyhaemoglobin, NO and, 452, 460–1
Oxypurinol, ischaemia–reperfusion injury and effects of, 506, 515
OxyR1 gene and protein, 63, 72, 73, 148, 159–67
membrane damage and the, 162–3
mutagenesis and the, 162–3
regulation, mechanisms, 163–7
Ozone, *see* O_3

p47 protein of NADPH oxidase, 412
p67 protein of NADPH oxidase, 412
P450-dependent monooxygenases, *see* Oxygenases
PAN, *see* Peroxyacetyl nitrate
Pancreatic islet cell damage by streptozotocin, 300–1
Papillomas, forestomach, phenolic antioxidants inducing, 259
Paracetamol, glutathione conjugation and, 172
Particulate pollutants
animals affected by, 39–44
anthropogenic sources, 8
catalysis involving, 24–5
Perfusion, *see* Hypoperfusion; Reperfusion
Peritonitis, zymosan-induced, 609
Peroxidase(s), *see also specific peroxidases*
eosinophil, 403
human plasma, 214–15
membranous, 323, 326, 327–9
neutrophil, 400
selenium-dependent, 327, 328–9, 553
xenobiotic/carcinogen activation involving, 282, 287–92 *passim*
Peroxidation
DNA, repair, 189

lipid, 61–2, 174, 188–9, 319–36, 389
aldehydes formed by β-cleavage reactions of alkoxyl radicals arising from, 337–69
in animals (in general), 61–2, 174, 188–9, 319–36
ascorbic acid and, 225, 227
cell cycle and, 354
glutathione conjugation and, 174, 188–9
inhibition, 188–9, 228
ischaemia–reperfusion injury and, 502
of low-density lipoproteins, *see* Lipoproteins
phenolic antioxidants and, 252, 258
photooxidative skin damage and, 567–8
in plants, 25, 28
repair, 188–9, 325–7
shock/trauma-related organ failure and the products of, 599–600
α-tocopherol and, 228
uric acid and, 227
xenobiotic/carcinogen formation and, 258, 284
Peroxides, *see also specific peroxides*
biomolecules reacting with, 22
endogenous, glutathione transferases and, 181–9
organic
free radical metabolites of, 305–7
as tumour promoters, 305–7
Peroxisome proliferator, 8-hydroxy-deoxyguanosine in DNA and effects of, 106–7
Peroxyacetyl nitrate (PAN)
biomolecules reacting with, 22
generation, 15
Peroxyl radicals
chemical activation enhanced by, 286
LDL oxidation and, 379, 380, 381
in membranes, 323
phenolic antioxidants and, 323
in plasma, 226, 227, 228
production, 228, 287
trapping of, 226, 227
Phaeomelanins, skin cancer and, 563
Phagocytes, 399–420, *see also specific types*
oxidants from activated, 230
and respiratory burst, 399–420
priming, 413–15
shock-related organ damage and, 601–4
Phase II enzymes, preneoplastic nodules and tumours and, 204, 206
o-Phenanthroline, lens epithelial cells and effects of, 538
Phenol
anticarcinogenic effects, 206
from benzene, generation of, 293–4
toxicity, 198
quinone reductase and, 199
Phenolic antioxidants, 245–73
adverse actions, 259–66
oxidative metabolism in relation to, 263–6
prooxidative activity in relation to, 263–6
in membranes, 323
protective actions, 246–59, 323
Phenoxyl radicals
chemical activation enhanced by, 286
phenolic antioxidants converted to, 262
Phenyl butyl nitrone in shock therapy, 610
Phorbol esters
ornithine decarboxylase induction by, antioxidants suppressing, 258
respiratory burst and, 410, 411, 414
Phosphatidylchloline hydroperoxide as substrate for phospholipid hydroperoxide glutathione peroxidase, 327–8
Phosphatidylinositol cycle, platelet, 425–6
3-Phosphoglycolaldehyde (PGA), repair of sites terminated with, 136–7
3-Phosphoglycolaldehyde (PGA) diesterase, yeast, 139
3′-Phosphoglycolate esters in DNA, 124
Phosphoinositide cycle, platelet, 439
Phospholipase(s)
fatty acid hydroperoxides released by, reduction of, 329
glutathione peroxidase plus, antiperoxidative action exerted by, 329

Phospholipase A_2
  LDL oxidative modifications involving, 381, 383
  phospholipid hydroperoxide elimination mediated by, 324–7
Phospholipase C
  platelet, 425–6
  respiratory burst and, 409, 411
Phospholipid(s)
  aldehydes bound to, role, 338
  membrane, hydroperoxidation and, 321–4
Phospholipid hydroperoxide(s), elimination, 323–9
Phospholipid hydroperoxide glutathione peroxidase, 323, 327–9
  glutathione peroxidase and, relationship between, 328–9
3′-Phosphomonoesters in DNA, 124
Photoallergy, drug-induced, 561
Photocarcinogenesis, 66, 563, 568, 574–5
Photochemical reactions, 9–10, 14–17, 45
  atmospheric, 9–10
  in lenses, 536–7, 542, 549
  melanin in, 562
  in plants, 25–9
  porphyrins in, 563
  primary, 9–10
Photochemotherapy, 565–6
Photodissociation, 15
Photolysis, 9–10
Photooxidants, 559–83
  main, 10
  skin and effects of, 559–83
Photosensitization, 561–6
  endogenous, 561–4
  exogenous, 564–5
Photosmog, 14–17
Photosynthesis, 25–9 *passim*, 31
  inhibition/depression, 27, 31
Photosystem I, oxygen reduction involving, 26
Phototoxicity, drug-induced, 561
Physiological processes in plants, pollutants affecting, 31–4
Phytoalexins, 34
Phytofluene, antioxidant role of, 229
  LDL oxidation and the, 390
Plants, 26–34
  alkaloids, 4–hydroxyalkenal production from, 340
  photochemical reactions in, 25–9
  pollutants damaging, 17, 23, 26–34
  surfaces of
    changes at, 29–31
    photooxidant formation near, 16
Plasma, endogenous antioxidant defences, 213–43
Platelet(s) (thrombocytes), 421–43
  activation and aggregation, 422, 423–9, 433–4, 438–9
    inhibition (=inactivation), 251, 428–9, 434–7, 439
  formation, 422
  function and physiological responses, 422–3, 424–5
  morphology, 423–4
  oxidative stress in, 421–43
  exogenous, 431–7, 439
    physiological, 430–1, 438–9, 440
  phenolic antioxidants and, 251
Platelet-activating factor
  respiratory burst and, 408, 414
  shock and, 601
PolA, DNA repair involving, 67
Pollution, air, oxygen radicals and, 3–55
Polyamines, antioxidative mechanisms in plants involving, 33
Polyamino acids as NO synthetase substrates, 458
Polychlorinated dioxins, sources, 9
Polycyclic hydrocarbons, 295–6
  oxidation to free radicals, 295
  quinone reductase induced by, 205
  toxicity/carcinogenicity, 40, 291–2, 295–6
Polyethylene glycol-conjugated superoxide dismutase, 513, 514–15
Polymorphonuclear neutrophils, *see* Neutrophils
Polytrauma, 596–7
Polyunsaturated fatty acids
  peroxidation, 319–36
    aldehydes formed by β-cleavage reactions of alkoxyl radicals arising from, 337–69
  shock/trauma-related organ failure and, 599–600

Porphyrins
haemato-, 566
photosensitization and, 563–4
singlet oxygen generation and the role of, 64–5
Preneoplastic nodules, quinone reductase activity in, 204
Probucol, protective effects of, 247–59 *passim*
Promotion, tumour, *see* Tumour
Promyelocyte granules, 400
Prooxidant activity
of endogenous plasma antioxidants, 224–5
of phenolic antioxidants, 263–6
at skin surface, of antioxidants, 577
Propyl gallate, *see* Gallic acid esters
Prostacyclin
endotoxins and, 587
NO and, 468
phenolic antioxidants affecting synthesis of, 251
Prostaglandin(s)
cutaneous, UV effects, 568
endotoxins and, 587
glutathione conjugation and, 173, 190
platelet, 427
α,β-unsaturated ketone-containing, glutathione conjugates of, 186–7
Prostaglandin endoperoxides, platelet, 427, 433
Prostaglandin H synthetase (PHS)
benzidine and, 301
xenobiotic activation and, 282–3
Prostaglandin synthetases, glutathione conjugation and, 173, 182
Prostanoids, *see* Arachidonic acid metabolism *and specific prostanoids*
Protease(s)
endotoxin-induced cytokine release and, 590, 591
neutrophil, 402
in lung, ischaemia–reperfusion injury and, 517
serine, 590
α-Protease inhibitor (antiprotease/antiproteinase)
endotoxin-induced cytokine release and, 590
inactivation, 36
protection against, 231
in shock and organ failure, 606
Protein(s)
antioxidant, 216–18, 226, 227, 228, 230
cationic, in eosinophils, 404
DNA crosslinking with, *see* DNA
lens
age-related changes, 534
disulphides in, *see* Disulphides
membrane and cytosolic, linkage, 545–7
UV effects, 58, 549
serum, 4-hydroxynonenal bound to, 354
Protein kinase C, 331, 401
phenolic/flavonoid antioxidants and, 248, 259
platelet, 426, 431
respiratory burst and, 409, 410, 411
Protein X, eosinophil, 404
Protoporphyria, erythropoietic, 564, 577
Pseudopodia, platelet, 425
Psoralens, 565–6
Pulmonary function/damage etc, *see* Lung
Pumps, ion, in lens, 548
Pyrogallol, superoxide release by, 463–4
Pyruvate, antioxidant activity, 222

Quercetin
adverse effects, 261, 264
protective effects of, 247–59 *passim*
Quinoid metabolites, formation from butylated hydroxyanisole, 265
Quinone(s), 199–205, 296–300
oxidation to free radicals, 296–300
toxicity, 296, 296–300
mechanisms, 197–8, 290, 296, 296–300
protection from, 196, 199–205, 290
Quinone reductase, protective roles, 195–211

RAD50 and *RAD52* locus in yeast, 144
Radiation, *see specific forms*
Radicals, free, *see* Free radicals *and specific (types of) radicals*

Reactive oxygen species, *see* Oxygen species
Reacylation of lysophospholipids, 326
RecA-dependent DNA repair, 62, 67
Receptors
  for chemotactic agonists, respiratory burst and, 408
  scavenger, 372, 381
Recombination
  enzymes used in, DNA repair via, 143–4
  intrachromosomal, in *xth*⁻ mutants, 137
Red blood cells, 234–6
  antioxidants, 215, 234–6
Redox cycling
  quinones/quinone reductase and, 197, 199–200, 201, 202, 300
  of butylated hydroquinone (TBHQ), 266
Redox-cycling agents, energy balance upset by, 148
Redox status
  in lenses, of glutathione, 539–40
  in platelets, perturbation in, 430
Redox endonucleases, 130
Reducing agents, LDL oxidative modifications and the role of, 386, 387
Reduction, oxygen, in the lens, 534–5
Relative electrophoretic mobility in measurements of LDL oxidation, 376
Renal tissue, *see* Kidney
Repair
  of DNA, *see* DNA
  of lipid peroxidation, 188–9, 325–7
  of membranes with oxidative damage, 325–7
Reperfusion, 495–519, *see also* Ischaemia–reperfusion
  arrhythmias induced by, 504–507
  benefits of (reversal of ischaemia by), 493, 495
  free radicals produced during, 496–503
  injury caused by, 493–6, 503–19
Respiration in plants, pollutants and, 32
Respiratory burst, 399–420
  activation, 407–11
  in neutrophils, 406–11, 414–15, 500–1
  phenolic antioxidants and the, 250–1
Respiratory chain
  ischaemia–reperfusion injury and the, 500
  NO binding to enzymes in, 446–7
  superoxide production, 500
Respiratory distress syndrome, adult, endotoxin and, 586
Respiratory system in animals, pollutants affecting, 35–44 *passim*
Reticulocytes, lipoxygenase, role, 330–1
Reticuloendothelial system, macrophages of, 405
Reticuloid, actinic, cell line, UV effects, 68–9
Riboflavin, lens, 536–7, 542
RNA
  8-Hydroxydeoxyguanosine triphosphate incorporation into, 108
  liver, 8-hydroxyguanine in, 107
  synthesis, hydrogen peroxide stress and, 166–7
RNA polymerase, hydrogen peroxide stress and, 166–7

Saccharomyces spp., *see* Yeast
Salmonella typhimurium
  hydrogen peroxide in, 155–69
  4-hydroxyalkenals as mutagens in, 357–8, 359
Scavenger receptors, macrophage, 372, 381
Scavenging of radicals, 447, 462–6, 476–7, 608–10
  at epidermal surface, 576
  ischaemia–reperfusion injury and the, 507–10
  by NO, 447, 462–6, 476–7, 478
  by NO donors, 476–7, 478
  by phenolic antioxidants, 250, 258–9
  by plasma antioxidants, 227
  shock/trauma-related organ failure and, 600, 608–10
  by superoxide dismutase, 202
Selenium
  cerebral ischaemia and effects of, 518
  lens cataract development with dietary deficiency of, 543–4
Selenium-dependent peroxidases, 327, 328–9, 553

Selenocompounds
lens oxidative insult prevention using, 552–4
shock and organ failure and effects of, 610
Semiquinones, 202
in cigarette smoke, 38
quinone reduction to, 197
Sepsis
lipid peroxidation and, 599–600
mediators, 585–93
Serine proteases, endotoxin-induced cytokine release and, 590
Serum
antioxidants in, 230
proteins, 4-hydroxynonenal bound to, 354
Shock, 585–616
endotoxic, 585–93
mediators of, 585–93
onset, 596–7
Signal transduction, transmembrane
phenolic antioxidants affecting, 248–50
respiratory burst and, 407–11
Silicates, 42–3
Singlet oxygen, *see* $^1O_2$
Sister chromatid exchange, 4-hydroxyalkenal genotoxicity and, 361–4
Skin, 65–6, 559–83
components, 560
disease, 570–4
glutathione in, 70
inflammation, 65–6, 568
photooxidative stress, 559–83
damage caused by, 567–70
tumours, *see* Tumours
UVA effects, *see* Ultraviolet
Smog, 13–17
Smoke, cigarette, *see* Cigarette smoke
Smooth muscle
LDL oxidative modifications involving cells of, 381
phenolic antioxidants inhibiting contraction in, 252
vascular, NO/oxyhaemoglobin/methaemoglobin and, 461
SO_2
atmospheric oxidation, 12–13, 14
and its derivatives, biomolecules reacting with, 18, 27, 33
emissions, 45
plant damage caused by, 27
$SO_3^{\bullet}$ radical, formation, 287
SO_3^{2-}, conversion to sulphur trioxide radical, 287
sodA and *sodB*, 71
Sodium nitroprusside, 472, 473, 475
Sodium-potassium ATPases in lens, 548
Soot particles
diesel, damage caused by, 39–41
sources and emissions, 8, 45
SOS response of DNA-damaged cells, 146
SoxR regulon, 148
Spanish oil syndrome, toxic, 350
Spin-trapping agents in shock therapy, 610
Squamous cell carcinomas, forestomach, phenolic antioxidants inducing, 259
Starch, heated, 8-hydroxydeoxyguanosine formation in DNA by, 104
Stomach cancer, fore-, phenolic antioxidants inducing, 259, 264
Stomata, pollutants and, 29, 31–2
Streptozotocin, free radicals from the oxidation of, 300–1
Stress, *see also specific stressors*
oxidative, definition, xv–xxii, 421
responses
in mammalian cells, UVA and, 70, 75
in plants, 34
Stroke, ischaemia–reperfusion injury and, 519
Sudan III [1-(4-phenylazophenylazo)-2-naphthol], chemoprotective effects of, 206
Sugar, *see also* Carbohydrates
cataracts induced by, 541
in DNA, radical-induced damage, 123–5
repair, 135–43
Sulphanilic acid diazotization, NO assay via, 453
Sulphenic acids, xenobiotics oxidized to, 88–9, 92–4

Sulphur-containing oxidants/inorganic compounds, *see specific formula*
Sulphur-containing xenobiotics, glutathione oxidation by, 88–94
Sulphur radicals, LDL oxidation and, 379–81, *see also specific formula*
Sulphydryl groups, *see* Thiol
Suntanning devices, carcinogenic risks, 574
Superoxide, *see* $O_2^{-\bullet}$
Superoxide dismutase, 200–2, 507–15, 578
  benzene and, 281, 294
  extracellular human, 214
  in ischaemia–reperfusion injury, 507–15, 515, 516, 518, 519
    dose–response, 511–13
    necrosis delayed by, 510–11
    post-treatment with, 514–15
    pretreatment with, 513–14
  LDL oxidation and, 387, 388
  in lens, 544
  in mammalian cells, 72, 75
  in microbes, 71
  NO and, 447, 466–7
  in plants, 27, 33
  quinone reductase and, 200–2
  in shock and organ failure, 601, 608–9
  in skin, 571, 575, 578
  UV radiation and, 71, 72, 575
  xanthine oxidase and, 601
  xeroderma pigmentosum and, 571
Superoxide:semiquinone oxidoreductase, superoxide dismutase as a, 202
Surfactant, lung, 607
Survival, DNA repair and, 145–8
Sydnonimines, 472, 473, 475–6, 477–8
  therapeutic uses, 477–8
Synergism between butylated hydroxyanisole and butylated hydroxytoluene, 266
Systemic lupus erythematosus, 572

Tar condensate in cigarette smoke, damage caused by, 38–9, 102, 107
Tension, oxygen, lenses and, 534
2(3)-*tert*-butylated compounds, *see under* Butyl
2,3,7,8-Tetrachlorodibenzo-*p*-dioxin (TCDD)
  carcinogenicity, 44
  emission, 9
12-*O*-Tetradecanoylphorbol-13-acetate, ornithine decarboxylase induction by, antioxidants suppressing, 258
Thiobarbituric acid reactive substance (TBARS)
  in LDL oxidation assays, 373–4, 378, 598
  photooxidative skin damage and, 567–8
  in shock/trauma-related organ failure, 598, 599
Thiocarbamides, oxidation, 92
6-Thioguanine resistant cells, 4-hydroxynonenal mutagenicity in, 358–9
Thiol (sulphydryl) groups, 88–94, 218–19, 545–8
  erythrocytic glutathione and, 235
  in lens, 541, 545–8
  NO and, 465–6
  oxidation, 431–2, 541, 545–8
    as antioxidant protection, 219, 226, 227, 228
    as oxidative damage, 219
  peroxyacetyl nitrate reacting with, 22
  platelet, 431–2, 436
  protein, 218–19
  xenobiotics bearing, glutathione oxidation by, 88–94
Thiol-oxidizing agents in platelet studies, 431–2
Thioredoxin/thioredoxin reductase in lens, 544
4-Thiouridine, UVA and, 73
Thiyl radicals (GS$^\bullet$), 219, 287
  formation, 219, 220, 287
3′-termini of DNA, repair of lesions at, 136–7, 139, 139–40, 142
Thrombocyte, *see* Platelet
Thromboxane, synthesis
  endotoxin-stimulated, 587
  phenolic antioxidants affecting, 251
  platelet, 427, 431, 438
Thromboxane synthase, platelet, 427
Thylakoid membranes, reactive oxygen species in, generation of/damage by/detoxification of, 26

Thymine in DNA
of lens, 549
radical-induced damage, 120, 121–2, 189
Thymine glycol (in DNA)
in lens, UVB-induced, 549
radical-induced formation, 120, 121, 147
Thymine glycol glycosylase, 128–31
Tissue macrophages, 405–6
Tobacco smoke, *see* Cigarette smoke
Tocopherol(s), *see also* Vitamin E
antioxidant role of, 222–3, 226–9, 231–3, 575, 576–7
LDL oxidation and the, 390–2
ascorbic acid and, interactions, 231–3
lipid peroxidation and, 225, 323
phenolic antioxidants and, 323
photoradiation and, 575, 576–7
as a prooxidant, 577
in shock/trauma-related organ failure, 598, 609–10
α-Tocopheryl radical, formation, 232
Total radical-trapping antioxidant parameter (TRAP) values, 226–7
Toxic Spanish oil syndrome, 350
Toxins, particle-bound, animals affected by, 39–44
Transcription of DNA, hydrogen peroxide stress and, 166–7
Transferrin, antioxidant activity, 216, 217, 218
Transformation, potentiation and inhibition by phenolic antioxidants, 260, 261
Transition metals, LDL oxidation involving, 378–80, 386, 389
Transmembrane signal transduction, *see* Signal transduction
TRAP values, 226–7
Trauma
lipid peroxidation and, 599
organ failure following, 595–616
poly-, 596–7
Tryptophan, photochemical degradation, 58
Tubular epithelium, renal, oxidative damage and the, 605
Tubulin, platelet, glutathione-oxidizing compounds and, 432
Tumour(s), *see also* Carcinogen; Carcinogenesis; Chemotherapeutic agents; Cytotoxic agents *and sites and specific types of tumours*
progression (from benign to malignant), 279–80
progressors, free radical metabolites of, 304–11
promoters, free radical metabolites of, 304–11
promotion, 278–9
phenolic antioxidants suppressing, 258–9
reactive oxygen species in, 258
quinone reductase activity in, 204
skin
haematoporphyrins in treatment of, 566
photochemical processes and, 563, 568
Tumour necrosis factor
endotoxins and, 458, 459, 588, 589–90, 606
NO and, 458, 459
Tumour necrosis factor-α
endotoxins and, 590
phagocyte priming by, 415
Tungsten, dietary, intestinal ischaemia–perfusion injury and effects of, 516, 517

U74006F
cerebral ischaemia and effects of, 519
shock and organ failure and effects of, 609
Ubiquinol-10, antioxidant activity, 224, 229
Ubiquinone in sun-exposed skin, 575
Ultraviolet (all/unspecified wavelengths)
carcinogenicity, 66, 563, 573–4
skin antioxidants and effects of, 574–6
Ultraviolet-A radiation, 57–83
skin and effects of, 65–6, 70, 560, 567, 569, 573, 575, 576
Ultraviolet-B radiation
lenses and effects of, 58, 549
skin and effects of, 65–6, 70, 567, 567–8, 569, 573–4, 575, 576, 577, 578

Ultraviolet-C radiation, skin and effects of, 567, 568
Umbilical vein endothelial cells, 4-hydroxyalkenal effects, 347, 351, 352
United States of America, anthropogenic emissions of main pollutants in, 7
Urea in DNA, radical-induced formation, 120, 121
Uric acid as an antioxidant, 221, 226, 227, 228, 230
UvrABC complex, 132–3

Vanadyl ion, LDL autoxidation accelerated by, 380, 385
Vascular smooth muscle, NO/oxyhaemoglobin/methaemoglobin and, 461
Vasodilation/vasorelaxation, 445–89
 drug-dependent, 472–6
 endogenous, 448–50
 by NO/EDRF, 448–50, 453–4
Vegetables, anticarcinogens in, 207
Vitamin C, *see* Ascorbic acid
Vitamin E (α-tocopherol), *see also* Tocopherols
 antioxidant role of, 222–3, 226–33, 542–3, 564, 576–7
  LDL oxidation and the, 390–2
  in lenses, 542–3
 ascorbic acid and, interactions, 231–3
 cerebral ischaemia and effects of, 518–19
 erythropoietic protoporphyria and, 564
 lipid peroxidation and, 225, 323
 phenolic antioxidants and, 323
 photoradiation and, 576–7
 shock/trauma-related organ failure and, 598
 ultraviolet and, 564
Volatile organic compounds (VOCs), 45
 natural, 5–6

Water-soluble antioxidants, 220–2, 226–7
 lipid-soluble and, interaction between, 231–3
White blood cells, *see* Leukocytes

X-radiation, 8–hydroxydeoxyguanosine formation in DNA induced by, 103
Xanthine dehydrogenase, conversion to xanthine oxidase, 497, 516, 517
Xanthine oxidase
 ischaemic–reperfused organs and free radical generation by, 496–8, 506–7, 515–16
 in shock, 600–1
 xenobiotic/carcinogen activation involving, 283
Xanthine oxidase inhibitors, 506–7, 610
 ischaemia–reperfusion injury and effects of, 506–7
 shock and organ failure and effects of, 610
Xenobiotics
 electrophilic, glutathione conjugation and, 178
 enzymatic activation, 280–4, 287–92 *passim*
 inflammatory cells activating, 287–92
 thiol-bearing, glutathione oxidation by, 88–94
Xeroderma pigmentosum, 571
Xth gene, DNA repair involving, 67, 133, 135–7

Yeast (Saccharomyces)
 apurinic/apyrimidinic (AP) endonuclease, 139–40, 142–3
 DNA repair in, 130, 139–40, 142–3, 144, 145
 3-phosphoglycolaldehyde diesterase, 139
 thymine glycol glycosylases, 130

Zeaxanthin, antioxidant role of, 229
 LDL oxidation and the, 390
Zymosan-induced peritonitis, 609